ROYAL BOTANIC GARDENS, KEW

Kew Bulletin Additional Series XI

FLORA OF THE CAYMAN ISLANDS

George R. Proctor

Former Head
Natural History Division, The Institute of Jamaica
and
Atkins Visiting Scholar, Harvard University

With a section on Environment and Plant Associations

by M.A. Brunt

Land Resources Development Centre
Overseas Development Administration

LONDON
HER MAJESTY'S STATIONERY OFFICE

First published 1984

This flora is dedicated to
Dr MEC Giglioli, 1927–1984,
for his contributions to science in
the Cayman Islands.

ISBN 0 11 242548 8

CONTENTS

PREFACE

This flora was written by Dr G.R. Proctor, former head of the Natural History Division of the Institute of Jamaica, lately Atkins Visiting Scholar at Harvard University, subsequently Supervisor of the National Botanical Garden Herbarium, Santo Domingo, the Dominican Republic, and now attached to the Department of Natural Resources, Puerta de Tierra, Puerto Rico.

The production of the Flora results from collaboration between the Land Resources Development Centre* (L.R.D.C.) of the U.K. Overseas Development Administration and the Cayman Island Government's Mosquito Research and Control Unit (M.R.C.U.).

In 1966 and 1967, Mr M.A. Brunt of L.R.D.C. undertook ecological studies to help the M.R.C.U. establish its programme of mosquito research and control. In 1967, at the request of Dr M.E.C. Giglioli, Director of the M.R.C.U., the Institute of Jamaica kindly sent Dr G.R. Proctor to the Cayman Islands to help with the botanical side of this work. Following this initial collaboration by Dr Proctor with L.R.D.C., Mr J.C. Cumber, then Administrator of the Cayman Islands, asked the Centre whether it could publish a Cayman flora based on the recent botanical and ecological studies and also on the earlier plant collections. This was subsequently agreed, with the backing and approval of Mr C. Bernard Lewis, at that time Director of the Institute of Jamaica, and Mr R. Ross, Keeper of Botany, British Museum (Natural History). The Institute of Jamaica had at one time contemplated the production of such a flora, and the British Museum (Natural History) had also intended compiling an annotated list of the plants of the Cayman Islands to supplement the *Flora of Jamaica* that they were publishing: the way in which the *Flora of the Cayman Islands* has been assembled from a number of collections and records is described in the Foreword by Mr C. Bernard Lewis and in the Introduction by the author.

As the time for publication approached, Her Majesty's Stationery Office, who publish L.R.D.C. reports, suggested that the Flora might be more appropriately issued as one of the Kew Bulletin Occasional Series rather than as an L.R.D.C. report. Those concerned readily gave their agreement, and this volume is the result.

The Flora was edited and prepared for publication by Mr W.J. Baulkwill of the Land Resources Development Centre. The printer's copy was prepared for Her Majesty's Stationery Office by the Ministry of Agriculture, Fisheries and Food Specialist Typing Service, Guildford, working on behalf of the Royal Botanic Gardens, Kew.

M A BRUNT
Land Resources Development Centre
January 1984.

---

*Formerly the Land Resources Division of The Directorate of Overseas Surveys, also of the Overseas Development Administration.

FOREWORD: BY C.B. LEWIS

It was in 1937 that all of this started. His Honour A.W. Cardinall (later Sir Alan Cardinall), Commissioner of the Cayman Islands, was anxious to learn as much as possible about his territory. He found that, apart from a collection of birds, virtually all scientific specimens from the three islands had been gathered by American collectors and were deposited in American museums and that the British Museum (Natural History) had practically no Cayman material. He proposed a Cayman Expedition which would make an addition to the British collections.

I had become interested in Jamaica in 1936 when I was a graduate student in Zoology at Johns Hopkins University. At that time it was the practice of botanical students of that institution to spend their third summer in Jamaica, and I had attempted to go on such a trip as an "extra". However, I had to forego this experience, as I had to pay my passage to England instead, having been awarded a Rhodes Scholarship to Oxford University during the year.

At Oxford, arrangements were made for me to go on the Cayman Expedition, in view of my deep interest in the Caribbean area. W.G. Alexander was in charge, and I was responsible for organizing the scientific work. I tried to get a botanist from Oxford with no success. On our departure from England at the end of March, 1938, we sailed from Liverpool on the S.S. Samala with the name of Wilfred Kings, Science Master of the Lawrence Sheriff School at Rugby, as a strong possibility to fill the botanical post.

Our first contact with anyone from the Cayman Islands was a young seaman named Berkeley Bush, employed on the "Samala", who was very well thought of by the captain and crew. Discussions of the islands with him ensued, and many of our questions brought the reply, "Ask Ernest". This turned out to be good advice, as Ernest Panton was not only his brother-in-law and also the Assistant Commissioner, but was also full of information on everything that the expedition required, as we learned when we reached Grand Cayman.

Our ship landed us at Kingston, Jamaica. Here, the family of J.S. Webster, shipowners, did much to fill us in. They had lived in Jamaica from early in the century, but had maintained close contact with their homeland, Grand Cayman.

After our short stay in Jamaica, we took the only regular transport, the "Cimboco", which visited each of the Cayman Islands about once a fortnight. We arrived at Georgetown, Grand Cayman, on April 17, 1938. Until it was definite that Kings would join us, I tried to collect plants, among many other things. However, he arrived in May, and quickly settled in as a member of the group.

Our initial investigations were carried out on Grand Cayman. We found the few roads on the island to be rough, and their condition somewhat restricted our work. None were paved and the surfaces were very poor, making transportation difficult. To supplement our efforts, we kept a "cat" boat at the Georgetown end of North Sound, from which we explored the coastline particularly to the east. At other times we used a truck to convey us around the island.

Part of the group visited the Lesser Caymans in late May and early June. Those who took part in this exercise included Kings, G. Thompson, and myself. On Cayman Brac we worked for about ten days, May 18–28, the interval between visits of the "Cimboco". We stayed with Mr & Mrs A.S. Rutty, who provided us with every facility. Although only the single main road on the Brac was kept in fairly good condition, our coverage of this island was good considering the time allowed.

On Little Cayman, where we worked from May 28 to June 10, we used facilities provided by Capt. Sam Bodden and his niece Miss Monica Bodden. This island was difficult to work because there were no roads. Much of the interior was virtually inaccessible. Further, the mangroves, which form a broad strip along a large part of the south coast, were dead in a zone up to about 300 yards wide and nearly five miles long, due to the ravages of the great 1932 hurricane. About 70 persons lived on Little Cayman when we were there, all but one family being settled near the southwest corner. A foot track crossed the island to the north coast from this area. We planned to use most of our time near the village and along this track. When we decided to journey eastward, we were unable to get any food or water from Capt. Bodden, who opposed our working at the east end of the island. We were later told that in all probability this was because he had cached somewhere in that area a treasure in gold that he had found in the Pedro Cays in 1903, and which for many years he kept in a safe in a house he had built on Little Cayman. This house was blown away in the 1932 hurricane, leaving the safe intact but with the combination corroded so that it could not again be opened. It was suspected that he had buried this safe near the east end of the island.

At any rate, we succeeded in obtaining two "cat" boats with the help of Capt. James Banks, and sailed one night at midnight to elude Capt. Bodden, eastward behind the reef, reaching the East End about daybreak. We planned to spend several days, and hoped to find water (or catch rain-water) as well as catch fish for food. However, the weather turned out to be extremely dry and hot, with temperatures well over a hundred, and it was all we could do to keep going. One of our men went temporarily berserk from the heat. Our chief food was lobsters, which were abundant. By the fourth or fifth day we set off on the return trip, during which it began to rain! Capt. Bodden was right on hand to give us a close inspection when we got back, but all we had were natural history specimens and no gold.

Kings worked very hard making collections of plants on all three of the islands, but except for a rather brief general report of his findings, never published, his specimens have remained unreported in detail up to the present. Only in the present Flora are they listed in their entirety.

For a long time I had planned to write up the flora myself, using as a basis not only the Kings collection, but also the published records of A.S. Hitchcock (1893) and C.F. Millspaugh (1900). Also to be incorporated were the brief records made earlier by William Fawcett. The list of plants collected by Kings was prepared at the British Museum (Natural History) with the help of specialists at that institution. Meanwhile, I had been appointed Curator at the Institute of Jamaica, and in 1950 was made Director. Up to that time I had still not found the time to begin the actual writing.

It was about this time that life rapidly changed for the Islands. The second half of the century was going to be a boom period. Land that in 1946 at North Side I was offered for £1 per acre, and on West Bay Beach for £6 per acre, suddenly rose enormously in price, and has moved to as much as $1500 per running foot in the latter area. The need for a published Flora also became more urgent. Such a book would be useful not only for identifying plants and as an educational tool, but also would serve as a permanent record of the vegetation as it existed before suffering the changes that result from the pressures of development. It is my hope that the delay in bringing about the publication of this Flora more than forty years after the original Oxford Expedition study will be beneficial in that a much more detailed and thorough treatment is now possible.

In 1951, George R. Proctor joined the staff of the Institute of Jamaica. He spent many hours on the taxonomic problems included in this study. It was possible for him to examine the Kings specimens, as well as much of the actual material collected by Fawcett, Hitchcock, and Millspaugh. He was able to spend three weeks on Grand Cayman in 1956, and subsequently visited all three islands on a number of occasions, collecting numerous specimens. In 1967 he met Martin Brunt of the (then) Land Resources Division, Ministry of Overseas Development (United Kingdom), and partly as a result of their discussions he resolved to write the present Flora.

In view of the inclusion of many drawings originally published in the Fawcett & Rendle *Flora of Jamaica*, it is worth noting that the illustrator for the first volume of that work, Miss Helen Wood, was a Caymanian from Bodden Town. She joined the Jamaican Government service in 1897, and beginning in 1912 was employed at the Institute of Jamaica in charge of its Museum for fifteen years until a few months before her death in November, 1927.

C. BERNARD LEWIS,
formerly Director of the
Institute of Jamaica,
16 August, 1978.

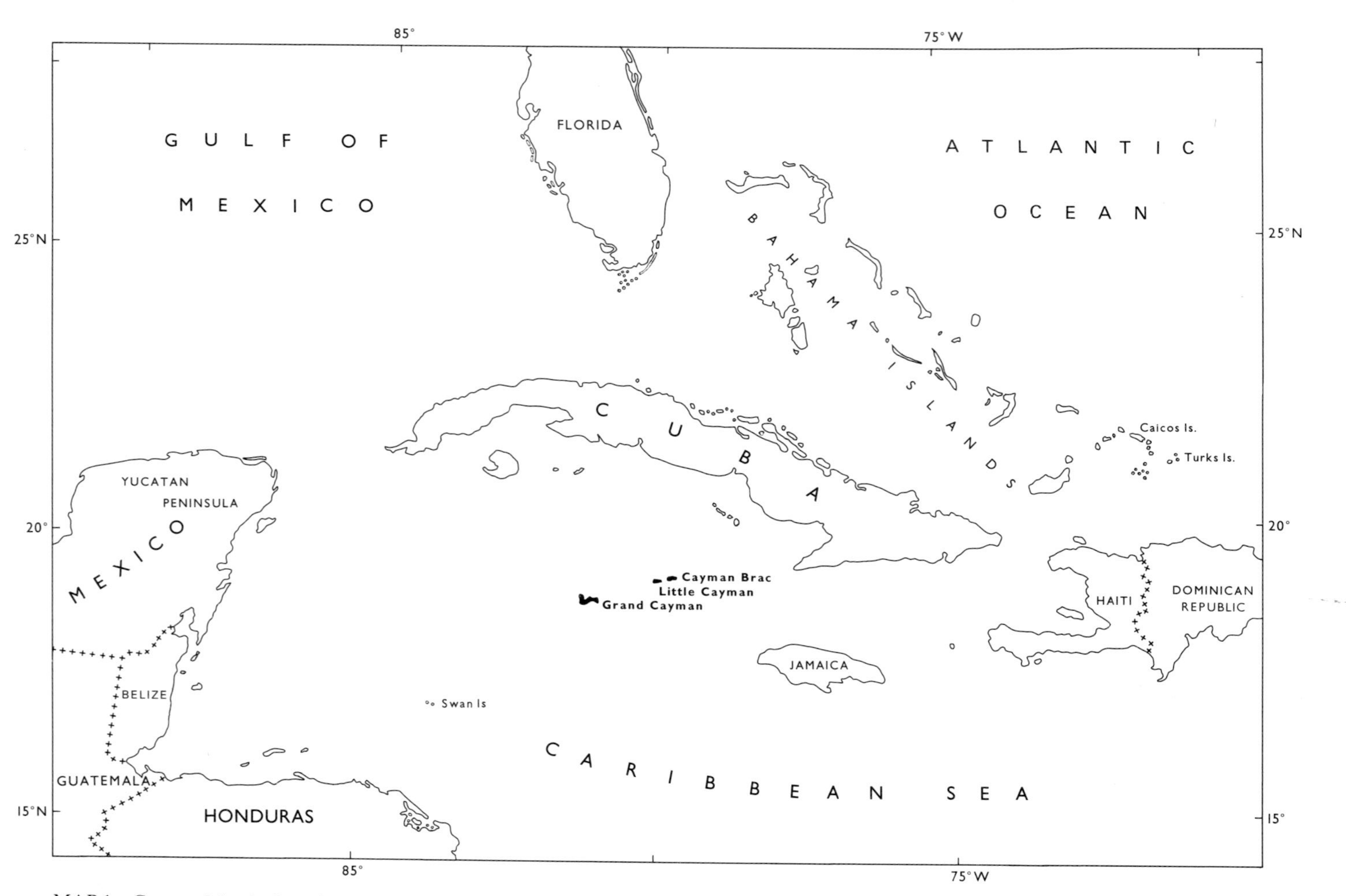

MAP 1 Cayman Islands: Location map

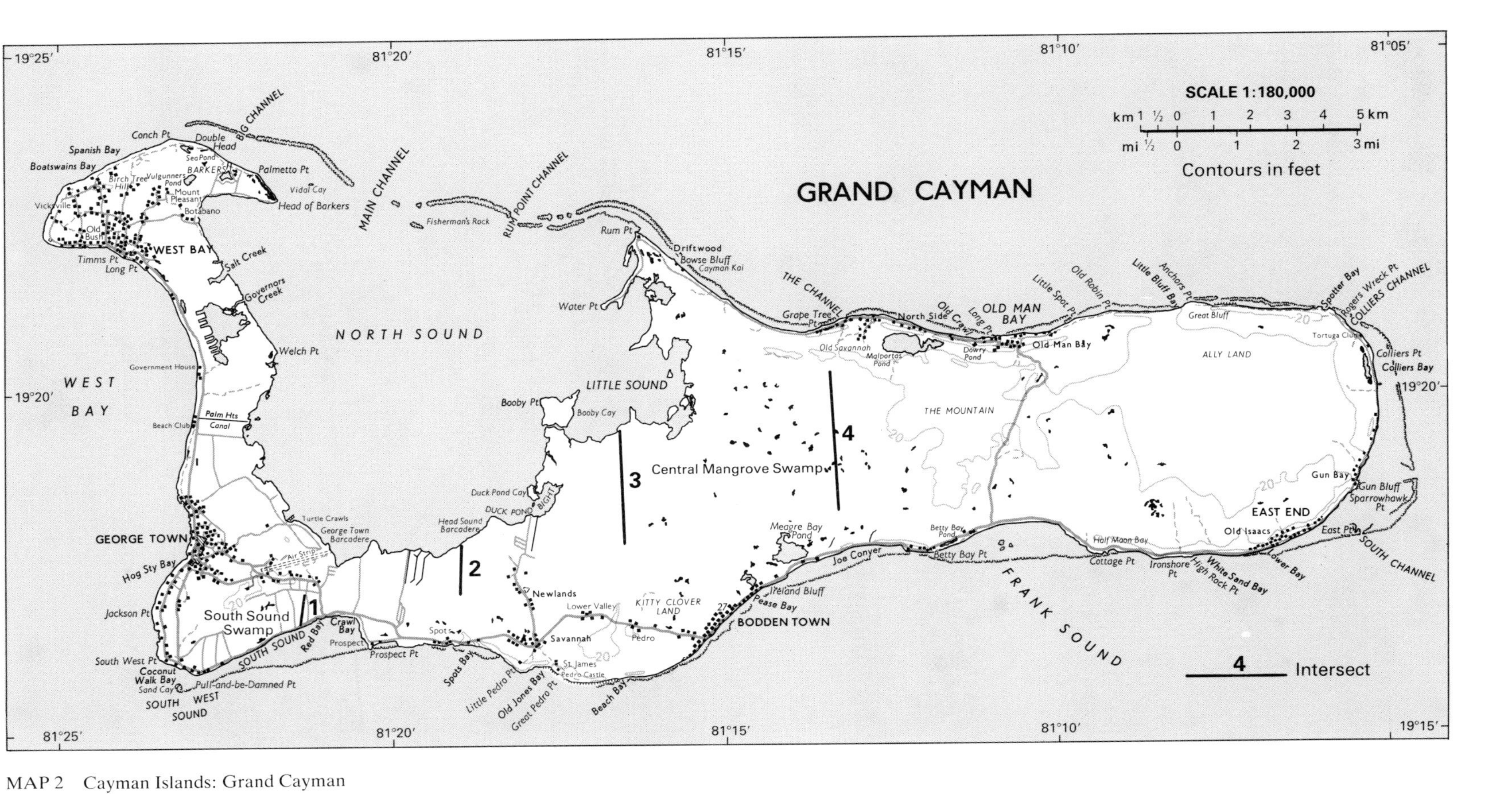

MAP 2 Cayman Islands: Grand Cayman

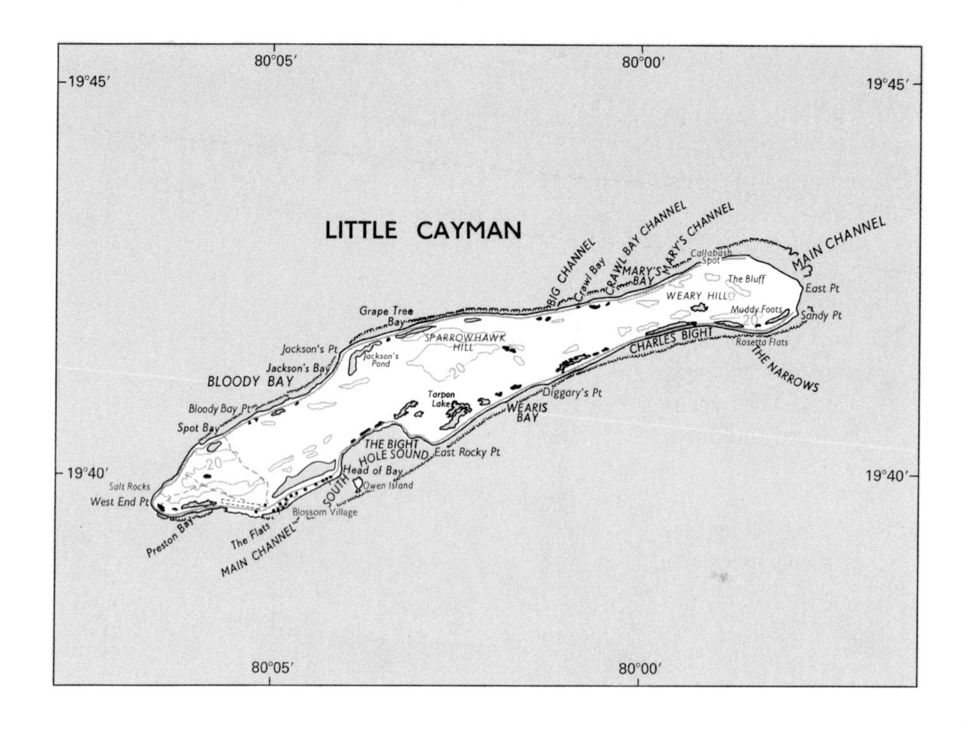

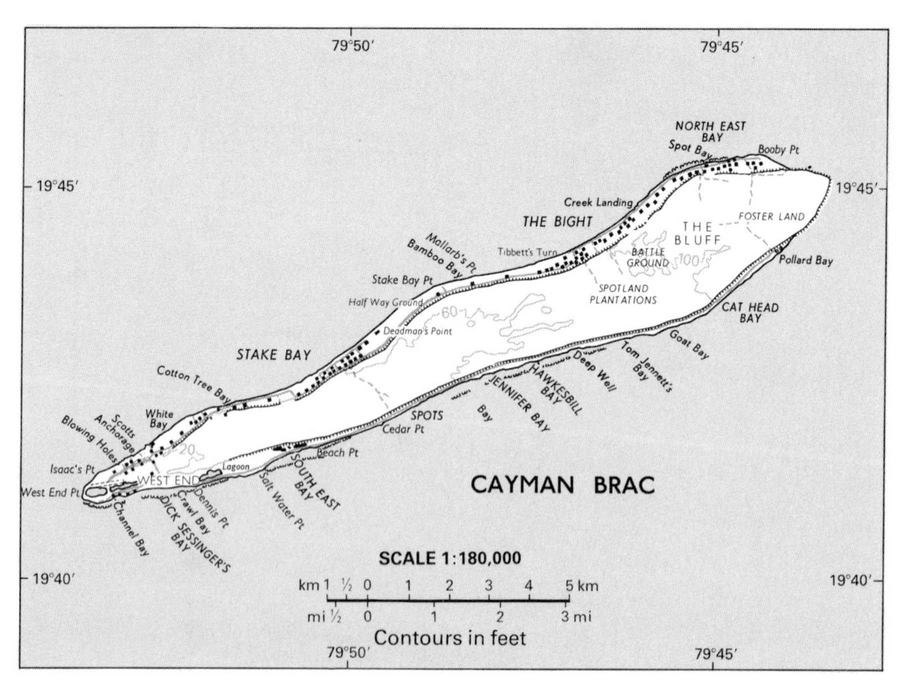

MAP 3 Cayman Islands: Little Cayman and Cayman Brac

INTRODUCTION

GEOGRAPHIC AND FLORISTIC BACKGROUND*

The Cayman Islands, consisting of Grand Cayman, Little Cayman, and Cayman Brac, are located in the Caribbean Sea between latitude 19° 15' and 19° 45'N., and longitude 79° 44' and 81° 27' W. Grand Cayman, much the largest of the three, is also the westernmost, and is separated from the other two by about 60 miles (97 km) of open sea. Altogether, their total area is about 102 square miles (264 km^2). They are composed of a hard core of massive white limestone of Oligocene-Miocene age, fringed by more recent consolidated reefs and other coastal formations. They are relatively flat and low-lying except for Cayman Brac, whose cliff-bordered interior rises gradually toward the precipitous eastern end to a height of 140 feet (43 m). Grand Cayman, located about 160 miles (257 km) W.N.W. of Jamaica, and about the same distance southwest of Cuba, is approximately 22 miles (35 km) long, with a breadth varying from less than 1 to 5 miles (1.6–8 km). The relatively narrow western end of this island, shaped like an arm bent at the elbow, encloses a large shallow bay called North Sound, 36 sq miles (93 km^2) in area, protected by a coral reef from the open sea to the north, and fringed by extensive mangrove thickets.

Little Cayman and Cayman Brac are both much smaller than Grand Cayman, both being about 10 miles (16 km) long and 1.5 miles (2.4 km) wide. They are adjacent to each other, both with their long axis oriented in a roughly east-west direction; the east end of Little Cayman is about 4.7 miles (7.5 km) northwest of the west end of Cayman Brac.

The climate of the Cayman Islands is subtropical and subject to distinct seasonal variations. The period from November to March is relatively cool and dry, with prevailing winds variable but often from the north, and an average temperature range between 65° and 75°F (18°-24°C). The remainder of the year averages about 10°F warmer or more, is subject to frequent showers and rather high humidity, and has prevailing easterly or southeasterly winds. Average annual rainfall is about 1400 mm or 55 inches.

Geologically, the Cayman Islands belong to the Greater Antilles, being emergent fragments of the otherwise submerged Cayman Ridge, which is continuous with the Sierra Maestra mountains of southeastern Cuba. This ridge, which extends in an east-west direction, is bordered by the Yucatan Basin to the north, while to the south the sea-bottom falls away abruptly to the very deep parallel Cayman Trench. There is no botanical evidence that the Cayman Islands have ever had a dry-land connection to any other land mass since they themselves emerged from the sea. If this is true, then it is necessary to conclude that their vascular flora, now known to include at least 601 indigenous or more recently

---

* The Section on Environment and plant communities, page 5, gives data on climate and soils and describes plant communities in detail.

naturalized species, has reached them by various means of dispersal over the sea. The position of these islands, together with the direction of the prevailing winds and sea-currents, suggests that the vegetation should be mostly Antillean in affinity, and this is in fact the case. Little or no relationship is shown with the mainland of Central America. Moreover, there exists an interesting overlapping of floral elements wherein numerous species of chiefly Cuban and Bahamian distribution grow in the Cayman Islands intermingled with others that are primarily or exclusively Jamaican. The local distribution of these species appears to be random among the three islands, i.e., Little Cayman and Cayman Brac, although closer to Cuba than Grand Cayman, show no special Cuban bias in their vegetation over that of Grand Cayman. The random distribution of Antillean elements in the Cayman flora can be interpreted as evidence of chance dispersal over the sea, most likely through the agency of hurricanes in the majority of cases. That this occurred sporadically over a very long period of time is suggested by the fact that local evolution of endemic species and varieties has had time to take place. All of the endemics show close affinities with Antillean congeners (with one exception), some of which also grow in the Cayman Islands along with their presumed derivative taxa.

SPECIES AND VARIETIES

The following 21 species and varieties are considered endemic to the Cayman Islands; the symbols "GC", "LC", and "CB" refer to Grand Cayman, Little Cayman, and Cayman Brac respectively; an asterisk (*) means that a close congener is also present:

1. *Aegiphila caymanensis* Moldenke (Verbenaceae) GC*
2. *Agalinis kingsii* Proctor (Scrophulariaceae) GC
3. *Allophylus cominia* var. *caymanensis* Proctor (Sapindaceae) GC, CB
4. *Argythamnia proctorii* Ingram (Euphorbiaceae) GC, LC, CB
5. *Caesalpinia caymanensis* Millsp. (Leguminosae-Caesalpinioideae) GC*
6. *Chamaesyce bruntii* Proctor (Euphorbiaceae) LC
7. *Chionanthus caymanensis* Stearn (Oleaceae) LC
8. *Coccothrinax proctorii* Read (Palmae) GC, LC, CB
9. *Cordia sebestena* var. *caymanensis* (Urban) Proctor (Boraginaceae) GC, LC, CB
10. *Crossopetalum caymanense* Proctor (Celastraceae) GC, CB*
11. *Dendropemon caymanensis* Proctor (Loranthaceae) LC
12. *Dendrophylax fawcettii* Rolfe (Orchidaceae) GC
13. *Epidendrum kingsii* Adams (Orchidaceae) LC
14. *Hohenbergia caymanensis* Britton ex L.B. Smith (Bromeliaceae) GC
15. *Oncidium caymanense* Moir (Orchidaceae) GC
16. *Pectis caymanensis* var. *robusta* Proctor (Compositae) GC*
17. *Phyllanthus caymanensis* Webster & Proctor (Euphorbiaceae) LC, CB
18. *Pleurothallis caymanensis* Adams (Orchidaceae) GC
19. *Salvia caymanensis* Millsp. & Uline (Labiatae) GC*
20. *Schomburgkia thomsoniana* Rchb.f. (Orchidaceae) GC (vars. LC, CB)
21. *Verbesina caymanensis* Proctor (Compositae) CB

The Jamaican element in the Cayman flora consists of 17 species and varieties. All of these are otherwise confined to Jamaica except for five that also occur on the Swan Islands, and one on Swan and Cozumel.

1. *Agave sobolifera* Salm-Dyck (Liliaceae-Agavoideae) GC
2. *Astrocasia tremula* (Griseb.) Webster (Euphorbiaceae) GC
3. *Bourreria venosa* (Miers) Stearn (Boraginaceae) GC, LC, CB (Swan)
4. *Casearia odorata* Macf. (Flacourtiaceae) GC*
5. *Cephalocereus swartzii* (Griseb.) Britton & Rose (Cactaceae) GC, LC, CB
6. *Cestrum diurnum* var. *venenatum* (Mill.) O.E.Sch. (Solanaceae) GC (Swan)
7. *Cordia brownei* (Friesen) I.M. Johnst. (Boraginaceae) GC, LC, CB
8. *Daphnopsis occidentalis* (Sw.) Krug & Urban (Thymelaeaceae) GC, CB
9. *Harrisia gracilis* (Mill.) Britton (Cactaceae) GC, CB (Swan ?)
10. *Opuntia spinosissima* Mill. (Cactaceae) CB
11. *Phyllanthus angustifolius* (Sw.) Sw. (Euphorbiaceae) GC, LC, CB (Swan)
12. *P. nutans* ssp. *nutans* (Euphorbiaceae) GC, LC, CB (Swan)
13. *Tabernaemontana laurifolia* L. (Apocynaceae) GC
14. *Tournefortia astrotricha* var. *astrotricha* (Boraginaceae) GC, CB (Swan)
15. *T. astrotricha* var. *subglabra* Stearn (Boraginaceae) GC
16. *Trichilia glabra* L. (Meliaceae) GC, LC, CB (Swan, Cozumel)
17. *Vernonia divaricata* Sw. (Compositae) GC, LC, CB

The strictly Cuban element in the Cayman flora consists of only 11 species and varieties, but this figure can be reinforced by numerous others of wider distribution that occur in Cuba but not in Jamaica. Cayman plants found otherwise only in Cuba are as follows:

1. *Cestrum diurnum* var. *marcianum* Proctor (Solanaceae) GC
2. *Chamaesyce torralbasii* (Urban) Millsp. (Euphorbiaceae) CB
3. *Clerodendrum aculeatum* var. *gracile* Griseb. ex Moldenke (Verbenaceae) GC
4. *Leptocereus leonis* Britton & Rose (Cactaceae) CB
5. *Malpighia cubensis* Kunth (Malpighiaceae) GC
6. *Pectis caymanensis* var. *caymanensis* (Compositae) GC, LC, CB
7. *Phyllanthus nutans* ssp. *grisebachianus* (Muell. Arg.) Webster (Euphorbiaceae) LC
8. *Polygala propinqua* (Britton) Blake (Polygalaceae) LC
9. *Portulaca tuberculata* Leon (Portulacaceae) LC, CB
10. *Roystonea regia* (Kunth) O.F. Cook (Palmae) GC
11. *Tillandsia fasciculata* var. *clavispica* Mez in DC. (Bromeliaceae) LC

The vegetation of the Cayman Islands shows virtually no direct affinity with that of Central America. Hence the discovery in 1975 of the herein-described *Phyllanthus caymanensis* Webster & Proctor on Little Cayman and Cayman Brac was quite surprising, as it is closely related to two Central American species, and represents the only known endemic Antillean element in its section of the genus.

There is, on the other hand, a total of 39 Cayman species whose distribution includes Cuba, the Bahamas, and in some cases Hispaniola (or in two cases – *Chascotheca domingensis* and *Cynanchum picardae* – Hispaniola alone), which do not occur in Jamaica.

A further group of 9 species represents plants endemic to the Greater Antilles which occur in **both** Cuba and Jamaica:

1. *Bunchosia swartziana* Griseb. (Malpighiaceae) GC
2. *Capparis ferruginea* L. (Capparidaceae) GC
3. *Cassia clarensis* (Britton) Howard (Leguminosae-Caesalpinioideae) GC (Note: this entity is not very distinct from a wider-ranging species)
4. *Celtis trinervia* Lam. (Ulmaceae) GC
5. *Cleome procumbens* Jacq. (Capparidaceae) GC
6. *Evolvulus arbuscula* Poir. (Convolvulaceae) LC (Note: closely related forms in the Bahamas may not be distinct)
7. *Paspalum distortum* Chase (Gramineae) GC
8. *Selenicereus grandiflorus* (L.) Britton & Rose (Cactaceae) GC
9. *Solanum havanense* Jacq. (Solanaceae) GC

This list can be augmented by at least 17 other species of primarily Greater Antillean distribution which extend slightly beyond this area (chiefly into the Bahamas), and occur in both Cuba and Jamaica, also usually Hispaniola, but not Puerto Rico. The great majority of these have been found only on Grand Cayman in our area.

The remainder of the species considered indigenous (i.e., those not introduced by human agency) consist chiefly of wide-ranging tropical American species, pan-tropical strand plants, and more or less pantropical weeds. An unknown number of these occurred in the Cayman Islands before the earliest human settlement, but it can be assumed that a certain percentage have introduced themselves since that time. A further 65 species, now more or less established or naturalized, owe their presence to intentional (or sometimes unintentional!) introduction by humans and this number will no doubt grow during the course of time.

The whole list of vascular plants now known to grow without cultivation in the Cayman Islands totals 601 species (20 pteridophytes, 1 gymnosperm, 141 monocotyledons 439 dicotyledons). It is quite probable that others will eventually be found, as a few relatively inaccessible portions of all three islands have still not been adequately investigated at all seasons. On the other hand, there is real danger that the building of new roads, hotels, and housing settlements will bring about the total extermination of many interesting or unique plant species unless enough representative habitats can be saved from the bulldozers.

ENVIRONMENT AND PLANT COMMUNITIES: BY M.A. BRUNT

This account of the vegetation of the Cayman Islands is largely based on field work by the author. It also draws on the work of others, namely Johnston (1975), Stoddart (1980), Brunt & Giglioli (1980), Woodroffe (1982) and Sauer (1982). The vegetation is similar to that found on other coralline islands and limestone formations in the West Indies, which has been widely described (see Bibliography: Environment and plant communities). These writers have mostly adopted Beard's system of classifying Tropical American vegetation, developed in a series of papers (1944, 1949a and 1955). It is again adopted here.

Under the Beard system floristically different plant associations are grouped together on the basis of similar structure and physiognomy into physiognomic units called formations; in turn, on the basis of habitat similarities, these are grouped into formation-series. Thus the association bears a floristic name, e.g. *Rhizophora mangle – Avicennia nitida* mangrove association; the formation a physiognomic name, e.g. Woodland or Thicket, and the formation-series a habitat name, e.g. Dry evergreen or Seasonal swamp.

The vegetation of the Cayman Islands falls into three of Beard's formation series:

The Dry evergreen formation-series
The Seasonal swamp formation-series
The Swamp formation-series

These are shown in Diagram 1, which includes some formations that do not occur in the Cayman Islands.*

Before the vegetation is described, the main features of the environment are noted.

**Note* Measurements are given in metres and centimetres and in feet and inches. The intersect diagrams however are scaled in feet alone. The relevant conversions are:

1 in	=	2.54	cm	1 cm	=	0.393	in
1 ft	=	30.48	cm	1 m	=	3.280	ft
	=	0.304	m	1 km	=	3 280.8	ft
1 mile	=	1 609.34	m		=	0.621	mi

The environment

The development of the different plant formations has been affected by the soil, the climate, the proximity of the islands to seed sources (mainly Jamaica and Cuba) and by ocean currents. (See Darbyshire, Bellamy and Jones (1976) on ocean currents, Guppy (1917) on seed dispersal, and Sauer (1982) on the evolution of the seashore vegetation). These formations have developed under sub-optimum conditions. Although tall tropical forest will grow under climatic conditions similar to those in the Cayman islands, the thin, relatively poor soils overlying limestone have largely prevented this happening. Even the mangrove swamps are dwarfed compared to swamps of the same species in the region (e.g. Guiana).

Climate

There is a good general description in Stoddart (1980a). The north-east trade winds are the dominant feature of the climate. This dominance is most marked in the winter, except for the occasional "northwester" of several days duration. These bring stormy weather, rainfall and lower temperatures; wind speeds will increase from the norm of about 5 m/sec to 13 m/sec (11.18 – 29.08 mi/h). In the summer south-easterly winds tend to occur.

There are two distinct seasons, a dry season from November to April, and a wet season from June to October. The average annual rainfall is about 1 400 mm (55.12 in), although there can be considerable variation between one year and the next, e.g. the annual rainfall recorded at Blossom Village during the years 1971–76 varied from 806 mm (31.73 in) to 1 872 mm (73.70 in). Temperatures are highest in the wet season and lowest in the dry season: August and January mean daily temperatures of 29.5°C (85.1°F) and 26.0°C (78.8°F) respectively, with a diurnal range of about 5.5°C (9.9°F). The absolute annual range is 12-33°C (54-91°F).

The associated tidal pattern peaks in August from a February low, the mean amplitude being 244 mm (9.61 in). The main climate and tidal features are summarised in Diagram 2, which also shows the associated flooding pattern in the mangrove swamps.

Hurricanes are also important, some 40 having been recorded since 1751; the last of these to hit Grand Cayman was in 1917. Their importance lies in the damage they do to the vegetation, especially the mangrove swamps. The complex zonation of the latter is due in part to such storms.

Soils

Each of the main geographical units of the islands – the uplands, the seasonal swamps and the marine swamps – have characteristic soils and vegetation. (See Baker (1974) and the FAO/UNESCO Soil Map of the World). The classification of these soils, and their relationship to the vegetation formations is shown in Table 1.

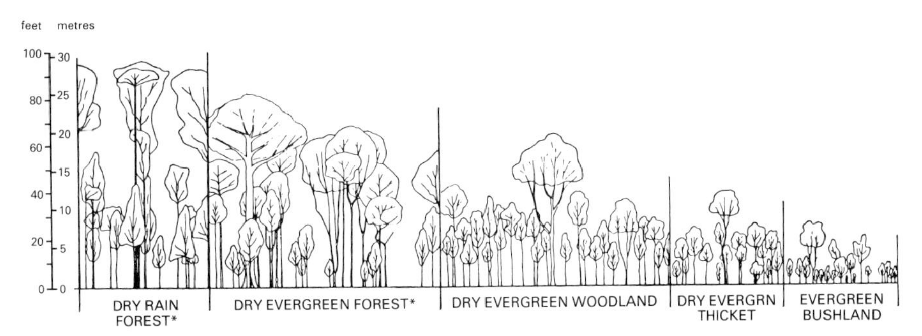

The Dry Evergreen Formation-Series. Dry Rain Forest redrawn from Fanshawe's diagram of "Wallaba Forest" (1952), Dry Evergreen Forest from Beard's Xerophytic Rain Forest in 1944b, the remainder imaginary, from descriptions.

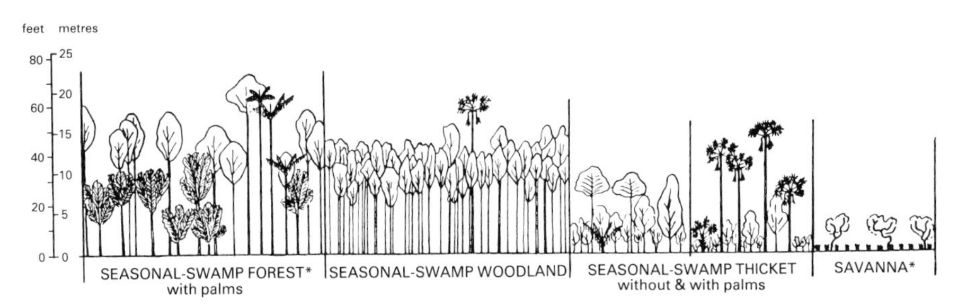

The Seasonal-Swamp Formation Series. Seasonal-Swamp Forests and Woodland redrawn from Fanshawe's diagrams (1952). Seasonal-Swamp Thickets after the Palm Marsh of Beard 1944a, Savanna diagrammatic.

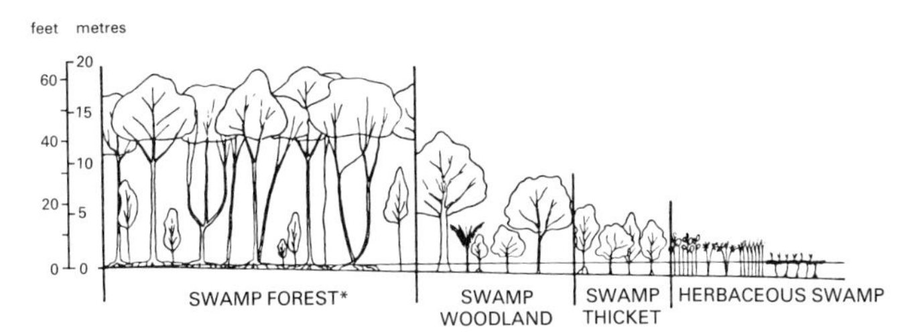

The Swamp Formation-Series. Swamp Forest after Beard, 1944a, the remainder imaginary and diagrammatic.

(From Beard (1955): The classification of Tropical American Vegetation-Types. Ecology 36(1) 89-100)

*Absent from the Cayman Islands

Diagram 1 Vegetation formation-series occurring in part in the Cayman Islands

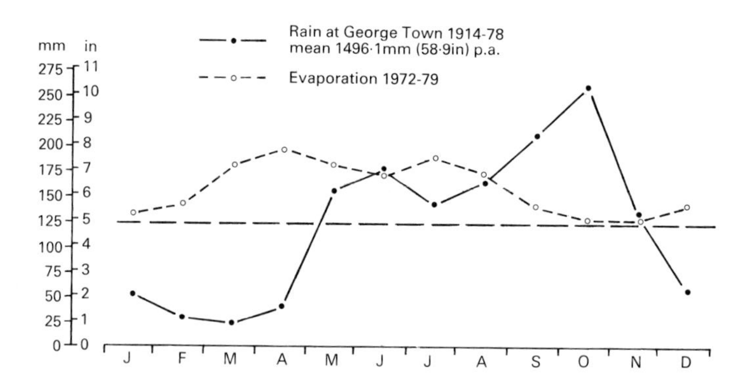

a Mean monthly rainfall and evaporation

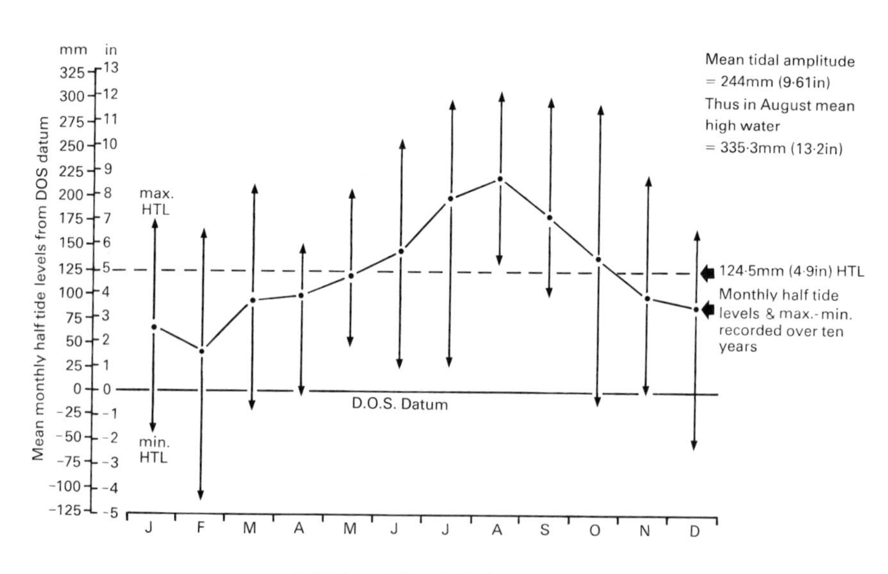

b Tides at South Sound 1970-82

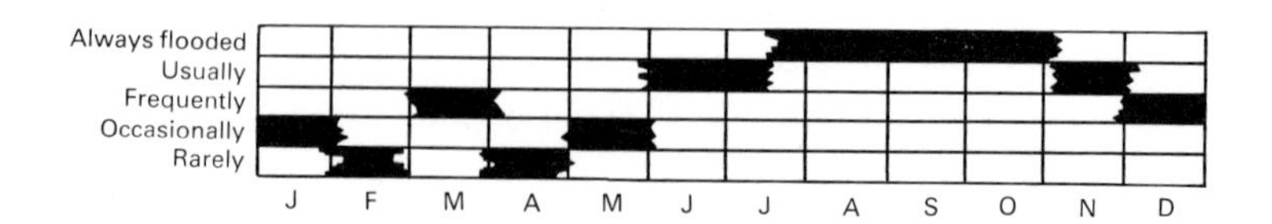

c Flood state of the mangrove swamps

(After Giglioli)

Diagram 2 Climatic and tidal features of Grand Cayman

The upland soils: developed from limestone parent material, vary in depth from a few centimetres over rock, to pockets of soil more than a metre deep. Large areas of the uplands have very thin soil cover. They vary in colour from dark brown (7.5 YR 4/4) to dark reddish-brown (2.5 YR 2/4) to dark red (10 R 3/6). They are clays or clay loams, with moderate to rapid internal drainage, and have low to moderate moisture-supplying capacities. They are neutral to mildly alkaline, and generally have a high nutrient status, but nutrient availability to plants is probably restricted. In ecological terms the shallow nature of these soils (which will exacerbate the effect of the dry season), and their nutritional limitations have certainly limited the development of the Evergreen formations.

The seasonal swamp soils: have developed in shallow hollows in the limestone which vary in size from a few square metres to several hectares (Halfway Pond, Georgetown), and in depth from 20 cm to over a metre. The parent material is limestone.

The soils vary from dark reddish brown (5 YR 2/2) to very dark greyish brown (10 YR 3/2); mottling may occur. They are loams to clay loams, with restricted internal drainage. They vary from mildly alkaline to mildly acid. Depending on the amount of organic matter incorporated in the profile, their fertility varies, and may be high, although the chemical constraints mentioned above also apply.

The marine swamp soils: support the extensive mangrove swamps. Where these border the sea the soils are saline. Towards the landward edge of the swamps the tidal influence is less and the salt is leached by the rainfall. The soils vary in depth from a few centimetres to over three metres and are peaty. Their colour varies from black (10 YR 2/1) to dark yellowish brown (10 YR 4/4) with dark reddish brown (5 YR 2/2) variants. They are neutral to mildly alkaline, and due to their organic nature have a very high exchange capacity, and a high nutrient status. The chemical constraints described for the other two soils will again apply.

Table 1 shows the island's major geographical units and their distinctive soils and vegetation. This simple pattern is however subject to variation. Site factors – for example the local landform – soil type, the degree of exposure to wind (salt laden or otherwise), or the availability, depth and salinity of the groundwater – may give rise to local floristic variation with a formation. Human settlement and farming have also resulted in changes to the Dry evergreen woodland formation. These range from its replacement by rough pastures and various kinds of thicket, to the establishment of the man-made floristically distinct Logwood woodland (*Haematoxylum campechianum*) once exported as a leather dye source. Since the Logwood is no longer cut, this woodland now has the appearance of a natural formation.

Table 1 Cayman vegetation formation-series, formations and soil classification

Geographical Unit	*Vegetation formation -series (f.s.) and formation*	*Soil Classification* U.S. Soil Taxonomy classification	F.A.O./U.N.E.S.C.O. Soil Map of the World legend	UWI Regional (Baker, 1974) and older names (*)
Uplands	Dry evergreen f.s. Woodland	Lithic Rendolls	Lithosols Shallow Rendzinas	Soils formed over hard limestone or bluff soils
	Thicket	Lithic Ustochrepts	Lithic Chromic Cambisols	Soils of Ironshore Formation Terra rossa
	Bushland	Lithic Rhodoxeralf	Lithic Chromic Luvisols	Red brown earths Brown forest soils
	Rock pavement vegetation			
Seasonal swamps	Seasonal swamp f.s. *Conocarpus* thicket *Conocarpus* bushland	Aquepts	Gleysols	Hydromorphic soils*
	Herbaceous swamp	Calciaquolls	Calcaric Gleysols	
Marine swamps	Marine swamp f.s. Mangrove swamp woodland Mangrove swamp thicket	Histosols Fibrists Hemists Saprists	Histosols	Bog soils* Organic peat soils

Classification of Cayman Islands vegetation

The composition of the Cayman Islands formation-series is set out below. In the notation adopted, the formation-series are distinguished by roman numerals. Formations*, i.e. physiognomic groupings within the formation-series, are shown by letters. Associations, i.e. floristic groupings within the formations, particularly site types and derived cultivation types, are shown by small roman numerals.

I	DRY EVERGREEN FORMATION SERIES
Ia	*Dry evergreen woodland*
Ia(i)	Dry evergreen littoral woodland
Ib	*Dry evergreen thicket*
Ib(i)	Dry evergreen littoral thicket and bushland littoral hedge
Ib(ii)	Dry evergreen logwood (Haematoxylum) thicket
Ic	*Herb grassland (rough pasture) derived from dry evergreen woodland and thicket*
Id	*Dry evergreen bushland*
Id(i)	Dry evergreen bushland littoral hedge
Id(ii)	Dry evergreen bushland on sandy and pebbly beach ridges and associated coastal communities
Id(iii)	Dry evergreen Coccothrinax palm bushland
Id(iv)	Dry evergreen manchineel (Hippomane) bushland
Ie	*Dry evergreen vegetation on rock pavements*
II	SEASONAL SWAMP FORMATION SERIES
IIa	*Seasonal Conocarpus (buttonwood) swamp thicket*
IIb	*Seasonal Conocarpus (buttonwood) swamp bushland*
IIc	*Seasonal herbaceous swamp*
IIc(i)	Seasonal herbaceous grassland swamp derived from Conocarpus (buttonwood) swamp thicket and bushland
IIc(ii)	Seasonal undifferentiated herbaceous swamp probably derived from Conocarpus (buttonwood) swamp woodland and thicket
IIc(iii)	Seasonal undifferentiated brackish herbaceous swamp
IIc(iv)	Seasonal Panicum grassland swamp
IIc(v)	Seasonal Eleocharis sedge swamp
IIc(vi)	Seasonal Typha bulrush swamp
III	SWAMP FORMATION SERIES: A: MARINE

(In the description of the Marine swamp-formation series *in the text*, the conventional order of description from optimum to pessimum is reversed, to make the narrative, which starts with the pioneering *Rhizophora* community, easier to follow. The text order is thus IIIc, IIIb, IIIa).

* A definition of the formations follows this classification.

IIIa	*Mangrove swamp woodland*
IIIa(i)	Avicennia-Laguncularia-Rhizophora mangrove swamp woodland
IIIa(ii)	Avicennia mangrove swamp woodland
IIIa(iii)	Rhizophora mangrove swamp woodland
IIIa(iv)	Laguncularia mangrove swamp woodland
IIIa(v)	Laguncularia-Avicennia, Laguncularia-Rhizophora, Laguncularia-Conocarpus mangrove swamp woodland
IIIb	*Mangrove swamp thicket*
IIIb(i)	Rhizophora mangrove swamp thicket
IIIb(ii)	Avicennia mangrove swamp thicket
IIIb(iii)	Laguncularia mangrove swamp thicket
IIIb(iv)	Conocarpus-Avicennia and Conocarpus-Laguncularia mangrove swamp thicket
IIIc	*Mangrove swamp bushland*
IIIc(i)	Rhizophora bushland
IIIc(ii)	Rhizophora-Conocarpus bushland
IIIc(iii)	Rhizophora-Avicennia bushland
IIIc(iv)	Laguncularia-Avicennia-Conocarpus bushland

III SWAMP FORMATION SERIES: B: FRESHWATER

IIId *Undifferentiated herbaceous swamp, including ponds*

DEFINITION OF FORMATIONS

The formations: (physiognomic groupings within the formation series) listed above were defined by Beard (1955) as follows:

Woodland: a two-storied formation with the canopy formed of densely packed, attenuated trees, not larger than 45-50 cm (18-20 in) in diameter and about 6-12 m (20-40 ft) high. There is a discontinuous emergent layer between 18-25 m (60-80 ft) high, made up of larger trees. The woodland is almost entirely evergreen.

Dry evergreen thicket; a two-storied formation with a low, open, or dense canopy, and a dense or sparse undergrowth. The canopy is between 6-12 m (20 and 40 ft) high, and trees have slender stems not larger than 15-20 cm (6-8 in) in diameter. There is an occasional larger emergent.

Dry evergreen bushland: low, dense impenetrable vegetation of few species, consisting of gnarled small trees and bushes 3-4 m (10-13 ft) high, with no readily definable stratification. There is sometimes an upper open stratum of scattered dominants and a lower story of shrubs. Stem diameters seldom exceed 15 cm (6 in). Moss and ferns absent, lianes few, but epiphytes are abundant. The bushland is often irregularly distributed with patches of bare soil or rock in between (Beard, 1949a; Fanshawe, 1952; D'Arcy, 1975; Loveless and Asprey, 1957; Loveless, 1960)

Dry evergreen vegetation on rock pavements: irregular and open growth of herbaceous and woody plants less than 2 m (9.8 ft) in height, growing in crevices or in mats of humus upon sheet rock, usually with much bare rock between.

Seasonal swamp woodland: a low woodland of small stemmed regularly spaced trees with scattered emergent ...... trees. The canopy lies between 9-15 m (30-50 ft) .... The shrub layer is virtually absent; the ground cover sparse or dense. The mean tree diameter is 10 cm (4 in); emergents if present are from 40-90 cm (16-36 in). The stocking is high ...... but the flora is restricted to about 20 tree species. Lianes are few, epiphytes rare. Leaves are evergreen, mesophyllous, simple. Stilt roots are a common feature (Fanshawe 1952).

Seasonal swamp thicket: a low dense formation up to 12 m (40 ft) high of small trees only a few centimetres in diameter, but often with rather spreading branches. The canopy will be more or less closed, and the undergrowth dense. The ground cover when present is predominantly of sedges.

Seasonal herbaceous grassland swamp: Land dominated by grasses and occasionally by other herbs, sometimes with widely scattered or grouped trees and shrubs, the canopy cover of which does not exceed 2% (Pratt *et al.* 1966), included are a number of floristic variants which may be seasonally flooded with fresh or brackish water.

Mangrove swamp woodland and thicket: Comprises one or a few species, and has a simple structure and is remarkably homogeneous. There is one tree stratum 3-12 m (10-40 ft) high, and the canopy is loose and open. A shrubby or herbaceous undergrowth may be present. Stilt and breathing roots (pneumatophores) are common. Leaves are evergreen, mesophyllous and thickly cutinised.

I DRY EVERGREEN FORMATION SERIES

Ia Dry evergreen woodland

This account is based upon the writer's work in Grand Cayman, supplemented by Stoddart's descriptions of Little Cayman (Stoddart, 1980). The latter also described a Dry evergreen forest south of Sparrow-hawk Hill in Little Cayman, noting its similarity to Dry evergreen woodland. In the writer's view forest in the generally accepted sense does not occur in the Cayman Islands; in consequence Dry evergreen forest is not here recognised. Nor in the following description has Dry evergreen woodland been separated from Dry evergreen thicket in Grand Cayman. Both formations certainly occur, but in a complex mosaic due to the exploitation of the woodland in the past; greater in Grand Cayman than Little Cayman.

For generations Caymanians depended on the local woodland for timber for house building, boat construction and for fuel, with the result that most of the larger trees have been felled. This appears to have resulted in the virtual destruction, at least in Grand Cayman, of the discontinuous 18-24 m (60-80 ft) emergent layer which distinguished Beard's woodland from thicket: both having a canopy between 6-12 m (20-40 ft) in height. Thus in Grand Cayman little true woodland remains, although there may be a few isolated patches in the east of the island and on Cayman Brac. Today therefore most of the upland supports Dry evergreen thicket.

Distribution: in Grand Cayman it occupies areas in the east and central part of the island, and to the north of South Sound. It has been virtually destroyed in the West Bay area. In Little Cayman it is found in the centre of the island south of Sparrowhawk Hill, at Paradise End, and inland from Jackson's Bay and Salt Rocks. In Cayman Brac it occupies areas throughout the island on the Bluff Limestone formation.

Structure and physiognomy: examples of this formation were not studied in detail on Grand Cayman. However a conspicuous group of 30 m (100 ft) high Royal Palms *(Roystonia regia)* should be noted near the trans-island road in the centre of the island; with the surrounding evergreen vegetation it could be described as woodland. Stoddart (1980) noted that the taller examples of the formation on Little Cayman (which he called forest) reach 12 to 20 metres (39-65 ft) in height, and the woodland 6 to 12 metres (19-39 ft); the emergent layer reaching 30 m and 20 m (98 and 65 ft) respectively; leaves being generally simple, evergreen, most stiff and fleshy, many with latex or essential oil; bark-shedding being a conspicuous feature.

Floristic composition: Stoddart (1980) noted floristic differences between the forest and woodland on Little Cayman which are detailed below

Dry evergreen formations on Little Cayman: floristic composition

Component	*Species*	*Forest*	*Woodland*
Trees	Bumelia salicifolia	x	
	Bursera simaruba	x	x
	Calyptranthes pallens	x	
	Canella winterana	x	x
	Chionanthus caymanensis	x	
	Citharexylum fruticosum		x
	Coccoloba uvifera		x
	Cordia gerascanthus		x
	C. sebestena var. caymanensis	x	
	Croton nitens	x	x
	Erythroxylum rotundifolium	x	
	Ficus aurea		x
	Guapira discolor		x
	Hypelate trifoliata	x	
	Metopium toxiferum		x
	Myrcianthes fragrans	x	
	Picrodendron baccatum		x
	Tabebuia heterophylla	x	x
	Trichilia glabra		x
	Swietenia mahagoni		x

Component	*Species*	*Forest*	*Woodland*
Shrubs	Adelia ricinella		x
	Allophylus cominia var. caymanensis	x	
	Amyris elemifera	x	
	Argythamnia proctorii	x	x
	Ateramnus lucidus		x
	Calyptranthes pallens		x
	Capparis flexuosa		x
	Chiococca alba	x	
	Coccothrinax proctorii		x
	Colubrina elliptica	x	x
	Croton lucidus	x	
	Jacquinia berterii	x	x
	Malvaviscus arboreus	x	
	M. arboreus var. cubensis		x
	Maytenus buxifolia	x	x
	Phyllanthus angustifolius		x
	Randia aculeata	x	
	Savia erythroxyloides		x
Herbs	Lasiacis divaricata	x	x
	Paspalum blodgettii	x	
Succulent	Agave sobolifera	x	
Epiphytes	Schomburgkia thomsoniana	x	
	Tillandsia spp.	x	
Vines	Capparis flexuosa		x
	Ipomoea indica var. acuminata		x
	I. violacea		x

The dominant big trees of the forest are *Bumelia salicifolia, Calyptranthes pallens* and *Chionanthus caymanensis* compared to the woodland in which the most conspicuous species is *Bursera simaruba*.

The ecology of the woodland on Cayman Brac has not been investigated.

Ia(i) Dry evergreen littoral woodland

Dry evergreen littoral woodland is probably restricted to Little Cayman. Stoddart (1980) has written: "The Spot Bay woodland is probably the last surviving remnant of Littoral Woodland on any of the three Cayman Islands, and is thus of considerable interest. It includes tall trees of *Swietenia mahagoni* (reaching 15 m or 4.9 ft), and *Bursera simaruba*, with *Dalbergia ecastaphyllum, Cordia sebestena* var. *caymanensis, Terminalia catappa, Coccoloba uvifera* and *Jacquinia berterii. Coccoloba* forms a hedge along the seaward side".

Ib *Dry evergreen thicket*

On Grand Cayman most of what must once have been Dry evergreen woodland has apparently been reduced to thicket status. Stoddart (1980) also notes that on Little Cayman the woodland is "a rather variable entity".

Distribution: on Grand Cayman the Dry evergreen woodland/thicket mosaic, the latter dominating, occupies the eastern part of the island; while remnants also occur in the Georgetown – Newlands vicinity. On Little Cayman it occupies a north-facing gentle slope some 200 m (650 ft) wide bordering the northern shore, and the Bluff limestone ridge further inland, the differences in aspect and exposure resulting in differences in floristic composition. On Cayman Brac the formation is mainly found on the higher ground, also of Bluff Limestone; see also Nathan (1975).

Structure and physiognomy: this two-storey formation has a canopy which varies in height from 6-10 m (20 to 35 ft); on exposed sites it may be lower. The canopy, usually discontinuous, may become closed with thin, occasional emergent trees. Stoddart noted three variants in Little Cayman, which are summarised in Table 2, with data on stem size and stems per acre relating to Grand Cayman only. The latter derived from limited sampling are therefore indicative.

Table 2 Formation Ib. Variations in structure (Stoddart, 1980)

		Grand Cayman	*Little Cayman*	
Characteristic		*Bluff Limestone (BL) bordering Central Mangrove Swamp, and BL patch reefs in swamp. Intersect 4*	*1. Seaward slopes of Bluff Limestone ridges*	*2. Dissected Bluff Limestone ridge 3. Inland slopes of Bluff Limestone ridge*
Canopy height (first storey)	(ft) (m)	20 – 35 sometimes + 40 6 – 10 " + 12	10 – 13 3 – 4	20 – 40 6 – 12
Emergent height (second storey)	(ft) (m)	40 – 50 12 – 15	not recorded	not recorded
Av. crown diameter	(ft) (m)	20 – 30 6 – 9	" "	" "
Av. stem size (d.b.h.)	(in) (cm)	4 – 6* 10 – 15	" "	" "
Stems	(per acre) (per ha)	435 979	" "	" "

* occasional trees of 16 in (40.6 cm) d.b.h. recorded

On Grand Cayman the occasional tree which emerges above the canopy is often the ubiquitous red birch (*Bursera simaruba*). The widespread distribution of the species, particularly where the thicket has been disturbed, is probably due to its fire resistance, its low timber-value (it is seldom cut), and because it is used for fence posts which root and grow into trees. The trees generally are thin boled and mostly thin stemmed, with a tendency to branch near the ground. Some trees, *Swietenia mahagoni* and *Picrodendron baccatum* occur with notably short, large-diameter stems 3.6-4.5 m (12-15 ft) and 40-50 cm (16-20 in) d.b.h. *Swietenia* and *Bursera* are truly deciduous, while some of the species lose a high proportion of their leaves in the dry season; the formation however is predominantly evergreen. Leaves are generally microphyllous and coriaceous. The formation occurs on both the deeply dissected Tertiary limestone, and the smoother surfaced Ironshore formation. On the former the trees have long-branched root systems with sprawl over the rocky surface and penetrate into the crevices, but leave much of the rock uncovered. On the Ironshore formation the trees mostly form a dense root mat up to 45 cm (18 in) thick which completely blankets the rock surface. Everywhere the soil is shallow. The ground layer is sparse and frequently absent; ferns may occur locally. Climbers including climbing cacti (*Selenicereus*) are well represented. Epiphytes including orchids (*Oncidium* and *Schomburgkia*) and bromeliads (*Tillandsia* spp.) may be locally abundant. Moss is rare, but occasionally a group of trees may be shrouded with the epiphytic lichen "Spanish moss" (*Ramelina* cf. *Usnea*).

The structure of this formation growing on one of the emergent patch reefs in the Bluff Limestone, crossed by Intersect 4, is shown in Diagram 3. *A key to the symbols employed in this and all subsequent intersect drawings is shown at Diagram 4.*

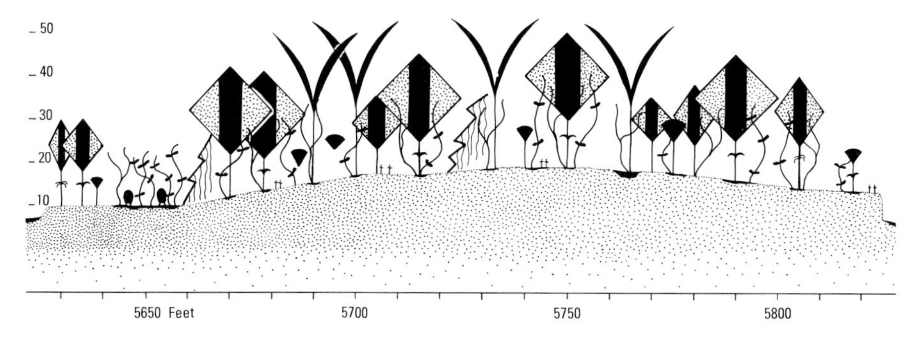

Diagram 3 Dry evergreen thicket on an emergent patch reef in the Bluff Limestone. Intersect 4. (Key: Diagram 4)

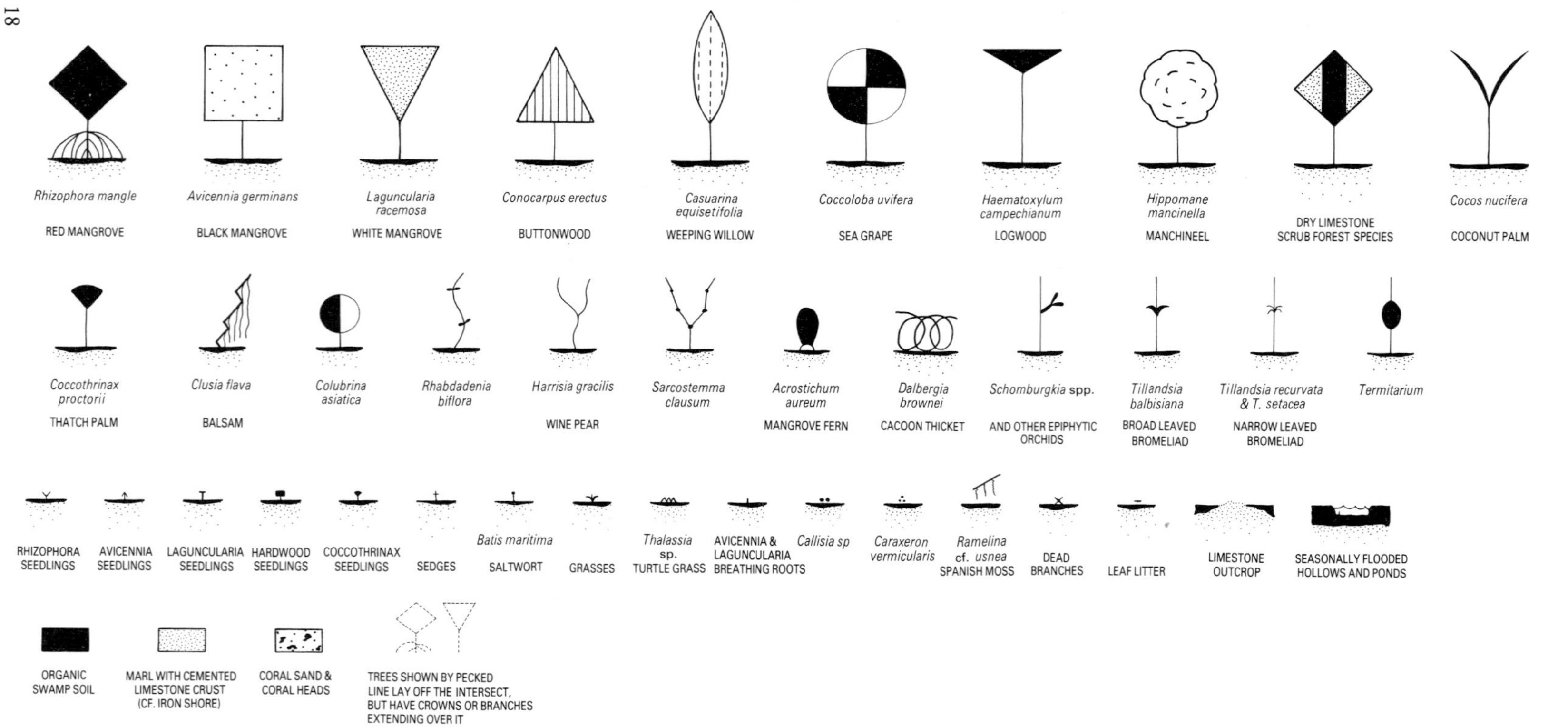

Diagram 4 Key to the symbols used in the intersects

It will be noted that coconut palms occur. Whether these were planted or grew from seed washed through the surrounding mangrove swamps on to these patch reef islands is not certain. The latter explanation seems rather unlikely.

Floristic composition: owing to the long history of cutting and burning, the composition of undisturbed evergreen thicket has not been established. In Grand Cayman (as in Jamaica: Loveless and Asprey, 1957) there does not appear to be a dominant species or group of species, although red birch (*Bursera simaruba*) and Palms, (*Coccothrinax proctorii* and *Thrinax radiata*) may be locally dominant. On Little Cayman, however, Stoddart (*op. cit.*) noted that the formation on the seaward slopes of Bluff Limestone ridges was characterised by three taller species: *Plumeria obtusa, Cephalocereus swartzii* and *Agave sobolifera*; *Guapira discolor* being a common tree, and *Savia erythroxyloides* the most abundant shrub. The details of the floristic composition of this formation on the three Little Cayman sites, and in Grand Cayman generally, are as follows:

Dry evergreen thicket, floristic composition

		Grand Cayman	*Little Cayman*		
Component	*Species*	*Bluff Limestone Ironshore formations*	*Seaward slopes Bluff Limestone ridges*	*Dissected Bluff Limestone ridge*	*Inland slopes of Bluff Limestone ridge*
Trees	Allophylus cominia				x
	Amyris elemifera				x
	Annona squamosa	x			
	Antirhea lucida				x
	Bumelia glomerata				x
	B. salicifolia				x
	Bunchosia media	x			
	Bursera simaruba	x		x	
	Calyptranthes pallens	x			
	Canella winterana	x		x	
	Capparis cynophallophora	x			
	Chionanthus caymanensis				x
	Clusia flava	x			
	Coccothrinax proctorii	x			
	C. gerascanthus			x	
	C. sebestena			x	
	Maclura tinctoria	x			
	Crescentia cujete	x			
	Croton nitens	x			
	Elaeodendron xylocarpum	x			

		Grand Cayman	Little Cayman		
Component	*Species*	*Bluff Limestone Ironshore formations*	*Seaward slopes Bluff Limestone ridges*	*Dissected Bluff Limestone ridge*	*Inland slopes of Bluff Limestone ridge*
Trees (Contd)	Erythroxylum areolatum	x			
	Eugenia axillaris	x			
	Ficus aurea	x			x
	Guapira discolor		x	x	x
	Guettarda elliptica		x	x	x
	Haematoxylum campechianum	x			
	Hippomane mancinella	x			
	Hypelate trifoliata			x	
	Jacquinia keyensis	x			
	Malvaviscus arboreus	x			
	M. v. cubensis	x			
	Myrcianthes fragrans	x	x	x	x
	Myrsine acrantha	x			
	Ocotea coriacea	x			
	Petitia domingensis	x			
	Picrodendron baccatum	x			
	Phyllanthus angustifolius	x			
	Plumeria obtusa		x	x	x
	Psidium guajava	x			
	Psychotria nervosa	x			
	Roystonia regia	x			
	Savia erythroxyloides	x			
	Schoepfia chrysophylloides			x	
	Swietenia mahagoni	x			
	Tabebuia heterophylla	x	x	x	x
	Thrinax radiata	x			
	Tournefortia astrotricha	x			
	Ximenia americana	x			
	Zuelania guidonia				x
Shrubs	Adelia ricinella	x			
	Alvaradoa amorphoides	x			
	Amyris elemifera		x		
	Argythamnia proctorii	x			
	Bauhinia divaricata	x	x	x	x
	Bourreria venosa				x
	Calyptranthes pallens		x	x	
	Capparis ferruginea	x			
	Colubrina asiatica				x
	Cordia brownei	x			
	Croton linearis		x		x
	C. lucidus		x		
	Dalbergia brownei	x			
	Erithalis fruticosa	x	x	x	
	Erythroxylum rotundifolium	x			

		Grand Cayman	*Little Cayman*		
Component	*Species*	*Bluff Limestone Ironshore formations*	*Seaward slopes Bluff Limestone ridges*	*Dissected Bluff Limestone ridge*	*Inland slopes of Bluff Limestone ridge*
Shrubs (Contd)	Evolvulus arbuscula		x		
	Gyminda latifolia				x
	Jacquinia berterii		x	x	x
	Phyllanthus angustifolius			x	
	P. nutans	x	x		x
	Piper amalago	x			
	Randia aculeata	x	x		x
	Euphorbia cassythoides	x			
	Rochefortia acanthorphora	x			
	Savia erythroxyloides		x	x	
	Solanum havanense	x			
	Strumpfia maritima		x		
	Tournefortia volubilis	x		x	
Climbers	Callisia repens (herb)	x			
	Capparis flexuosa	x			
	Chiococca alba	x			
	Peperomia magnoliifolia	x			
	Philodendron sp.	x			
	Smilax havanensis	x			
Succulents	Agave sobolifera	x	x	x	x
	Cephalocereus swartzii		x	x	
	Selenicereus grandiflorus	x			
Epiphytes	Oncidium calochilum	x			
	O. caymanense	x			
	Schomburgkia thomsoniana var. thomsoniana	x			
	S. thomsoniana var. minor			x	x
	Tillandsia balbisiana	x		x	x
	T. recurvata	x		x	x
	T. tenuifolia	x		x	x
	Ramelina (cf. Usnea)	x			
Herbs	Bromelia pinguin	x			
	Fimbristylis dichotoma	x			
	Paspalum blodgettii				x
	Lasiacis divaricata (grass)	x			
	Psilotum nudum	x			
	Rivina humilis	x			
	Salvia occidentalis				x
	Sansevieria hyacinthoides	x			
	Scleria lithosperma	x			
	Spiranthes squamulosa	x			
	Solanum bahamense	x			
Ferns	Polypodium heterophyllum	x			
	P. polypodioides	x			

Ib(i) Dry evergreen littoral thicket and bushland littoral hedge

Beard (1955) retained the term littoral to differentiate dry evergreen vegetation under the influence of the sea. See also Asprey and Loveless (1958). As the littoral thicket usually grades into the bushland littoral hedge Id(i), the latter is included here for ease of reference. This littoral thicket is extensively developed along the eastern portion of the north shore of Grand Cayman. Stoddart (1980) noted its occurrence on the north shore of Little Cayman – calling it *Coccoloba* hedge. Sauer (1982) has recorded a number of occurrences, including an example from the foot of the cliffs at the east end of Cayman Brac.

Structure and physiognomy: the height of this thicket varies from about a metre (2-3 ft) at the seaward edge to 4.5-6.1 m (15-20 ft) at its apex, and has a streamlined form in the direction of the prevailing wind. It seldom exceeds 32 m (150 ft) in width, the shorter part of the formation being bushland. The shrubs are usually dense, interlaced (often impenetrable), distorted in form, with a tendency to stagheadedness and dead branchlet tips.

Floristic composition: on Grand Cayman the dominant and often the sole species present is the sea grape, *Coccoloba uvifera*. *Thespesia populnea, Morinda royoc, Caesalpinia bonduc* and *Jacquinia keyensis* also occur occasionally. The ground surface is usually bare or covered with drying *Coccoloba* leaves. In Little Cayman, the littoral thicket is again virtually monospecific. The seaward edge however may be occupied by a zone, up to 100 m (326 ft) wide, of the shrub *Suriana maritima*. This may be 2 m (6.5 ft) tall. Other shrubs: *Tournefortia* sp., and *Scaevola plumieri* also occur occasionally. Further associated species noted by Sauer (1982) include *Chrysobalanus icaco, Coccothrinax proctorii, Cyperus planifolius, Ficus aurea, Hymenocallis caribaea, Ipomoea violacea* and *Plumeria obtusa*.

Ib(ii) Dry evergreen logwood (Haematoxylum) thicket

Logwood *(Haematoxylum campechianum)* was probably introduced to the Cayman Islands in the late 18th or early 19th century. How extensively it was planted is unknown as it spreads naturally by stolons. Today it covers many acres of land in the central part of Grand Cayman. It was last cut as a source of dye in the depression of the late 1920's. Where it occurs it has largely replaced the Dry evergreen thicket. This formation has not been recorded from either Little Cayman or Cayman Brac, although it may occur on the latter.

Ia Dry evergreen woodland. The Bluff, Cayman Brac (p.13)

Ia Dry evergreen woodland, with emergent Royal Palms *(Roystonia regia)*. Trans-Island Road, Grand Cayman (p.14)

PLATE 1(A) Dry evergreen formation series I

Ib Dry evergreen thicket, with *Cephalocereus swartzii.* Eastern end of Grand Cayman (pp.16,19)

Id(ii) Dry evergreen Cocothrinax palm bushland, with emergent *C. proctorii* "Thatch" palms. Eastern end of Grand Cayman (p.31)

PLATE 1(B) Dry evergreen formation series I

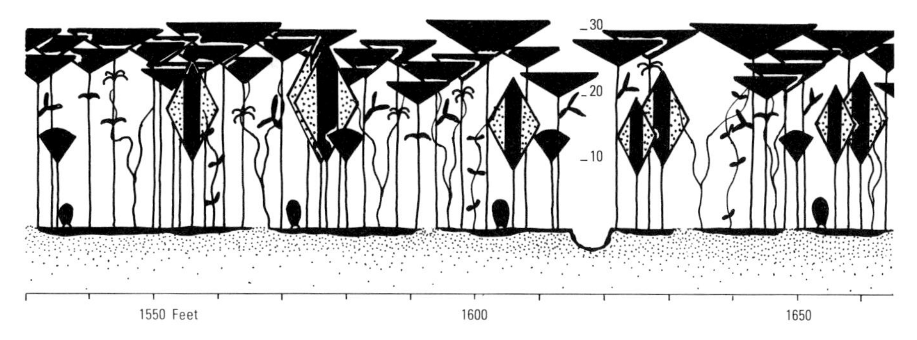

Diagram 5 Dry evergreen logwood *Haematoxylum* thicket. Intersect 2 (Key: Diagram 4)

Structure and physiognomy: the canopy which averages between 6-9 m (20 and 30 ft) in height, is composed almost entirely of logwood (*Haematoxylum campechianum*). The light, obconic tree crowns interfinger, but as their foliage is not dense the thicket floor is only partially shaded. In places a second discontinuous layer composed of either 3.6-4.5 m (12-15 ft) high *Coccothrinax* palms or 1.8-3.0 m (6-10 ft) high shrubs may occur. Creepers, epiphytes including orchids, bromeliads and lichen (Spanish moss) may be locally abundant. The formation is illustrated in Diagram 5, representing an intersect through a section of Logwood thicket encountered on Intersect 2, south of North Sound.

Floristic composition:

Trees:

Capparis cynophallophora
Clusia flava
Coccothrinax proctorii
Elaeodendron xylocarpum
Haematoxylum campechianum
Myrcianthes fragrans
Myrsine acrantha
Picrodendron baccatum
Psidium guajava
Psychotria nervosa
Roystonia regia (seedlings)

Shrubs:

Bauhinia divaricata
Capparis ferruginea

Climbers:

Capparis flexuosa
Chiococca alba
Selenicereus grandiflorus
Smilax havanensis

Herbs:

Bromelia pinguin
Callisia repens
Rivina humilis
Solanum bahamense

Epiphytes:

Schomburgkia thomsoniana	T. tenuifolia
Tillandsia balbisiana	Ramelina sp. (cf. Usnea)
T. recurvata	

Ferns:

Polypodium polypodioides

This list is not exhaustive, being the commonly occurring species found in the Logwood thicket; many of the Dry evergreen thicket species will also occur occasionally.

Ic *Herb grassland (rough pasture) derived from Dry evergreen woodland and thicket*

These rough pastures have been largely established on land cleared of Dry evergreen thicket on the Ironshore formation, in the West Bay, Georgetown, Bodden Town and Trans-Island road areas of Grand Cayman. They are not extensive on Little Cayman, and have not been studied on Cayman Brac. The deeply dissected Tertiary limestone of the eastern part of Grand Cayman is not suitable for pasture establishment. In the past the pastures were regularly burnt to encourage the growth of grasses, and to try and control the invasion of unwanted woody species such as maiden plum *(Comocladia dentata)* and manchineel *(Hippomane mancinella)*. Chemical sprays are now being used for this purpose. Prickly pear *(Opuntia dillenii)* occurs commonly in some of the pastures.

Structure and physiognomy: as the grasslands are grazed for most of the year, the vegetation seldom grows to its maximum height. The majority of the dominant grasses are tufted, with an average height of 45.7-60.9 cm (18-24 in) *(Panicum maximum, Andropogon glomeratus, A. pertusus, Sporobolus domingensis)*, and are interspersed with a large proportion of herbs (many creepers) and woody shrubs. The latter vary from being dwarfed and prostrate to shrubs or even small trees on average 1.2-3.0 m (4-10 ft) high, while some specimens of *Hippomane* and *Canella* may exceed 6.1 m (20 ft), and may in places become so closely spaced as to form small patches of bushland often dominated by one species (e.g. Manchineel or *Coccothrinax* palm bushland). Swabey and Lewis (1947) noted that Guinea grass *(Panicum maximum)* was a rapid coloniser after burning and Seymour grass *(Andropogon pertusus)* "made its appearance (since 1938), and has occupied large areas of pastureland particularly between Georgetown and the Great Sound and in the centre of the island, along the road between Frank Sound and Old Man Bay". During the dry season the pastures become very dry and brown in colour.

Floristic composition:

Grasses and sedges:

Andropogon glomeratus
A. pertusus
Chloris barbata
C. inflata
C. petraea
Cyperus ligularis (sedge)
Eleusine indica
Panicum maximum
Paspalum blodgettii
P. millegrana
Setaria geniculata
Sporobolus domingensis

Herbs:

Argemone mexicana
Bastardia viscosa
Bauhinia divaricata
Bidens alba var. radiata
Blechum brownei
Capraria biflora
Cassia occidentalis
C. nictitans var. aspera
Catharanthus roseus
Chamaesyce hypericifolia
Corchorus siliquosus
Conyza canadensis
Crotalaria retusa
C. verrucosa
Desmodium incanum
D. tortuosum
Emilia fosbergii
Eupatorium villosum
Euphorbia cyathorphora
E. heterophylla
Eustoma exaltatum
Heliotropium angiospermum
H. curassavicum
Lippia nodiflora
Lithophila muscoides
Malachra alceifolia
Melochia pyramidata
Parthenium hysterophorus
Phaseolus lathyroides
Pluchea odorata
Porophyllum ruderale
Portulaca rubricaulis
Priva lappulacea
Ruellia tuberosa
Spermacoce assurgens
Spilanthes urens
Stachytarpheta jamaicensis
Stylosanthes hamata
Synedrella nodiflora
Tridax procumbens
Turnera ulmifolia
Vernonia cinerea
Waltheria indica
Zephyranthes citrina
Z. rosea
Z. tubispatha

Creepers:

Abrus precatorius
Cissus trifoliata (on shrubs)
Citrullus vulgaris
Clitoria ternatea
Cucumis anguria
Ipomoea indica var. acuminata (on shrubs)
I. tiliacea
I. triloba
Jasminum fluminense
Merremia dissecta
Momordica charantia
Passiflora suberosa
Sarcostemma clausum (on shrubs)
Urechites lutea

Shrubs and small trees:

Acacia farnesiana
Calyptranthes pallens
Canella winterana
Casearia aculeata
Cordia globosa
Croton humilis
C. linearis
Eugenia axillaris
Ernodia littoralis
Forestiera segregata
Gossypium barbadensis
Hippomane mancinella
Lawsonia inermis
Leucaena leucocephala
Malpighia cubensis
Morinda royoc
Savia erythroxyloides
Vernonia divaricata

Id *Dry evergreen bushland*

This formation is well developed in the eastern coastal area of Grand Cayman, where the inland thicket formation grades into bushland. The change is probably due to drier conditions and exposure to salty winds. The drier conditions arise from a lower total rainfall, and from higher evaporation due to exposure to the prevailing north-east trade winds. There is little soil; what there is is found in the solution hollows of the bare limestone. On Little Cayman the formation covers much of the western end of the island on moderately dissected terrain (Stoddart 1980). The formation has not been studied on Cayman Brac.

Structure and physiognomy: in Grand Cayman much of this formation is no higher than 1.5-1.8 m (5-6 ft). This may in part be due to coppicing as a result of past clearing for cultivation, or cutting for firewood. Where the bushland has clearly established itself on old agricultural land, occasional trees (mostly birch, *Bursera simaruba*) rise above the canopy level to a height of 4.5-6.1 m (15 or 20 ft). In some cases these are probably old fence posts which have grown. Climbers and epiphytes are frequent. The herb layer is generally not well developed, but herbs and grass do occur in openings in the bushland. Stoddart (1980) noted that the bushland on Little Cayman had also been much modified by grazing and cultivation, *Plumeria, Agave* and *Bursera* being present; shrubs noted included *Jacquinia berterii, Evolvulus arbuscula* and *Strumpfia maritima,* with clumps of the fern *Acrostichum aureum* in solution hollows. The formation is only some 2-3 m (6-10 ft) above sea level on Little Cayman, compared with Grand Cayman where it is over 9.1 m (30 ft) a.m.s.l.

Floristic composition: the following notes refer to Grand Cayman only.

Trees:

Antirhea lucida
Annona glabra
Bourreria venosa
Bursera simaruba
Calyptranthes pallens
Canella winterana
Elaeodendron xylocarpum
Euphorbia tirucalli
Guettarda eliptica
Haematoxylon campechianum
Hypelate trifoliata
Savia erythroxyloides
Spondias purpurea
S. dulcis
Thrinax radiata

Shrubs:

Amyris elemifera	Erythroxylum areolatum
Caesalpinia bonduc	Helicteres jamaicensis
Catesbaea parviflora	Lantana involucrata
Cordia gerascanthus	Phyllanthus nutans
Colubrina cubensis	P. angustifolius
Capparis flexuosa	Schaefferia frutescens
C. cynophallophora	Salmea petrobioides
Dalbergia brownei	Solanum bahamense
Ernodia littoralis	Trema lamarckianum
Erithalis fruticosa	Turnera ulmifolia

Creepers:

Echites umbellata (on trees)	Morinda royoc
Jacquemontia havanensis	Tournefortia volubilis

Epiphytes:

Schomburgkia thomsoniana	Tillandsia setacea
Tillandisa flexuosa	T. utriculata

Herbs:

Eupatorium villosum	Melochia tomentosa
Euphorbia cyathophora	Physalis angulata
Heliotropium angiospermum	Portulaca rubricaulis
Lasiacis divaricata (grass)	Rivina humilis

Id(i) Dry evergreen bushland littoral hedge (See section Ib(i))

Id(ii) Dry evergreen bushland on sandy and pebbly beach ridges and coastal sand areas, and associated coastal communities

This formation, although similar to Dry evergreen littoral thicket is shorter, is not wind sheared, and is therefore recognised separately. It is often associated with one or more minor communities occurring to seaward or landward. These have not all been separated on Grand Cayman. Stoddart (1980) however included them in a purely empirical scheme specific to Little Cayman:

Strand community on sandy beach
Strand community on cobble beach
Coastal sand flats
Rock pavement community
Steep rocky coast community

The first of these includes "Littoral Hedge" which is here included in the description of Dry evergreen littoral thicket: Ib(i). Rock pavement vegetation and the steep rocky coast community are discussed at Ie. Data on the other three communities for Little Cayman are included in Table 3 at the end of this section for ease of reference.

Dry evergreen bushland occurs along certain parts of the coast of Grand Cayman, where behind the present day strandline there is a sandy beach ridge, formed by past storm action, which may include a fair proportion of pebbles and old coral heads. The bushland growing on these ridges is distinctive and may be dominated by the seagrape-almond association (*Coccoloba uvifera* – *Terminalia catappa*). The latter although an introduced species is now firmly established around the coast, and must be regarded as a component of the natural vegetation. Other important species noted by Sauer (1982) include *Caesalpinia bonduc*, the creeper *Canavalia maritima* and the two shrub's *Chrysobalanus icaco*, and *Colubrina asiatica*, which are not found in other beach habitats. As these ridges are narrow, and as much of the coastal area is being developed and the ridges are excavated for building material, this community has been somewhat disturbed. Behind these relict beach ridges there is usually a further area of sand, which may give way to the Ironshore formation, be backed by mangrove swamp, or be replaced by the Bluff Limestone formation. This coastal sand area is nowhere very extensive, the land behind West Bay Beach being the largest area. Much of it now supports a mosaic of scattered patches of bushland interspersed with areas of herbaceous scrub land.

Structure and physiognomy: the Dry evergreen bushland formation has a 3.0-3.9 m (10-13 ft) high canopy composed predominantly of *Terminalia* and *Coccoloba*, which may include other species such as *Bursera* and *Calyptranthes*. Sauer (1982) notes that this formation on the beach ridges is usually 1-2 m (3.2-6.5 ft) taller than the same formation growing on limestone. This canopy is now seldom seen intact. Emergents above the canopy include *Terminalia* and *Casuarina*; the latter is also an introduced species which has been both planted and has established itself in numbers along the sandy part of the coast. The broken nature of much of this formation has resulted in the shrub layer usually being better developed than is normal in evergreen thicket formations. The edges of the formation, and areas within which lack tree or shrub cover usually have a ground cover of herbs, many of them creepers, and grass.

Floristic composition:

Trees:

Bourreria venosa
Bursera simaruba
Casearia aculeata
Casuarina equisetifolia
Calyptranthes pallens
Coccoloba uvifera
Ficus sp.
Eugenia axillaris
Gyminda latifolia
Leucaena leucocephala
Morinda citrifolia
Pouteria campechiana
Terminalia catappa
Trema lamarckianum
Thespesia populnea
Thrinax radiata
Yucca aloifolia

Shrubs:

Agave sp.
Argusia gnaphalodes
Bourreria cf. venosa
Cassia clarensis
Chrysobalanus icaco
Colubrina asiatica
Crossopetalum rhacoma
Cryptostegia grandiflora
Erithalis fruticosa
Heliotropium ternatum
Iva cheiranthifolia
Scaevola plumierii
Suriana maritima
Turnera ulmifolia

Creepers:

Caesalpinia bonduc
Canavalia rosea*
Ipomoea pes-caprae
  ssp. brasiliensis*
I. triloba*
I. violacea*
Spilanthes urens*
Tribulus cistoides*

(* = herbaceous species)

Herbs and grasses:

Arundo donax
Anthephora hermaphrodita
Boerhavia erecta
Capraria biflora
Cenchrus echinatus
C. tribuloides
Chamaesyce hirta
C. mesembrianthemifolia
Chloris petraea
Commelina elegans
Cyperus brunneus
C. peruvianus
Desmodium tortuosum
Eragrostis ciliaris
E. domingensis
E. tenella
Hymenocallis latifolia
Malvastrum coromandelianum
Melochia tomentosa
Panicum maximum
Paspalum blodgettii
Pectis caymanensis
Pennisetum purpureum
Portulaca oleracea
Ruellia brittoniana
Russelia equisetiformis
Salvia serotina
Sansevieria hyacinthoides
Sida stipularis
Spigelia anthelmia
Stemodia maritima
Stenotaphrum secundatum
Stylosanthes hamata
Vigna luteola

Many of the herbaceous creepers and shrubs occur on the margins of the formation, and on the adjacent sandy and pebbly areas, as sparse scrub or intermittent ground cover.

The details of three minor communities on Little Cayman mentioned at the beginning of this section are summarised in Table 3:

Table 3 Association Id(ii). Floristic composition of three minor communities on Little Cayman

Vegetation community sites and site details		
Strand community on sandy beaches (excluding "Littoral hedge": see Ib(i)	*Strand community on cobble beaches*	*Coastal sand flats*
Low, narrow beaches protected by offshore reefs	*Frequently protected from the sea by more recent sandy beach deposits*	*50-100 m average width, never more than 200 m; less than 2.5 m a.m.s.l. the flats extend inland from beach ridge crest: on N and S shores*
Pioneer species: (a) *shrubs* Scaevola, Argusia, Suriana, Caesalpinia, Cordia (occasionally); (b) *herbs/creepers* Cakile lanceolata, Sesuvium portulacastrum, Sporobolus virginicus; and further inland: Ipomoea violacea, Chamaesyce mesembrianthemifolia, Hymenocallis latifolia and Cenchrus *Climax species: shrubs* (1-2 m (3.2-6.5 ft) tall) Suriana maritima (dominant) Argusia sp. (occasional) Scaevola plumieri (occasional) These communities often occupy the beach between the "Littoral hedge" and the strand line	*Seaward sandy deposits*: (a) *shrubs* Suriana maritima; (b) *herbs/creepers* Ambrosia hispida Ipomoea violacea Sesuvium portulacastrum Hymenocallis latifolia Borrichia arborescens *Cobble beaches: shrubs* Coccoloba uvifera Morinda citrifolia Thespesia populnea Hymenocallis latifolia	The *climax vegetation* was Littoral woodland (see Ia(i) for details), now largely replaced by 2-5 m (6.5-16.4 ft) high secondary scrub: *Trees and shrubs* Caesalpinia Chrysobalanus icaco Coccoloba uvifera Colubrina asiatica Comocladia dentata Conocarpus erectus Cordia Lantana *Herbs and grasses* occupy more opened ground: Abrus precatorius Ambrosia hispida Bryophyllum pinnatum Canavalia rosea Cassytha filiformis Cyperus spp. Ernodea littoralis Ipomoea pes-caprae Portulaca oleracea Sporobolus virginicus Stachytarpheta

Id(iii) Dry evergreen Coccothrinax palm bushland

At the eastern end of Grand Cayman the ordinary dry evergreen bushland differs in having a layer of emergent *Coccothrinax proctorii* palms, 4.5-9.1 m (15-30 ft) high. Elsewhere the dry evergreen bushland has been cleared, but the thatch palm *(Coccothrinax proctorii)* has been left, as the leaf fibre was used for rope making. The remaining palms which seem to be fire resistant have multiplied to form patches of dense palm-dominant bushland.

Id(iv) Dry evergreen manchineel (Hippomane) bushland

Manchineel *(Hippomane mancinella)* is a component of both the evergreen thicket and bushland formations. It has a very caustic sap, and when land is cleared it alone is often left growing. It appears to be fire resistant, and thrives in the open. Thus left in the pastures, manchineel seeds effectively and patches of pure manchineel bushland develop. The ground cover within these patches is indistinguishable from the rough pastures derived from the evergreen woodland and thicket, to which section reference should be made for details.

Ie *Dry evergreen vegetation on rock pavements*

Around part of the coast of Grand Cayman and Little Cayman, and most of the coast of Cayman Brac there is a raised beach called the Ironshore formation. This consists of coral marl with a fissured cemented surface. It is windswept and exposed to salt spray. The vegetation which grows under these inhospitable conditions is an open community of herbs and specialised dwarf shrubs varying in height from 10 cm (4 in) to 0.6-0.9 m (2-3 ft). Asprey and Loveless (1958) described a similar community in Jamaica and noted the dwarfing effects of the habitat: Seagrape *(Coccoloba)* and buttonwood *(Conocarpus)* which will both grow to over 9 m (30 ft) under optimum conditions, seldom exceed 30 cm (12 in) in this environment, forming a "prostrate mat of sprawling branches". Leaves are generally small, leathery and evergreen. There is no structure in the normally accepted sense of the word; the plants grow in cracks and fissures in the rock surface where they can establish themselves, leaving at least 25% of the surface uncovered. The seaward edge of this community is usually dominanted by the herb *Sesuvium portulacastrum*, or the shrubs *Rhachicallis americana* and *Borrichia arborescens* (Sauer 1982).

Floristic composition:

Shrubs:

Atriplex sp.
Borrichia arborescens
Jacquinia keyensis
Rhachicallis americana

Shrubs: (Contd)

Coccoloba uvifera
Conocarpus erectus
Erithalis fruticosa
Sansevieria hyacinthoides
Strumpfia maritima
Suriana maritima

Herbs:

Chamaesyce mesembrianthemifolia
Caraxeron vermicularis
Pilea microphylla
Sesuvium portulacastrum

Grass:

Cenchrus tribuloides
Paspalum distichum
Sporobolus virginicus

The *Pilea* is seldom found on the true rock pavement, but is common in the cracks of the limestone cliffs on the north-east coast of Grand Cayman. Similar rock pavements (the same geological formation) occur inland shielded from wind and salt spray, *Coccoloba* and *Conocarpus* are therefore absent. The following species were noted: *Borrichia arborescens, Crinum* sp., *Hippeastrum puniceum, Jacquinia keyensis, Paspalum blodgettii, P. vaginatum.*

Stoddart (1980) noted the following zonation from sea to land on a 50 m (164 ft) broad pavement standing about 2.5 m (8.2 ft) a.m.s.l. at West End Point, Little Cayman.

Outer zone 0-15/20 m: (0-49/65 ft)	(1)	Scattered patches of *Sesuvium, Caraxeron, Cenchrus* and *Sporobolus* in depressions
Inner zone 15/20-50 m:	(2)	Fleshy mat of *Sesuvium*
(49/65-164 ft)	(3)	Dwarfed prostrate shrubs: *Conocarpus, Borrichia, Rhachicallis americana, Strumpfia maritima*
	(4)	Transitional zone with the above shrubs plus *Suriana maritima*
	(5)	Inner zone of 3-4 m (9.8-13.1 ft) tall thick bushy *Suriana maritima*
	(6)	A hedge of dense *Coccoloba uvifera* at the inner edge of the pavement, where it passes beneath sand or cobble beach

He also described (Table 4) a related community of four zones growing on the steep, deeply dissected, rocky Bluff Limestone coast at the east end of Little Cayman.

Table 4 Formation Ie, Little Cayman: vegetation communities on the steep rocky East coast (Stoddart, 1980)

Zone	*Height a.m.s.l. m (ft)*	*Distance from the shore m (ft)*	*Vegetation community*
1	0-2/3 (0-6/9.8)	0-3 (0-9.8)	Lower splash zone: only patches of *Sesuvium* in potholes
2	2/3-6 (6/9-19)	3-15 (9.8-49.0)	Prostrate scrub of dwarf shrubs: *Conocarpus, Borrichia, Strumpfia maritima*
3	6-7 (19-22)	15-30 (49.0-98.4)	*Suriana maritima* extensive, with *Capparis flexuosa, Erithalis fruticosa* and *Phyllanthus angustifolius*
4	+ 7 (+ 22)	30-70/80 (98.4-229/262.4)	Low *Coccoloba uvifera* scrub, with trees of *Ficus aureus* growing in potholes
5	+ 7 (+ 22)	+ 70/80 (+229/262.4)	Bushland of *Bursera, Thespesia, Cordia, Plumeria, Thrinax, Cephalocereus*. Emergent trees of *Bursera* deformed into an L shape

For further details of these rock pavement communities see Sauer's (1982) analysis of transect data for his Output Zone on Limestone.

II SEASONAL SWAMP FORMATION SERIES*

These swamps which are seasonally flooded vary from brackish swamps dominated by buttonwood *(Conocarpus)* to various freshwater swamps. In Grand Cayman the landward fringe of the mangrove swamp is often dry for part of the dry season, and could have been included with the seaonal swamp formations. However on the one hand the limit of seasonal flooding *within* the mangrove swamps is difficult to determine and varies from year to year, and on the other hand the centre of some of the *Conocarpus* swamps may remain flooded throughout wet years. On practical grounds the mangrove swamps have been excluded from the seasonal swamp formation series.

---

* Readers interested in the detailed distribution of these swamp formations should refer to the maps by Brunt and Giglioli (1980)

IIa *Seasonal Conocarpus (buttonwood) swamp thicket*

Seasonal *Conocarpus* swamp thicket occupies a number of extensive inland sites, where there are depressions in the ground (limestone) surface. The water in these swamps is brackish to fresh, and although they may be inland they, along with other swamps, presumably have subterranean connections with the sea. There are good examples about one mile to the south east and east of Old Man Village near the north coast of Grand Cayman. Elsewhere *Conocarpus* often occurs on the landward fringe of the mangrove swamps, or it may occur in patches within the mangrove swamps, where there is a slight elevation in the swamp sufficient to reduce the frequency of salt water flooding, enabling *Conocarpus* to establish itself. A good example of this from Intersect 2 is shown in Diagram 6.

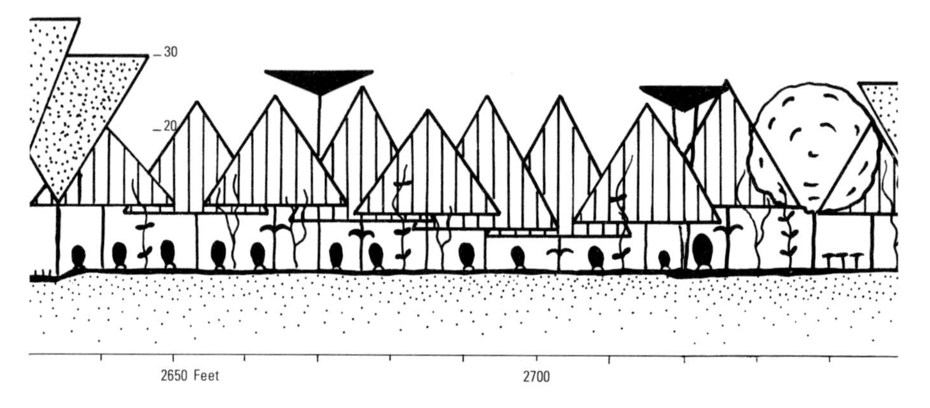

Diagram 6 Seasonal *Conocarpus* swamp thicket. Intersect 2 (Key: Diagram 4)

This section of Seasonal *Conocarpus* swamp thicket abuts *Avicennia* Swamp Woodland; separating the two is a narrow band, often no more than a tree or two wide of *Laguncularia racemosa*.

At the eastern end of Grand Cayman, further areas of Seasonal *Conocarpus* swamp thicket occur at an elevation above high tide level. They depend on rainfall and are not brackish.

Structure and physiognomy: canopy height varies between 6-9 m (20 and 30 ft); average crown diameter is 1.5-6.0 m (15-20 ft). Stem size averages 10-15 cm (4-6 in) d.b.h., although multiple stem trees are not uncommon, with three or more stems averaging 5-7 cm (2-3 in) d.b.h.; stems per ha average 978 (436 per ac). Emergents seldom occur, although well established mangrove species may occasionally tower above the canopy. There is no shrub layer, but a well developed herbaceous layer, usually composed of the giant mangrove fern *(Acrostichum aureum)*, is often present. These may be 1.5-1.8 m (5-6 ft) in height with a 1.8 m (6 ft) spread, growing on mounds 30-45 cm (12-18 in) high, and occurring every 0.6-1.2 m (2-4 ft). The swamp floor is also usually littered in places with dead branches, and some *Conocarpus* trunks may grow parallel to the ground for as much as 6 m (20 ft) before sending up vertical shoots. Climbers which ascend into the canopy, and bromeliads, are common.

Floristic composition: The composition is very restricted, *Conocarpus erectus* having total dominance; very occasionally an odd tree from the Dry evergreen woodland manages to establish itself.

Trees:

Conocarpus erectus	Hippomane mancinella
Haematoxylum campechianum	Laguncularia racemosa

Climbers:

Rhabdadenia biflora	Selenicereus grandiflorus

Epiphytes:

Tillandsia balbisiana	T. recurvata

Ferns:

Acrostichum aureum

IIb *Seasonal Conocarpus (buttonwood) swamp bushland*

Seasonal Conocarpus swamp thicket may become dwarfed to form bushland. This often happens towards the middle of the swamp. The bushland may occur on patches of higher ground which are less frequently flooded – for example in the South Sound Swamp, see Diagram 7, where it is surrounded by broken *Laguncularia* woodland.

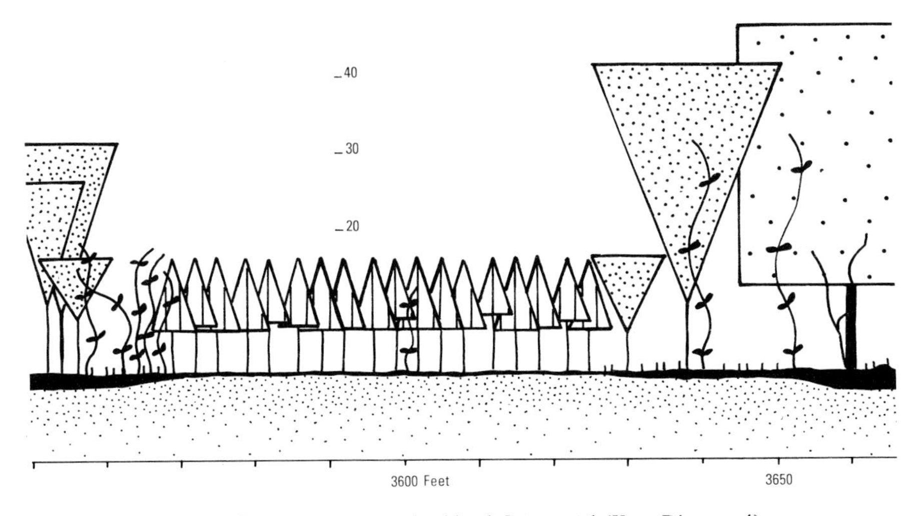

Diagram 7 Seasonal *Conocarpus* swamp bushland. Intersect 1 (Key: Diagram 4)

Small patches may surprisingly also occur in the middle of *Rhizophora* mangrove swamp growing on swamp soil more than 3 m (10 ft) deep, suggesting that a change in the flooding regime in this part of the swamp has occurred, allowing the establishment of *Conocarpus*. This is illustrated in Diagram 8, of part of the Central Mangrove Swamp of Grand Cayman.

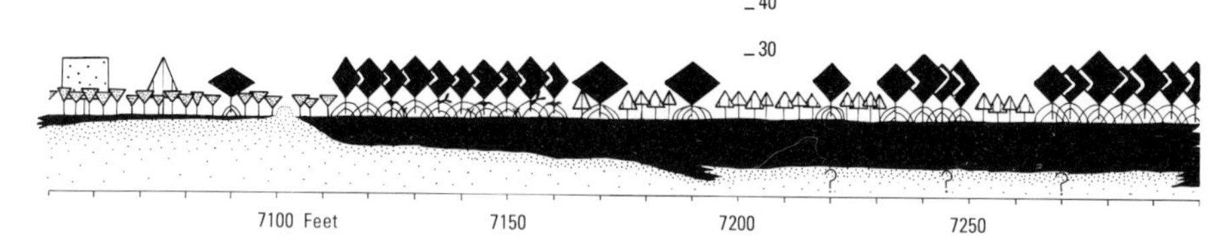

Diagram 8 Pockets of Seasonal *Conocarpus* swamp bushland in *Rhizophora* mangrove swamp. Intersect 4 (Key: Diagram 4)

A further variant (Diagram 9) which also grows on deep soil in the same swamp, consists of 1.8 m (6 ft) high *Conocarpus* bushland, with emergent 3.6-4.5 m (12-15 ft) high *Avicennia* trees scattered singly, 6-15 m (20-50 ft) apart, through the swamp. Sedges and ferns may be locally common, climbers are frequent, and there is usually an appreciable layer of leaf litter.

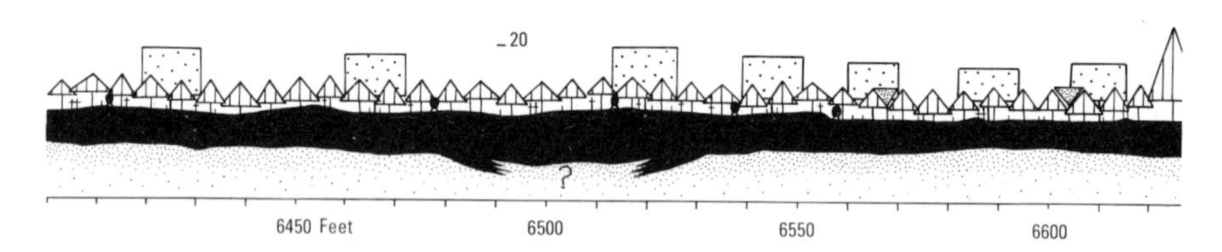

Diagram 9 Seasonal *Conocarpus* Bushland with scattered emergent *Avicennia*. Intersect 4 (Key: Diagram 4)

Stoddart (1980) also noted large areas of dense, twisted *Conocarpus* and *Laguncularia* on low-lying Bluff Limestone in the interior of Little Cayman, with *Erithalis* and *Strumpfia* occurring in the localities marginal to these swamps.

Structure and physiognomy: canopy height varies on average between 1.8-3.0 m (6 and 10 ft), although well developed bushland may reach 4.5 m (15 ft). Average crown diameter is small and falls between 1.2 and 1.8 m (4 and 6 ft), the branches of each crown intricately intergrown with its neighbours. Stems are short in length, seldom exceeding a metre in height, or over 5.0 cm (2 in) in diameter, and often very twisted. The stem population averages 2 700 per ha (1 200 per ac) although cases of 6 075 per ha (2 700 per ac) have been recorded. Individual bushes may grow out of low mounds standing some 30 cm (12 in) above the average level of the swamp floor. The bushes also tend to have a distinct cone of "prop roots", extending 45-60 cm (18-24 in) up the trunk, and occupying a similar diameter at ground level. In drier parts of the swamp 0.9 m (3 ft) high clumps of the sedge *Cladium jamaicense* with flowering spikes reaching 1.8 m (6 ft) or more may be locally common, and associated with 0.9 m

(3 ft) high clumps of the fern *Nephrolepis multiflora*. Occasional clumps of the mangrove fern *Acrostichum aureum* also occur.

Floristic composition: The canopy is virtually monospecific, *Avicennia* or *Laguncularia* only occurring occasionally.

Trees:

Avicennia nitida

Shrubs:

Conocarpus erectus
Laguncularia racemosa
Myrica cerifera

Creepers:

Rhabdadenia biflora
Sarcostemma clausum

Herbs:

Cladium jamaicense
Eleocharis sp. aff. minima
Spigelia anthelmia

Ferns:

Acrostichum aureum
Blechnum serrulatum
Nephrolepis multiflora

IIc *Seasonal herbaceous swamp*

Six herbaceous swamp types have been distinguished, two of which occupy land that was probably originally *Conocarpus* swamp. The others are either grassland or bulrush swamp. All occupy seasonally flooded depressions in the limestone surface, and some occur in zonal succession, reflecting varying lengths of seasonal flooding. Some of the swamps are flooded with brackish or salt water.

IIc(i) Seasonal herbaceous grassland swamp derived from Conocarpus (buttonwood) swamp thicket and bushland

This community often occurs next to *Conocarpus* swamp thicket, and *Conocarpus* stumps are frequently found in these swamps. The main sites, saucer-shaped depressions in the limestone, occur in the West Bay area of Grand Cayman, and without exception have shallow soil. Salt flush may also occur.

Structure and physiognomy: short grass-sedge-herb swamp with occasional shrubs and taller herbs. The dominant ground cover consists of short sedges (usually *Eleocharis mutata*) with grass (*Paspalum vaginatum*) and herbs, often the 10-15 cm (4-6 in) high prostrate *Lippia* and *Philoxera*; the latter may be dominant over large areas. The sedge/grass components may grow to 91 cm (36 in) or more, but, due to grazing and/or trampling, this seldom happens. Isolated 3.0-3.6 m (10-12 ft) high *Conocarpus* bushes, or their stumps and/or clumps of the mangrove fern (*Acrostichum aureum*) may occur.

Creepers sometimes trail over the swamp surface. Clumps of the tall bulrush (*Typha*) and the associated herb (*Sagittaria*) have also been recorded.

Floristic composition:

Shrubs:

Conocarpus erectus

Creepers:

Rhabdadenia biflora
Sarcostemma clausum

Herbs:

Centella asiatica
Lippia nodiflora
Caraxeron vermicularis
Portulaca oleracea
Sagittaria lancifolia

Grasses and sedges:

Eleocharis mutata
Paspalum vaginatum
Typha domingensis

Ferns:

Acrostichum aureum

IIc(ii) Seasonal undifferentiated herbaceous swamps, probably derived from Conocarpus (buttonwood) swamp woodland and thicket

These swamps nearly all occur in shallow depressions in the limestone (usually Ironshore formation) surface. On floristic grounds they seem to occupy a place between the brackish *Conocarpus* (buttonwood) swamps and the fresh water *Typha* (bulrush) swamps.

Structure and physiognomy: Short grass-herb-sedge swamps, with occasional shrubs and taller herbs. The dominant ground cover consists of short sedges (*Eleocharis mutata*), grass (*Paspalum vaginatum*) and herbs, again the prostrate *Lippia* and *Caraxeron* are common. Clumps of the mangrove fern (*Acrostichum aureum*) are uncommon, but creepers (*Rhabdadenia* and *Sarcostemma*) on the swamp surface are common. The swamps very often border ponds which in dry weather may evaporate, exposing the muddy bottom. Where the swamps fringe *Typha* swamps, the transition is often marked by a zone of the sedge *Eleocharis mutata*.

Floristic composition:

Sedges:

Fimbristylis ferruginea
Eleocharis mutata
Cyperus brunneus and planifolius
Rynchospora stellata

Grasses:

Sporobolus virginicus
Paspalum vaginatum
Stenotaphrum secundatum
Leptochloa fascicularis
Panicum purpurascens

Herbs:

Lippia nodiflora
Caraxeron vermicularis
Wedelia trilobata
Vigna luteola
Ammannia latifolia
Sesuvium sp.
Spilanthes urens
Ipomoea sp.
Sternodia maritima
Ruellia tuberosa
Phaseolus lathryoides
Portulaca oleracea

Creepers:

Rhabdadenia biflora
Sarcostemma clausum

Ferns:

Acrostichum aureum

The richness of this flora compared to that of the herbaceous swamps derived from *Conocarpus* will be noted. Marginal to these swamps, 1.2-1.5 m (4-5 ft) bushes of *Iva cheiranthifolia* will often be found. A number of the commoner species *Paspalum, Lippia* and *Caraxeron* also occur, though less frequently than in the rough pastures surrounding these seasonally flooded depressions.

IIc(iii) Seasonal undifferentiated brackish herbaceous swamps

This unit includes divergent communities, which have been grouped together for convenience. The first three which occur on the landward side of mangrove swamps near Barkers are seasonally flooded by brackish water. They lie to the east of Sea Pond, next to Vulgunners Pond and west of Sea Pond respectively. These ponds dry out in the dry season. The first two consist of the 30-38 cm (12-15 in) high, succulent herb *Batis maritima*. In the third, the *Batis maritima* zone which fringes a discontinuous belt of *Avicennia* mangrove and *Conocarpus*, gives way to a series of discrete monospecific herbaceous zones:

Sesuvium portulacastrum 10 cm (4 in) high succulent herb
Fimbristylis cymosa ssp. *spathacea* 30 cm (12 in) high sedge
Andropogon pertusus 45 cm (18 in) high grass

This swamp is flooded by rain water, with subterranean contamination by salt water.

The fourth site NW of Vicksville, is a shallow depression in the limestone surface, surrounded by Dry evergreen thicket. The centre is a mud bottomed pond, which dries up and cracks in the dry season. The mud surface is covered by an algal film. The pond is surrounded firstly by a seasonally flooded 76 cm (30 in) high grass zone of *Distichlis spicata*; secondly by occasional 1.2 m (4 ft) high clumps of the mangrove fern *Acrostichum aureum*, the woody creeper *Rhabdadenia biflora* and 0.9 m (3 ft) high clumps of the shrub *Iva cheiranthifolia*.

IIc(iv-vi) Seasonal Panicum grassland, Eleocharis sedge and Typha bulrush swamps

There are a number of seasonally flooded fresh water herbaceous swamps. The swamp centres, frequently dominated by the bulrush *Typha domingensis*, are often surrounded successively by zones of the sedge *Eleocharis mutata*, followed by the grass *Panicum purpurascens*. These are described below.

IIc(iv) Seasonal Panicum Grass Swamp

These swamps occupying shallow depressions in the limestone surface are flooded by rainfall and runoff from the surrounding land. They occur in the West Bay area, near Georgetown, the Trans-Island Road, and along the south coast. The drier depressions are occupied by *Panicum purpurascens* alone, which if ungrazed can grow to a height of 3 m (10 ft). The centre may be a pond, which dries up in the dry season. Increasingly moist conditions are reflected by the presence of herbaceous species in the swamp centre, including the 1.5 m (5 ft) high *Ludwigia erecta* and 1.8 m

IIa Seasonal Conocarpus (Buttonwood) swamp thicket. Near Bodden Town, Grand Cayman (p.34)

IIc(vi) Seasonal Typha bulrush swamp. Georgetown area, Grand Cayman (p.41)

PLATE 2 Seasonal swamp formation series II

(6 ft) *Typha domingensis*, with *Panicum purpurascens* being progressively replaced by the sedge *Eleocharis mutata*. The *Typha* may dominate the swamp centre, while the introduced water hyacinth (*Eichhornia crassipes*) has become established at the centre of one swamp.

IIc(v) Seasonal Eleocharis sedge swamp

These swamps also occur in shallow depressions in the limestone surface, mainly in the West Bay and Georgetown areas. Here the 45-60 cm (18-24 in) high sedge *Eleocharis mutata* occurs as the dominant of either (a) the whole depression, with or without the prostrate creeper *Lippia nodiflora*, tangles of the creeper *Sarcostemma clausum*, and occasional clumps of the grass *Paspalum vaginatum*; or (b) a zone intermediate between the *Typha* bulrush swamps, which frequently occupy the centre of these shallow depressions, and the surrounding outer zone. The outer zone is frequently occupied by *Panicum* grassland swamp; by the 30 cm (12 in) high sedge *Rynchospora stellata* and the fleshy herb *Centella asiatica* occurring as co-dominants, or by the 7-12 cm (3-5 in) high *Lippia nodiflora*.

Eleocharis can also withstand considerable flooding, and may occur in ponds which seldom if ever dry out. It also grows in association with *Conocarpus erectus, Sarcostemma clausum* and *Rhabdadenia biflora* under brackish conditions.

IIc(vi) Seasonal Typha bulrush swamp

These swamps found in depressions in the Ironshore formation are restricted to the West Bay, Georgetown and Trans Island road areas of Grand Cayman; they do not occur in Little Cayman.

They vary from small, roughly circular depressions, 6-9 m (20-30 ft) in diameter, to elongated depressions measuring 91 x 251 m (300 ft x 825 ft), with the swamp surface no more than 45-60 cm (18-24 in) below the level of the surrounding land. The depth of soil averages 30 cm (12 in), although 91 cm (36 in) deep pockets occur occasionally (see Diagram 10). They are mostly waterlogged throughout the year. In the dry season the swamp surface may become dry and salt flushed.

The *Typha* bulrush*, average height of 3.0-3.6 m (10-12 ft) may cover large areas, with dead *Typha* plants forming a tangle some 1.2 m (4 ft) high, which completely covers the swamp surface. The herb *Sagittaria lancifolia* often grows in this tangle. Patches of 1.8-3.0 m (6-10 ft) high mangrove fern (*Acrostichum aureum*), or 45 cm (18 in) high *Eleocharis* sedge, or the 1.2 m (4 ft) high herb *Polygonum glabrum*, or tangles of the creeper

---

*Also known as "Cat-tail" and "Rush"

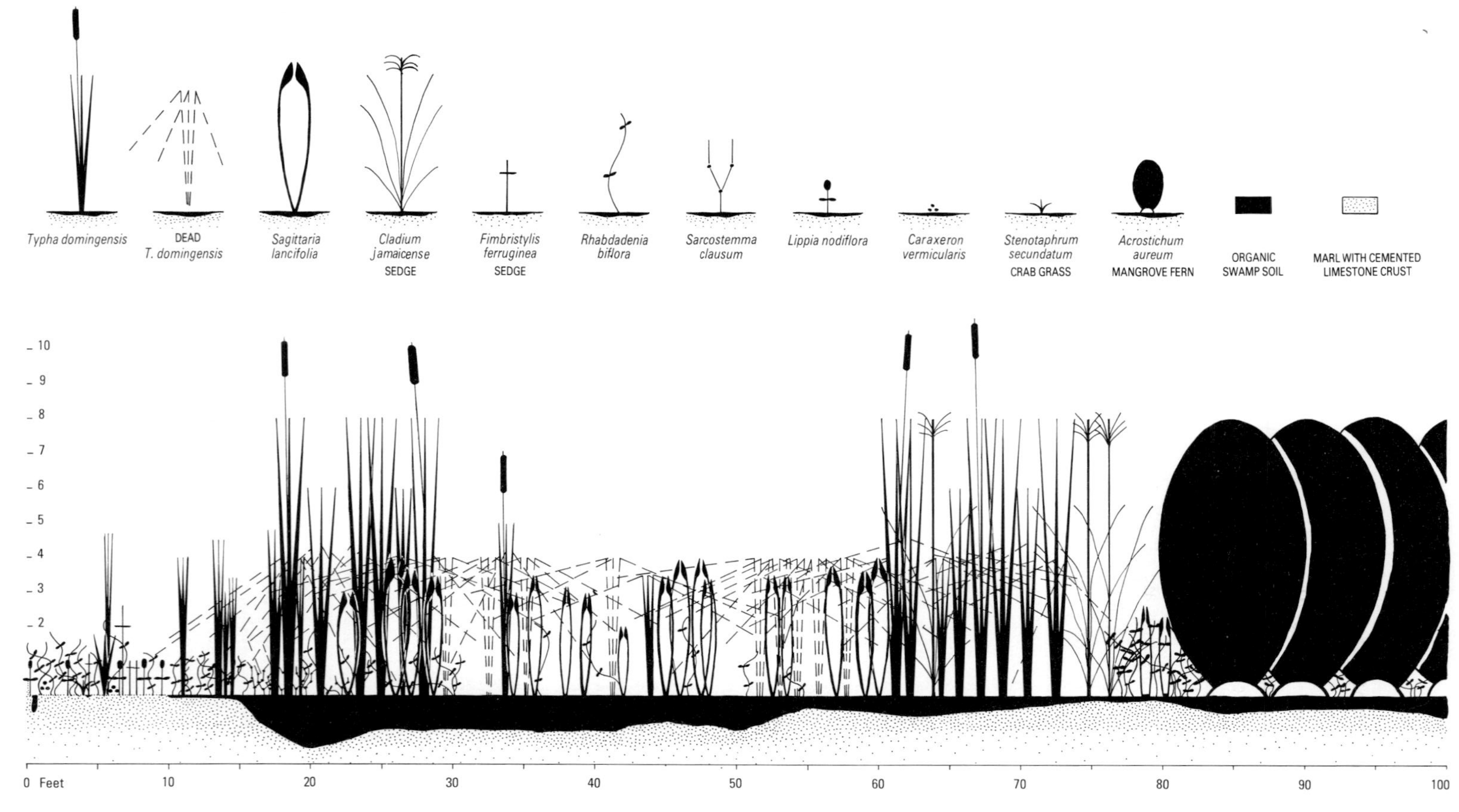

Diagram 10 Seasonal *Typha* bulrush swamp. Measured intersect from the dryland edge towards the centre of the swamp: Half Way Pond, Georgetown.

Sarcostemma clausum, or the 15 cm (6 in) high prostrate herb *Caraxeron vermicularis* may replace the *Typha*. There may also be a grass ground layer of 15-20 cm (6-8 in) high *Paspalum vaginatum* or *Panicum purpurascens*. Creepers, both *Sarcostemma* and *Rhabdadenia* are common. In places the mud surface may be exposed, and is often covered by an algal scum.

These swamps are of simple composition, large areas being dominated by one or two species.

Floristic composition:

Bulrush:

Typha domingensis

Sedges:

Cladium jamaicense	Eleocharis mutata
Cyperus brunneus or planifolius	Fimbristylis ferruginea
Cyperus sp.	

Ferns:

Acrostichum aureum	Thelypteris interrupta

Herbs:

Centella asiatica	Polygonum glabrum
Lippia nodiflora	Sagittaria lancifolia
Caraxeron vermicularis	

Grasses:

Panicum purpurascens	Paspalum vaginatum

Creepers:

Ipomoea hederifolia	Sarcostemma clausum
Rhabdadenia biflora	

III SWAMP FORMATION SERIES A: MARINE*

The marine swamps in the Cayman Islands consist principally of mangrove species. They occupy two-thirds of Grand Cayman, but are less widespread on the other two islands. Owing to the small tidal variation in sea level, the gradual build-up of the swamp floor levels by leaf fall, and the extent of these formations, the landward portions of the mangrove swamps are in effect seasonal swamps. They have been included in the swamp formation for convenience as already noted.

*Readers interested in the detailed distribution of these swamp formations should refer to the maps by Brunt and Giglioli (1980).

The development of the swamps seems to have been a complex process. Towards the end of the Pleistocene the western part of Grand Cayman was probably a large lagoon, with a maximum depth of about 30 ft, fringed by a growing reef, and containing small patch reefs. A fall in sea level at the end of the Pleistocene, which reduced the size of the lagoon to the present North Sound, also it seems allowed the pioneer *Rhizophora* mangrove to establish itself progressively in the shallowing water. Thereafter the colonisation does not seem to have followed an easily discernable pattern. Hurricanes altering the silting pattern, and the effects of the patch reefs which would have emerged as islands, and the possible seepage of fresh water through the limestone to emerge as springs in the swamps, all appear to have contributed to the present complex distribution of the different kinds of mangrove swamp. Further information on the evolution of the West Bay mangrove swamps will be found in Woodruffe (1982).

In the following description of the marine swamp formations, the conventional order of description from optimum to pessimum is reversed, as it is probably easier to follow a narrative which starts with the pioneering community – *Rhizophora* bushland – even though the path to the optimum may be difficult to follow, and it may appear contradictory to describe a formation which grows in the sea as bushland. *The text is therefore in the order IIIc, IIIb, IIIa.* We will leave it to other writers to propose improvements to what is admittedly an imperfect system. It should be remarked however that Stoddart's term "scrub" is not acceptable, as it is applied to formations which under the Beard system, which he accepts, would be classified as bushland.

IIIc *Mangrove swamp bushland*

There are a number of floristically different mangrove bushland formations, ranging from the monospecific pioneer *Rhizophora* formation which establishes itself in shallow water along sheltered shores, to various combinations of the main mangrove species: *Laguncularia* and *Avicennia*, with or without the brackish/fresh water species *Conocarpus*, which also occurs alone in association with *Rhizophora*. *Conocarpus* dominant bushland with scattered *Avicennia* has been described in Section IIb. In the former case *Rhizophora* (normally found near the sea with its prop root system immersed in water) grows on deep peat soil which is only seasonally inundated. The reasons for the distribution of this curious dwarfed formation are not fully understood.

IIIc(i) Rhizophora bushland

Rhizophora bushland occurs as a narrow band on the seaward edge of the eastern and southern sides of the North Sound mangrove swamps of Grand Cayman. It also occurs in Little Cayman to a very limited extent along the north shore of the Bight in South Hole Sound and on the nearby leeward shore of Owen Island. It is absent from Cayman Brac.

Rhizophora is the pioneer mangrove species, and this formation represents the colonising edge of the mangrove swamps. Air photograph evidence suggests that colonisation has been imperceptible in the last thirty years or so; the swamps seem to be in equilibrium with the environment.

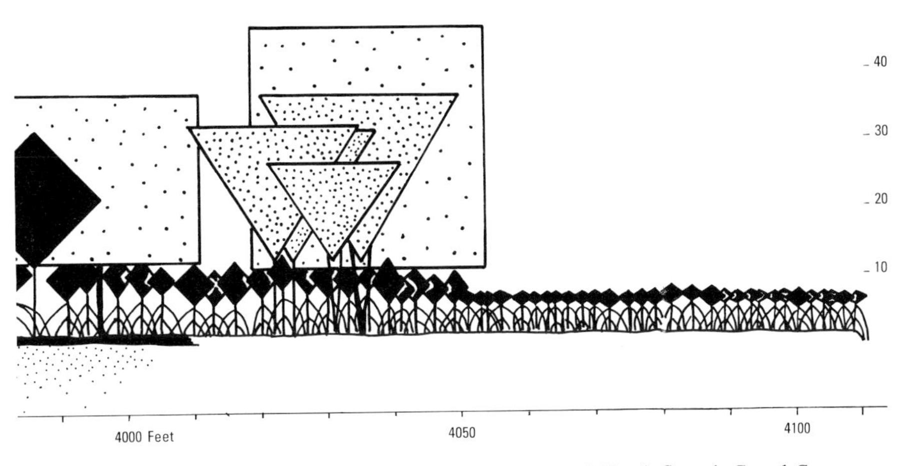

Diagram 11 Pioneer *Rhizophora* bushland, southern shore of North Sound, Grand Cayman. Intersect 2 (Key: Diagram 4)

Diagram 11 shows the formation in North Sound and its relationship with the adjacent *Avicennia – Laguncularia* woodland, into which the *Rhizophora* continues as an understorey.

Rhizophora bushland also occurs surprisingly as an isolated inland community on three sites in the western part of Little Cayman. These are surrounded by Bluff Limestone which rises to 6-13 m (19-42 ft) above sea level. The swamp surface however lies close to sea level (Stoddart, 1980).

Structure and physiognomy: the canopy height of the pioneer *Rhizophora* is 2.4-2.7 m (8-9 ft) above the bed of the sea, and some 1.8 m (6 ft) above sea level. The average crown diameter is 1.2 m (4 ft). The prop roots extend up the trunk to 1.2 m (4 ft) above sea bed level. The formation is very dense, and stem numbers may exceed 22 500 per ha (10 000 per acre). There are not emergents.

The *Rhizophora* bushland on Little Cayman was described by Stoddart (1980) as varying in height from 1.5-1.9 m (4.9-6.2 ft), with the lowest foliage 65-90 cm (25-35 in) above the ground. The prop root system is very dense averaging 80 per m^2 with a mean root diameter 50 cm (19 in) above ground level of 2.3 cm (0.9 in). The leaves are small, the mean length and breadth of a sample of 50 leaves measured 10.1 (3.98 in) and 3.8 cm (1.50 in) respectively, compared to the largest leaf which measured 12.2 (4.0 in) and 4.6 cm (1.6 in).

Floristic composition: the formation is virtually monospecific; there are no epiphytes or climbers. The Little Cayman inland swamps have a few scattered *Conocarpus* bushes in them.

IIIc(ii) Rhizophora – Conocarpus bushland (Diagram 12)

There is a good example, in the centre of the Grand Cayman Central Mangrove Swamp on Intersect 4, of this variant in which *Conocarpus* occurs more frequently. Here the ground surface, covered in deep leaf litter, is broken by 76 cm (30 in) deep depressions, and termitaria occur.

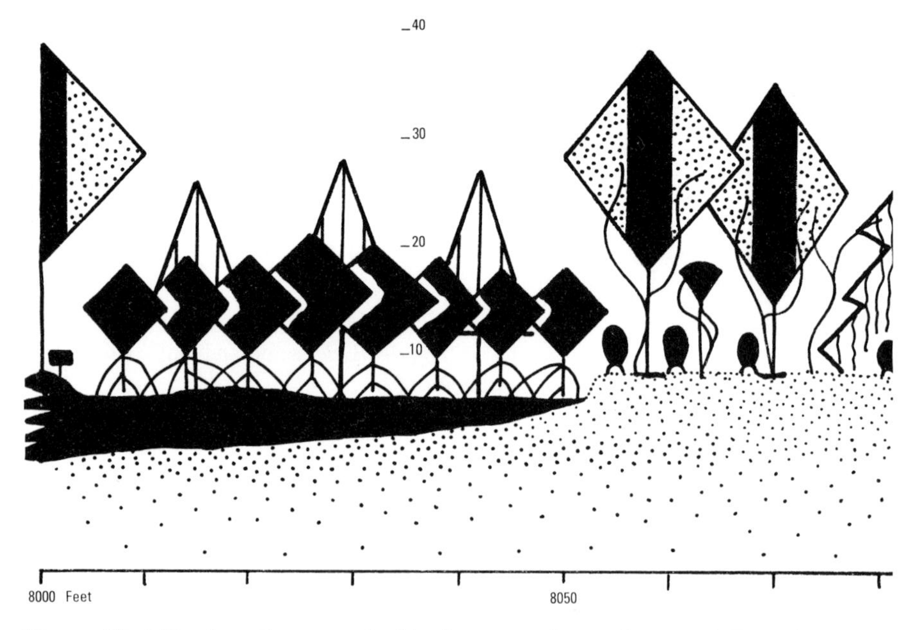

Diagram 12 *Rhizophora-Conocarpus* bushland, next to a dry cay. Intersect 4 (Key: Diagram 4)

Structure and physiognomy: the *Rhizophora* canopy varies in height between 1.8 and 3.6 m (6 and 12 ft); the trunk is virtually absent, the prop roots growing into the 0.9 m (3 ft) deep crowns with the result that the formation is very dense. *Conocarpus* bushes 4.5-9.1 m (15-30 ft) high emerge above this canopy, spaced 3.0-3.6 m (10-12 ft) apart. Bromeliads occur to within 3.0 cm (12 in) of ground level, suggesting this is the maximum water level when the swamps are flooded. Dense patches of creeper cover the ground surface and grow into the trees in places. *Rhizophora* seedings occur occasionally.

Floristic composition: *Rhizophora mangle* is the dominant species.

Shrubs:

Conocarpus erectus
Rhizophora mangle

Creepers:

Rhabdadenia biflora

Epiphytes:

Schomburgkia thomsoniana
Tillandsia balbisiana
T. recurvata
T. setacea

This formation presumably represents a transition from saline to brackish swamp conditions, *Conocarpus* having invaded the *Rhizophora*. It will be seen from Diagram 12 that the formation occurs close to a dry cay in the swamp with dry evergreen thicket cover. It is not clear however why the *Rhizophora* is dwarfed, and why other species – *Avicennia* and *Languncularia* – are virtually absent. As noted the closely related formation in which *Conocarpus* is dominant has been described in section IIb.

IIIc(iii) Rhizophora – Avicennia bushland

Rhizophora bushland is also associated with *Avicennia* in the Central Mangrove Swamp on Grand Cayman; this formation occurs between two stands of *Avicennia* woodland on Intersect 4, see Diagram 13. It continues beyond this but the *Rhizophora* increases in height, and becomes a thicket.

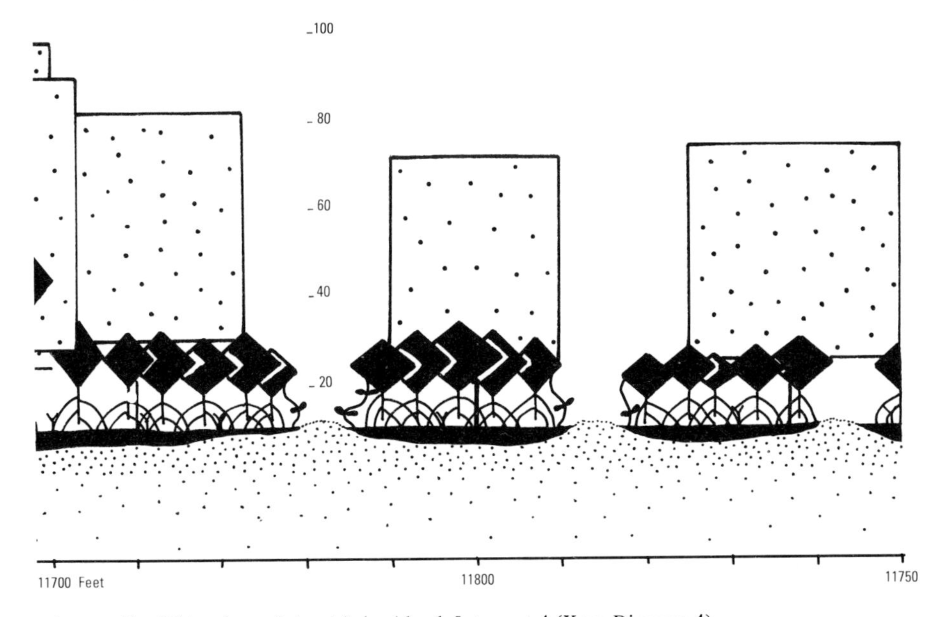

Diagram 13 *Rhizophora-Avicennia* bushland. Intersect 4 (Key: Diagram 4)

Structure and physiognomy: there is a continuous dense stratum of *Rhizophora*, the canopy height above swamp level varying between 1.8 and 3.6 m (6 and 12 ft). Growing above this are large 9-12 m (30-40 ft) high *Avicennia* trees, with crowns up to 9 m (30 ft) in diameter, spaced about 6 m (20 ft) apart. *Rhizophora* seedlings and *Avicennia* pneumatophores characterise the ground surface.

Floristic composition: Rhizophora is the dominant species, *Avicennia* being subdominant. The creeper *Rhabdadenia biflora* and the orchid *Schomburgkia* occur locally.

IIIc(iv) Laguncularia – Avicennia – Conocarpus bushland

The above formation also occurs in the centre of the main Grand Cayman mangrove swamp, as shown in Diagram 14 of Intersect 4. It grows on deep peat soil – over 3.0 m (10 ft) in places; elsewhere islands of limestone (old patch reefs) stand above swamp level, on which Dry evergreen thicket grows. In other places a lace-like pattern of the bushland grows on the banks separating very large numbers of small, shallow ponds, which dry out in the dry season.

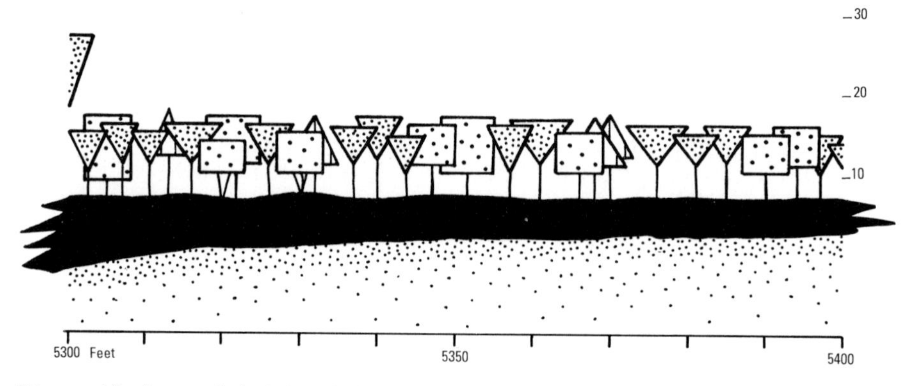

Diagram 14 *Laguncularia-Avicennia-Conocarpus* bushland. Intersect 4 (Key: Diagram 4)

Structure and physiognomy: the canopy varies in height from 1.8 to 3.0 m (6 to 10 ft), the average crown diameter being 2.4 m (8 ft). Stem sizes vary between 1.2 and 7.6 cm (½ and 3 in) d.b.h., the number of stems per hectare averaging 7 200 (3 200 per ac). The *Laguncularia* bushes are frequently multi-stemmed, individual stems measuring 1.2-2.5 (½-1 in) in diameter. There are no emergents. The swamp floor is often densely covered, with short *Laguncularia* and *Avicennia* pneumatophores.

Floristic composition: the main canopy components are *Laguncularia*, *Avicennia* and *Conocarpus*. *Laguncularia* is usually more or less dominant, but the ratio, between the three species (in the above order) varies from site to site, e.g. 4:2:1, 7:1:5.

In the drier parts of the bushland occasional clumps of the 45 cm (18 in) high sedge *Fimbristylis ferruginea* and the 1.2 m (4 ft) high mangrove fern *Acrostichum aureum* may occur occasionally.

IIIb *Mangrove swamp thicket*

Mangrove swamp thicket may be monospecific or a mixture of two species e.g. *Avicennia* – *Conocarpus*. It has only one stratum 3-12 m (10-40 ft) in height.

IIIb(i) Rhizophora mangrove swamp thicket

There are appreciable areas of this formation in the Barker's Peninsula and West Bay Peninsula (Salt Creek and Governor's Creek) swamps of Grand Cayman, where it is subject to daily tidal inundation. It also occurs in the Central Mangrove Swamp of that island where it is only seasonally flooded. There are small areas of it in the Tarpon Lake swamp on Little Cayman.

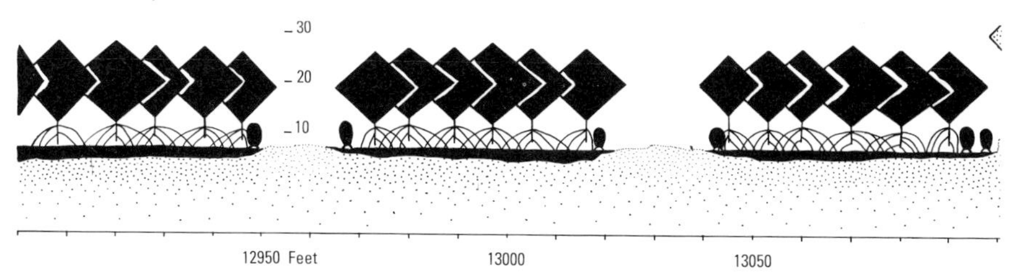

Diagram 15 *Rhizophora* mangrove swamp thicket. Intersect 4 (Key: Diagram 4)

Structure and physiognomy: the average height of the canopy is 3.6 m (12 ft), but it may reach 6.1 m (20 ft) in places. It is not always continuous and there may be 3.0-4.5 m (10-15 ft) wide gaps in it. Crown diameter ranges from 2.1 to 4.5 m (7 to 15 ft), and averages 3.0 m (10 ft). Stem diameters vary from 3.0 to 8.5 cm (1.2 to 3.4 in), averaging 5.0 cm (2.0 in). The number of stems per hectare averages 4 050 (1 800 per ac). The formation may grow in peat soil varying in depth from 0.3-2.4 m (1-8 ft), crab holes occurring occasionally. There are good examples of the formation on Intersect 4: Diagram 15.

Floristic composition: the formation is monospecific. The mangrove fern *Acrostichum aureum* may occur occasionally on drier sites.

IIIb(ii) Avicennia mangrove swamp thicket (Diagram 16)

There are extensive areas of *Avicennia* woodland in the West Bay peninsula and in the Central Mangrove Swamp on Grand Cayman. In places, particularly near its margin, this formation reduces in height and passes into thicket. The formation does not occur on Little Cayman or Cayman Brac.

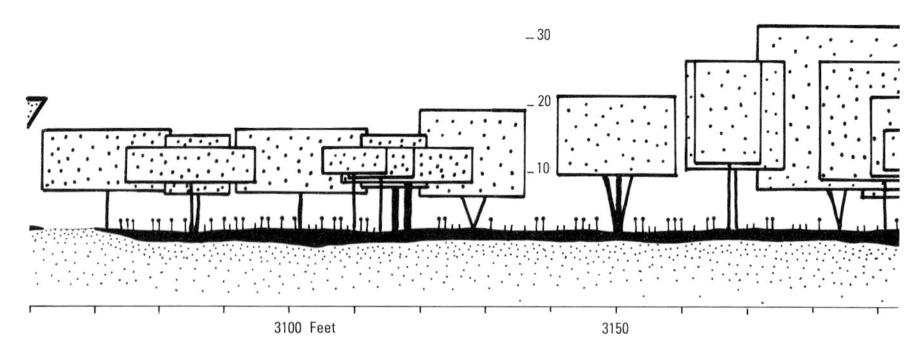

Diagram 16 *Avicennia* mangrove swamp thicket. Intersect 2 (Key: Diagram 4)

Structure and physiognomy: the canopy height varies between 3.6 and 5.4 m (12 and 18 ft), averaging 4.2 m (14 ft). Crown diameters range from 1.8 to 6.1 m (6 to 20 ft) and average 4.2 m (14 ft). Stem size varies from 5.0 to 30.4 cm (2 to 12 in) d.b.h. and averages 13.9 cm (5.5 in); stocking being about 8 775 stems per ha (3 900 stems per ac). A few of the *Avicennias* have multiple stems: two or three. The ground is densely covered with *Avicennia* pneumatophores.

Floristic composition: the thicket consists only of *Avicennia*. The ground is frequently covered by a dense tangle of the succulent herb *Batis maritima*, 30.4-38.1 cm (12-15 in) in height.

IIIb(iii) Laguncularia mangrove swamp thicket

Laguncularia is the species which colonises mangrove swamps after clearing. Uniform stands of *Laguncularia* are therefore frequently encountered on cleared swamp land that has been abandoned – this accounts for their presence in the main swamps, as for example in South Sound Swamp, Grand Cayman: Diagram 17. This example, however, is atypical in having a low stocking.

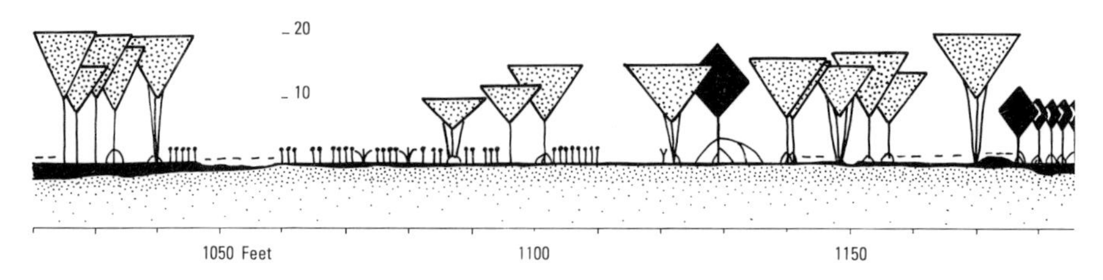

Diagram 17 *Laguncularia* swamp thicket, South Sound Swamp. Intersect 1 (Key: Diagram 4)

Structure and physiognomy: the canopy height averages (4.5 m) 15 ft, but may reach 6.0 m (20 ft) or more. Crown diameters range from 2.4 to 5.4 m (8 to 18 ft), averaging 3.6 m (12 ft). Stem diameters range from 2.5 to 10.1 cm (1 to 4 in) d.b.h., averaging 4.75 cm (1.9 in); stocking ranges from 10 125 to 23 625 stems per ha (4 500 to 10 500 stems per ac). A feature of this thicket is the dense stocking of slender saplings, resulting from the high survival rate and rapid growth of large numbers of *Laguncularia* seedlings. Many trees have four or more stems; while some stems grow parallel to the ground, anchored by lateral roots, with branches growing vertically to form part of the canopy. The root crown of individual trees will frequently stand (30.4 cm) 12 in or more above ground level, and will have a 'cone' of small prop-like roots about it. There is usually an accumulation of leaf litter 2-3 in deep.

Floristic composition: *Laguncularia* dominates the formation. Patches of *Batis maritima* may occur.

IIIb(iv) Conocarpus – Avicennia and Conocarpus – Laguncularia mangrove swamp thicket

These two floristic variants of mangrove swamp thicket are considered together as they are almost certainly both examples of the invasion of the swamp by *Conocarpus*, occasioned by the development of drier conditions. The two examples considered are both from the Central Mangrove Swamp on Grand Cayman, Intersect 4.

Conocarpus – Laguncularia occurs near an outcrop of Bluff Limestone with Dry evergreen woodland cover, and probably results from previous disturbance to the swamp. *Conocarpus-Avicennia* swamp thicket (Diagram 18) grows between shallow ponds, with *Avicennia* bordering the ponds.

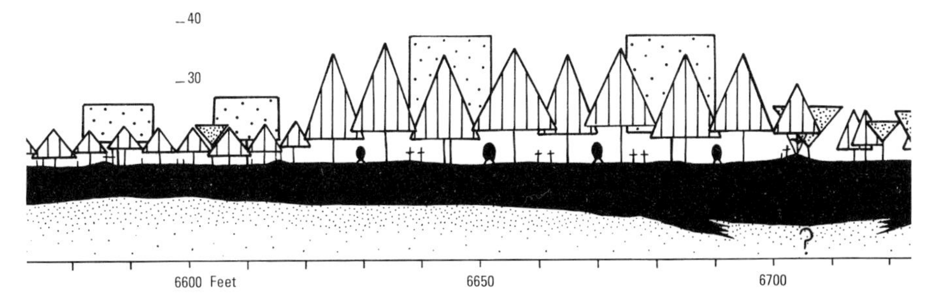

Diagram 18 *Conocarpus-Avicennia* swamp thicket. Intersect 4 (Key: Diagram 4)

Structure and physiognomy: in both cases the canopy height is about 7.6 m (25 ft), the average crown size is 3.6 m (12 ft), and stem diameters vary from 5 to 10 cm (2 to 4 in). The depth of the peat soil may exceed 3 m (10 ft).

Floristic composition: in addition to the canopy species, sedges and ferns may occur occasionally including *Cladium jamaicense, Fimbristylis ferruginea* (sedges), *Acrostichum aureum*, *Nephrolepis multiflora* (ferns).

IIIa *Mangrove swamp woodland* (Diagram 19)

Mangrove swamp woodland, like the equivalent thicket may be mono-specific, or a mixture of two or three of the main mangrove species, With drier conditions on the landward side of the swamps *Conocarpus* may occur in the woodland.

The floristically richest and tallest *Avicennia – Laguncularia – Rhizophora* mangrove swamp woodland occurs on the coastal edge of the swamps, usually flanked on the seaward side by pioneer *Rhizophora* bushland (Diagram 11), and on the landward side by *Avicennia* woodland: Diagram 19 shows swamps on the southern side of North Sound, Grand Cayman.

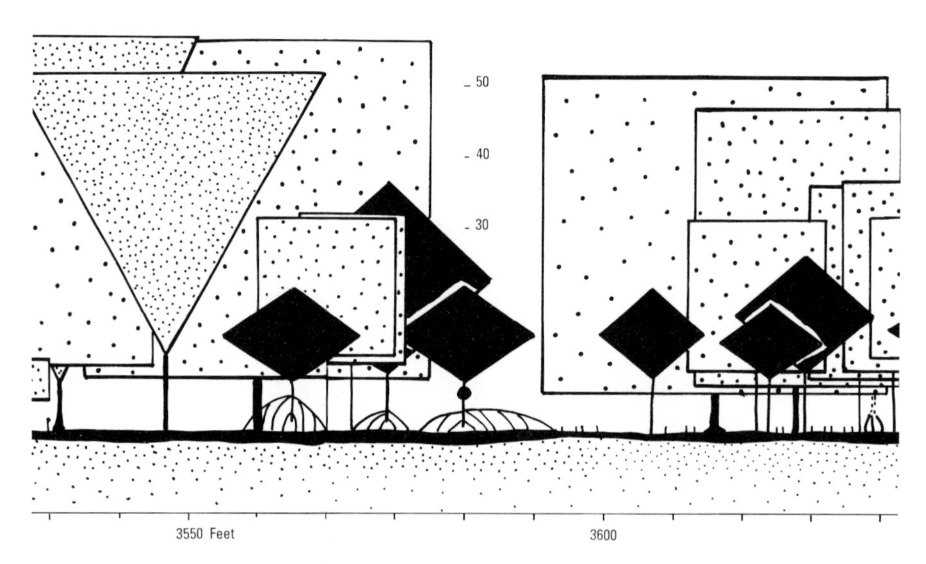

Diagram 19 *Avicennia-Laguncularia-Rhizophora* mangrove swamp woodland. Intersect 2 (Key: Diagram 4)

Rhizophora woodland occurs less frequently, while *Laguncularia* woodland probably results from recolonisation of cleared mangrove swamp as previously described. Varying mixtures of any two of the main species are found in the middle of the Central Mangrove Swamp of Grand Cayman.

Mangrove swamp woodland occurs extensively on Grand Cayman, to a lesser extent on Little Cayman, but is absent from Cayman Brac.

IIIc(i) Rhizophora bushland (pioneer) and IIIa(v) Laguncularia-Avicennia mangrove swamp woodland. North Sound, Grand Cayman (pp.44and56)

IIIa(iii) Rhizophora mangrove swamp woodland, proproot system. South Sound Swamp, Grand Cayman (p.55)

PLATE 3(A) Swamp formation series III: A: Marine

IIIa(ii) Avicennia mangrove swamp woodland with *Batis maritima* ground cover. West Bay area, Grand Cayman (p.54)

IIId Undifferentiated herbaceous swamp including ponds: freshwater pond with *Nymphoides indica* Water Lily. Old Man village, Grand Cayman (p.58)

PLATE 3(B) Swamp formation series III: A: Marine (above) and B: Freshwater

IIIa(i) Avicennia – Laguncularia – Rhizophora mangrove swamp woodland

In general terms this formation occurs as a broad zone round the southern and western shore of North Sound, Grand Cayman. It also surrounds Tarpon Lake on Little Cayman, although the structure is slightly different. It is the formation that usually develops following the establishment of a belt of pioneer *Rhizophora* mangrove.

Structure and physiognomy: in the example illustrated in Diagram 20. the canopy is dominated by *Avicennia* and is about 15 m (50 ft) high. Individual *Avicennias* vary in height from 3.0 to 16.7 m (10 to 55 ft), averaging 11.5 m (38 ft), with crown diameters ranging from 4.5 to 15.2 m (15 to 50 ft), averaging 8.5 m (28 ft). *Avicennia* stem diameters vary from 3.8 to 45.7 cm (1½ to 18 in) d.b.h., some trees have two stems. The larger stems may have small buttresses, extending 30 cm (12 in) up the trunk, and up to 6.1 cm (24 in) outwards from the trunk; branching may not start until 5.4-6.1 m (18-20 ft) from the ground. There are far fewer *Laguncularias*, about one for every three *Avicennias*. In height they vary from 9.1 to 16.7 m (30 to 55 ft), averaging 12.8 m (42 ft). Crown diameters range from 5.4 to 13.7 m (18 to 45 ft), averaging 9.1 m (30 ft). Stem diameters vary from 7.6 to 25.4 cm (3 to 10 in), and average 16.5 cm (6.5 in) d.b.h. The *Rhizophora* forms a lower stratum in the woodland which may not be continuous. It represents the pioneer shore-line *Rhizophora* bushland, in which trees of *Avicennia* and *Laguncularia* have established themselves. It is not clear why the *Rhizophora* has not also grown higher to form part of the canopy, as is the case in the stand of this woodland surrounding Tarpon Lake in Little Cayman. Individual *Rhizophora* trees vary in height from 3.6 to 10.6 m (12 to 35 ft), averaging 5.7 m (19 ft). The crown diameter varies from 3.0 to 9.1 m (10 to 30 ft), but averages 4.2 m (14 ft). Crowns are usually interlocked to form a dense continuous understorey layer. Stem diameters vary from 2.5 to 7.6 cm (1-3 in) and average 5.7 cm (2¼ in) d.b.h. The stocking of all three species is 353 stems per ha (157 stems/ac).

The lower part of the swamp comprises a dense tangle of interlocking *Rhizophora* prop roots 0.9-1.5 m (3-5 ft) high, with the ground covered by *Avicennia* pneumatophores. Occasionally 61 cm (24 in) high termitaria occur.

Floristic composition: in addition to the three mangrove species orchids (*Schomburgkia*) and bromeliads (*Tillandsia* ssp.) occur. There may also be occasional patches of the succulent herb *Batis maritima*. Seedlings are uncommon.

IIIa(ii) Avicennia mangrove swamp woodland

This formation, which is only found in Grand Cayman, frequently occurs on the inland side of the *Avicennia* – *Laguncularia* – *Rhizophora* woodland, and forms a broad if discontinuous belt parallel to the North Sound shore. Here the surface of the swamp tends to be a little lower, in consequence the *Avicennia* swamp remains flooded for longer periods than the rest of the swamp, and is a prime breeding ground of the Mangrove Mosquito *Aedes taeniorynchus*. (See Diagram 20).

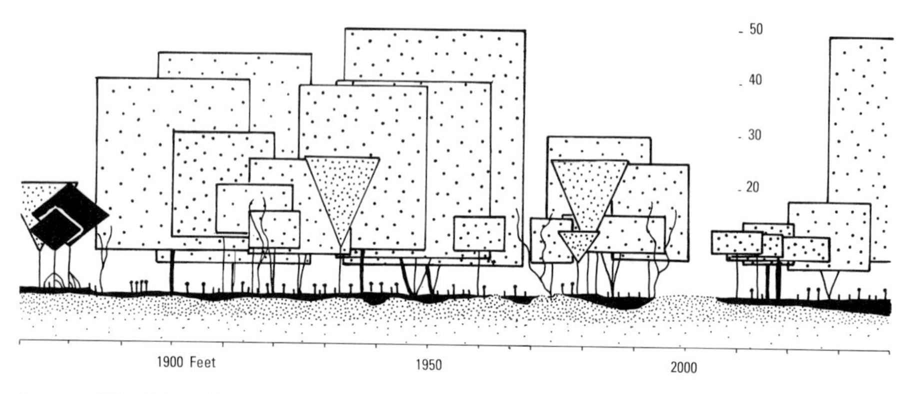

Diagram 20 *Avicennia* mangrove swamp woodland. Intersect 2 (Key: Diagram 4)

Structure and physiognomy: the height of the formation is somewhat variable the average canopy rising from 7.0 to 13.1 m (23 to 43 ft) as the community grades from thicket into woodland. Individual trees may reach 15.2 m (50 ft) in height. Crown diameters also vary from as little as 0.9 to 15.2 m (3 ft to 50 ft), canopy crowns average 8.5 m (28 ft). Stem diameters range from 6.3 to 16.5 cm (2½ to 6½ in) and average 11.4 (4½ in) d.b.h. Stocking ranges from 967 to 3 375 stems per ha) 430 to 1 500 stems per ac), the latter occurring near the border with *Avicennia* thicket. The ground is covered by a very dense mat of *Avicennia* pneumatopores.

Floristic composition: the woodland is dominated by *Avicennia* although occasional trees of *Laguncularia* may occur. The ground is frequently covered by a dense tangle of the succulent herb *Batis maritima* 38-61 cm (15-24 in) high. Epiphytic orchids (*Schomburgkia thomsoniana*) and bromelliads (*Tillandsia* spp.) occur occasionally, as do the climbers – the spiny *Selenicereus grandiflorus* and *Ipomoea violacea*

IIIa(iii) Rhizophora mangrove swamp woodland

The formation is only found in Grand Cayman. Stands of it occur on the western shore of North Sound, in South Sound, and south and south-east of Booby Cay in the Central Mangrove swamp.

In the case of South Sound, this swamp must once have been open to the sea, but was closed by a beach ridge, Diagram 21. New colonisation by *Rhizophora* is on the seaward side of the ridge (see Sauer, 1983).

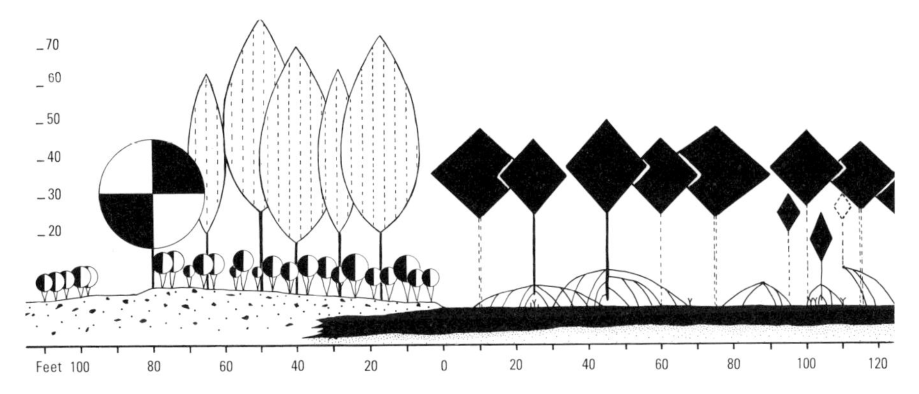

Diagram 21 *Rhizophora* mangrove swamp woodland. Intersect 1 (Key: Diagram 4)

Structure and physiognomy: the average canopy height is 15.2 m (50 ft), although it may in places reach 21 m (70 ft). Crown diameters vary from 5 to 50 ft and average 30 ft. Stem diameters vary from 2.5 to 25.4 cm (1 to 10 in) and average 15.2 cm (6 in) d.b.h. Stocking averages 200 stems per acre. The *Rhizophora* prop root system may extend up the trunks to 3 m (10 ft) above ground level resulting in an impenetrable mass of interlocking roots. Other roots may descend from branches 6.1 m (20 ft) above the ground. A second stratum is often absent.

Floristic composition: the swamp is dominated by *Rhizophora*; occasional trees of *Avicennia* or *Laguncularia* may occur. There are some patches of *Batis maritima* 45.7-60.9 cm (18-24 in) high, but the swamp floor, 5.0-7.6 cm (2-3 in) deep in leaf litter, is usually without vegetation. Ferns occur occasionally: *Acrostichum aureum* and *Nephrolepis biserrata* and *multiflora*, climbers (*Ipomoea violacea* and *Rhabdadenia biflora*) and the bromeliad *Tillandsia balbisiana* may be locally common.

IIIa(iv) Laguncularia mangrove swamp woodland

The colonising role of *Laguncularia* has already been discussed (*Laguncularia* mangrove swamp thicket; IIIb(iii)). Many of these colonising thickets will grow to become woodland, which is here described. It is found in small pockets throughout the Grand Cayman mangrove swamps.

Structure and physiognomy: the average canopy height is 9.1 m (30 ft); it may reach 15.2 m (50 ft) in places. Crown diameter varies from 3 to 12 m (10 to 40 ft) and averages 6.7 (22 ft). The average size is 8.9 cm (3.5 in) d.b.h., but varies from 2.5 to 30.4 cm (1 to 12 in). Stocking is also variable and ranges from 1 170 to over 3 555 stems per ha (520 to over 8 000 per ac), the average is 7 650 per ha (3 400 per ac). The emergent stratum is not very strongly developed; similarly the shrub layer is virtually absent. Ground cover is sparse, although creeper tangles occur occasionally. These characteristics are well illustrated in Diagram 22.

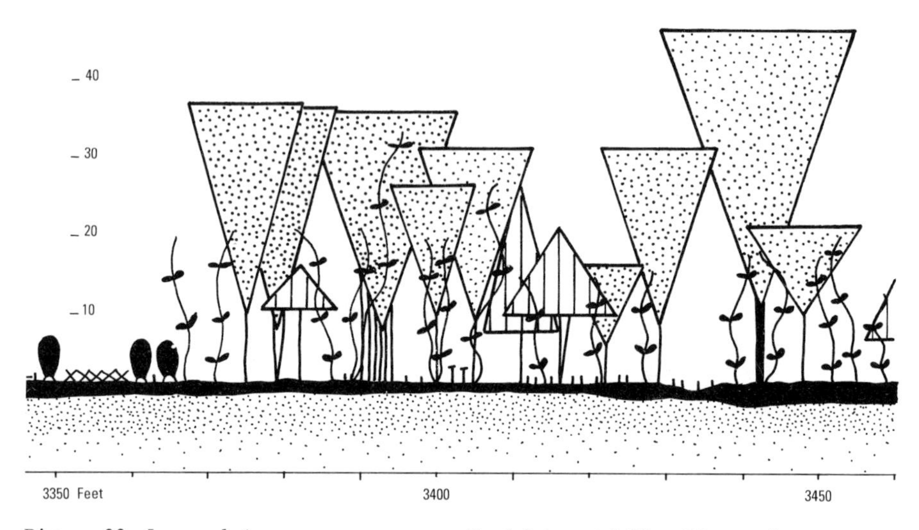

Diagram 22 *Laguncularia* mangrove swamp woodland. Intersect 1 (Key: Diagram 4)

Floristic composition: *Laguncularia* is the dominant component of the canopy and shrub layer. *Conocarpus* may occur occasionally as a tree or shrub. The ground cover is sparse although 1.2 m (4 ft) high clumps of the fern *Acrostichum aureum* may occur locally, or patches of (15.2 cm) 6 in high *Caraxeron vermicularis*. Individual tree crowns may be festooned with the creeper *Rhabdadenia biflora*; epiphytic orchids (*Schomburgkia thomsoniana*) are locally common.

IIIa(v) Laguncularia – Avicennia, Laguncularia – Rhizophora, Laguncularia – Conocarpus mangrove swamp woodland

The main mangrove species grow in different proportions, resulting in a number of floristically distinct mangrove swamp woodland communities. The above variants are illustrated below in Diagrams 23-25 from which their major characteristics can be appreciated. They will not therefore be described in detail, particularly as the variation that occurs between one site and another is considerable. In general terms these floristically distinct types conform structurally with the earlier examples of mangrove swamp woodland. The major components, climbers, epiphytes etc are also similar.

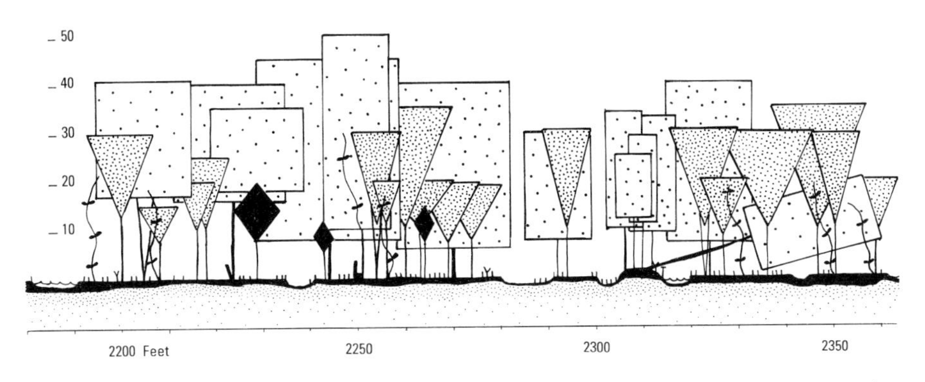

Diagram 23 *Laguncularia-Avicennia* mangrove swamp woodland. Intersect 1 (Key: Diagram 4)

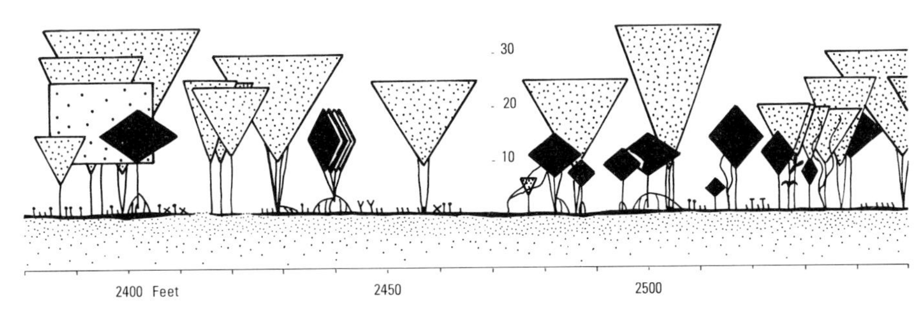

Diagram 24 *Laguncularia-Rhizophora* mangrove swamp woodland. Intersect 2 (Key: Diagram 4)

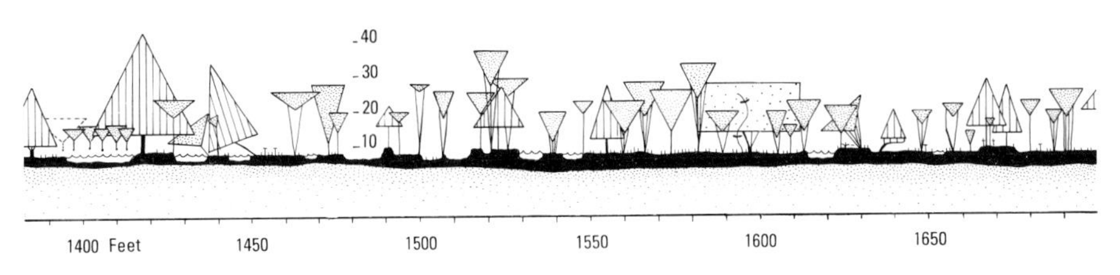

Diagram 25 *Laguncularia-Conocarpus* mangrove swamp woodland. Intersect 4 (Key: Diagram 4)

These variants occur throughout the main mangrove swamps on Grand Cayman, and to a lesser extent on Little Cayman. Their distribution can be appreciated from the Cayman Island swamp maps (Brunt and Giglioli, 1980).

III SWAMP FORMATION SERIES: B: FRESHWATER

IIId *Undifferentiated herbaceous swamp, including ponds*

There are a number of ponds in the islands, which have swamp margins. The principal ones investigated include those near Brinkleys, and those south of Old Man village near the north coast of Grand Cayman.

Comparable ponds – if they exist – on Little Cayman and Cayman Brac have not been investigated.

Two species occur in the ponds: the water lily *Nymphoides indica*, and the floating herb *Eichhornia crassipes*, which may grow to 50.8 cm (20 in) in height above the pond surface.

The surrounding herbaceous swamp is characterised by the following species, which grow to an average height of 30.4-38.1 cm (12-15 in), with 60.9 cm (24 in) clumps of *Panicum* grass.

Herbs:

Eclipta prostrata
Ludwigia erecta

Grasses:

Leptochloa fascicularis
Panicum purpurascens

Sedges:

Eleocharis geniculata
E. interstincta
E. mutata

For the sake of convenience the flora found on the walls of old wells and water holes is also recorded here:

Shrub:

Petitia domingensis

Ferns:

Adiantum tenerum and melanoleucum
Nephrolepis multiflora
Tectaria incisa
Thelypteris kunthii
T. reptans

Sedge:

Cyperus swartzii

HISTORY OF THE BOTANICAL COLLECTIONS

Botanical investigation began late in the Cayman Islands, and has involved comparatively few people. The earliest known plant collections* were made in May, 1888, by William Fawcett, then Director of Public Gardens and Plantations in Jamaica, and later co-author of *Flora of Jamaica*. His visit, of but a few days' duration, resulted in the publication of a short report dealing with the natural and agricultural resources of the islands, and also reporting on such subjects as the disease of coconut palms (presumably what is now called Lethal Yellowing) then ravaging Grand Cayman. Added to this report was a list of 112 plants that he had collected, including both indigenous and introduced species. The specimens on which this list was based were deposited in the herbarium of the Royal Botanic Gardens, Kew, England. Among Fawcett's discoveries was the endemic orchid *Dendrophylax fawcettii*, described by Rolfe in the *Gardener's Chronicle* of Nov. 10, 1888. Alluding to this plant, Fawcett later commented, "As I saw but a small portion of the Islands, and that chiefly on the sea shore, I feel little doubt that a complete collection of the plants would be of very great interest, and that perhaps other endemic species would be found.".

Following Fawcett, several American botanists paid short visits to the Cayman Islands during the 1890's. Of these, the most important were Albert S. Hitchcock (in 1891), who later became a leading authority on grasses, and Charles F. Millspaugh (in 1899). Hitchcock's specimens (mostly miserable scraps, unfortunately) were collected while he was a member of a party of naturalists led by J.T. Rothrock; they are preserved in the herbarium of the Missouri Botanical Garden in St. Louis. Millspaugh made his collections while a guest of Allison V. Armour (the Chicago meat-packing millionaire) on a West Indian cruise of the yacht "Utowana"; they visited the Cayman Islands during February, 1899. The chief set of Millspaugh's specimens is in the herbarium of the Field Museum of Natural History in Chicago. Both Hitchcock and Millspaugh published lists of their collections.

After Millspaugh, there came an interval of 39 years during which there was virtually no botanical activity in the Cayman Islands. A few plants only were gathered on Cayman Brac by the geologist C.A. Matley in 1924.

The Oxford University Biological Expedition to the Cayman Islands, a party of five under the leadership of W.G. Alexander, was in the field from April 17 to August 27, 1938. The primary objects of attention were plants, insects, reptiles, and fishes, but nearly all animal taxa received some attention. The official botanist of this group was Wilfred W. Kings, who joined the expedition about a month later than the others; he had been especially recruited from Lawrence Sheriff School,

---

*The oldest reference found to a Cayman plant appears on p. 214 of William Herbert's monograph of Amaryllidaceae (1837), where he described *Hymenocallis caymanensis*. The author noted that this species came from Grand Cayman, but cited no collector.

Rugby, because Oxford had no available botanist at that time. Prior to his arrival, some plant-collecting was done by C. Bernard Lewis, whose interests were otherwise chiefly zoological. Kings gathered a large collection of material from all three islands; until recently, these excellent specimens constituted the major basis of our knowledge of the Cayman flora. The main set of the Kings collection is deposited at the British Museum (Natural History) in London, while duplicate material is to be found in several other herbaria.

Lewis, then an Oxford student (a Rhodes Scholar from the United States), later became Director of the Institute of Jamaica in Kingston; he collected further Cayman plant specimens during the 1940's. His continued interest in the Cayman Islands has been a constant source of encouragement during the writing of this book.

Coincident with the Oxford Expedition was a brief visit during June, 1938, of the Cap Pilar Expedition. The "Cap Pilar" was a French sailing vessel (a "barquentine") that made a 2-year trip around the world beginning in September, 1936, under the captaincy of Adrian Seligman, then of Wimbledon, England. The purpose of the trip was partly adventure and partly to collect plants for the Royal Botanic Gardens, Kew. Originally, the botanist of this expedition was A.F. Roper, but when the ship reached South Africa on its outward journey, his place was taken (for unexplained reasons) by C.M. Maggs, then a horticulturist at the Kirstenbosch Botanical Garden. The "Cap Pilar" visited Australia, and after making collections on various Pacific islands (including the Galapagos) and at Panama, a final stop was made at Grand Cayman, where a few plants were collected.

The present writer first visited Grand Cayman on April 19, 1948, while a member of the Catherwood-Chaplin West Indies Expedition of the Academy of Natural Sciences of Philadelphia. This one-day stop was followed eight years later by a longer collecting trip on behalf of the Institute of Jamaica and the British Museum (Natural History), traveling to the islands as a guest of the British survey ship H.M.S. Vidal. Nearly three weeks were spent on Grand Cayman at that time, followed by a one-day visit to Cayman Brac on May 2, 1956. His next Cayman trip was made during June, 1967, at the invitation of Dr. M.E.C. Giglioli, Director of the Cayman Government Mosquito Research and Control Unit (M.R.C.U.), following a suggestion by Martin Brunt. The latter, an ecologist on the staff of the (now) Land Resources Development Centre of the British Overseas Development Administration (O.D.A.), was at the time making ecological studies of the Cayman swamps for M.R.C.U. He had himself already collected numerous Cayman plant specimens, but he and his colleagues were having difficulties with identification of these, partly because no descriptive Flora of these islands had ever been written. The stimulating collaboration with Brunt in the field at that time led to the author's agreement to write this Flora at the request of Mr J.A. Cumber, then Administrator of the Cayman Islands. It has been Brunt's continued interest that has been largely responsible for making its publication possible.

Later, short collecting trips were made by the writer to Little Cayman (July 7-12, 1967, and August 2-12, 1975), Cayman Brac (August 6-10 and November 9-11, 1968), and again to Grand Cayman (September 6-7, 1969), in order to augment the available botanical information about these islands. Particular mention should be made of the hospitality of the late Dr Logan Robertson, which made possible the earlier work on Little Cayman. The second trip to that island in 1975 was sponsored by the Cayman Islands Government as part of their joint study of the island with the Royal Society of London. This complemented the O.D.A.-funded Natural Resources Study of all the Cayman Islands. I am indebted to Dr David Stoddart of Cambridge University for being invited to participate in this program.

The prime set of the writer's collections is preserved in the herbarium of the Institute of Jamaica; most of this material is duplicated at the British Museum (Natural History), and smaller sets have been distributed to other herbaria.

In recent years an unknown number of American collectors have visited the Cayman Islands (usually only Grand Cayman), but for the most part it has not been possible to take their material into account, except for the collection of Marie-Helène Sachet (1958), some of the specimens collected by Jonathan Sauer (1962 and 1967), and the collection of Donovan S. and Helen B. Correll (1979). Sauer's gatherings totalled 243 numbers, according to lists he kindly supplied, but only a small fraction of these were actually seen in connection with the present study. A single species, *Atriplex pentandra*, is recorded from the Cayman Islands solely on the basis of a Sauer specimen. Sauer's material is deposited in the herbarium of the University of Wisconsin at Madison; duplicate material is filed at the Field Museum of Natural History in Chicago.

The Corrells' specimens, 65 numbers in all, are deposited in the herbarium of the Fairchild Tropical Garden, Miami. Included in the collection are the first examples of *Corchorus hirsutus* and *Gomphrena globosa* from the Cayman Islands, and the first of *Citharexylum fruticosum*, *Passiflora cupraea*, and *Phoradendron quadrangulare* from Grand Cayman.

Table 5 summarizes the Cayman Islands specimens seen by the present writer, listed chronologically by dates of collection. The herbarium symbols signify the following: A, Arnold Arboretum; BM, British Museum (Natural History); F, Field Museum of Natural History; FTG, Fairchild Tropical Garden; GH, Gray Herbarium of Harvard University; IJ, Institute of Jamaica; K, Royal Botanic Gardens, Kew; MO, Missouri Botanical Garden; US, U.S. National Herbarium, Washington, D.C.

Table 5 Collectors of Cayman Islands Plants, in chronologic order

	Name	*Dates*	*Herbaria*	*No. of specimens*
1.	William Fawcett	May 1888	K	16
2.	John T. Rothrock	"Winter of 1890-1891"	F	10
3.	A.S. Hitchcock	Jan. 1891	MO, F	166
4.	Charles F. Millspaugh	Feb. 1899	F	216
5.	C.A. Matley	Jan. 1924	K	3
6.	Wilfrid W. Kings	May-Aug., 1938	BM, GH, MO	645
7.	C. Bernard Lewis	Apr. 1938 Dec. 1944 Mar. 1945 Dec. 1945	BM, IJ	43
8.	C.M. Maggs	June 1938	K	18
9.	George R. Proctor	Feb. 1948 Apr.-May 1956 June-July 1967 Aug. 1968 Nov. 1968 Sept. 1969 Aug. 1975	IJ, BM, GH	1007
10.	Marie-Helène Sachet	Sept. 1958	US	95
11.	Robert A. Dressler	May 1964	IJ	2
12.	Richard A. Howard & B. Wagenknecht	Jan. 1969	A	11
13.	Martin Brunt	May-June 1967	BM, IJ	503
14.	Jonathan Sauer	June 1967	F	15
15.	John Popenoe	Apr. 1968	FTG, IJ	1
16.	Donovan S. & Helen B. Correll	Nov. 1979	FTG	65
			Total	2816

The writer visited all the cited institutions during the preparation of the present book, and wishes to thank the relevant officials, curators, and staff for much help and many kindnesses received while studying the Cayman plant specimens under their care.

BIBLIOGRAPHY

General

The literature pertaining directly to Cayman Islands vegetation is scanty, and consists chiefly of lists or short reports. Other literature, dealing with aspects of the geology, zoology, and history of these islands is more extensive, but not relevant to the present book. The following botanical references have been consulted in addition to general taxonomic works and special papers or monographs on particular groups of plants:

Fawcett, W. (1888) Cayman Islands. Kew Bull. 1888, 160-163.
______(1889) Report by the Director of Public Gardens and Plantations on the Cayman Islands. Bull. Bot. Jamaica no. 11, 6-7.
Guppy, H.B. (1917) Plants, seeds and currents in the West Indies and Azores. London: Williams and Norgate.
Hitchcock, A.S. (1893) List of plants collected in the Bahamas, Jamaica and Grand Cayman. Ann. Rep. Missouri Bot. Gard. 4, 47-179.
______(1898) List of cryptogams ... Ibid., 9, 111-120.
Millspaugh, C.F. (1900) Plantae Utowanae. Field Mus. Bot. Ser. 2, 3-133.
Proctor, G.R. (1980). Checklist of the plants of Little Cayman. Atoll Res. Bull. No. 241, 71-80.
Savage English, T.M. (1913) Some notes on the natural history of Grand Cayman. Handbook of Jamaica for 1912, 598-600.
______(1913) Some notes from a West Indian coral island. Kew Bull. 1913, 367-372.
Swaby, C. and Lewis, C.B. (1946) Forestry in the Cayman Islands. Development and welfare in the West Indies, Bull. no. 23. Barbados.

Floras

The following standard West Indian floras were frequently consulted:

Adams, C.D. (1972) Flowering plants of Jamaica. University of the West Indies, Mona, 1-848.
Alain, H. (1962) Flora de Cuba, Vol. 5. Rio Piedras, Puerto Rico Univ.
Britton, N.L. & Millspaugh, C.F. (1920) Bahama flora. New York, 1-695.
Britton, N.L. & Wilson, P. (1923-1930) Botany of Porto Rico and the Virgin Islands. Sci. Surv. of Porto Rico and the Virgin Islands, Vols. 5-6. New York, Acad. Sci.
Fawcett, W. & Rendle, A.B. (1910-36) Flora of Jamaica. Vols. 1, 3, 4, 5, & 7 (all publ.) British Museum (Natural History), London.
Gooding, E.G.B., Loveless, A.R., & Proctor, G,R. (1965) Flora of Barbados. London: H.M.S.O. 1-486.
Grisebach, A.H.R. (1859-64) Flora of the British West Indian Islands. London, Lovell Reeve.
León, H. (1946) Flora de Cuba. Vol. 1, Havana, Cuba, Cultural S.A.
______& Alain, H. (1951) Flora de Cuba. Vol. 2, Havana, Cuba, P. Fernandez.
______& Alain (1953) Flora de Cuba. Vol. 3, Havana, Cuba, P. Fernandez.
______& Alain (1957) Flora de Cuba. Vol. 4, Havana, Cuba, P. Fernandez.
Urban, I. (1898-1928) Symbolae Antillanae. Vols. 1-9, London, Williams and Norgate.

Environment and plant communities

The following list of references relates to the section by M A Brunt on the environmental background and the description of the formation series.

Asprey G.F. and Loveless A.R. (1958) The dry evergreen formations of Jamaica. II. The raised coral beaches of the north coast. J. Ecol. 46, 457-570.

Asprey G.F. and Robbins R.G. (1953) The vegetation of Jamaica. Ecol. Monogr. 23, 359-412.
Baker R.J. (1974) Cayman Islands: Soil and land use surveys No 26. Department of Soil Science, Univ. West Indies, Trinidad.
Beard J.S. (1944) Climax vegetation in tropical America. Ecology, 25, 127-158.
______(1949a) The natural vegetation of the Windward and Leeward Islands. Oxford Forestry Mem. 21, 1-192.
______(1949b) Ecological studies upon a physiognomic basis. Actas del 2º Congreso Sudamericano de Botanica, Lilloa 20,45-53.
______(1955) The classification of tropical American vegetation-types. Ecology, 36, 89-100.
Brunt M.A. and Giglioli M.E.C. (1980) Cayman Islands Swamp Maps, Sheets 1-3, 1:25 000 scale. Land Resources Development Centre, UK. Overseas Development Administration.
Darbyshire J., Bellamy I. and Jones B. (1976) Cayman Islands Natural Resources Study. Part III Results of the investigations into the physical oceanography. Ministry of Overseas Development, London.
D'Arcy W.G. (1971) The island of Anegada and its flora. Atoll Res. Bull. 139, 1-21.
______(1975) Anegada Island: vegetation and flora. Atoll Res. Bull. 188, 1-40.
Davies J.H. (1942) The ecology of the vegetation and topography of the sand keys of Florida. Pap. Tortugas Lab. 33, 113-195.
Fanshaw D.B. (1952) The vegetation of British Guiana. Imp. For. Inst. Paper No 29, Oxford.
F.A.O. (1975) FAO-Unesco Soil map of the world 1 5 000 000. Vol. III Mexico and Central America. Paris: Unesco.
Harris D.R. (1965) Plants, animals, and man in the Outer Leeward Islands, West Indies: an ecological study of Antigua, Barbuda, and Anguilla. Univ. Calif. Publs Geog. 18, i-ix, 1-164.
Howard R.A. and Briggs W.R. (1953) The vegetation of coastal dogtooth limestone in southern Cuba. J Arnold Arb. 34, 88-96.
Johnston D.W. (1975) Ecological analysis of the Cayman Island avifauna. Bull. Fla St. Mus. biol. Sci. 19, 235-300.
Loveless A.R. (1960) The vegetation of Antigua, West Indies, J. Ecol. 48, 495-527.
Loveless A.R. and Asprey G.F. (1957) The dry evergreen formations of Jamaica. I. The limestone hills of the south coast. J. Ecol. 45, 799-822.
Nathan M.B. 1975 The mosquitoes of Cayman Brac and Little Cayman, with particular reference to the ecology and control of *Aedes egypti*. PhD thesis, University of London, 1-262.
Poggie J.J. 1962 Coastal pioneer plants and habitat in the Tampico region, Mexico. Cstl Stud. Inst., La St. Univ., Tech. Rept. 17A, 1-62.
Pratt D.J., Greenway P.J. and Gwynne M.D. (1966) A classification of East African rangeland, with an appendix on terminology. J. appl. Ecol. 3, 369-382.
Sauer J.D. (1959) Coastal pioneer plants of the Caribbean and Gulf of Mexico. Madison: University of Wisconsin, Departments of Botany and Geography.

Sauer J.D. (1967) Geographic reconnaissance of seashore vegetation along the Mexican Gulf Coast. Cstl Stud. Inst. La St. Univ. Cstl Stud. Ser. 21, i-x, 1-59.
_______(1976) Problems and prospects of vegetational research in coastal environments. Geoscience and Man, 14, 1-16.
_______(1982) Cayman Island Seashore Vegetation. University of California Publications in Geography; vol 25, 1-137
Seifriz W. (1943) The plant life of Cuba. Ecol. Monogr. 13, 375-426.
Stoddart D.R. (1980) Vegetation of Little Cayman. Atoll Research Bulletin No 241, 53-70.
Swabey C. and Lewis C.B. (1946) Forestry in the Cayman Islands, Development and Welfare in the West Indies. Barbados: Government Printer, Bull. 23, 1-31.
Woodroffe C.D. (1982) Geomorphology and development of mangrove swamps, Grand Cayman. Bull. Marine Sci. 32(2), 381-398.

PLAN OF THE FLORA

The development of plant classification since the time of Linnaeus has been marked by a series of more or less competing systems, all of which were intended to express contemporary ideas about species and their relationships, and (since the acceptance of Darwin's theory of evolution) to arrange groups of species in hierarchies according to their supposed phylogeny. It has been customary for writers of Floras to use whatever sequence of taxa was considered orthodox for their particular region or nationality, and this has often tended to delay and restrict the spread of new taxonomic ideas. For example, it has long been customary for Floras in the North American and West Indian region to be arranged more or less according to the system of Engler & Prantl, first proposed in Germany in the late nineteenth century. However, in recent years the accumulation of much new information has led to profound changes in taxonomic concepts. In particular (though not exclusively), the views of such writers as Cronquist and Thorne in the U.S.A. and Takhtajan in the U.S.S.R., reinforced by the work of numerous monographers, have convincingly demonstrated that many details of the older systems have become obsolete. Unfortunately, while these facts are usually taught to modern university students, the writing of floristic books has lagged behind. In an effort to redress this situation in one small area, it was decided to arrange the families of flowering plants in this book according to Arthur Cronquist's *The Evolution and Classification of Flowering Plants* (1968). Even though modifications of this system will undoubtedly occur as knowledge improves, at the present time it reflects much current thinking about plant classification, and therefore deserves to be exercised at the practical level.

The chief exceptions to the original Cronquist system in the present book are (1), the monocotyledon series is placed before the dicotyledons, more out of habit than conviction, as there seems no doubt that the monocotyledons as a group were derived from primitive dicotyledonous ancestors; (2), the Agavaceae are reduced to a subfamily of Liliaceae; (3), the Viscaceae are separated as a family from the

Loranthaceae; (4), the name Surianaceae is used in place of Stylobasiaceae (on the basis of being a *nomen conservandum*); and (5), *Picrodendron* is included in the Euphorbiaceae.

Unlike Engler & Prantl, Cronquist does not deal with classification below the family level, so in this regard the writer has simply followed his own preferences. For the most part, a broad concept of genus has been followed, in the belief that the binomial system of nomenclature works better as a means of recognition and information-retrieval when the number of generic names is kept to a minimum consistent with what is known or believed about natural relationships. It is the writer's opinion that the trend toward more and more splitting at the generic and family levels results in an unbalanced taxonomy that unduly exaggerates the significance of differences at the expense of obscuring resemblances, and at the same time vitiates the usefulness of names as a means of recognition.

In the text, descriptions of families, genera and species are provided, together with keys for the identification of taxa at each level. The identification keys are dichotomous; users should note that characters used in the keys are frequently not repeated in the relevant descriptions. *Thus the keys should be referred to as an integral part of the descriptions.*

All taxa believed to be not indigenous to the Cayman Islands (especially those supposedly or certainly introduced by human agency) are enclosed in square brackets: [...........].

NOTES ON THE ILLUSTRATIONS

The illustrations have been obtained from a number of sources as indicated by bracketed initials inserted at the end of each caption. The explanation of these initials is as follows:

(D.E.), drawn by Derek Erasmus; (F. & R.), reproduced with the permission of the Trustees of the British Museum (Natural History) from Fawcett & Rendle, *Flora of Jamaica*; (G.), drawn by Miss V Goaman; (H.), reproduced with the permission of the Office of the Secretary, United States Department of Agriculture, from Hitchcock, *Manual of the grasses of the West Indies* and collateral publications, the original drawings kindly made available by the Rachel McMasters Miller Hunt Botanical Library, Carnegie-Mellon University, Pittsburgh; (J.C.W.), drawn by Mrs Derek Erasmus; (St.), made available by Dr William T Stearn from drawings prepared for volume 6 of *Flora of Jamaica* (unpublished) – these drawings are held in the *Flora of Jamaica* collection in the Department of Botany at the British Museum; (R.R.I.), a drawing of *Pennisetum purpureum* is by W E Trevithick in a book by R Rose-Innes*; (H.U.), a drawing of *Epidendrum cochleatum* is

---

*Rose-Innes R, (1977) *A manual of Ghana grasses.* Surbiton, Surrey, England: Land Resources Division.

reproduced with the permission of L.G. Garay, Curator of the Orchid Herbarium of Oake Ames Botanical Museum, Harvard University.

It should be noted that, although many of the drawings have been published in other works, a significant number (116) represent original illustrations of species either not depicted before in any publication, or else seldom or never before drawn in as much detail. The high quality of the drawings by Mr and Mrs Erasmus and by Miss Goaman provide this book with a value beyond that which the text alone could support. The important role of Dr William T Stearn and Mr A C Jermy (ferns) in supervising or monitoring the preparation of these drawings is also acknowledged.

Mrs Neil Cruikshank, of the Mosquito Research and Control Unit, Cayman Islands, drew the diagrams in the section on 'Environment and plant communities'.

RECENT NOMENCLATURAL CHANGES

1. (Page 641, 646) The plant named *Ipomoea stolonifera* in this Flora should be *Ipomoea imperati* (Vahl) Griseb., Cat. Pl. Cub. 203. 1866; La Valva & Sabato in Taxon 32: 110-132. 1983.

2. (Page 530, 544) Dr. Grady L. Webster, a leading specialist on Euphorbiaceae, has made *Ateramnus* P.Br. a synonym of *Sapium*, thus restoring the name *Gymnanthes* for the genus named *Ateramnus* in this Flora. See Taxon 32: 304-305. 1983.

ACKNOWLEDGEMENTS

In the course of this Introduction the author has already thanked the many organisations and individuals who have assisted the work of compiling the Flora. He reiterates his grateful acknowledgement of these contributions and especially of those made by the Cayman Islands authorities, the United Kingdom Overseas Development Administration and the Institute of Jamaica, who supported the project as described in the Preface, and by the custodians of plant materials and illustrations: among the latter the author particularly thanks the Trustees of the British Museum (Natural History) and the Rachel McMasters Miller Hunt Botanical Library for their kind permission to use the important drawings already described. Grateful acknowledgement is also made to the Overseas Development Administration for commissioning the original drawings and arranging for permission to use the others; Mr Martin Brunt of the Administration's Land Resources Development Centre has been largely responsible for the success of these arrangements.

As explained in the Preface, the publication of the Flora by Her Majesty's Stationery Office resulted from a joint effort by the Land Resources Development Centre, the Royal Botanic Gardens, Kew, and the Ministry of Agriculture, Fisheries and Food: special thanks are due to the Ministry's Specialist Typing Service, Guildford, for their care in preparing the printer's copy, in particular to the supervisor Mrs M. Willis and the typist Miss Linda Salmon.

DIVISION I. PTERIDOPHYTA

Vascular plants without flowers, fruits or seeds, having a life-cycle of two distinct, independent phases, called sporophyte and gametophyte, which are normally produced in an alternating sequence. The sporophyte, nearly always the more conspicuous phase, has parts differentiated into roots, stem (rhizome or caudex) and leaf (frond in the ferns). These structures always contain special hardened (vascular) tissues for the conduction of water and various water-soluble substances. In leafy tissues the vascular strands are normally evident as veins. The sporophyte moreover produces dust-like asexual spores (reproductive granules) which may be all similar (plant homosporous) or unlike in size and form (plant heterosporous). (All Cayman pteridophytes are homosporous). Spores are borne in very small or minute capsules (sporangia), these isolated or variously clumped on the sporophyte leaves or in their axils. A germinating spore grows by normal cell division into a gametophyte (prothallus) characteristic in form according to taxonomic group; those of true ferns usually scale-like or semi-filamentous, green, and lacking vascular tissue except in rare cases. The prothalli of homosporous species are all alike in general appearance, but may be either monoecious (organs of both sexes on the same individual), or dioecious (male and female organs on separate individuals). The male reproductive cells (gametes or spermatozoids) of all pteridophytes are motile by means of coiled cilia. Fertilization of the egg cell (female gamete or archegone) by a spermatozoid requires the presence of a droplet of water in order to reach its destination; the resulting zygote immediately initiates the growth of a new sporophyte plant, which soon assumes the characteristic appearance of a fern or other pteridophyte. Both the sporophyte and the gametophyte in many cases are also capable of reproduction by vegetative means.

A world-wide group of more than 12,000 species, classified in a number of families whose relationship in many cases is remote. Only two of these families occur in the Cayman Islands.

KEY TO FAMILIES

1\. Plant without true roots or leaves, the plant body consisting chiefly of a dichotomously branched system of erect green stems set with scattered minute pointed scales, the fertile branches bearing isolated 3-celled sporangial capsules: **1. Psilotaceae**

1\. Plant with true roots, stems and leaves, these clearly differentiated; sporangia minute, variously clustered on the backs or margins of fertile leaves: **2. Polypodiaceae**

Family 1

PSILOTACEAE

Plants mostly epiphytic, the stems either simple and provided with small 2-ranked leaves *(Tmesipteris)* or many times dichotomous and appearing leafless, the leaves being remote, minute, and scale-like. Sporangia 2-celled *(Tmesipteris)* or 3-celled and dehiscing vertically, attached on adaxial base of minute bifid sporophylls. Spores reniform, all similar. Gametophytes subterranean or embedded in humus, tuberous.

A very primitive plant-group, with 2 genera, *Tmesipteris* of the South Pacific islands, and *Psilotum*, pantropical.

Genus 1. **PSILOTUM** Sw.

Plants epiphytic or on old masonry, sometimes terrestrial, the rhizomatous part of the stem short-creeping, beset with small brownish hairs. Aerial stems loosely clustered or solitary, the lower unbranched part more or less elongate, dichotomous above into several to numerous 2- or 3-angled divisions. Scale-like leaves few, scattered, minutely awl-shaped, alternate and distichous or else 3-ranked. Sporangia depressed-globose, sessile, 3-lobed, 3-celled; spores hyaline.

A pantropical genus of 2 species.

1. **Psilotum nudum** (L.) Beauv., Prodr. Fam. Aethéog. 112. 1805.

Mostly erect, 20–60 cm high, the main stalk 2–4 mm thick. Branches more or less 3-angled. Scale-like leaves remote, mostly 1–2 mm long, rarely to 3 mm. Sporangia c. 2 mm thick, yellow or brownish.

GRAND CAYMAN: *Brunt 1920; Kings GC 174, GC 408.*
– Pantropical.

Family 2

POLYPODIACEAE

Small to rather large leafy plants of various habit, terrestrial or epiphytic, the usually hairy or scaly rhizome creeping to erect, producing few or many roots and sometimes also stoloniferous. Fronds usually stalked (stipitate), uniform or dimorphic, the lamina (blade) simple to several-times pinnately compound, or rarely digitately- or palmately-divided. Sporangia long-stalked, dehiscing by means of an annulus of elastic cells interrupted by thin-walled cells at one side, and arranged in lines or clusters (sori), or completely covering the underside of fronds or frond-divisions (pinnae or segments). Sori naked or protected by scale-like indusia which develop either from the veins or from the margins of the lamina (rarely both). Spores bilateral to globose or tetrahedral, often surrounded by a husk-like outer covering called the perispore, this smooth or variously sculptured or even spinulose, the form usually characteristic for genera or often for species. Gametophytes green, flat, glabrous or hairy, sometimes glandular.

More that 170 genera (up to 240 or more recognized by some authors) and nearly 10,000 species, as here defined. Some specialists subdivide the group into numerous smaller families, but there is wide divergence of opinion as to whether and how this should be done. The present writer does not favour fragmentation of the ferns at the family level, preferring a concept of subfamilies. In any case, the relatively very few Cayman fern genera are more conveniently arranged under a single family heading. A summary of the habitats of the Cayman ferns is given in Figure 2A, page 77.

KEY TO GENERA

1. Sporangia completely covering the backs of fertile pinnae; large, coarse ferns of brackish swamps: **6. Acrostichum**

1. Sporangia clustered in linear or roundish sori, never completely covering the lamina tissue; small to medium-sized ferns of various habitats, but not occurring in brackish or saline situations:

  2. Sori linear or marginal (or both):

    3. Sori marginal:

      4. Stipes wiry, black or nearly so, glossy:

        5. Ultimate frond-divisions mostly sessile and less than 3 mm broad; sporangia arising from the lamina, protected by a recurved marginal indusium: **1. Cheilanthes**

        5. Ultimate frond-divisions distinctly black-stalked and up to 10 mm broad; sporangia arising from underside of the indusial flap: **2. Adiantum**

      4. Stipes neither black, wiry, nor particularly glossy:

        6. Rhizome elongate, without scales; fronds scattered, long-stalked, the lamina deltate in outline and 3-4-pinnate: **3. Pteridium**

6. Rhizome short-creeping, densely scaly; fronds clustered, oblanceolate in outline, 1-pinnate: **4. Pteris**

3. Sori not marginal, elongate and parallel to midveins of pinnae: **5. Blechnum**

2. Sori round or nearly so, not marginal:

7. Lamina fully pinnate, at least at the base (i.e., at least the lowest pinnae attached to the rhachis by a junction much narrower than the width of the pinna); indusium present or apparently absent:

8. Veins all free, none joined or connivent in the tissue; pinnae entire to serrate, jointed to the rhachis: **7. Nephrolepis**

8. At least some veins joined or connivent in the tissue; pinnae more or less lobed, and not jointed to the rhachis:

9. Pinnae few (3-6 pairs), 1-6 cm broad with entire margins, the lowest with basal lobes; veins reticulate: **8. Tectaria**

9. Pinnae numerous (7-25 or more pairs), 0.5-1.5 cm broad with margins lobed throughout; only basal veins joined or connivent: **9. Thelypteris**

7. Lamina simple or pinnatisect (i.e., joined to the rhachis by a broad base, as broad as the segment itself); sori without any trace of indusium: **10. Polypodium**

Genus 1. **CHEILANTHES** Sw.

Small terrestrial or epipetric ferns. Rhizomes short-creeping, scaly. Fronds clustered, erect or spreading, often glandular, hairy or scaly (but not in the single Cayman species), the lamina 1-4-pinnate, or 1-pinnate and variously pinnatifid, the ultimate divisions in some species minute. Rhachis and costae terete, black; veins free, simple or forked. Sori marginal, arising from the enlarged tips of the veins, usually numerous and narrowly confluent, more or less protected by the reflexed, membranous, indusioid margin. Spores globose or tetrahedral and smooth, granulose, or sometimes corrugated.

A world-wide genus of more than 125 species, many of them characteristically occurring in the crevices of cliffs or among rocks.

1. Cheilanthes microphylla (Sw.) Sw., Syn. Fil. 127. 1806. **FIG. 1.**

Rhizome clothed with soft, narrowly linear scales, these pale yellow or reddish-brown. Fronds mostly 8-20 cm long (becoming larger in other regions), the wiry black stipes much shorter than or equalling the lamina, and bearing a few hair-like scales near the base. Lamina oblong-lanceolate, seldom exceeding 2 cm in width in Cayman plants, 2-3-pinnate; rhachis and costae sparsely clothed with lax, jointed,

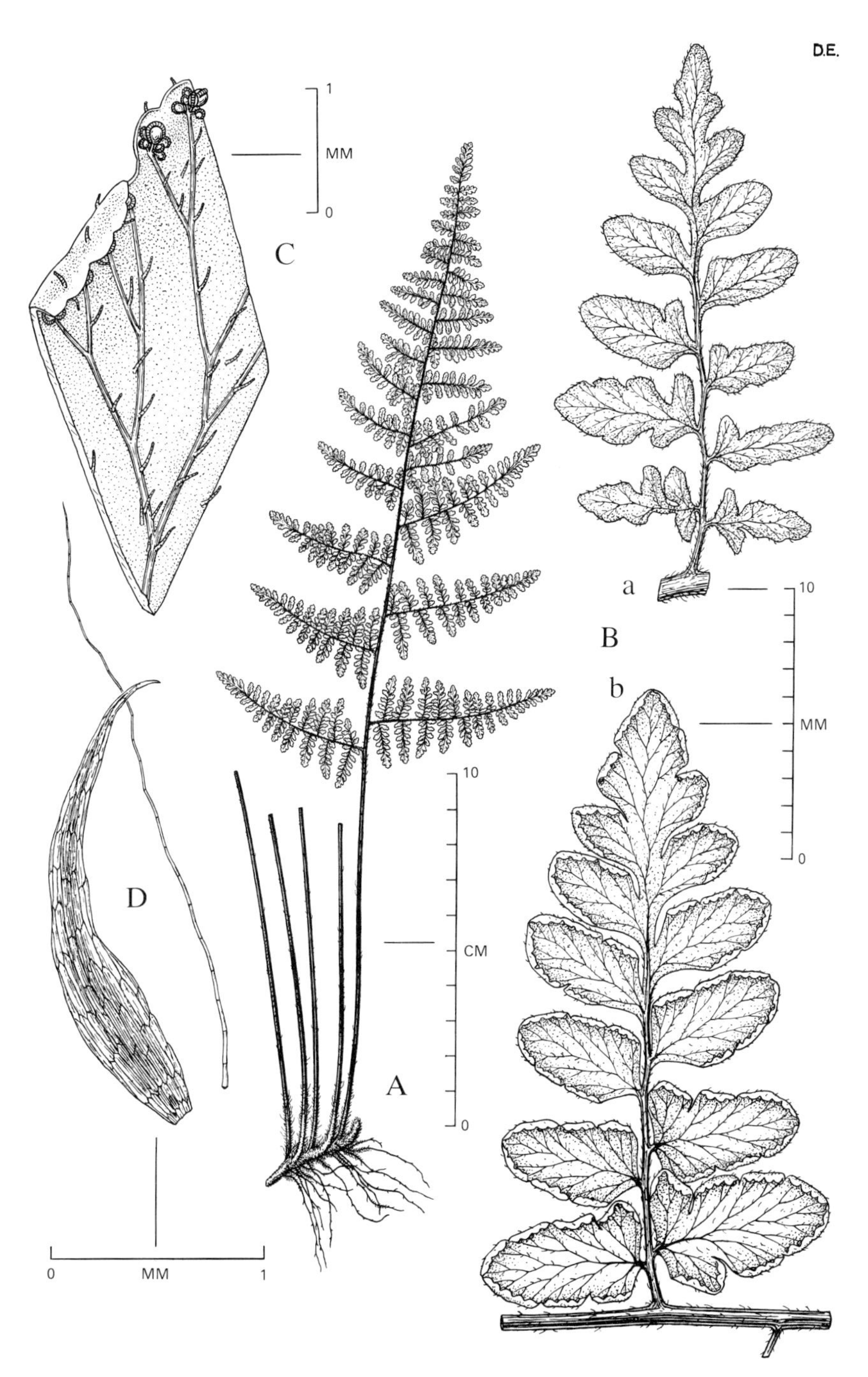

FIG. 1 **Cheilanthes microphylla.** A, habit. B, single pinna; a, upper side; b, lower side. C, portion of fertile pinnule. D, scale and hair of rhizome. (D.E.)

whitish hairs. Pinnules simple, rarely lobed or pinnate, the ultimate divisions always more or less oval or rounded-oblong.

CAYMAN BRAC: *Proctor 29007.*
– Southern United States and Mexico, the West Indies (common), and northern South America, usually on shaded limestone ledges or similar habitats. The Cayman Brac plants are frequent in certain places along the northern side of The Bluff.

Genus 2. **ADIANTUM** L.

Terrestrial ferns of forest floors, shady ravines and rocky banks or cliffs. Rhizomes slender and wide-creeping to short and sub-erect, bearing numerous narrow scales chiefly near the apex. Fronds usually distichous, apart or clustered with firm, dark to black, usually highly glossy stipes. Lamina 1-5-pinnate (rarely simple), of various patterns of dissection, most often glabrous or apparently so, sometimes glaucous beneath. Ultimate divisions sessile or stalked, often articulate and deciduous; veins free, often flabellately-branched. Sori borne on the underside of the reflexed margin (or marginal lobe), the sporangia arising along the ends of (and sometimes between) the ultimate veinlets. Spores most often tetrahedral, dark, smooth.

A large pantropical genus of about 200 species, most numerous in South America; a few also occur in temperate regions. Often called "maidenhair ferns".

KEY TO SPECIES

1. Lamina 1-pinnate, or (if 2–3-pinnate) with an elongate terminal pinna essentially like the lateral ones; ultimate divisions mostly curved-oblong with one entire side, the stalks not articulate: **1. A. melanoleucum**

1. Lamina compoundly dissected without a distinct terminal pinna; ultimate divisions more or less rhombic or flabellate-cuneate with 2 entire sides, the stalks articulate at the apex: **2. A. tenerum**

1. Adiantum melanoleucum Willd. in L., Sp. Pl. 5: 443. 1810. **FIG. 2.**

Rhizome short-creeping, densely clothed with dull brownish, concolorous scales. Fronds closely distichous, in Cayman plants seldom over 30 cm long, the stipes lustrous purple-black, scabrous. Lamina variable, in juvenile or depauperate forms linear, 1-pinnate, and mostly less than 20 cm long; those of well-developed mature plants 2-pinnate (or 3-pinnate at the base); pinnules or ultimate divisions oblong, close, dimidiate. Sori oblong-lunate or linear and deeply curved, occurring only along the distal margins.

GRAND CAYMAN: *Kings F 24.*
– Florida, Bahamas, and the Greater Antilles, on shaded limestone ledges or cliffs. The Grand Cayman plants were found on the sides of a "well".

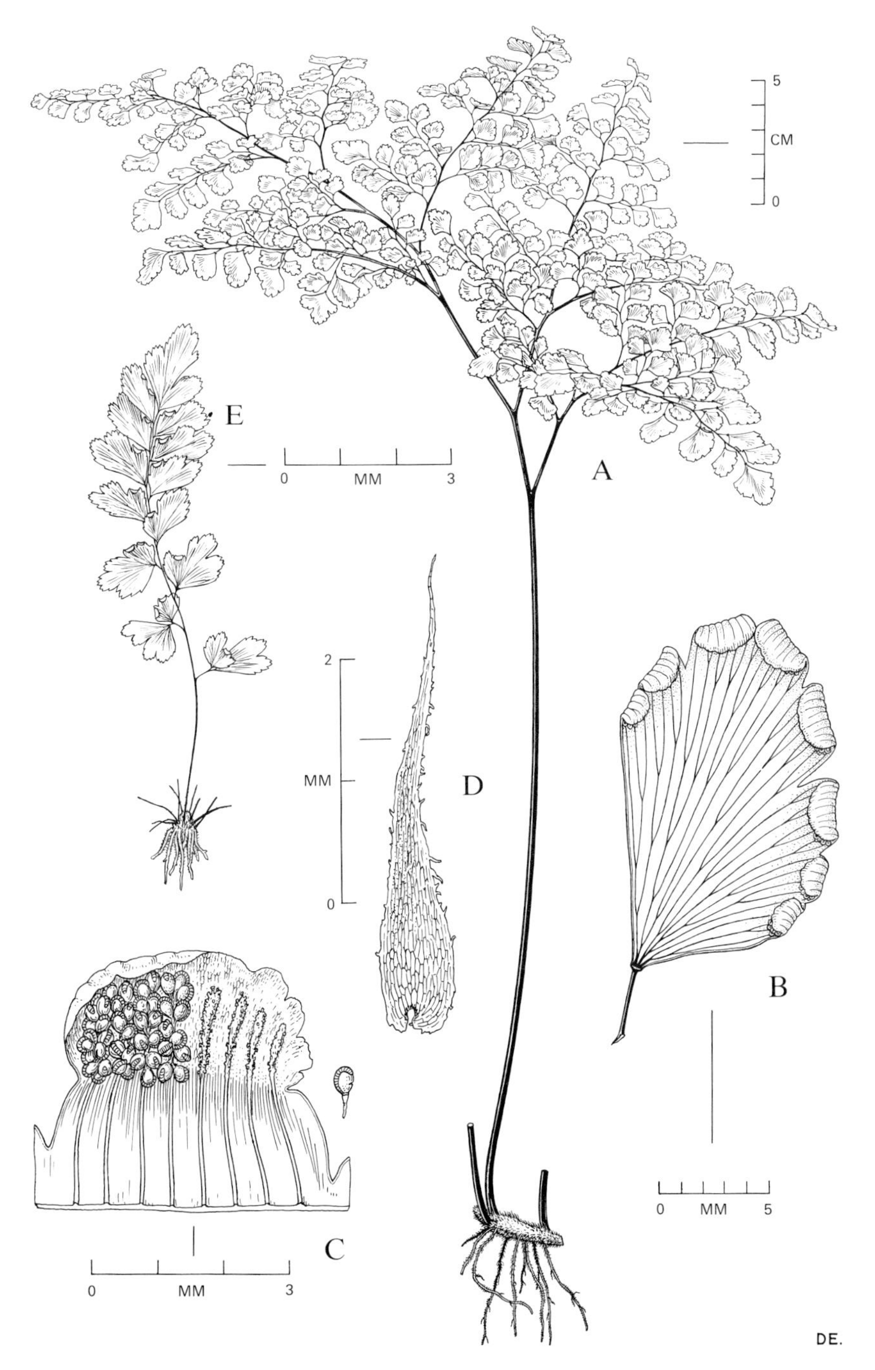

FIG. 2 **Adiantum.** A, **A. tenerum,** habit. B, pinnule. C, fertile lobe (dissected). D, scale of rhizome. E, **A. melanoleucum**, depauperate plant. (D.E.)

2. **Adiantum tenerum** Sw., Nov. Gen. & Sp. Pl. 135. 1788. **FIG. 2.**

Rhizome short-creeping, densely clothed with lustrous dark brown scales having pale, lacerate margins. Fronds few, distichous, erect and spreading or sometimes pendent, reaching 70 cm long or more, the stipes lustrous black or purple-black, smooth. Lamina deltate-ovate, subpentagonal, nearly as broad as long, 3-5-pinnate at the base; pinnae alternate, stalked; ultimate divisions more or less rhombic-oblong to flabellate-cuneate, on 2-4 mm stalks, these distinctly jointed at the apex. Sori retuse-oblong, borne in pairs on each of the lightly bifid lobes of the distal margin.

GRAND CAYMAN: *Kings F 40; Proctor 15274.*

– Southern United States, the West Indies, and Mexico to northern South America. Cayman plants occur on shaded limestone banks and ledges, and in moist, shaded "wells", not common.

Genus 3. **PTERIDIUM** Gled. ex Scop., nom. cons.

Terrestrial ferns; rhizomes deeply subterranean, extensively creeping, repeatedly branched, lightly clothed toward the apex with reddish-brown, pluricellular, articulate hairs; scales lacking. Fronds of coarse, hard texture, tripartite and mostly 3-4-pinnate, the lower pinnae with glandular so-called "nectaries" at base when young; ultimate segments variable in size and shape, but always with revolute margins; veins free. Sori marginal, usually continuous, the sporangia borne between the outer reflexed margin and an inner indusial flange, on a vascular strand connecting the vein-ends. Spores globose-tetrahedral, very finely spinulose.

Usually considered to consist of a single, variable, world-wide species, a concept which unfortunately results in a cumbersome hierarchy of trinomial and quadrinomial names. In many countries this fern is (or these ferns are) known as "Bracken", and is considered a pernicious weed in most places where it grows.

1. **Pteridium aquilinum** (L.)Kuhn in v.d.Decken, Reisen in Ost-Afrika 33: 11. 1879. var. **caudatum** (L.) Sadebeck in Jahrb. Hamb. Wiss. Anst. 14, Beiheft 3: 5. 1897.

Fronds often 1 m tall or more, the greenish to straw-coloured stipes nearly as long as the deltate lamina, glabrous throughout except for the lamina tissue beneath. Ultimate segments mostly linear, 1-5 cm long, 1-4 mm broad, distant and distinct.

GRAND CAYMAN: *Kings F 39.*

– Florida, the West Indies, Mexico and Central America (this variety). Apparently rare in the Cayman Islands; it occurs chiefly in old fields and secondary thickets.

HABITAT	GENUS NUMBER AND SPECIES NAME	OCCURRENCE (INCLUDING ISLANDS WHERE IDENTIFIED)
Mangrove (*Rhizophora* or *Laguncularia*) swamp woodland and thicket	6. Acrostichum aureum*	More or less brackish swamps (and wet pockets of limestone, see below) (GC, LC)
	6. Acrostichum danaeifolium*	” ” ” ” ” (GC)
	7. Nephrolepis biserrata* (R)	Only in *Rhizophora* swamp woodland (GC)
	7. Nephrolepis multiflora* (R)	” ” ” ” ” (GC)
Seasonal *Conocarpus* swamp thicket and bushland	6. Acrostichum aureum*	Brackish to freshwater swamps (GC, LC)
	6. Acrostichum danaeifolium*	” ” ” ” (GC)
	5. Blechnum serrulatum*	In the drier parts of the swamp associated with sedges (GC)
	7. Nephrolepis biserrata* (R)	An epiphyte on swamp trees or on the ground (GC)
	7. Nephrolepis multiflora* (R)	Brackish to fresh water swamps (GC, CB)
Seasonal *Typha* bulrush swamp	9. Thelypteris interrupta	Fresh water ponds and boggy edges of *Typha* swamps (GC)
	6. Acrostichum aureum*	Sometimes co-dominant with *Typha* (GC)
Seasonal herbaceous and grassland swamp	5. Blechnum serrulatum*	In fresh water grassland swamps and wet ditches (GC)
Moist marly swales	9. Thelypteris kunthii*	Moist marly swales and depressions (also ledges and “wells”, see below) (GC, CB)
Shady and/or moist limestone ledges, crevices, “wells”	1. Cheilanthes microphylla	Shaded limestone cliff ledges (CB)
	2. Adiantum melanoleucum	Shaded limestone cliff ledges/“wells” (GC)
	2. Adiantum tenerum (R)	” ” ” ” ” (GC)
	7. Nephrolepis multiflora (R)	” ” ” ” ” (GC, CB)
	8. Tectaria incisa (R)	Moist shaded “wells” (GC)
	9. Thelypteris reptans	Shaded limestone ledges (GC)
	4. Pteris longifolia var. bahamensis (R)	Moist limestone ledges, pits (CB, GC)
	9. Thelypteris kunthii*	Ledges, pits, wells (and swales, see above) (GC, CB)
	9. Thelypteris augescens (R)	Moist ledges (GC)
	10. Polypodium dispersum	More or less shaded limestone rock outcrops (GC)
	10. Polypodium heterophyllum (R)*	Sterile plants sometimes on limestone rocks (main habitat woodland, see below) (GC)
	6. Acrostichum aureum*, A. danaeifolium*	Sometimes on wet pockets in the limestone (main habitat in swamps, see above) (GC, LC)
Dry evergreen woodland and thicket	10. Polypodium polypodioides (R)	Tree trunks in sheltered woodland (GC, CB)
	10. Polypodium heterophyllum (R)	Climbing, in dense woodland (GC)
	10. Polypodium phyllitidis (R)	Tree trunks in sheltered woodland (GC)
Cultivated areas, old fields and secondary thickets	3. Pteridium aquilinum var. caudatum (also various species of 7. Nephrolepis (see text))	Old fields and secondary thickets (GC)
R rare or not common.	*listed twice, under different habitats. GC Grand Cayman.	LC Little Cayman. CB Cayman Brac.

FIG. 2A. Habitats of the Cayman ferns (Polypodiaceae). See also the detailed descriptions of species and the classification of Cayman vegetation described in the Introduction

Genus 4. **PTERIS** L.

Terrestrial ferns; rhizomes often stout and woody, creeping to erect, clothed with scales at least near the apex. Fronds various in outline, the veins free or areolate. Sori marginal, usually linear, the sporangia borne in a continuous line on a slender inframarginal vein connecting the ultimate vein-ends; paraphyses usually present; indusium linear, formed by the modified reflexed margin, opening inwardly. Spores tetrahedral (rarely bilateral), and smooth, tuberculate or otherwise sculptured.

A large, worldwide genus of at least 280 species, most abundant in tropical regions.

1. Pteris longifolia L., Sp. Pl. 2: 1074. 1753.
var. **bahamensis** (Ag.)Hieron. in Hedwigia 54: 289. 1914. **FIG. 3.**

Pteris bahamensis (Ag.) Fée. 1852.

Rhizome decumbent, densely clothed with yellowish to rusty-brown hair-pointed scales. Fronds ascending or spreading (forming an open rosette), usually less than 50 cm long including the short stipe; stipes 5–15 cm long, bearing soft scales which on falling leave a smooth surface; lamina glabrous except for a few scattered minute hairs and glands along the rhachis, oblanceolate or rarely elliptic-oblong, 3–12 cm broad, 1-pinnate, terminating abruptly in a linear segment; pinnae narrowly linear, ascending, 2-7 mm broad, the sterile ones with finely crenulate margins; veins simple or once-forked.

GRAND CAYMAN: *Proctor 15304.* CAYMAN BRAC: *Kings CB 97; Proctor 29363.*
– Florida, Bahamas, and Cuba, the Cayman population occurring on moist limestone ledges, the sides of old excavations, and on talus slopes below cliffs, not common. Differs from "typical" *P. longifolia* in its smaller fronds with smooth stipes and ascending pinnae which are truncate or nearly so at the base. Grand Cayman plants show a diploid chromosome count as contrasted with the tetraploid Jamaican population of *P. longifolia.* (T. Walker, personal communication.)

Genus 5. **BLECHNUM** L.

Mostly rather coarse terrestrial ferns; rhizomes ascending to erect and often stout or even trunk-like, or else elongate and scandent, densely scaly, sometimes stoloniferous. Fronds 1-pinnate (rarely simple), usually glabrous, all similar (as in the Cayman species) or dimorphic; veins of sterile pinnae all free. Sori elongate-linear, usually continuous, borne near or against the pinna midvein (costa) on an elongate transverse veinlet connecting the main veins and parallel to the costa; indusium narrowly linear, continuous or nearly so, firm, opening toward the costa.

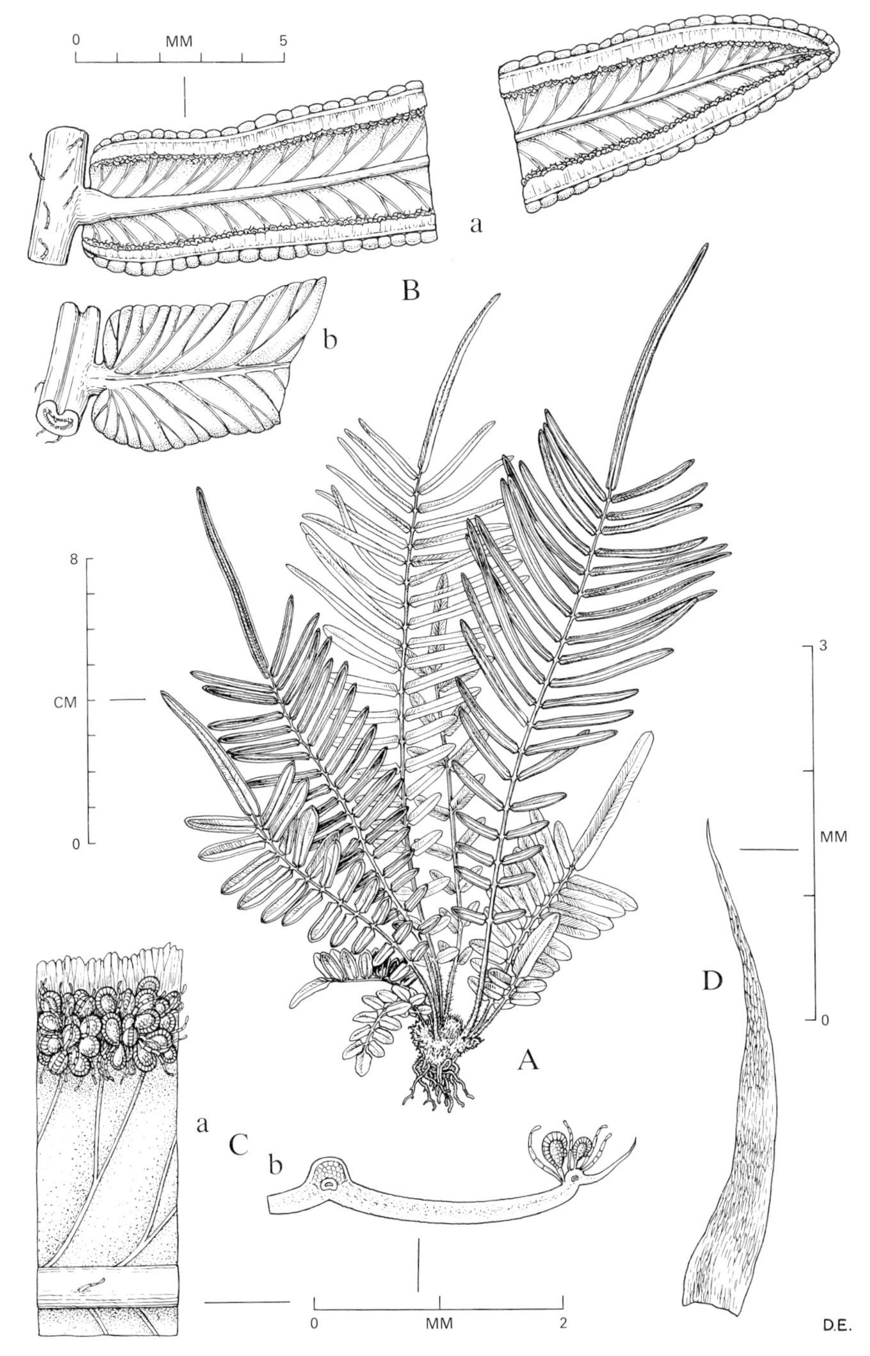

FIG. 3 **Pteris longifolia** var. **bahamensis.** A, habit. B, pinna; a, lower side; b, upper side. C, details of fertile pinna; a, with indusium removed; b, section. D, scale of rhizome. (D.E.)

Spores bilateral, usually smooth.

A world-wide genus of at least 180 species, the majority in the southern parts of both hemispheres.

1. Blechnum serrulatum L.C.Rich. in Act. Soc. Hist. Nat. Paris 1: 114. 1792. **FIG. 4.**

Rhizome subterranean, wide-creeping and stolon-like, with erect, subwoody branches bearing the fronds, and clothed with small, rigid, dark brown scales. Fronds rigidly erect, up to 1 m long or more, the stipes shorter than the lamina; lamina narrowly to broadly oblong, mostly 10–20 cm broad, terminating in an apical pinna similar to the lateral ones; pinnae linear-ligulate, 5–15 mm broad, sessile and articulate at the base, eventually deciduous; midvein (costa) bearing a few small yellowish scales beneath; veins close, 1-3-forked; tissue of hard texture. Sori as described for the genus; indusium becoming minutely and irregularly lacerate.

GRAND CAYMAN: *Brunt 1951; Kings F 37, F 43; Lewis P 39; Proctor 3241.* – Florida, the West Indies, and Mexico to South America; Cayman plants form tufts in grassy swamplands and in roadside ditches.

Genus 6. **ACROSTICHUM** L.

Large, coarse ferns of brackish or saline swamps; rhizomes stout, woody, erect, scaly at the apex. Fronds erect, 1-pinnate with large, simple pinnae; venation closely areolate without included free veinlets. Sporangia densely covering the under-surface of fertile pinnae, copiously mingled with sterile paraphyses; indusium lacking. Spores tetrahedral, minutely tuberculate.

A small pantropical genus with several species of wide distribution.

KEY TO SPECIES

1. Fertile fronds with only the upper pinnae bearing sporangia; paraphyses capitate-stellate, dark glistening brown: **1. A. aureum**

1. Fertile fronds with (usually) all the pinnae bearing sporangia, the sterile and fertile fronds thus completely dimorphic; paraphyses sausage-shaped, pale brown: **2. A. danaeifolium**

1. Acrostichum aureum L., Sp. Pl. 2: 1069. 1753. **FIG. 5.**

Rhizome bearing at apex a dense tuft of rigid linear, dark brown scales c. 1.5 cm long or less. Fronds up to 3 m long (usually much less); stipes much shorter than the lamina, subterete, light brown, bearing at base a cluster of ovate, thinly papery

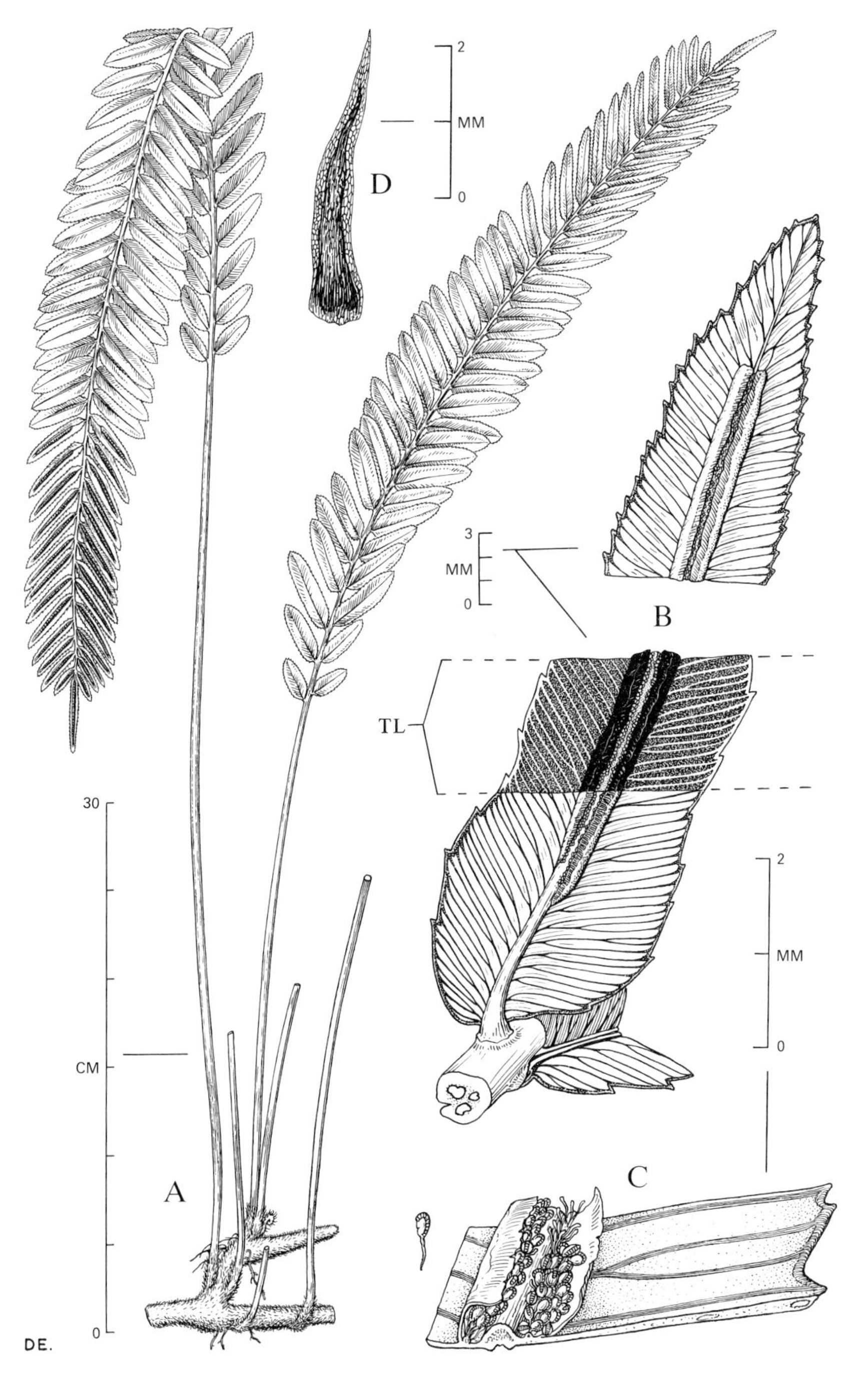

FIG. 4 **Blechnum serrulatum.** A, habit. B, pinna (TL, by transmitted light). C, enlarged portion of fertile pinna. D, scale of rhizome. (D.E.)

scales, these pale brown with a blackish midrib, the margins minutely lacerate, up to 1.5 cm long; upward the stipes bearing several alternate hard spurs. Lamina oblong, 20-40 cm broad, with 10-14 ascending pinnae on each side and a terminal one; tissue of hard texture, glabrous; fertile pinnae (when present) 1-4 on a side.

GRAND CAYMAN: *Correll & Correll 51049; Maggs II 66; Proctor 15270.* LITTLE CAYMAN: *Kings LC 54; Proctor 28090.*
– Pantropical; common in more less brackish swamps, also occasionally in wet pockets of limestone rock.

2. Acrostichum danaeifolium Langsd. & Fisch., Ic. Fil. 1: 5, t. 1. 1810. **FIG. 5.**

Rhizome massive, clothed at apex with rigid, linear, dark brown scales up to 2 cm long. Fronds up to 4 m tall; stipes very stout, grooved, dark brown, bearing at base a cluster of broadly linear scales up to 2.5 cm long, these with a dark brown longitudinal band in the middle and paler, retrorsely fibrillose margins. Sterile fronds erect to somewhat spreading, the fertile ones taller and rigidly erect; pinnae very numerous, 20-40 or more on a side and a terminal one; tissue of sterile pinnae sometimes finely pubescent beneath; fertile pinnae of somewhat fleshy texture.

GRAND CAYMAN: *Proctor 15269.*
– Western Hemisphere tropics, in the same habitats as the preceding, but usually much less common.

Genus 7. **NEPHROLEPIS** Schott

Plants terrestrial or epiphytic, the rhizomes mostly erect and short, scaly and usually extensively stoloniferous. Fronds clustered, crowded, persistent, with relatively short stipes; lamina usually oblong-linear, more or less elongate, the apex of slow growth (thus often appearing indeterminate), normally 1-pinnate (some horticultural forms are more finely dissected). Pinnae articulate to the rhachis and eventually deciduous; margins entire to more or less serrate; veins 1-4-forked, all but the fertile branches reaching nearly to the margins, with enlarged tips often secreting a small white calcareous scale on the upper (adaxial) side of the pinna. Sori terminal on the first distal vein-branches, medial to submarginal in relative position, in a single row on either side of the midvein, each protected by an orbicular to lunate indusium which is attached at the sinus. Spores bilateral.

A pantropical genus of about 30 species, some of them variable and difficult to define. One indigenous and one naturalized species occur outside of cultivation in the Cayman Islands. Several horticultural forms of other species are sometimes grown for ornament.

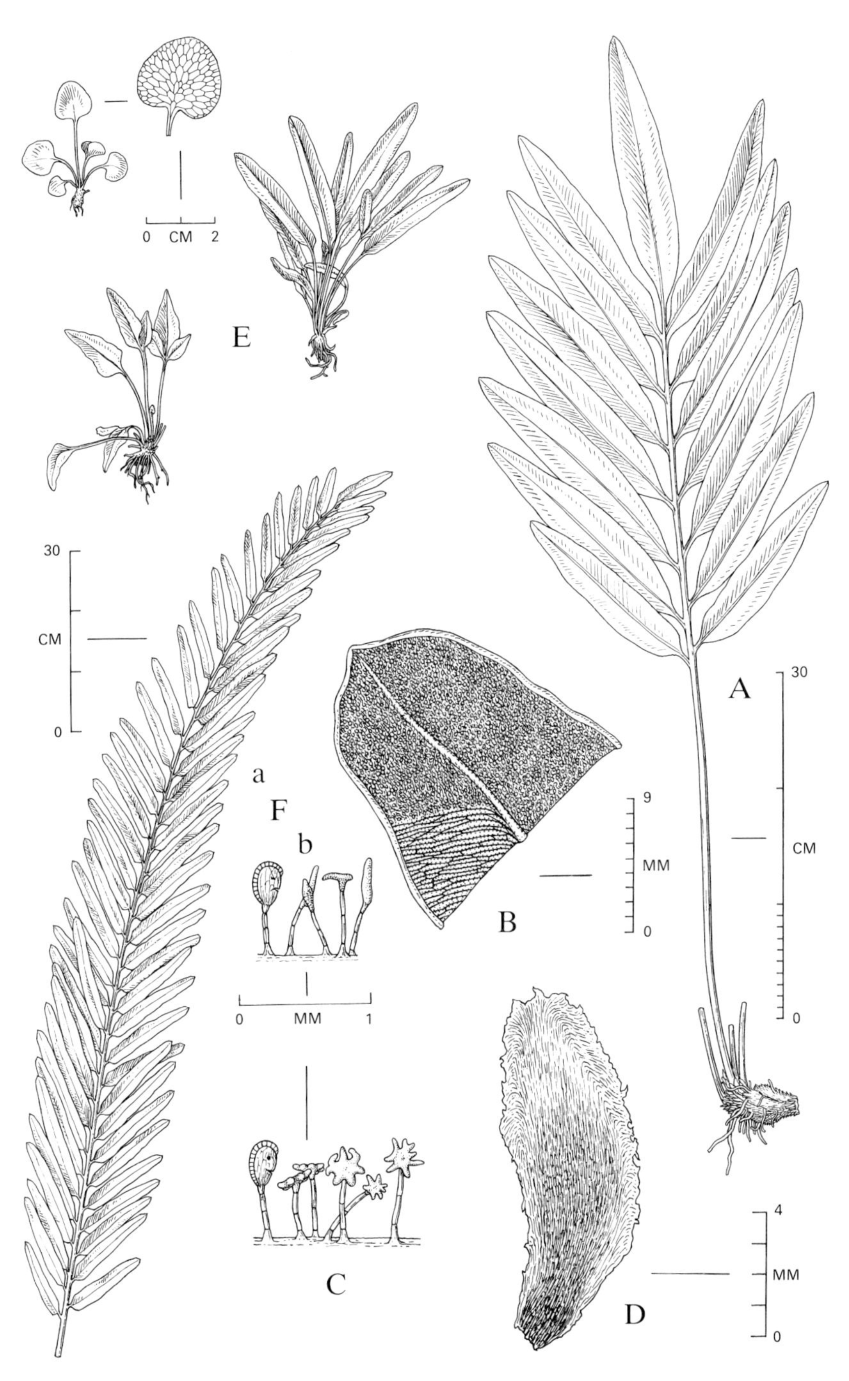

FIG. 5 **Acrostichum.** A, **A. aureum,** habit. B, tip of fertile pinna, lower side. C, sporangium and paraphyses. D, scale of rhizome. E, juvenile sporophytes. F, **A. danaeifolium**; a, frond; b, sporangium and paraphyses. (D.E.)

KEY TO SPECIES

1. Midvein of pinnae glabrous on upper side, or else bearing a few scattered flexuous hairs 0.5–1 mm long; scales of stipe spreading and rather loose; pinnae broadly cuneate at base, some of them short-stalked: **1. N. biserrata**

1. Midvein of pinnae on upper side densely clothed with very short, straight, pale brown pluricellular hairs 0.2–0.3 mm long; scales of stipe (or many of them) closely appressed and blackish-brown with pale minutely fibrillose margins; pinnae subcordate and auricled at the sessile base: **[2. N. multiflora]**

1. **Nephrolepis biserrata** (Sw.)Schott, Gen. Fil. under t. 3. 1834. **FIG. 6.**

Rhizome woody, densely clothed at apex with light brown, sparingly ciliate and denticulate, hair-pointed scales; stolons numerous, slender, elongate, scaly. Fronds 1-5 m long, often arching or pendent, with fibrillose to nearly naked rhachis; pinnae 1.2-2.5 cm broad, the margins varying from finely dentate-serrulate in sterile pinnae to bicrenate in fertile ones, the crenations often minutely toothed.

GRAND CAYMAN: *Kings F-33; Millspaugh 1376.*
– Pantropical. Rare in the Cayman Islands, where its precise habitat is not recorded. Elsewhere, it occurs both on the ground, on rocky ledges, and as an epiphyte on trees especially in swamps.

[2. **Nephrolepis multiflora** (Roxb.) Jarrett ex Morton in Contr. U.S. Nat. Herb. 38: 309. 1974.

Rhizome short, woody, erect, clothed especially at the apex with lance-attenuate, mahogany- to dark-brown adpressed scales, these with pale, finely fimbriate-ciliate margins; similar scales on stipe and rhachis, becoming progressively smaller above; stolons numerous, wiry. Fronds 15–120 cm long, the stipes relatively very short. Lamina narrowed at base; pinnae 2-7.5 cm long, 0.5–1.2 cm broad above the auricled base, the base sessile and subcordate with the acroscopic auricle often overlapping the rhachis, the margins bluntly serrulate to finely crenate. Sori submarginal, with nearly orbicular indusium.

GRAND CAYMAN: *Brunt 1854, 1864, 2034; Proctor 15285.* CAYMAN BRAC: *Proctor 29364.*
– Native to India, this species has become widely naturalized in tropical areas, including southern Florida and the West Indies. Although elsewhere often a road-side weed, Cayman plants occur on shaded limestone ledges or on the moist sheltered sides of "wells", and are not common.]

The following may be found under cultivation:

(a) *Nephrolepis cordifolia* (L.) Presl, with narrow fronds, the stolons usually bearing small, scaly tubers. Tuberless clones are also known.

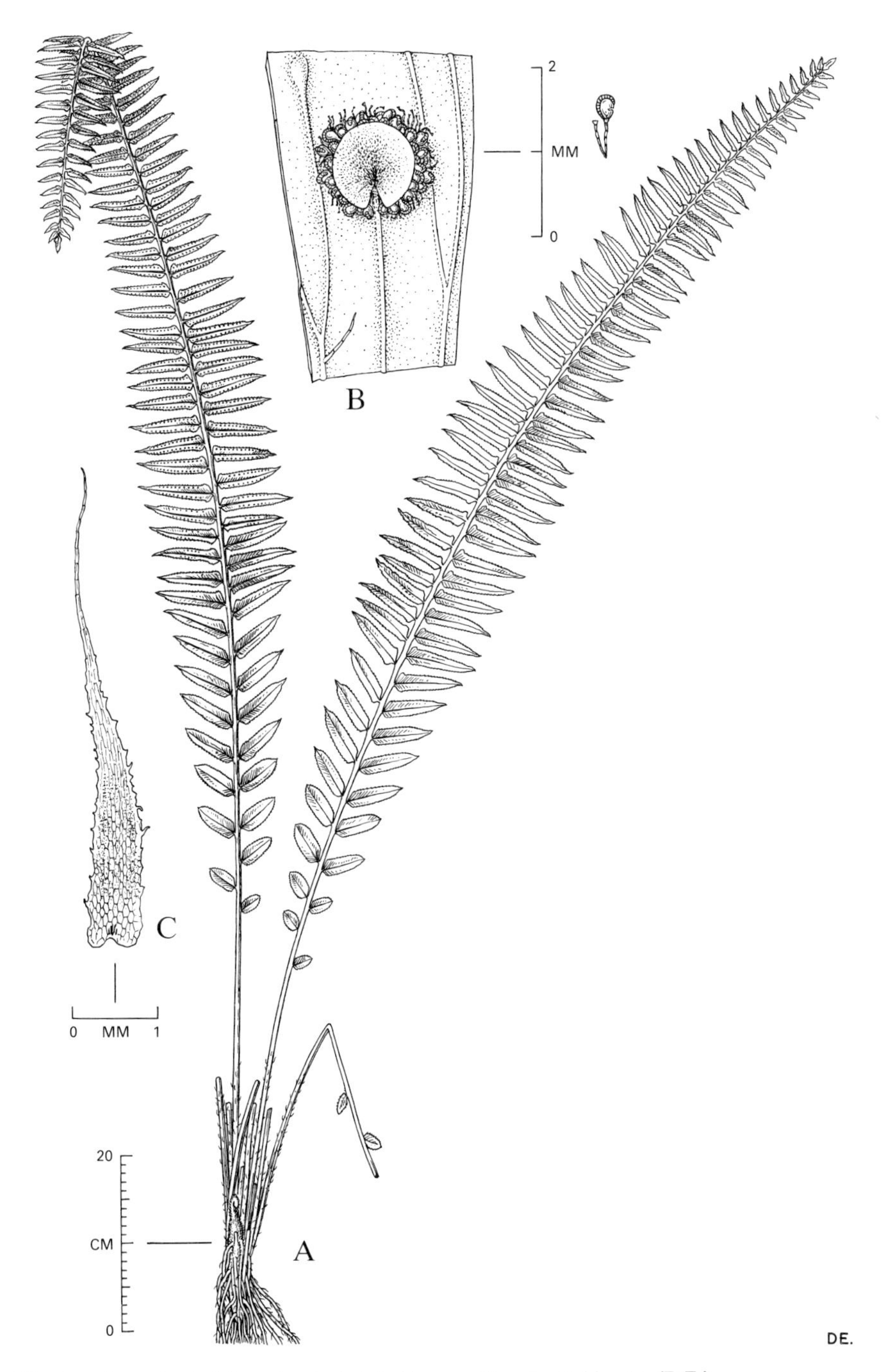

FIG. 6 **Nephrolepis biserrata.** A, habit. B, sorus. C, scale of rhizome. **(D.E.)**

(b) *N. cordifolia* cv. 'Duffii' (*N. cordifolia* f. *duffii* (T.Moore) Proctor), with small orbicular or doubly-orbicular pinnae and repeatedly forked rhachises, the stolons without tubers. This form is nearly always sterile. Its alleged relationship with *N. cordifolia* may be incorrect.

(c) *N. exaltata* (L.) Schott cv. 'Boston Fern' (*N. exaltata* var. *bostoniensis* Davenp.), a variable, more or less finely dissected form which occasionally reverts to the typical 1-pinnate condition. A more extreme form is cv. 'Verona', the most finely dissected of all.

(d) *N. falcata* (Cav.) C.Chr. cv. 'Furcans' (*N. falcata* f. *furcans* (T.Moore in Nicholson) Proctor), the "fishtail fern" often cultivated in the West Indies, with pinnae and sometimes the rhachis forked 1 to several times. It is often incorrectly called a form of *N. biserrata.* The ancestral stock of cv. 'Furcans' apparently originated in Australia or New Guinea.

(e) *N. hirsutula* (Forst.)Presl cv. 'Superba', a hairy, always sterile form with deeply and irregularly lobed pinnae. Typical *N. hirsutula* is native to tropical Asia and islands of the western Pacific Ocean. It has been reported as occasionally adventive in the American tropics.

Genus 8. **TECTARIA** Cav.

Medium to rather large terrestrial ferns; rhizomes woody, short-creeping to erect, scaly at the apex. Fronds clustered, simple to 1-pinnate or pinnately dissected, the divisions usually broad with entire or sparingly incised margins; venation reticulate, the areoles with or without free included veinlets. Sori mostly round (sometimes somewhat elongate or irregular), scattered or borne in open rows; indusium peltate, reniform, or sometimes lacking. Spores bilateral with thick perispore, becoming tuberculate or spinulose.

A pantropical genus of perhaps nearly 100 species, many of them poorly understood.

1. **Tectaria incisa** Cav., Descr. Pl. 249. 1802. **FIG. 7.**

Rhizome woody, erect, at maturity 1.5–3 cm thick, the apex clothed with dark brown, lance-deltate, denticulate-fimbriate scales up to 7 mm long. Fronds scarcely more than 50 cm long in Cayman plants, elsewhere up to 1.5 m long, the grooved stipe about as long as the lamina; lamina oblong or ovate-oblong, 1-pinnate with basal pinna deeply 2-lobed, the lower division shorter than the upper; undivided pinnae narrowly or broadly oblong with long-acuminate apex, usually several cm broad, the terminal pinna confluent with the uppermost lateral ones; tissue normally glabrous; vein-areoles often with a free included veinlet. Sori round; indusium round-reniform, persistent, often appearing peltate by the overlapping of the basal lobes.

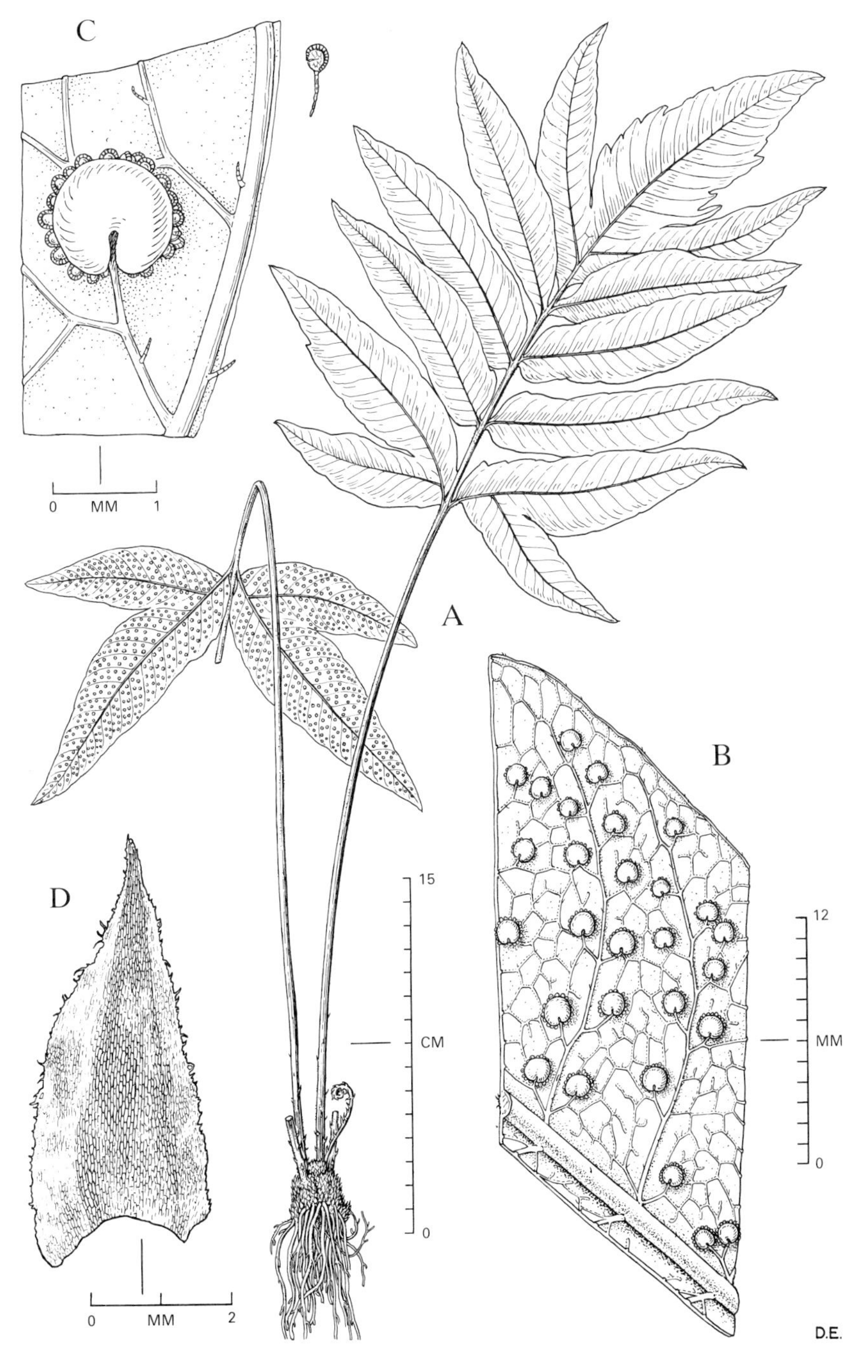

FIG. 7 **Tectaria incisa.** A, habit. B, portion of fertile pinna, lower side. C, sorus. D, scale of rhizome. (D.E.)

GRAND CAYMAN: *Kings F 46.*

– West Indies and Mexico to South America. Rare in the Cayman Islands, where it occurs on the moist, shaded sides of "wells".

Genus 9. **THELYPTERIS** Schmid.

Small to moderately large terrestrial ferns. Rhizomes slender and wide-creeping to thick and erect, clothed with scales. Fronds scattered or clustered, the lamina usually 1-pinnate (rarely simple), the pinnae with margins variously lobed or pinnatifid, glabrous or commonly pubescent, the hairs simple, forked, or stellate. Veins free, or the lowermost connivent at the sinuses of lobes, or adjacent pairs joined in the tissue, at their junction producing a short excurrent veinlet; in some species (not occurring in the Cayman Islands) all the adjacent veins are joined, the venation thus being completely reticulate. Sori roundish or elliptic, dorsal on veins, with or without indusium; indusium (if present) usually roundish-reniform, attached at the sinus. Spores bilateral, monolete, with variously wrinkled or sculptured perispore.

One of the largest genera of ferns, worldwide in general distribution, with probably at least 900 species. Some authors have preferred to subdivide the group into several or many smaller genera, a procedure not favoured by the present writer. However, in areas where many species occur, it is convenient to arrange them in groups called subgenera which serve to emphasize lines of closer afinity within the genus. It is hardly necessary to do this in describing the few Cayman Islands species.

KEY TO SPECIES

1. Lowest veins of adjacent lobes free or connivent to a membrane at the sinus, not truly joined in the tissue:

2. Median pinnae with mostly less than 25 lobes on a side (on mature fronds), these mostly longer than broad, the basal ones of lower pinnae usually enlarged and overlapping the rhachis: **1. T. kunthii**

2. Median pinnae with up to 40 lobes on a side, these mostly as broad as long and distinctly triangular, the basal ones of lower pinnae, if slightly enlarged, not overlapping the rhachis: **2. T. augescens**

1. Lowest veins of adjacent lobes clearly joined in the tissue:

3. Hairs always simple; fronds never arching and rooting at the tip; plant of swampy or boggy habitat, with wide-creeping blackish rhizome: **3. T. interrupta**

3. Hairs both branched and simple (stellate hairs abundant on vascular parts); sterile fronds often elongate, arching, and rooting at the tip; plant of limestone ledges, with short decumbent rhizome: **4. T. reptans**

1. Thelypteris kunthii (Desv.) Morton in Contr. U.S. Nat. Herb. 38: 53. 1967. **FIG. 8.**

Thelypteris normalis (C.Chr.) Moxley, 1920.

Rhizomes creeping, clothed at the apex with narrow, pale brown, ciliate scales. Fronds erect, with glabrate or sparsely pubescent stipes shorter than the lamina; lamina lance-oblong to broadly ovate-oblong, up to 50 cm long or more and 25 cm broad (often less), the lowest pinnae often deflexed at an angle to the plane of the lamina; vascular parts obliquely fine-hairy above, the underside finely whitish-hairy and also minutely capitate-glandular. Pinnae linear with attenuate apex, usually not over 2 cm broad; segments more or less oblong-falcate with 8–10 pairs of veins. Sori medial, with persistent hairy indusium.

GRAND CAYMAN: *Brunt 1821, 1925; Kings F 10, F 17, F 18, F 20, F 22, F 28; Proctor 15271.* CAYMAN BRAC: *Proctor 29365.*
– Southeastern United States, the West Indies, and Mexico to Central America; rare in the Lesser Antilles and northern South America. Cayman plants occur in moist marly swales and depressions, on ledges and the sides of excavations, and in "wells"; frequent.

2. Thelypteris augescens (Link) Munz & Johnston in Amer. Fern J. 12: 75. 1922.

Rhizome creeping, the apex clothed with lance-attenuate, yellow-brown, ciliate scales. Fronds few, spaced 1–3.5 cm apart, erect to arching, with quadrangular glabrous stipes equalling or a little shorter than the lamina; lamina broadly oblong to roundish-ovate, up to 60 cm long or more (usually less in Cayman plants), mostly 20–40 cm broad, rather abruptly ending in a hastate-pinnatifid apex or else a somewhat distinct terminal pinna wider than the lateral ones and 7–17 cm long; underside finely and rather densely pubescent (hairs mostly 0.2–0.4 mm long). Pinnae linear with attenuate apex, usually not over 1 cm broad and of firm texture, incised ½ or a little more to the costa; segments oblique, deltate or subfalcate, with about 5–8 pairs of veins. Costae bearing small, linear, ciliate scales beneath. Sori slightly supramedial, the persistent indusium pubescent.

GRAND CAYMAN: *Proctor 15272.*
– Florida, Bahamas, Cuba, and Guatemala. Cayman plants were found on moist limestone ledges overhanging an open depression, and were seen only at this one locality.

3. Thelypteris interrupta (Willd.) Iwatsuki in Jour. Jap. Bot. 38(10): 314. 1963.

Thelypteris gongylodes (sometimes spelled *"goggilodus"*) (Schkuhr) Small, 1938.

T. totta (Thunb.) Schelpe, 1963.

Rhizome wide-creeping, branched, black, 2–5 mm thick, bearing scattered, lanceolate, dark purple-brown, sparsely ciliate scales. Fronds erect, distant, with glabrous stipes equalling or longer (sometimes much longer) than the lamina; lamina narrowly oblong, up to 50 cm long or more and 20 cm broad (commonly less), with abruptly acuminate apex, glabrous on the upper (adaxial) side, sparsely pubescent beneath. Pinnae linear with bluntly acuminate apex, 5–15 mm broad, lobed less than halfway to the costa; costae bearing a few small brown scales beneath, the costules minutely resinous-glandular; lobes deltate or rounded, with 7–15 pairs of veins, the basal adjacent ones united in the tissue. Sori medial, with glabrous persistent indusium.

GRAND CAYMAN: *Brunt 1975.*
– Pantropical, usually in mats of floating vegetation on small freshwater ponds or in boggy marshes; rare in the Cayman Islands.

4. Thelypteris reptans (J.F.Gmel.)Morton in Fieldiana (Bot.) 28: 12. 1951; Amer. Fern Jour. 41: 87. 1951. **FIG. 9.**

Rhizome short, decumbent, at the apex bearing lance-attenuate, brown, stellate-pubescent scales. Fronds clustered, variable in form, procumbent and often attenuate to a proliferous tip to shorter and laxly ascending or erect, the sterile and fertile ones often of different form; stipes much shorter than the lamina. Lamina oblong to linear-attenuate, mostly 10–30 cm long, 2–5 cm broad, pinnate throughout or pinnatifid toward the apex; rhachis stellate-pubescent and usually often bearing longer simple hairs. Pinnae few or numerous, close or distant, linear to oval, the apex rounded to acute, the margins more or less crenately lobed, minutely stellate-pubescent on both sides; veins 2–7 pairs per lobe, the basal ones usually united and sending an excurrent veinlet to the sinus between the lobes. Sori inframedial in relation to the midveins of the lobes; indusium minute, bearing numerous long, simple or forked hairs.

GRAND CAYMAN: *Brunt 1820; Kings F 47; Proctor 15273.*
– Florida, the West Indies, and Mexico to Venezuela. Cayman plants occur along the shaded bases of limestone ledges and are frequent.

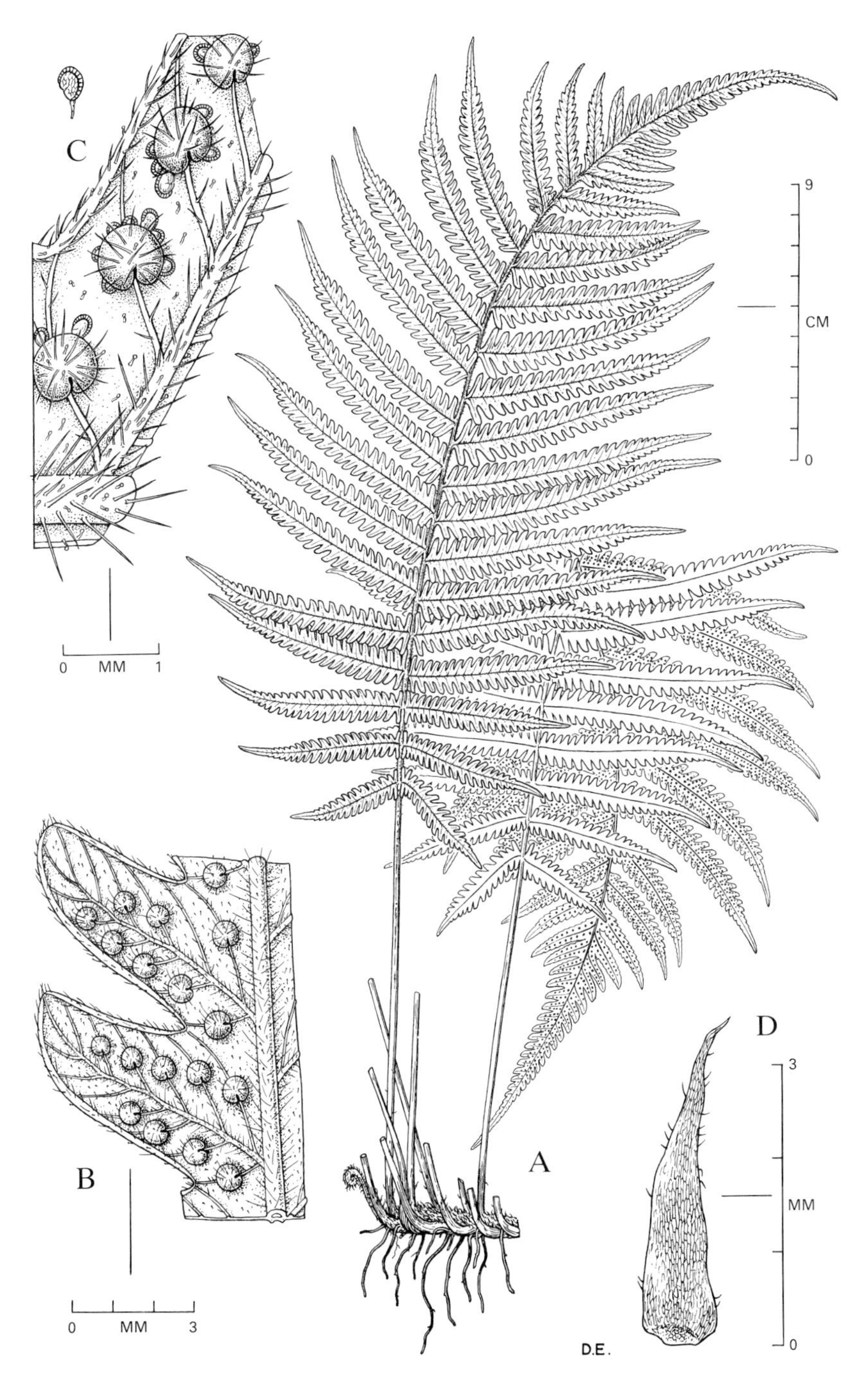

FIG. 8 **Thelypteris kunthii.** A, habit. B, portion of fertile pinna, lower side. C, enlargement of same. D, scale of rhizome. (D.E.)

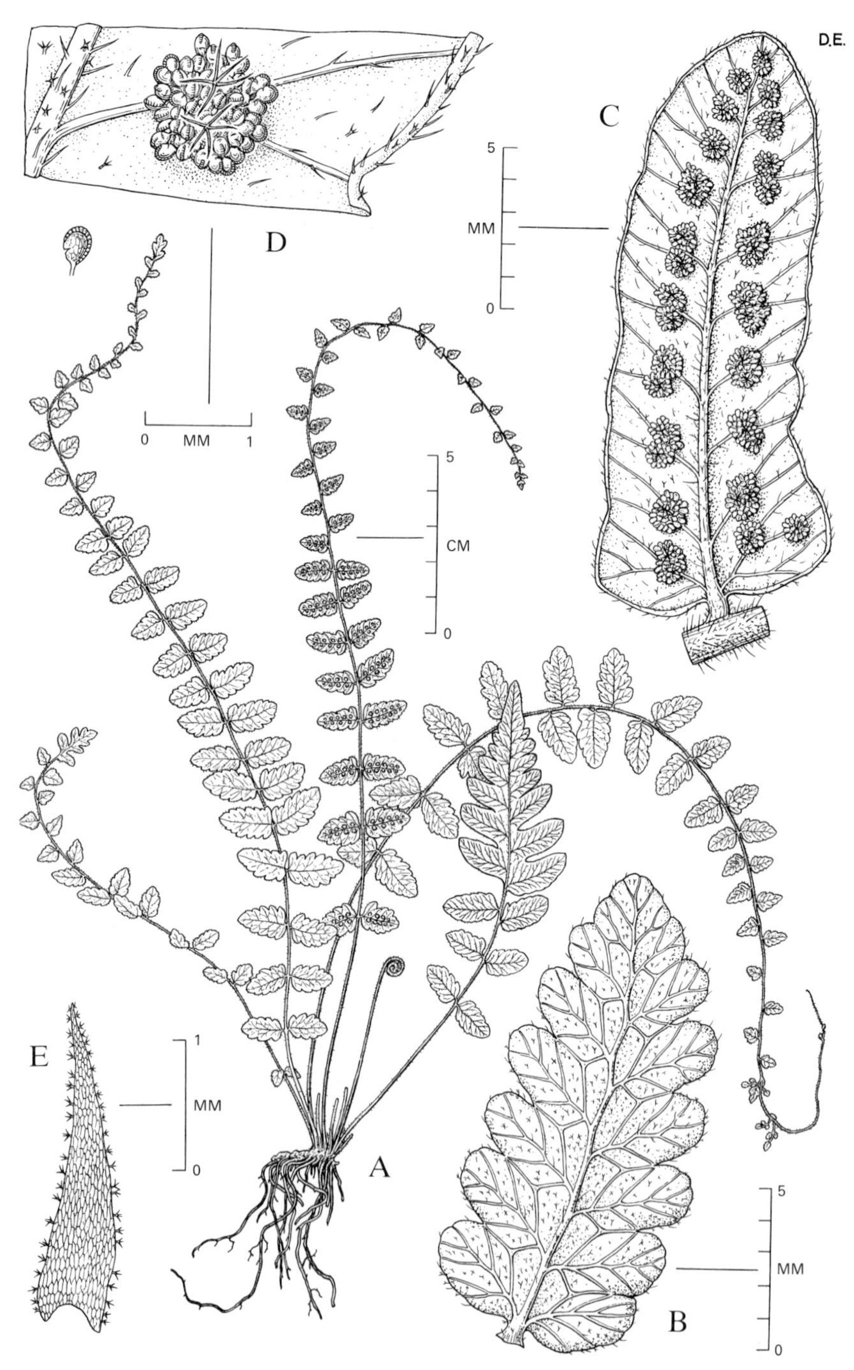

FIG. 9 **Thelypteris reptans.** A, habit. B, pinna, upper side. C, fertile pinna, lower side. D, sorus. E, scale of rhizome. (D.E.)

Genus 10. **POLYPODIUM** L.

Small to moderately large, chiefly epiphytic or epipetric ferns of varied habit. Rhizomes creeping, clothed with often clathrate scales, rarely nearly naked. Fronds articulate to the rhizome, all similar or somewhat dimorphic, glabrous or variously hairy or scaly; lamina simple to pinnatisect or 1-pinnate, sometimes further subdivided, the margins entire or rarely crenate-toothed; veins branched, free or variously reticulate, the areoles often with free included veinlets. Sori round or oval (rarely oblong to linear), terminal on vein-branchlets, not marginal, usually in 1 or more regular rows, always lacking an indusium. Spores bilateral, with smooth to tuberculate perispore.

A world-wide genus of about 225 species, often subdivided into smaller genera on the basis of frond-outline and variations in the pattern of venation.

KEY TO SPECIES

1. Lamina pinnatisect:

    2. Fronds essentially naked except for scattered minute scales along the rhachis beneath; veins all free: **1. P. dispersum**

    2. Fronds densely scaly, especially beneath; veins more or less reticulate: **2. P. polypodioides**

1. Lamina simple:

    3. Rhizome elongate, less than 1.5 mm thick, bearing scattered fronds less than 12 cm long, of thin texture; sori uniseriate: **3. P. heterophyllum**

    3. Rhizome short-creeping, 5-8 mm thick, embedded in a mass of roots, bearing clustered fronds often more than 20 cm long (to 50 cm or more), of stiff, somewhat cartilaginous texture; sori multiseriate: **4. P. phyllitidis**

1. Polypodium dispersum A.M.Evans in Amer. Fern Jour. 58: 173, t. 27. 1968 (1969). **FIG. 10.**

Rhizome short-creeping, 4-8 mm thick, densely clothed with narrow, attenuate, dark brown, denticulate scales to 3 mm long. Fronds elastic-herbaceous, with blackish puberulous stipes much shorter than the lamina; lamina narrowly elliptic-oblong, ovate-oblong, or linear-oblong, 12–40 cm long, mostly 4–6 cm broad; rhachis black, bearing scattered small brown scales beneath, these flat with denticulate margins; segments numerous, close, horizontal, ligulate, with ciliolate margins, the underside minutely septate-puberulous; veins 2-forked, obscure. Sori terminal on distal vein-forks, superficial.

GRAND CAYMAN: *Kings F 3, F 34.*

– Florida, the West Indies, and continental tropical America, usually on more or less shaded limestone rocks. This species is apogamous, i.e., it produces diploid spores and gametophytes that give rise to new sporophyte plants vegetatively, without the intervention of a sexual process.

2. Polypodium polypodioides (L.)Watt in Canadian Nat. II, 13: 158. 1867. **FIG. 10.**

Rhizome wide-creeping, 1-2 mm thick, clothed with lance-subulate, dark brown scales with pale denticulate-ciliolate margins. Fronds scattered, hygroscopic (curling up in dry weather, opening out after rain), the densely appressed-scaly stipes shorter than the lamina; lamina chiefly oblong or linear-oblong, 4-15 cm long, 1.5-6 cm broad, densely clothed beneath with roundish or deltate-ovate, peltate scales that entirely conceal the tissue, on the upper side bearing a few scattered, slender, pale, fimbriate scales; segments 2.5-5 mm broad, rounded at the apex, dilated at the base; veins obscure. Sori supramedial, protruding from pockets in the lamina tissue.

GRAND CAYMAN: *Brunt 2146; Kings F 32, F 32-a, F 36, F 45; Proctor 15034.* CAYMAN BRAC: *Proctor 29084.*

– Widespread in the warmer parts of the Western Hemisphere from Ohio and Virginia to Argentina; common in the West Indies. Cayman plants occur on mossy logs and tree-trunks in sheltered woodlands, and are rather rare.

3. Polypodium heterophyllum L., Sp. Pl. 2: 1083. 1753. **FIG. 11.**

Rhizome elongate, densely clothed with linear-attenuate, tawny to reddish-brown, denticulate scales, these peltately attached far above the narrowed base. Fronds distant, variable in form, glabrous, the stipes very short or nearly lacking; sterile lamina oval or elliptic and 1-3 cm long, or lanceolate to linear and 3-12 cm long, narrowed at both ends, with margins undulate, sinuate, or (rarely) irregularly crenate to incised; fertile lamina mostly narrower than the sterile and somewhat longer; veins reticulate, forming a row of areoles on either side of the costa, producing numerous short free excurrent veinlets near the margins, and a single free or closed veinlet within each areole. Sori borne singly on the infra-areolar veinlets, forming a single row on either side of the costa; sporangia accompanied by brown hairlike paraphyses.

GRAND CAYMAN: *Brunt 2174.*

– Florida and the West Indies, often climbing on the stems of shrubs or small trees in dense woodlands; sterile plants sometimes grow on limestone rocks. Rare in the Cayman Islands.

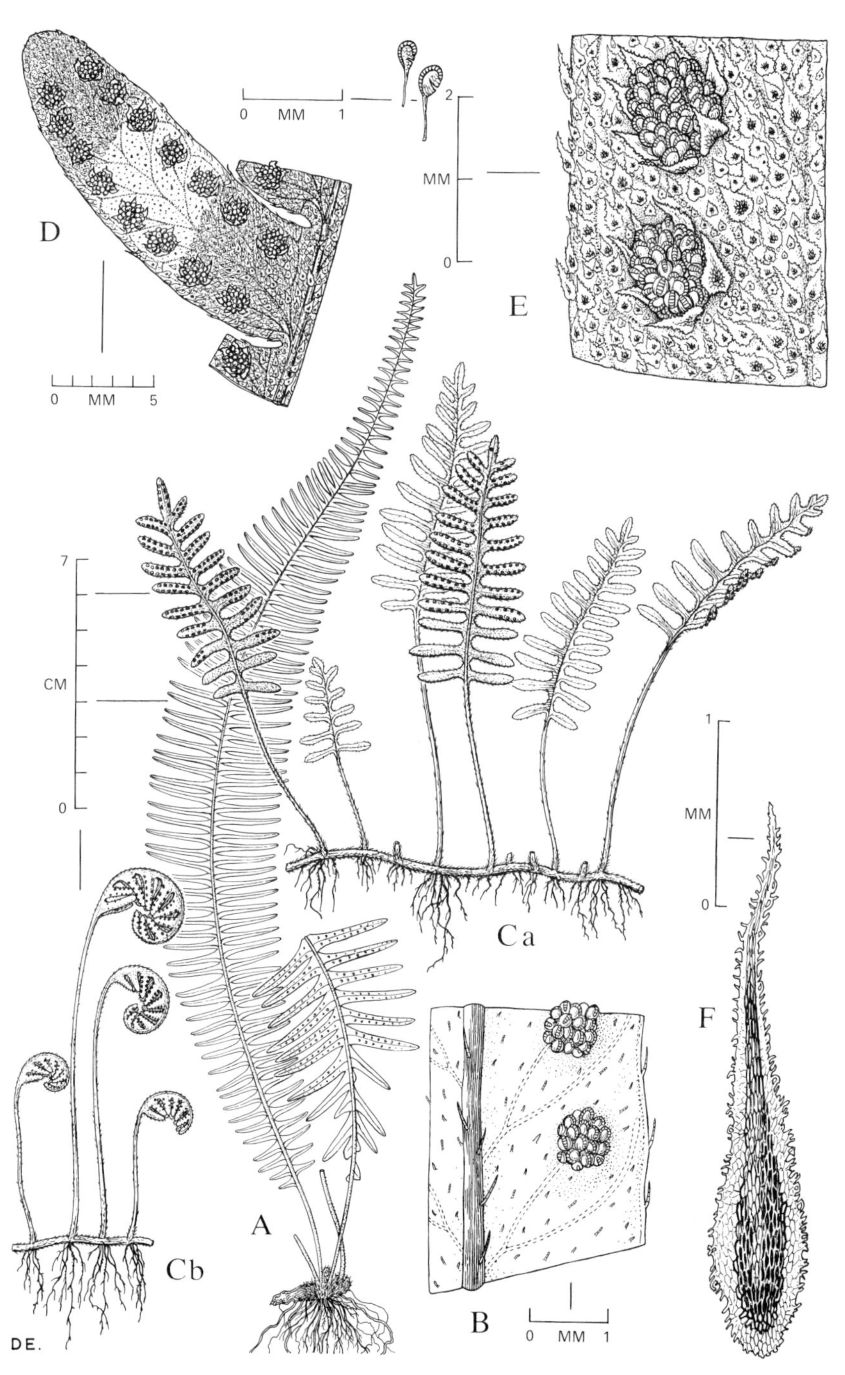

FIG. 10 **Polypodium.** A, **P. dispersum,** habit. B, portion of fertile pinna (lower side). C, **P. polypodioides**; a, habit (expanded after rain); b, habit during dry weather. D, fertile segment (lower side). E, same enlarged. F, scale of rhizome. (D.E.)

4. Polypodium phyllitidis L., Sp. Pl. 2: 1083. 1753. **FIG. 11.**

Rhizome short-creeping, subwoody, usually enveloped in a mass of brown-tomentose rootlets, bearing at the apex a few appressed-imbricate, more or less ovate scales, these 2–6 mm long, acute to acuminate, gray-brown, clathrate, glabrous. Fronds with stipes very short or nearly lacking, passing gradually upwardly into the long-decurrent lamina; lamina naked, glabrous, narrowly lance-oblong, 4–8 cm broad near the middle; principal veins ascending, joined by arching cross-veins, the areoles mostly with 2 or 3 included free excurrent veinlets, one of these sometimes prolonged and joined to the next cross-vein above. Sori in 2 rows between the main lateral veins.

GRAND CAYMAN: *Brunt 1865; Kings F 2, F 2-a, F 26, F 41, F 42; Proctor 27973.*

– Florida, the West Indies, and continental tropical America, on trunks or mossy bases of trees in sheltered woodlands, uncommon in the Cayman Islands.

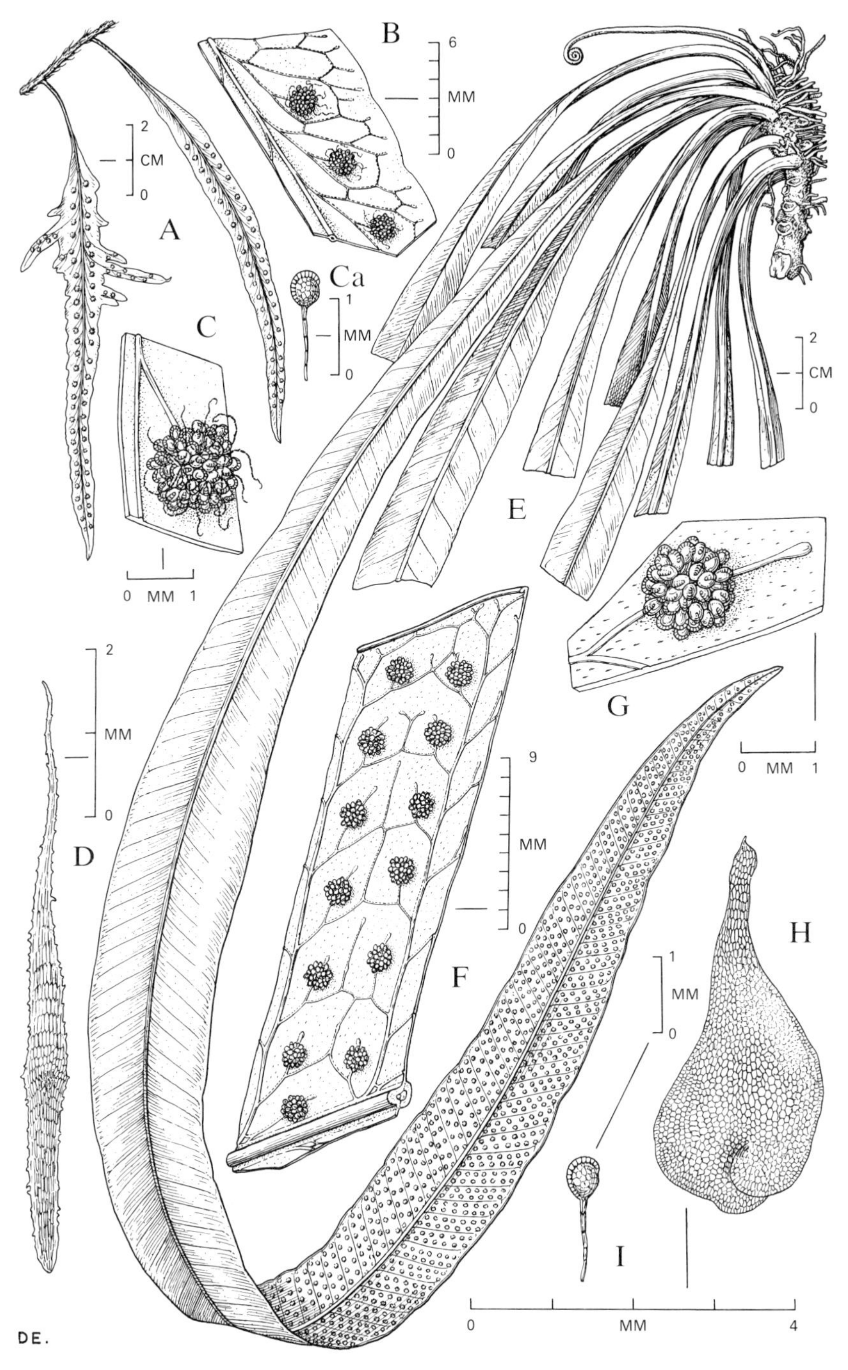

FIG. 11 **Polypodium.** A, **P. heterophyllum,** habit. B, portion of fertile frond (lower side). C, sorus; a, sporangium. D, scale of rhizome. E, **P. phyllitidis,** habit. F, portion of fertile frond (lower side). G, sorus. H, scale of rhizome. I, sporangium. (D.E.)

DIVISION II. SPERMATOPHYTA

Vascular plants that produce seeds containing an embryo, i.e., a rudimentary plant which remains in a dormant condition until germination. An alternation of sporophyte and gametophyte generations (cf. discussion above under Division I, Pteridophyta) occurs, but in this case the gametophyte is much reduced and wholly dependent on the sporophyte for its nutrition. The sporophylls (equivalent to the sporangium-bearing structures of the Pteridophyta) are arranged in groups (cones or strobili in Gymnosperms, or flowers in Angiosperms) of definite or indefinite number and are heterosporous, those bearing microsporangia (anther-sacs) termed stamens, and those producing megasporangia termed pistils. The megasporangium itself (nucellus) is enclosed along with the embryo-sac inside an integument (testa), the whole structure termed an ovule. Each ovule, when fertilized, becomes a seed. The megasporangiate sporophyll, containing one to many ovules (open and often scale-like in Gymnosperms, closed in Angiosperms) is called a carpel. The female gametophyte is confined within the megasporangium, where its egg-cell is fertilized by a non-motile (motile only in Cycadaceae) male gamete introduced by the growth of a pollen-tube. The cellular contents of a pollen-grain together with the pollen-tube constitute the male gametophyte. Pollen is normally transmitted from an anther-sac to the receptive point or surface (stigma) of a carpel or fused group of carpels by the agency of wind, insects, birds, bats, or other means. The evolution of flowering plants is closely associated with that of various active agents of pollination. After the fertilization of an ovule, the young sporophyte begins its development, forming a seed, while still attached to the sporophyte of the preceding generation. The production of seeds, allowing a period of dormancy during times unfavourable for growth, and also providing a wide variety of devices for dispersal, gives seed-bearing plants a competitive advantage over the pteridophytes, whose spores and gametophytes must fend for themselves without any special protective or dispersal mechanism. This is the chief reason why the seed-plants now dominate the world's land vegetation.

Seed-bearing plants, numbering an estimated total of more than 250,000 species, are subdivided into two classes, the Gymnospermae and the Angiospermae.

Class 1. GYMNOSPERMAE

More or less woody plants, monoecious or dioecious, the ovules not enclosed in an ovary, typically borne on scale-like carpels which are arranged in a cone or strobilus, or else sometimes terminal on naked or bracteate stalks. Stamens solitary or clustered on the scales of a strobilus, likewise the ovules. Male and female strobili separate and dissimilar. Pollen transferral always by means of wind. Seeds nut-like, winged or unwinged, or else berry-like or resembling a drupe; endosperm present. The endosperm of gymnosperm seeds is residual female-gametophyte (embryo-sac) tissue (cf. Angiospermae, for different origin of endosperm).

The vascular tissues of gymnosperms differ from those of most angiosperms in lacking xylem vessels.

A world-wide taxon of about 700 species, mostly trees or shrubs (a few vines, a few others with subterranean stems or almost acaulescent), which represent the remnant of a group more abundant in past geologic periods. Included are many trees of great economic value, such as various species of pines (*Pinus*). Only one of the several gymnosperm families occurs naturally in the Cayman Islands.

Family 1

CYCADACEAE

Woody or subwoody plants with erect trunks, or the caudex sometimes short and partly or wholly buried in the ground, growing only from the apex and thus often unbranched; leaves rather large and of leathery texture, pinnately compound without a terminal leaflet, forming a fern-like or palm-like apical crown. Dioecious, the stamens and carpels in terminal strobili or on modified leaves. Scales of staminate strobili bearing several anther-sacs on the lower (abaxial) side; pollen distributed by wind but fertilization effected by means of motile antherozoids. Scales of carpellate strobili bearing 2 or more naked ovules. Seeds drupe-like or nut-like.

A tropical and subtropical family of 9 genera and about 75 or more species, occurring chiefly in America, Australia and Africa.

Genus 1. **ZAMIA** L.

Caudex subwoody or toughly succulent, more or less starchy, wholly or partly buried in the ground; leaves stiff and coriaceous, the segments parallel-veined and entire or toothed, often serrulate at the apex; petiole unarmed in our species. Strobili dense, many-scaled, oblong-cylindric to subglobose, the female larger and thicker than the male; scales stalked, peltate, more or less hexagonal, eventually deciduous. Anther-sacs several per scale, sessile. Ovules 2 per scale, sessile. Seeds drupe-like, angled.

A tropical American genus of several variable species. West Indian forms have been used as a source of starchy flour, from which a poisonous stubstance must be extracted before it is edible.

1. **Zamia pumila** L., Sp. Pl. ed. 2, 2:1659. 1763; Eckenwalder in Jour. Arnold Arb. 61:715-719. 1980. **FIG. 12.**

Zamia media var. *commeliniana* Schuster in Engler & Diels, 1932.

"Bulrush", "Bull Rush"

Caudex stout, up to 20 cm long or more and 6 cm thick, usually completely buried in the ground, clothed at the apex with silky-villous, narrowly acuminate scales 2-3.5 cm long. Leaves usually 4-6 in a crown, erect-arching, up to 75 cm long; leaflets 12-20 on a side, alternate or subopposite, linear to narrowly oblanceolate, 9-14 cm long, 0.7-2 cm broad, serrulate at the obtuse apex. Peduncles villous-pubescent, 1-3 cm long; strobili with dark red tomentose scales. Male strobili oblong, 3.5-7.5 cm long at anthesis. Carpellate strobili broadly oblong-ellipsoid, 6-9 cm long. Seeds red, angled, c. 2 cm long.

GRAND CAYMAN: *Brunt 1868, 1876; Kings GC 256; Proctor 15008.* LITTLE CAYMAN: *Proctor 35122.* CAYMAN BRAC: *Fawcett; Kings PCB 1; Proctor 15325.*

– Florida, Bahamas, and the Greater Antilles, in rocky woodlands at low elevations. Many local variants have been given names as species or varieties, but their claims to recognition have not survived taxonomic analysis. The description given above is based on the Cayman Islands population, and does not necessarily apply in various details to plants from other areas.

[*Cycas revoluta* Thunb., "Sago", occurs under cultivation in Grand Cayman *(Kings PGC 258).*]

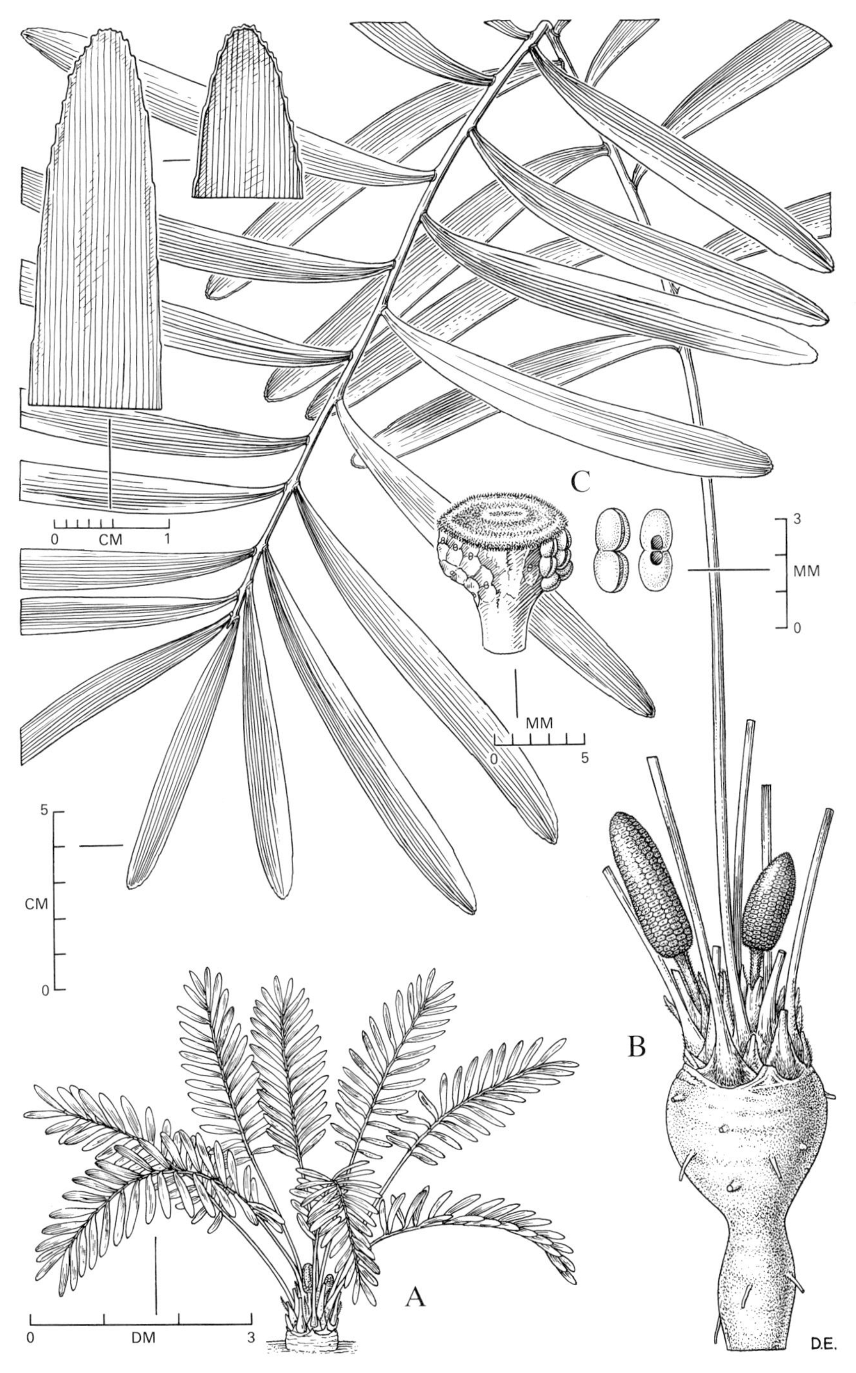

FIG. 12 **Zamia pumila.** A, habit. B, male plant with details of leaf-tips. C, scale of male cone and pollen-sacs. (D.E.)

Class 2. ANGIOSPERMAE

Plants of very diverse habit, structure, form, size, and habitat; ovules (and seeds) borne enclosed in carpels located in the center of flowers, these carpels interpreted as modified fertile leaves (megasporophylls) infolded along a median line or zone so that the margins form a more or less firmly sealed ventral (adaxial) suture. Carpels either free or often several united into a compound pistil, the ovule-bearing portion (the ovary) ripening into a fruit. Fertilization is usually a double process in which (1), the egg-cell is fertilized by a male gamete reaching it by means of the pollen-tube, and (2), other female gametophyte cells within the ovule are united with additional pollen-tube nuclei, and subsequently often develop into a triploid or polyploid nutritive tissue called endosperm which is closely associated with the embryo in the seed.

The flowering plants dominate the land vegetation of most regions of the world, and in all form a vast array of probably nearly 250,000 species, including the majority of plants having economic value, and nearly all those used as a source of food or fiber. They are customarily divided into two subclasses, as follows:

KEY TO THE SUBCLASSES

1. Stem (if present) with vascular bundles scattered through a large solid or spongy pith, or sometimes hollow with the bundles scattered in the surrounding cylinder; leaves usually parallel-veined (reticulate in Araceae and Smilacaceae); parts of the flower nearly always in 3's; embryo with 1 cotyledon:
   Subclass 1. MONOCOTYLEDONES

1. Stem with a single continuous or interrupted vascular cylinder, in woody members this cylinder increasing in diameter by growth of a cambium, the pith (if any) usually small and with no vascular bundles scattered through it; leaves usually reticulate-veined; parts of the flower most often in 5's or 4's, sometimes more numerous, seldom in 3's; embryo with 2 cotyledons (rarely more);
   Subclass 2. DICOTYLEDONES

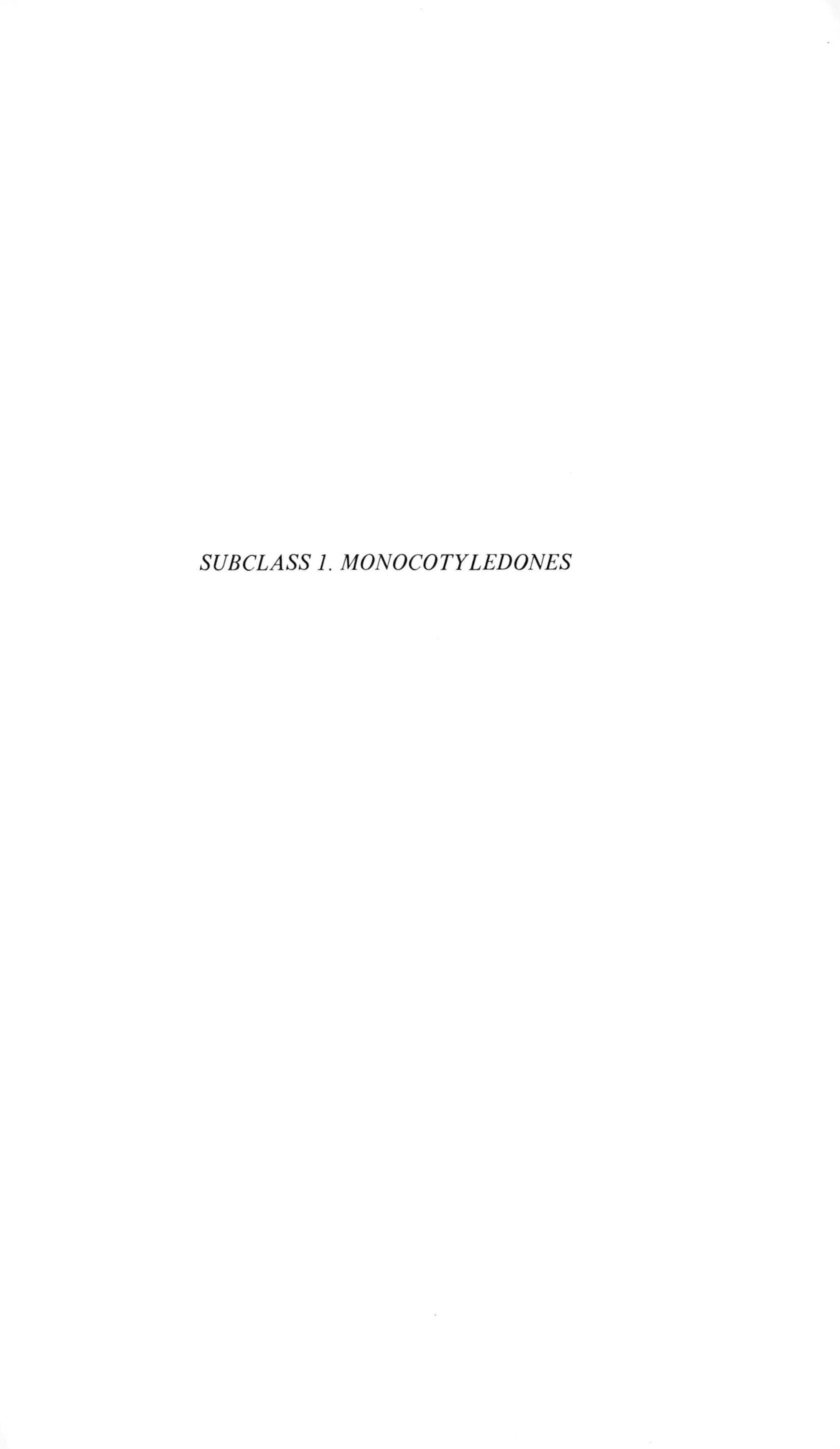

SUBCLASS 1. MONOCOTYLEDONES

SUBCLASS 1. MONOCOTYLEDONES

The monocotyledonous families are arranged in the following sequence, in accordance with the taxonomic sequence of Cronquist:

ORDER 1. ALISMATALES

Family 1. Alismataceae

ORDER 2. HYDROCHARITALES

Family 2. Hydrocharitaceae

ORDER 3. NAJADALES

Family 3. Ruppiaceae

4. Zosteraceae

ORDER 4. COMMELINALES

Family 5. Commelinaceae

ORDER 5. CYPERALES

Family 6. Cyperaceae

7. Gramineae

ORDER 6. TYPHALES

Family 8. Typhaceae

ORDER 7. BROMELIALES

Family 9. Bromeliaceae

ORDER 8. ARECALES

Family 10. Palmae

ORDER 9. ARALES

Family 11. Araceae

12. Lemnaceae

ORDER 10. LILIALES

Family 13. Pontederiaceae

14. Liliaceae

15. Smilacaceae

ORDER 11. ORCHIDALES

Family 16. Orchidaceae

KEY TO THE MONOCOTYLEDONOUS FAMILIES

1. Aquatic plants, at least the base normally growing in water:

2. Plants entirely submerged:

3. Plants of brackish water; stems trailing loosely in the water, not subterranean: **3. Ruppiaceae**

3. Plants of sea water; stems ("rootstocks") wide-creeping, subterranean:

4. Leaves straplike, c. 10 mm wide: **2. Hydrocharitaceae**

4. Leaves narrowly linear, less than 1.5 mm wide: **4. Zosteraceae**

2. Plants floating or else rooted in mud, often with emergent leaves:

5. Plants floating on fresh water pools or tanks, minute, lacking leaves, the plant-body a flat disk usually less than 4 mm long, multiplying by budding: **12. Lemnaceae**

5. Plants usually rooted in mud and always with conspicuous leaves:

6. Each flower with numerous separate carpels; petals 3, white, conspicuous: **1. Alismataceae**

6. Each flower with 1 carpel, or if more than one, these joined in a common ovary; flowers not white:

7. Flowers minute, unisexual, aggregated into dense brown cylindric spikes: **8. Typhaceae**

7. Flowers not as above:

8. Leaves with inflated petioles; flowers with showy 6-parted perianth: **13. Pontederiaceae**

8. Leaves linear and lacking petioles; flowers without a perianth, or this structure represented by minute bristles or scales:

9. Leaves (when present) 3-ranked, the basal sheaths tubular and not split; aerial stems usually 3-angled, not hollow; anthers attached at the base: **6. Cyperaceae**

9. Leaves 2-ranked, the basal sheaths usually split to the point of attachment; aerial stems usually cylindric and hollow; anthers attached by the middle: **7. Gramineae**

1. Terrestrial or epiphytic plants, not normally growing in water:

10. Ovary superior or naked:

11. Perianth lacking or represented by minute bristles or scales; flowers aggregated in spikes or spikelets:

12. Leaves linear, without differentiated petiole and blade; veins parallel:

13. Leaves (when present) 3-ranked, the basal sheaths tubular and not split; aerial stems usually 3-angled, not hollow; anthers attached at the base: **6. Cyperaceae**

13. Leaves 2-ranked, the basal sheaths usually split to the point of attachment; aerial stems usually cylindric and hollow; anthers attached by the middle: **7. Gramineae**

12. Leaves differentiated into petiole and flat expanded blade; veins reticulate: **11. Araceae**

11. Perianth obviously present:

14. Plants tree-like with solid woody trunks; leaves large, pinnately compound or palmately divided into numerous segments: **10. Palmae**

14. Plants herbaceous or vinelike, the leaves always simple:

15. Perianth segments all petal-like, equal:

16. Woody vine with unisexual flowers, the plants dioecious; fruit a berry: **15. Smilacaceae**

16. Terrestrial herbs with bisexual flowers; fruit a dehiscent or indehiscent capsule: **14-a. Liliaceae**

15. Perianth segments in 2 series, the inner petal-like, the outer sepal-like and usually green:

17. Plants epiphytic or terrestrial; leaves often clothed with very minute scales, and usually forming a stiff rosette (except in *Tillandsia recurvata*) **9. Bromeliaceae**

17. Plants always terrestrial; leaves naked, alternate on elongate trailing or ascending stems (forming a rosette in *Rhoea spathacea,* with leaves red-purple beneath): **5. Commelinaceae**

10. Ovary inferior:

18. Flowers regular (actinomorphic) or nearly so; stamens 6, free from the style; seeds with endosperm; always terrestrial:

19. Leaves with entire margins and lacking an apical spine; bulbous herbs with solitary or umbellate flowers: **14-b. Liliaceae**

19. Leaves with prickly margins or at least with an apical spine; plants coarse and rigid, not bulbous at base:

20. Perianth segments all similar, petaloid, united below to form a tube; flowers in compound panicles much exceeding the leaves: **14-c. Liliaceae**

20. Perianth segments of two kinds, consisting of dissimilar sepals and petals; flowers in simple or compound headlike inflorescences shorter than or barely equalling the leaves: **9. Bromeliaceae**

18. Flowers irregular (zygomorphic); stamens 1 or 2, adnate to the style; seeds powdery, without endosperm; epiphytic or sometimes terrestrial: **16. Orchidaceae**

Family 1

ALISMATACEAE

Annual or perennial acaulescent herbs, chiefly of marshes or wet places. Leaves basal, the elongate petioles sheathing the base, the blades flat and several-ribbed. Scapes erect, simple or branched. Flowers perfect or unisexual, regular, whorled, borne in terminal racemes or panicles. Calyx of 3 persistent green sepals. Petals 3, white, soon falling. Stamens 6 or more, the filaments distinct, the anthers 2-celled. Carpels distinct and free, few or many, capitate in our species, each with usually 1 ovule; style usually persistent, appearing as a beak on the fruit. Fruit (in the Caymanian species) a head of achenes; seeds curved, the embryo horseshoe-shaped.

A widely distributed family of about 12 genera and 75 species.

Genus 1. **SAGITTARIA** L.

Perennial herbs growing in shallow fresh water or wet soil, arising from thick, somewhat tuberous horizontal rhizomes. Leaves erect, with spongy petioles. Scapes simple or branched, the flowers monoecious or dioecious, in whorls of 3, the flowers of the upper whorls usually staminate, the lower pistillate. Stamens few or many; carpels very numerous, crowded on a headlike receptacle.

A cosmopolitan genus of about 20 species; one species is recorded from the Cayman Islands.

1. **Sagittaria lancifolia** L., Pl. Jam. Pug. 27. 1759. **FIG. 13.**

Plants coarse, glabrous, up to 1 m tall or sometimes more. Leaves with lance-linear to elliptic blades usually 12–35 cm long, 3–11 cm broad, acute or acuminate at apex, acuminate at base, conspicuously nerved. Scapes simple or often branched, with flowers on long, slender, spreading peduncles. Flowers 2–4 cm broad with 3 delicate pure white petals. Fruiting heads 1–1.5 cm in diameter; achenes more or less obovate, dorsally winged, with a short horizontal beak at the apex.

GRAND CAYMAN: *Brunt 1658, 2158; Kings GC 189; Lewis 4289; Proctor 27963, 31024.* Chiefly occurring in the Georgetown area.
– S. United States, Mexico and Central America, West Indies, S. America.

Family 2

HYDROCHARITACEAE

Aquatic herbs, floating or submerged, with short to elongate leafy stem. Leaves linear or strap-shaped, sessile or stalked. Flowers regular, unisexual or rarely perfect, usually solitary within a stalked, tubular, 2-lipped spathe or else subtended by 2 overlapping bracts. Calyx of 3 sepals; corolla of 3 thin petals or lacking. Stamens 3–12, with linear 2-celled anthers. Carpels 3–15, united, the ovary 1–15-celled with numerous ovules; styles or stigmas as many as the ovary-cells. Fruit a berry or capsule.

A widely-distributed family of about 16 genera and 80 species.

Genus 1. **THALASSIA** Solander

Marine herbs with thick, creeping rootstocks, the strap-like leaves tufted on short erect branches at the nodes. Scapes elongate, arising singly among the leaves, terminating in a one-flowered inflorescence enclosed in a tubular, 2-cleft spathe. Staminate flowers stalked, consisting of 3 petaloid sepals and 8–13 distinct stamens with very short filaments. Pistillate flowers nearly sessile within the spathe, consisting of an ellipsoid inferior ovary terminated by 12–18 linear-spindle-shaped, pilose stigmas, these always arranged in pairs. Ovary incompletely 6-12 celled by the projection of irregular placental lobes. Pollination entirely submarine. Fruit a stalked capsule with warty surface, opening by valves.

A genus of 2 species, one occurring in the Pacific and Indian Oceans, the other Caribbean.

1. **Thalassia testudinum** Konig ex Konig & Sims in Ann. Bot. 2: 96. 1805. **FIG.14.**

"Turtle grass".

Plants forming extensive colonies; leaves 2-5 in a cluster, sheathing at base, up to 30 cm long or more, 6-11 mm wide, rounded and minutely denticulate at the apex. Flowers seldom observed; sepals 10-12 mm long; anthers linear, 8 mm long; stigmas 10 mm long. Fruit ellipsoid or spindle-shaped, short-stalked and short-beaked.

GRAND CAYMAN: *Kings GC 20, GC 158, GC 160, GC 184; Proctor 15133; Sachet 435.* LITTLE CAYMAN: *Kings LC 62a, LC 98, LC 121; Proctor 28109.* CAYMAN BRAC: *Proctor 29373.*
– Florida to northern South America, commonly occurring in shallow sandy bays and lagoons, and especially in areas protected by reefs.

Family 3

RUPPIACEAE

Submerged aquatic herbs with long, thread-like, forking stems; leaves alternate, narrowly linear, l-veined, and sheathing at the base. Flowers perfect, without bracts, in 2-flowered spikes on short peduncles which elongate after flowering; perianth lacking. Stamens 4, the anthers sessile, 2-celled, soon deciduous. Carpels 4, each with 1 pendulous ovule, sessile at first, at length each carpel long-stalked; stigma minutely peltate. Fruit of long-stalked drupelets each crowned with the persistent stigma.

One genus of 2 somewhat variable species, widely distributed.

Genus 1. **RUPPIA** L.

Characters as given for the family; both species occur in the Cayman Islands. They are not easy to differentiate.

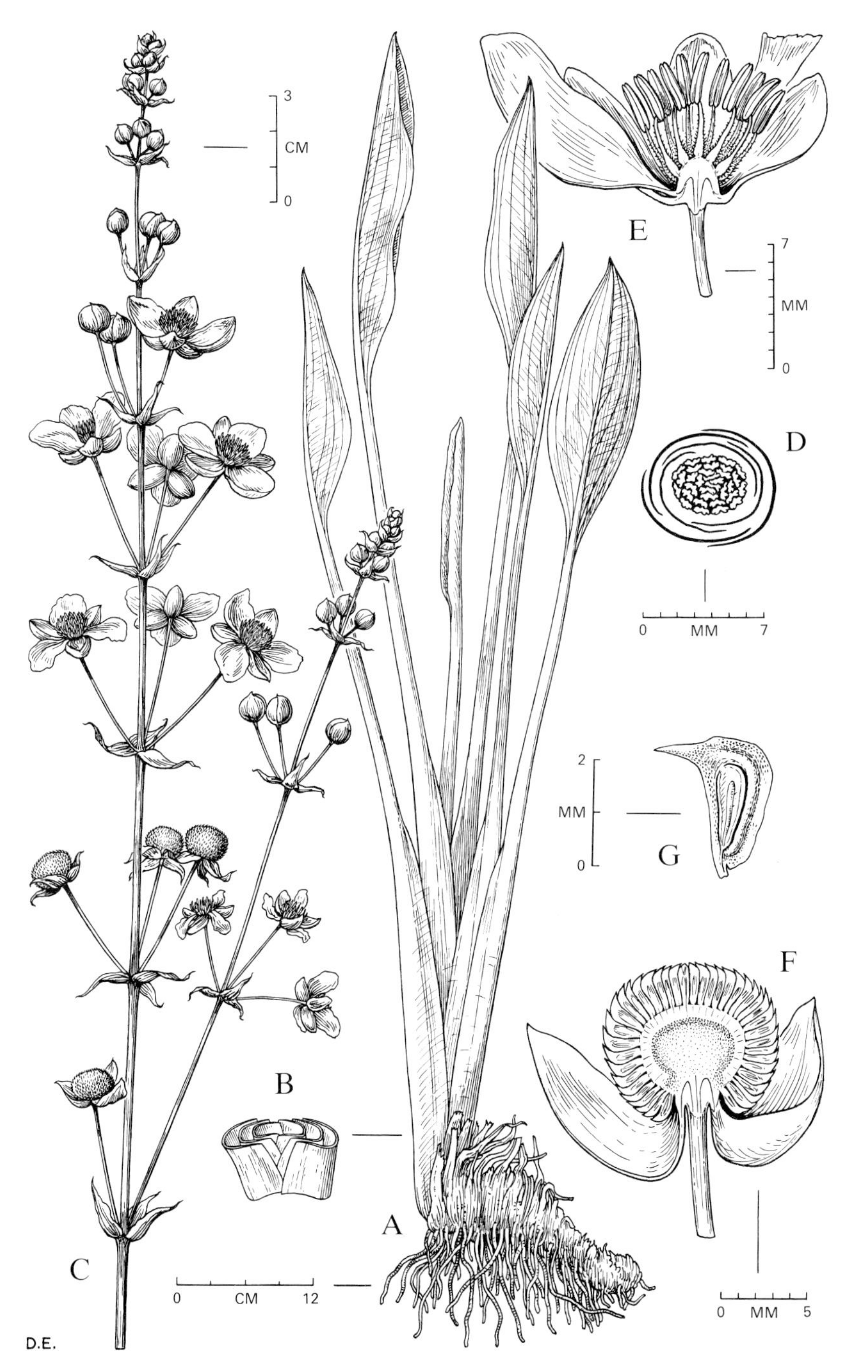

FIG. 13 **Sagittaria lancifolia.** A, habit. B, x-section showing imbricate leaf-bases. C, inflorescence. D, floral diagram. E, staminate flower, long. section. F, pistillate flower, long. section. G, achene, long. section. (D.E.)

KEY TO SPECIES

1. Peduncle becoming elongate in fruit and often spiral; leaves of juvenile plants obtuse or rounded: **1. R. cirrhosa**

1. Peduncle not much longer in fruit than in flower; leaves always sharply acute: **2. R. maritima**

1. Ruppia cirrhosa (Petagna) Grande in Bull. Orto Bot. Univ. Nap. 5: 58. 1918. **FIG. 15.**

Ruppia spiralis L. ex Dumort., 1827.

Stems flaccid, slender, repeatedly forked, up to 1 m long or more, forming loose, intricate, submerged masses. Leaves linear-capillary, up to 12 cm long, 0.3–0.8 mm wide, broadly sheathing at the base, the sheaths 6–12 mm long. Fruiting peduncles capillary, elongate, and usually curved, flexuous or spiral and up to 10 cm long or more. Drupelets ovoid, 2-3 mm long, on slender stalks (carpophores) 1-3 cm long.

GRAND CAYMAN: *Kings GC 405.*
– Cosmopolitan in suitable habitats. The Cayman plants were found "submerged in pond at Battle Ground, Batabano".

2. Ruppia maritima L., Sp. Pl. 1: 127. 1753.

Similar in habit to the preceding species, differentiated by the key characters.

CAYMAN BRAC: *Proctor 29129.*
– Cosmopolitan. The Cayman plants were found "in warm shallow brackish pools, South Side".

Family 4

ZOSTERACEAE

Submerged marine perennial herbs with slender creeping stems ("rootstocks"). Leaves linear with sheathing bases, tufted at nodes on the rootstock. Flowers unisexual or (rarely) perfect, solitary or clustered, usually naked, rarely with minute bracts; perianth lacking. Staminate flower consisting of 1 sessile or 2 long-stalked, 2-celled anthers bearing thread-like pollen. Pistillate flower of 1 or 2 fused carpels, sessile or stalked, with hairlike style. Fruit a 1-seeded drupelet.

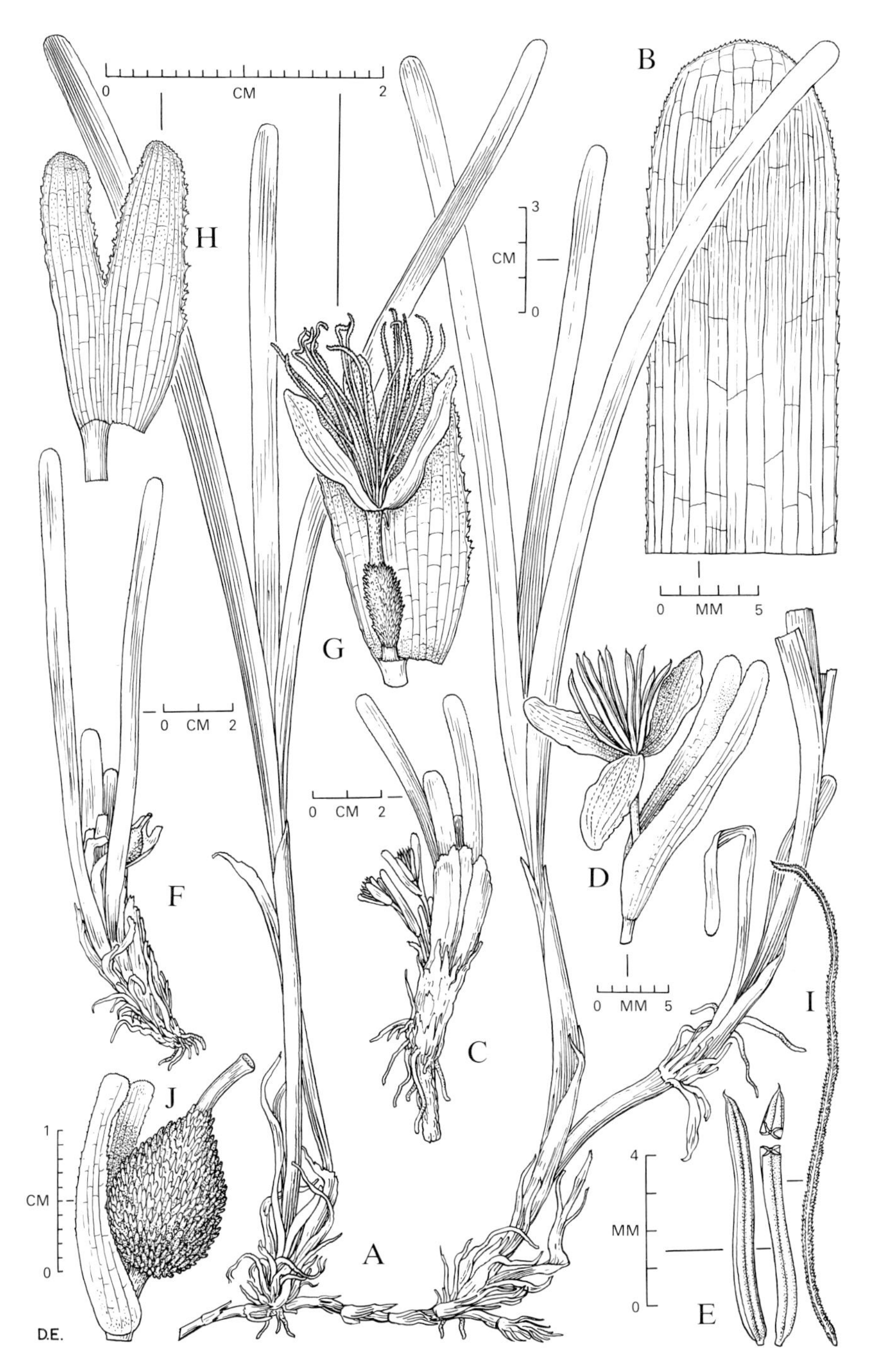

FIG. 14 **Thalassia testudinum.** A, habit. B, leaf-tip. C, plant with staminate flowers. D, staminate flower. E, anthers. F, plant with pistillate flower. G, pistillate flower. H, spathe of pistillate flower. I, stigma. J, fruit. (D.E.)

Four genera with about 22 species, widely distributed in tropical, subtropical and temperate regions.

KEY TO GENERA

1. Leaves flat; stigma 1: **1. Halodule**

1. Leaves terete; stigmas 2: **2. Syringodium**

Genus 1. **HALODULE** Endl.

Dioecious herbs with creeping jointed rootstocks, rooting at the nodes. Leaves linear, grass-like, flat, 2- or 3-toothed at the apex, two or several tufted on short spurs at nodes of the rootstock, each tuft protected at the base by a membranous, ligulate sheath. Flowers solitary, unisexual. Staminate flower consisting of 2 anthers attached at different levels at the end of a long pedicel, the pedicel sheathed at the base. Pistillate flower consisting of a single naked carpel; style short, ending in a slender stigma. Fruit a small globose drupelet.

Four species are attributed to the Caribbean area (and Bermuda) in a recent monographic study (Den Hartog in Blumea 12(2): 289-313. 1964). This publication did not ascribe any species to the Cayman Islands, but the same author's later book "The sea-grasses of the world" (1970) cites *Proctor 15134* as *H. wrightii.*

1. Halodule wrightii Aschers. in Sitz. -ber. Ges. Nat. Freunde Berlin. 1868: 19. 1868; Bot. Zeit. 26: 511. 1868.

Diplanthera wrightii (Aschers.) Aschers., 1897.

"Eel Grass"

Leaves 0.3-0.8 mm wide, 2-toothed at the apex without secondary projections on the teeth, the inner side of these teeth concave. Flowers rarely observed; anthers 2-celled, about 6 mm long; carpel c. 3 mm long. Mature fruit black.

GRAND CAYMAN: *Kings GC156, GC159; Proctor 15134.* LITTLE CAYMAN: *Kings LC 120a; Proctor 28108.*
– Coasts of Florida and West Indian islands to Puerto Rico; apparently absent from Jamaica, where it is allegedly replaced by *H. beaudettei* (Den Hartog) Den Hartog, said to differ in having broader leaves with 3-toothed tips. None of the Cayman collections cited above is fertile. They were found in sandy salt-water lagoons and bays.

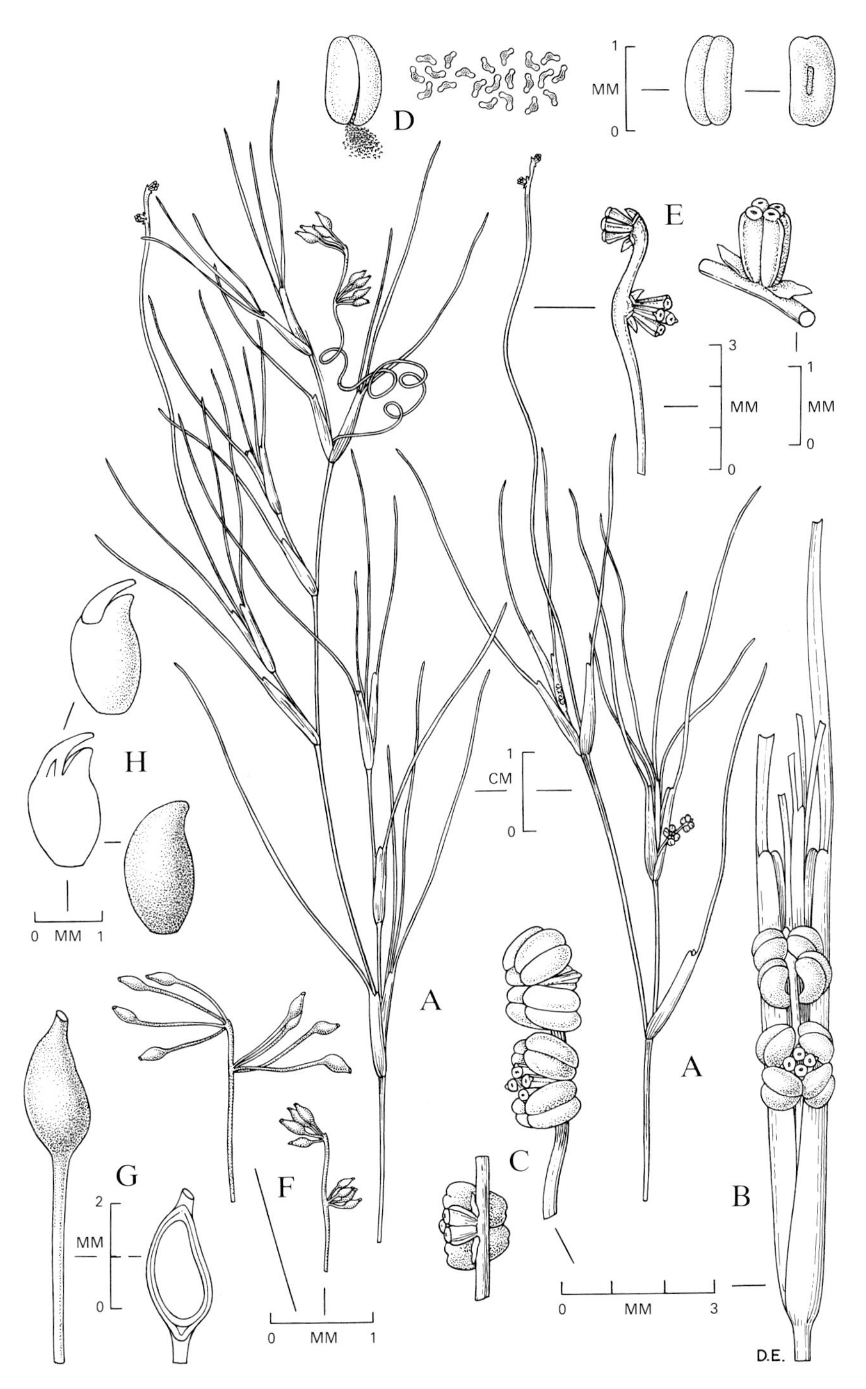

FIG. 15 **Ruppia cirrhosa.** A, habit. B, inflorescence in situ. C, inflorescence showing position of anthers. D, anther and pollen. E, inflorescence with early development of fruits. F, stages in development of fruits. G, single fruit. H, germination of seed. (D.E.)

Genus 2. **SYRINGODIUM** Kütz.

Dioecious herbs with creeping jointed rootstocks, rooting at nodes. Leaves linear, terete, pointed, 2 or several tufted on small erect branchlets at the nodes, each tuft protected at the base by an auricled membranous sheath. Flowers solitary or in cymose clusters arising from the leaf-axils. Staminate flowers consisting of 2 anthers attached at the same level on a slender stalk. Pistillate flowers of 2 fused carpels; pistil with a slender style terminating in 2 stigmas.

A genus of 2 species, one occurring in the Indian and western Pacific Oceans, the other in the Caribbean region.

1. Syringodium filiforme Kütz. in Hohenack, Alg. Marin. Sico. IX, no. 426. 1860.

Cymodocea manatorum Aschers., 1868. For discussion of this species, see Dandy & Tandy in Jour. Bot. Lond. 77: 116. 1939.

"Manatee Grass"

Rootstock red-brown, wide-creeping; leaves mostly 10–25 cm long, 0.5–1 mm thick, in tufts of 2 or 3 separated by internodes 1–5 cm long; basal sheaths 2.5–6 cm long. Flowers rarely observed. Fruit ellipsoid, flattened, 5–6.5 mm long, 2.5 mm broad, beaked by the persistent style; seed 2.5–3 mm long.

GRAND CAYMAN: *Kings GC 157; Sachet 437.* LITTLE CAYMAN: *Kings LC 120, MA 93a; Proctor 28107.* Undoubtedly occurs at Cayman Brac.
– Southern coasts of the United States; Bermuda and West Indian islands, in sandy salt-water lagoons and around reefs.

Family 5

COMMELINACEAE

Annual or perennial herbs, often rather succulent, with alternate entire leaves, these usually with sheathing or clasping bases. Flowers perfect, regular or irregular, in cymes or umbels usually subtended by spathe-like or leafy bracts. Sepals 3, free and distinct, usually green and herbaceous. Petals usually 3, delicate and soon withering, free or united into a tube. Stamens typically 6, in two whorls of 3 each, but in some genera dimorphic or reduced in number; anthers 2-celled. Ovary superior, 2- or 3-celled; ovules 1 to several in each cell; sometimes only 1 or 2 cells fertile. Fruit a 2- or 3-valved capsule, or indehiscent.

A tropical and warm-temperate family of about 25 genera and 600 species.

KEY TO GENERA

1. Plants trailing or suberect, with elongate slender stems; leaves less than 10 cm long:

 2. Inflorescences pedunculate; corolla irregular, one petal much smaller than the other two: **1. Commelina**

 2. Inflorescences sessile in leaf axils; corolla regular, of 3 equal petals: **2. Callisia**

1. Plants with short erect stem c. 1.5 cm thick; leaves usually 20–35 cm long, red-purple beneath: [**3. Rhoeo**]

Genus 1. **COMMELINA** L.

Annual or perennial branching herbs, the stems trailing or ascending, somewhat succulent. Leaves sessile or short-petioled. Inflorescence a small pedunculate cyme or group of cymules, more or less enclosed by a folded, heart-shaped spathe-like bract; flowers blue or white. Sepals unequal, 2 usually being united toward the base. Petals markedly unequal, 2 of them being much longer than the third, clawed. Fertile stamens 3, two with oblong anthers, the other incurved with a longer anther; sterile stamens 3, smaller than the fertile ones, bearing small, cross-shaped, non-functional anthers; filaments glabrous. Ovary 3-celled, one of the cells often reduced or abortive; fertile cells with 1 or 2 ovules. Seeds smooth, rough, or reticulate.

A large genus of about 230 tropical or warm-temperate species, several of them pantropical weeds; one species occurs in the Cayman Islands.

1. Commelina elegans Kunth in H.B.K., Nov. Gen. 1: 259. 1816. **FIG. 16.**

"Water grass"

Stems more or less decumbent, glabrous. Leaves lanceolate to narrowly elliptic or oblong-lanceolate, 3–8 cm long, acute or acuminate at apex; sheaths minutely ciliate on margins. Spathes 1–2 cm long, glabrous or minutely pubescent, acute, the margins joined near the base (in the closely related pantropical weed *C. longicaulis* Jacq., which may yet be found in the Cayman Islands, the spathe-margins are not joined at all). Petals blue, pale bluish, or sometimes white. Capsule 4–5 mm long, 3-seeded; seeds smooth, 3–3.5 mm long (in *C. longicaulis* the seeds are reticulated and 2–3 mm long).

GRAND CAYMAN: *Brunt 2089; Kings GC 312, GC 313, GC 366; Proctor 27994.* CAYMAN BRAC: *Proctor 29055.*
– Widespread throughout tropical America, common along damp roadside ditches and in other low moist situations.

Genus 2. **CALLISIA** Loefl.

Creeping, trailing or ascending herbs. Flowers in sessile axillary glomerules, or umbellate on filiform exserted peduncles; sepals 2 or 3; petals 2 or 3, equal; stamens 1-3, with glabrous filaments, the anther-cells separated by a broad, hastate connective. Ovary 2- or 3-celled, compressed or angled, with 2 ovules in each cell; stigma more or less deeply 3-lobed. Capsule membranous, 2- or 3-valved.

A tropical American genus of 12 species, one occurring in the Cayman Islands.

1. **Callisia repens** (Jacq.) L., Sp. Pl. 2, 1: 62. 1762. **FIG. 17.**

Stems prostrate, creeping, glabrous, with short erect flowering branches. Leaves ovate, 1–4 cm long, 0.6–1.3 cm wide, rather fleshy or succulent, often minutely speckled, apex sharply acuminate, base clasping and margins minutely ciliate. Flowers in a dense, sessile, axillary glomerule more or less enclosed by the clasping, spathe-like leaf base; sepals 3–4 mm long, pilose on back; petals 3, translucent white, shorter than the sepals. Ovary pilose at the apex; stigma trifid, long-pilose. Capsule oblong, 1.5 mm long; seeds dark brown, wrinkled, 1 mm long.

GRAND CAYMAN: *Brunt 1775, 1777; Kings GC 201, GC 341; Proctor 27974.*
– West Indies, Central & South America, in partly shaded, often rocky situations. Various populations of this species are somewhat different in appearance, but the variations seem to be merely clonal.

[Genus 3. **RHOEO** Hance

Erect herbs with a relatively short to somewhat elongate, thick, glabrous stem. Leaves somewhat fleshy, attached in a dense close spiral to form a rosette-like cluster much longer than the inflorescences. Peduncles axillary, short; flowers in contracted scorpioid clusters enclosed within an erect, firm, boat-shaped spathe formed by 2 overlapping bracts. Sepals free, somewhat petaloid; petals 3, equal, free; stamens 6, subequal, with hairy filaments. Ovary 3-celled, with a single ovule in each locule. Fruit a 3-valved capsule; seeds rugose.

A single Central American species, widely cultivated and naturalized throughout the West Indies.

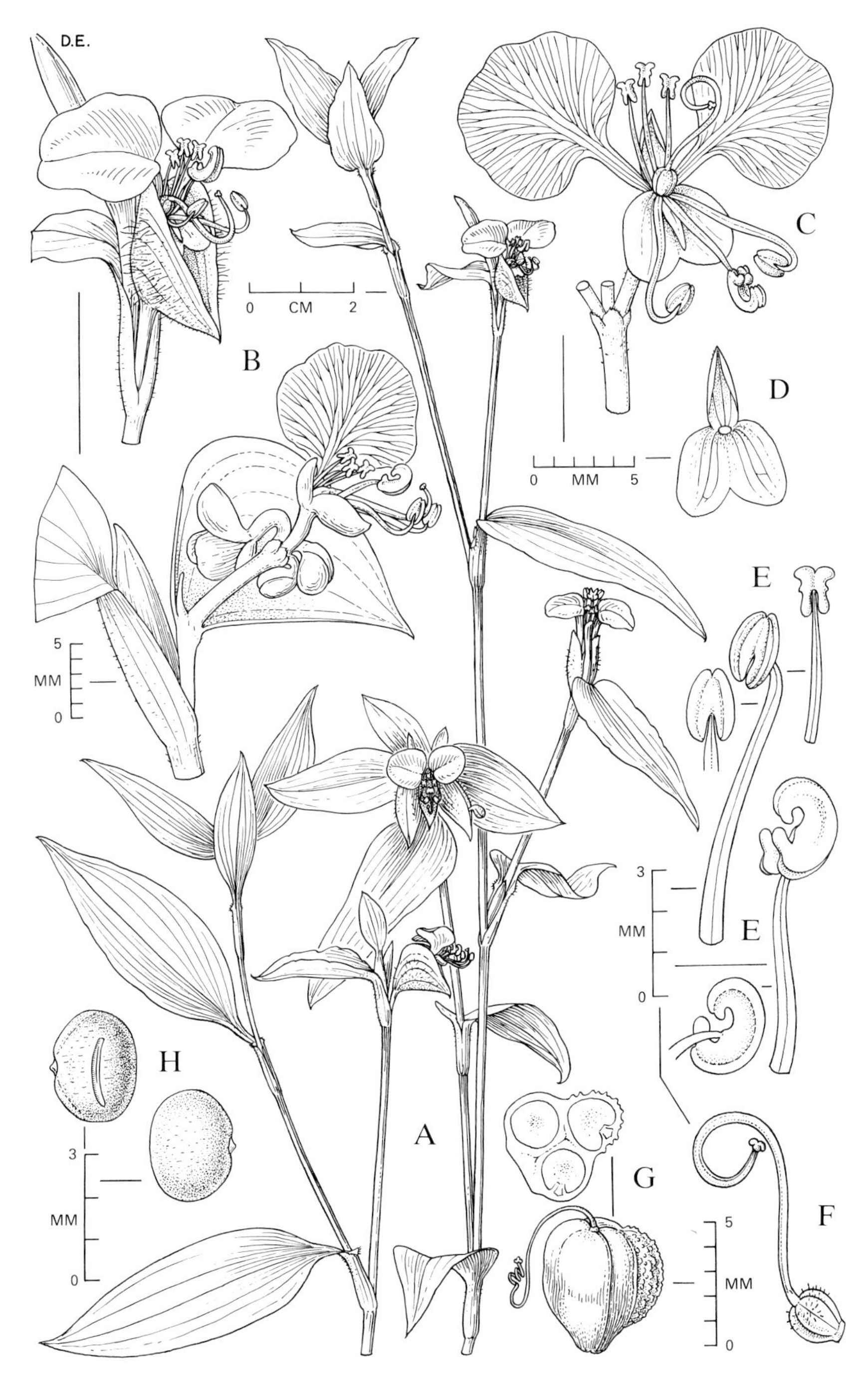

FIG. 16 **Commelina elegans.** A, habit. B, inflorescence. C, flower. D, sepals. E, stamens. F, pistil. G, capsule. H, seeds. (D.E.)

1. **Rhoeo spathacea** (Sw.) Stearn in Baileya 5: 195. 1957.

Rhoeo discolor (L'Herit.) Hance, 1853.

"Boat Lily"

Plants often forming colonies; stems up to 20 cm tall, or old plants more elongate, becoming procumbent-ascending, c. 1.5 cm thick. Leaves nearly erect, oblong-lanceolate, easily broken, mostly 20–35 cm long, 3–5 cm broad, long-acuminate at apex, dark green above and red-purple beneath in the common form. Peduncles 2–4 cm long; inflorescences several- to rather many-flowered; pedicels c. 1 cm long. Petals white, 5–8 mm long. Capsule 4–4.5 mm long; seeds 3–4 mm long.

GRAND CAYMAN: *Kings GC 188.* CAYMAN BRAC: *Proctor (sight),* cult. and escaping.
– Type from Nicaragua; widely cultivated and naturalized.]

Family 6

CYPERACEAE

Grass-like or rush-like herbs (rarely somewhat woody), annual or perennial, often with more or less elongate rhizomes. Culms usually erect, solid (rarely hollow), more or less 3-angled (rarely terete). Leaves linear or nearly so, 3-ranked, with closed sheaths at base, this lacking a ligule or the ligule very small; sometimes lacking a blade, the entire leaf then consisting of a tubular sheath. Inflorescences simple in heads, or variously compound; flowers bi-sexual or unisexual, arranged in dense or loose spikelets, the individual flowers always solitary in the axils of deciduous or persistent papery scale-like glumes. Spikelets 1- to many-flowered, clustered or numerous, rarely solitary. Perianth composed of bristles or small scales, rarely calyx-like, sometimes lacking. Stamens usually 1-3, the anthers 2-celled. Ovary superior, 1-celled, with a single erect ovule; style usually 2- or 3-cleft (rarely more divided). Fruit an achene.

A large, world-wide family of about 48 genera and over 3800 species, very few of which are of any economic importance. Members of this family are often mistaken for grasses. All are wind-pollinated with the exception of a few white-bracted species of *Rhynchospora*; one of these *(R. stellata)* commonly occurs in Grand Cayman.

KEY TO GENERA

1. Culms terminating in a single spikelet:

 2. Leaves all reduced to bladeless sheaths; spikelets conic-cylindric, the glumes spirally arranged: **2. Eleocharis**

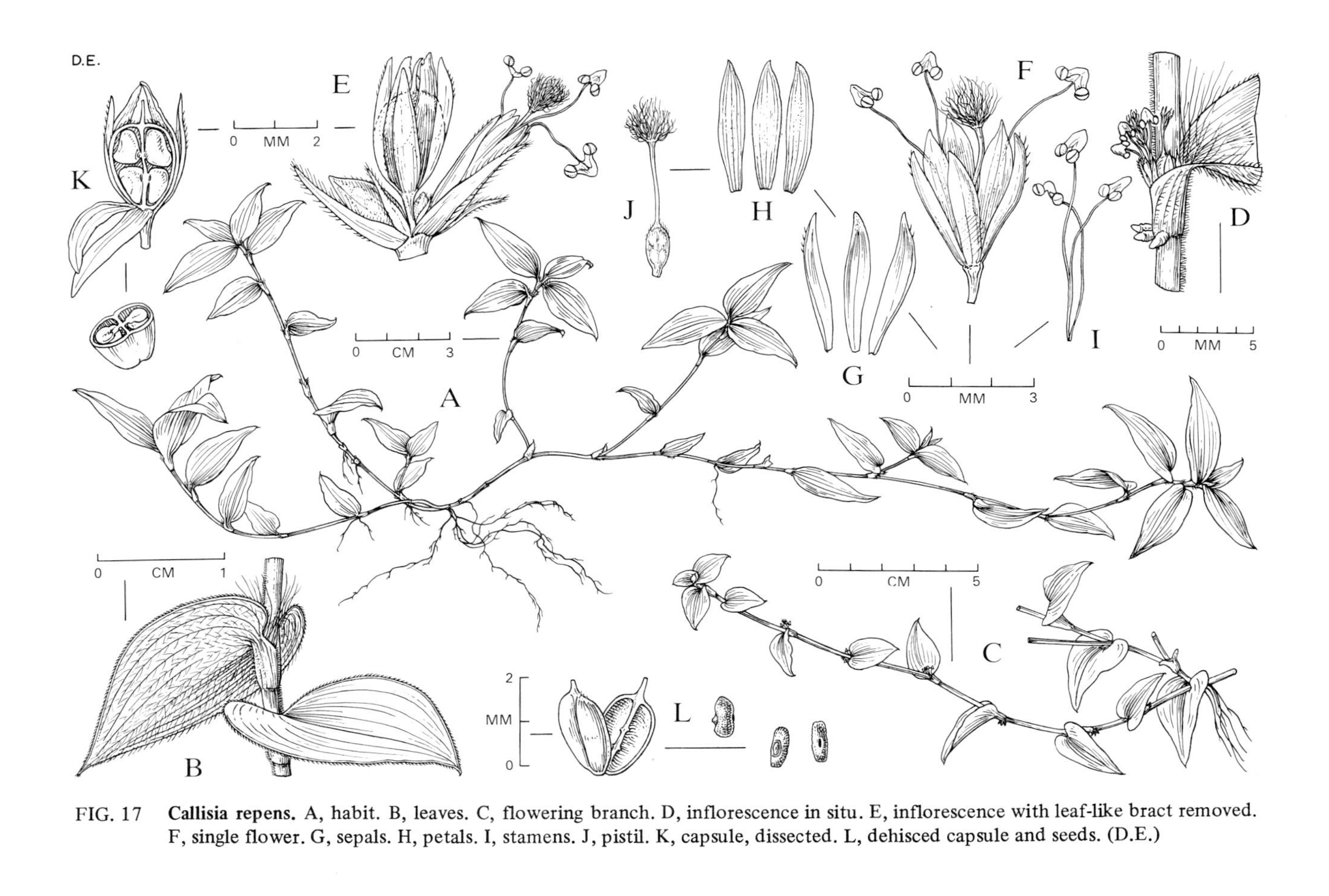

FIG. 17 **Callisia repens.** A, habit. B, leaves. C, flowering branch. D, inflorescence in situ. E, inflorescence with leaf-like bract removed. F, single flower. G, sepals. H, petals. I, stamens. J, pistil. K, capsule, dissected. L, dehisced capsule and seeds. (D.E.)

2. Leaves with free blades present; spikelets flattened, the glumes 2-ranked: **3. Abildgaardia**

1. Culms terminating in more than one spikelet, the inflorescence compound or condensed:

3. Glumes 2-ranked, the spikelets thus evidently flattened: **1. Cyperus**

3. Glumes spirally imbricate:

4. Bracts below the inflorescence white at base; spikelets condensed into a dense, headlike cluster: **6. Rhynchospora**

4. Bracts below the inflorescence green throughout, or else minute or lacking:

5. Flowers all unisexual (plants monoecious); achenes white, shining, enamel-like: **9. Scleria**

5. At least the fertile flowers bisexual; achenes not white and enamel-like:

6. Empty glumes at base of spikelet 1 or 2:

7. Base of style swollen; perianth bristles absent: **4. Fimbristylis**

7. Base of style not swollen; perianth bristles present: **5. Scirpus**

6. Empty glumes at base of spikelet 3 or more:

8. Fertile flower lateral in the spikelet; spikelets in loose panicles; culms erect, usually 1–3 m tall, distantly leafy: **7. Cladium**

8. Fertile flower terminal in the spikelet; spikelets condensed into a dense, headlike cluster; culms decumbent or ascending, less than 30 cm long, densely leafy: **8. Remirea**

Genus 1. **CYPERUS** L.

Annuals or perennials, the leaves and culms solitary or tufted, sometimes with corm-like thickenings at the base, often proliferating by slender creeping subterranean rhizomes which in some cases may bear tuber-like nodules, all the underground parts frequently aromatic. Culms simple, erect, leafy near the base (the leaves shorter than or exceeding the culms), at apex bearing a capitate or branched umbelliform inflorescence, subtended by 1 to many more or less leafy bracts. Spikelets flattened or angular, few to many in dense or loose spikes, or, by a shortening of the rhachis, in a capitate cluster. Glumes 2-ranked, deciduous or persistent. Flowers bisexual, without perianth, sometimes only 1 or 2 of a spikelet fertile. Stamens 1-3; style 2- or 3-cleft. Achenes flattened or 3-angled, neither beaked not tuberculate.

A cosmopolitan genus of about 820 species, most numerous in tropical and warm-temperate regions. The five major subdivisions *(Kyllinga, Pycreus, Cyperus, Mariscus and Torulinium)* by some authors are treated as separate genera, whereas the present writer prefers to view them as subgenera. All five are represented in the Cayman Islands. In the following treatment, the correct assignment with respect to these taxa is indicated in the synonymy of each species.

Few species of *Cyperus* have any economic value. Among the Cayman representatives, *C. compressus* and *C. rotundus* have found minor usefulness elsewhere, the former as a source of fiber, the latter for perfume and medicine.

KEY TO SPECIES

1. Styles 2; achenes flattened:

  2. Spikelets aggregated into a single dense ovoid or globose head; rhachillas jointed at base, the whole spikelet falling when ripe:

    3. Leaves reduced to bladeless sheaths; bracts not exceeding the inflorescence: **1. C. peruvianus**

    3. Leaves with foliaceous blades; bracts leaf-like, much exceeding the inflorescence: **2. C. brevifolius**

  2. Spikelets fasciculate, forming penicillate clusters either on a branched or contracted lobed or headlike inflorescence; rhachillas not jointed at base, persistent after the achenes and glumes have fallen: **3. C. polystachyos**

1. Styles 3; achenes 3-angled:

  4. Rhachillas not jointed at all, persistent after the achenes and glumes have fallen:

    5. Culms mostly solitary; spikelets red-brown; rhizomes elongate, slender, bearing aromatic nut-like tubers: **4. C. rotundus**

    5. Culms tufted; spikelets green; rhizomes absent or very short:

      6. Plants annual with fibrous roots only; culms and leaves neither viscid nor septate-nodulose; glumes appressed, 3-3.5 mm long: **5. C. compressus**

6. Plants perennial with short rhizomes; culms and leaves viscid, the latter septate-nodulose; glumes spreading, mucronate, 2–2.5 mm long: **6. C. elegans**

4. Rhachillas jointed at least at the base:

7. Rhachillas jointed at base only, the entire spikelets falling as a unit at maturity:

8. Spikelets compressed, more than 4 mm long:

9. Spikes cylindric, the spikelets not over 6 mm long: **7. C. ligularis**

9. Spikes flattened, the spikelets more than 6 mm long (often much more):

10. Spikelets 1.5–1.7 mm wide; glumes acutely keeled, c. 1 mm wide, usually light brown; achenes narrowly obovate in outline: **8. C. planifolius**

10. Spikelets 2.5–2.8 mm wide; glumes obtusely keeled, 1.5–1.8 mm wide, deep red-brown; achenes broadly elliptic in outline: **9. C. brunneus**

8. Spikelets inflated, less than 3 mm long: **10. C. swartzii**

7. Rhachillas jointed both at base and between flowers, the spikelets breaking into internode-segments when falling:

11. Leaves 4–13 mm wide; culms 2–5 mm thick; annual with fibrous roots only: **11. C. odoratus**

11. Leaves not over 2 mm wide; culms 0.3–0.5 mm thick; perennials with knotted rhizomes:

12. Leaves 1–2 mm wide; culms up to 35 cm tall; spikelets mostly 9–20 mm long, the glumes 2.5–3 mm long: **12. C. filiformis**

12. Leaves 0.6–0.9 mm wide; culms less than 9 cm tall; spikelets not over 8 mm long, the glumes 1.8–2 mm long: **13. C. floridanus**

1. Cyperus peruvianus (Lam.) F.N. Williams in Bull. Herb. Boiss. ser. 2, 7: 90. 1907.

Kyllinga peruviana Lam., 1789.

Perennial by short creeping rhizomes covered with conspicuous brown ovate scales. Leaves reduced to tubular green sheaths, membranous at the apex. Culms mostly 20–50 cm tall or sometimes taller, appearing naked. Bracts triangular, inconspicuous. Spikelets c. 4 mm long, arranged in a dense globular head c. 1 cm in diameter, each spikelet with about 3 glumes, only 1 of these fertile. Achenes flattened, yellowish-brown, c. 1 mm long.

GRAND CAYMAN: *Brunt 2085; Proctor 15286; Sachet 453.*
– Widespread in tropical America and Africa, in wet soils, not common.

2. **Cyperus brevifolius** (Rottb.) Radlk, ex Hassk., Cat. Pl. Hort. Bogor. 24. 1844.

Kyllinga brevifolia Rottb., 1773.

Perennial by creeping rhizomes; leaves shorter than the culms, mostly 2–3 mm wide. Culms slender, mostly 10–30 cm tall. Inflorescence a single small (5–10 mm long), ovoid or globose head of densely aggregated spikelets, subtended by 3 elongate but unequal leafy bracts. Spikelets each with 3–4 glumes, only 1 of these fertile. Achenes flattened, brown, c. 1 mm long.

GRAND CAYMAN: *Kings GC 182.*
– Pantropical, also extending somewhat into warm-temperate regions, in damp soils, often a weed of lawns and gardens.

3. **Cyperus polystachyos** Rottb., Descr. Pl. 21. 1772.

Pycreus polystachyos (Rottb.) Beauv., 1807.

Cyperus odoratus sensu Britton & Wilson, 1923, not L., 1753.

Annual, or short-lived perennial with short rhizomes. Culms tufted, slender, mostly 20–60 cm tall, usually nearly equalled by the basal leaves. Inflorescence compound-umbellate (rarely capitate), subtended by several spreading, elongate, leafy bracts. Spikelets linear-lanceolate, chiefly 8–12 mm long, 1–1.5 mm wide, gray- or yellow-brown, densely fasciculate. Achenes flattened, brown to black, 1 mm long.

GRAND CAYMAN: *Brunt 2092; Kings GC 182b, GC 194, GC 425; Lewis GC 29, GC 33; Proctor 15241, 27950, 31036.*
– Pantropical, often a weed of roadside ditches.

4. Cyperus rotundus L., Sp. Pl. 1: 45. 1753.

"Nut Grass"

Perennial by long, slender, fragile, deeply subterranean rhizomes bearing aromatic, nut-like tubers at wide intervals. Culms mostly 15–30 cm tall (rarely taller). Inflorescence compound with 2 or 3 erect, simple rays subtended by shorter leafy bracts. Spikelets few, 1–2 cm long (rarely much longer), loosely spreading. Achenes 3-angled, shining black, 1.5–2 mm long.

GRAND CAYMAN: *Proctor 31798.*

– Pantropical, also extending into many warm-temperate regions. A persistent weed of lawns, gardens and open waste ground, this species easily regenerates from any tubers left in the ground. These tubers contain an aromatic oil that has been used in some countries in perfume, and in medicines as a remedy for digestive disorders.

5. Cyperus compressus L., Sp. Pl. 1: 46. 1753.

Annual, with fibrous roots only; leaves 1–3 mm wide. Culms tufted, 10–40 cm tall, wiry. Inflorescence a simple capitate umbel of spikelets or more often compound, having 2 or 3 slender rays up to 1.5 cm long, each bearing an umbel of spikelets at the apex, the whole inflorescence subtended by 2 or 3 bracts, one of these longer than the inflorescence. Spikelets 3–10 in a cluster, 8–25 mm long, 3–5 mm wide, much flattened; glumes light green with a yellowish band along each side, deciduous beginning from base of the spikelet. Achenes 3-angled, brown to black, shining, 1–1.3 mm long.

CAYMAN BRAC: *Proctor 29130.*

– Pantropical. The culms contain fibers that have been used in India for weaving small mats.

6. Cyperus elegans L., Sp. Pl. 1: 45. 1753.

Cyperus viscosus Sw., 1788.

Perennial by short tangled rhizomes. Culms tufted, mostly 30–60 cm long, sometimes overtopped by the more or less convolute wiry basal leaves. Inflorescence decompound-umbellate, subtended and exceeded by several narrow, elongate bracts; rays up to 15 cm long. Spikelets mostly 4–10 mm long, up to 3 mm wide, borne in numerous small heads; glumes greenish-brown, sharply mucronate. Achenes 3-angled, nearly black, 1.5 mm long.

GRAND CAYMAN: *Brunt 1848, 2040; Proctor 15065, 15300.* CAYMAN BRAC: *Proctor 29002.*
– Florida, West Indies, and continental tropical America, chiefly growing in moist subsaline or brackish soils near seacoasts.

7. **Cyperus ligularis** L., Syst. Nat. ed. 10, 2: 867. 1759.

Mariscus ligularis (L.) Urban, 1900.

"Cutting Grass"

Perennial, forming large, dense clumps. Leaves conspicuous, pale green, often 1 cm wide or more, with harsh, rough margins. Culms stout, up to 1 m tall or more. Inflorescence compound, subtended by long leafy scabrous bracts. Spikelets mostly 4-6 mm long, c. 1 mm wide, borne in dense cylindric spikes up to 4 cm long and 1 cm thick, these often aggregated closely; glumes light brown. Achenes 3-angled, brown, 1.2-1.4 mm long.

GRAND CAYMAN: *Brunt 1796; Kings GC 168, GC 278; Maggs II 62; Proctor 15104.* LITTLE CAYMAN: *Proctor 28072.* CAYMAN BRAC: *Proctor 28983.*
– Tropical America and Africa, usually at low elevations near the sea.

8. **Cyperus planifolius** L.C. Rich. in Act. Soc. Hist. Nat. Paris 1: 106. 1792. **FIG. 18.**

Mariscus planifolius (L.C. Rich.) Urban, 1900.

"Cutting Grass"

Perennial, densely tufted from short rhizomes. Leaves many, glaucous green and of stiff texture, much longer than the culms and 8-10 mm wide. Culms 60-90 cm tall, 3-4 mm thick toward base. Inflorescence compound, subtended and much exceeded by 5-8 leafy bracts; rays very unequal, bearing corymbose clusters of spikes, these ovoid, 1.5-3 cm long, densely composed of many spikelets. Spikelets linear, 6-16 mm long, 1.5-1.7 mm wide, light brown sometimes tinged with red, 8- to 14-flowered; glumes rather loosely overlapping. Achenes 3-angled with shallowly concave sides, 1.2-1.5 mm long.

GRAND CAYMAN: *Sauer 4231.*
– West Indies and French Guiana, usually in coastal sands.

9. Cyperus brunneus Sw., Fl. Ind. Occ. 1: 116. 1797.

Cyperus brizaeus Vahl, 1806.

Mariscus brunneus (Sw.) C.B. Clarke in Urban, 1900.

"Cutting Grass"

Perennial, densely tufted from short rhizomes. Leaves many, glaucous green and of rather stiff texture, shorter than to equalling the culms and 3–8 mm wide. Culms 20–80 cm tall, 1.5–3 mm thick toward base. Inflorescence compound, either open with elongate rays or congested in one headlike lobed cluster, much overtopped by 3–5 leafy bracts; rays (when developed) 1–7 cm long, bearing 1–4 spikes, these broadly ovoid, up to 3 cm long and 2.5 cm thick. Spikelets divergent, linear-oblong or lance-oblong, 10–20 mm long, mostly 2.5–2.8 mm wide (rarely less), compressed, usually deep red-brown, 6- to 17-flowered; glumes rather densely overlapping. Achenes 3-angled with convex sides, 1.2–1.5 mm long.

GRAND CAYMAN: *Proctor 15130.*
– Florida, the West Indies, and southern Mexico, frequent in coastal sands.

The preceding two species have customarily been combined under the name *Cyperus planifolius* because of their close similarity, and only recently has their distinctness been recognized (T. Koyama in Howard, Flora of the Lesser Antilles 3: 268–269. 1979). It has not been possible to re-study the numerous specimens of this complex in order to cite them under their correct names. They can be listed as follows: GRAND CAYMAN: *Brunt 1705; Kings GC 28, GC 267; Lewis GC 40; Millspaugh 1248; Sachet 387a, 387b.* LITTLE CAYMAN: *Kings LC 3, LC 92, LC 101; Proctor 28071.* CAYMAN BRAC: *Brunt 1675; Kings CB 10, CB 62; Proctor 28924.*

10. Cyperus swartzii (Dietr.) Boeck, ex Kük. in Fedde Repert. 23: 186. 1926.

Mariscus gracilis Vahl, 1806.

Cyperus caymanensis Millsp., 1900.

Perennial with short, tangled, woody rhizomes, forming dense tufts. Leaves few, or sometimes apparently lacking, shorter than the culms, 1–4 mm wide, slightly rough on the margins; basal sheaths red-brown. Culms slender and wiry, usually 15–40 cm tall or more. Inflorescence capitate, consisting of up to 3 densely aggregated spikes each bearing 30–40 crowded spikelets, the whole subtended by 3–5 unequally elongate bracts, these 1–2 mm wide. Spikelets 1–2.2 mm long, asymmetrically inflated, 1-flowered; fertile glumes pale greenish, longitudinally 11- to 15-nerved. Achenes 3-angled with concave sides, purplish-brown, 1-1.2 mm long.

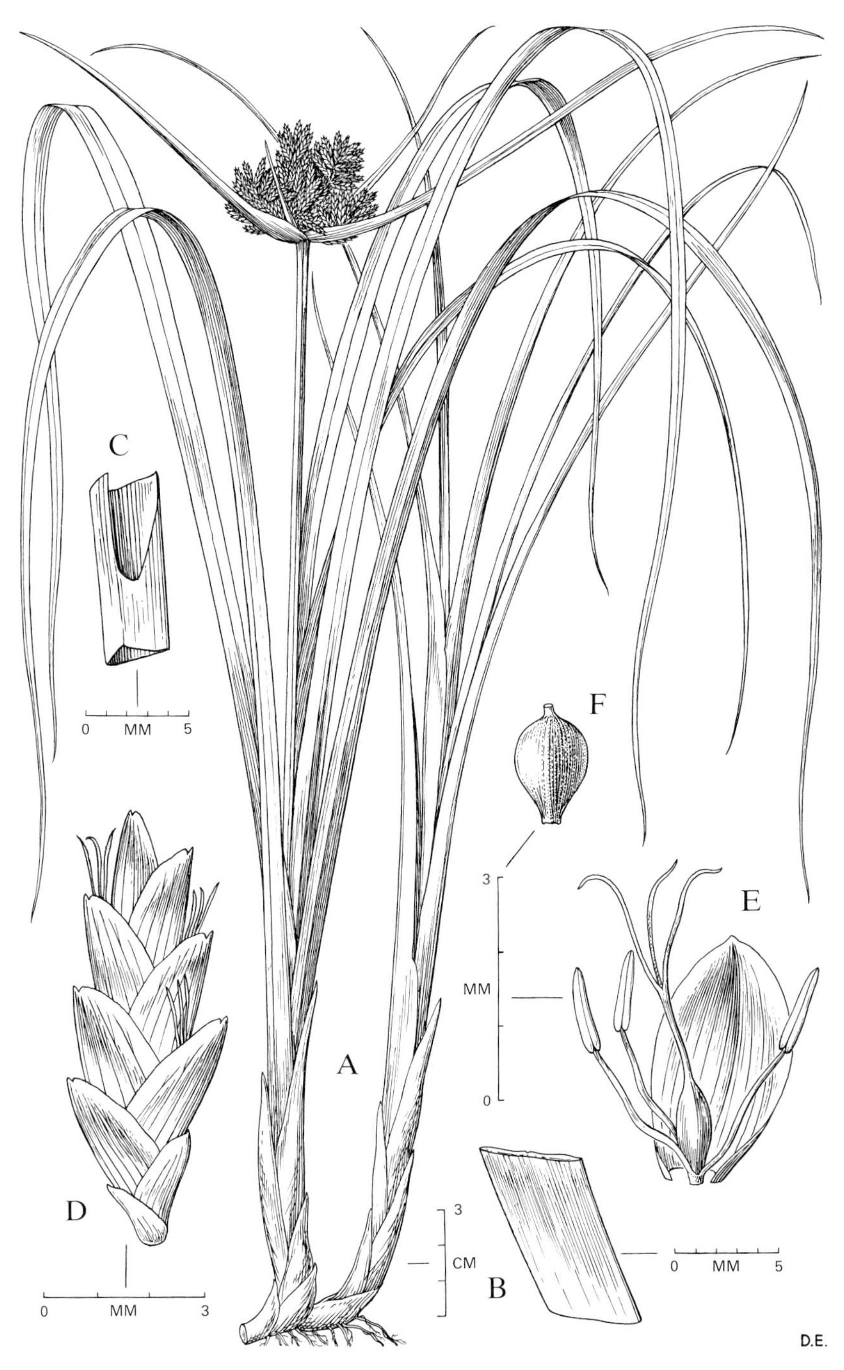

FIG. 18 **Cyperus planifolius.** A, habit. B, section of leaf. C, section of stem. D, spikelet. E, floret. F, achene. (D.E.)

GRAND CAYMAN: *Brunt 1822; Kings GC 177, GC 286; Millspaugh 1334* (type of *C. caymanensis*); *Proctor 27964.*

– Greater and Lesser Antilles, also reported from southern Mexico, often growing in damp shaded glades.

11. Cyperus odoratus L., Sp. Pl. 1: 46. 1753.

Cyperus ferax L.C. Rich., 1792.

Torulinium odoratum (L.) Hooper, 1972.

Annual or perhaps sometimes a short-lived perennial, solitary or loosely tufted, with fibrous roots. Leaves shorter than the culms, up to 13 mm wide. Culms stout, smooth, up to 1 m tall. Inflorescence loosely compound, ample, subtended and overtopped by several long leafy bracts; rays 5-12, up to 10 cm long or more (rarely to 20 cm), bearing loose oblong-cylindric spikes 2-3 cm long, with 20–40 spikelets. Spikelets linear, 10-20 mm long, less than 1 mm wide, the glumes light yellowish-brown, 2 mm long or more, only slightly overlapping; at maturity each rhachilla separating into 1-fruited joints. Achenes 3-angled, brown to black, 1-1.5 mm long.

GRAND CAYMAN: *Brunt 1845.*

– Widespread in tropical and warm-temperate regions, commonly a weed of roadside ditches, damp fields and marshes.

12. Cyperus filiformis Sw., Nov. Gen. & Sp. Pl. 20. 1788.

Torulinium filiforme (Sw.) C.B. Clarke in Urban, 1900.

Perennial, densely tufted, the rhizomes short and knotted. Leaves few (1-3 per culm), half or less as long as the culms, filiform, 0.5-1.5 mm wide, slightly rough-margined, the sheaths red-brown. Culms usually 20-35 cm tall, 0.3-0.5 mm thick. Inflorescence a solitary, lateral-appearing, loose cluster of 2-8 spikelets of unequal length, subtended by 2 or 3 bracts one of which is much longer than the others and appears like an extension of the culm. Spikelets 9-20 mm long, 0.7-1 mm wide; glumes yellow-brown, 2.5-3 mm long, appressed. Achenes 3-angled, brown to black, 2 mm long.

GRAND CAYMAN: *Kings GC 384.* CAYMAN BRAC: *Proctor 28936.*

– Bahamas, Greater Antilles and the Virgin Islands, in shaded sandy or rocky soils. Reports from Florida pertain to the next species.

13. Cyperus floridanus Britton in Small, Fl. Southeastern U.S. 170, 1327. 1903. **FIG. 19.**

Mariscus floridensis C.B. Clarke, 1908, illegit. (*C. floridanus* Brit. has not so far been validly transferred to *Torulinium* with its relatives).

Cyperus filiformis var. *densiceps* Kük., 1926.

C. kingsii C.D. Adams, 1977.

Perennial with short knotted aromatic rhizomes; leaves and culms densely tufted. Leaves numerous, filiform, often exceeding the culms, 0.6–0.9 mm wide. Culms 4–8 cm tall, 3-angled and more slender than the leaves. Inflorescence a solitary, lateral-appearing loose cluster of usually 10–16 spikelets, the aggregate 1 cm in diameter or less. Spikelets 6–8 mm long, 0.6–0.8 mm wide, 3- to 7-flowered; glumes yellow-brown, striate, broadly green-ribbed, 1.8–2 mm long, mucronate. Achenes oblong-ellipsoid, 3-angled, tawny to black, 1–1.1 mm long.

GRAND CAYMAN: *Kings GC 410* (type of *C. kingsii*). LITTLE CAYMAN: *Proctor 35098, 35205.* CAYMAN BRAC: *Proctor 28910.*
– Florida Keys, Bahamas, Cuba, Jamaica and Hispaniola (Isla Saona), growing in pockets of flat limestone, in loose dry calcareous sand, or in thin bauxitic soil of pastured slopes.

Genus 2. **ELEOCHARIS** R. Br.

Annual or perennial herbs of watery habitats or wet soil, solitary or tufted, often with long rhizomes. Leaves reduced to bladeless sheaths enclosing base of culm. Culms angled, flattened, grooved or terete, sometimes hollow and septate. Inflorescence a single spikelet; subtending bracts minute or absent. Spikelet terminal, erect, few- or many-flowered; glumes spirally imbricate, deciduous. Flowers bisexual, with a perianth of normally 6 bristles, these usually retrorsely barbed; stamens 1–2; style 2–3-cleft, the expanded base persisting on apex of ripe achene as a cap-like "tubercle". Achene flat, 3-angled or turgid.

A world-wide genus of about 200 species.

KEY TO SPECIES

1. Spikelet about same diameter as the culm or only a little thicker, 1–5 cm long; culms 1–5 mm thick; plants perennial:

 2. Culms nodose-septate: **1. E. interstincta**

 2. Culms not nodose-septate:

3. Culms terete or nearly so: **2. E. cellulosa**

3. Culms rather sharply 3-angled: **3. E. mutata**

1. Spikelet obviously much thicker than the culm, less (usually much less) than 0.6 cm long; culms slender, 0.5 mm thick or less; plants annual:

4. Achenes jet black, smooth and shining, more or less flattened; glumes ovate, faintly bicolorous:

5. Perianth-bristles brown; achenes 1 mm long; culms mostly 5-20 cm tall: **4. E. geniculata**

5. Perianth-bristles nearly white; achenes 0.5 mm long; culms mostly 2-7 cm tall: **5. E. atropurpurea**

4. Achenes whitish to pale brown, reticulate-striate, acutely 3-angled; glumes ovate-lanceolate, strongly bicolorous: **6. E. minima**

1. Eleocharis interstincta (Vahl) Roem. & Schult. in L., Syst. Veg. ed. nov. 2: 149. 1817. **FIG. 20.**

Perennial, spreading by stout rhizomes. Culms mostly 40–100 cm tall, 4–5 mm thick, terete, hollow and nodose-septate; basal sheaths membranous, oblique at apex. Spikelets cylindric, up to 4 cm long; glumes rigid, oblong or ovate, often acute, pale yellowish or greenish, striate. Bristles 6, retrorsely barbed, about as long as the achene. Style 2- or 3-cleft. Achenes 1.5-2 mm long, brown or yellowish-brown, with minute transverse ridges; tubercle conic, acute.

GRAND CAYMAN: *Brunt 1794, 1842; Kings GC 191, GC 192.*
– Southern United States, West Indies, Central and South America.

2. Eleocharis cellulosa Torr. in Ann. Lyc. N. Y. 3:298. 1836.

Perennial, with deep-seated creeping rhizomes, often forming dense colonies. Culms mostly 30-70 cm tall, 1-3.5 mm thick, terete in upper part, obscurely 3-angled toward base; sheaths membranous, oblique at apex, usually purple. Spikelets cylindric, 1.5–4 cm long, thicker than the culm; glumes rigid, orbicular or obovate, obtuse, yellowish with brown border and whitish membranous margins, striate. Bristles about 6, not barbed, about as long as the achenes. Style 3-cleft. Achenes 2–2.5 mm long, olive-brown, nearly smooth; tubercle broadly conic, whitish, tipped by the blackish style-base.

FIG. 19 **Cyperus floridanus.** A, habit. B, spikelet. C, achene. (D.E.)

CAYMAN BRAC: *Proctor 29130.* The single colony seen was growing densely in brackish sandy mud.

– Southern United States, Bermuda, West Indies, Mexico and Belize.

3. Eleocharis mutata (L.) Roem. & Schult. in L., Syst. Veg. ed. nov. 2:155. 1817. **FIG. 21.**

Perennial, spreading by stout rhizomes. Culms up to 1 m tall or more, sharply 3- or 4-angled; sheaths membranous, oblique at apex. Spikelets cylindric, 2–5 cm long; glumes firm, ovate or obovate, pale yellowish. Bristles 6, brown, retrorsely barbed, about as long as the achene. Style 3-cleft. Achenes 1.5–2.3 mm long, yellowish-brown, lustrous, obscurely longitudinally striate; tubercle a low, annular depressed cap surmounted from the middle by the acuminate style-base.

GRAND CAYMAN: *Brunt 1799, 1967, 2093; Kings GC 102; Proctor 27947.*

– Widespread in tropical America; also in Africa.

4. Eleocharis geniculata (L.) Roem. & Schult. in L., Syst. Veg. ed. nov. 2:150. 1817. **FIG. 22.**

Eleocharis caribaea (Rottb.) Blake, 1918.

E. capitata R.Br., 1810.

Annual, with fibrous roots. Culms densely tufted, mostly 5–20 cm tall; basal sheaths firm, oblique at apex. Spikelets ovoid, obtuse, 2–6 mm long; glumes ovate, obtuse, yellowish or pale brown. Bristles 5–8, brown, longer than the achene. Style 2-cleft. Achenes c. 1 mm long, shining black, with a short whitish tubercle.

GRAND CAYMAN: *Brunt 1965, 2162; Kings GC 206.*

– Pantropical, extending also into temperate regions. Usually a weed of wet ditches and damp open ground.

5. Eleocharis atropurpurea (Retz.) Kunth, Enum. Pl. 2:151. 1837.

A miniature annual. Culms densely tufted, very slender, 2–7 cm tall. Spikelets ovoid, obtuse or subacute, 2.5–4 mm long; glumes ovate-oblong, light purple-brown with green midvein and narrow scarious margins. Bristles 2–4, whitish, retrorsely hairy, nearly as long as the achene. Style 2- or 3-cleft. Achenes 0.5 mm long, shining black, with minute whitish tubercle constricted at base.

GRAND CAYMAN: *Kings GC 248.*

– Of scattered but wide distribution in the warmer parts of both hemispheres.

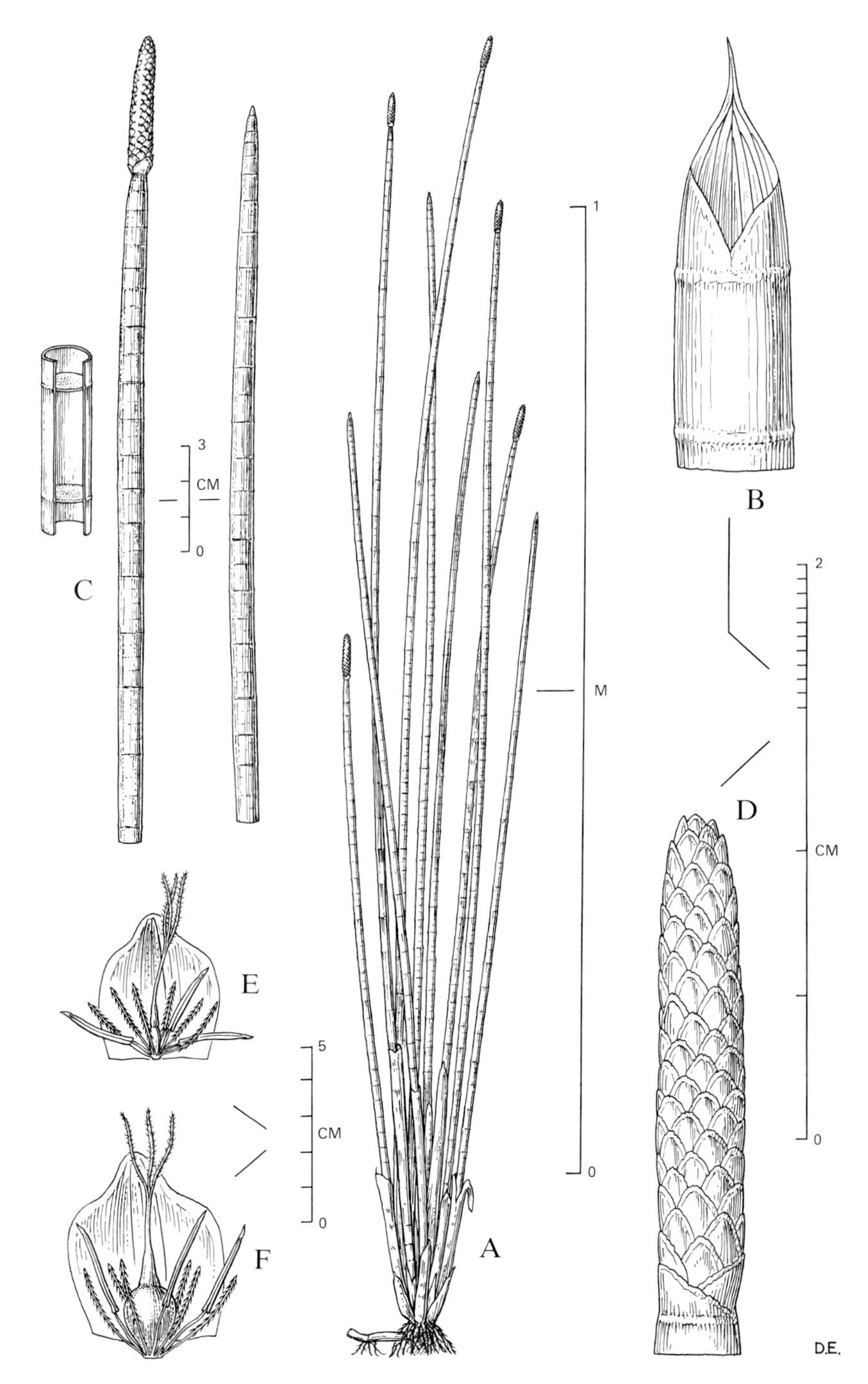

FIG. 20 **Eleocharis interstincta.** A, habit. B, basal sheath. C, sterile and fertile culms (upper end). D, spikelet. E, young floret. F, fruiting floret. (D.E.)

6. Eleocharis minima Kunth, Enum. Pl. 2:139. 1837.

A miniature annual. Culms tufted, capillary, 2–7 cm tall, longitudinally grooved; sheaths short, conspicuous, with inflated apex. Spikelets lanceolate or lance-ovate, 3–7 mm long; glumes ovate-lanceolate, usually acute, dark brown with greenish midvein and broad white hyaline margins. Bristles several, white, obscurely toothed, much shorter than the achene. Style 3-cleft. Achenes 0.7–1 mm long, narrowed at both ends; tubercle short-pyramidal, brownish or gray.

GRAND CAYMAN: *Brunt 1954.*
– Widespread in tropical America but in the West Indies known otherwise only from Cuba.

Genus 3. **ABILDGAARDIA** Vahl

Glabrous tufted perennials. Leaves basal, flat, sharp-pointed. Culms slender, numerous in a tuft. Inflorescence a solitary spikelet (rarely more than 1), subtended by 1 or 2 very small inconspicuous bracts. Spikelet few-to many-flowered, the deciduous scales imbricated in two rows. Bristles lacking. Stamens 1-3. Style pubescent, deciduous, with enlarged base; stigmas 3. Achenes 3-angled.

A small genus of about 15 species, chiefly occurring in the Old World tropics.

1. Abildgaardia ovata (Burm.f.) Kral in Sida 4:72. 1971.

Abildgaardia monostachya (L.) Vahl, 1805.

Fimbristylis monostachya (L.) Hassk., 1848.

A low, tufted plant with a hard, knotted base. Leaves up to 20 cm long (usually less), 0.5–1.5 mm wide, with roughened, minutely bristly margins. Culms wiry, about twice as long as the leaves. Bracts much shorter than the pale, solitary spikelets. Spikelets ovate or ovate-lanceolate, 5–15 mm long, up to 5 cm broad, flattened; glumes ovate, keeled, mucronate, pale greenish-brown with whitish margins. Achenes 2–2.5 mm long, constricted near the base, yellowish, tuberculate.

LITTLE CAYMAN: *Proctor 28190, 35133.* CAYMAN BRAC: *Proctor 35153.*
– Pantropical, chiefly at low to medium elevations. The Cayman specimens were collected in pockets of soil on flat, shaded limestone pavements.

Genus 4. **FIMBRISTYLIS** Vahl

Annual or perennial plants, usually tufted, with culms leafy toward base. Inflorescence a simple or compound cyme (often appearing umbellate by reduction), subtended by 1 to several narrow bracts. Spikelets loosely or densely aggregated, cylindric, the glumes spirally imbricate and deciduous. Flowers perfect (but apparently sometimes functionally unisexual and monoecious), without

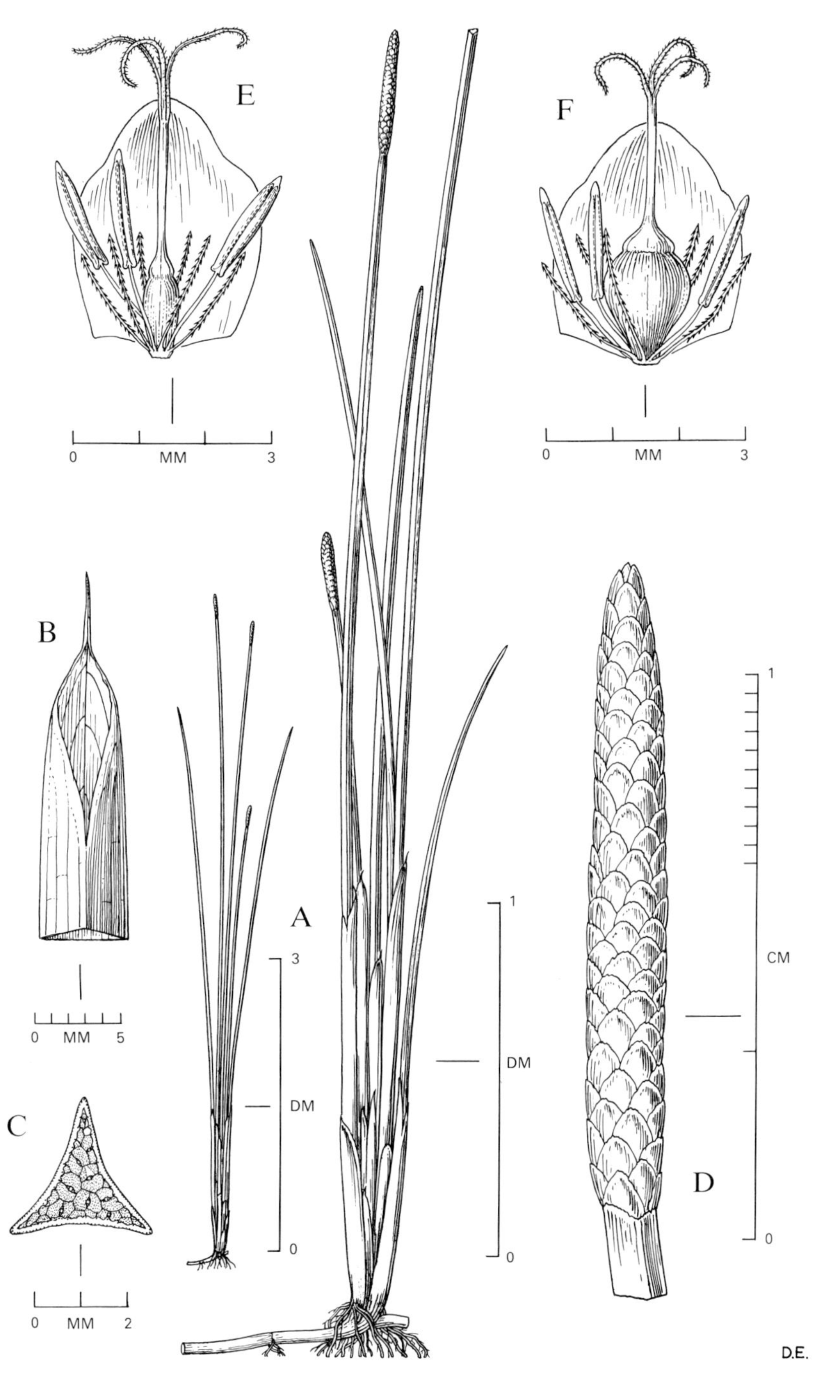

FIG. 21 **Eleocharis mutata.** A, habit. B, basal sheath. C, x-section of culm. D, spikelet. E, young floret. F, fruiting floret. (D.E.)

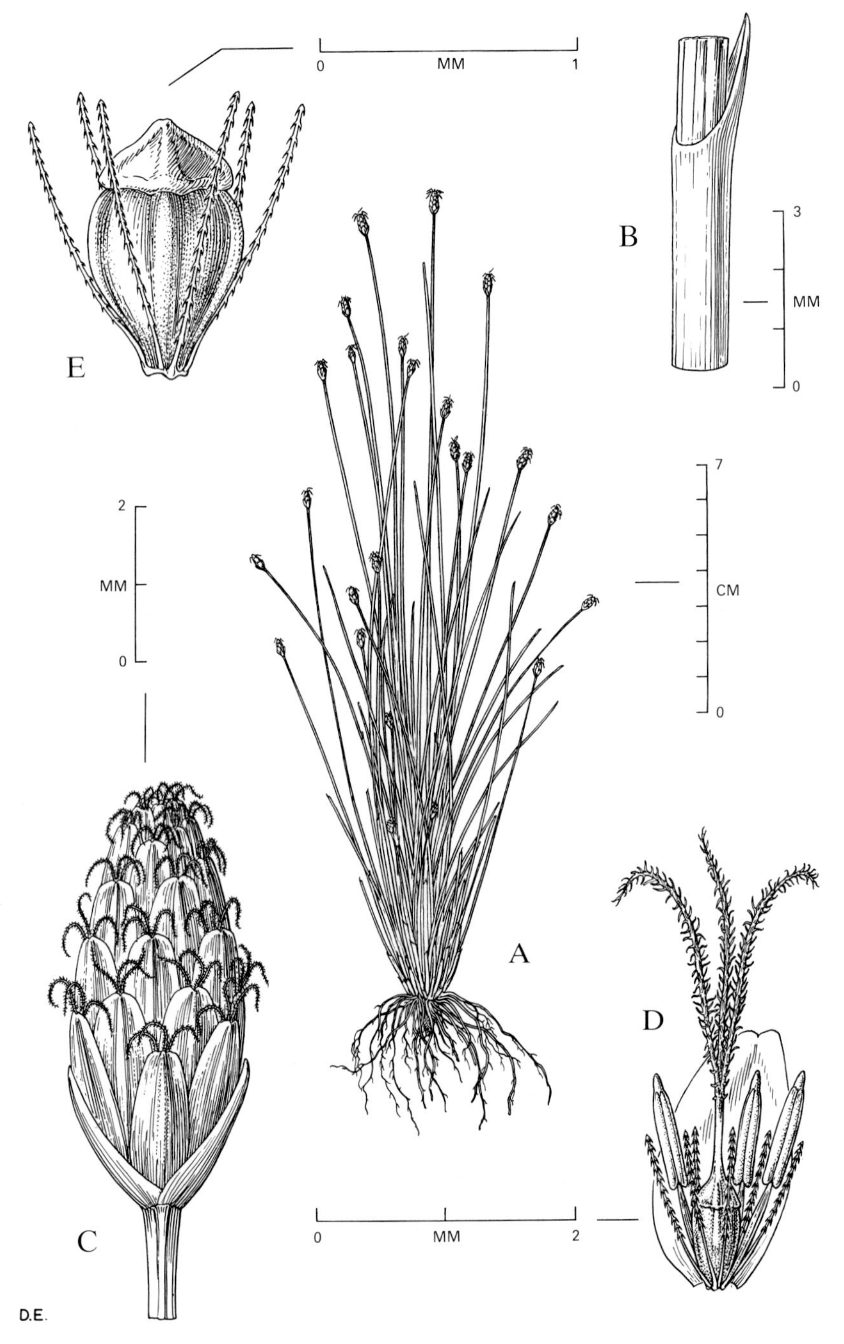

FIG. 22 **Eleocharis geniculata.** A, habit. B, basal sheath. C, spikelet. D, floret. E, achene with bristles. (D.E.)

perianth. Stamens 1-3. Style 2- or 3-cleft, often pubescent near the enlarged base, completely deciduous. Achenes flattened or 3-angled, often finely reticulate or striate.

A large genus of about 150 tropical or subtropical to warm-temperate species, most abundant in the Indonesian-Australian region.

KEY TO SPECIES

1. Spikelets more than 5 mm long (up to 15 mm), in loose inflorescences (or contracted in *F. ferruginea*); achenes various but not granular:

  2. Glumes glabrous; achenes not smooth:

    3. Achenes minutely reticulate in lines but not ribbed; sheaths at base of culm dark brown to nearly black:

      4. Spikelets narrowly oblong- to lance-cylindric; longest bract of inflorescence equalling or longer than the entire inflorescence: **1. F. spadicea**

      4. Spikelets ovoid or broadly ellipsoid; longest bract of inflorescence much shorter than the entire inflorescence: **2. F. castanea**

    3. Achenes longitudinally ribbed as well as minutely reticulate; sheaths at base of culm green or pale brown, inconspicuous: **3. F. dichotoma**

  2. Glumes minutely pubescent; achenes appearing smooth: **4. F. ferruginea**

1. Spikelets usually less than 4 mm long, aggregated in dense heads; achenes with granular surface: **5. F. cymosa** subsp. **spathacea**

1. Fimbristylis spadicea (L.) Vahl, Enum. Pl. 2:294. 1805.

Perennial, densely tufted, spreading by elongate stolons. Leaves shorter than or nearly equalling the culms, erect, more or less linear-involute with roughened margins. Culms mostly 30-70 cm tall, wiry, the broad basal sheaths dark brown to black. Inflorescences with 3-8 unequal rays, these less than 1 cm to 3 cm long, often branched. Spikelets usually 1-1.5 cm long; glumes ovate, obtuse to apiculate, shiny dark brown to nearly black with pale margins. Stamens 2 or 3, with dark brown subulate anthers 1.5-2 mm long. Achenes flattened-obovoid, brown, 1.5-1.8 mm long, the surface finely reticulate due to regular rows of deeply pitted quadrangular cells.

GRAND CAYMAN: *Brunt 1642; Proctor 15242.*
– Widespread in tropical and temperate America, occurring as far north as southern Canada. Many variants occur, and the following species is often combined with it. In the West Indies, *F. spadicea* occurs chiefly in damp brackish soils at low elevations.

2. **Fimbristylis castanea** Vahl, Enum. Pl. 2:292. 1805.

Perennial, rather loosely tufted, the bases deeply set in the substratum, the outer and older leaf-bases persistent as imbricated scales. Leaves and culms similar to those of *F. spadicea.* Inflorescences with 3–6 slender rays, these 2–9 cm long and often unbranched. Spikelets relatively fewer than in *F. spadicea,* and rarely more than 1 cm long; glumes broadly ovate with rounded apex, dull brown, Stamens as in *F. spadicea,* likewise the achenes.

GRAND CAYMAN: *Kings GC 193.*
– Seacoasts of eastern North America from New York to the Yucatan peninsula; also the Bahamas and Cuba, always in coastal marshes or swales.

3. **Fimbristylis dichotoma** (L.) Vahl, Enum. Pl. 2:287. 1805.

Fimbristylis diphylla (Retz.) Vahl. 1805.

F. annua (All.) Roem. & Schult., 1817.

Annual or short-lived perennial, tufted. Leaves shorter than the culms, flat, to 3 mm wide. Culms slender, 10–60 cm tall, glabrous or pubescent. Inflorescence loose, compound-umbellate with slender unequal rays, subtended by several very short rough-margined bracts, these ciliate at the sheathing base. Spikelets usually numerous, oblong to ovoid, mostly 5–10 mm long, c. 2 mm thick or less, solitary or in clusters of 2 or 3; glumes ovate, 2–3 mm long, brown with green midrib, acute or apiculate. Stamens 1 or 2; style 2-forked. Achenes flattened-obovoid, whitish to pale yellow-brown, 1 mm long, longitudinally ribbed, the ribs minutely tuberculate.

GRAND CAYMAN: *Brunt 2175; Kings GC–193; Proctor 27949.*
– Cosmopolitan; one of the most widely distributed plants in the world, but often known by other names. One authority estimates that about 400 synonyms apply to this species.

4. **Fimbristylis ferruginea** (L.) Vahl, Enum. Pl. 2:291. 1805. **FIG. 23.**

Perennial, in slender tufts. Leaves inconspicuous, much shorter than the culms or often lacking. Culms 20–70 cm tall, narrowly brown-sheathed at base. Inflo-

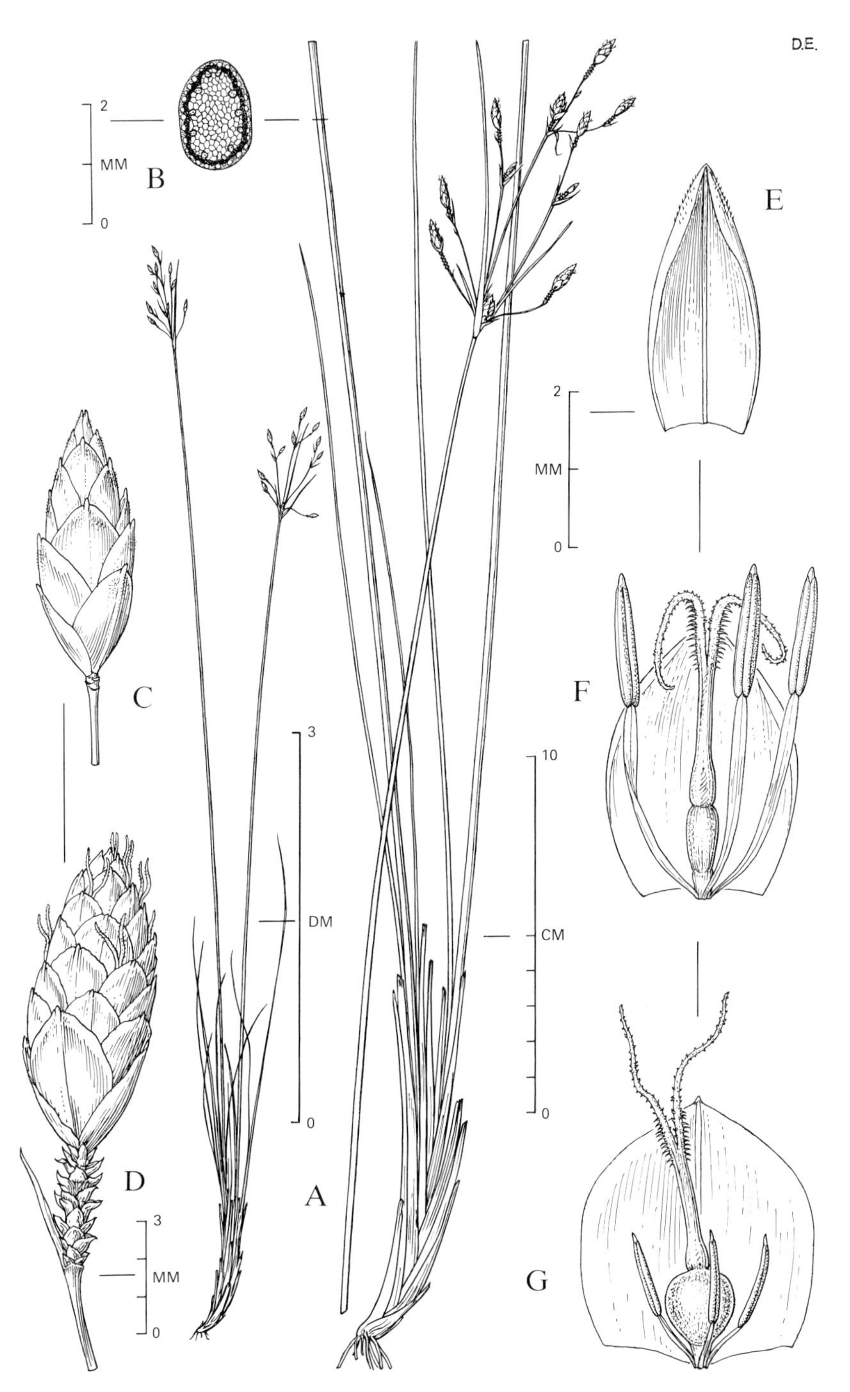

FIG. 23 **Fimbristylis ferruginea.** A, habit. B, x-section of culm. C, young spikelet. D, mature spikelet. E, bract. F, functionally staminate floret. G, fruiting floret. (D.E.)

rescence rays 0.5–2 cm long, forming a small rather compact cluster. Spikelets 5–10 (rarely 1) per inflorescence, ovoid-oblong, 8–20 mm long; glumes ovate, minutely apiculate, appressed puberulous toward the apex. Stamens 3, anthers 1–1.5 mm long. Achenes flattened-obovoid, 1.5 mm long, the surface very minutely reticulate (seen under high magnification), ordinarily appearing smooth.

GRAND CAYMAN: *Brunt 1774, 1922, 1945; Kings GC 169, GC 356; Proctor 15059.* LITTLE CAYMAN: *Kings LC 74.* CAYMAN BRAC: *Proctor 29131.* – Pantropical.

5. **Fimbristylis cymosa** R. Br. subsp. **spathacea** (Roth) T. Koyama in Micronesica 1:83. 1964.

Fimbristylis spathacea Roth, 1821.

F. cymosa sensu Adams, 1972, not R. Br., 1810.

Perennial, forming dense tufts often aggregated in a turf. Leaves much shorter than the culms, stiff, flat, spreading, 1.5–3 mm wide, mucronate at tip. Culms 10–40 cm talk, stiffly erect. Inflorescence a dense head of small spikelets, subtended by short glabrous entire bracts. Spikelets ellipsoid to short-cylindric, 2–3(–6) mm long, 1.5–2 mm thick; glumes ovate, 1–1.3 mm long, light to dark brown, broadly keeled and obtuse to emarginate. Stamens 1 or 2, the anthers 0.6 mm long; style 2-forked (very rarely 3-forked). Achenes dark brown, c. 0.75 mm long.

GRAND CAYMAN: *Brunt 1851, 2024; Proctor 15287.* LITTLE CAYMAN: *Proctor 35177.* CAYMAN BRAC: *Proctor (sight).*
– Pantropical, chiefly near sea-coasts in brackish or subsaline soils. The similar *F. cymosa* subsp. *cymosa* differs chiefly in having 3-branched styles and slightly different achenes; it does not occur in the West Indies. The two are maintained as separate species by some authors, but the differentiating characters are reported not to be constant.

Genus 5. **SCIRPUS** L.

Usually perennials, spreading by stolons or sometimes caespitose. Culms terete or angulate, leafy or the leaves reduced to tubular sheaths. Inflorescence umbellate, paniculate, or reduced to 1 spikelet or a small cluster of sessile spikelets, subtended by 1 to many bracts. Spikelets terminal or lateral, sessile or pedunculate, 3- to many-flowered; glumes spirally imbricate, all fertile or the lowest sometimes empty. Flowers bisexual, with perianth of 1–6 bristles, or bristles sometimes lacking. Stamens 2–3; style 2- or 3-cleft, not enlarged at base, usually deciduous. Achenes flattened or 3-angled.

A cosmopolitan genus of about 300 species, usually of wet habitats. A single species is reported from the Cayman Islands.

1. **Scirpus validus** Vahl, Enum. Pl. 2:268. 1805. **FIG. 24.**

Rhizomes elongate, creeping, c. 1 cm in diameter, covered with brown papery scales. Culms erect, up to 2 m tall (often less), terete, reddish- or brownish-sheathed at base, the sheaths terminating in a narrow free blade 0.5–10 cm long. Inflorescence compound-umbellate, subtended by a narrow rigid green bract 1–7 cm long. Spikelets ovoid, 5–10 mm long, 4–5 mm thick, numerous, in clusters of 2 to 4 on scabrous-margined peduncles; glumes mucronate; bristles 6, reddish, tortuous, retrorsely barbed. Style 2-cleft. Achenes flattened, 2.5 mm long, dark gray when ripe.

GRAND CAYMAN: *Kings GC 411.*
– Widely distributed in chiefly temperate regions; also common in the Greater Antilles, usually in open marshes or swales at low elevations.

Genus 6. **RHYNCHOSPORA** Vahl

Perennial sedges, spreading by means of creeping rhizomes. Culms usually solitary, leafy below. Inflorescence (in the only Caymanian species) a terminal head of sessile spikelets, subtended by long leafy bracts white toward base. Spikelets somewhat compressed; scales spirally arranged in 2 or 3 rows, several usually empty or with abortive flowers. Flowers perfect, with or without perianth. Stamens 3. Style 2-cleft. Achenes flattened, transversely rugose, capped with the beaklike persistent base of the style.

A nearly cosmopolitan genus of more than 250 species, best represented in the tropics. The Cayman species belongs to a group that is often maintained as a separate genus *Dichromena*, and the generic description given above applies more especially to this taxon.

1. **Rhynchospora stellata** (Lam.) Griseb. in Gött. Abh. 7:271. 1857. **FIG. 25.**

Dichromena colorata (L.) Hitchc., 1893.

Rhizomes elongate (to 30 cm) and very slender, stolon-like. Leaves several, shorter than the culm, 1-3 mm wide, usually flat. Culms 20–50 cm tall. Bracts 4–6, leafy, 1–12 cm long, spreading or reflexed, white at base, glabrous. Spikelets numerous, 6–8 mm long; scales broadly ovate-lanceolate, 3–4 mm long. Achenes c. 0.8 mm long, with a flat tubercle.

GRAND CAYMAN: *Brunt 1646, 1921; Kings GC 100; Proctor 15288.*
– Southeastern United States; Mexico to Belize; Bermuda and throughout the West Indies to Martinique. One of the few sedges pollinated by insects.

Genus 7. **CLADIUM** R. Br.

Tall coarse perennials with thick rhizomes. Leaves long and narrow, with rough margins. Culms tall, the inflorescence a compound terminal panicle with numerous small clusters of sessile spikelets, or a series of axillary panicles. Spikelets mostly 1- to 3-flowered with usually only the lowest flower fertile; scales imbricate, the lower ones empty. Flowers perfect, with perianth of bristles or lacking. Stamens 2-3. Style 3-cleft, the thickened base often persistent on apex of the achene as a tubercle. Achenes somewhat flattened or 3-angled, smooth or rugose.

A genus of 4 species of pantropical and warm-temperate distribution. A single species occurs in the Cayman Islands.

1. **Cladium jamaicense** Crantz, Inst. R. Herb. 1:362. 1766. **FIG. 26.**

Mariscus jamaicensis (Crantz) Britton, 1913.

"Cutting Grass", "Sawgrass"

Rhizomes elongate, woody, 4–6 mm thick, clothed with brown papery overlapping scales. Culms 1–3 m tall, up to 2.5 cm thick near base, obscurely 3-angled, leafy; cauline leaves numerous, up to 1 m long or more, 6–20 mm wide, of hard texture, the margins very scabrous. Panicles several in leaf-axils along upper part of the culms, densely or laxly compound-umbelliform, mostly 2–12 cm long. Spikelets narrowly ovoid, 4–5 mm long, brown, acute. Perianth lacking. Stamens 2, anthers 3 mm long. Achenes ovoid, 2 mm long, acute, brown and somewhat rugose.

GRAND CAYMAN: *Brunt 1690, 1863, 1948; Kings GC 256, GC 363.*
– Pantropical as a species, subsp. *jamaicense* confined to the American tropics. The tough leaves are used in Jamaica for weaving baskets.

Genus 8. **REMIREA** Aubl.

A perennial with elongate rhizomes and solitary or clustered densely leafy culms. Leaves rigid, linear-lanceolate, spinulose-tipped, the basal sheaths imbricate. Inflorescence a dense solitary head of 1-flowered spikelets; rhachilla jointed above the base. Flowers bisexual, without perianth. Stamens 3; style 3-cleft. Achenes smooth, sessile.

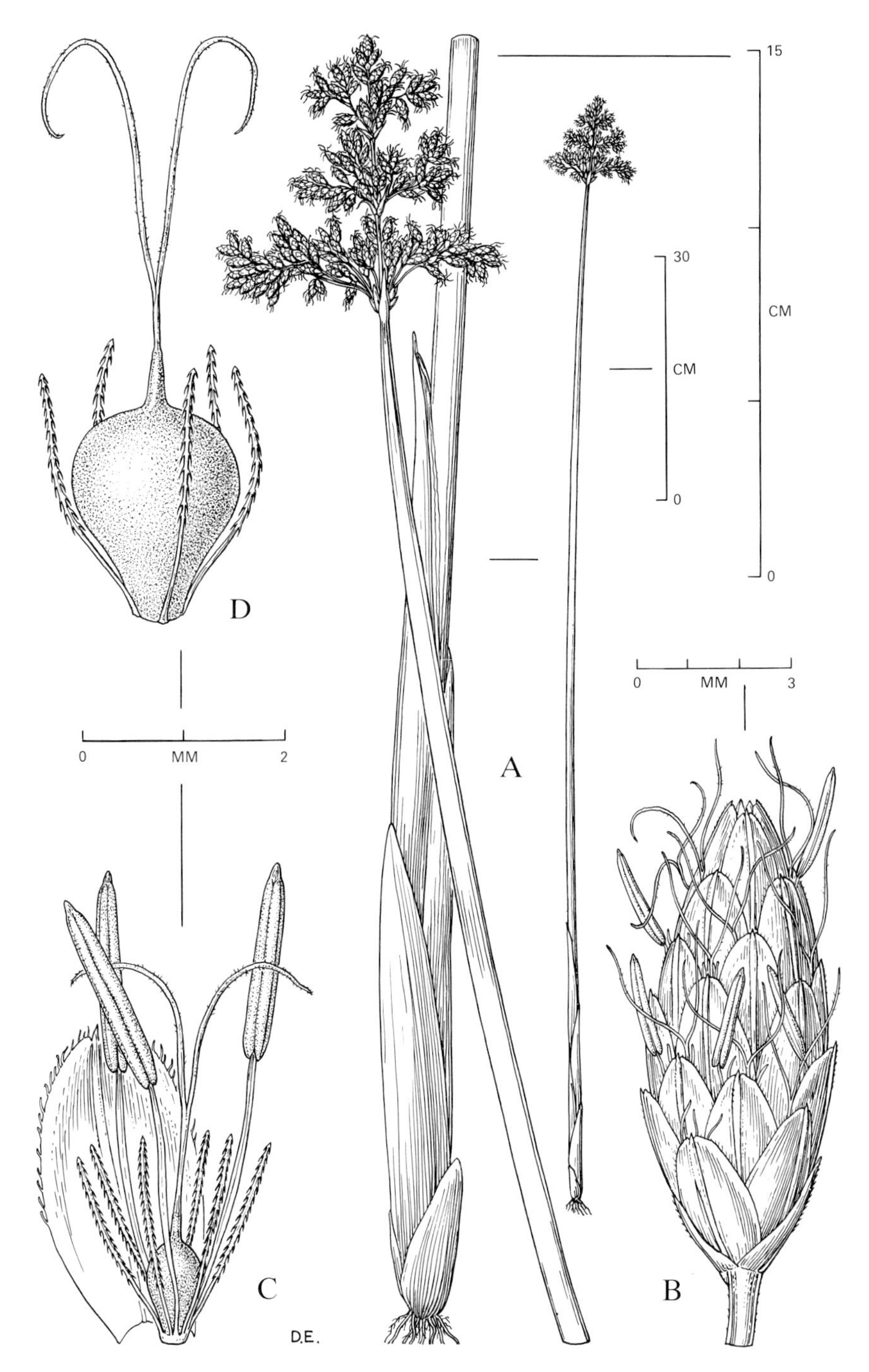

FIG. 24 **Scirpus validus.** A, habit. B, spikelet. C, floret. D, achene with bristle and persistent style. (D.E.)

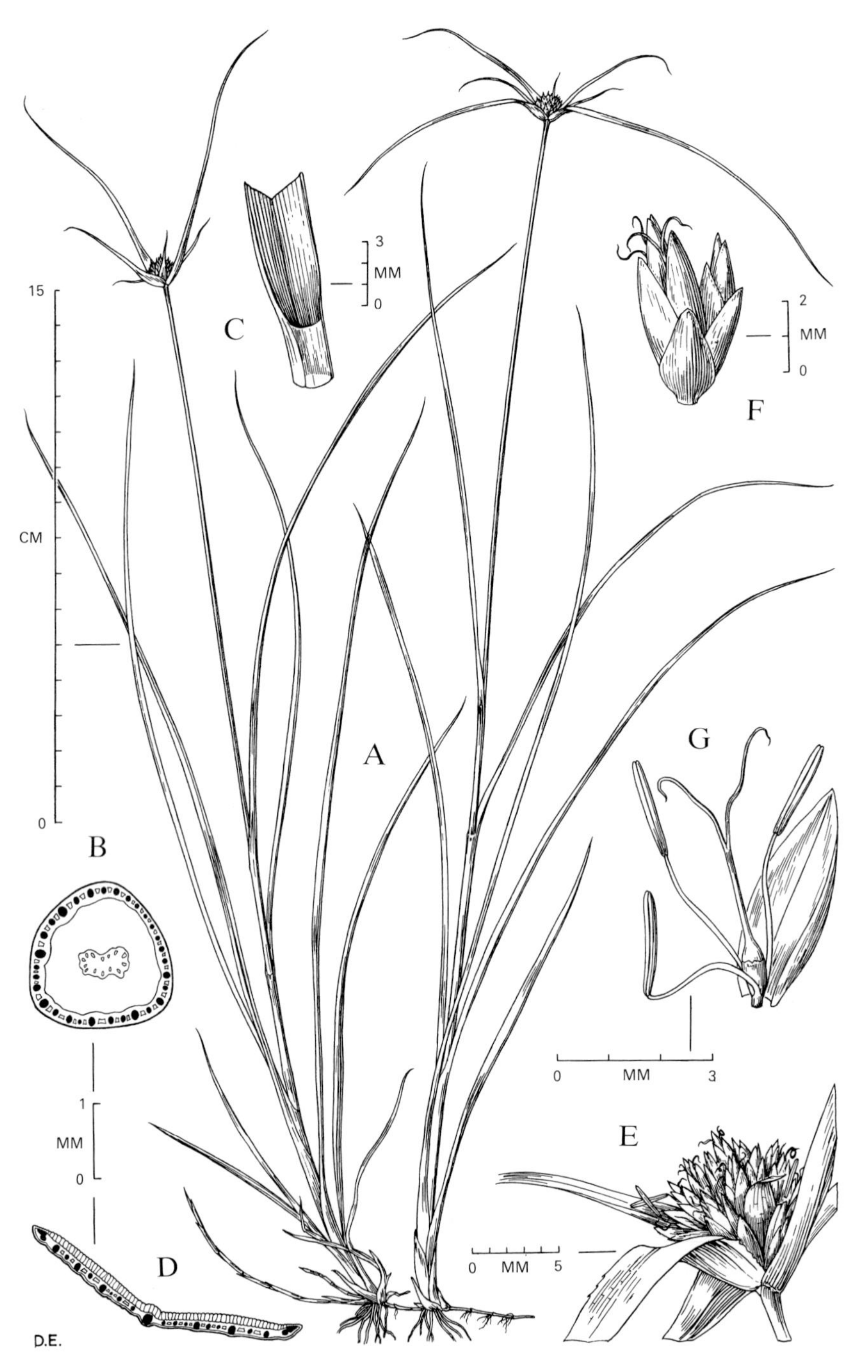

FIG. 25 **Rhynchospora stellata.** A, habit. B, x-section of culm. C, junction of leaf-blade with sheath. D, x-section of leaf-blade. E, inflorescence. F, two spikelets. G, floret with scale. (D.E.)

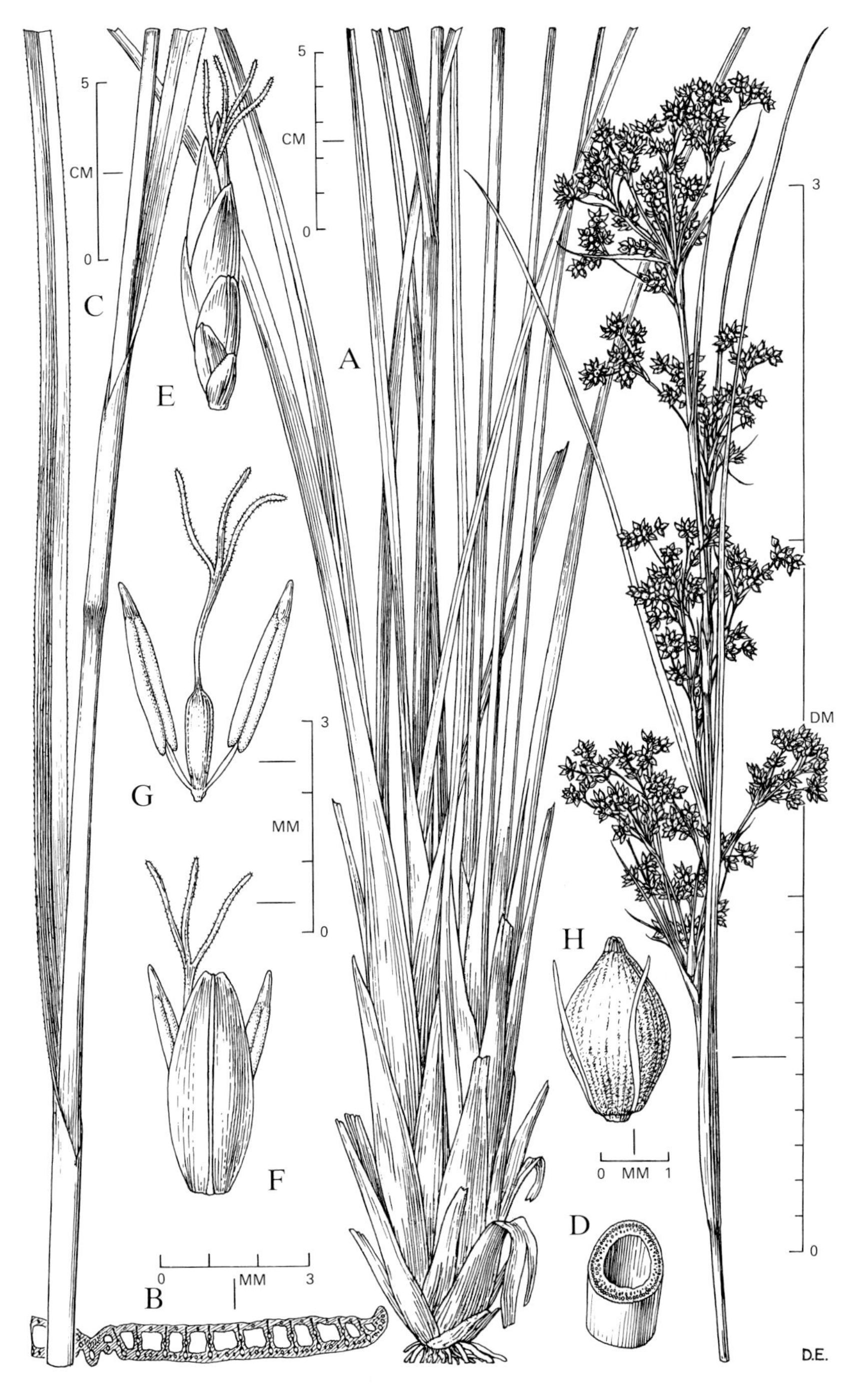

FIG. 26 **Cladium jamaicense.** A, base of plant. B, x-section of leaf. C, lower part of culm. D, x-section of culm. E, spikelet. F, floret. G, floret with glume removed. H, achene.

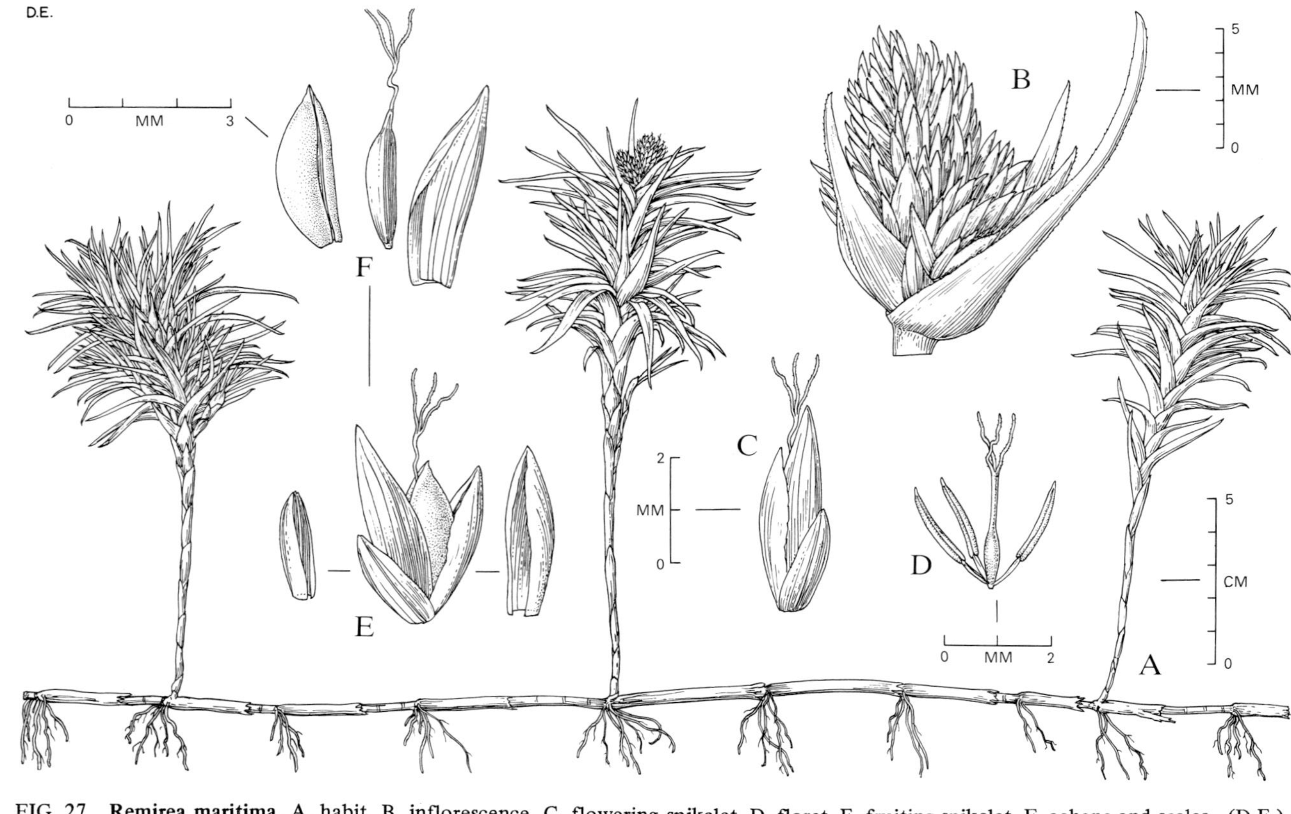

FIG. 27 **Remirea maritima.** A, habit. B, inflorescence. C, flowering spikelet. D, floret. E, fruiting spikelet. F, achene and scales. (D.E.)

A pantropical genus of 1 species. Some recent authors have included it in *Cyperus* or its segregate *Mariscus* on narrow technical grounds, despite its uniquely distinctive habit and other features.

1. **Remirea maritima** Aubl., Pl. Guian. 1:45. 1775. **FIG. 27.**

Remirea pedunculata R. Br., 1810.

Cyperus pedunculatus (R. Br.) Kern, 1958; Adams, 1972.

Mariscus pedunculatus (R. Br.) T. Koyama, 1977.

Culms 2-30 cm long, decumbent to erect, leafy throughout. Leaves flat, 2-8 cm long, 2-7 mm wide near base. Heads 1-2 cm long, subtended by leaf-like bracts. Spikelets many, ovoid, brownish, 3-5 mm long, each with 4 glumes. Glumes ovate, several-veined, 3-4.5 mm long. Achene trigonous, grayish-brown, 2.5 mm long.

GRAND CAYMAN: *Proctor 15216.* LITTLE CAYMAN: *Kings LC 81; Proctor 35084.*
– Of wide but sporadic distribution, usually found in loose sand at the top of sea-beaches.

Genus 9. **SCLERIA** Bergius

Chiefly perennial, often with rhizomes, and with sharply 3-angled more or less leafy culms, these short and erect or to tall and erect, or sometimes greatly elongate and vinelike. Spikelets small, few-flowered, clustered in terminal or axillary fascicles, or in interrupted spikes; glumes spirally attached. Flowers monoecious, without perianth; staminate spikelets several-flowered; pistillate ones 1-flowered. Stamens 1-3. Ovary sometimes supported on a disc (hypogynium); stigmas 3. Achenes globose to ovoid, usually white, with gleaming, enamel-like surface.

A pantropical genus of 200 species, a few extending into temperate regions.

1. **Scleria lithosperma** (L.) Sw., Nov. Gen. & Sp. Pl. 18. 1788. **FIG. 28.**

Perennial, the culms 20-60 cm tall, glabrous, more or less clustered from short, hard, knotted rhizomes. Leaves involute, 10-20 cm long, 0.5-3 mm broad, the lowest ones reduced to finely pubescent, papery sheaths. Inflorescence of several small, distant clusters of spikelets, each subtended by a more or less filiform bract. Hypogynium lacking. Achenes obovoid-ellipsoid, c. 2 mm long.

GRAND CAYMAN: *Brunt 2175; Kings GC 235, GC 245.* LITTLE CAYMAN: *Kings LC 58; Proctor 28120.* CAYMAN BRAC: *Proctor 28998.*
– Pantropical, mostly found in rather dry woodlands.

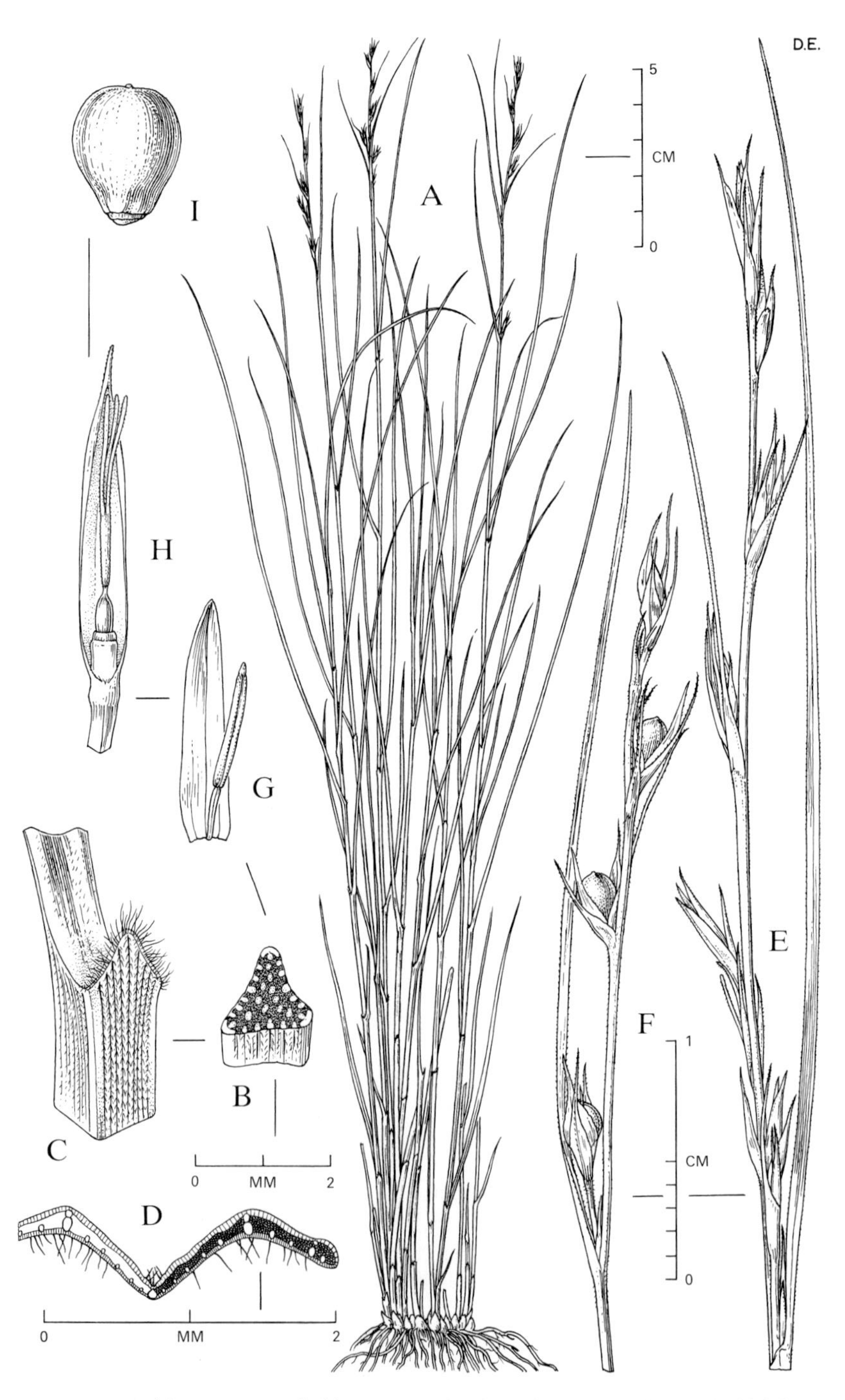

FIG. 28 **Scleria lithosperma.** A, habit. B, x-section of culm. C, junction of leaf-blade and sheath. D, x-section of leaf. E, staminate inflorescence. F, pistillate inflorescence. G, staminate floret and scale. H, pistillate floret and scale. I, achene. (D.E.)

Family 7

GRAMINEAE

Annual or perennial herbs, rarely woody or tree-like. Stems (culms) usually terete and hollow (rarely solid), with closed nodes. Leaves 2-ranked, usually elongate, with a distinct basal sheath split on one side to the point of attachment, and a small appendage (the ligule) at the junction of sheath and blade. Inflorescence a spike, raceme or panicle of densely or loosely aggregated spikelets, each spikelet consisting of a rhachilla bearing 1 to several flowers subtended by small bractlets; the lowest 2 bractlets (glumes) sterile; each succeeding bractlet (lemma) enclosing a flower, with another bractlet (palea) subtending the flower on the upper side. Flowers perfect or unisexual. Perianth none, or reduced to 2 or 3 minute scales (lodicules). Stamens 1–6 (usually 3), separate; anthers 2-celled, attached at the middle. Ovary superior, 1-celled with 1 ovule; styles 1–3 (usually 2), more or less plumose. Fruit a seed-like grain (caryopsis); endosperm copious, starchy.

A large, world-wide family of about 620 genera and more than 8000 species (some authorities estimate as many as 10,000). This is economically the most important family of plants. Because of the complexity of classification, it is customary to arrange grasses in groups called tribes. Eight of these occur in the Cayman Islands, and can be characterised as follows:

TRIBE I. BAMBUSEAE. Culms woody, perennial, usually hollow; spikelets 1- to many-flowered in panicles or racemes, or in close heads or fascicles; often 1 or more sterile lemmas at base of spikelet; lemmas usually awnless; blades usually articulated with the sheath. **Genus 1.**

TRIBE II. FESTUCEAE. Spikelets usually several-flowered, in open, narrow, or sometimes spike-like panicles (rarely in racemes); lemmas awnless or awned from the tip, rarely from between the teeth of a bifid apex; rhachilla usually disarticulating above the glumes and between the florets. **Genera 2–4.**

TRIBE III. AGROSTEAE. Spikelets 1-flowered, usually perfect, in open, contracted, or spike-like panicles, but not in true spikes nor in one-sided racemes. **Genus 5.**

TRIBE IV. ZOYSIEAE. Spikelets subsessile in short spikes of 2–5, each spike falling entire from the continuous axis, usually 1-flowered; flowers all perfect, or perfect and staminate together in the same spike; glumes usually firmer than the lamma and palea, sometimes awned, the lemma awnless. **Genus 6.**

TRIBE V. CHLORIDEAE. Spikelets 1- to several-flowered, in 2 rows on one side of a continuous rhachis, forming one-sided spikes or spike-like racemes, these solitary, digitate, or racemose along the main axis. **Genera 7–12.**

TRIBE VI. PANICEAE. Spikelets with one perfect terminal floret, and below this a sterile floret and two glumes; fertile lemma and palea indurate or at least firmer than the glumes and sterile lemma, a lunate line of thinner texture at the back just above the base, the rootlet protruding through this at germination; articulation occurring below the spikelet. **Genera 13–23.**

TRIBE VII. ANDROPOGONEAE. Spikelets in pairs along a rhachis, the usual arrangement being one of the pair sessile and fertile, the other pedicellate and staminate or neuter, rarely lacking with only the pedicel present; fertile spikelet consisting of one perfect terminal floret and, below this, a staminate or neuter floret, the lemmas thin or hyaline, and two awnless glumes, one or usually both firm or indurate. **Genera 24–27.**

TRIBE VIII. MAYDEAE. Spikelets unisexual, the staminate in pairs (sometimes in 3's), 2-flowered, the pistillate usually single, 2-flowered, the lower floret sterile, embedded in hollows of the thickened articulate axis and falling attached to the joints, or enclosed in a thickened involucre or sheath; glumes membranaceous or thick and rigid, awnless; lemmas and palea hyaline, awnless; plants monoecious. **Genus 28.**

KEY TO GENERA

1. Spikelets 1- to many-flowered, the reduced florets (if any) above the perfect ones (sterile lemmas below as well as above in *Bambusa*); articulations usually above the glumes; spikelets usually more or less laterally compressed:

2. Plants woody, the culms perennial: **[1. Bambusa]**

2. Plants herbaceous (except in *Arundo*), the culms annual:

3. Spikelets 1-flowered (or the staminate 2-flowered) in groups (short spikes) of 2–5, these groups racemose along a main axis, each one falling entire; glumes thick and leathery: **6. Anthephora**

3. Spikelets not as above:

4. Spikelets sessile on a usually continuous rhachis (short-pedicellate in *Leptochloa*):

5. Spikes racemosely arranged:

6. Spikelets articulate above the glumes, 2- to several-flowered: **7. Leptochloa**

6. Spikelets articulate below the glumes, 1-flowered: **11. Spartina**

5. Spikes digitately arranged, or nearly so:

7. Spikelets with 2 or more perfect florets:

8. Rhachis not extending beyond the spikelets: **8. Eleusine**

8. Rhachis extending beyond the spikelets: **9. Dactyloctenium**

7. Spikelets with 1 perfect floret only:

9. Sterile floret lacking; plants with elongate rhizomes, and strongly stoloniferous: **10. Cynodon**

9. Sterile floret present; plants tufted or weakly stoloniferous: **12. Chloris**

4. Spikelets pedicelled, in open or contracted (sometimes spike-like) panicles, rarely racemose:

10. Spikelets 1-flowered: **5. Sporobolus**

10. Spikelets 2- to many-flowered:

11. Spikelets with copious long silky hairs; tall woody reeds with large plume-like panicles: **[2. Arundo]**

11. Spikelets not long-hairy; herbaceous grasses with culms not at all woody:

12. Lemmas 3-nerved; plants tufted, without elongate rhizomes: **3. Eragrostis**

12. Lemmas 5- or more nerved; plants spreading by elongate scaly rhizomes: **4. Distichlis**

1. Spikelets with one perfect terminal floret (disregarding staminate or neuter spikelets) and a sterile or staminate floret below, usually represented by a sterile lemma only, one glume sometimes (rarely both glumes) lacking; articulation below the spikelets, these falling entire, singly, in groups, or together with joints of the rhachis; spikelets (or at least the fruits) more or less dorsally compressed:

13. Glumes membranous, the sterile lemma like the glumes in texture:

14. Inflorescence-axis thickened and corky, the spikelets sunk in cavities in its joints, these disarticulating at maturity: **14. Stenotaphrum**

14. Inflorescence-axis not thickened, nor the spikelets sunk in cavities:

15. Spikelets subtended or surrounded by one to many bristles or spines, these distinct or more or less connate at the base, forming a false involucre:

16. Bristles persistent, hairlike: **20. Setaria**

16. Bristles falling with the spikelets at maturity, hairlike or spine-like:

17. Bristles not united at the base, usually slender, often plumose: **[21. Pennisetum]**

17. Bristles more or less united at the base, forming a bur: **22. Cenchrus**

15. Spikelets not subtended nor surrounded by bristles:

18. Fruit cartilaginous, not rigid, papillose, usually dark-coloured, the lemma with white hyaline margins, these not inrolled; spikelets with short pubescence or glabrous; fruit elliptic; racemes digitate or subdigitately arranged: **13. Digitaria**

18. Fruit indurate, rigid:

19. First glume typically lacking; spikelets plano-convex, subsessile in spike-like racemes: **15. Paspalum**

19. First glume present; spikelets usually biconvex to globose, in panicles (spike-like racemes in *Panicum geminatum*):

20. Glumes and lemmas awnless:

21. Culms neither woody nor bamboo-like: **16. Panicum**

21. Culms woody, scrambling, bamboo-like: **17. Lasiacis**

20. Glumes or lemmas, or both, awned or at least sharp-pointed:

22. Spikelets scabrous or stiffly hispid but not silky, arranged in short dense racemes: **18. Echinochloa**

22. Spikelets long-silky (the hairs reddish), arranged in loose panicles: **[19. Tricholaena]**

13. Glumes indurate; fertile lemma and palea hyaline or membranous, the sterile lemma like the fertile in texture:

23. Spikelets in pairs, one sessile and perfect, the other pedicelled and usually staminate or neuter, in mixed panicles or racemes:

24. Spikelets all perfect, awnless:

25. Rachis continuous; culms slender (less than 5 mm thick), not juicy, seldom over 1 m tall; wild plants: **23. Imperata**

25. Rhachis disjointing; culms thick (to 5 cm or more), solid and juicy, up to 3 mm tall or more; cultivated plants: **[24. Saccharum]**

24. Spikelets not all perfect:

26. Spikelets in evident racemes, these digitate or numerous; awns conspicuous, not deciduous: **25. Andropogon**

26. Spikelets in reduced racemes of 1–5 (rarely 7) joints, these peduncled in open or compact panicles; awns, if present, usually deciduous; cultivated plants: **[26. Sorghum]**

23. Spikelets unisexual, the pistillate below, in axillary sheathed "cobs", the staminate in large terminal panicles; cultivated plants: **[27. Zea]**

[*TRIBE I. BAMBUSEAE*

Genus 1. **BAMBUSA** Schrad.

Large woody perennials with conspicuously jointed, hollow culms. Spikelets arranged in branched leafy or leafless panicles or in panicled spikes, often oblong or ovate, 1- to many-flowered; paleas 2 -keeled. Stamens 6.

A primarily Old World genus of about 70 species, a few of them now widely naturalized. "Bamboo".

1. **Bambusa vulgaris** Schrad. ex Wendl., Collect. Pl. 2:26, t. 47. 1810.

Bambusa sieberi Griseb., 1864.

A giant bamboo up to 16 m tall or more (usually much less in the Cayman Islands), forming large open clumps from branched, creeping, woody rhizomes. Culms 10–12 cm in diameter, bright green or yellow. Leaf-blades linear-lanceolate, acuminate, 10–25 cm long, mostly 1–3 cm broad, usually glabrous at least on the surfaces. Flowers seldom seen. Spikelets 15–20 mm long, closely 6–10-flowered.

GRAND CAYMAN: *Brunt 1686; Kings GC 163.*
– Pantropical in cultivation; probably of Asiatic origin but now naturalized in many countries of both hemispheres. The culms form durable poles and have many uses. The very young shoots 15–30 cm long are edible both cooked or pickled, and are considered a delicacy by many people.]

TRIBE II. FESTUCEAE

[Genus 2. **ARUNDO** L.

Tall perennial reeds, with broad linear leaf-blades and large plume-like terminal panicles. Spikelets several-flowered, the florets successively smaller, the summits of all about equal; rhachillas glabrous, disarticulating above the glumes and between the florets; glumes somewhat unequal, membranaceous, 3-nerved, narrow, tapering into a slender point, about as long as the spikelet; lemmas thin, 3-nerved, densely long-pilose, gradually narrowed at the apex, the nerves ending in slender teeth, the middle tooth extending into a straight awn.

A genus of 12 temperate and tropical species of wide distribution.

1. **Arundo donax** L., Sp. Pl. 1:81. 1753. **FIG. 29.**

Culms 2–6 m tall, growing in large clumps, sparingly branched from thick, knotty rhizomes. Leaf-blades numerous, elongate, 3–5 cm broad or more, with a characteristic light convoluted area at the junction of the sheath, and slightly pubescent there; sheath glabrous; ligule 1 mm long, with toothed margin. Inflorescence a large, tawny panicle to 70 cm long; spikelets 1 cm long, 3-flowered; glumes with 3 prominent nerves.

GRAND CAYMAN: *Brunt 1712.*
– Indigenous to the Mediterranean region and widely distributed in warmer parts of the Old World; cultivated chiefly for ornament in America, and occurring as an escape in the southern United States through the West Indies to South America. Grows well in sandy soils. The hard, hollow culms are used for making such things as fishing-rods, flutes, and bagpipes.]

Genus 3. **ERAGROSTIS** Beauv.

Mostly tufted annuals or perennials with open or contracted panicles. Spikelets few- to many-flowered, the florets usually imbricate, the rhachilla disarticulating above the glumes and between the florets, or continuous, the lemmas deciduous, the paleas persistent; glumes somewhat unequal, acute or acuminate, 1-nerved, or

FIG. 29 **Arundo donax.** A, inflorescence, x $^1/_3$. B, culm with leaves, x $^1/_3$. C, rhizome. x $^1/_3$. D, spikelet, x3. E, floret, x3. (H.)

the second rarely 3-nerved; lemmas acute or acuminate, keeled or rounded on the back, 3-nerved, the nerves usually prominent; paleas 2-nerved, the keels sometimes ciliate.

A cosmopolitan genus of about 300 species, the majority occurring in subtropical regions.

KEY TO SPECIES

1. Annuals; paleas ciliate on the keels, the cilia conspicuous, usually as long as the width of the lemma; culms mostly 10–30 cm tall:

  2. Pedicels as long as the spikelets or longer; panicle open, oblong: [**1. E. tenella**]

  2. Pedicels very short; panicle spike-like, more or less interrupted: **2. E. ciliaris**

1. Perennials; paleas scaberulous on the keels but not prominently long-ciliate; culms 40–150 cm tall: **3. E. domingensis**

[**1. Eragrostis tenella** (L.) Beauv. ex Roem. & Schult. in L., Syst. Veg. ed. nov. 2:576. 1817.

Eragrostis amabilis (L.) Wight & Arn., 1841.

Small tufted annual usually 15 cm tall or less. Leaf-blades 3–8 cm long, c. 2 mm broad, glabrous, with inconspicuous ligule. Panicle open and rather delicate, c. 6 cm long or sometimes longer. Spikelets 2 mm long.

GRAND CAYMAN; *Correll & Correll 51056.* LITTLE CAYMAN: *Proctor 28192.* CAYMAN BRAC: *Proctor 15310.*
– Native of the Old World tropics, originally described from India. This little weed now occurs throughout the West Indies.]

1. Eragrostis ciliaris (L.) R. Br. in Tuckey, Narr. Exped. Riv. Zaire 478. 1818. **FIG. 30.**

Small tufted annual seldom much over 20 cm tall. Leaf-blades average 6 cm long, 5 mm broad, glabrous except for a tuft of long hairs (3 mm long) just above the junction with the sheath; sheath glabrous except for a ring of hairs on the back just below where the 3 mm hairs occur; basal sheath purplish ligule reduced to a trace. Inflorescence a ragged spike-like panicle averaging 7 cm long. Spikelets 8-flowered; cilia of the paleas with a characteristic comb-like appearance when viewed through a hand-lens.

FIG. 30 **Eragrostis ciliaris.** A, plant, x½. B, spikelet, x5. C, floret, x10. (H.)

GRAND CAYMAN: *Brunt 1708, 2179; Proctor 11965.* LITTLE CAYMAN: *Proctor 28155.* CAYMAN BRAC: *Millspaugh 1190; Proctor 29060.*
– A pantropical weed.

3. Eragrostis domingensis (Pers.) Steud., Syn. Pl. Glum. 1:278. 1854.

Tufted perennial with strong erect culms up to 1.5 m tall. Leaf-blades elongate, 2–5 mm broad; sheaths glabrous or sparsely pilose at summit. Panicles 25–50 cm long, loosely-flowered. Spikelets short-pedicellate, 5–10 mm long, 10–25-flowered; lemmas 1.5–1.8 mm long.

GRAND CAYMAN: *Brunt 1660, 1704, 2067, 2178; Hitchcock; Millspaugh 1240; Proctor 11966.* CAYMAN BRAC: *Proctor 29003.*
– West Indies.

Genus 4. **DISTICHLIS** Raf.

Dioecious perennials with extensively creeping scaly rhizomes. Culms erect, stiff, leafy, terminating in a small dense panicle. Spikelets several- to many-flowered, the rhachilla of the pistillate ones disarticulating above the glumes and between the florets. Glumes unequal, broad, acute, keeled, mostly 3-nerved; lemmas closely imbricate, firm, 3-nerved with intermediate striations; pistillate palea enclosing the grain.

A genus of 4 North American, 8 South American, and 1 Australian species.

1. Distichlis spicata (L.) Greene in Bull. Calif. Acad. Sci. 2:415. 1887. **FIG. 31.**

Sometimes forming extensive colonies of sterile plants. Leaves strongly distichous, with stiff, glabrous, involute blades. Spikelets c. 1 cm long, aggregated in a small, compact panicle.

GRAND CAYMAN: *Brunt 1824, 1866, 1886; Kings GC 424.*
– United States to Mexico and the northern West Indies, in saline soils.

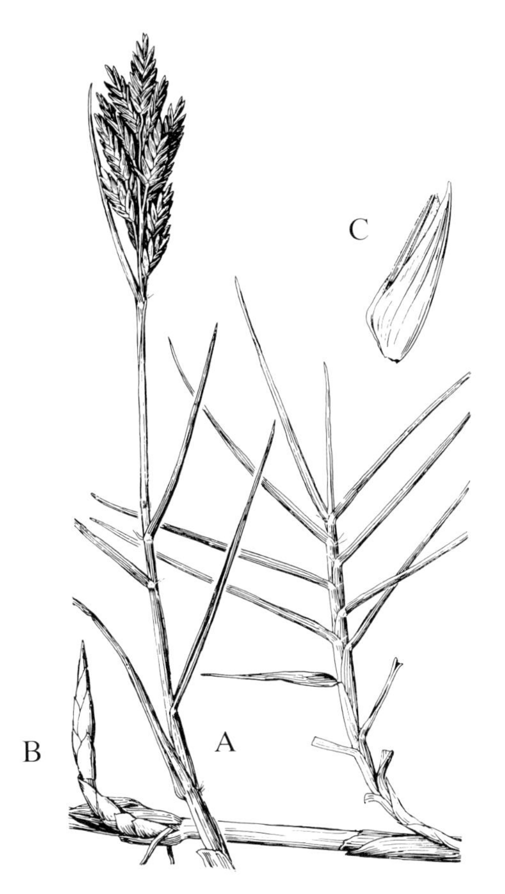

FIG. 31 **Distichlis spicata.** A, plant with sterile shoot, x1. B, fertile shoot with inflorescence, x1. C, floret, x5. (H.)

TRIBE III. AGROSTEAE

Genus 5. **SPOROBOLUS** R. Br.

Perennials (or some species annual) with narrow to open panicles having numerous very small spikelets. Spikelets 1-flowered, the rhachilla disarticulating above the glumes; glumes usually unequal, the second often as long as the spikelet; lemma membranous, 1-nerved, awnless; palea usually prominent and as long as the lemma or longer. Seed often free from the pericarp.

A genus of about 150 species, widespread in tropical and warm-temperate regions.

KEY TO SPECIES

1. Plants tufted, without creeping rhizomes; leaves loosely or rather densely arranged from base of culm:

   2. Glumes nearly equal, both much shorter than the spikelet: **1. S. jacquemontii**

   2. Glumes very unequal, the second about as long as the spikelet: **2. S. domingense**

1. Plants with creeping rhizomes; leaves stiffly distichous: **3. S. virginicus**

1. Sporobolus jacquemontii Kunth, Révis. Gram. 2:427, t. 127. 1831.

Sporobolus indicus of Hitchc., 1936, etc., not (L.) R. Br., 1810.

Tufted perennial; culms 40–100 cm tall. Leaves elongate, to 50 cm long, mostly 1–4 mm wide, flat or somewhat involute, glabrous, the midrib dorsally marked as a white line; ligule very small, ciliate. Inflorescence a dense narrow panicle 15–30 cm long. Spikelets a little over 1 mm long.

GRAND CAYMAN: *Proctor 15070, 15258.*
– West Indies; Mexico to Brazil, common in open waste places.

2. Sporobolus domingensis (Trin.) Kunth, Révis. Gram. 2:427. 1831. **FIG. 32.**

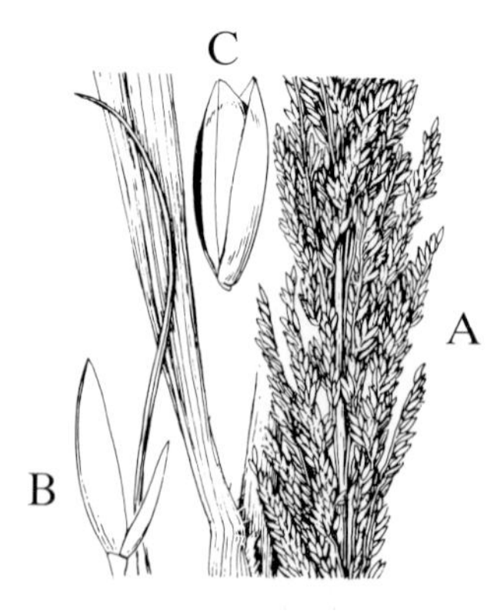

FIG. 32 **Sporobolus domingensis.** A, portion of leaf and inflorescence, x1. B, glumes, x10. C, floret x10. (H.)

Tufted glabrous perennial; culms 30-80 cm tall. Leaves much shorter than the culms, to 20 cm long, 3-8 mm wide, more or less involute. Panicles contracted, densely flowered, 7-15 cm long. Spikelets c. 2 mm long or slightly less.

GRAND CAYMAN: *Brunt 1683, 1726.*
– Florida, Bahamas, and the Greater Antilles, along sandy borders of salinas and open waste places near the sea.

3. **Sporobolus virginicus** (L.) Kunth, Révis. Gram 1:67. 1829. **FIG. 33.**

Culms stiffly erect, 15-50 cm tall or more, from extensively creeping tough scaly perennial rhizomes. Leaves stiff, much shorter than the culms, 1-5 mm wide, involute on drying; ligule minute, ciliate. Panicles spike-like, usually 2-10 cm long; spikelets glabrous, 1.8-3.2 mm long, the glumes unequal or nearly equal.

GRAND CAYMAN: *Brunt 1624, 1641, 1861, 1910, 2105, 2120; Hitchcock; Proctor 27944.* LITTLE CAYMAN: *Kings LC 84; Proctor 28081.* CAYMAN BRAC: *Brunt 1682; Kings CB 42; Proctor 28928.*
– Eastern United States south to Brazil; common in saline soil or coastal sands throughout the West Indies. Often forms extensive colonies with mostly or entirely sterile culms, the flowering narrowly seasonal.

TRIBE IV. ZOYSIEAE

Genus 6. **ANTHEPHORA** Schreb.

Loosely tufted annuals, the inflorescence narrow and spike-like. Spikelets in small clusters of 4, the indurate first glumes united at base, forming a pitcher-shaped pseudo-involucre, the clusters subsessile and erect on a slender flexuous continuous axis; glumes rigid, acute, or produced into short awns.

A small, pantropical genus of 20 species.

1. **Anthephora hermaphrodita** (L.) Ktze., Révis. Gen. Pl. 2:759. 1891. **FIG. 34.**

Culms leafy, branched, ascending or decumbent; leaf-blades flat, up to 8 mm broad, velvety-pubescent on the upper surface, the margins more or less wavy; sheaths glabrous with ciliate margins; ligule prominent, 1.5 mm long. Inflorescence with numerous approximate, deciduous groups of spikelets, each enclosed by leathery lower glumes.

FIG. 33 **Sporobolus virginicus**, x½. (H.)

FIG. 34 **Anthephora hermaphrodita**, x½. (H.)

GRAND CAYMAN: *Brunt 2066; Proctor 15219.* LITTLE CAYMAN: *Proctor 28156.* CAYMAN BRAC: *Proctor 28920.*

– Throughout the West Indies and other parts of tropical America; common at lower elevations.

TRIBE V. CHLORIDEAE

Genus 7. **LEPTOCHLOA** Beauv.

Annuals or perennials with flat leaf-blades and numerous narrow racemes scattered along a common axis, forming a rather ample panicle. Spikelets 2- to several-flowered, sessile or short-stalked, approximate or somewhat distant along one side of a slender rhachis; rhachilla disarticulating above the glumes and between the florets; glumes unequal or nearly equal, awnless or mucronate, 1-nerved, usually shorter than the first lemma; lemmas obtuse or acute, sometimes 2-toothed and mucronate or short-awned from between the teeth, 3-nerved, the nerves sometimes pubescent.

A genus of 27 tropical or warm-temperate species, the majority in the Western Hemisphere.

KEY TO SPECIES

1. Plants annual:

  2. Spikelets 1–2 mm long, the lemmas awnless, 1.5 mm long; leaf-sheaths papillose-hispid: **1. L. filiformis**

  2. Spikelets 7–12 mm long, the lemmas distinctly awned, 4–5 mm long; leaf-sheaths glabrous or scabrous: **2. L. fascicularis**

1. Plants perennial; lemmas 1.5–2mm long, awnless or short-awned; leaf-sheaths glabrous (usually somewhat glaucous): **3. L. virgata**

1. Leptochloa filiformis (Lam.) Beauv., Ess. Nouv. Agrost. 166. 1812. **FIG. 35.**

Tufted annual with flat leaf-blades 4–10 mm broad; ligule long and membranaceous. Inflorescence up to 25 cm long, usually lax and open with numerous very slender racemes, slightly viscid. Spikelets 3–4 flowered; glumes acuminate, longer than the first floret; lemmas pubescent on the nerves.

FIG. 35 **Leptochloa filiformis.** A, plant, x½. B, inflorescence, x½. C, spikelet, x10. D, floret, x10. (H.)

GRAND CAYMAN: *Brunt 2176.*
– Virginia to California, south to South America. A common weed of fields and cultivated ground throughout the West Indies.

2. Leptochloa fascicularis (Lam.) A. Gray, Man. 588. 1848. **FIG. 36.**

Somewhat succulent annual with freely branching culms 30-100 cm long, erect to spreading or prostrate. Leaf-blades flat to loosely involute, 1-3 mm broad; sheaths often purple. Inflorescence often partly enclosed by the upper leaves, the several to numerous racemes ascending or appressed, or at maturity spreading. Spikelets overlapping, 6-12-flowered; awn of lemma up to as long as the lemma.

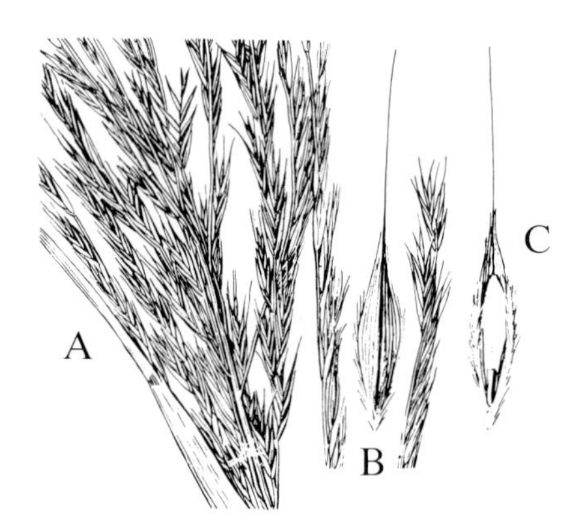

FIG. 36 **Leptochloa fascicularis.** A, portion of inflorescence, x1. B, C, floret (two views), x10. (H.)

GRAND CAYMAN: *Brunt 1631, 1966.*
– United States, Mexico, and the West Indies, mostly in ditches and wet places.

3. Leptochloa virgata (L.) Beauv., Ess. Nouv. Agrost. 166. 1812.

Tufted perennial with wiry culms 50-100 cm tall. Inflorescence ample, the slender racemes several to many, often laxly spreading, the lower distant. Spikelets 3-5-flowered.

GRAND CAYMAN: *Hitchcock.*
– Mexico and the West Indies to South America, common.

Genus 8. **ELEUSINE** Gaertn.

Annuals, with two to several rather stout spikes digitately arranged at summit of the culms, rarely with one or two (or a whorl) a short distance below, or a single terminal spike. Spikelets few- to many-flowered, compressed, sessile and closely imbricate, in two rows along one side of a rather broad rhachis, the latter not prolonged beyond the spikelets; rhachilla disarticulating above the glumes and between the florets; glumes unequal, rather broad, acute, 1-nerved, shorter than the first lemma; lemmas acute, with 3 stong green nerves close together forming a keel, the upper floret somewhat reduced. Seed dark brown, loosely enclosed in the thin pericarp.

A small genus of 9 tropical or subtropical species, occurring in both hemispheres. Several are useful fodder plants, and one *(E. coracana)* is cultivated as a cereal and for an alcoholic beverage in Ceylon, India, and Africa.

1. **Eleusine indica** (L.) Gaertn., Fruct. & Sem. Pl. 1:8. 1788. **FIG. 37.**

Loosely tufted, branching at base, the compressed culms ascending or prostrate, less than 50 cm long. Leaf-blades usually flat (sometimes folded), 3-8 mm broad. Spikes mostly 2-6 in number, 4-15 cm long.

GRAND CAYMAN: *Brunt 2037; Millspaugh 1270; Proctor 15097; Sachet 391.* LITTLE CAYMAN: *Proctor 28154.* CAYMAN BRAC: *Proctor 29106.*
– A common pantropical weed.

Genus 9. **DACTYLOCTENIUM** Willd.

Annuals or perennials, with flat leaf-blades and two to several short thick spikes digitate and widely spreading at the summit of the culms. Spikelets 3-5-flowered, compressed, sessile and closely imbricate, in two rows along one side of the rather narrow flat rhachis, the end projecting in a point beyond the spikelets; rhachilla disarticulating above the first glume and between the florets; glumes somewhat unequal, broad, 1-nerved, the first persistent upon the rhachis, the second mucronate or short-awned below the tip, deciduous; lemmas firm, broad, keeled, acuminate or short-awned, 3 -nerved, the lateral nerves indistinct, the upper floret reduced; palea about as long as the lemma. Seed subglobose, ridged or wrinkled, enclosed in a thin, early-disappearing pericarp.

A small warm-climate genus of 10 species, chiefly African.

1. **Dactyloctenium aegyptium** (L.) Beauv., Ess. Nouv. Agrost. Expl. Planch. 10. 1812. **FIG. 38.**

Creeping stoloniferous annual, up to 15 cm tall. Leaf-blades 6–14 cm long, to 5 mm broad, hairy on both sides; sheath sometimes glabrous, sometimes hairy; ligule 1 mm long, ciliate. Inflorescence of 3–4 spikes arranged digitately on the end of the culm; spikes to 4 cm long with rhachis extending beyond the spikelets. Spikelets 4–5 flowered, c. 3 mm long and broad.

GRAND CAYMAN: *Millspaugh 1267; Proctor 15067.*
– Pantropical, in various habitats.

Genus 10. **CYNODON** L.C. Rich.

Perennials, with creeping stolons or rhizomes, short leaf-blades, and several slender digitate spikes. Spikelets 1-flowered, awnless, sessile in two rows along one side of a slender continuous rhachis, the rhachilla disarticulating above the glumes and prolonged behind the palea as a slender naked bristle, this sometimes bearing a rudimentary lemma; glumes narrow, acuminate, 1-nerved, about equal, shorter than the floret; lemma strongly compressed, pubescent on the keel, firm, 3-nerved, the lateral nerves close to the margins.

About 10 tropical and subtropical species.

1. **Cynodon dactylon** (L.) Pers., Synops. Pl. 1:85. 1805. **FIG. 39.**

"Bahama grass"; "Bermuda grass"

Stoloniferous perennial with wiry creeping rhizomes. Flowering culms erect, usually 10–30 cm tall; leaf-blades 2–5 cm long or more, 1.5–2.5 mm broad, glabrous or finely pubescent; ligule in some strains with a double row of hairs, one row being shorter than the other, other strains with only a single line of hairs. Inflorescence of 3–5 spikes digitate at apex of culm; spikes 2.5–5 cm long. Spikelets 2–2.5 mm long.

GRAND CAYMAN: *Hitchcock; Proctor 15132.* CAYMAN BRAC: *Proctor 29099.*
– Widespread in the warmer parts of both hemispheres. Commonly used for lawns and pastures throughout the West Indies; it especially thrives in sandy soils.

Genus 11. **SPARTINA** Schreb.

Stout perennials with tough leaves and 2 to many spikes racemose on a main axis. Spikelets 1-flowered, much flattened laterally, sessile and usually closely

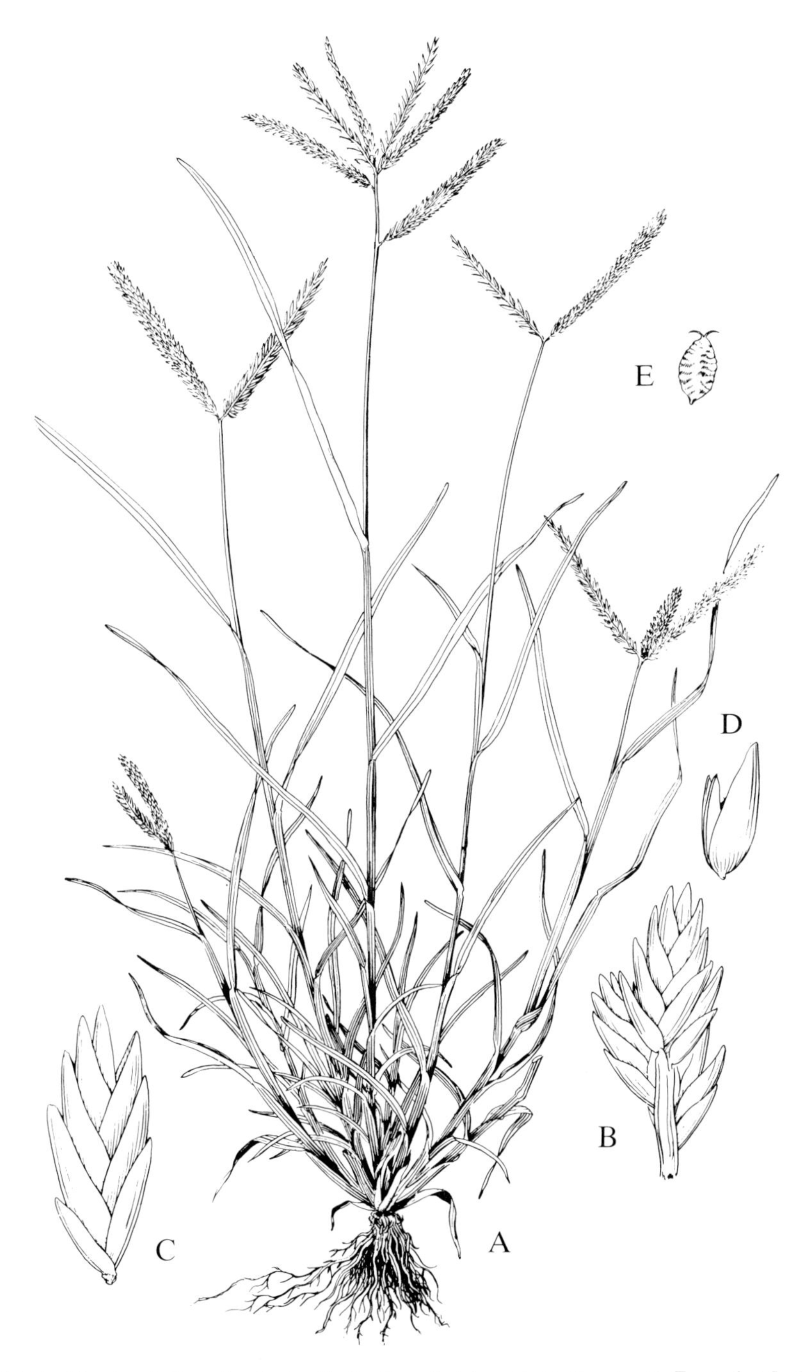

FIG. 37 **Eleusine indica.** A, plant, x½. B, C, spikelets, x5. D, floret, x5. E, seed, x5. (H.)

FIG. 38 **Dactyloctenium aegyptium.** A, plant, x½. B, spikelet, x5. C, floret, x5. D, anthers, x5. E, second glume, x5. (H.)

FIG. 39 **Cynodon dactylon.** A, plant x½. B, spikelet, x5. C, floret, x5. (H.)

imbricate on one side of a continuous rhachis, disarticulating below the glumes, the rhachilla not produced beyond the floret. Glumes keeled, 1-nerved, acute or short-awned, the first shorter, the second often exceeding the lemma; lemma firm, keeled, the lateral nerves obscure, narrowed to a rather obscure point; palea 2-nerved, keeled and flattened, the keel between or at one side of the nerves.

A chiefly temperate-climate genus of 16 species, all halophytes.

1. **Spartina patens** (Ait.) Muhl., Descr. Gram. 55. 1817.

Plants forming loose (often sterile) colonies by means of hard, scaly, long-creeping, deep-seated rhizomes. Culms erect, tough and wiry, up to 1 m tall or more. Leaf-blades hard, involute, extending into a long fine point. Spikes mostly 2–8, ascending or spreading, remote, 4–6 cm long; spikelets closely imbricate, 6–8 mm long, the second glume scabrous on the keel.

GRAND CAYMAN: *Brunt 2065; Proctor 31045.* LITTLE CAYMAN: *Kings LC 20; Proctor 28194.*
– Southeastern United States and the West Indies, on sandy sea-beaches and salt marshes.

Genus 12. **CHLORIS** Sw.

Perennials or annuals, with flat or folded leaf-blades and 2 to many often showy digitate spikes. Spikelets with 1 perfect floret, sessile in 2 rows along one side of a continuous rhachis, the rhachilla disarticulating above the glumes, produced beyond the perfect floret and bearing 1 to several reduced florets consisting of empty lemmas, these often truncate, and, if more than one, the smaller ones enclosed in the lower, forming a usually club-shaped rudiment; glumes somewhat unequal, the first shorter, narrow, acute; lemma keeled, usually broad, 1–5-nerved, often villous on the callus and villous or long-ciliate on the keel or marginal nerves, awned from between the short teeth of a bifid apex, the awn slender or sometimes reduced to a mucro, the sterile lemmas awned or awnless.

About 40 species of tropical and warm temperate regions. Several are useful pasture grasses.

KEY TO SPECIES

1. Lemmas awnless; spikes dark brown: **1. C. petraea**

1. Lemmas awned; spikes pale or purplish:

2. Plants perennial, the culms commonly more than 1 m tall; spikes flexuous, 8-10 cm long: **2. C. barbata**

2. Plants annual, the culms usually less than 75 cm tall; spikes less than 8 cm long: **3. C. inflata**

1. Chloris petraea Sw., Nov. Gen. & Sp. Pl. 25. 1788. **FIG. 40.**

Eustachys petraea (Sw.) Desv., 1810.

Stoloniferous perennial, glabrous and glaucous. Culms flat, averaging about 30 cm tall; leaf-blades 6-20 cm long, 3-6 mm broad, folded lengthwise and markedly keeled; ligule very reduced, difficult to distinguish. Inflorescence of 4-7 spikes,

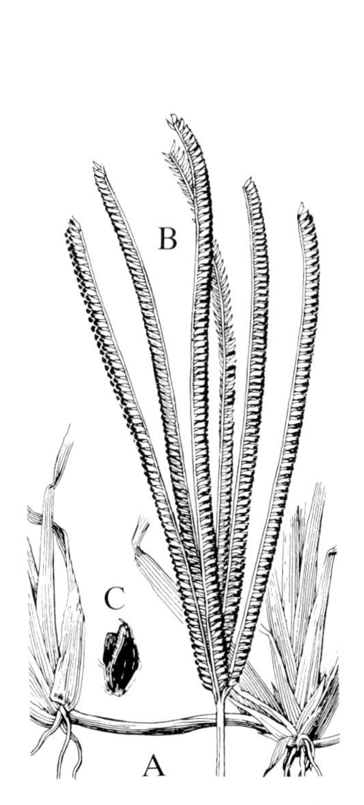

FIG. 40 **Chloris petraea.** A, portion of plant with stolons, xl. B, inflorescence, xl. C, florets, x5. (H.)

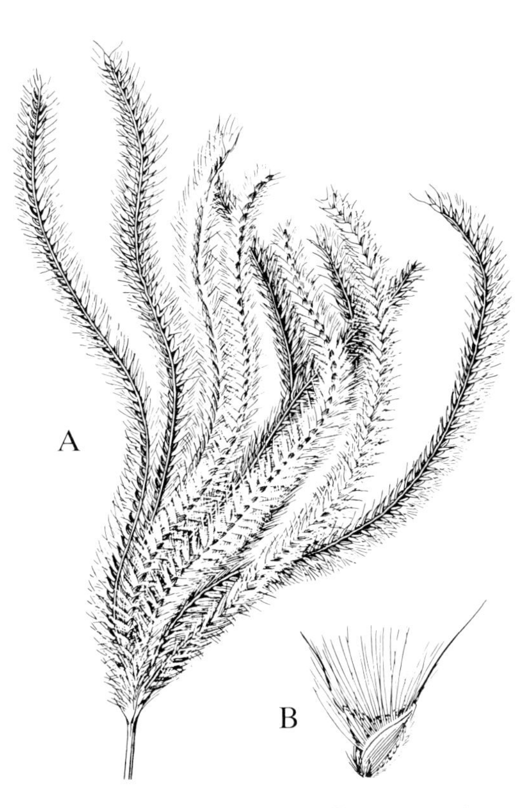

FIG. 41 **Chloris barbata.** A, inflorescence, x1. B, florets, x5. (H.)

these 5–10 cm long, erect or ascending; spikelets 2 mm long, 2-flowered, only one of the florets maturing; glumes green, lemmas dark brown.

GRAND CAYMAN: *Brunt 1802, 2039; Hitchcock; Kings GC 266; Lewis 3849, GC 26a; Millspaugh 1255; Proctor 15160; Sachet 388b.* LITTLE CAYMAN: *Kings LC 103; Proctor 28083.* CAYMAN BRAC: *Millspaugh 1181.*
– Southern United States, West Indies, and eastern Mexico south to Panama and Trinidad; common on sandy beaches and similar inland habitats.

2. Chloris barbata (L.) Sw., Fl. Ind. Occ. 1:200. 1797; Fosberg in Taxon 25(1): 176–178. 1976. **FIG. 41.**

Andropogon barbatus L., 1759.

Chloris polydactyla Sw., 1788, illegit.

C. dandyana C.D. Adams, 1971.

"Bitter Grass"

Culms rather stout; leaf-blades up to 15 mm broad, scabrous toward the long-acuminate tip. Spikes 5–10, pale; spikelets closely imbricate, silky; first lemma c. 2.5 mm long, pubescent on the keel, long-ciliate on the margins, the hairs 2 mm long, the awn about as long as the lemmas; second lemma shorter and narrower than the first, the awn c. 2 mm long.

GRAND CAYMAN: *Brunt 2036; Hitchcock; Kings GC 262; Millspaugh 1271; Proctor 15129.*
– Florida and the West Indies south to Brazil, chiefly in grassy fields.

3. Chloris inflata Link, Enum. Pl. 1:105. 1821.

Chloris barbata Sw., 1797, based on *Andropogon barbatus* L., 1771, not *A. barbatus* L., 1759.

Tufted annual, with compressed culms mostly 30–75 cm tall; leaf-blades up to 20 cm long or more, 2–5 mm broad, glabrous except at the junction of the sheath where a few long scattered hairs occur; sheath glabrous and markedly keeled; ligule small, ciliate. Inflorescence of mostly 5–8 reddish-purple spikes, 4–6 cm long. Spikelets 3-flowered with one fertile and two sterile florets; lemmas of the three florets each with an awn c. 7mm long.

GRAND CAYMAN: *Brunt 1785, 2035; Correll & Correll 51055; Maggs II 61; Proctor 15057.* CAYMAN BRAC: *Proctor 29090.*
– Mexico and the West Indies south to Argentina. A common weed of roadsides, pastures, and sandy waste ground.

TRIBE VI. PANICEAE

Genus 13. **DIGITARIA** Heist. ex Fabr.

Erect or prostrate annuals or perennials, the slender racemes glabrous to silky-pubescent, digitate, or if paniculate then aggregated along the upper part of the culms. Spikelets solitary or in 2's or 3's, subsessile or short-stalked, alternate in two rows on one side of a 3-winged or wingless rhachis, lanceolate or elliptic, plano-convex; first glume minute or lacking; second glume equalling the sterile lemma or shorter, glabrous or silky-hairy; fertile lemma cartilaginous with pale hyaline margins.

A large pantropical genus of nearly 400 species, many extending into warm-temperate regions.

KEY TO SPECIES

1. Racemes paniculately arranged, silky-pubescent; fruit lance-acuminate: **1. D. insularis**

1. Racemes digitately or subdigitately arranged, glabrous or with inconspicuous short pubescence; fruit elliptic: **2. D. horizontalis**

1. Digitaria insularis (L.) Mez ex Ekman in Ark. f. Bot. 11(4):17. 1912. **FIG. 42.**

Panicum insulare (L.) Meyer, 1818.

Trichachne insularis (L.) Nees, 1829.

Valota insularis (L.) Chase, 1906.

Culms tufted, coarse, erect, 50–150 cm tall; leaf-blades flat, usually scabrous, up to 25 cm long and 15 mm broad; sheaths sparsely silky-hairy chiefly along the margins; ligule distinct, 1 mm long. Inflorescence a greenish to pale tawny panicle mostly 15–20 cm long; spikelets 4–5 mm long, silky-hairy.

GRAND CAYMAN: *Brunt 2167; Hitchcock; Kings GC 252; Proctor 15284.* LITTLE CAYMAN: *Kings LC 97; Proctor 28158.* CAYMAN BRAC: *Millspaugh 1153; Proctor 29081.*
– Widely distributed in the warmer parts of America, chiefly occurring in open ground and waste places throughout the West Indies at low elevations.

FIG. 42 **Digitaria insularis.** A, B, C, parts of plant x½. D, spikelet, x10. E, floret, x10 (H.)

2. Digitaria horizontalis Willd., Enum. Hort. Berol. 92. 1809. **FIG. 43.**

Stoloniferous annual with decumbent branching culms; leaves flat, pubescent, the sheaths pilose. Inflorescence of 5-15 very slender, lax, sparsely hairy racemes up to 8 cm long, subdigitate or in fascicles along a slender axis. Spikelets narrow, c. 2.5 mm long; first glume obsolete or lacking.

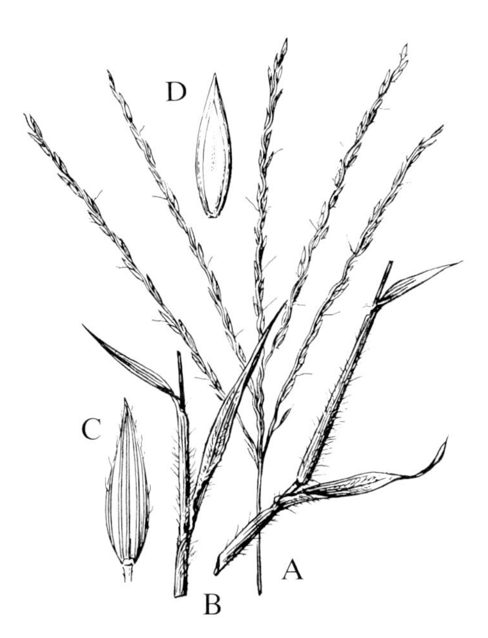

FIG. 43 **Digitaria horizontalis.** A, portions of plant, x1. B, spikelet, x10. C, spikelet, x10. D, spikelet, x10. (H.)

GRAND CAYMAN: *Hitchcock; Kings GC 344.* LITTLE CAYMAN: *Proctor 28152.*

– Pantropical, common in fields, open ground, and waste places. A closely related species, *D. ciliaris* (Retz.) Koeler, probably will be found in the Cayman Islands. It can be distinguished from *D. horizontalis* by the glabrous rhachises of the inflorescence branches, the larger spikelets (c. 3 mm long), and by the small but evident first glumes. *D. bicornis* (Lam.) R. & S. may likewise occur, as it is also common in the West Indies. In this genus, the species are often not easy to distinguish.

Genus 14. **STENOTAPHRUM** Trin.

Stoloniferous perennials with short flowering culms, rather broad obtuse leaf blades, and flat terminal and axillary racemes. Spikelets enbedded in one side of an enlarged flattened corky rhachis disarticulating at maturity, the spikelets remaining attached. First glume small; second glumes and sterile lemma about equal, the latter with a palea or staminate flower; fertile lemma chartaceous.

A small tropical and subtropical genus of 7 species.

1. **Stenotaphrum secundatum** (Walt.) Ktze., Rèvis. Gen. Pl. 2:794. 1891. **FIG. 44.**

Plants extensively creeping, the flat stolons with long internodes and short erect leafy branches. Flowering culms 10-30 cm tall; leaf-blades mostly 10-15 cm long, up to 8 mm wide, glabrous except for a few short hairs on the margins just above junction with the sheath; ligule represented by a line of hairs. Racemes 5-10 cm long; spikelets remote, 4-6 mm long.

GRAND CAYMAN: *Brunt 1768; Hitchcock; Kings GC 214; Proctor 15038.* LITTLE CAYMAN: *Kings LC 44.* CAYMAN BRAC: *Brunt 1680; Kings CB 20.* – Southern United States south to Argentina, common at low to medium elevations. An excellent pasture grass, it is also sometimes used for lawns, especially in poor sandy soils.

Genus 15. **PASPALUM** L.

Annuals or perennials with 1 to many spike-like racemes, these single, paired at the summit of the culms, or racemosely arranged along the main axis. Spikelets plano-convex, usually obtuse, subsessile, solitary or in pairs, in two rows on one side of a narrow or dilated rhachis, the back of the fertile lemma toward it; first glume usually lacking; second glume and sterile lemma commonly about equal, the former rarely lacking; fertile lemma usually obtuse, chartaceous-indurate, the margins inrolled.

A nearly cosmopolitan genus of 250 species, especially common in tropical regions.

KEY TO SPECIES

1. Racemes 2, conjugate or nearly so at the summit of the culm; plants with creeping stolons and rhizomes, in more or less saline soils: **1. P. vaginatum**

1. Racemes 1 to many, if more than one then racemose or fascicled on the axis, not conjugate at the apex; plants tufted, not stoloniferous:

2. Plants perennial; spikelets with entire margins:

FIG. 44 **Stenotaphrum secundatum.** A, plant, x1. B, portion of spike. C, D, two views of spikelet. (H.)

3. Racemes solitary; spikelets solitary (not paired) on the rhachis: **5. P. distortum**

3. Racemes 2 to many:

4. Plants slender, the culms usually less than 1 m tall; spikelets less than 2 mm long, pubescent:

5. Racemes many, usually 10 or more; spikelets subhemispheric, 1.3-1.5 mm long, pubescent but the hairs not glandular: **2. P. paniculatum**

5. Racemes few, usually 3-8; spikelets elliptic or oval:

6. Spikelets 1.3 mm long, oval, glandular-pubescent: **3. P. blodgettii**

6. Spikelets 1.5-1.8 mm long, elliptic, sparsely appressed-pubescent or nearly glabrous: **4. P. caespitosum**

4. Plants robust, the culms usually 1-2 m tall; spikelets 2-2.4 mm long, glabrous: **6. P. millegrana**

2. Plants annual; spikelets with a broad firm notched margin: **7. P. fimbriatum**

1. Paspalum vaginatum Sw., Nov. Gen. & Sp. Pl. 21. 1788.

Extensively creeping perennial with horizontal rhizomes and long leafy stolons, often forming large colonies. Flowering culms 8-60 cm tall; leaf-blades 1.5-15 cm long, 3-8 mm wide, glabrous, flat or folded with involute margins; ligule very short, truncate. Inflorescence usually a pair of racemes at apex of culm; racemes 1-7.5 cm long; spikelets 3-4.5 mm long, pale, glabrous.

GRAND CAYMAN: *Brunt 1825, 1858, 1862, 1867, 1961, 2041; Proctor 15265; Sachet 396.* LITTLE CAYMAN: *Kings LC 84a; Proctor 28067.* CAYMAN BRAC: *Kings CB 99a.*
– Pantropical, extending along seacoasts into warm-temperate areas; common in the West Indies. The closely related *P. distichum* L., which may also occur in the Cayman Islands, can be distinguished by the appressed-pubescent second glume and lemma of the lower floret, and by hispid pubescence on the lower culm-nodes. This species is less salt-tolerant than *P. vaginatum.*

2. Paspalum paniculatum L., Syst. Nat. ed. 10, 2:855. 1759.

Culms erect or decumbent at base, mostly 50-100 cm tall; leaf-blades flat, mostly 10-25 cm long, 10-20 mm broad, more or less hairy; sheaths papillose-hispid. Panicle usually 8-20 cm long, of several to many arched-spreading, somewhat fascicled racemes, the lower ones 4-12 cm long; spikelets c. 1.3 mm long, subhemispheric, pubescent.

GRAND CAYMAN: *Millspaugh 1406* (syntype of var. *minor* Scribn.).
– Mexico and the West Indies to Argentina; a weed of cultivated and waste places, also occurring in ditches and moist open ground.

3. Paspalum blodgettii Chapm., Fl. South. U.S. 571. 1860. **FIG. 45.**

Culms erect, 40-100 cm tall; leaf-blades flat, 5-25 cm long, 5-10 mm wide; sheaths pubescent. Racemes 3-8, remote (or the upper approximate), 2-8 cm long; spikelets c. 1.3 mm long, bearing numerous small gland-tipped hairs.

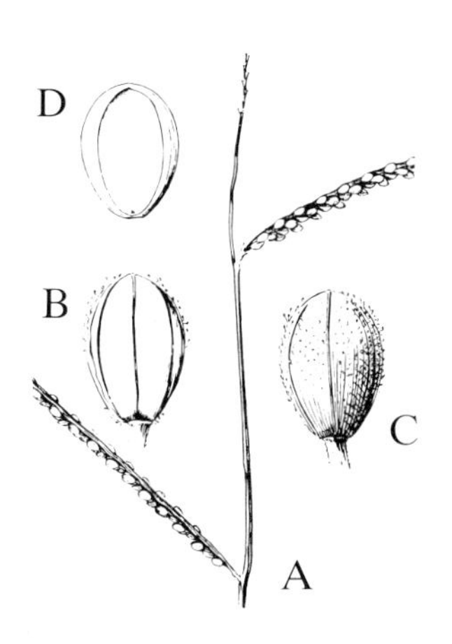

FIG. 45 **Paspalum blodgettii.** A, portion of inflorescence, x1. B, C, two views of spikelet, x10. D, floret, x10. (H.)

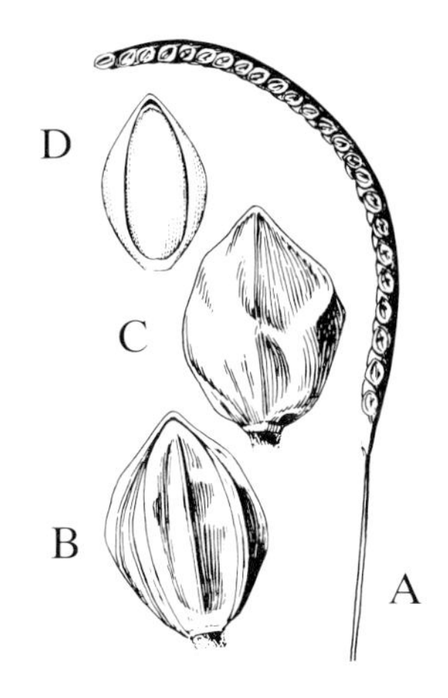

FIG. 46 **Paspalum distortum.** A, inflorescence, x1. B, C, two views of spikelet, x10. D, floret, x10. (H.)

GRAND CAYMAN: *Brunt 1918, 1953, 2038; Kings GC 136; Proctor 15068, 15257, 15265.* LITTLE CAYMAN: *Proctor 28082.* CAYMAN BRAC: *Kings CB 72; Proctor 28992.*
– Florida, Central America, and the West Indies, in pastures, sandy clearings, and in soil pockets or rocky thickets and cultivated ground.

4. Paspalum caespitosum Flügge, Monogr. Pasp. 161. 1810.

Densely tufted, the culms erect, slender, 30-60 cm tall, the base hard and slightly enlarged; leaf-blades flat or commonly folded or involute, 5-20 cm long, 4-10 mm wide. Racemes 3-5, remote, rather thick, ascending or somewhat spreading, usually 1.5-4 cm long; spikelets crowded, c. 1.5 mm long or a little more.

GRAND CAYMAN: *Hitchcock*
– Southern Florida, Central America and the West Indies, usually in partly shaded humus in pockets of limestone rock.

5. Paspalum distortum Chase in Contr. U.S. Nat. Herb. 28:142, f. 35. 1929. **FIG. 46.**

Culms slender, 15-50 cm tall; leaf-blades 15-40 cm long, 10-15 mm broad, involute, often somewhat tortuous. Inflorescence a solitary raceme 2.5-6 cm long, arcuate; spikelets 2 mm long, ovate to somewhat rhomboid, the glume and sterile lemma irregularly crumpled, glabrous.

GRAND CAYMAN: *Kings GC 234; Proctor 15255.*
– Greater Antilles, in grassy clearings and savannas.

6. Paspalum millegrana Schrad. in Schult., Mant. 2:175. 1824. **FIG. 47.**

Robust tufted perennial; culms 1-2 m tall; leaf-blades elongate, 7-15 mm broad, glabrous but with sharply scabrous margins; ligule 3 mm long, with a tuft of hairs just behind. Racemes numerous (usually 12-25), rather thick, 6-16 cm long; spikelets 2-2.4 mm long, suborbicular, flattened, pale to leaden-purplish.

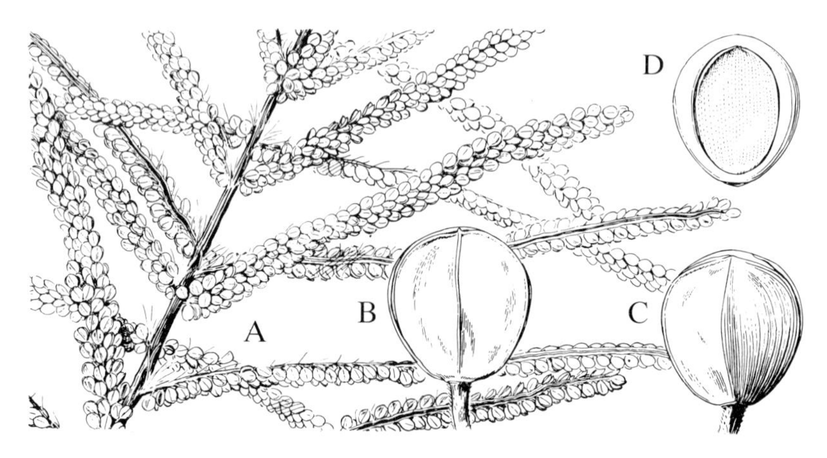

FIG. 47 **Paspalum millegrana.** A, portion of inflorescence, x1. B, C, two views of spikelet, x10. D, floret, x10 (H.)

GRAND CAYMAN: *Brunt 1897.*
– Central America and the West Indies to Brazil, in moist savannas and along ditches.

7. Paspalum fimbriatum Kunth in H.B.K., Nov. Gen. 1:93, t. 28. 1816. **FIG. 48.**

Loosely tufted annual, the erect culms 25-100 cm tall; leaf-blades flat, mostly 10-20 cm long, 5-12 mm broad. Racemes 3-8, ascending or spreading, 2.5-8 cm long; spikelets c. 2.3 mm long, with a broad, firm, notched wing, the wing and spikelets together c. 3 mm long and wide.

GRAND CAYMAN: *Kings GC 25; Proctor 15071.*
– West Indies, Panama, and northern South America, a common weed of open waste ground.

Genus 16. **PANICUM** L.

Annuals or perennials of various habit. Inflorescence paniculate, rarely racemose; spikelets more or less compressed dorsiventrally; glumes herbaceous, nerved, usually very unequal, the first often minute, the second typically equalling the sterile lemma, the latter of the same texture and simulating a third glume, bearing in its axil a membranous or hyaline palea and sometimes a staminate flower, or rarely the palea lacking; fertile lemma chartaceous-indurate, typically obtuse, the nerves obsolete, the margins inrolled over an enclosed palea of the same texture.

A large genus of about 500 species, widely distributed in tropical and warm-temperate regions. Several, known collectively as "millets", are important cereal crops, especially in India and southern Europe. Others are important pasture grasses. The genus has been subject to considerable taxonomic fragmentation in recent years, into groups which the present writer prefers to consider subgenera.

KEY TO SPECIES

1. Spikelets arranged in spike-like racemes along one side of the panicle-branches:

   2. Nodes glabrous; racemes erect, 0.5-3 cm long: **1. P. geminatum**

   2. Nodes bearded; racemes ascending, 3-9 cm long: **2. P. purpurascens**

1. Spikelets in open or condensed panicles, the branches not raceme-like:

   3. Panicles open, diffuse, 20-50 cm long, mostly 6-20 cm broad; spikelets 3 mm long, the glumes neither keeled nor scabrous: **[3. P. maximum]**

   3. Panicles (or panicle-branches) narrow and condensed, 10-25 cm long, mostly less than 5 cm broad; spikelets 2.5 mm long, the glumes with a scabrous keel: **4. P. rigidulum**

1. **Panicum geminatum** Forsk., Fl. Aegypt.-Arab. LX, 18. 1775. **FIG. 49.**

Paspalidium geminatum (Forsk.) Stapf in Prain, 1920.

Stoloniferous perennial; culms 25–80 cm tall, spreading from a decumbent base; leaf-blades 10–20 cm long, 3–8 mm wide, glabrous; ligule represented by a line of hairs just under 1 mm long. Inflorescence a narrow panicle, the main axis bearing 12–18 erect, subdistant racemes less than 3 cm long; spikelets 2-3 mm long, glabrous.

GRAND CAYMAN: *Brunt 2090, 2099; Proctor 15283.*
– Pantropical, occurring in swamps and damp places at low elevations.

2. **Panicum purpurascens** Raddi, Agrost. Bras. 47. 1823.

Brachiaria purpurascens (Raddi) Heur., 1940.

Panicum muticum sensu Adams, 1972, not Forsk., 1775.

"Para Grass"

Widely creeping stoloniferous perennial; culms decumbent at base, rooting at the lower nodes, 2–6 m long, sometimes clambering on bushes or trees; leaf-blades 15–25 cm long, 10–15 mm wide, glabrous with scabrous margins; sheaths pubescent; nodes densely villous; ligule c. 1 mm long, membranous or represented by a line of hairs. Inflorescence an open panicle; spikelets 3 mm long, glabrous.

GRAND CAYMAN: *Brunt 1795, 1841, 1846, 1855, 1903.*
– Widespread in tropical America, especially in shallow ponds and moist places; also in Africa. It is a valuable forage grass.

[3. **Panicum maximum** Jacq., Collect. 1:76. 1787. **FIG. 50.**

"Guinea Grass"

Robust tufted perennial; culms erect, 1-2.5 m tall, glabrous except for hairy nodes; leaf-blades flat, to 1 m long or more, 10–35 mm wide, usually glabrous or nearly so and with scabrous margins; sheath with scattered long hairs toward base. Inflorescence an open, diffuse panicle, the lower branches whorled; spikelets 3 mm long, oblong, acute.

GRAND CAYMAN: *Brunt 1747; Kings GC 251; Proctor 15222; Sachet 375.* LITTLE CAYMAN: *Proctor 28151.* CAYMAN BRAC: *Proctor 29337.*
– Native of Africa, now cultivated and naturalized in most tropical countries. It is an important pasture grass, or often cut and fed green.]

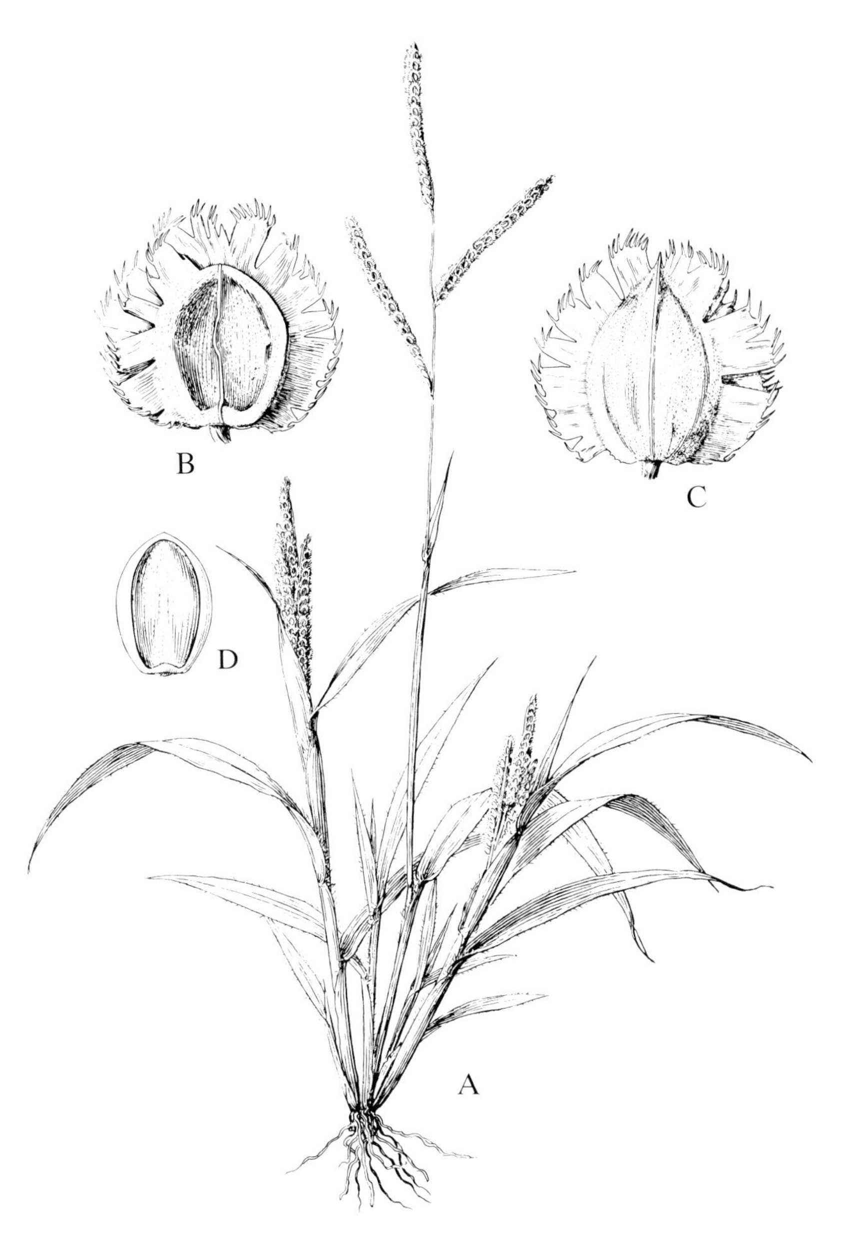

FIG. 48 **Paspalum fimbriatum.** A, plant, x½. B, C, two views of spikelet, x10. D, floret, x10. (H.)

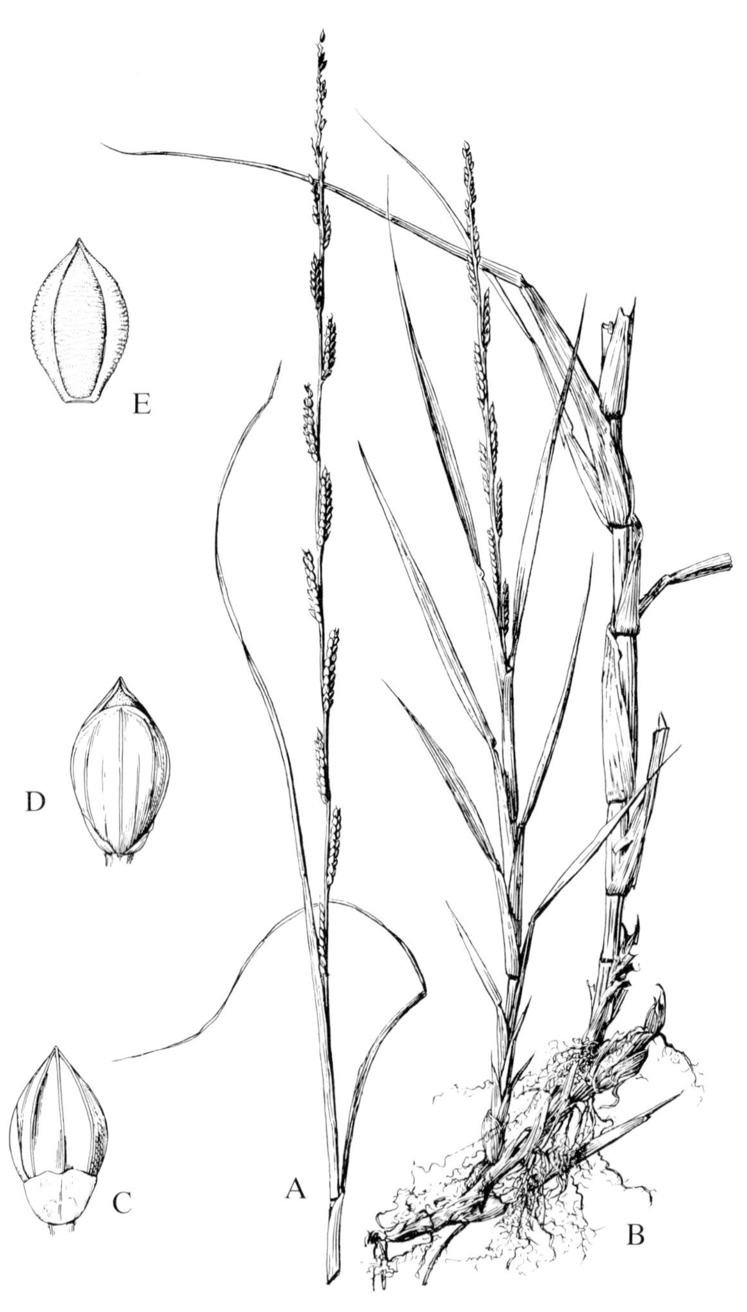

FIG. 49 **Panicum geminatum.** A, B, inflorescence and portion of plant, x½. C, D, two views of spikelet, x10. E, floret, x10. (H.)

FIG. 50 **Panicum maximum.** A, B, C, vegetative portions of plant, x½. D, inflorescence, x½. E, F, two views of spikelet, x10. G, floret, x10. (H.)

4. Panicum rigidulum Nees, Agrost. Bras. 163. 1829. **FIG. 51.**

Panicum condensum Nash in Small, 1903; Adams, 1972.

Tufted perennial, essentially glabrous; culms erect, 0.5–2 m tall; leaf-blades flat from a folded base, 15–30 cm long, 3–10 mm wide, margins slightly scabrous; sheaths (esp. the lower ones) compressed-keeled. Panicles terminal and axillary, bearing appressed branchlets with crowded spikelets; spikelets c. 2.5 mm long, narrowly oblong, acuminate, on scabrous pedicels.

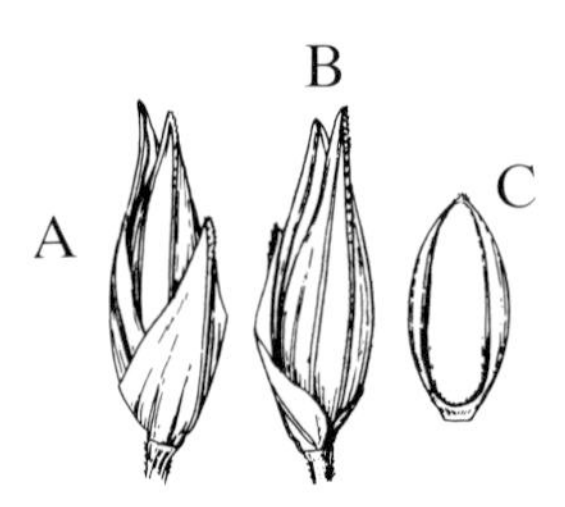

FIG. 51 **Panicum rigidulum.** A, B, two views of spikelet, x10. C, floret, x10. (H.)

GRAND CAYMAN: *Proctor 15282.* CAYMAN BRAC: *Proctor 29131.*
– Eastern United States, West Indies and Central America south to Brazil, in marshes and wet places.

Genus 17. LASIACIS (Griseb.) Hitchc.

Large branching perennials with woody culms, often clambering, the mostly firm leaf-blades narrowed into a minute petiole, and with open or somewhat compact panicles. Spikelets subglobose, placed obliquely on their pedicels; first glume broad, somewhat inflated-ventricose, usually not more than one-third the length of the spikelet; second glume and sterile lemma about equal, broad, abruptly apiculate, papery-chartaceous, shining, many-nerved, glabrous or lanose at the apex only, the lemma enclosing a membranous palea and sometimes a staminate flower; fertile lemma white, bony-indurate, obovoid, obtuse, this and the palea of the same texture bearing at the apex in a slight crateriform depression a tuft of woolly hairs, the palea concave below, gibbous above, the apex often free at maturity.

About 30 species of tropical and subtropical America.

5. Lasiacis divaricata (L.) Hitchc. in Contr. U.S. Nat. Herb. 15:16. 1910. **FIG. 52.**

Woody perennial climber, usually glabrous throughout (except margins of sheaths); culms much-branched, reaching 3 m or more in length; leaf-blades

FIG. 52 **Lasiacis divaricata.** A, portion of plant, x½. B, spikelet, x10. C, floret, x10. (H.)

narrowly lanceolate, 5–12 cm long, 5–15 mm broad, with scabrous margins; ligule inconspicuous, ciliate. Panicles terminating the main culm and fertile branches, 5–20 cm long, loosely flowered; spikelets ovoid, c. 4 mm long.

GRAND CAYMAN: *Brunt 1694, 2031, 2191; Kings GC 166, GC 166a; Lewis 3859, Sachet 372.* LITTLE CAYMAN: *Kings LC 94.* CAYMAN BRAC: *Kings CB 44; Millspaugh 1172, 1226; Proctor 28962.*

– Florida, West Indies, Central and South America, common in thickets and woodlands at low to middle elevations.

Genus 18. **ECHINOCHLOA** Beauv.

Coarse, often succulent, annuals or perennials, with compressed sheaths, linear flat leaf-blades, and rather compact panicles composed of short, densely-flowered racemes along the main axis. Spikelets plano-convex, often stiffly hispid, subsessile, solitary or in irregular clusters on one side of the panicle-branches; first glume about half the length of the spikelet, pointed; second glume and sterile lemma equal, pointed, mucronate, or the glume short-awned and the lemma long-awned, sometimes conspicuously so, enclosing a membranous palea and sometimes a staminate flower; fertile lemma plano-convex, smooth and shining, acuminate-pointed, the margins inrolled below, flat above, the apex of the palea not enclosed.

A genus of about 30 species occurring widely in tropical and subtropical regions.

KEY TO SPECIES

1. Racemes simple, 1–2 cm long; awn of the sterile lemma reduced to a short point; leaves 3–6 mm broad: **1. E. colonum**

1. Racemes often branched, more than 2 cm long; awn of the sterile lemma 1–2 cm long, conspicuous; leaves more than 10 mm broad: **2. E. walteri**

1. Echinochloa colonum (L.) Link, Hort. Bot. Berol. 2:209. 1833. **FIG. 53.**

Weedy annual, usually much branched at base; culms decumbent-ascending or erect, 20–40 cm long, glabrous; leaf-blades rather lax, 5–10 cm long, 3–6 (–10) mm broad, somewhat scabrous on the margins; sheaths glabrous; ligule lacking. Inflorescence a compressed panicle of 4–6 appressed or ascending racemes; spikelets crowded, 3 mm long, in 4 rows.

GRAND CAYMAN: *Brunt 2098; Kings GC 399.*

– Pantropical, often in roadside ditches and moist fields.

FIG. 53 **Echinochloa colonum,** x1. (H.)

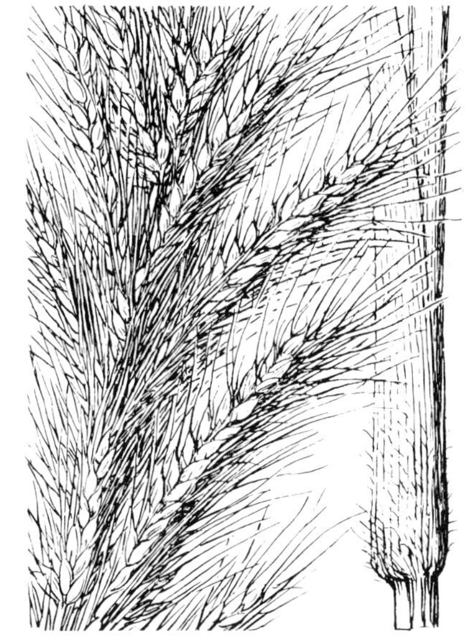

FIG. 54 **Echinochloa walteri,** x1. (H.)

2. Echinochloa walteri (Pursh) Heller, Cat. N. Amer. Pl. ed. 2, 21. 1900. **FIG. 54.**

Annual; culms erect, often succulent and rooting at the lower nodes, 1–2 m tall, up to 2.5 cm thick at base; leaf-blades 10–15 (-30) mm broad, scabrous on both sides; sheaths papillose-hairy or papillose only, rarely glabrous; ligule lacking or represented by a line of hairs. Panicle dense, up to 30 cm long, erect or nodding; racemes appressed or ascending, up to 10 cm long; spikelets close, long-awned, often purple, c. 3 mm long.

GRAND CAYMAN: *Brunt 1659, 1772; Kings GC 279; Proctor 15021.*
– Eastern United States, Cuba, Jamaica, and Hispaniola, in marshes, lagoons and wet places.

[Genus 19. **TRICHOLAENA** Schrader in Schultes

Perennials or annuals, with rather open panicles of silky spikelets. Spikelets on short capillary pedicels; first glume minute; second glume and sterile lemma equal, raised on a stipe above the first glume, emarginate or slightly lobed, short-awned, covered (except toward the apex) with long silky hairs, the palea of the sterile lemma well-developed; fertile lemma shorter than the spikelet, cartilaginous, smooth, boat-shaped, obtuse, the margins thin, not inrolled, the palea not enclosed.

A genus of 37 or more species occurring chiefly in tropical Africa, Madagascar, and from Arabia to Viet Nam, the following widely naturalized elsewhere.

1. **Tricholaena rosea** Nees, Cat. Sem. Hort. Vratisl. 1836; Fl. Austr. 17, 1841.

Rhynchelytrum roseum (Nees) Stapf & Hubb. ex Bews, 1929.

R. repens sensu C.E. Hubb., 1934, in part, not *Saccharum repens* Willd., 1797.

Tricholaena repens sensu Hitchc., 1936, not *S. repens* Willd., 1797.

Tufted annual; culms often decumbent at base, rooting freely at the lower nodes, 30–100 cm tall; leaf-blades flat, 5–15 cm long, 2–7 mm broad, nearly glabrous; sheath hairy especially just below junction with the blade and on the nodes; ligule represented by a line of hairs 1 mm long. Inflorescence a showy red-purple panicle 10–15 cm long; spikelets c. 5 mm long, covered with long silky hairs.

GRAND CAYMAN: *Brunt 2131; Proctor 15306.* LITTLE CAYMAN: *Proctor 28143.* CAYMAN BRAC: *Proctor 28995.*
– Native of Africa; introduced in the West Indies, now widespread and common along roadsides and in fields. A somewhat useful pasture grass on poor soils, it is also cultivated in some countries to make dry bouquets.]

Genus 20. **SETARIA** Beauv.

Annuals or perennials, with narrow (often spike-like) terminal panicles. Spikelets subtended by 1 to several bristles (sterile branchlets), when ripe falling free from the persistent bristles, awnless; first glume broad, usually less than half the length of the spikelet, 3–5 nerved; second glume and sterile lemma equal, or the glume shorter, several-nerved; fertile lemma coriaceous-indurate, smooth or rugose.

A genus of 140 species widely occurring in tropical and warm-temperate regions. *S. italica* (L.) Beauv., "Italian millet", is cultivated as a cereal from southern Europe to Japan.

1. Setaria geniculata (Lam.) Beauv., Ess. Nouv. Agrost. 178. 1812. **FIG. 55A.**

Chaetochloa geniculata (Lam.) Millsp. & Chase, 1903.

Perennial with knotty branching rhizomes up to 4 cm long; culms usually erect, 30–100 cm tall, hard and wiry at base; leaf-blades flat, 10–25 cm long, 4-9 mm broad, nearly glabrous or somewhat hairy toward base on upper surface; sheath glabrous; ligule represented by a conspicuous line of hairs. Inflorescence a cylindric spike-like panicle 1-10 cm long (usually about 5 cm), 4–8 mm thick (excluding the bristles); spikelets mostly 2–2.5 mm long, subtended by bristles 1 to 3 times longer than the spikelets.

FIG. 55A **Setaria geniculata**, x1. (H.)

GRAND CAYMAN: *Brunt 1640, 1896, 1927; Hitchcock; Kings GC 167, GC 325; Lewis GC 31; Proctor 15254.*

– United States to Argentina, common throughout the West Indies in pastures, cultivated areas, and moist ground. Hitchcock notes that "this species is exceedingly variable in general appearance due to the variation in the colour and length of the bristles . . . .".

[Genus 21. **PENNISETUM** L.C. Rich.

Annuals or perennials, with usually flat leaf-blades and dense spike-like panicles. Spikelets solitary or in groups of 2 or 3, surrounded by an involucre of bristles, these not united except at the very base, often plumose, falling attached to the spikelets; first glume shorter than the spikelet, sometimes minute or lacking; second glume shorter than or equalling the sterile lemma; fertile lemma chartaceous, smooth, the margin thin, enclosing the palea.

A genus of 130 species widely distributed in warm regions (chiefly Old World). *P. typhoīdeum* L.C. Rich., "Pearl millet", is extensively cultivated in India.

1. **Pennisetum purpureum** Schum. in K. Dansk. Vid. Selsk. Naturv. & Math. Afh. 3:64. 1828. **FIG. 55B.**

"Elephant grass"; "Napier grass"

Robust tufted leafy perennial 2–4 m tall; leaf-blades elongate, 20–30 mm broad, the lower ones glabrous, the upper ones with long hairs on dorsal surface; sheath with scattered hairs; ligule a line of hairs 3–4 mm long; nodes conspicuously hairy. Inflorescence a tawny spike-like panicle up to 30 cm long; spikelets 7 mm long, unequally pedicelled in fascicles, these subtended by long hair-like bristles.

GRAND CAYMAN: *Brunt 1735; Kings GC 354; Proctor 15289.*
– Native of tropical Africa; introduced in the West Indies and other areas as a forage grass and becoming naturalized.]

Genus 22. **CENCHRUS** L.

Annuals or perennials, with terminal racemes of spiny burs. Spikelets sessile, 1 to several together, permanently enclosed in a bristly or spiny involucre or bur, composed of more or less coalesced sterile branchlets; burs sessile or nearly so on a slender, compressed or angled axis, its apex produced into a short point beyond the uppermost bur, the burs falling entire, the grains germinating within them; involucre somewhat oblique, its body irregularly cleft, the lobes rigid, in most species resembling the spines, the cleft on the side of the bur next to the axis reaching to the tapering, abruptly narrowed or truncate base, the bristles or spines barbed, at least toward the summit; spikelets mostly glabrous or nearly so; first glume 1-nerved, usually narrow, sometimes lacking; second glume and sterile lemma 3–5-nerved, the lemma enclosing a well developed palea and usually a staminate flower. Fruit usually turgid, indurate, the lemma acuminate, the nerves visible toward the summit, the margins thin, flat.

About 25 tropical and warm-temperate species; most are troublesome weeds because of their prickly burs.

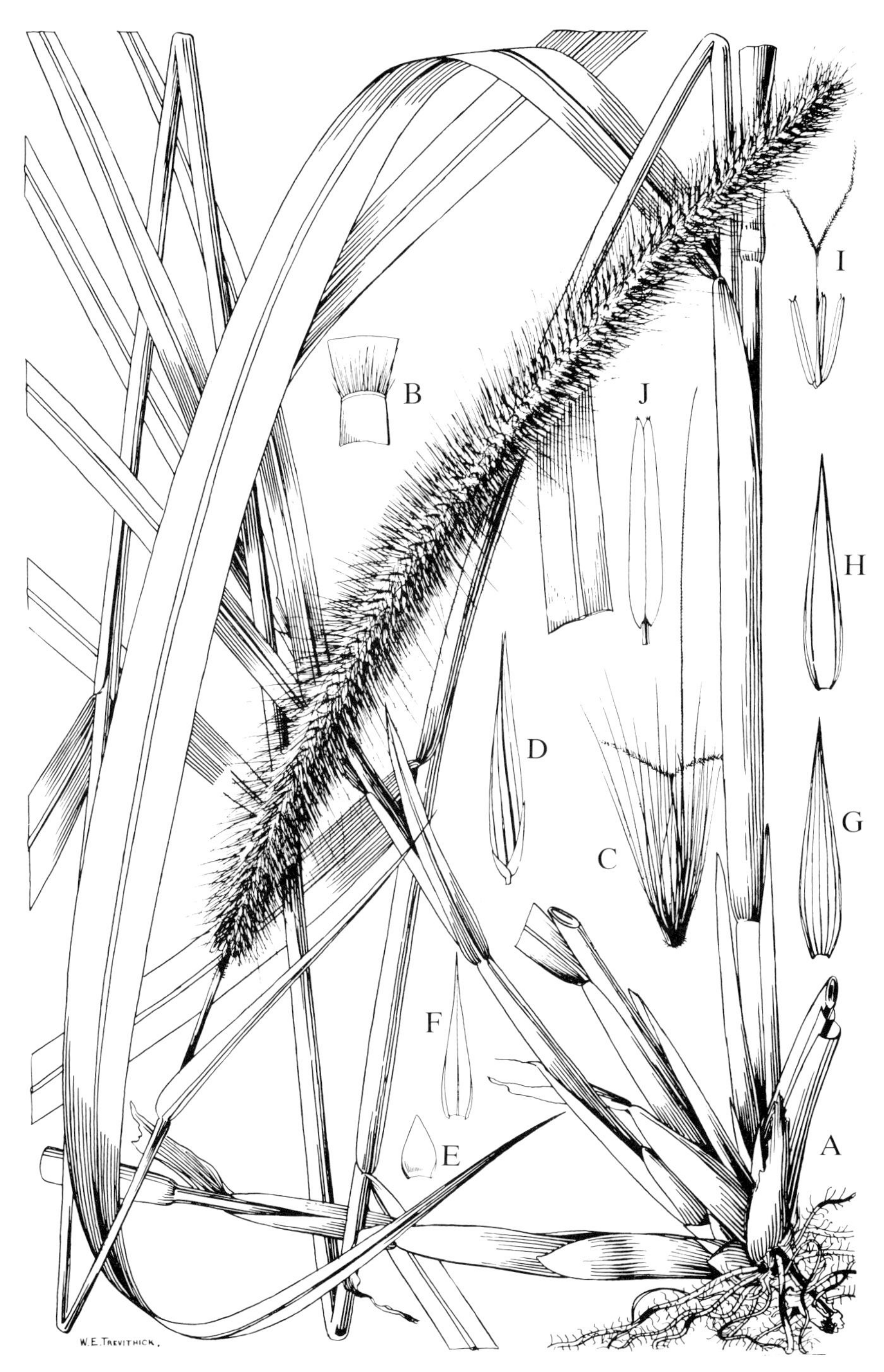

FIG. 55B **Pennisetum purpureum.** A, habit, x½. B, ligule, x1. C, spikelet surrounded by bristles, x3. D, spikelet, x5. E, lower glume, x5. F, upper glume, x5. G, upper lemma, x5. H, palea, x5. I, flower, x5. J, stamen, x12. (R.R.I.)

KEY TO SPECIES

1. Plants perennial; burs glabrous except at base of spines: **3. C. gracillimus**

1. Plants annual; burs more or less pubescent (the hairs sometimes visible only under a strong lens):

 2. Involucre with a ring of slender bristles at base:

  3. Burs, excluding the bristles, not more than 4 mm wide, numerous, crowded in a long spike; lobes of involucre interlocking, not spine-like: **1. C. brownii**

  3. Burs, excluding the bristles, c. 5.5 mm wide, not densely crowded; lobes of the involucre erect or nearly so (or rarely one or two lobes loosely interlocking), the tips spine-like: **2. C. echinatus**

 2. Involucre beset with flattened spreading spines, no ring of slender bristles at base:

  4. Burs, including spines, 7–8 mm wide, finely and minutely pubescent; **4. C. incertus**

  4. Burs, including spines, 10–12 mm wide, usually densely woolly-pubescent, some hairs as long as the width of the spines near the base: **5. C. tribuloides**

1. Cenchrus brownii Roem. & Schult. in L., Syst. Veg. 2:258. 1817. **FIG. 56.**

Culms erect from a sparingly-branched, more or less geniculate base, 30–100 cm tall; leaf-blades flat, thin, mostly 10–30 cm long, 6–12 mm broad. Spike 4–10 cm long, dense; burs with numerous slender outer bristles, the inner usually exceeding the bur, erect or spreading; lobes of the bur usually 6–8, interlocking at maturity; spikelets usually 3.

GRAND CAYMAN: *Hitchcock; Millspaugh 1268; Proctor 15174.* CAYMAN BRAC: *Kings CB 36.*

– Florida and Mexico to Brazil; a common weed of fields and waste places throughout the West Indies.

2. Cenchrus echinatus L., Sp. Pl. 2:1050. 1753. **FIG. 57.**

"Soft Bur"

Culms ascending from a branched decumbent base, often rooting at the lower nodes, 25–60 cm long or longer. Leaf-blades 6–20 cm long, 3–9 mm broad; sheath

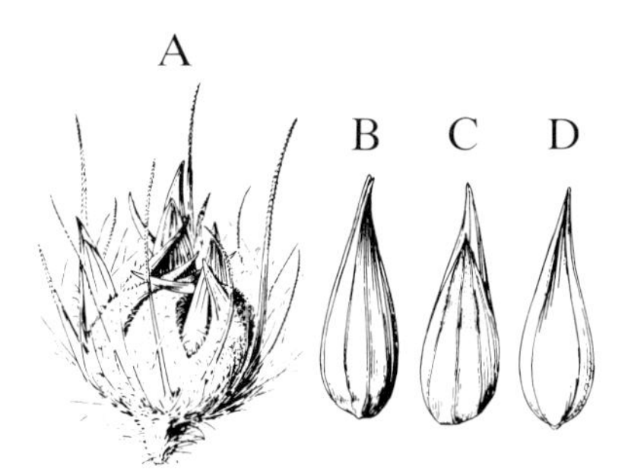

FIG. 56 **Cenchrus brownii.** A, bur, x5. B, C, two views of spikelet, x5. D, floret, x5. (H.)

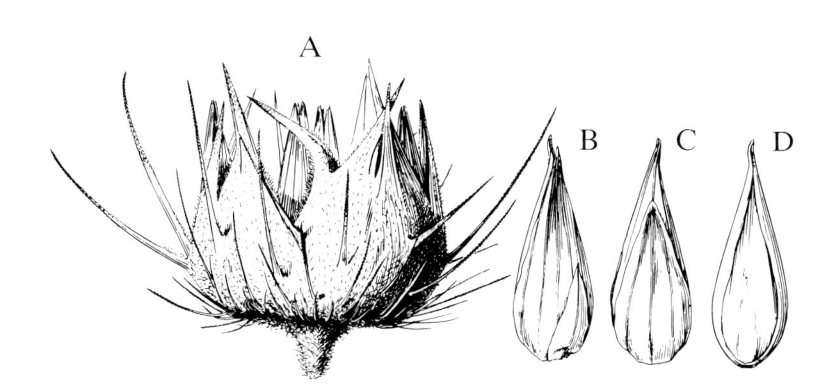

FIG. 57 **Cenchrus echinatus.** A, bur, x5. B, C, two views of spikelet, x5. D, floret, x5. (H.)

with marginal silky hairs. Spikes 3-7 (-10) cm long; burs truncate at base, the body 4-7 mm high and as wide or wider; longest bristles about equalling the lobes of the bur in length; lobes of the bur about 10, erect or bent inward, the tips hard and spine-like; spikelets about 4.

GRAND CAYMAN: *Brunt 1707; Hitchcock; Kings GC 170; Proctor 15256.* LITTLE CAYMAN: *Proctor 28153.* CAYMAN BRAC: *Proctor 29098.*
– Common throughout tropical and subtropical America, a weed of fields and open waste ground.

3. Cenchrus gracillimus Nash in Bull. Torr. Bot. Club 22:299. 1895. **FIG. 58.**

Plants tufted, at length forming dense clumps, glabrous throughout. Culms erect or ascending, 20-80 cm tall, slender and wiry; leaf-blades 5-20 cm long, 2-3 mm broad, usually folded. Spikes 2-6 cm long, the burs not crowded; burs 3.5-5 mm wide (excluding the spines), somewhat tapering at the base; spines spreading or reflexed, all flat and broadened at base, retrorsely scabrous toward apex; lobes of the bur 5-6 mm long; spikelets 2 or 3, 5.5-7 mm long.

GRAND CAYMAN: *Lewis 3853; Proctor 15146, 31046.*
– Florida, Cuba, Jamaica and Hispaniola, on sandy open ground.

4. Cenchrus incertus M.A. Curtis in Boston J. Nat. Hist. 1:135. 1837. **FIG. 59.**

Cenchrus pauciflorus Benth., 1840.

Plants sometimes forming large mats; culms 20-90 cm long or more, spreading to ascending from a branched decumbent base. Leaf-blades 3-15 cm long, 2-7 mm broad, flat or folded. Spikes 3-10 cm long; burs mostly 4-6 mm wide, often densely (but very minutely) pubescent, rarely nearly glabrous; spines numerous, spreading or reflexed, flat and broadened at base, some of the upper ones 5 mm long; lobes of the bur about 8, rigid and spine-like; spikelets usually 2.

GRAND CAYMAN: *Sachet 384.* LITTLE CAYMAN: *Proctor 28157.* CAYMAN BRAC: *Proctor 28991.*
– United States to Argentina, in the West Indies common in sandy open ground.

5. Cenchrus tribuloides L., Sp. Pl. 2:1050. 1753. **FIG. 60.**

Plants with stout, branching, radiate-decumbent culms 15-60 cm long or more, rooting at the nodes and with numerous ascending branches 10-30 cm tall; leaf-sheaths usually overlapping, sharply keeled, those below the spikes inflated. Spikes

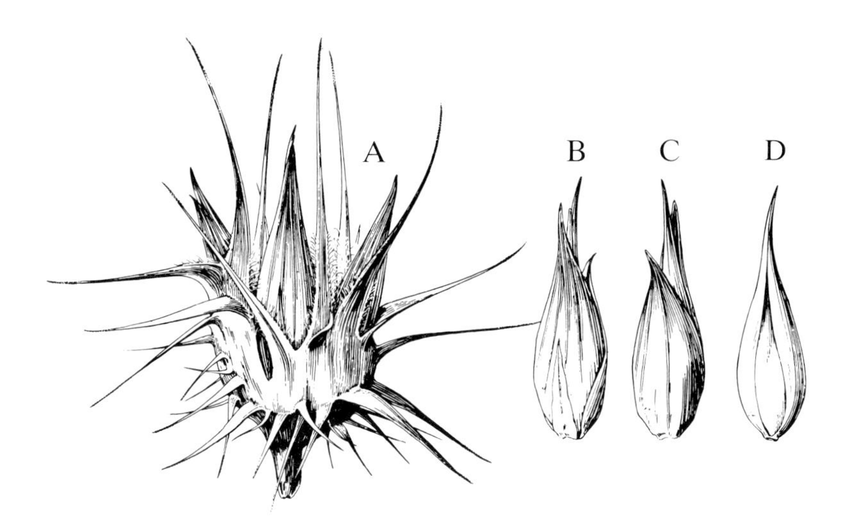

FIG. 58 **Cenchrus gracillimus.** A, bur, x5. B, C, two views of spikelet, x5. D, floret, x5. (H.)

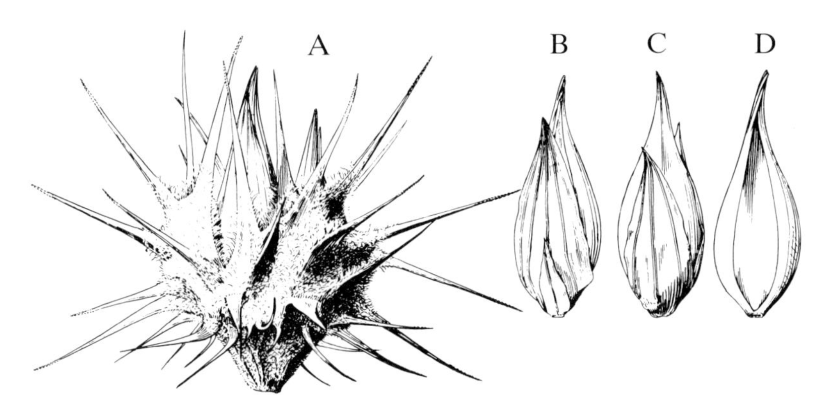

FIG. 59 **Cenchrus incertus.** A, bur, x5. B, C, two views of spikelet, x5. D, floret, x5. (H.)

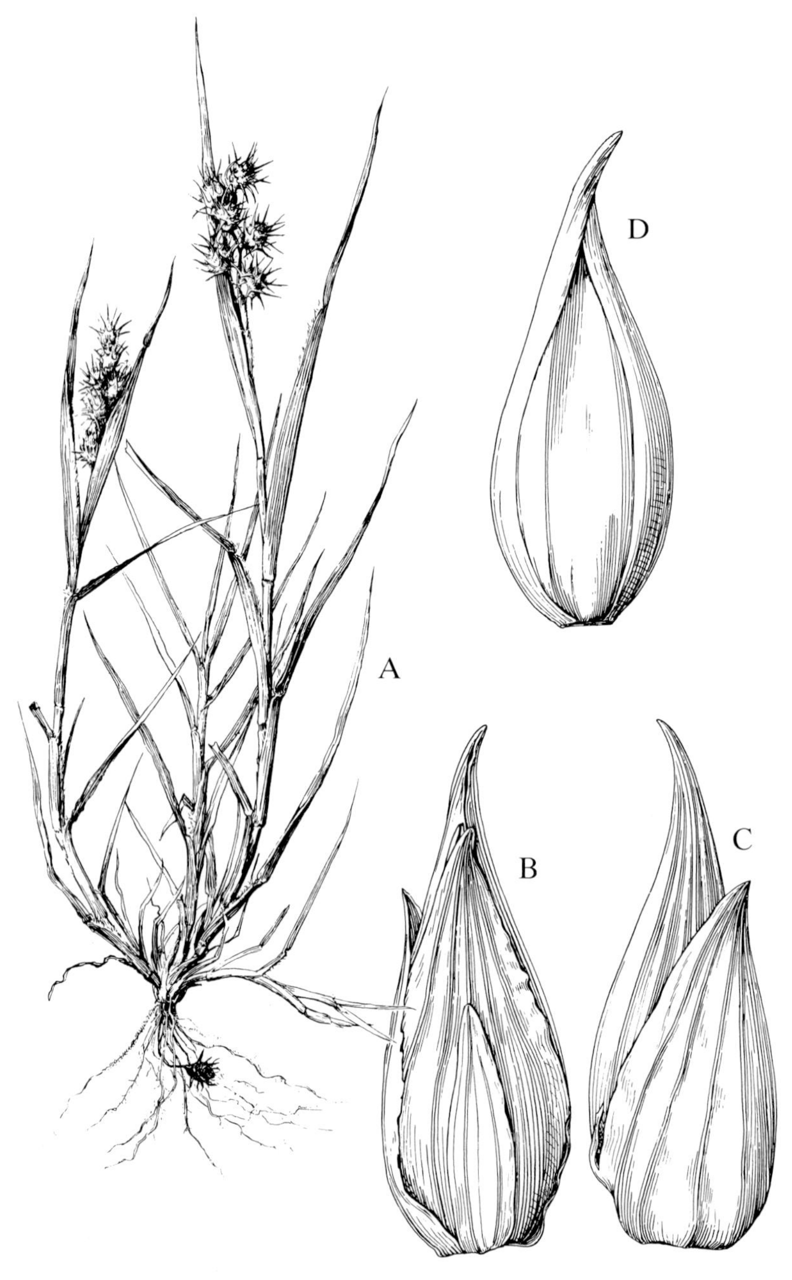

FIG. 60 **Cenchrus tribuloides.** A, plant, x½. B, C, two views of spikelet, x10. D, floret, x10. (H.)

3-9 cm long; burs 5-6 mm wide and 8-9 mm high (excluding the spines); spines more or less spreading, flat, the upper sometimes as much as 3 mm broad at base; spikelets usually 2, 7-8 mm long.

GRAND CAYMAN: *Brunt 1741; Kings GC 26; Millspaugh 1249; Proctor 15069.* LITTLE CAYMAN: *Kings LC 107.* CAYMAN BRAC: *Kings CB 39; Millspaugh 1152.*
– Eastern United States, the West Indies, and Brazil, in loose coastal sands.

TRIBE VII. ANDROPOGONEAE

Genus 23. **IMPERATA** Cyrillo

Slender erect perennials with terminal narrow silky panicles. Spikelets all alike, awnless, in pairs, unequally pedicellate on a slender continous rhachis, surrounded by long silky hairs; glumes about equal, membranous; sterile lemma, fertile lemma, and palea thin and hyaline.

A small genus of 10 tropical and subtropical species.

1. **Imperata contracta** (Kunth) Hitchc. in Rep. Mo. Bot. Gard. 4:146. 1893.

Tufted, spreading by scaly rhizomes; culms erect, often 1 m tall or more, leafy; leaf-blades flat, elongate, 6-10 mm broad, glabrous but with sharply scabrous margins. Inflorescence a narrow whitish panicle up to 40 cm long; spikelets 3 mm long, densely silky-hairy from the base, the hairs much longer than the spikelet.

GRAND CAYMAN: "Hitchcock in 1891", as recorded in *Manual of the Grasses of the West Indies,* p. 379. No Cayman material seen in preparing the present Flora.
– Southern Mexico and the West Indies to Brazil, in swamps and moist open ground.

[Genus 24. **SACCHARUM** L.

Tall, coarse perennial grasses with large plume-like inflorescences of panicled racemes. Culms solid, juicy (not hollow as in most grasses). Spikelets all alike, in pairs, one sessile, the other stalked, surrounded at base with long silky hairs, the rhachis readily disarticulating below the spikelets. Glumes firm, 1-3-nerved, acute or acuminate; sterile lemma similar to the glumes but hyaline; fertile lemma shorter than the glumes, hyaline, awnless, sometimes lacking.

5 or more species of the Old World tropics.

1. Saccharum officinarum L., Sp. Pl. 1:54. 1753.

"Sugar Cane"

Loosely tufted, with solid nodose leafy culms 2–3 m tall or more; leaf-blades 1.5–2 m long, 30–50 mm broad, with scabrous margins; sheaths deciduous from lower part of the culm; ligule 2 mm long, membranous, with tuft of hairs just behind. Panicles conspicuous, plume-like, silvery-white, with abundant woolly hairs, the whole inflorescence 50 cm long or more.

Cayman specimens not collected; sometimes cultivated and persisting.
– Cultivated in nearly all tropical and subtropical countries; origin somewhere in southeastern Asia.]

Genus 25. **ANDROPOGON** L.

Annuals or more often coarse perennials with solid leafy culms; spikelets arranged in racemes, these solitary, paired, clustered, or in panicles, the common peduncles often enclosed in or exserted from a spathe-like sheath, these sheaths often aggregated in a loose or dense, often silky, compound inflorescence. Spikelets paired at each node of a disarticulating rhachis, one sessile and perfect, the other pedicellate and staminate or neuter (often very much reduced), the rhachis and sterile pedicels usually densely ciliate or villous. Glumes of sessile spikelet indurate, several-nerved, the median nerve obscure or lacking, the margins keeled toward the apex; sterile lemma hyaline; fertile lemma hyaline, narrow, shorter than the glumes, awnless or awned from the apex or from between lobes, the awn straight or geniculate. Pedicellate spikelet either as large as the sessile, or else more or less reduced, sometimes only the pedicel present.

A pantropical and warm-temperate genus of nearly 200 species, by some authors split into several smaller genera.

KEY TO SPECIES

1. Robust erect tufted grass over 1 m tall; inflorescence dense, feathery, club-shaped, with numerous hairy racemes in pairs, each pair subtended by a spathe-like sheath; glumes not pitted: **1. A. glomeratus**

1. Low laxly tufted stoloniferous grass with ascending culms less than 1 m tall, inflorescence a loose cluster of 2–8 racemes not subtended by a spathe-like sheath; first glume with a pit or pinhole in the middle of the back: **[2. A. pertusus]**

1. Andropogon glomeratus (Walt.) B.S.P., Prel. Cat. N.Y. 67. 1888. **FIG. 61.**

Densely tufted perennial; culms compressed, 1-1.5 m tall. Leaves with flat or folded blades 3-5 mm broad; lower sheaths crowded, keeled. Inflorescence densely corymbose, villous, with paired racemes 1.5-3 cm long; rhachis slender, flexuous; sessile spikelet 3-4 mm long, with awn 1.5 cm long.

GRAND CAYMAN: *Brunt 1656, 1974; Hitchcock; Kings GC 164; Lewis GC 21.* – Southeastern United States, Mexico and the West Indies to northern South America. Frequent in moist open ground. The very similar *A. bicornis* (not yet recorded from the Cayman Islands) can be distinguished by having all its spikelets awnless.

FIG. 61 **Andropogon glomeratus,** x1. (H.)

[2. **Andropogon pertusus** (L.) Willd. in L., Sp. Pl. 4:922. 1806.

Amphilophis pertusa (L.) Stapf, 1917.

Bothriochloa pertusa (L.) A. Camus, 1931.

Rather low, loosely tufted perennial with branched, ascending leafy culms 20-100 cm tall, the nodes bearded. Leaves usually pubescent, 1-4 mm broad, variable in length from 2 to 20 cm. Racemes few, flabellately clustered on a short naked axis, 2-6 cm long, villous; awn of sessile spikelet twice-geniculate, brown, c. 15 mm long.

GRAND CAYMAN: *Brunt 1839, 1852, 1894a; Proctor 15069.* LITTLE CAYMAN: *Proctor 35200.* CAYMAN BRAC: *Proctor 29338.*
– Pantropical in present distribution, but indigenous to tropical Asia and introduced elsewhere. Common in open fields and along roadsides, quickly springing from dormant roots after rains and tending to crowd out other grasses. Eaten by stock, but not very palatable.]

[Genus 26. **SORGHUM** Moench

Annuals or perennials, with elongate leaves and open, often large panicles of short racemes. Spikelets in pairs, one sessile and fertile, the other stalked and usually staminate, these pairs attached at the nodes of a tardily disarticulating rhachis of a short, few-jointed raceme, the terminal sessile spikelet with 2 pedicellate spikelets. Glumes of the fertile spikelet indurate, the first rounded but somewhat keeled at the summit; fertile lemma awnless or with a short, usually geniculate, twisted awn. Pedicellate spikelets herbaceous, lanceolate, the first glume several-nerved, 2-keeled in the upper half.

A genus of about 35 species, chiefly found in tropical Africa.

KEY TO SPECIES

1. Plants perennial, with creeping rhizomes; panicles loose, open; leaves less than 2 cm broad: **1. S. halepense**

1. Plants annual, tufted (creeping rhizomes lacking); panicles usually dense, heavy; leaves up to 5 cm broad: **2. S. saccharatum**

1. Sorghum halepense (L.) Pers., Syn. Pl. 1:101. 1805.

"Johnson Grass"

Perennial with numerous strong rhizomes. Culms erect, usually 1-1.5 m tall, with appressed-pubescent nodes. Leaves elongate, mostly 1-1.5 cm broad, whitish-scabrous on the margins; sheaths shorter than the internodes, glabrous; ligule membranous, ciliate, c. 2 mm long. Panicles 15-30 cm long; spikelets 5 mm long, acute, entirely pale or often with a large red spot at base, the awn (when present) 1-1.5 cm long, deciduous.

GRAND CAYMAN: *Hitchcock.*
– Widely naturalized in all warm countries, introduced from the Mediterranean region. A weed of fields and waste places.

2. **Sorghum saccharatum** (L.) Moench, Meth. Pl. 207. 1794.

Sorghum vulgare Pers., 1805.

"Sorghum", "Guinea Corn"

Tufted annual, often 2 m tall or more (but dwarf races exist). Culms coarse, erect, the ample leaves with broad flat blades to 5 cm broad, the midrib white; ligule membranous, ciliate, 4 mm long. Panicles commonly dense and compact, with numerous turgid persistent spikelets.

CAYMAN BRAC: *Millspaugh 1224.* Also occasionally cultivated on Grand Cayman, but no preserved specimens have been seen.
– Pantropical in cultivation, with numerous varieties, some very different in appearance; occasionally escaping or persisting after cultivation. The species probably originated in Africa.]

TRIBE VIII. MAYDEAE

[Genus 27. **ZEA** L.

A coarse annual with broad drooping leaf-blades. Spikelets unisexual; staminate spikelets 2-flowered, in pairs on one side of a continuous rhachis, one flower nearly sessile, the other stalked; glumes membranous, acute; pistillate spikelets sessile, in pairs, consisting of one fertile and one sterile floret, the latter sometimes developed as a second fertile floret; glumes broad, rounded or emarginate at apex; sterile lemma similar to the fertile, with palea present; style very long and slender, stigmatic along both sides well toward the base.

A single variable species, existing almost wholly as a cultivated plant.

1. **Zea mays** L., Sp. Pl. 2:971. 1753.

"Corn", "Maize"

A tall annual, extremely variable in size of plant and character of the pistillate inflorescences. Leaves conspicuously distichous, the blades up to 7 cm broad or more. Inflorescences monoecious, the staminate flowers in spikelike racemes, these numerous, forming large spreading panicles terminating the culms; pistillate inflorescences in the axils of the leaves, the spikelets in 8–16 (or even to 30) rows on a thickened, almost woody axis (cob), the whole enclosed in numerous large foliaceous bracts (husks), the long styles (silk) protruding from the top as a silky mass of threads.

No specimens have been collected in the Cayman Islands, but it is commonly cultivated.
– Grown in all tropical and temperate regions; apparently of tropical American origin.]

Family 8

TYPHACEAE

Erect, perennial, gregarious herbs with creeping rhizomes, erect terete flowering stems, and long linear leaves sheathing at the base. Flowers unisexual, densely crowded in elongate, cylindric, terminal spikes, the staminate above, the pistillate below. Perianth lacking. Staminate flowers composed of 1-7 stamens (usually 3), subtended by hairs; filaments very short, connate; anthers linear, 4-celled. Pistillate flowers consisting of a 1-celled ovary elevated on a short hairy stalk, containing 1 anatropous ovule, and terminated by a short style and a linear to spatulate stigma. Fruit a linear or fusiform achene, elevated on a long hairy stalk, and terminating in the persistent but fragile, greatly elongated style.

A small but cosmopolitan family with a single genus.

Genus 1. **TYPHA** L.

Characters as given for the family. It should be noted that the pistillate flower-spikes contain numerous sterile flowers mingled with the fertile ones; these enlarge after anthesis and become dilated at the apex.

A genus of about 10 species, all marsh plants, widely distributed in tempearate and tropical regions. In some countries, the fibrous stems and leaves are used for many purposes, such as for thatch, soft matting, ropes and baskets. The downy wool of the inflorescence can be applied like cotton to wounds and ulcers. The young shoots are edible, and are said to taste like asparagus. The pollen can be used as flour to make bread, and is also highly inflammable.

1. **Typha domingensis** Pers. Syn. Pl. 2:532. 1807. **FIG. 62.**

"Cat-tail", "Rush", "Bulrush", "Bull Rush"

Stems 1-3 m tall; leaves to 1.5 m long or more, 3-20 mm broad, nearly flat, glabrous. Spikes pale brown, the staminate and pistillate portions separated, each 10-40 cm long; hairs of pistillate pedicels usually minutely club-shaped at apex; pollen-grains 1-celled.

GRAND CAYMAN: *Brunt 1689; Hitchcock; Kings CG 216.*
– Southern United States and throughout the West Indies, south to southern South America, chiefly in coastal marshes.

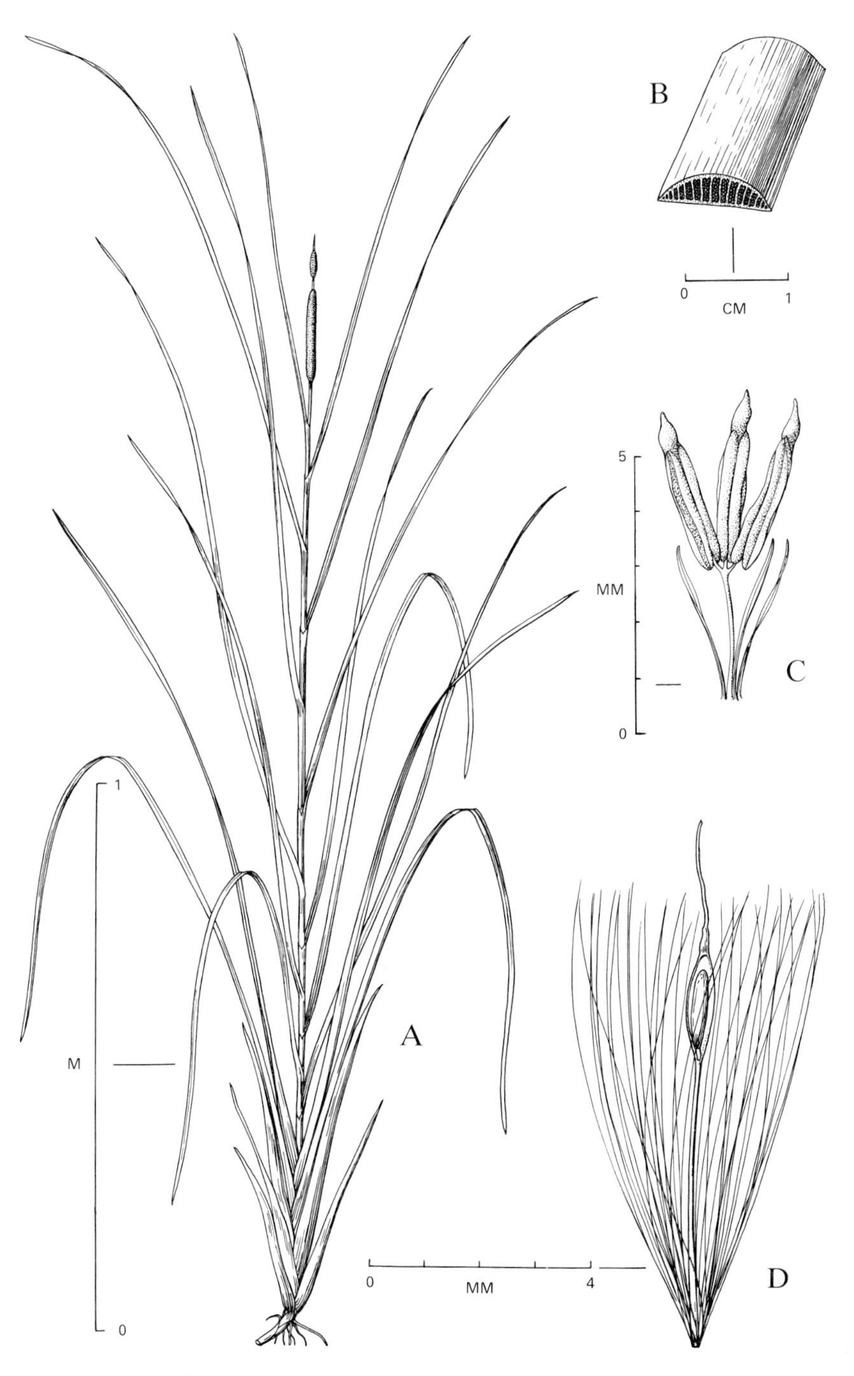

FIG. 62 **Typha domingensis.** A, habit. B, partial x-section of stem. C, staminate flower. D, pistillate flower. (D.E.)

Family 9

BROMELIACEAE

Mostly perennial herbs, epiphytic or terrestrial. Leaves spirally arranged, usually forming a basal tuft, dilated-sheathing toward the base, simple, the margins entire or spiny, the surface nearly always bearing minute peltate scales at least when young. Flowers perfect or functionally dioecious; inflorescence simple or paniculate, often with brightly coloured bracts. Sepals and petals dissimilar, free or connate; stamens 6, in 2 series; filaments free, or joined to the petals or to each other; ovary superior to inferior, 3-celled; stigmas 3 or style 3-parted. Fruit a capsule or berry; seeds often plumose or winged, and containing plentiful mealy endosperm.

A chiefly tropical American family of about 50 genera and more than 1,500 species. The pineapple *(Ananas comosus)* is the most important economic plant of this family. Many of the species of this family are commonly called "Wild Pine".

KEY TO GENERA

1. Leaves with entire margins; ovary superior or nearly so; fruit a dry capsule; seeds with plumose appendage: **1. Tillandsia**

1. Leaves with spiny margins; ovary wholly inferior; fruit a berry; seeds naked:

  2. Petals naked, 3 cm long; ovaries free from each other; leaves bearing long curved spines on the margins: **2. Bromelia**

  2. Petals appendaged, less than 1.5 cm long:

    3. Ovaries remaining distinct from each other in fruit; leaves 8–12 cm broad with minutely spiny margins; inflorescence not foliaceous at apex: **3. Hohenbergia**

    3. Ovaries fused with each other and with the fleshy bracts to form a headlike syncarp; leaves less than 4 cm broad with densely spiny margins; inflorescence crowned by a tuft of leaves: **[4. Ananas]**

Genus 1. **TILLANDSIA** L.

Caulescent or acaulescent herbs of varied habit, usually epiphytic. Leaves tufted or distributed along a stem, linear, ligulate, or narrowly triangular. Inflorescence on a distinct scape, simple or compound, usually of one or more distichous-flowered spikes or sometimes reduced to a single polystichous-flowered spike by the reduction of lateral spikes to single flowers, or rarely the whole inflorescence reduced to a single flower. Flowers perfect; sepals usually symmetric, free, or equally or posteriorly connate; petals free or joined at the base, unappendaged. Stamens of various lengths relative to the petals and pistil. Ovary superior, glabrous; ovules usually many, caudate. Capsule septicidal. Seeds erect, narrowly cylindric or fusiform, the plumose appendage basal, straight, white.

A large genus of about 350 species widely distributed in the warmer parts of the Western Hemisphere.

KEY TO SPECIES

1. Stems elongate, loosely clothed with scattered, gray, linear leaves; inflorescence 1–2-flowered; stamens shorter than the petals: **1. T. recurvata**

1. Stems very short, the leaves tufted, often with more or less broadly sheathing bases; inflorescence with several to numerous flowers; stamens longer than the petals:

  2. Floral bracts much less than twice the length of the internodes:

    3. Flowers appressed to the rhachis; inflorescence up to 1 m long, 2–3-pinnate with long slender branches; leaves numerous, not twisted or marked with transverse stripes: **2. T. utriculata**

    3. Flowers more or less spreading; inflorescence less than 40 cm long, simple or with 2 or 3 short simple branches; leaves few, twisted and marked with transverse stripes: **3. T. flexuosa**

  2. Floral bracts twice the internodes or more:

    4. Leaf-blades linear-setiform, straight, less than 1 mm in diameter for most of their length, the sheaths not inflated: **4. T. setacea**

    4. Leaf-blades flat or definitely if narrowly triangular, more than 1 mm broad (often much more), the basal sheaths broadly expanded, erect, and more or less clasping:

      5. Inflorescence many-branched, the branches (spikes) 15–30 cm long; leaves with basal sheaths essentially flat, not recurved-inflated: **5. T. fasciculata**

      5. Inflorescence simple or few-branched, the branches less than 10 cm long; leaves with basal sheaths more or less recurved-inflated:

6. Basal leaves equalling or exceeding the inflorescence; plants not silvery-canescent:

7. Swollen base of plant ellipsoid or spindle-shaped; leaf-sheaths loosely imbricate, gradually narrowing into the blades; floral bracts coriaceous: **6. T. balbisiana**

7. Swollen base of plant ovoid or bulbous; leaf-sheaths tightly imbricate, abruptly narrowing into the inrolled blades; floral bracts thin and papery: **7. T. bulbosa**

6. Basal leaves shorter than the inflorescence; plants densely silvery-canescent throughout: **8. T. circinnata**

1. Tillandsia recurvata (L.) L., Sp. Pl. ed. 2, 1:410. 1762. **FIG. 63.**

"Old Man's Beard"

Stems several or many in a clump, simple or few-branched, 1–10 cm long, mostly shorter than the leaves; leaves distichous, 3–17 cm long, densely and minutely pruinose-scaly, the thin, many-nerved sheaths overlapping each other and concealing the stem, the blades recurved, linear, terete, 0.5–2 mm in diameter and of soft or lax texture. Scape terminal, up to 13 cm long, c. 0.5 mm thick, bearing 1 or rarely 2 narrow scape-bracts immediately below the inflorescence (or sometimes one of them remote); inflorescence usually 1- or 2-flowered; floral bracts resembling the scape-bracts; sepals lanceolate, 4–9 mm long; petals narrow, pale violet or white; stamens deeply included but longer than the pistil. Capsule up to 3 cm long.

GRAND CAYMAN: *Kings GC 108; Proctor 15172.* LITTLE CAYMAN: *Kings LC 76; Proctor 28134.*
– Southern United States south to Argentina and Chile, always epiphytic, or in some countries (notably Jamaica) often growing on electric and telephone wires.

2. Tillandsia utriculata L., Sp. Pl. 1:286. 1753.

Plants stemless; leaves many in a dense rosette, up to 1 m long, densely and finely pale-appressed-scaly throughout; sheaths rather large, subovate; blades linear-triangular, long-acuminate, 2–7 cm broad at the base, the outer ones usually recurving. Scape erect, equalling or often exceeding the leaves,8–14 mm thick, glabrous; scape-bracts erect, tubular-involute, barely overlapping or the uppermost sometimes remote. Inflorescence 2–3-pinnate (rarely simple) with lax branches, glabrous; primary bracts like the upper scape-bracts, not over 4 cm long; racemes up to 35 cm long with an elongate sterile base bearing several bracts, the rhachis

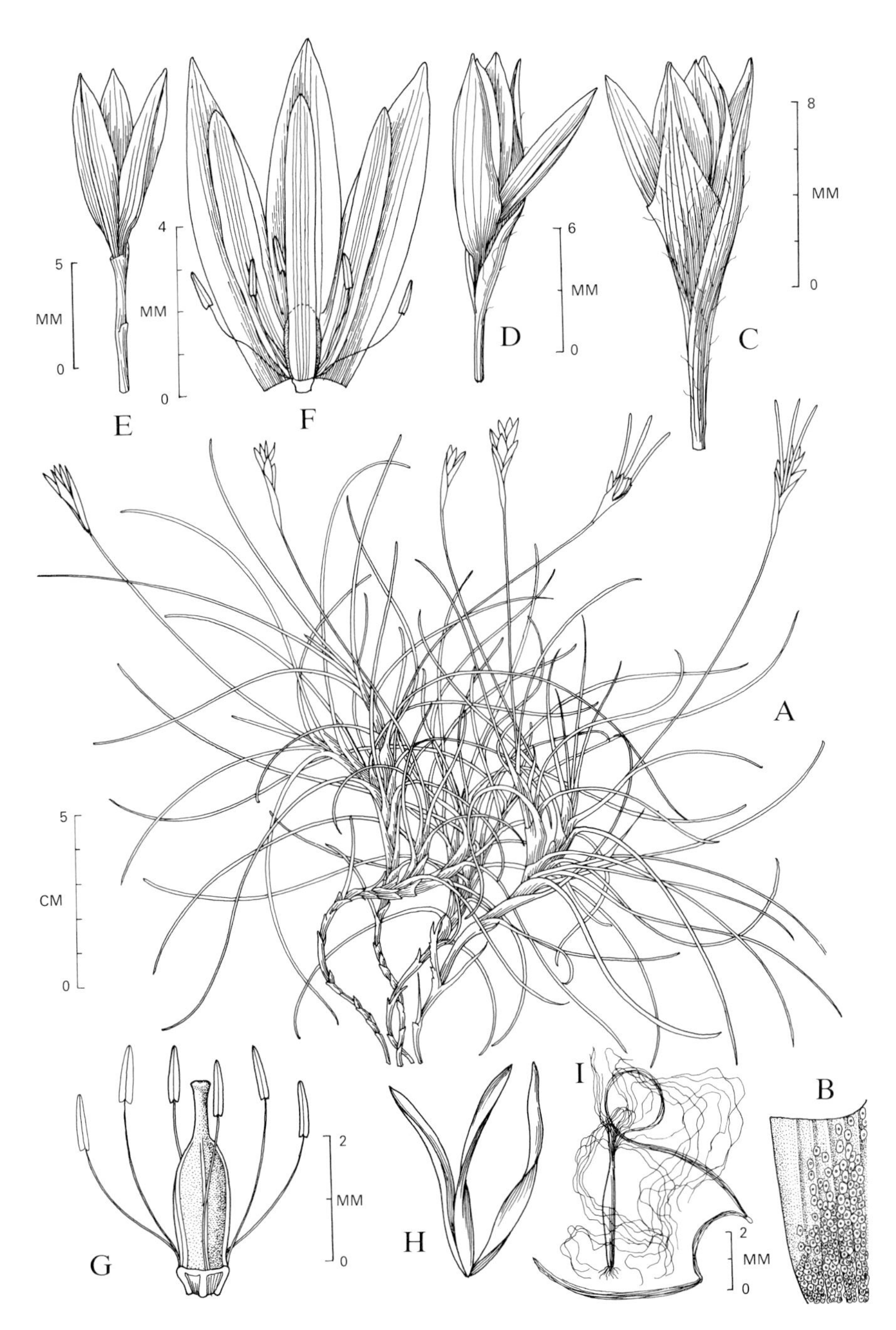

FIG. 63 **Tillandsia recurvata.** A, habit. B, portion of leaf-base greatly enlarged, showing scales. C, flower. D, flower with one sepal removed. E, flower with all sepals removed. F, flower spread out to show the stamens. G, stamens and pistil alone. H, fruit after dehiscence. I, seed with plumose fibrils. (G.)

slender, undulate, sulcate, and strongly flattened next to the flowers; floral bracts erect, enfolding the base of the flower but very little of the rhachis, much exceeded by the sepals, and equalling or shorter than the internodes, closely and prominently nerved, the margin often purple. Flowers appressed to the rhachis, the pedicels up to 5 mm long; sepals obtuse, 14-18 mm long; petals linear, 3-4 cm long, white; stamens and pistil exserted. Capsule 4 cm long, narrowly cylindric, acute.

GRAND CAYMAN: *Brunt 1832, 1833, 1834; Kings GC 407; Proctor 15004.* LITTLE CAYMAN: *Kings LC 52. Proctor 35079, 35192.* CAYMAN BRAC: *Kings CB 34; Proctor 29345.*
– Florida, West Indies, Mexico, Belize and Venezuela, common on trees at low elevations.

3. **Tillandsia flexuosa** Sw., Nov. Gen. & Sp. Pl. 56. 1788.

Tillandsia aloifolia Hook., 1827.

Plant stemless; leaves 10 or more in a dense, twisted, often semi-bulbous rosette, 15-30 cm long or more, densely pale-appressed-scaly and usually marked with broad pale transverse stripes, the sheaths ovate, very large but passing into the blade without clear distinction, long-acuminate, stiff and curved at the apex. Scape erect, slender, glabrous; scape-bracts erect, tubular-involute, at least the upper ones shorter than the internodes. Inflorescence simple or with few branches; primary bracts like the upper scape-bracts; racemes very laxly-flowered; rhachis slender, flexuous, sharply angled, glabrous; floral bracts 2-3 cm long, equalling or shorter than the sepals and about equalling the internodes. Flowers with pedicels to 7 mm long; sepals 2-3 cm long; petals tubular-erect, linear, to 4 cm long, usually pale rose; stamens exserted; capsule narrowly cylindric, up to 7 cm long.

GRAND CAYMAN: *Brunt 1835; Proctor 31026.* LITTLE CAYMAN: *Kings LC 116; Proctor 28132.* CAYMAN BRAC: *Kings CB 32; Millspaugh 1219; Proctor 15317.*
– Florida, Bahamas and Greater Antilles, and Panama to Guayana.

4. **Tillandsia setacea** Sw., Fl. Ind. Occ. 1:593. 1797.

Tillandsia tenuifolia L., 1753, in part, not as to type.

Plant stemless; leaves numerous in a dense fasciculate rosette, 15-30 cm long or more, often longer than the inflorescence, densely and finely lepidote throughout; sheaths c. 2 cm long, often keeled, brownish; blades setiform, usually less than 1 mm thick. Scape erect, very slender, sparsely lepidote; scape-bracts erect, closely involute, densely overlapping. Inflorescence simple and distichous-flowered or shortly-branched, mostly 2.5-5 cm long; primary bracts like the upper scape-bracts; spikes subsessile; floral bracts densely overlapping, mostly 8-14 mm long, longer than the sepals. Flowers subsessile; sepals 7-12 mm long, glabrous; petals tubular-erect, linear, 2 cm long, violet; stamens and pistil exserted. Capsule narrowly cylindric, 2.5 cm long.

GRAND CAYMAN: *Brunt 1836; Hitchcock; Kings GC 180, GC 197; Proctor 15003.*
– Florida, Greater Antilles, northern Central America, Venezuela, Brazil.

5. Tillandsia fasciculata Sw., Nov. Gen. & Sp. Pl. 56. 1788.
var. **clavispica** Mez in DC., Monog. Phan. 9:682. 1896.

Plant stemless; leaves numerous, up to 70 cm long, forming a stiff, rather dense rosette; sheaths ovate, nearly flat, dark brown near the base; blades narrowly triangular, long-acuminate, rigid, not more than 3 cm broad, finely and densely brown-appressed-lepidote throughout (but the surface appearing smooth and rather lustrous if examined without a lens). Scape erect, stout, shorter than the leaves, clothed with densely overlapping bracts, the lower of these leaf-like. Inflorescence compound with simple branches, overtopping the leaves; primary bracts like the upper scape-bracts; spikes ascending, clavate with an elongate sterile base, the apical fertile portion densely-flowered. Floral bracts 2-2.8 cm long, bright pink or red at anthesis. Flowers erect, subsessile, with violet petals, the stamens and pistil exserted. Capsule acuminate, 4 cm long.

LITTLE CAYMAN: *Proctor 35171.*
– Cuba. In Little Cayman this taxon grows on the ground or on the lower parts of trees in very dense woodland not near the sea. Other variants of *T. fasciculata* are widespread in the West Indies.

6. Tillandsia balbisiana Schult. f. in R. & S., Syst. Veg. 7, pt. 2:1212. 1830. **FIG. 64.**

Plant nearly stemless, sometimes several together in a loose clump; leaves many in a dense bulbous rosette, often longer than the inflorescence if extended (but typically recurved), densely and minutely pale-appressed-lepidote throughout; sheaths ovate, large, inflated, forming a more or less ellipsoid "pseudobulb" up to 12 cm long, the blades abruptly spreading or recurved from the tops of the sheaths, of hard texture and with involute margins, long-attenuate at the apex. Scape erect, slender, subglabrous; scape-bracts erect, overlapping, with long, linear, spreading or reflexed-contorted apices. Inflorescense simple and distichous-flowered or more often densely pinnate, 6-20 cm long; primary bracts like the upper scape-bracts, at least their sheaths shorter than the axillary spikes. Spikes sessile, strict, linear, 3-12 cm long, c. 1 cm wide at anthesis; floral bracts overlapping, 1.5-2.2 cm long, longer than the sepals, often bright red at anthesis. Flowers erect, subsessile; sepals coriaceous; petals 3-4.5 cm long, violet or purple; stamens and pistil exserted. Capsule narrowly cylindric, 4 cm long.

GRAND CAYMAN: *Brunt 1873; Kings GC 110, GC 195, GC 381.* LITTLE CAYMAN: *Kings LC 2, LC 113, LC 115; Proctor 28131.* CAYMAN BRAC: *Kings CB 31; Proctor 15316.*
– Florida, Bahamas, Greater Antilles except Puerto Rico, and from Mexico south to Colombia and Venezuela, always epiphytic (occasionally on electric or telephone wires).

7. **Tillandsia bulbosa** Hook., Exot. Fl. 3:t. 173. 1826.

Plants stemless, solitary or few in a clump; leaves 8–15, often exceeding the inflorescence, clothed with minute, closely adhering, cinereous scales; sheaths orbicular, inflated, forming a dense ovoid pseudobulb 2–5 cm long, green or greenish-white, abruptly contracted into narrow, contorted, involute blades, these up to 30 cm long and 2–7 mm thick. Scape erect; scape-bracts foliaceous with elongate blades exceeding the inflorescence. Inflorescence simple or subdigitate with few spikes, red or green; primary bracts ovate with foliaceous blades; spikes spreading, 2–5 cm long, 2-8-flowered; floral bracts overlapping, 15 mm long, longer than the sepals, 2 or 3 times as long as the internodes, keeled. Flowers sessile; sepals 13 mm long, glabrous; petals linear, 3–4 cm long, blue or violet; stamens and pistil exserted. Capsule cylindric, to 4 cm long.

LITTLE CAYMAN: *Kings LC 114.*
– West Indies and Mexico to Colombia and Brazil.

8. **Tillandsia circinnata** Schlecht. in Linnaea 18:430. 1844.

Plants stemless, solitary or few in a clump; leaves mostly 8–10, much shorter than the inflorescence, covered throughout with coarse, closely appressed, cinereous scales; sheaths broadly ovate, inflated, forming a narrowly ovoid or ellipsoid pseudobulb 5–15 cm long, gradually contracted into the blades, the outer ones reduced and bladeless; blades involute-subulate, shorter than or not much exceeding the sheaths, 3–7 mm thick, curved, contorted or coiled. Scape erect or decurved; scape-bracts overlapping, with curved, rigid, foliaceous blades. Inflorescence simple or digitately or pinnately compound, the spikes few; primary bracts like the scape bracts. Spikes 5–12 cm long, 2–10 flowered; floral bracts overlapping, pink, 2–3 cm long, longer than the sepals, 2 or 3 times as long as the internodes, not keeled. Flowers sessile; sepals c. 20 mm long, glabrous; petals linear, 4 cm long, purple; stamens and pistil exserted. Capsule narrowly cylindric, 4 cm long.

LITTLE CAYMAN: *Kings LC 114a; Proctor 28133.*
– Florida, Bahamas, Cuba, Hispaniola and Mexico.

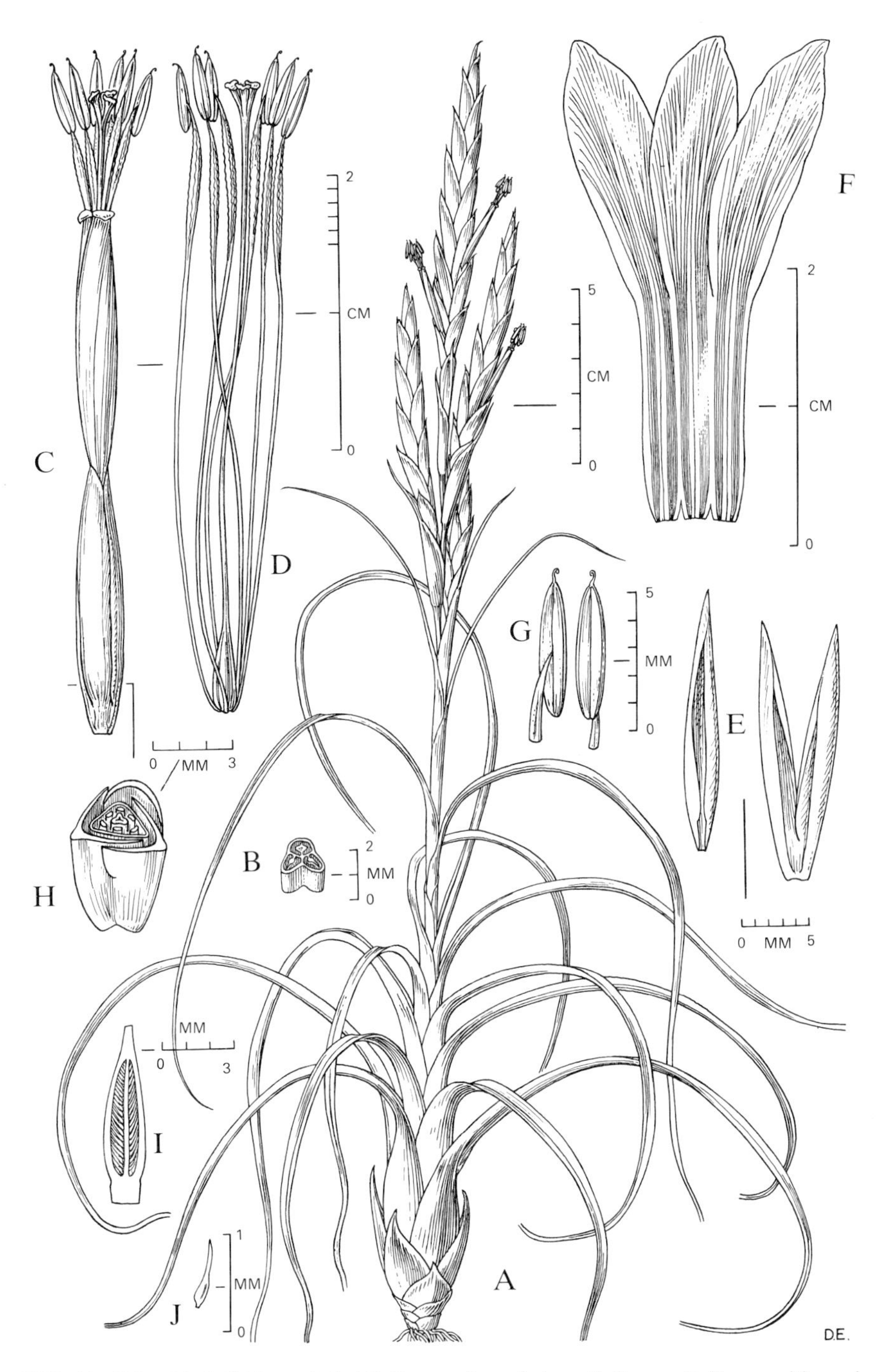

FIG. 64 **Tillandsia balbisiana.** A, habit. B, x-section of stem. C, flower. D, flower with sepals and petals removed. E, sepals. F, corolla spread to show petals joined in the basal half. G, anthers. H, x-section of ovary. I, long.-section of ovary. J, seed. (D.E.)

Genus 2. **BROMELIA** L.

Coarse terrestrial herbs, spreading by underground stolons. Leaves arranged in a stiff, ascending rosette; margins armed with large curved spines. Inflorescence scapose or sessile, paniculate. Flowers pedicellate; sepals free or rarely somewhat united, obtuse or acute, rarely mucronulate; petals rather fleshy, with narrow free blades, the basal claws centrally united to the filament-tube but with free margins, unappendaged; stamens included, the anthers narrow and acute; ovary passing gradually into the pedicel, the epigynous tube conspicuous to nearly lacking. Berry succulent, relatively large; seeds few to many, flattened, naked.

A genus of nearly 50 species, widely distributed from Mexico and the West Indies to Argentina.

1. **Bromelia pinguin** L., Sp. Pl. 1:285. 1753. **FIG. 65.**

"Pingwing"

Leaves numerous, usually 1–2 m tall, with broad, coarsely tomentose-lepidote sheaths at the base; blades linear, acuminate, up to 5 cm broad, armed with stout curved teeth up to 1 cm long. Scape stout, white-farinose; scape-bracts foliaceous with red, subinflated sheaths. Inflorescence many-flowered, narrowly pyramidal, white-farinose; primary bracts like the scape-bracts but the upper ones entire; branches up to 12-flowered; floral bracts linear-subulate from a short broad base, 3 cm long. Flowers up to 6 cm long, distinctly stalked; sepals 15–30 mm long; petals 3 cm long, rose with white base and margins, densely white-tomentose at the apex; ovary 2 cm long. Berry ovoid, c. 3.5 cm long, yellowish, edible but very acid.

GRAND CAYMAN: *Brunt 1957; Kings GC 185; Proctor 27992.*
– Mexico and the West Indies to Guyana and Ecuador.

Genus 3. **HOHENBERGIA** Schult.f.

Stemless, coarse, usually epiphytic herbs, sometimes growing on rocks. Leaves rosulate, the sheaths tightly overlapping and holding water, the blades broadly ligulate with more or less spiny margins. Scape well developed. Inflorescence bipinnate or tripinnate, composed of dense conelike spikes. Flowers perfect, sessile, compressed; sepals nearly or quite free, distinctly asymmetric, mucronate; petals nearly or quite free, the claw bearing 2 scales on the inner surface; stamens shorter than the petals, the first series free, the second adnate to the petals; pollen with 2 or 4 pores. Ovary wholly inferior, compressed and more or less winged, usually changing little in fruit; ovules several in each locule.

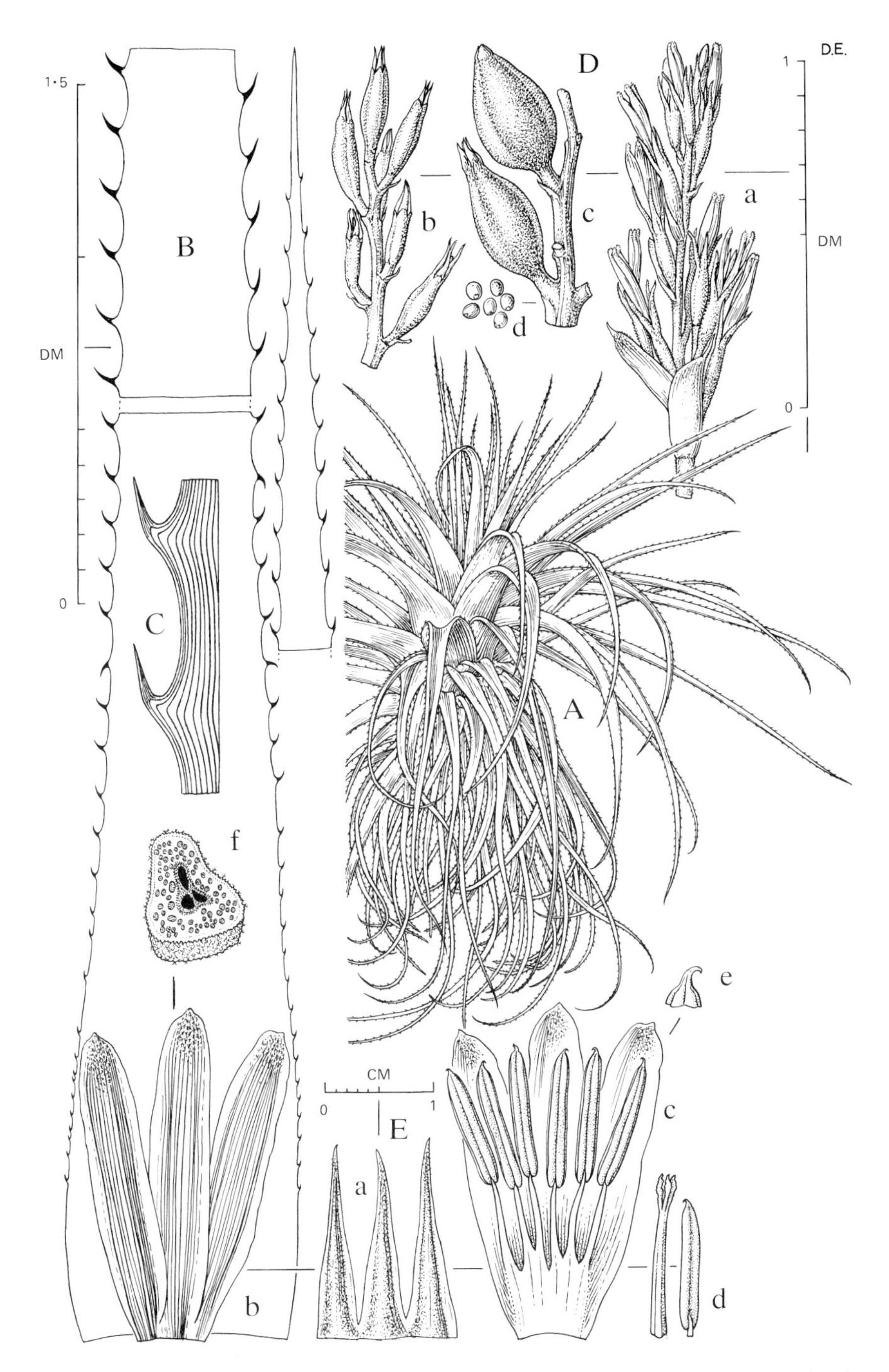

FIG. 65 **Bromelia pinguin.** A, habit, much reduced. B, outline of leaf segments showing marginal spines. C, marginal spines, enlarged. D, portions of inflorescence; a, with flowers; b, after fall of perianths; c, fruits; d, seeds. E, floral details; a, floral bracts; b, sepals; c, petals with attached stamens; d, style and anther; e, apex of anther; f, x-section of ovary.

A genus of about 35 species, concentrated chiefly in Jamaica and eastern Brazil, with a few single species scattered elsewhere in the West Indies, Central America and northern South America.

1. **Hohenbergia caymanensis** Britton ex L.B. Smith in Proc. Amer. Acad. 70:150. 1935.

Leaves large, 1 m long or more and up to 13 cm broad, minutely brown-punctulate throughout, broadly rounded and apiculate at the apex, and with finely serrulate margins (the teeth less than 1 mm long). Scape 8–12 mm in diameter, densely and minutely brown-lepidote; scape-bracts overlapping, linear-lanceolate, minutely serrulate. Inflorescence rather laxly bipinnate, c. 40 cm long; primary bracts like the scape-bracts, the lower ones much longer than the spikes; spikes densely ellipsoid, 2.5–4 cm long, stalked; floral bracts 12 mm long, with a mucro as long as the triangular-ovate base, strongly nerved. Flowers 14 mm long; sepals 6 mm long; petals greenish-white, 10 mm long; stamens included.

GRAND CAYMAN: *Rothrock 495 (type) F!, Kings GC 196; Proctor 31031.* – Endemic, apparently confined to a small area of rocky woodland about a half mile southeast of Georgetown. The plants mostly grow on rocks, but also occasionally on the lower parts of tree-trunks. This species is not very different from *H. penduliflora* of Cuba and Jamaica.

[Genus 4. **ANANAS** Mill.

A terrestrial herb with numerous narrow, stiff leaves in a rosette, their margins plentifully armed with sharp, curved spines. Scape erect, stout, spiny-bracted. Inflorescence dense, headlike, crowned with a tuft of sterile leafy bracts. Flowers sessile, usually violet; sepals free, 5–7 mm long, slightly asymmetric; petals free, c. 15 mm long, each bearing 2 slender scales; stamens included; pollen-grains ellipsoid with 2 pores. Ovaries coalescing with each other and with the bracts and axis to form a fleshy compound fruit. Berry sterile in the cultivated varieties, no seeds being formed.

A monotypic genus probably of South American origin.

1. **Ananas comosus** (L.) Merrill, Interpr. Rumph. Amb. 133. 1917.

"Pineapple"

Characters of the genus. Leaves up to 1 m long, 1–3 cm broad, the marginal prickles c. 2 mm long. Flowering spikes 4–10 cm long, enlarging in fruit, the fruits variable in shape according to the variety.

Often cultivated and sometimes persisting after cultivation; no Cayman specimens examined.]

[The Zingiberales are represented by cultivated plants of several families, notably Musaceae. *Musa sapientum* L., the banana, and *Musa paradisiaca* L., the plantain, are often grown.]

Family 10

PALMAE

Mostly trees with unbranched, erect trunks and a terminal bud. Leaves usually large, pinnately or palmately divided, forming a crown. Flowers perfect or unisexual; if the latter, then either on the same or on different plants. Inflorescence (spadix) paniculate, subtended or at first enclosed by a spathe usually of 2 valves, one of these usually much longer than the other. Sepals and petals 3 each, free or connate. Stamens 6–12; filaments distinct or joined toward the base; in pistillate flowers the stamens may be reduced to staminodes or lacking. Ovary usually 1–3-celled, each cell with a single ovule; style short or lacking. Fruit a 1-seeded berry (drupe); seed containing a horny or cartilaginous endosperm frequently rich in oil.

A large, economically important family of more than 200 genera and 2,500 species, represented in nearly all tropical and many warm-temperate regions.

KEY TO GENERA

1. Leaves pinnate:

  2. Petioles forming a smooth green crownshaft; inflorescences produced at base of crownshaft: **1. Roystonea**

  2. Petioles not forming a crownshaft; inflorescences produced among the leaves:

    3. Lower pinnae slender, stiff, and spinelike; trunk with persistent leaf-bases or conspicuous geometric scars; suckers produced at base of trunk; fruit less than 7 cm long: **[2. Phoenix]**

    3. Lower pinnae not spinelike; leaf-bases not persistent; trunk with rather smooth ring-like scars; no suckers produced; fruit up to 30 cm long: **[3. Cocos]**

1. Leaves palmate:

4. Leaves green on both sides; fresh ripe fruits white; seed smooth: **4. Thrinax**

4. Leaves silvery beneath; fresh ripe fruits black; seed grooved and fissured: **5. Coccothrinax**

Genus 1. **ROYSTONEA** O.F. Cook

Tall, erect, simple, unarmed, monoecious trees with columnar, light-coloured trunks often bulged or swollen. Leaves with tubular green petioles forming a crownshaft at top of the trunk, this 1–2 m long, surmounted by a spreading crown of large pinnate leaf-blades; pinnae (leaflets) 1- or 2-seriate. Inflorescences springing from base of the crownshaft, each protected in bud by a pointed, glabrous, long-fusiform spathe; spathe-parts unequal, deciduous at anthesis. Spadix a whitish compound panicle, erect or ascending in flower, deflexed in fruit; flowers superficial on the rhachillae, unisexual, the staminate and pistillate ones randomly scattered on the same rhachillae. Sepals of staminate flowers imbricate and much shorter than the petals; stamens 6, bearing conspicuous anthers; pistillate flowers much smaller than the staminate, with staminodes represented by scales or a cup or ring; pistil of 3 carpels but usually only 1 developing, forming a 1-seeded fruit with stigmatic scar on frontal side near the base. Fruit drupelike, less than 2 cm long, oblong to globular, sessile; seed with raphe making a branched or lacerated pattern on its face; micropyle and embryo basal; albumen hard.

A chiefly Antillean genus of less than 10 species; several of these are prized as ornamentals and are planted in all tropical countries because of their tall, stately habit.

1. Roystonea regia (Kunth in H.B.K.) O.F. Cook in Science ser. 2, 12:479. 1900; Bull. Torr. Bot. Club 28:554. 1901. **FIG. 66.**

Oreodoxa regia Kunth in H.B.K., 1815.

Tree to 18 m tall or more, with trunk usually tapered at both ends, slightly bulged toward the middle. Leaf-blades rather lax and down-curved; leaflets biseriate and standing irregularly in two planes, giving the whole leaf a somewhat shaggy, plumose appearance, usually less than 1 m long and c. 2–3 cm broad. Spadix 1 m long or more at anthesis, glabrous; staminate flowers with petals 5 mm long, the anthers concealed in bud; pistillate flowers c. 3 mm long. Fruit oblong-globose, 8–13 mm long and 8–10 mm thick, dark purplish at maturity.

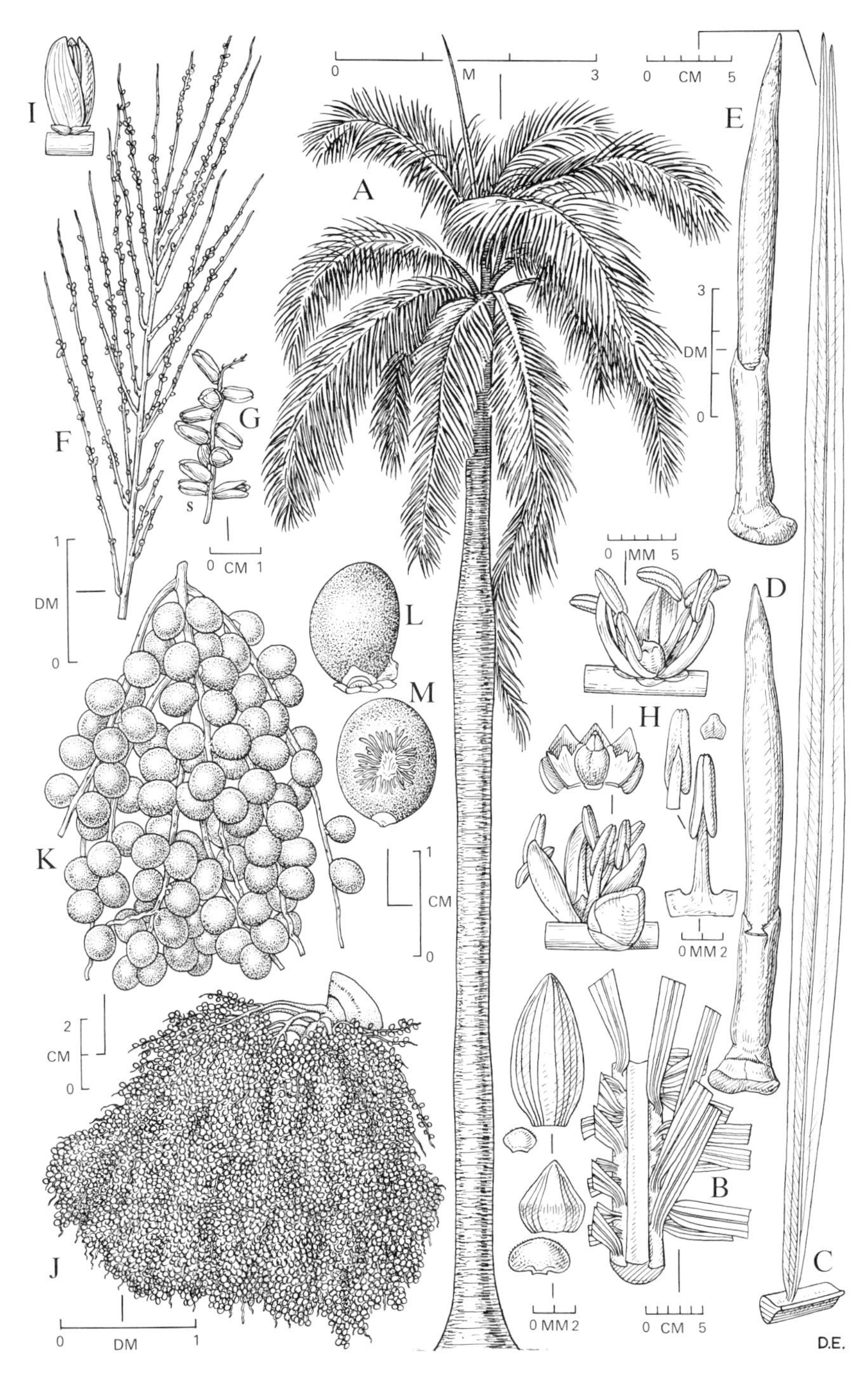

FIG. 66 **Roystonea regia**. A, habit. B, portion of leaf-rhachis. C, single pinna. D, E, two views of inflorescence-bud. F, portion of expanded inflorescence. G, portion of single rhachilla; s = staminate flowers (buds only). H, staminate flowers and anthers in various aspects. I, pistillate flower. J, fruiting inflorescence. K, enlarged portion of same. L, single fruit. M, seed showing branched raphe. (D.E.)

GRAND CAYMAN: *Brunt 1753, 1947; Kings PGC 259; Proctor (photo).* – Florida and Cuba; also widely cultivated. The Grand Cayman population has the appearance of being indigenous; the trees are scattered or in small groups in swampy woodlands and clearings in the central part of the island.

[Genus 2. **PHOENIX** L.

The date-palm, *Phoenix dactylifera* L., has long been cultivated in the Cayman Islands, presumably just as an ornamental, as it seldom if ever produces fruit here. It is not common, but being long-lived tends to persist near the sites of former dwellings. This species, which is native to North Africa and Arabia, has been cultivated for many thousands of years, especially in desert oases and in the Tigris-Euphrates valley. It is an important source of food and revenue in these and some other regions.

GRAND CAYMAN: *Proctor (sight).* LITTLE CAYMAN: *Kings LC 104a.*]

[Genus 3. **COCOS** L.

The coconut-palm, *Cocos nucifera* L., is economically the most important of all cultivated palms, but unfortunately was largely wiped out in the Cayman Islands by a mycoplasmic disease. It survives here chiefly as an occasional ornamental, but the introduction of disease-resistant varieties could perhaps restore the species as a significant local source of food and drink. In other tropical parts of the world, coconuts are often grown as a commercial crop. The oil extracted from the "copra" or dried endosperm of the large seeds has many culinary and industrial uses.

GRAND CAYMAN: *Proctor (sight).* LITTLE CAYMAN: *Proctor (sight).* CAYMAN BRAC: *Kings CB 52, CB 53.*]

Genus 4. **THRINAX** Sw.

Mostly rather slender, unarmed trees. Leaves long-petiolate, the chiefly orbicular blades palmately cleft with narrow lobes, these 2-parted at the apex and obliquely folded; hastula thick, concave, pointed, often tomentose within when young. Spadix interfoliar, paniculate, partly enclosed toward the base by numerous tubular tomentose spathes. Flowers perfect, subequally 6-parted; stamens mostly 6, the filaments connate at the base; ovary superior, the stigma flat or concave; ovule solitary with lateral micropyle. Fruit 1-seeded, globose, usually white at maturity; seed with smooth endosperm.

A chiefly West Indian genus of 8 or 10 species.

1. **Thrinax radiata** Lodd. ex J.A. & J.H. Schultes in L., Syst. Veg. 7(2):1301. 1830. **FIG. 67.**

Thrinax multiflora Mart., 1838.

T. excelsa sensu L.H. Bailey, 1938, not Griseb., 1864.

A small tree rarely more than 8 m tall (usually much shorter), the trunk 8-13 cm in diameter; leaf-blades 120–160 cm broad, the segments broadest at the point of fusion, beneath bearing scattered small fimbriate scales each having a conspicuous translucent central portion. Inflorescence and flowers creamy-white at anthesis, the axes glabrous. Fruits white and smooth at maturity, 7–8 mm in diameter.

GRAND CAYMAN: *Brunt 1905, 1943; Kings PGC 255, PGC 257, PGC 257a; Proctor 27966.* LITTLE CAYMAN: *Kings LC 47, LC 104.*
– Florida, Bahamas, Greater Antilles except Puerto Rico, Yucatan and Belize; frequent in rocky coastal thickets.

Genus 5. **COCCOTHRINAX** Sarg.

Slender unarmed trees. Leaves long-petiolate and palmately cleft, with segments bifid at the apex and obliquely folded; hastula free, thin, erect, concave, and usually pointed. Spadix interfoliar, paniculate, bearing numerous papery 2-cleft spathes. Flowers perfect, 6-parted; stamens 9–12, with subulate filaments barely united at the base. Ovary superior, the stigma funnel-shaped; ovule solitary with sub-basal micropyle. Fruit 1-seeded, globose, black or dark brown at maturity; seed with deeply grooved endosperm.

A chiefly West Indian genus of about 30 species.

1. **Coccothrinax proctorii** R.W. Read in Phytologia 46 (5):285. 1980. **FIG. 67.**

"Thatch", "Silver Thatch"

Slender tree, the trunk 2-5 (–10) m tall. Leaf-blades orbicular in outline, with 39–48 narrowly trullate segments, these mostly 61–80 cm long and 3.2–4.2 cm wide at the widest, tapering to a very slightly bifid apex; in unexpanded blades the segment-tips are connected by a threadlike strand; adaxial surface dark green; abaxial (lower) surface appearing silvery to golden, covered with a dense indumentum of persistent irregularly shaped and interlocked fimbriate scales, the stalk or central portion of each scale conspicuous as a dark-colored dot; hastula variable in outline and width (2–3.3 cm wide), its free adaxial extension 0.5–1 cm long, conspicuously ciliate; petiole (32–) 75–80 cm long, the abaxial surface usually densely covered with white scales that are soon lost. Inflorescence with 5–7 or more primary branches; flowers stalked, fragrant, white at anthesis, soon turning creamy; stamens about 10 in number, about equalling the pistil. Fruit purple-black at maturity, the mature fruiting pedicels mostly 1-4 mm long; seeds mostly 5-5.4 mm in diameter.

GRAND CAYMAN: *Brunt 1906; Proctor 27991 (type); Sachet 380; Sauer 4091 (WIS, cited by Read).* LITTLE CAYMAN: *Proctor 28033, 35082.* CAYMAN BRAC: *Brunt 1674; Proctor 29045.*
– Endemic, in rocky thickets and woodlands. The manufacture of rope from the leaves of the Thatch Palm was formerly an important cottage industry on Grand Cayman and Cayman Brac. This rope was especially prized by fishermen, both locally and in Jamaica, because of its durability in contact with sea-water. Synthetic substitutes have now largely replaced this commodity, and increased prosperity among the Cayman people has reduced the incentive to do this type of rather arduous and relatively low-paying work.

Family 11

ARACEAE

Terrestrial, epiphytic, or aquatic herbs with thick, fleshy, tuberous rhizomes, or corms, or sometimes rather slender creeping or climbing stems. Leaves alternate, petiolate, simple or compound, usually with reticulate venation. Inflorescence a fleshy cylindric or clavate spike (spadix) usually subtended by a conspicuous flat or hooded bract (spathe) which may be green, white, or otherwise colored. Spadix densely beset with minute usually unisexual flowers, the staminate on the apical portion and the pistillate toward the base, or the plants dioecious; a few genera with bisexual flower. Perianth, when present, minute, of 4-6 or more free or connate segments. Stamens 4-6, mostly without filaments or sometimes with broad flat filaments. Ovary 1- to several-celled; ovules 1 to several in each cell. Fruit a berry with one to many seeds; seeds usually with endosperm.

A chiefly tropical and warm-temperate family of more than 100 genera and 2,000 species. Many species find a useful place in horticulture, and a few are commonly used for food. *Caladium bicolor* (Aiton) Vent. is a commonly cultivated ornamental with spotted or variegated leaves; another is *Dieffenbachia maculata* (Lodd.) G. Don, the "Dumb Cane". *Colocasia esculenta* (L.) Schott, "Dasheen", and *Xanthosoma sagittifolium* (L.) Schott, "Coco", are widely cultivated in the West Indies for their edible starchy tubers. The latter is recorded as being grown on Grand Cayman *(Kings GC-84).*

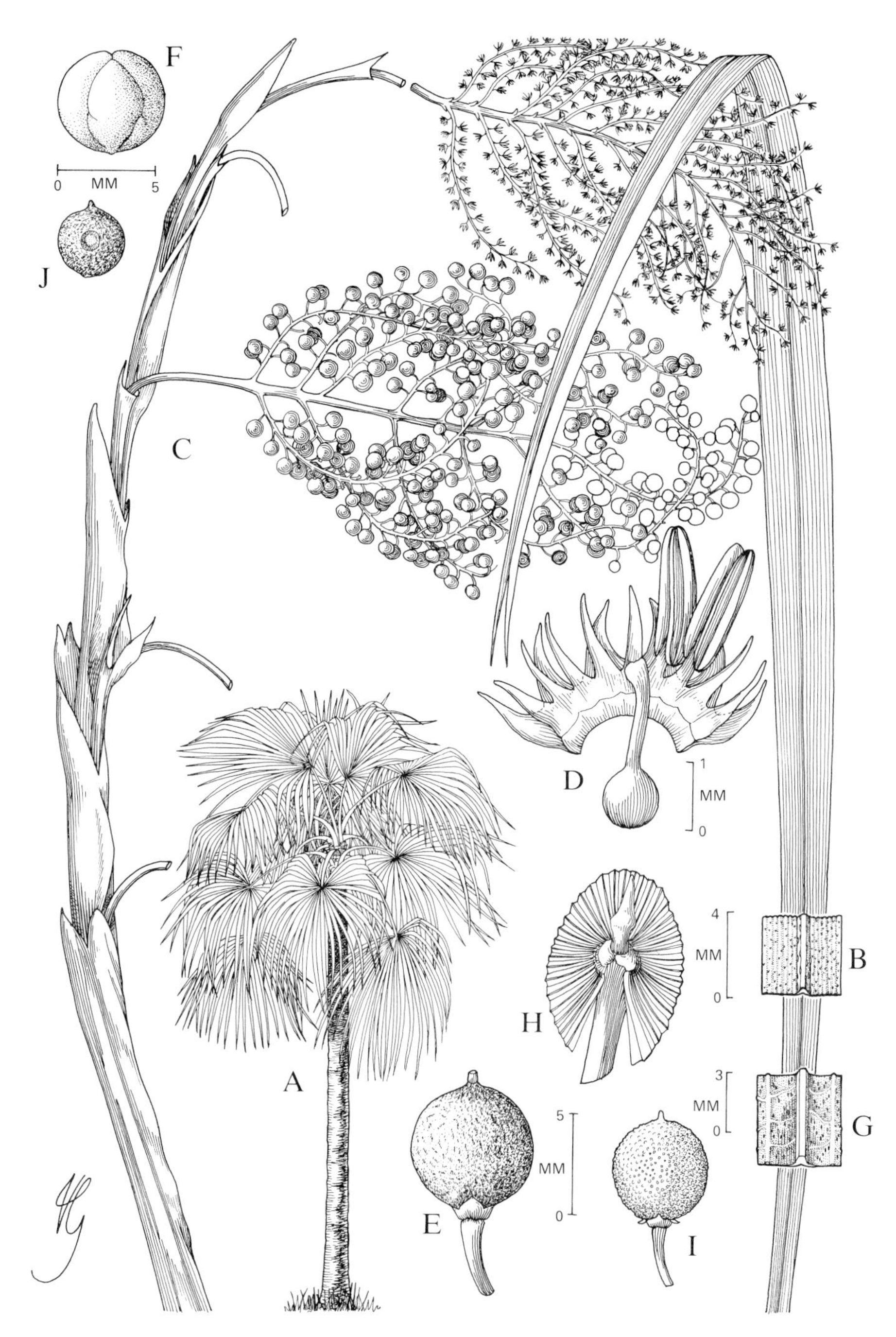

FIG. 67 **Coccothrinax proctorii.** A, habit (from photograph). B, section underside of leaf. C, portion of inflorescence with flowers and young fruits. D, dissected flower. E, fruit. F, seed showing grooved endosperm. **Thrinax radiata.** G. section, underside of leaf. H. attachment of leaf-blade to petiole, showing hastula. I, fruit. J, seed with smooth endosperm. (G.)

Genus 1. **PHILODENDRON** Schott

Mostly scandent or vine-like, often high-climbing and emitting long aerial roots; internodes mostly elongate. Leaves with terete or vaginate petioles; blades entire or variously lobed or parted, the lateral nerves all parallel. Peduncles usually short; spathe persistent, convoluted into a tube at base and closely surrounding the pistillate portion of the spadix, the apical part more or less hooded, whitish or green. Spadix erect, nearly equalling the spathe, sessile or short-stipitate, the basal pistillate portion cylindric and densely many-flowered, becoming fleshy in fruit, the apical staminate portion sterile toward base, fertile above, withering and recurved in fruit. Flowers unisexual, naked, the staminate with 2–6 stamens, these obpyramidal-prismatic, truncate at apex, the anthers oblong or linear; ovary of the pistillate flowers 2- to several-celled, ovules 1-several; stigma sessile. Fruits crowded, 1- to several-seeded.

A large genus of about 275 species, widely distributed in warm countries.

1. Philodendron aff. **scandens** C. Koch & H. Sello, Ind. Sem. Hort. Berol. 1853: App. 4. 1853.

A vine with slender, terete stems c. 5 mm in diameter. Leaves with slender terete petioles up to 16 cm long; blades triangular-hastate, up to 11 cm long from attachment of the petiole to the shortly acuminate or merely acute apex, and mostly 5–9 cm broad, narrowly to broadly cordate at base with rather open sinus, the rounded lobes somewhat divergent. Inflorescence unknown.

GRAND CAYMAN: *Brunt 2170; Kings GC 259; Proctor 27998.*
– Probably an endemic, undescribed species related to *P. scandens,* but its precise identity cannot be established as long as only sterile material is known. The Kings specimen was gathered at Forest Glen, the Brunt and Proctor material in dense woodlands about a mile due north of Breakers; plants from the latter locality were grown in Jamaica for several years without showing any sign of flowering.

Family 12

LEMNACEAE

Monoecious, very rarely dioecious, small to minute aquatic leafless annual herbs, floating on or submerged in fresh water. Plant body (frond) either solitary or connected in small groups, symmetric or asymmetric, flat or inflated, varying in shape from reniform, round, elliptic, lanceolate and linear to globose, usually green but red or brown pigments sometimes also present; roots several, one, or none. Vegetative propagation from budding pouches; when there is one budding pouch there is also a flowering cavity bearing a spatheless inflorescence (except

Wolffiopsis with 2 flowering cavities); when there are two budding pouches one of these may give rise to an inflorescence surrounded by a spathe. Inflorescence consisting of 1 pistillate and 1 or 2 staminate flowers; perianth none. Staminate flower consisting of a single stamen, the anther 1- or 2-locular; pistillate flower consisting of 1 sessile, globular, 1-celled ovary with a short apical style; ovules 1-4. Fruit a 1-4-seeded utricle ; seeds ribbed or smooth.

A small world-wide family of 6 genera and about 29 species, representing the world's smallest flowering plants. The present treatment is based on the synopsis presented by den Hartog & van der Plas in Blumea 18(2): 355-368. 1970.

Genus 1. **LEMNA** L.

Small water plants, floating on the surface or submerged (in which case they rise to the surface during flowering periods). Fronds solitary or in small connected groups of 2-10, symmetric or slightly asymmetric, round, elliptic, oblong, obovate or lanceolate, flat or slightly swollen, often with a median row of papillae on the dorsal side; tissues containing raphides; stomata present on dorsal side of floating plants only; margin entire or rarely denticulate; nerves 1-3 (rarely 5); root 1 per frond (rarely absent), terminated by a root cap of distinctive shape. Budding pouch opening at margin of frond, rarely slightly dorsal or ventral in position. Seeds longitudinally ribbed, rarely smooth.

A cosmopolitan genus of about 9 species.

1. Lemna aequinoctialis Welw. in Anaes Conselho Ultramar. 55:543. 1858.

Lemna perpusilla sensu Adams, 1972, not Torrey, 1843.

"Duck Weed"

Fronds solitary or 2-5 connected, obovate to elliptic, slightly asymmetric, 1.5-3.3 mm long, 0.8-2.5 mm wide, the upper surface minutely keeled with a median line of small papillae, with a more prominent papilla opposite the attachment of the root and often another at the apex; lower surface flat to slightly convex; root sheath with lateral wings; nerves usually 3 but often indistinct. Ovary with 1 ovule. Seed oblong-ovoid, prominently ribbed.

GRAND CAYMAN: *Kings GC 215.*
– Of cosmopolitan distribution, occurring widely in both tropical and temperate regions. The single Cayman collection was gathered at a "small cow well 1 mile S.E. of Georgetown" on July 6, 1938.

Family 13

[PONTEDERIACEAE

Perennial aquatic herbs, rooting in mud or sometimes floating. Leaves chiefly basal, often with more or less succulent petioles, the blades with finely parallel venation. Flowers bisexual, irregular (zygomorphic), solitary or in spikes, racemes or panicles subtended by leaf-like bracts (spathes). Perianth corolla-like, 6-parted, the lobes usually joined toward the base. Stamens 3 or 6, the filaments adnate to the perianth-tube; anthers 2-celled, sometimes dimorphic. Ovary superior, 1-3-celled; ovules 1-several; style 1; stigma 3-6-lobed. Fruit a capsule with several seeds, or achene-like with 1 seed; seeds with mealy endosperm.

A small family of 8 genera and about 25 species, chiefly of tropical and warm-temperate regions.

Genus 1. **EICHHORNIA** Kunth

Plants floating or rooted in mud. Leaves with slender, stout, or inflated petioles and flat blades, or sometimes completely narrow and grasslike when submerged. Flowers solitary to numerous; perianth conspicuous, somewhat 2-lipped, the 6 parts in 2 series, all united into a tube below. Stamens 6, 3 included and 3 exserted. Ovary 3-celled; ovules numerous. Fruit a many-seeded capsule.

About 5 species, natives of tropical America.

1. **Eichhornia crassipes** (Mart.) Solms in A. & C. DC., Monog. Pha. 4:527. 1883. **FIG. 68.**

Piaropus crassipes (Mart.) Britton, 1923.

"Water Hyacinth"

Plants floating or the rhizomes rooting in mud. Leaves 5-40 cm long, in a basal tuft, glabrous, the petioles usually inflated, the blades ovate to orbicular or reniform, faintly many-nerved, 4-12 cm broad. Inflorescence a contracted panicle 4-15 cm long, subtended by a pair of unequal spathelike bracts. Perianth 5-7 cm broad when fully expanded, lilac or rarely white, the upper lobe bearing a central violet blotch with yellow center. Stigma 3-lobed, glandular-hairy.

GRAND CAYMAN: *Brunt 1840; Kings GC 230.*
– Originally from Brazil, but introduced in most warm countries and widely naturalized, often becoming a pest of waterways and fresh-water lakes.]

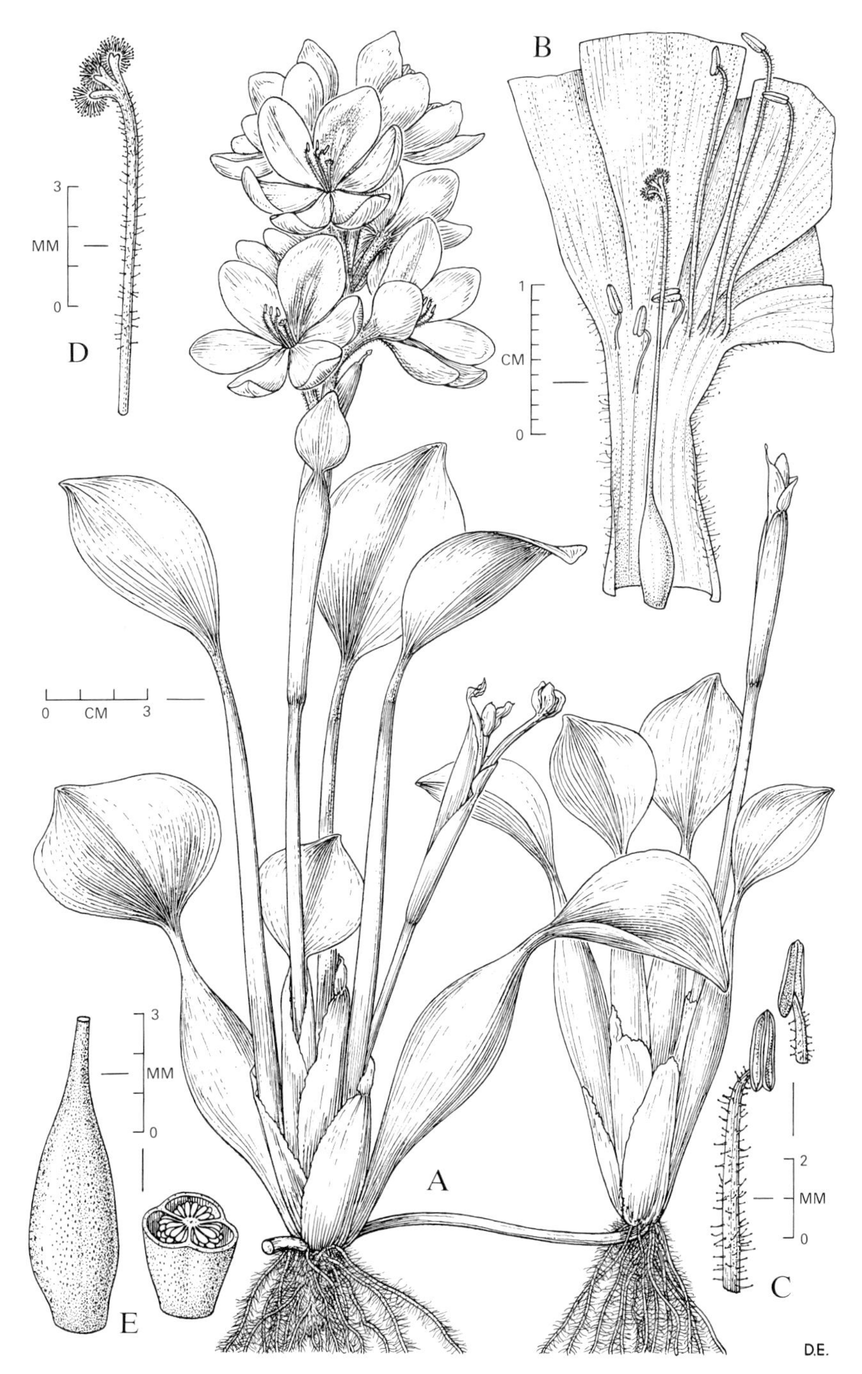

FIG. 68 **Eichhornia crassipes.** A, habit. B, dissected flower. C, stamen with two views of an anther. D, stigma and upper portion of style. E, ovary, whole and in x-section. (D.E.)

Family 14

LILIACEAE

Herbs, vines, or somewhat woody plants, usually perennial, of widely diverse habit and structure. Flowers bisexual and either solitary, umbellate, or in spikes, racemes or panicles; perianth 6-parted or 6-lobed, regular or slightly zygomorphic, petaloid; stamens usually 6, the filaments free or adnate to the perianth or to each other; ovary superior or inferior, 3-celled; styles 3, more or less united into 1 (rarely lacking); stigmas 3, separate or more or less united; ovules 1 or few to many. Fruit a 3-valved capsule or an indehiscent berry; seeds with fleshy endosperm.

As here defined, a very large family of more than 300 genera and 4,900 species, represented all over the world. The diversity of structure had led to segregation of several groups as independent families, but the overlapping of basic characters makes arbitrary division very difficult. In the present work the family is treated in a broad sense, subdivided into 3 subfamilies.

KEY TO SUBFAMILIES

1. Ovary superior; plants stemless or nearly so, with tufted leaves, spreading by subterranean stolons; flowers in racemes or narrow panicles: *SUBFAM. 1. LILIOIDEAE*

1. Ovary inferior, or if superior the plant with erect subwoody stem; plants not or somewhat stoloniferous; flowers solitary, umbellate, or in very large panicles:

  2. Bulbous herbs; flowers solitary or umbellate, subtended by a spathe; leaves entire, not spine-tipped: *SUBFAM. 2. AMARYLLIDOIDEAE*

  2. Coarse herbs of hard or fleshy texture, not bulbous, sometimes developing a trunk; flowers amply paniculate; leaves with prickly margins or at least with spiny tips: *SUBFAM. 3. AGAVOIDEAE*

[*SUBFAMILY 1. LILIOIDEAE*

KEY TO GENERA

1. Leaves green with white mottling, leathery, with entire margins; flowers greenish-white: **1. Sansevieria**

1. Leaves plain green without mottling, succulent, with toothed margins; flowers yellow: **2. Aloe**

Genus 1. **SANSEVIERIA** Thunb.

Perennial herbs forming large colonies by means of creeping stolons. Leaves entire, leathery, erect, clustered, flat or terete. Flowers greenish-white, the inflorescence a raceme or panicle on an unbranched nearly naked scape. Perianth with slender tube and narrow spreading lobes; stamens 6, the filaments inserted at base of perianth-lobes; anthers oblong or linear. Ovary 3-celled, each cell with a single, erect ovule; style 1, long and filiform; stigma capitate. Fruit a thin-walled capsule; seeds 1-3, fleshy.

A genus of more than 50 species, indigenous to Africa and Asia. The leaves of many species yield valuable fibres; some are also widely cultivated for the ornamental value of their curious, sword-like leaves.

1. Sansevieria hyacinthoides (L.) Druce in Rep. Bot. Exch. Cl. Brit. Isles 1913 (3): 423. 1914.

"Bowstring Hemp", "Lion's Tongue"

Leaves lanceolate, nearly flat, mostly 30–100 cm long, 5–9 cm broad at the middle, narrowed at both ends, dark green mottled with white, the margin bordered with a fine red line. Inflorescence about as long as the leaves or slightly longer; flowers in clusters of 2 or 3 along the main axis; pedicels mostly less than 5 mm long; perianth greenish-white, the lobes about 1.5 cm long.

GRAND CAYMAN: *Brunt 1942, 1956; Kings GC 82; Proctor 15206.* LITTLE CAYMAN: *Proctor (sight).* CAYMAN BRAC: *Kings CB 90; Proctor 29086.*
– The original habitat of this species was somewhere in Africa. It is rather widely naturalized in the West Indies, and can readily be distinguished from the similar *S. trifasciata* Prain (not yet recorded from the Cayman Islands, but may be present) by the red-lined margins of the leaves. The two species have often been confused under the name *S. guineensis.* The somewhat fleshy leaves of these and other related species contain numerous fine, white, silky fibers that can be used for many purposes; their strength is said to be about the same as those in Sisal *(Agave sisalana).*

Genus 2. **ALOE** L.

Succulent plants with fleshy, tufted leaves and bitter sap; leaves with toothed, serrate, or roughened margins. Inflorescence an elongate raceme with numerous nodding flowers. Perianth subcylindric, the lobes more or less coherent with tips slightly spreading; stamens 6, with slender free filaments and oblong anthers; ovary sessile, 3-celled, 3-angled, the single style tipped by a very small stigma; ovules many in each cell of the ovary. Capsule leathery; seeds numerous, black.

A genus of more than 300 species, chiefly occurring in Africa and Madagascar.

1. **Aloe vera** (L.) Burm.f., Fl. Ind. 83. 1768.

Aloe barbadensis Mill., 1768.

A. vulgaris Lam., 1783.

"Bitter Aloes", "Sempervirens"

Stemless or with a very short upright stem, spreading by creeping stolons. Leaves narrowly deltate-lanceolate, 30–60 cm long, acuminate, turgid and watery within, pale glaucuous-green; marginal teeth usually less than 2 cm apart. Scape 60–120 cm tall, bearing distant, broad, acute scales; raceme dense, 10–30 cm long; bracts longer than the short pedicels. Flowers yellow, about 2.5 cm long; stamens about as long as the perianth, the style longer.

GRAND CAYMAN: *Kings GC 122.*
– Native of the Mediterranean region; widely naturalized in Florida, the West Indies, and Central America. In addition to various medicinal uses, the plant can produce a valuable natural dye, and also fibre.]

SUBFAMILY 2. AMARYLLIDOIDEAE

KEY TO GENERA

1. Flowers solitary: **3. Zephyranthes**

1. Flowers usually 2 or more in an umbel:

2. Filaments free to the base:

3. Flowers red; perianth-tube with scales or a corona at the throat; seeds black: **4. Hippeastrum**

3. Flowers white (or white with rose stripes); perianth-tube lacking scales or a corona at the throat; seeds green, fleshy: **[5. Crinum]**

2. Filaments joined below by a cup-like membrane; flowers pure white: **6. Hymenocallis**

Genus 3. **ZEPHYRANTHES** Herbert

Bulbous herbs with narrow, rather grass-like, glabrous leaves. Flowers solitary at the apex of a leafless scape, the pedicel subtended by a papery bract (spathe). Perianth funnel-shaped, erect or inclined, of various colours, with 6 subequal lobes. Stamens 6, the filaments adnate to the throat of the perianth-tube. Ovary 3-celled, with numerous ovules in 2 rows in each cell; style filiform, with a 3-lobed or nearly capitate stigma. Capsule 3-lobed, 3-celled, 3-valved; seeds blackish, usually flattened.

A genus of about 35 species occurring widely in the warmer parts of America. Many of them are popular in cultivation, and as they tend to escape and become naturalized in fields and along roadsides, it is often hard to determine if a given species is truly indigenous to a particular area.

KEY TO SPECIES

1. Perianth yellow: **1. Z. citrina**

1. Perianth pink: **2. Z. rosea**

1. Perianth white: **3. Z. tubispatha**

1. Zephyranthes citrina Baker in Curt., Bot. Mag. 108: t. 6605. 1882.

Zephyranthes eggersiana Urban, 1907.

"Yellow crocus"

Bulb ovoid-globose, about 2-3 cm in diameter. Leaves mostly 15-30 cm long, 1.5-2.5 mm broad, narrowly linear. Scape 15-25 cm tall; spathe 2-2.5 cm long, about half as long as the pedicel or more. Perianth yellow, 3-4 cm long, the lobes elliptic; stamens shorter than the perianth; style 2 cm long, with a slightly trifid stigma.

GRAND CAYMAN: *Brunt 2123; Kings GC 414b; Proctor 27933.*
– West Indies.

2. Zephyranthes rosea Lindl., Bot. Reg. t. 821. 1824. **FIG. 69.**

Bulb subglobose, 2-2.5 cm in diameter. Leaves linear, 10-25 cm long, 2-5.5 mm broad. Scape 10-25 cm tall, usually longer than the leaves; pedicel longer than the 2 cm spathe. Perianth bright pink, 3-4 cm long; stamens much shorter than the perianth; style nearly equalling the perianth, with a deeply trifid stigma.

GRAND CAYMAN: *Brunt 2124; Kings GC 414; Proctor 27962.*
– Cuba; naturalized in Bermuda, the Bahamas, and Jamaica.

3. Zephyranthes tubispatha (L'Herit.) Herbert in Edw., Bot. Reg. 7, App. 36. 1821.

Bulb subglobose, 1.5-2.5 cm in diameter. Leaves narrowly linear, 16-30 cm long, 2-7 mm broad. Scape 15-25 cm tall (rarely less), shorter than or equalling the leaves; spathe 2-3.5 cm long, shorter than the pedicel. Perianth white, tinged greenish toward base, 3.5-5 cm long; stamens shorter than the perianth; style with deeply trifid stigma, overtopping the anthers.

GRAND CAYMAN: *Brunt 2125; Proctor 27961.*
– West Indies and northern South America; often cultivated.

Genus 4. **HIPPEASTRUM** Herbert

Bulbous plants with more less strap-shaped, flat, entire leaves. Flowers large, showy, usually 2 or more in an umbellate cluster at the apex of a hollow, leafless scape. Perianth funnel-shaped, more or less nodding, the lobes nearly equal, the throat with scales or a corona; stamens inserted in the throat, the filaments separate; anthers linear or linear-oblong; ovary inferior, 3-celled; ovules many; style long, with a capitate or trifid stigma. Capsule globose, 3-valved; seeds flattened, with a thick black testa.

A genus of about 75 species, widely distributed in tropical and subtropical America.

1. Hippeastrum puniceum (Lam.) Kuntze, Rev. Gen. Pl. 2:703. 1891 (err. *purpureum*).

Hippeastrum equestre Herbert, 1821.

"Red lily"

Bulb globose or ovoid-globose, stoloniferous, 4-5 cm long, the outer papery coats brown. Leaves mostly 25-50 cm long, 2.5-5 cm broad, gradually narrowed to the blunt apex, absent at flowering time. Scape terete, glaucous, 25-60 cm tall; umbel 2-4-flowered, subtended by papery, lanceolate bracts; pedicels 3.5-7 cm long. Perianth about 9 cm long or sometimes less, about equal in width when expanded, the segments bright red or rose-red with green at base; stamens shorter than the perianth-segments.

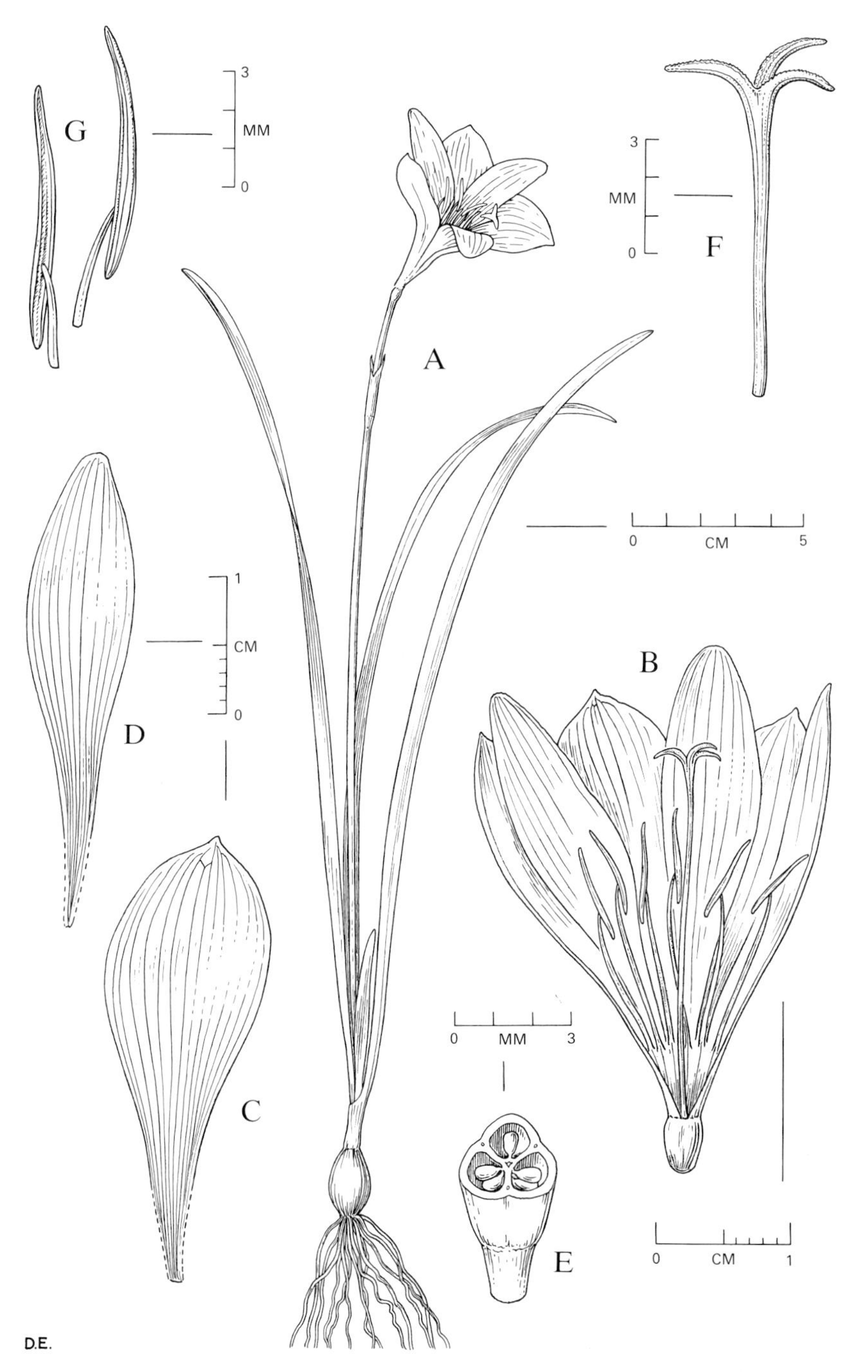

FIG. 69 **Zephyranthes rosea.** A, habit. B, dissected flower. C, D, outer and inner perianth-segments. E, x-section of ovary. F, stigmas and upper portion of style. G, anthers. (D.E. & G.)

GRAND CAYMAN: *Brunt 2061.*
– West Indies, Central and South America, often cultivated and persisting after cultivation.

[Genus 5. **CRINUM** L.

Plants arising from bulbs narrowed at the apex into a short or long neck; leaves narrowly or broadly strap-shaped with entire, toothed, or wavy margins. Flowers rather large, white or striped pink, in an umbellate cluster at the apex of a solid leafless scape; pedicels short or none. Perianth with subequal lobes broadly spreading or connivent-funnel-shaped, from a long, slender tube; stamens inserted in the throat of the perianth-tube, the filaments long and filiform, the anthers linear; ovary 3-celled with few (sometimes only 2) ovules in each cell; style long and filiform with a minute capitate stigma. Capsule irregular in shape, tardily dehiscent; seeds large, green and bulb-like, with very thick endosperm.

An imperfectly known genus of about 100 species, at least a few occurring in nearly all warm countries. Three species are known to occur in the Cayman Islands; all have been introduced for their fragrant, ornamental flowers, and all tend to persist or become naturalized. It is not certain that any of them are here correctly named, as the preserved material is scanty, and a few discrepancies with published descriptions have not been resolved. However, until the genus as a whole is better understood, greater precision in identification probably is not possible. It should be noted that many Crinums hybridize readily, and it is likely that many plants labelled with a particular name may in fact represent a mixture of two or more natural species. The descriptions given here are based primarily on Cayman specimens, reinforced by observations on what appear to be the same species in Jamaica.

KEY TO SPECIES

1. Bulb with a long, stalk-like massive neck rising 30 cm or more above ground; leaves 9–15 cm broad; flowers numerous, 18 or more per umbel: **1. C. amabile**

1. Bulb with neck, if any, subterranean; leaves less than 8 cm broad; flowers few, 8 or less per umbel:

  2. Perianth-lobes white or pale pink with pink median stripe, connivent and c. 2 cm broad; leaf-margins entire but undulate: **2. C. zeylanicum**

  2. Perianth-lobes pure white, divergent and c. 1 cm broad; leaf-margins roughened or minutely toothed: **3. C. americanum**

1. **Crinum amabile** Donn, Hort. Cantab. ed. 6, 83. 1811.

"Giant Lily"

Bulb with a rather massive, stem-like neck rising 30 cm or more above ground. Leaves numerous, spreading, narrowly lanceolate, entire, 70–100 cm long, 9–15 cm broad, clasping at the slightly narrowed base and gradually tapering to the apex. Scape 40–80 cm tall, 2-keeled. Spathe-valves deltate, reddish, 12–18 cm long. Umbel of 18–30 pedicellate flowers; perianth with slender cylindric deep rose or red tube 7–12 cm long; lobes c. 2 cm broad, equalling or exceeding the tube, pink outside, and whitish with deep pink median stripe within. Fruit not seen.

GRAND CAYMAN: *Proctor 15208,* collected in sandy soil along the seacoast near Gun Bay.
– Native of Sumatra; widely cultivated, and perhaps of hybrid origin.

2. **Crinum zeylanicum** (L.) L., Syst. Nat. ed. 12, 1:236. 1767.

Bulb said to be subglobose, 10–15 cm in diameter, the neck short (but similar plants naturalized in Jamaica have a slender neck up to 10 cm long). Leaves 6–10, strap-shaped, 30–60 cm long, 3–5 cm broad, the margins entire but more or less undulate. Scape 30–50 cm tall, not keeled, often tinged red. Spathe-valves deltate, thinly papery, 6–8 cm long. Umbel of 4–8 subsessile or short-pedicellate flowers; perianth with slender curved greenish or reddish tube 7–14 cm long; lobes c. 2 cm broad or more, oblong-lanceolate with partly connivent margins, white with a pink to deep rose median stripe. Fruit not seen.

GRAND CAYMAN: *Brunt 2062,* growing by roadside in shallow soil over limestone pavement near Red Bay; *Sachet 445.* LITTLE CAYMAN: *Proctor (sight).*
– Native of tropical Africa and Asia.

3. **Crinum americanum** L., Sp. Pl. 1:292. 1753.

"Seven Sisters"

Bulb soft-succulent, cylindric, 7–11 cm thick, subterranean and stoloniferous. Leaves several, strap-shaped, 50–90 cm long, 4–7 cm broad (rarely more), the margins minutely roughened or irregularly denticulate. Scape 30–40 cm tall, not keeled. Spathe-valves narrowly to broadly deltate, papery, 5–7 cm long. Umbel of 3–8 sessile fragrant flowers; perianth pure white with slender cylindric tube 11–15 cm long, and wide-spreading narrow lobes c. 1 cm broad; filaments rose. Fruit not seen, said to be irregularly subglobose and 4–6 cm thick.

GRAND CAYMAN: *Proctor 15307*, from sandy yards around Georgetown. – Southeastern United States and Jamaica; cultivated in Cuba. Wild plants of this species customarily grow in swamps.]

Genus 6. **HYMENOCALLIS** Salisb.

Bulbous plants with usually linear or strap-shaped leaves, sometimes contracted below into a petiole. Flowers pure white (or greenish in some continental species), few or many in an umbel at the apex of a solid, leafless scape, commonly fragrant chiefly at night; pedicels lacking or short. Perianth with an elongate narrow tube and equal, spreading or recurved narrow lobes; stamens inserted in the throat of the perianth-tube, united in their lower part to form a rather conspicuous membranous cup, usually with an entire or bifid process between each 2 filaments; anthers linear. Ovary 3-celled, with 1 or 2 ovules in each cell; style long and filiform with a small capitate stigma. Capsule fleshy and somewhat irregular, splitting open as the seeds ripen; seeds large, fleshy, green.

A tropical American genus of about 30 species, doubtfully separable from the Old World genus *Pancratium*. "Spider lilies"

1. **Hymenocallis latifolia** (Mill.) M.J. Roem., Syn. Monogr. 4:168. 1847. **FIG. 70.**

Pancratium latifolium Mill., 1768.

Hymenocallis caymanensis Herbert, 1837.

"Lily", "Wild White lily", "Easter Lily"

Bulb ovoid-conic, up to 15 cm long or more and 6-8 cm thick, clothed with brownish papery-membranous epidermis. Leaves strap-shaped, 20-80 cm long, 3.5-6 cm broad, acute at the apex. Peduncle up to 60 cm long; flowers 6-15 in a cluster, greenish in bud, white at anthesis. Perianth tube 10-15 cm long, the linear-lobes 9-12 cm long, somewhat recurved; staminal cup mostly 2-3 cm long; anthers 1-1.3 cm long. Ovary sessile. Capsule 2-4 cm in diameter.

GRAND CAYMAN: *Brunt 1713, 2181; Kings GC 111.* LITTLE CAYMAN: *Kings LC 105; Proctor 28079.* CAYMAN BRAC: *Kings CB 89; Proctor 28023.* – Florida, Cuba, Jamaica and Hispaniola, chiefly in sandy clearings near the sea.

SUBFAMILY 3. AGAVOIDEAE

KEY TO GENERA

1. Flowers white; ovary superior; leaf-margins minutely roughened but not spiny: **7. Yucca**

1. Flowers yellow or greenish; ovary inferior; leaf-margins armed with sharp, recurved spines: **8. Agave**

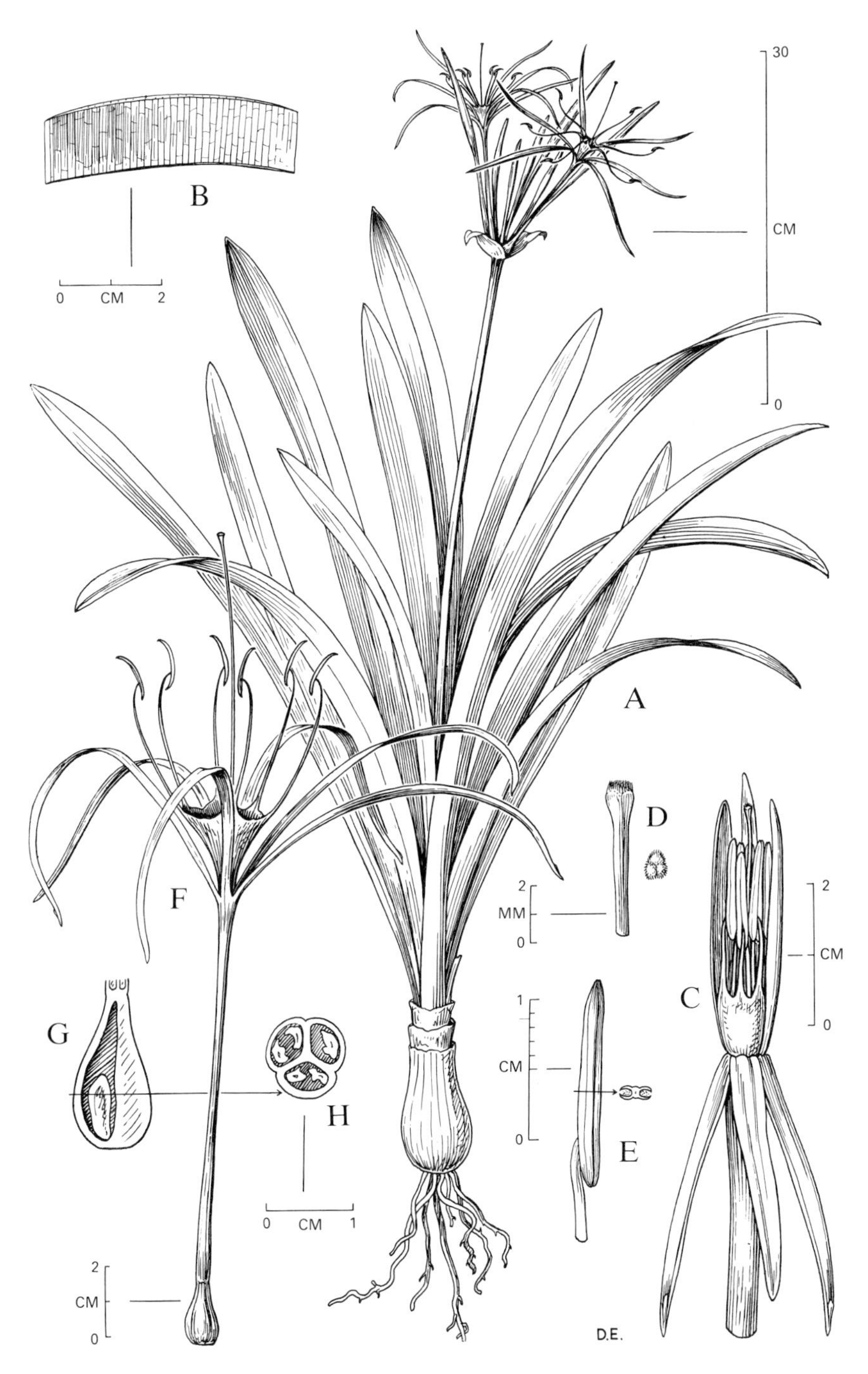

FIG. 70 **Hymenocallis latifolia.** A, habit. B, section across leaf. C, mature bud with three perianth-segments pulled down. D, apex of style and stigma. E, anther. F, flower. G, H, sections through ovary. (D.E.)

Genus 7. **YUCCA** L.

Large coarse plants, the majority with a woody trunk, often tall and treelike, simple or branched. Leaves crowded toward the apex of the trunk or its branches, linear-lanceolate, thick and rigid, usually spine-tipped, the margins entire or fibrous. Flowers white or creamy, in large terminal panicles; perianth long-persistent after withering, the segments separate or nearly so but more or less connivent, rather fleshy. Stamens 6, hypogynous, free, much shorter than the perianth, with thick filaments and small sessile anthers. Ovary sessile or rarely stipitate, 3-celled, the cells incompletely partitioned in two; ovules numerous; style short, stout, divided at apex into 3 or 6 stigmatic lobes. Fruit either indehiscent and pulpy or spongy within, or dry and splitting open by 6 valves; seeds black, flat.

A genus of more than 25 species, most of these occurring in the southwestern United States and Mexico, a single one being found in the West Indies.

1. **Yucca aloifolia** L., Sp. Pl. 1:319. 1753.

A shrub or small tree up to 4 m tall, unbranched or with a few short branches, often growing in clumps or colonies. Leaves numerous, dark green, rigid, mostly 30-60 cm long 2-3.5 cm broad, minutely roughened on the margins and terminating in a sharp brown spine. Panicle compact, erect, up to 60 cm long; flowers numerous, nodding, creamy-white (rarely tinged purple), 3-5 cm long; ovary short-stipitate. Fruit an indehiscent pulpy capsule 7-9 cm long, the pulp dark purple; seeds roundish-flattened, up to 7 mm in greatest diameter.

GRAND CAYMAN: *Brunt 1998.* LITTLE CAYMAN: *Kings LC 62.*
– Southeastern United States and eastern Mexico, Bermuda, Bahamas, and the Greater Antilles. Often cultivated, so that its true natural distribution is hard to determine; the Cayman records may represent escapes from cultivation. The flowers are edible, either raw (in salads) or cooked as a vegetable. Those of *Y. elephantipes* in Guatemala are often dipped in egg batter and fried.

Another species of *Yucca,* not *aloifolia* but not yet identified, has been planted in Georgetown, Great Cayman.

Genus 8. **AGAVE** L.

Large, slow-growing, fleshy herbs with massive leaves forming a basal rosette; rarely developing a short trunk. Leaves stiff, persistent, armed with a sharp spine at the apex and usually prickly along the margins. Flowers paniculate on tall scapes (or spicate in a few continental species). Perianth 6-parted, more or less funnel-shaped, rather fleshy. Stamens 6, exserted. Ovary 3-celled, with numerous ovules forming 2 rows in each cell; style 1; stigma capitately 3-lobed. Capsule oblong to globose, many-seeded; seeds flat, thin, black. Reproduction often by means of small vegetative "bulbils" produced in large numbers on the inflorescence after flowering. The parent plant always dies after flowering, whether or not seeds or bulbils are formed.

A genus of perhaps 300 species distributed from the southern United States to tropical South America, especially numerous in Mexico. About 50 indigenous species are recorded from the West Indies, but this number may be excessive, as many are distinguished on the basis of very small differences. These plants are characteristic of rocky or arid habitats, and are often abundant. Several Mexican species, notably *A. sisalana* (Sisal) and *A. fourcroydes* (Henequen) are economically important for their useful fibres; these and others yield a copious flow of sap from their cut young inflorescences, which can be fermented to form "pulque", a Mexican national drink. Distillation of pulque yields "mescal" and "tequila", which are highly intoxicating. In some countries, *Agave* species are often planted to form living fences, and several, especially variegated forms of *A. americana,* find a wide use in horticulture.

KEY TO SPECIES

1. Leaves curved at base; flowers yellow; indigenous species: **1. A. sobolifera**

1. Leaves straight; flowers greenish or greenish-yellow, the filaments and style maroon-dotted; cultivated or persistent after cultivation: **[2. A. sisalana]**

1. Agave sobolifera Salm-Dyck, Hort. Dyck. 8:307. 1834. **FIG. 71.**

"Corato"

Leaves massive, dark green, narrowly lanceolate, up to 1.5 m long, acuminate at the spine-tipped apex; marginal prickles 5-15 mm apart, glossy dark brown, curved or reflexed-triangular (rarely straight), 1-4 mm long, often growing from the tops of green prominences of the margin, the intervening margin more or less concave. Inflorescence up to 6 m tall or more, paniculate; flowers yellow, on pedicels 5-10 mm long; ovary 15-25 mm long, narrowly spindle-shaped, longer than the perianth. Perianth-segments erect, 12-19 mm long; filaments 30-35 mm long. Capsules of Cayman plants not seen; in Jamaican material they are narrowly oblong, 45-50 mm long, 13-20 mm thick, turbinately narrowed at base and shortly beaked at the apex; seeds 4-5 mm × 7 mm.

GRAND CAYMAN: *Brunt 1765; Proctor 15160.* LITTLE CAYMAN: *Proctor (sight).* CAYMAN BRAC: *Kings CB 79, CB 88.*
– Jamaica. Like most of its relatives, this species grows in dry, rocky, exposed situations.

There is some evidence that another indigenous species occurs in the Cayman Islands, but this entity is only known in an immature condition and cannot be identified. Also, a juvenile individual of the Little Cayman population identified as *A. sobolifera* was transplanted to a garden in Kingston, Jamaica, in 1975; already, this plant shows vegetative differences from Jamaican *A. sobolifera* in that the leaves are much broader in proportion to their length. There is no doubt that *Agave* in the Cayman Islands requires further study.

[2. **Agave sisalana** Perrine in House Rep. Doc. 564:8. 1838.

"Sisal"

Leaves at first somewhat glaucous, eventually green and rather glossy, linear-lanceolate, nearly flat, mostly up to c. 1.5 m long and 10 cm broad near the base; apical spine dark brown, straight or slightly recurved, 2-2.5 cm long; marginal prickles often absent, when present slender and widely spaced. Inflorescence up to 6 m tall, the upper half loosely oblong-paniculate; flowers greenish or greenish-yellow with maroon speckles on the filaments and style; pedicels 5-10 mm long; ovary 20-25 mm long, shorter than the perianth. Perianth-segments erect, 15-20 mm long. Capsules rarely produced, oblong, c. 60 mm long; seeds 7 × 10 mm.

GRAND CAYMAN: *Proctor 15207.* CAYMAN BRAC: *Proctor (sight).*
– Originally from Yucatan; often cultivated in the hot drier parts of tropical countries, and tending to persist after cultivation.

Another *Agave,* not identified, occurs as an escape from cultivation on Little Cayman, but no material has been available for study.]

Family 15

SMILACACEAE

Shrubs or commonly vines, the latter climbing with the aid of petiolar tendrils. Leaves usually alternate and of leathery or hard texture, with 3 or more longitudinal nerves and reticulate venation between the nerves. Flowers usually dioecious, small, in axillary umbels, racemes or spikes. Perianth regular, 6-parted, the lobes all similar. Stamens usually 6, with confluent anther-cells. Ovary superior, 3-celled; ovules 1 or 2 in each cell. Fruit a berry; seeds with hard endosperm and small embryo.

A widely-distributed family in tropical and temperate regions, with 4 genera and about 375 species.

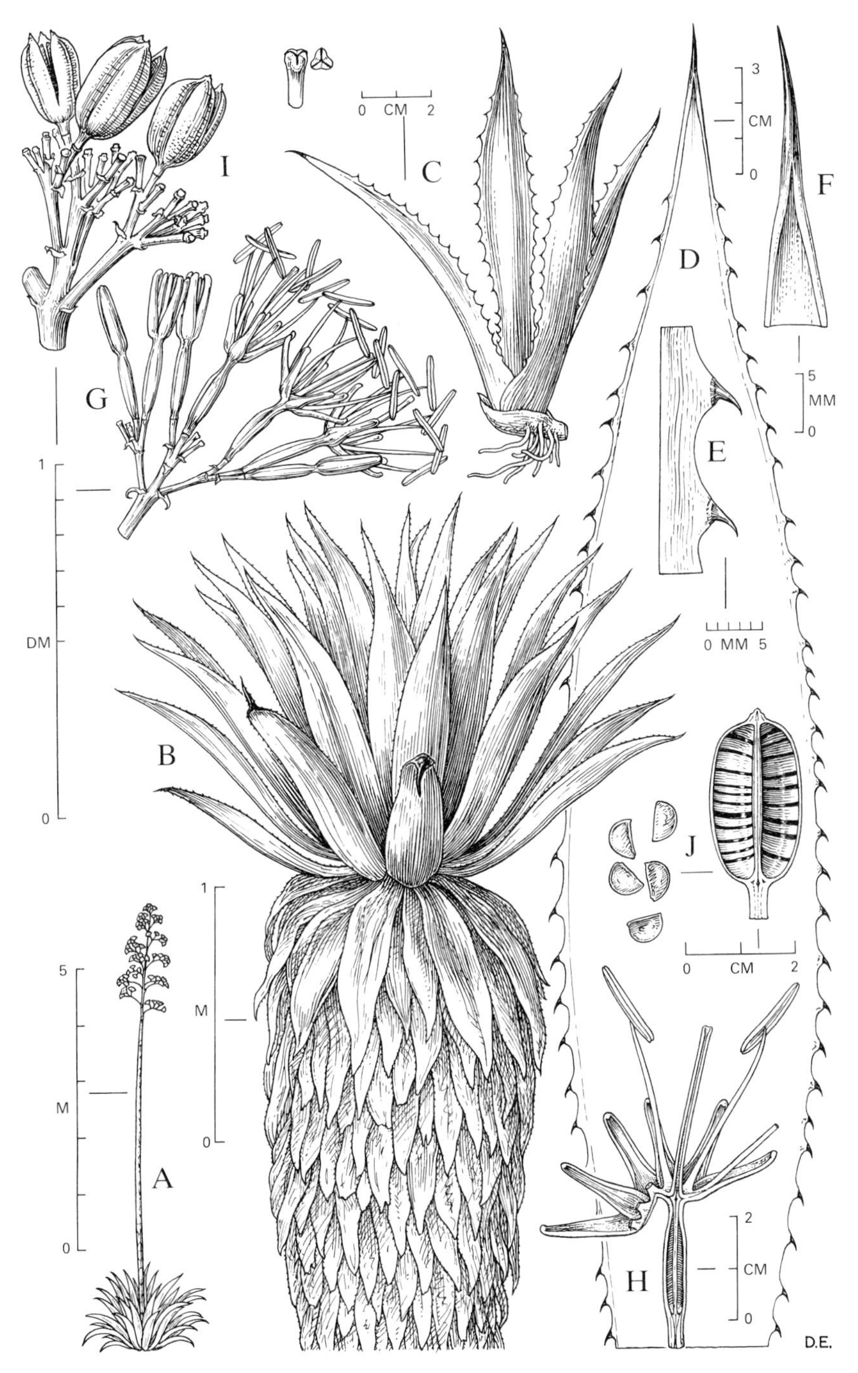

FIG. 71 **Agave sobolifera.** A, general habit. B, single very old plant. C, young plant growing from bulbil. D, outline of leaf. E, marginal spines. F, apical spine. G, portion of inflorescence, H, dissected flower. I, portion of fruiting inflorescence. J, long.-section of capsule and seeds. (D.E.)

Genus 1. **SMILAX** L.

More or less woody vines, often armed with prickles on stems and leaves, growing from woody or fleshy tubers or long creeping rhizomes. Leaves petiolate, with entire, lobed, or prickly margins; petioles sheathing at base, and bearing a pair of coiling tendrils. Flowers umbellate on a globose or convex receptacle, terminating an axillary peduncle; dioecious, usually small and greenish; perianth-lobes separate to the base. Pistillate flowers usually smaller than the staminate. Fruit a red, blue, or black berry.

A widespread genus of about 350 species. The drug known as Sarsaparilla, formerly believed to have medicinal value, is obtained from the roots of various tropical species of Smilax. The roots of several species found in southeastern United States yield a refreshing jelly-like condiment.

1. **Smilax havanensis** Jacq., Enum. Syst. Pl. Carib. 33. 1760. **FIG. 72.**

"Wire wiss"

Trailing or climbing slender woody vine, armed with short hooked prickles, or sometimes nearly unarmed, up to 8 m long, the branches striate or angled, often zigzag. Leaves coriaceous, narrowly lanceolate or lance-elliptic to broadly ovate or suborbicular, 2–10 cm long, 0.6–4.5 cm broad or more, 3–7-nerved, the margins usually spinulose or rarely entire, the base rounded or subcordate. Peduncles about equalling the petioles; inflorescence 4-30-flowered; pedicels c. 4 mm long; flowers green, 2-3 mm broad. Fruit black, 4–6 mm in diameter.

GRAND CAYMAN: *Brunt 1788, 2084; Correll & Correll 51040; Kings GC 187; Proctor 15022, 15281; Sachet 379.* Frequent in thickets and woodlands.
– Florida, Bahamas, Cuba, and Hispaniola.

Family 16

ORCHIDACEAE

Perennial herbs of diverse habit, terrestrial or epiphytic, with tuberous, fleshy or otherwise specialized roots. Leaves simple, entire, often leathery or fleshy, sometimes reduced to scales; stipules absent. Flowers perfect, irregular, bracted, solitary or variously arranged in spikes, racemes, or panicles. Perianth of 6 segments, the outer 3 (sepals) similar or nearly so, 2 of the inner ones (petals) lateral, alike; the third inner one (lip) dissimilar to the other 2, usually larger, often spurred or otherwise modified, in position superior, or inferior by twisting of the ovary or pedicel. Stamens much reduced, usually 1, more or less united with the style on an

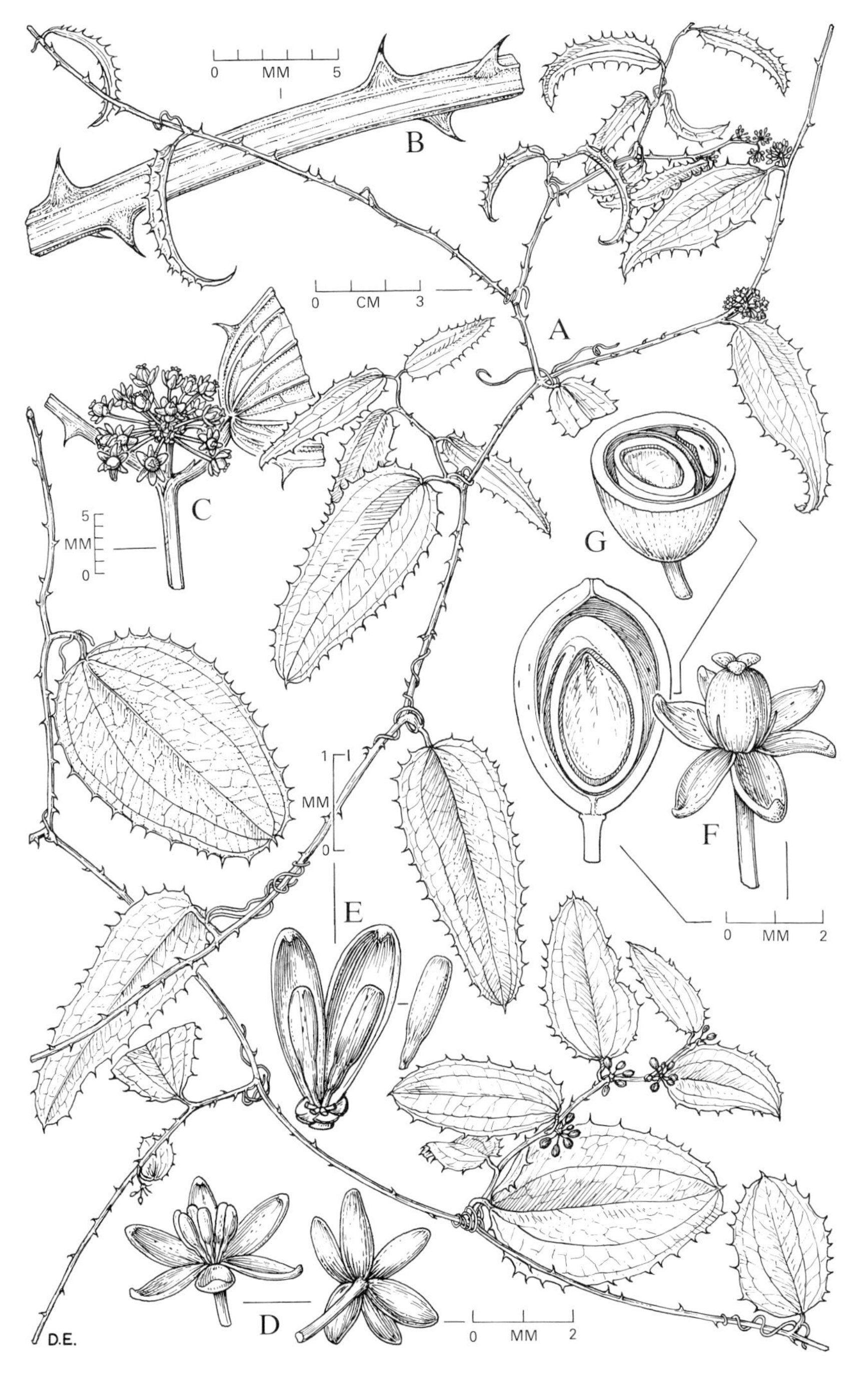

FIG. 72 **Smilax havanensis.** A, habit. B, enlarged portion of stem. C, inflorescence (pistillate). D, two views of a staminate flower. E, dissected portion of a staminate flower. F, pistillate flower. G, fruit, x- and long.-sections. (D.E.)

elongation of the floral axis known as the column, with a single anther; pollen aggregated in pollinia. Stigmas 3, one of them sterile and forming a structure known as the rostellum, the other 2 united to form a viscid surface below the anther or between its sacs, or sometimes at the end of the column. Ovary inferior, 1-celled; ovules numerous, on 3 parietal placentas. Fruit a 3-valved capsule containing numerous powdery seeds, these lacking endosperm. After seed germination, growth of the plant dependent on association with symbiotic fungi called mycorrhizae, which inhabit specialized root tissues.

Probably the largest family of flowering plants, with up to 800 genera and probably more than 25,000 species. They are best represented in the moister parts of tropical countries, but a few species occur even in subarctic regions.

KEY TO GENERA

1. Plants climbing, vinelike, with elongate green fleshy stems: **1. Vanilla**

1. Plants not vinelike:

2. Plants terrestrial, rooting in soil:

3. Base of plant a hard, bulb-like corm bearing a lateral inflorescence; flowers purple or rose: **[5. Bletia]**

3. Base of plant not bulb-like, the leaves springing from a cluster of fleshy roots, the inflorescence central:

4. Leaves reduced to scales (thus apparently absent); flowers dark purple; anther deciduous: **2. Triphora**

4. Leaves ample (sometimes absent at flowering time), 1–several in a basal tuft; flowers not purple; anther persistent:

5. Lip on lower side of flower, and more than 20 mm long: **3. Spiranthes**

5. Lip on upper side of flower, and less than 5 mm long: **4. Prescottia**

2. Plants epiphytic, or rarely on rocks, never rooted in soil:

6. Plants with evident leaves:

7. Plants with pseudobulbs:

8. Pseudobulbs conspicuous, all leaves being attached to their tips:

9. Pseudobulbs continuous, not jointed transversely; anthers 2-celled; pollinia 4: **7. Epidendrum**

9. Pseudobulbs apparently 2- to several-jointed; anthers imperfectly 8-celled; pollinia 8: **8. Schomburgkia**

8. Pseudobulbs concealed within the leaf-clusters, themselves usually leafless, very small: **9. Ionopsis**

7. Plants without pseudobulbs:

10. Leaves scattered alternately along an erect stem; flowers sessile in a narrow spike; pollinia 4: **7. Epidendrum**

10. Leaves solitary or else all basal; flowers pedicellate, often in panicles; pollinia 2:

11. Leaves solitary on short erect stems, the blades not over 1.5 cm long; lip entire, c. 2 mm long: **6. Pleurothallis**

11. Leaves several, basal, the blades more than 2 cm long; lip 3-lobed, the terminal lobe much larger than the lateral ones: **10. Oncidium**

6. Plant leafless, consisting of a conspicuous cluster of elongate roots and a short naked scape bearing 1 or 2 showy cream-white flowers, each with an elongate spur at base: **11. Dendrophylax**

Genus 1. **VANILLA** Sw.

Fleshy green-stemmed vines, often high-climbing, with or without leaves, often producing roots at the nodes. Leaves of various textures from membranous to leathery. Flowers in short axillary racemes or spikes, rather large, subtended by ovate bracts; sepals about equal, free, spreading; petals similar to the petals; lip and claw adnate to the column, the limb broad, concave, embracing the column at the base. Column long, without a foot; stigma transverse under the short rostellum; clinandrium short or obliquely raised. Anther attached to the margin of the clinandrium, with separate cells; pollinia powdery-granular, free or at length sessile on the rostellum. Capsule usually elongate, fleshy, not or but slightly dehiscent.

A pantropical genus of about 90 species. The spice called "vanilla" is obtained from the fermented pods of *V. planifolia* and (to a lesser extent) *V. pompona.*

1. **Vanilla claviculata** (Sw.) Sw. in Nov. Act. Upsal. 6:66. 1799.

Main stem leafless, long-trailing or climbing, 1-2 cm thick when fresh (shrinking by about one half when dried); internodes c. 10 cm long; adventitious roots often twisting like tendrils. Leaves (when present) 2-8 cm long, linear-lanceolate, rigid, recurved acuminate at the apex, half clasping at base, chiefly occurring on terminal shoots. Raceme with 8 or more sessile flowers, appearing stalked by the long narrow ovary; bracts reflexed, 2-10 mm long. Flowers fragrant, the perianth c. 4 cm long or more, glaucous green, the lip c. 5 cm long, its apex rounded, white and pink, more or less curled or ruffled on the upper margin; column 3 cm long. Capsule ellipsoid-cylindric, 8-10 cm long.

GRAND CAYMAN: *Brunt 2186; Kings GC 412.*
– Bahamas and Greater Antilles, rather common.

Genus 2. **TRIPHORA** Nutt.

Small terrestrial and leafy (or saprophytic and leafless) herbs, growing from a subterranean cormlike tuber, and spreading by delicate stolons. Leaves 2-4, alternate on the stem, relatively small or even reduced to mere sheaths. Inflorescence a several-flowered raceme or corymb. Flowers small, axillary, pedicellate, erect or nodding; sepals and petals similar, free or somewhat connivent; lip sessile or clawed, parallel to the column, almost entire to 3-lobed, 3-crested, decurved at apex; column slender, straight, apically entire or lobed; anther rigidly attached to top of column. Capsule ellipsoid.

A genus of about 10 temperate and tropical American species.

1. **Triphora gentianoides** (Sw.) Ames & Schltr., Orchidaceae 7:5. 1922.

Plant with erect glabrous dark red stem 8-15 cm tall or more; leaves reduced to rounded clasping sheaths. Flowers 6-8 in a short terminal corymb, on filiform pedicels up to 4 cm long or more, erect or somewhat spreading; bracts ovate, acuminate, concave, 5-9 mm long. Sepals narrowly oblong, 9 mm long, dark red-purple, 3-nerved; petals nearly as long as the sepals, white; lip nearly equalling the petals, 3-lobed. Capsule c. 1 cm long, ellipsoid with 6 narrow keels.

GRAND CAYMAN: *Kings GC 242.* Recorded only from the Forest Glen area.
– Florida and Mexico southward at scattered localities to Venezuela and Ecuador, rare or uncommon.

Genus 3. **SPIRANTHES** L.C. Rich.

Rather small terrestrial herbs with fleshy or tuberous roots. Stems erect, simple, more or less concealed by leaf-sheaths. Leaves basal, cauline, or both, often absent at flowering time when basal. Flowers in an erect spike or raceme, often spirally arranged; sepals free, the dorsal one usually erect and forming a galea with the petals, the lateral ones erect or spreading, more or less decurrent on the ovary to form a free or usually adnate chin-like structure. Petals narrow and usually adherent to the dorsal sepal. Lip sessile or clawed, simple or lobed, adherent to the column in most species. Column terete; clinandrium more or less membranous and conspicuous, often continued into the rostellum; rostellum various. Anther dorsal, erect, sessile or stipitate; pollinia 2, powdery or granular, usually attenuated at one end.

A very complex world-wide genus of at least 200 species, by many authors subdivided into smaller genera. The two Cayman species have often been placed in the genera *Pelexia* (or *Eltroplectris*) and *Stenorrhynchus* (or *Sacoila*) respectively, but for our purposes their resemblances seem more important than their differences. In this respect the present treatment follows that of Ames & Correll in *Orchids of Guatemala* (1952).

KEY TO SPECIES

1. Spur enclosing the elongated base of the lip; flowers greenish-white, the divisions long-attenuate: **1. S. calcarata**

1. Spur not enclosing base of lip; flowers brick-red, salmon or light buff, the divisions merely acute: **2. S. squamulosa**

1. Spiranthes calcarata (Sw.) Jiménez in Phytologia 8:326. 1962.

Pelexia setacea Lindl., 1840.

Eltroplectris calcarata (Sw.) Garay & Sweet, 1972.

Stems glabrous or minutely glandular-puberulous upwardly, 30–55 cm tall, sparingly clothed with a few appressed membranous sheaths having attenuate tips, and accompanied from the base by a single leaf (rarely 2) with petiole 8–15 cm long and lance-elliptic acute blade 10–15 cm long. Flowers 2–8 in a loose spike; bracts lance-attenuate, exceeding the ovary. Sepals pale green, linear-lanceolate, 2.5–2.8 cm long, the lateral ones connate at base into a spur-like appendage; petals like the sepals, attached their whole length to the median sepal; lip white, oblong, many-nerved, the margins fimbriate in the median part, attached at extreme base to inside of the sepal-spur. Capsule broadly ellipsoid, 2 cm long, with 6 narrow keels.

GRAND CAYMAN: *Kings GC 274.*
– Florida, Bahamas, Greater Antilles and northern South America, usually growing in shaded humus.

2. **Spiranthes squamulosa** (Kunth) Leon, Fl. Cuba 1:357. 1946.

Stenorrhynchus squamulosus (Kunth) Spreng., 1826.

Spiranthes orchioides of recent authors, in part, not A. Rich., 1850.

Sacoila squamulosa (Kunth) Garay, 1980.

Stems erect, minutely whitish-scurfy, 30–60 cm tall, clothed with about 8 acuminate sheaths 2–4 cm long. Basal leaves several, broadly lanceolate to elliptic, up to 30 cm long, absent at flowering time. Flowers more or less scurfy like the stem, up to 14 or more in a loose or somewhat crowded spike; bracts narrowly lanceolate, exceeding the ovary. Sepals brick red, salmon, or light buff, lance-acuminate, 5-nerved, the lateral ones very oblique and 2–2.35 cm long, prolonged at base into a chin-like spur; median sepal 1.5–1.75 cm long; petals coloured like the sepals, about as long as the adnate median sepal, subacute; lip 2–2.35 cm long, dilated and saccate about the middle, lance-acuminate above and linear-convolute below, more or less pubescent. Capsule scurfy, 1.1–1.3 cm long, 3-keeled.

GRAND CAYMAN: *Brunt 2190; Kings GC 244.*
– Florida and Mexico south to South America; in the West Indies chiefly occurring in the Greater Antilles. A more slender species, *S. lanceolata*, is found in some of the same countries but more especially in the Lesser Antilles. Some authors unite the two and characterize the combined population as "variable".

Genus 4. **PRESCOTTIA** Lindl.

Terrestrial herbs with clustered fibrous or fleshy roots. Leaves basal, membranous, sessile or petiolate. Inflorescence a slender erect spike of numerous small flowers. Sepals membranous, connate at base to form a short cup or tube, spreading or revolute at the apex; petals narrow, adnate to the sepal-cup. Lip on upper side of the flower, with claw adnate to the sepal-cup, auriculate at base, the apical part entire, deeply concave or galeate and often enclosing the column. Column very short, adnate to the sepal-cup. Pollinia 4, granular or powdery. Capsule small, suberect, ovoid or ellipsoid.

A genus of about 35 species found in tropical and subtropical America.

1. Prescottia oligantha (Sw.) Lindl., Gen. & Sp. Orch. Pl. 454. 1840.

Roots thick and fleshy in a small cluster; stems slender, erect, mostly 15-35 cm tall, glabrous, bearing several narrow, tubular sheaths. Basal leaves 2 or 3, 1.5-13 cm long including the petiole. Spike very slender, densely flowered, 2-13 mm long, 1.5-5 mm thick; bracts narrow, 2-3 mm long. Flowers minute, pinkish or whitish, less than 2.5 mm long. Lateral sepals arising from a broad connate base, 1-2.2 mm long; dorsal sepal ovate, usually obtuse, 1-2 mm long; petals more or less linear, 1-1.5 mm long; lip erect, suborbicular, concave-saccate, 1-1.5 mm long. Column laterally winged near the apex. Capsule ellipsoid, c. 4 mm long, narrowly 6-keeled.

GRAND CAYMAN: *Kings GC 427.*
– Florida, Mexico and the West Indies south to Brazil, Argentina and Peru; recently recorded from the Galapagos Islands, rather common and widespread, growing in the humus of rocky woodlands. Although the Cayman specimen is sterile, there is no doubt of its identity. The description was drawn up with the aid of Jamaican material.

[Genus 5. **BLETIA** Ruiz & Pav.

Terrestrial erect herbs growing from more or less globose corms, the leaves growing from top of corm, lanceolate, rather long and grasslike, usually withering at flowering time. Scape slender, arising from side of corm, bearing small distant scale-like leaves, the flowering racemes simple or branched; flowers showy, rather distant, long-stalked, with small ovate bracts. Sepals nearly alike, separate, ovate or oblong; petals similar to the sepals but slightly broader; lip oblong, longitudinally ridged or crested, 3-lobed at apex. Column elongate, arcuate, winged at apex, auriculate at base, footless. Capsule ellipsoid, erect.

A tropical American genus of less than 50 species, some of them very common and widespread, often attracting the attention of travellers because they frequently grow on roadside banks.

A species of *Bletia* which is probably *B. florida* (Salisb.) R. Br. is cultivated in Georgetown, Grand Cayman. It has deep purple flowers on branched inflorescences up to nearly 1 m tall, and was probably introduced from Jamaica. The terrestrial habit, ovoid pseudobulbs, and broad, somewhat grasslike leaves render identification easy.]

Genus 6. **PLEUROTHALLIS** R.Br.

Rather small epiphytic herbs, with stems clustered or else branched from a creeping primary stem or rhizome; pseudobulbs absent. Leaves usually solitary on a stem, often subtended by one to several sheaths. Inflorescence terminal (rarely lateral), fasciculate, racemose, or sometimes bearing a solitary flower. Flowers small, subtended by inconspicuous tubular bracts. Sepals equal or nearly so, erect or spreading, the median free or very shortly connate with the lateral; lateral sepals slightly or often completely connate. Petals shorter or narrower than the sepals, sometimes minute; lip shorter or rarely a little longer than the petals, simple or 3-lobed, usually contracted at base and jointed with the base of the column. Column equalling or shorter than the lip, winged or wingless, at base often with a small foot, this nearly obsolete to nearly as long as the column. Anther terminal, operculate, incumbent, 1- or 2-celled; pollinia 2 or 4, waxy. Capsule subglobose or ellipsoid.

A very large genus of about 600 species, confined to the Western Hemisphere, and most commonly found in the moist montane regions of the tropics and subtropics. It is rather surprising to find a species occurring near sea-level in the Cayman Islands.

1. Pleurothallis caymanensis C.D. Adams in Orquideologia 6:146, figs. A-C. 1971. **FIG. 73A.**

A small, dense epiphyte, the slender branched rhizomes concealed by the deflexed leaves; secondary stems with a single internode 2-3 mm long. Leaves thick, fleshy, mottled gray-green and minutely white-dotted, elliptic-oblanceolate, 6-15 mm long, sharply acuminate. Flowering scape solitary, filiform, 3-5-articulate, 20-35 mm long, bearing 1 to 3 flowers. Flowers greenish-white or yellowish with purple veins; median sepal oblong-oblanceolate, 7 mm long; lateral sepals almost wholly connate, 6 mm long, the free apices acute; petals broadly obovate, 2.1 mm long; lip ovate-oblong with broadly acute apex, entire, c. 2 mm long; column deeply curved. Ovary rugulose; capsule obliquely obovoid, c. 7 mm long.

GRAND CAYMAN: *Kings GC 250; Proctor 27983 (type).*
– Endemic; apparently related to *P. sertularioides* of Cuba and Jamaica. Confined to dense woodlands in the central part of Grand Cayman.

Genus 7. **EPIDENDRUM** Sw.

Plants epiphytic, on rocks, or rarely terrestrial, extremely diverse in size, erect or creeping and with or without a conspicuous rhizome. Stems either thickened into a cylindric to subglobose or compressed "pseudobulb" bearing leaves only at the apex (subgenus *Encyclia*), or slender and more or less leafy, simple or branched (subgenus *Epidendrum* mostly). Leaves 1 to numerous, terete or flattened, varying in outline from linear to oval. Inflorescence usually terminal (rarely lateral from a leafy stem), simple to diffusely paniculate, 1- to many-flowered; flowers minute to rather large. Perianth-segments spreading; petals usually much narrower than the

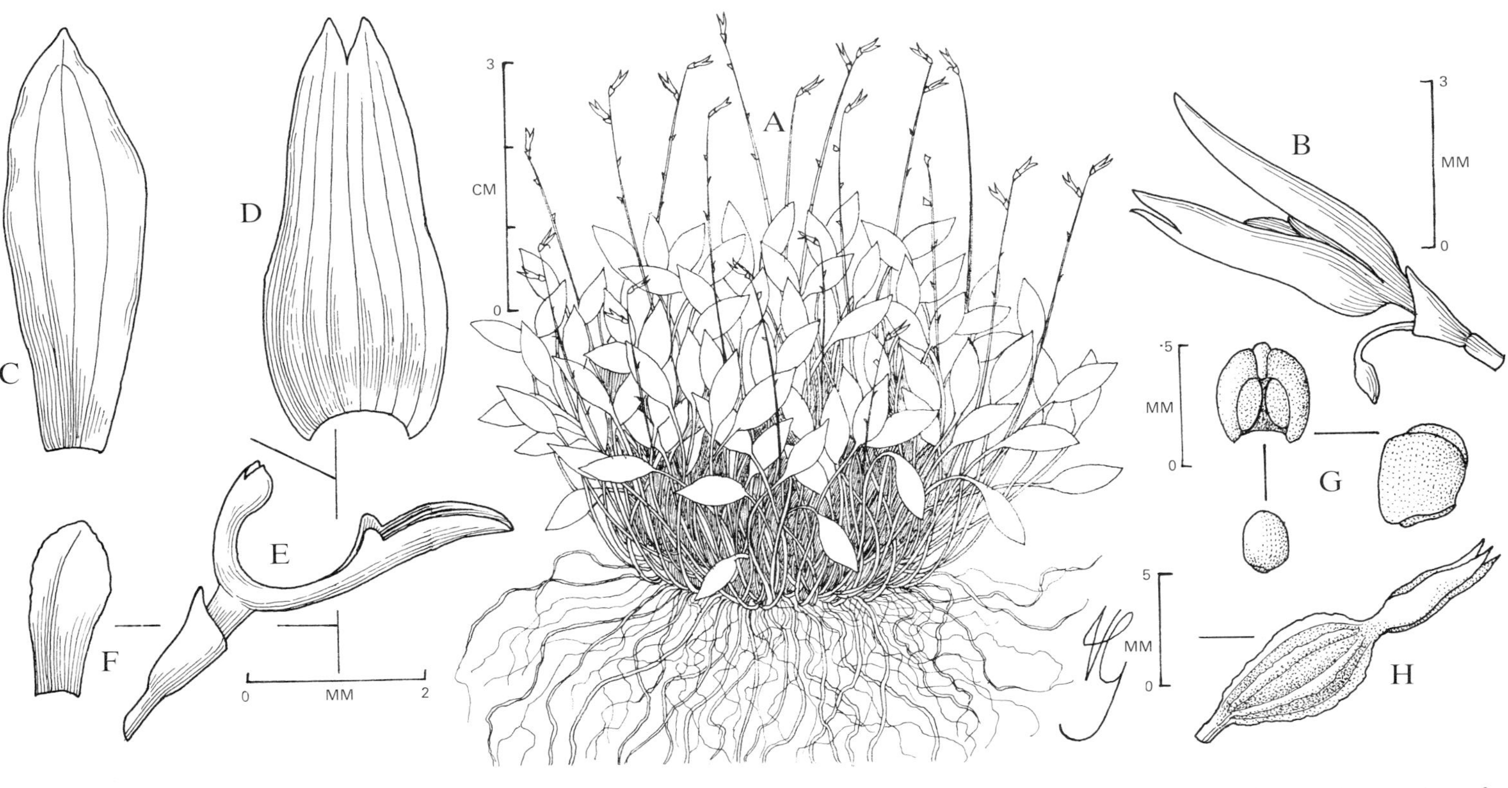

FIG. 73A **Pleurothallis caymanensis.** A, habit. B, single flower subtended by a small bud. C, D, perianth segments. E, column and lip. F, apex of lip. G, views of anther with pollinia. H, fruit. (G.)

sepals. Lip more or less adnate to the column, simple or 3-lobed, smooth or callose, often conspicuous. Column short to elongate, wingless, winged or auricled. Anther terminal, operculate, incumbent, 2-celled; pollinia 4, waxy, equal, more or less flattened. Capsule usually ellipsoid.

As here defined, this is the largest genus of neotropical orchids, with about 800 species. Some authors separate the pseudobulbous species as a genus *Encyclia.*

KEY TO SPECIES

1. Plants without pseudobulbs, the erect stems bearing 2–8 leaves; inflorescence a spike: **1. E. rigidum**

1. Plants with pseudobulbs:

2. Leaves 1 to a pseudobulb; inflorescence a panicle: **2. E. kingsii**

2. Leaves 2 to a pseudobulb; inflorescence a raceme:

3. Pseudobulbs orbicular, flattened and 2-edged, c. 2.5 cm long: **3. E. boothianum**

3. Pseudobulbs ovoid-conic or somewhat elongate, not 2-edged, 3–10 cm long or more:

4. Sepals and petals directed downward; lip c. 2 cm long, entire, strongly concave, and acuminate at the apex: **4. E. cochleatum**

4. Sepals and petals directed upward or spreading; lip 2.5–3.5 cm long, 3-lobed, essentially flat or slightly convex, notched-cuspidate at apex: **5. E. plicatum (?)**

1. Epidendrum rigidum Jacq., Enum. Pl. Carib. 29. 1760.

Epiphytic, with creeping, branched, compressed rhizome that gives rise to scattered erect stems, these covered by the leaf-sheaths and mostly 10–20 cm long. Leaves distichous, the blades articulate to the sheathlike clasping bases, usually 5–8 per stem, the lower ones reduced to mere sheaths, the upper with oblong or elliptic-oblong blades mostly 3–6 cm long, obliquely notched at the apex. Inflorescence with 2–7 scattered, yellowish-green, sessile flowers, the base of each partly enclosed by a folded greenish bract 9–15 mm long. Sepals ovate to oblong-elliptic, 4.5–10 mm long; petals linear to linear-oblanceolate, obtuse, denticulate on the margins, 4–9 mm long; lip adnate to the column, the free apex broadly rounded, crenulate-denticulate, 2.5-6 mm long, 3-5.5 mm wide; column 2-3 mm long, dentate at the apex. Capsule ellipsoid, beaked, mostly 1-1.5 cm long.

GRAND CAYMAN: *Dressler 2906 (IJ)*, collected in May, 1964.
– Widely distributed in tropical America from Florida south to Brazil and Bolivia. The sole Cayman record came from rocky woodland near Georgetown.

2. **Epidendrum kingsii** C.D. Adams in Orquideologia 6:145, Figs. A-C. 1971.

Epiphytic, with short rhizome; pseudobulbs subconic, 5-6 cm long and c. 1.5 cm thick. Leaves linear-oblong, 27-34 cm long, 1-1.5 cm broad, plicate toward the base. Inflorescence terminal on pseudobulbs, up to 80 cm long with branches to 12 cm long; bracts acute, c. 1 cm long. Flowers brownish-yellow, 1 to 8 on a branch, pedicellate; median sepal oblong-oblanceolate, 10 mm long, obtuse and minutely emarginate, with 5 free veins; lateral sepals similar but apex entire; petals oblanceolate, 10 mm long, the apex rounded, the veins reticulate. Lip free from the column, c. 9 mm long and broad, 3-lobed, the median lobe broadly rounded with emarginate-apiculate apex, the lateral ones with truncate apex; column winged at the apex. Ovary rugulose; capsule with pedicel up to 11 mm long, elongate-ellipsoid (immature), more or less verrucose.

LITTLE CAYMAN: *Kings LC 117A (type).* Reported to occur also on Cayman Brac.
– Endemic; related to *E. sintenisii* of the Greater Antilles.

3. **Epidendrum boothianum** Lindl. in Edw., Bot. Reg. 24:Misc. 5. 1838.

Encyclia boothiana (Lindl.) Dressler, 1961.

Epiphytic, clustered or solitary, with pseudobulbs less than 1 cm thick, subtended by ovate, scarious scales 1-2 cm long. Leaves oblong-oblanceolate, obtuse or acutish, 6-12 (-15) cm long, 1-2.5 cm broad. Flowering scape simple, usually longer than the leaves, subtended by an elongate spathe-like sheath. Flowers 1 to several; pedicels 3-5 mm long; sepals oblong to elliptic-oblanceolate, c. 13 mm long, acute at the apex, yellow mottled with brown; petals spatulate, about the same length and colour as the sepals; lip yellowish, c. 7 mm long, subrhombic; column without a foot, partly adnate to the lip. Capsule oblong ellipsoid, 2-3 cm long, 3-winged.

GRAND CAYMAN: *Kings GC 281.* Also reported from Cayman Brac.
– Florida, Bahamas, Cuba, Haiti, Mexico and Belize, on trees.

4. Epidendrum cochleatum L., Sp. Pl. ed. 2, 2:1351. 1763. **FIG. 73B.**

Encyclia cochleata (L.) Dressler, 1961.

Epiphytic or sometimes on rocks; pseudobulbs often more than 5 cm long and 2.5-5 cm broad. Leaves oblong-lanceolate, 20-30 cm long, 2-3.5 cm wide, acute at the apex. Raceme 20-50 cm long; flowers indefinite in number, distinctly pedicellate; sepals linear-lanceolate, 2.5-4 cm long, light green at first, turning yellow; petals similar to the sepals but slightly smaller; lip broadly ovate-deltate, subcordate, the concave side dusky purple or blackish purple with yellow lines. Capsules on pedicels 1.5 cm long, 3-winged, 3.5-4.5 cm long and 2.5 cm broad, the adaxial wing narrower than the others.

CAYMAN BRAC: *Mrs. Rena Reid s.n. (IJ).* The specimen consists of but a single flower which is perfectly characteristic of the species.

– Florida, Bahamas, Greater Antilles, Dominica and Venezuela, in many places very common. This was one of the earliest tropical orchids to be introduced into cultivation under glass in England.

5. Epidendrum aff. **plicatum** Lindl., Bot. Reg. 33: under t. 10; t. 35. 1847.

Encyclia plicata (Lindl.) Schltr., 1914.

Epiphytic or on logs; pseudobulbs clustered, 3-8 cm long. Leaves ligulate, dark green, 15-40 cm long, 1.5-3.5 cm wide, acute at the apex. Inflorescence usually a simple raceme or sometimes with one or two branches, much longer than the leaves, up to 75 cm long, bearing up to about 20 flowers. Sepals and petals brownish-yellow or olive, oblanceolate and sharply acuminate, 25-35 mm long, 5-11 mm wide above the middle. Lip about as long as the petals, light mauve, deeply 3-lobed, the middle lobe broader than long, cordate at base, crispate-margined, notched and cuspidate at the apex, the lateral lobes oblong-lanceolate, obtuse, upwardly bent and loosely enfolding the column. Column about 15 mm long. Capsule not seen.

CAYMAN BRAC: *Mrs. Rena Reid, photographs.* The identity of this species needs to be further substantiated with herbarium specimens. The relationship of Mrs. Reid's plant with *E. plicatum* seems likely, but they do not necessarily represent the same species.

– Cuba, so far as typical *E. plicatum* is concerned. The plant described by Britton & Millspaugh (1920) from the Bahamas is not the same as the Cuban plant illustrated by Lindley. Plants belonging to this group of orchids are frequently misidentified.

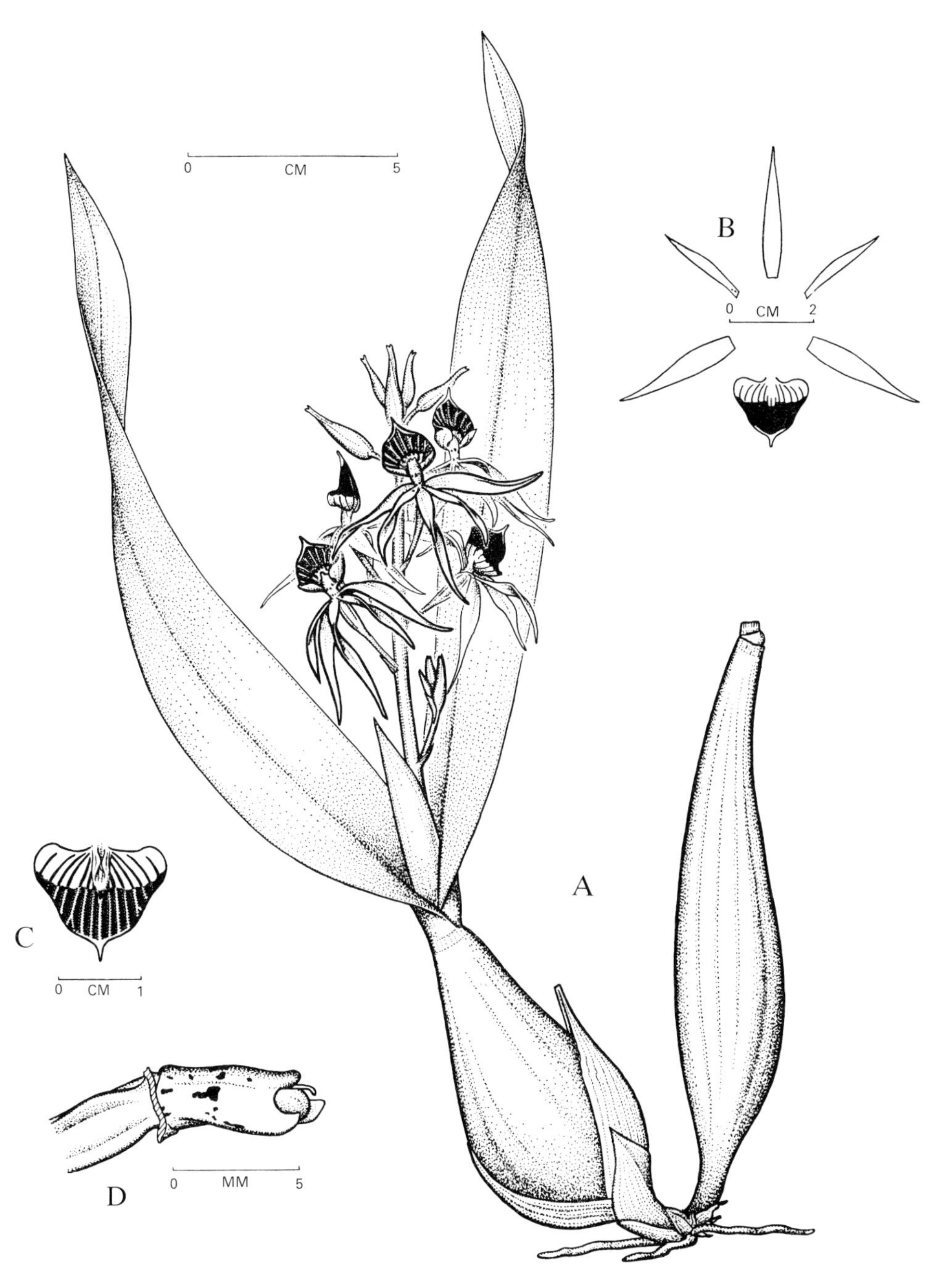

FIG. 73B **Epidendrum cochleatum.** A, habit. B, perianth-segments showing underside of lip. C, upperside of lip. D, column. (H.U.)

Genus 8. **SCHOMBURGKIA** Lindl.

Epiphytic, the pseudobulbs large and cylindric or conic, bearing several scarious or fibrous sheaths, these when fallen leaving joint-like horizontal scars at the lines of attachment; leaves 2–5 at the apex of the pseudobulb, more or less oblong, thick and leathery. Flowers stalked, in loose racemes or panicles on a terminal, elongate, sheathed peduncle; bracts persistent. Sepals almost equal, free, spreading, and more or less undulate; petals similar to the sepals. Lip very shortly connate at the base with the column, the sides loosely embracing it. Column straight or incurved, winged; clinandrium lobed. Anther attached to the apex of the median tooth, incumbent, imperfectly 8-celled; pollinia 8, waxy, superposed in pairs, broadly ovate, compressed, connected by a slender granular appendage. Capsule ovoid to ellipsoid.

A small genus of about a dozen tropical American species, sometimes included in the genus *Laelia*. Some of the species are noted for harbouring nests of stinging ants in their hollowed-out pseudobulbs.

1. **Schomburgkia thomsoniana** Rchb.f. in Gard. Chron. ser. 3, 2:38. 1887; 9:615. 1891; also in Veitch, Man. Orch. Pl. pt. 2:102. 1887.

S. thomsoniana var. *atropurpurea* Hook.f., 1902.

Laelia thomsoniana (Rchb.f.) L.O. Wms., 1941.

Schomburgkia brysiana var. *thomsoniana* (Hook.f.) H.G. Jones, 1963

S. brysiana var. *atropurpurea* (Hook.f.) H.G. Jones, 1963.

"Wild Banana"

Plants gregarious; pseudobulbs several, curved-ascending, up to 20 cm long or more and 3.5 cm thick. Leaves usually 4 or 5, oblong-elliptic, to 15 cm long or more and 5 cm broad. Peduncle to 1 m long or more (rarely 2 m) and 3–7 mm thick; inflorescence most often 6–20-flowered (but one recorded with 58 flowers!); perianth variously white, cream, or yellowish; lip with apex variously purple to nearly black, slightly bifid. Capsules ellipsoid, 4–4.5 cm long, longitudinally 6-grooved.

This species has sometimes been associated taxonomically with *Schomburgkia brysiana* of Central America, but that species, known only from Honduras and Guatemala, has perianth-segments that are orange to burnt-orange in colour, and a lip that is not only somewhat different in shape, but also is bright yellow with deep orange flecks on the lower part; it is probably more nearly related to the Central American *S. tibicinis* than to *S. thomsoniana.*

S. thomsoniana can be classified in two rather distinctive varieties:

1. Perianth white or cream, the lobes 3-3.5 cm long; lip purple-black at the apex and scarcely recurved; inflorescence usually a panicle; found on Grand Cayman only: **1a.** var. **thomsoniana**

1. Perianth yellow, the lobes 2-2.5 cm long; lip light purple at the reflexed apex; inflorescence usually a raceme; chiefly occurring on Little Cayman and Cayman Brac: **1b.** var. **minor**

1a. Schomburgkia thomsoniana var. **thomsoniana**

GRAND CAYMAN: *Brunt 1838; Hitchcock; Kings GC 15, GC 200, GC 292; Maggs II 64; Proctor 27957.*
– Endemic.

1b. Schomburgkia thomsoniana var. **minor** Hook.f. in Bot. Mag. 128: t. 7815. 1902. **FIG. 74.**

S. brysiana var. *minor* (Hook.f.) H.G. Jones, 1966.

This variety tends to be smaller and more slender than var. *thomsoniana,* the inflorescence averaging fewer flowers.

LITTLE CAYMAN: *Kings LC 7.* CAYMAN BRAC: *Kings CB 18; Proctor 15319.*
– Cuba?

Genus 9. **IONOPSIS** Kunth in H.B.K.

Epiphytic, with short, creeping rhizomes usually bearing 1 or 2 small pseudobulbs more or less concealed by the leaf-sheaths. Leaves few, tufted, narrow, leathery, with persistent, 2-ranked, overlapping sheaths. Inflorescences 1-3, lateral or subterminal, racemose or paniculate, with long peduncles. Flowers stalked; sepals nearly equal, erect or spreading at the apex, the median free, the lateral connate at the base, forming a small spurlike sac below the lip; petals like the median sepal but wider; lip adnate to base of the column, relatively large, clawed, the claw bearing 1 or 2 pairs of flat calli. Column short, erect, thick, not winged and lacking a foot. Anther terminal, operculate, incumbent, 1-celled or imperfectly 2-celled; pollinia 2, waxy, attached to linear stalk. Capsule ovoid to ellipsoid.

A small genus of about 10 species occurring in the warmer parts of the Western Hemisphere.

1. Ionopsis utricularioides (Sw.) Lindl., Coll. Bot. t. 39-A. 1821.

Pseudobulbs ellipsoid-conical, c. 1 cm long or a little more, usually leafless, often concealed, rarely absent. Leaves 2–4 in a cluster, lance-linear or lance-oblong, 3–15 cm long, acute at apex, leathery. Inflorescence lateral, from base of pseudobulb, paniculate, laxly few- to many-flowered. Flowers long-stalked, pale lilac, bluish or nearly white (rarely otherwise); sepals 3–6 mm long; petals similar to the sepals, 6–7 mm long. Lip 7–16 mm long, 7–18 mm wide toward the apex, broadly obcordate, tapering to a narrow stalklike base ("claw") bearing 2 small thin calli. Column c. 2 mm long. Capsule long-beaked, 1.2–1.6 cm long.

GRAND CAYMAN: *Kings GC 239.*
– Widely distributed throughout the American tropics, especially in drier regions.

Genus 10. **ONCIDIUM** Sw.

Chiefly epiphytic or growing on rocks, with or without pseudobulbs, the stems very short and leafy. Leaves flat, 3-edged or terete, usually equitant. Inflorescence a raceme or panicle, its peduncle attached laterally from the base of a pseudobulb or in the axil of a leaf. Flowers of various colours, often showy; sepals usually subequal, spreading or reflexed, all free or with the lateral ones connate. Petals similar to the median sepal or larger. Lip adnate to base of the column, the lower free part clawed and commonly crested or tuberculate, the apical part often variously lobed, the median lobe much larger than the lateral ones. Column short, thick, and often bearing petaloid wings on each side at the apex; footless. Anther 1-celled or imperfectly 2-celled, terminal, operculate, incumbent; pollinia 2, waxy, deeply grooved, attached to a flat stalk. Capsule beaked, ovoid to ellipsoid or fusiform.

A large genus of more than 450 species, confined to the American tropics and subtropics. Many of the species are prized in cultivation.

KEY TO SPECIES

1. Leaves terete, less than 2 mm thick; flowers yellow, the median lobe of the lip ovate-deltate: **1. O. calochilum**

1. Leaves flat, folded, mostly 5–10 mm broad when unfolded; flowers pink, the median lobe of the lip transversely reniform: **2. O. caymanense**

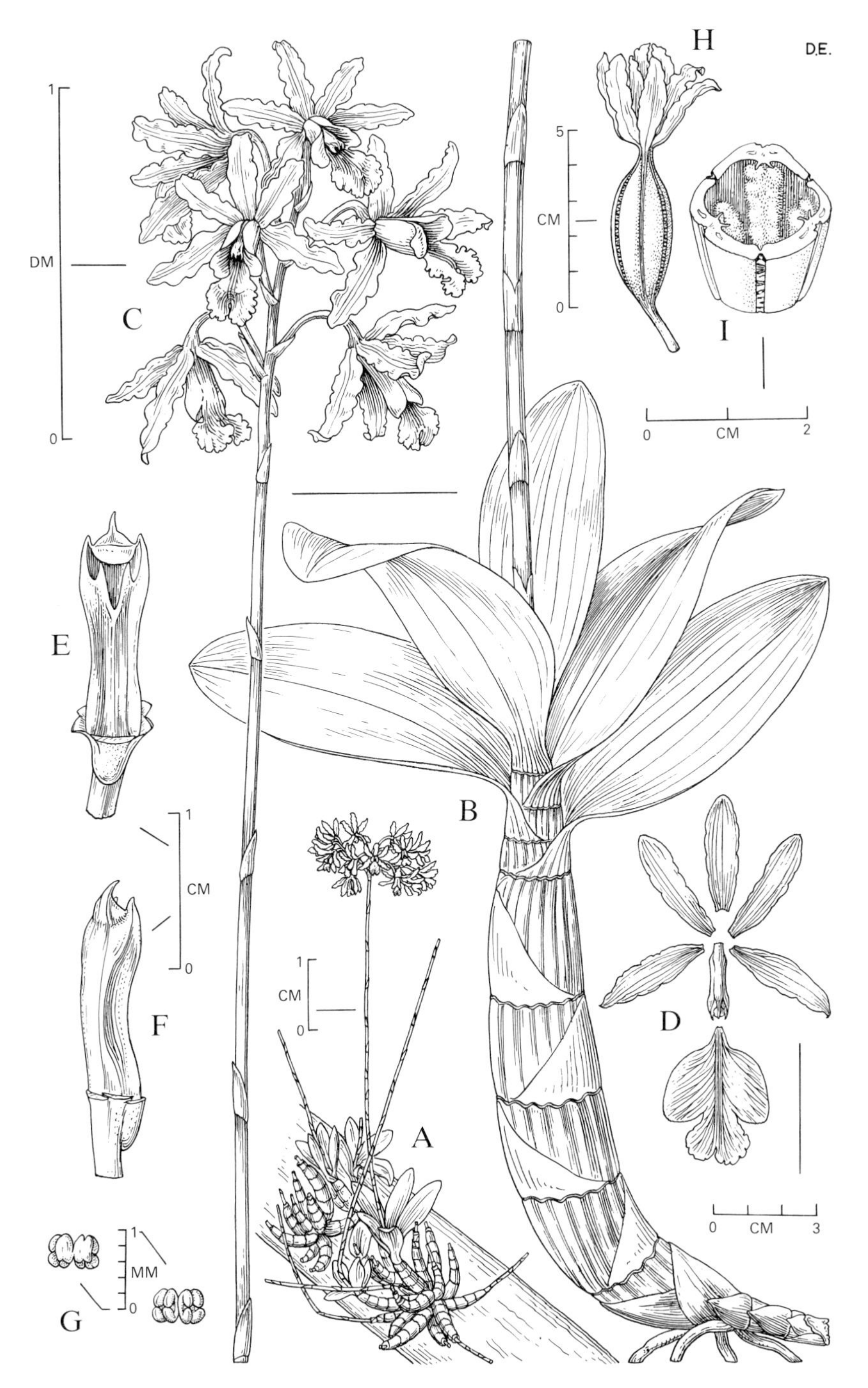

FIG. 74 **Schomburgkia thomsoniana** var. **minor.** A, habit. B, pseudobulb, leaves, and base of peduncle. C, inflorescence. D, flower-parts; a, lip; b, column. E, F, two views of column. G, two views of pollinia. H, capsule with withered flower attached. I, x-section of capsule. (D.E.)

1. **Oncidium calochilum** Cogn. in Urban, Symb. Ant. 6:660. 1910.

Epiphytic, forming small loose tufts; leaves terete, linear-subulate, mostly 3–10 cm long and 1–2 mm thick, their bases enclosed by loose, scarious, acuminate sheaths up to 1 cm long. Inflorescence 1-6-flowered on a peduncle 4–17 cm long. Flowers on delicate pedicels often 10 mm long or more, bright buttercup yellow; sepals 8–12 mm long, linear-acuminate, the lateral ones partly connate; petals c. 10 mm long, wider than the sepals. Lip 15–20 mm long, short-clawed and elongate-created at base, 3-lobed, the lateral lobes minute and inconspicuous, the apical one very large, ovate-deltate, shortly acuminate at the apex and with fimbriate margins. Column straight, 6 mm long. Capsule not seen.

GRAND CAYMAN: *Brunt 2169; = Proctor 28000,* both collected June 11, 1967.

– Cuba and Hispaniola, rare. In Grand Cayman this species is confined to dense woodland near the center of the island.

2. **Oncidium caymanense** Moir, Phytologia 17(6):427. 1968 (incomplete); ibid. 19:53. 1969.

Epiphytic, forming small tufts; leaves flat, folded, recurved, 2–10 cm long and up 10 mm broad (unfolded), the margins denticulate, the bases partly enclosed by loose scarious sheaths. Inflorescence racemose or paniculate, up to 14-flowered (usually less), on strong peduncles up to 22 cm long; bracts thin, keeled, c. 3 mm long. Flowers on delicate pedicels c. 10 mm long, chiefly light pink, the median sepal and petals each with a red spot at the base, and the claw and crested base of the lip yellow with brick-red speckles. Sepals narrowly oblanceolate, 5–8 mm long, the lateral ones partly connate; petals obovate, 6–7 mm long, 3–5 mm broad. Lip up to 10 mm long, clawed and crested at the base, 3-lobed at the apex, the lateral lobes rounded with fimbriate margins, the apical one much larger and transversely reniform (up to 11 mm broad), concave-apiculate on the outer edge, the margins entire. Column short, broadly 2-winged. Capsule ellipsoid, nearly 2 cm long.

GRAND CAYMAN: *Brunt 2168; = Proctor 27999,* collected June 11, 1967. This species was first described on the basis of cultivated material from the same collection cited above (after passing through several hands), but a type specimen was not cited with the original description of 1968. Originally, a number of live plants (part of *Proctor 27999*) were taken to Jamaica and given to Mr. Noel Gauntlett for propagation. According to Moir, "The plants were sent to Mr. Oris Russell in Nassau, Bahamas, who in turn gave them to the late Mr. Stanley Smith, of Nassau, who sent one speciman (sic) in flower to Hawaii, blooming during May – June." Subsequently, some fragments of this cultivated material were sent to the Ames Orchid Herbarium at Harvard University and designated the holotype.

– Endemic; related to *Oncidium variegatum* of the Greater Antilles. The collectors found it associated with *O. calochilum* in dense woodland near the center of Grand Cayman, where it is frequent in a very limited area. This species will probably be exterminated by real-estate development, if not already wiped out by orchid fanciers.

Genus 11. **DENDROPHYLAX** Rchb.f.

Epiphytic or epipetric leafless herbs with numerous elongated roots, a very short stem (or virtually stemless), and no pseudobulbs. Inflorescence a slender simple scape, the small to rather large flowers solitary or few in a short raceme. Sepals about equal, free, spreading; petals similar to the sepals. Lip sessile at base of the column, produced basally into an elongate hollow spur dilated toward the mouth and continuous with the column; apex of lip entire or 2-lobed, often large and showy. Column very short, thick, and footless. Anther terminal, operculate, incumbent, indistinctly 2-celled; pollinia 2, waxy, each attached to a simple or divided stalk. Capsule oblong or elongate.

A West Indian endemic genus of about 6 species.

1. **Dendrophylax fawcettii** Rolfe in Gard. Chron. ser. 3, 4:533. 1888. **FIG. 75.**

Roots 20-50 cm long, 2-3 mm in diameter. Peduncle 5-7 cm long, bearing below the middle one or two tubular bracts 4-4.5 mm long. Flowers solitary on pedicels c. 3 cm long; sepals and petals cream, narrowly elliptic-oblong, 2-2.5 cm long. Lip white, 3-4 cm long, broadly 2-lobed, the lobes divergent; spur slender, 11-15 cm long or more. Capsule cylindric, c. 8 cm long and 0.6 cm thick, longitudinally ribbed.

GRAND CAYMAN: *Dressler 2904; Fawcett (type); Kings GC 19, GC 19a.* – Endemic; grows on trees and rocks in sheltered situations.

This species is rare and should be given legal protection, otherwise it is likely to become extinct as a wild plant.

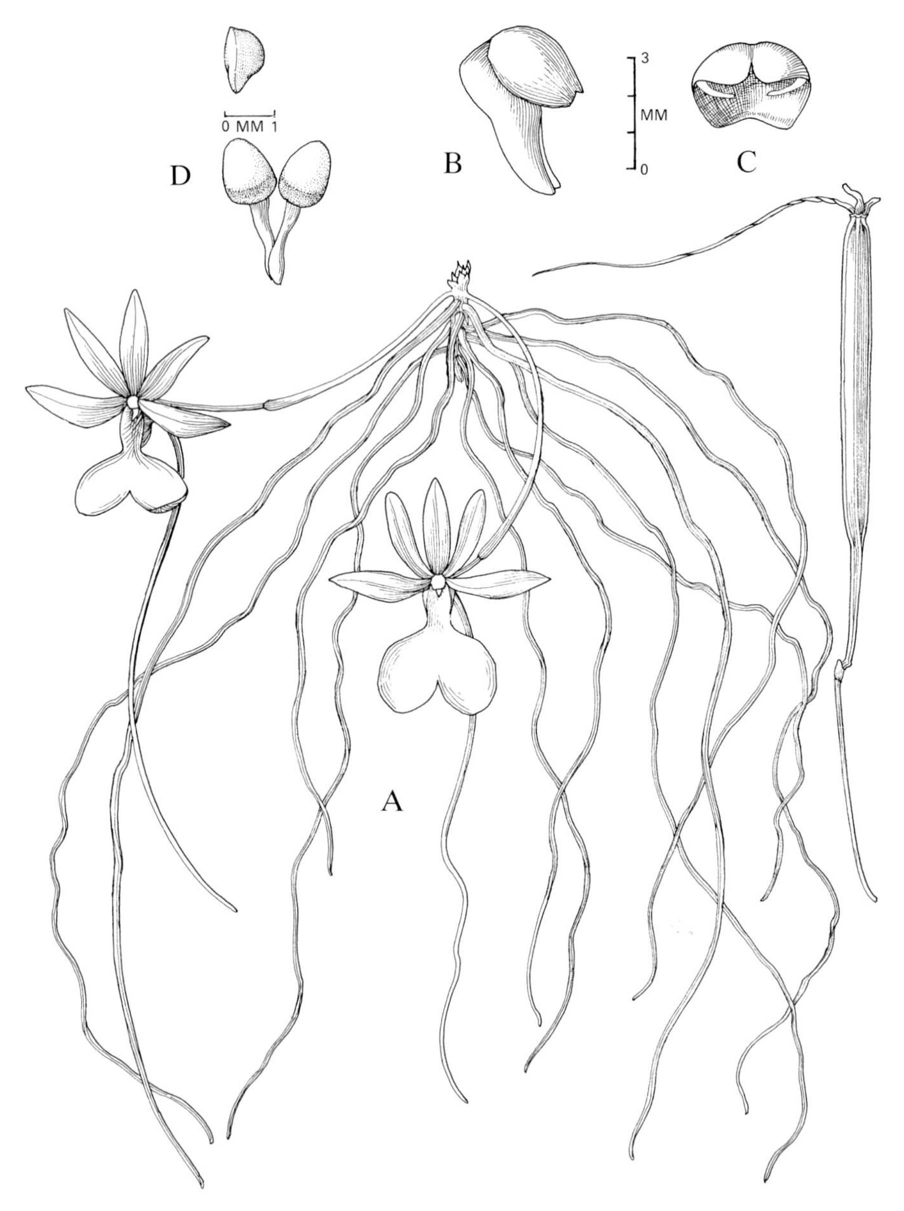

FIG. 75 **Dendrophylax fawcettii.** A, habit. B, column with terminal anther. C, front view of column with pollinia removed. D, pollinia. E, fruit. (G.)

SUBCLASS 2. DICOTYLEDONES

SUBCLASS 2. DICOTYLEDONES

The dicotyledonous families are arranged in the following sequence, in accordance with the taxonomic system of Cronquist:

ORDER 12. MAGNOLIALES

Family 17. Annonaceae

18. Canellaceae

19. Lauraceae

ORDER 13. PIPERALES

Family 20. Piperaceae

ORDER 14. NYMPHAEALES

Family 21. Nymphaeaceae

ORDER 15. RANUNCULALES

Family 22. Menispermaceae

ORDER 16. PAPAVERALES

Family 23. Papaveraceae

ORDER 17. URTICALES

Family 24. Ulmaceae

25. Moraceae

26. Urticaceae

ORDER 18. MYRICALES

Family 27. Myricaceae

ORDER 19. CASUARINALES

Family 28. Casuarinaceae

ORDER 20. CARYOPHYLLALES

Family 29. Phytolaccaceae

30. Nyctaginaceae

31. Cactaceae

32. Aizoaceae

33. Portulacaceae

34. Basellaceae

35. Chenopodiaceae

36. Amaranthaceae

ORDER 21. BATALES

Family 37. Bataceae

ORDER 22. POLYGONALES

Family 38. Polygonaceae

ORDER 23. THEALES

Family 39. Clusiaceae

ORDER 24. MALVALES

Family 40. Tiliaceae

41. Sterculiaceae

42. Malvaceae

ORDER 25. VIOLALES

Family 43. Flacourtiaceae

44. Turneraceae

45. Passifloraceae

46. Caricaceae

47. Cucurbitaceae

ORDER 26. CAPPARALES

Family 48. Capparaceae

49. Cruciferae

50. Moringaceae

ORDER 27. EBENALES

Family 51. Sapotaceae

ORDER 28. PRIMULALES

Family 52. Theophrastaceae

53. Myrsinaceae

ORDER 29. ROSALES

Family 54. Crassulaceae

55. Chrysobalanaceae

56. Leguminosae

ORDER 30. MYRTALES

Family 57. Lythraceae

58. Thymelaeaceae

59. Myrtaceae

60. Onagraceae

61. Combretaceae

ORDER 31. CORNALES

Family 62. Rhizophoraceae

ORDER 32. SANTALALES

Family 63. Olacaceae

64. Loranthaceae

65. Viscaceae

ORDER 33. CELASTRALES

Family 66. Celastraceae

ORDER 34. EUPHORBIALES

Family 67. Buxaceae

68. Euphorbiaceae

ORDER 35. RHAMNALES

Family 69. Rhamnaceae

70. Vitaceae

ORDER 36. SAPINDALES

Family 71. Surianaceae

72. Sapindaceae

73. Burseraceae

74. Anacardiaceae

75. Simaroubaceae

76. Rutaceae

77. Meliaceae

78. Zygophyllaceae

ORDER 37. LINALES

Family 79. Erythroxylaceae

ORDER 38. POLYGALALES

Family 80. Malpighiaceae

81. Polygalaceae

ORDER 39. UMBELLALES

Family 82. Umbelliferae

ORDER 40. GENTIANALES

Family 83. Loganiaceae

84. Gentianaceae

85. Apocynaceae

86. Asclepiadaceae

ORDER 41. POLEMONIALES

Family 87. Solanaceae

88. Convolvulaceae

89. Menyanthaceae

90. Hydrophyllaceae

ORDER 42. LAMIALES

Family 91. Boraginaceae

92. Verbenaceae

93. Avicenniaceae

94. Labiatae

ORDER 43. SCROPHULARIALES

Family 95. Oleaceae

96. Scrophulariaceae

97. Myoporaceae

98. Bignoniaceae

99. Acanthaceae

ORDER 44. CAMPANULALES

Family 100. Goodeniaceae

ORDER 45. RUBIALES

Family 101. Rubiaceae

ORDER 46. ASTERALES

Family 102. Compositae

KEY TO THE DICOTYLEDONOUS FAMILIES

1. Plants evidently parasitic:

  2. Stems twining and leafless: **19. Lauraceae**

  2. Stems shrubby and bearing green leaves:

    3. Flowers bisexual, borne in racemes: **64. Loranthaceae**

    3. Flowers unisexual, borne in rows partly embedded in internodes of articulate spikes: **65. Viscaceae**

1. Plants not, or not evidently, parasitic:

  4. Trees with leaves reduced to minute scales forming whorls at joints of thin, green, ribbed twigs; fruits in small, hard, globose heads: **[28. Casuarinaceae]**

  4. Not as above:

    5. Aquatic herbs with roundish, usually floating leaves:

      6. Leaves reticulate-ribbed beneath; flowers with numerous separate glabrous petals arranged spirally, the inner ones passing gradually into the many stamens: **21. Nymphaeaceae**

      6. Leaves smooth beneath; flowers with 5-lobed, fringed-hairy corolla and 5 distinct stamens: **89. Menyanthaceae**

    5. Terrestrial plants, or if aquatic the leaves aerial:

      7. Perianth absent; flowers often small or minute:

        8. Flowers bisexual, in dense, slender but fleshy spikes; ovary 1-celled: **20. Piperaceae**

        8. Flowers unisexual; inflorescence various:

          9. Leaves gland-dotted:

            10. Leaves glandular on both sides: **27. Myricaceae**

            10. Leaves glandular beneath only: **95. Oleaceae**

          9. Leaves not gland-dotted; ovary 2-4-celled:

11. Leaves opposite, fleshy, subterete; inflorescence cone-like; ovary 4-celled, each cell with solitary basal ovule: **37. Bataceae**

11. Leaves alternate (or if opposite, the plants with latex); inflorescence not cone-like; ovary usually 2–3-celled, each cell with 1 or 2 pendulous ovules: **68. Euphorbiaceae**

7. Perianth present:

12. Plants sometimes with latex; flowers unisexual, plants monoecious or dioecious; fruit a capsule, splitting into as many cocci as there are ovary-cells, or else a drupe: **68. Euphorbiaceae**

12. Plants not with combined characters as given above:

13. Perianth in 1 series, or apparently so:

14. Leaves opposite:

15. Anthers united into a ring or tube, the filaments free; flowers in heads, subtended by an involucre of bracts: **102. Compositae**

15. Anthers free:

16. Stipules present: **26. Urticaceae**

16. Stipules absent:

17. Ovary 1-celled:

18. Perianth petaloid, more or less tubular; fruit an anthocarp: **30. Nyctaginaceae**

18. Perianth dry, scarious, neither petaloid nor tubular; fruit an utricle: **36. Amaranthaceae**

17. Ovary 2- or more-celled:

19. Shrubs; flowers unisexual:

20. Monoecious; stems erect, bearing flat leathery leaves; fruit a 3-horned capsule: **67. Buxaceae**

20. Dioecious; stems arching, bearing fleshy subterete leaves; flowers minute, in small conelike spikes: **37. Bataceae**

19. Herbs; flowers bisexual; capsules not horned:

21. Trailing succulent herbs with terete stems: **32. Aizoaceae**

21. Erect herbs with 4-angled stems: **57. Lythraceae**

14. Leaves alternate:

22. Anthers united into a ring or tube, the filaments free; flowers in heads, subtended by an involucre of bracts: **102. Compositae**

22. Anthers free:

23. Fruits united in a headlike or hollow syncarp; plants with latex: **25. Moraceae**

23. Fruits not syncarpous; latex absent:

24. Ovary inferior or nearly so:

25. Herbs; flowers in loose umbels: **82. Umbelliferae**

25. Shrubs or trees; flowers in spikes, panicles or dense heads: **61. Combretaceae**

24. Ovary superior or nearly so:

26. Stipules present:

27. Stipules united to form a membranous or papery sheath closely surrounding the stem above each node: **38. Polygonaceae**

27. Stipules free, often minute and inconspicuous:

28. Flowers bisexual: **29. Phytolaccaceae**

28. Flowers unisexual or polygamous:

29. Stamens 8 or more: **43. Flacourtiaceae**

29. Stamens less than 8: **24. Ulmaceae**

26. Stipules absent or apparently so:

30. Ovary 3–6-celled; viscid shrub; fruit a papery-winged capsule: **72. Sapindaceae**

30. Ovary 1-celled; fruit not a papery-winged capsule:

31. Shrubs or small trees:

32. Flowers bisexual; perianth 6-parted; stamens 9, the anthers 4-celled; fruit a black or green drupe: **19. Lauraceae**

32. Flowers unisexual, plants monoecious; perianth 4-parted; stamens 8, the anthers 2-celled; fruit a white drupe: **58. Thymelaeaceae**

31. Herbs (rarely somewhat shrubby) or vines:

33. Stamens same in number as the perianth-segments:

34. Slender twining herbaceous vine without tendrils: **34. Basellaceae**

34. Plants erect or prostrate, not twining:

35. Inflorescence a fleshy spike with minute flowers embedded, or else a cluster with bracts becoming hard and cristate-toothed in fruit: **35. Chenopodiaceae**

35. Inflorescence neither fleshy nor the bracts becoming cristate-toothed: **36. Amaranthaceae**

33. Stamens more numerous than the perianth-segments; vine with tendrils and pink flowers: **38. Polygonaceae**

13. Perianth in 2 series, or apparently so:

36. Segments of the inner perianth-series separate, or united only at the extreme base:

37. Petals more than 10; plants succulent, leafless or with rudimentary leaves; plant-body more or less spiny: **31. Cactaceae**

37. Petals less than 10; plants obviously leafy:

38. Fruit a legume (i.e., a pod-like capsule formed from a single free carpel which splits along both ventral and dorsal edges into two halves), or an indehiscent or fragmenting modification of this type: **56. Leguminosae**

38. Fruit not a legume:

39. Stamens more than twice as many as the petals:

40. Leaves opposite:

41. Ovary superior: **39. Clusiaceae**

41. Ovary more or less inferior: **59. Myrtaceae**

40. Leaves alternate:

42. Flowers 3-parted (or with 6 petals); leaves simple; ovary with carpels not completely united (becoming more or less fused in aggregate fruits): **17. Annonaceae**

42. Flowers 4–5-parted; ovary syncarpous or of 1 carpel:

43. Leaves compound, with numerous glandular-pellucid dots: **76. Rutaceae**

43. Leaves simple:

44. Stipules, stipular glands, or stipular hairs present:

45. Ovary at least partly inferior:

46. Succulent herbs: **33. Portulacaceae**

46. Shrub or small tree: **55. Chrysobalanaceae**

45. Ovary superior:

47. Ovary 1-celled, exserted on a stalk (gynophore): **48. Capparaceae**

47. Ovary 2- or more-celled, sessile:

48. Filaments free except at base: **40. Tiliaceae**

48. Filaments more or less united into a tube or sheath:

49. Petals 5, equal and regular; hairs often stellate: **42. Malvaceae**

49. Petals 3, unequal (2 upper ones and a boat-shaped keel); hairs simple: **81. Polygalaceae**

44. Stipules absent:

50. Prickly herb with yellow sap: **23. Papaveraceae**

50. Shrub or small tree without prickles; sap colorless; foliage aromatic: **18. Canellaceae**

39. Stamens twice as many as the petals or less:

51. Ovary more or less inferior:

52. Plants herbaceous:

53. Vines with tendrils: **47. Cucurbitaceae**

53. Not vines; tendrils absent:

54. Flowers in umbels; ovary 2-celled with 1 ovule in each cavity; leaves sheathing at the base: **82. Umbelliferae**

54. Not as above:

55. Plants succulent, stems prostrate or ascending; ovary 1-celled; capsule splitting transversely: **33. Portulacaceae**

55. Plants not succulent, stems erect; ovary 4–6-celled; capsule splitting lengthwise: **60. Onagraceae**

52. Plants woody (shrubs or trees):

56. Leaves opposite:

57. Stilt-roots present; inflorescence 1–3-flowered; petals hairy: **62. Rhizophoraceae**

57. Stilt-roots absent; inflorescence many-flowered; petals glabrous: **61. Combretaceae**

56. Leaves alternate: **69. Rhamnaceae**

51. Ovary more or less superior, at least in flower:

58. Ovary half-immersed in the disk, which increases in size as the ovary ripens until it almost completely envelops the fruit; shrub or small tree: **63. Olacaceae**

58. Ovary completely superior, or if partly immersed in a disk, this not enlarging in fruit:

59. Leaves simple:

60. Stipules present (these small, or sometimes conspicuous):

61. Vines with tendrils:

62. Ovary stalked; flowers with persistent petals and conspicuous filamentous corona: **45. Passifloraceae**

62. Ovary sessile; flowers with minute petals, these soon falling; corona absent: **70. Vitaceae**

61. Not vines:

63. Leaves opposite:

64. Calyx bearing prominent sessile glands: **80. Malpighiaceae**

64. Calyx not glandular:

65. Flowers in terminal panicles; stamens 8: **57. Lythraceae**

65. Flowers in axillary cymes; stamens 4 or 5: **66. Celastraceae**

63. Leaves alternate:

66. At least some flowers unisexual; fruit a 2–3-seeded leathery capsule or dry drupe: **66. Celastraceae**

66. All flowers bisexual:

67. Leaves with 2 glands at base of blade; flowers yellow; ovary 1-celled: **44. Turneraceae**

67. Leaves without glands; flowers not yellow; ovary 3- or 5-celled:

68. Ovary 5-celled; hairs often stellate: **41. Sterculiaceae**

68. Ovary 3-celled; hairs never stellate:

69. Pedicels terete; stamens free to the base; fruit a 3-seeded capsule: **69. Rhamnaceae**

69. Pedicels angled; stamens united below; fruit a 1-seeded drupe: **79. Erythroxylaceae**

60. Stipules absent:

70. Ovary lobed or the carpels more or less distinct:

71. Twining vine; flowers unisexual; plants dioecious: **22. Menispermaceae**

71. Erect shrub; flowers bisexual with yellow petals: **71. Surianaceae**

70. Ovary entire, with fused carpels:

72. Treelike, usually unbranched giant herb, with thick hollow stem and large palmately-lobed leaves: **[46. Caricaceae]**

72. Not as above:

73. Small seaside herb with 4-parted white flowers and 6 stamens (4 long and 2 short); fruit a spindle-shaped indehiscent capsule: **49. Cruciferae**

73. Plants woody (trees):

74. Stamens 1–5; fruit a large 1-seeded drupe; leaves not gland-dotted: **74. Anacardiaceae**

74. Stamens 20 or more; fruit a large juicy berry with aromatic rind; leaves with pellucid glandular dots: **76. Rutaceae**

59. Leaves compound:

75. Leaves 1-pinnate (or 1-palmate):

76. Leaves opposite; stipules present; flowers yellow or blue: **78. Zygophyllaceae**

76. Leaves alternate; stipules absent; flowers neither yellow nor blue:

77. Plants herbaceous; leaves palmately divided: **48. Capparaceae**

77. Plants woody (shrubs or trees); leaves pinnate or 3-foliolate:

78. Leaflets more than 18 per leaf (up to 51), small, whitish beneath; flowers unisexual, plants dioecious: **75. Simaroubaceae**

78. Leaflets 3–17 per leaf; flowers bisexual or plants polygamo-dioecious:

79. Stamens united to form a tube; fruit a dehiscent capsule: **77. Meliaceae**

79. Stamens free or united only at the base; fruit a capsule or drupe:

80. Ovary 1-celled or the carpels free; fruit a drupe or of drupelets:

81. Foliage pellucid-dotted, or if not the stems spiny: **76. Rutaceae**

81. Foliage not pellucid-dotted; stems not spiny; sap irritating to the skin: **76. Anacardiaceae**

80. Ovary 3–5 celled:

82. Fruit a capsule:

83. Capsule 3-valved; stamens 8–10:

84. Capsule usually more than 6 cm long; stamens 8: **72. Sapindaceae**

84. Capsule less than 1.3 cm long; stamens 10: **73. Burseraceae**

83. Capsule 5-valved; stamens 4–6: **77. Meliaceae**

82. Fruit a drupe:

85. Drupe thin-fleshed, less than 3 cm in diameter, with dry leathery rind; flowers white or greenish: **72. Sapindaceae**

85. Drupe thick-fleshed, more than 3 cm in diameter, with thin epidermis; flowers red-purple: **74. Anacardiaceae**

75. Leaves 2-pinnate or more divided:

86. Ovary 1-celled; small tree with long triangular capsules containing many 3-winged seeds: **[50. Moringaceae]**

86. Ovary 3–6-celled:

87. Vines with tendrils; stamens 8, free; fruit a membranous inflated capsule: **72. Sapindaceae**

87. Small tree; stamens 10–12, united into a tube; fruit a small drupe: **77. Meliaceae**

36. Segments of the inner perianth-series united into one more or less tubular structure (corolla) at more than the extreme base:

88. Ovary inferior, at least during flowering:

89. Stipules present and often conspicuous, interpetiolar between opposite leaves: **101. Rubiaceae**

89. Stipules absent; leaves alternate or opposite:

90. Herbaceous vines with tendrils: **47. Cucurbitaceae**

90. Not vines; tendrils absent:

91. Flowers in heads subtended by an involucre of bracts; calyx highly modified in the form of scales, bristle or awns: **102. Compositae**

91. Flowers few and separate in stalked axillary dichasia; small fleshy seaside shrub; fruit a small black drupe: **100. Goodeniaceae**

88. Ovary not inferior at time of flowering:

92. Ovary apparently inferior in fruit:

93. Plants herbaceous with opposite leaves and funnel-shaped corolla-like perianth: **30. Nyctaginaceae**

93. Plants woody with alternate leaves; ovary half-immersed in the disk, which enlarges as the ovary ripens until it almost completely envelopes the fruit: **63. Olacaceae**

92. Ovary superior in flower and fruit:

94. Stamens twice as many as the corolla-lobes:

95. Plants woody (branched shrub or small tree); all leaves simple and entire; corolla bearded within: **63. Olacaceae**

95. Plants herbaceous (if treelike in stature, the stem hollow and usually unbranched, the leaves palmately lobed or compound):

96. Corolla 5-lobed; treelike unbranched giant herb with thick hollow stem and large palmate leaves: **[46. Caricaceae]**

96. Corolla 4-lobed; succulent herbs with simple or pinnately divided leaves: **54. Crassulaceae**

94. Stamens as many as or fewer than the corolla-lobes:

97. Stamens opposite to and equal in number with the corolla-lobes:

98. Plants with latex; ovary 4–14-celled: **51. Sapotaceae**

98. Plants without latex; ovary 1-celled:

99. Flowers in terminal racemes, the parts in 5's; leaves of stiff hard texture: **52. Theophrastaceae**

99. Flowers in small lateral clusters among or below the leaves (often at leafless nodes), the parts in 4's; leaves of thin, herbaceous texture: **53. Myrsinaceae**

97. Stamens alternating with the corolla-lobes and equal in number or fewer:

100. Corolla regular (actinomorphic), the divisions of equal size and similar in shape:

101. Leaves all opposite, or rarely whorled:

102. Stipules present: **83. Loganiaceae**

102. Stipules absent:

103. Stamens connected with the stigma to form a central column of intricate structure; plants with latex: **86. Asclepiadaceae**

103. Stamens separate:

104. Stamens 2; fruit a drupe or berry: **95. Oleaceae**

104. Stamens 4 or 5; fruit a capsule or follicle:

105. Corolla strongly twisted in bud:

106. Ovary 1-celled; fruit a capsule with numerous naked seeds; plants without latex: **84. Gentianaceae**

106. Ovary of 2 separate carpels initially connected by a common style; fruit of 2 follicles, the seeds with or without hairs or other appendages; plants usually with latex: **85. Apocynaceae**

105. Corolla not twisted in bud; latex absent:

107. Ovary 2-celled with numerous ovules in each cell; flowers solitary or paired: **96. Scrophulariaceae**

107. Ovary 1-4-celled with 1-4 ovules in each cell; flowers in cymes, spikes or heads:

108. Ovary 1-celled, the ovules pendulous from a free basal placenta; shrub or tree of saline habitats, the roots bearing upright aerial pneumatophores: **93. Avicenniaceae**

108. Ovary 2–4-celled; herbs, shrubs or trees not of saline habitats; pneumatophores lacking: **92. Verbenaceae**

101. Leaves all or mostly alternate (rarely some of them opposite or whorled):

109. Ovules 1 or 2 in each cavity:

110. Ovary 2- or 4-celled, entire; corolla usually trumpet- or salver-shaped, strongly twisted in bud; mostly trailing or twining herbaceous vines, sometimes with latex, rarely a small erect shrub: **88. Convolvulaceae**

110. Ovary 4-celled, usually 4-lobed; flowers in cymes, the corolla not or but slightly twisted in bud; herbs, shrubs, woody vines, or trees, never with latex: **91. Boraginaceae**

109. Ovules several to many in each cavity:

111. Stamens 4, the 5th a short staminode or absent; fruit a 2-celled, 2- or 4-valved capsule; ovules numerous, borne on swollen axile placentas: **96. Scrophulariaceae**

111. Stamens 5, all functional:

112. Ovary of 2 separate carpels united initially by the style; fruit of 2 follicles containing numerous winged seeds; latex copious: **85. Apocynaceae**

112. Ovary simple with fused carpels; seeds not winged; latex absent:

113. Ovary 1-celled with 2 parietal placentas; prostrate or decumbent annual herb with solitary or paired axillary flowers; fruit a small capsule: **90. Hydrophyllaceae**

113. Ovary 2- or 4-celled with thick axile placentas; erect herbs or shrubs, or woody vines, the flowers solitary or in cymose clusters or racemes: **87. Solanaceae**

100. Corolla irregular or oblique (zygomorphic), at least one of the divisions differing in size or shape from the others:

114. Leaves compound: **98. Bignoniaceae**

114. Leaves simple:

115. Flowers solitary or rarely several in a cluster:

116. Corolla 4 cm long or more, broadly bell-shaped, glabrous; ovary 1-celled with numerous ovules; fruit a large globose berry with a hard shell: **98. Bignoniaceae**

116. Corolla 2 cm long, narrowly cylindric with 2-lipped limb, the middle lobe of the lower lip densely bearded; ovary 2-celled with 4 ovules in each cavity; fruit a small drupe: **97. Myoporaceae**

115. Flowers in distinct inflorescences:

117. Ovules solitary in each cavity; fruit a drupe, or else a schizocarp of nutlets; plants often aromatic:

118. Style arising from the base of the ovary; ovary 4-celled, 4-lobed: **94. Labiatae**

118. Style terminal on the ovary; ovary 1–4-celled, entire: **92. Verbenaceae**

117. Ovules 2 or more in each cavity; fruit a capsule; plants not aromatic:

119. Fruit a beaked, 2-valved capsule, splitting elastically to the very base, flinging the seeds out by a sling action: **99. Acanthaceae**

119. Fruit a 2–4-valved capsule opening or splitting at the top but never all the way to the base, the segments not elastic: **96. Scrophulariaceae**

Family 17

ANNONACEAE

Trees or shrubs with alternate, entire leaves; stipules lacking. Flowers mostly perfect and 3-parted; sepals usually 3, valvate or imbricate; petals commonly 6 in two series, valvate or imbricate, the inner ones often rudimentary or absent. Stamens numerous, the anther-cells adnate; carpels numerous (rarely few), generally free, with 1 or more ovules in each cell, in fruit free or united to form a fleshy multiple fruit. Seeds with or without an aril, with copious ruminate endosperm and a minute embryo.

A pantropical family of about 75 genera or more, and more than 2000 species, the majority in the Old World.

Genus 1. **ANNONA** L.

Trees or shrubs, glabrous or with simple or stellate pubescence. Flowers solitary or in few-flowered clusters, these terminal, opposite the leaves, or apparently internodal; sepals 3, small, valvate; petals 6, free or connate at base, biseriate, the inner ones small or lacking, the outer valvate, fleshy, and usually concave. Stamens extrorse, the connective produced above the cells into a disk. Carpels often connate, containing solitary, basal, erect ovules. Fruit a fleshy aggregate, often edible.

A chiefly tropical American genus of about 120 species. In addition to those occurring in the Cayman Islands, a number of others are important for their edible fruits.

KEY TO SPECIES

1. Flowers globose or broadly pyramidal in bud; fruits smooth or covered with soft spines:

 2. Leaves more or less rounded at base and widest at or below the middle; pedicels glabrous; fruits smooth: **1. A. glabra**

 2. Leaves more or less acute at base and widest above the middle; pedicels sericeous; fruits soft-spiny: **2. A. muricata**

1. Flowers oblong or narrowly oblong in bud; fruits covered with rounded tubercles: **3. A. squamosa**

1. Annona glabra L., Sp. Pl. 1:537. 1753.

"Pond-apple".

A shrub or small tree to 6 m tall or more, the trunk often enlarged or buttressed at base. Leaves of stiff texture, petiolate, ovate-elliptic to oblong-elliptic, 7–14 cm long, 3–8 cm broad, acute at apex and rounded or obtuse at base. Flowers solitary, internodal, on pedicels 1.5–2 cm long; sepals 3–5 mm long, rounded and apiculate; petals 6, the outer ones ovate, 2.5–3 cm long, the inner ones smaller. Fruit globose-ovid, 5-12 cm long, smooth, yellowish at maturity, edible but insipid; seeds brown.

GRAND CAYMAN: *Brunt 1837, 2046.*
– Common at or near sea-level in suitable habitats throughout tropical America; also occurs in western Africa. This species grows in swamps or wet thickets, or often near mangroves.

2. **Annona muricata** L., Sp. Pl. 1:536. 1753.

"Soursop".

A small tree up to 8 m tall, the young branchlets clothed with minute reddish-brown hairs, soon becoming glabrate. Leaves lustrous, petiolate, oblong-obovate, 8-15 cm long, 3-6 cm broad, minutely sericeous beneath when young and with persistent domatia in the nerve-axils, abruptly acute at apex and less abruptly so at base. Flowers solitary, terminal or opposite leaves, on pedicels 1.5-2 cm long; sepals c. 3 mm long, broadly deltate; petals 6, the outer ones ovate-acuminate, thick, 2.5-3.5 cm long, the inner ones smaller. Fruit asymmetrically oblong-ovoid, up to 20 cm long or more, green at maturity and covered with curved, flexible spines, edible; seeds black.

CAYMAN BRAC: *Proctor 29107.* Probably occurs on Grand Cayman, but no specimens were seen during the preparation of the present work.
– Apparently indigenous to the West Indies, but widely cultivated and becoming naturalized in many tropical countries.

3. **Annona squamosa** L., Sp. Pl. 1:537. 1753. **FIG. 76.**

"Sweetsop".

A shrub or small tree up to 6 m tall with a rounded or spreading crown, the young branchlets grayish-sericeous. Leaves petiolate, elliptic or lance-elliptic, 5-11 cm long, 2-5 cm broad, acute at both ends, usually paler beneath and bearing small, scattered, soft hairs. Flowers greenish, solitary or several in small clusters opposite leaves, on pedicels 1-2 cm long; sepals 1.5-2 mm long, rounded-deltate; petals 6, the outer linear-oblong and 1.5-3 cm long, the inner ones rudimentary. Fruit globose or ovoid, usually 7-9 cm in diameter, glaucous, the carpels incompletely fused and projecting as smooth rounded tubercles, edible; seeds brown.

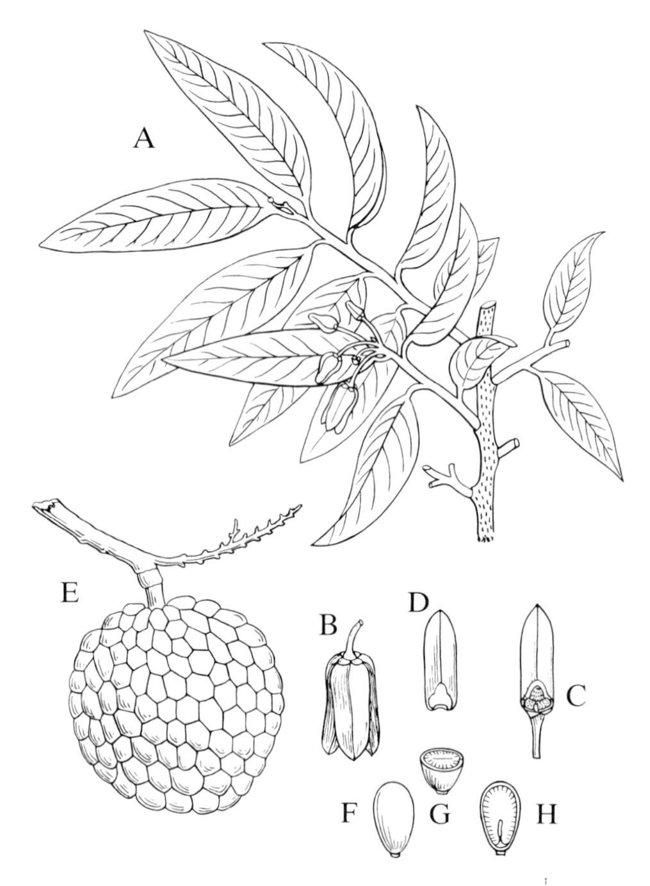

FIG. 76 **Annona squamosa.** A, habit, x⅓. B, flower, x⅔. C, receptacle and petal, x⅔. D, petal, x⅔. E, fruit, x⅓. F, seed, x⅔. G, H, x- and long. sections of seed. (F. & R.)

GRAND CAYMAN: *Brunt 2196; Kings GC 144; Proctor 15250.* CAYMAN BRAC: *Proctor 28964.*

– Apparently indigenous to the West Indies, but widely cultivated in tropical America and easily becoming naturalized.

The Custard-apple, *Annona reticulata* L., has been reported from Grand Cayman but no specimens have been seen by the present writer.

Family 18

CANELLACEAE

Small trees or shrubs with aromatic bark and alternate, entire leaves; stipules lacking. Flowers perfect, in axillary or terminal cymes or corymbs, rarely solitary; sepals 3, imbricate; petals 4 or 5, usually free, imbricate, sometimes alternating with petaloid scales; stamens rather numerous, the filaments united into a tube, the anthers attached closely together outside the tube and opening longitudinally; ovary free, 1-celled, with 2 or more ovules; style short, thick, with a 2-6-lobed stigma. Fruit a berry with 2 or more seeds, these smooth and hard; endosperm copious, fleshy, the embryo short.

A small family of 5 genera and about 16 species widely scattered in tropical America, eastern Africa and Madagascar.

Genus 1. **CANELLA** P.Br.

Evergreen glabrous trees or shrubs, the bark and leaves pleasantly aromatic. Flowers purple, red or violet in terminal, bracteolate corymbs; petals 5, petaloid scales lacking; stamens 10-20, the anthers contiguous on the filament-tube; ovary with 2 or 3 parietal placentas, each bearing 2 ovules. Berry globose, the gelatinous pulp enclosing few, obovoid to globose seeds.

A genus of 2 species, one occurring in Florida and the West Indies, the other in Colombia.

1. **Canella winterana** (L.) Gaertn., Fruct. & Sem. Pl. 1:373. 1788. **FIG. 77.**

"Pepper cinnamon".

A shrub or small tree rarely to 15 m tall, leaves oblanceolate or spatulate, 3-12 cm long, dark green and rather glossy above, the tissue minutely and closely gland-dotted, on petioles c. 1 cm long. Inflorescence several-to-many-flowered, the flowers on slender pedicels; sepals rounded, c. 3 mm long, ciliolate; petals blood-red, obovate, nearly twice as long as the sepals. Staminal tube 3-4 mm long, bearing yellow anthers. Berries deep orimson, 8-10 mm in diameter; seeds black.

GRAND CAYMAN: *Brunt 1990, 2143, 2151; Correll & Correll 51041; Hitchcock; Proctor 15186.* LITTLE CAYMAN: *Kings LC 15; Proctor 28117, 28177.* CAYMAN BRAC: *Kings CB 48; Matley; Proctor 28955, 29052.*
– Florida and the West Indies south to Barbados. Common in rocky woodlands and thickets, chiefly on limestone and at lower to middle elevations. Canella bark was formerly used in medicine as an aromatic stimulant and tonic. The berries,

when eaten by pigeons, impart to the flesh a distinct and pleasing flavour. This species has horticultural value as an ornamental, though apparently not often planted.

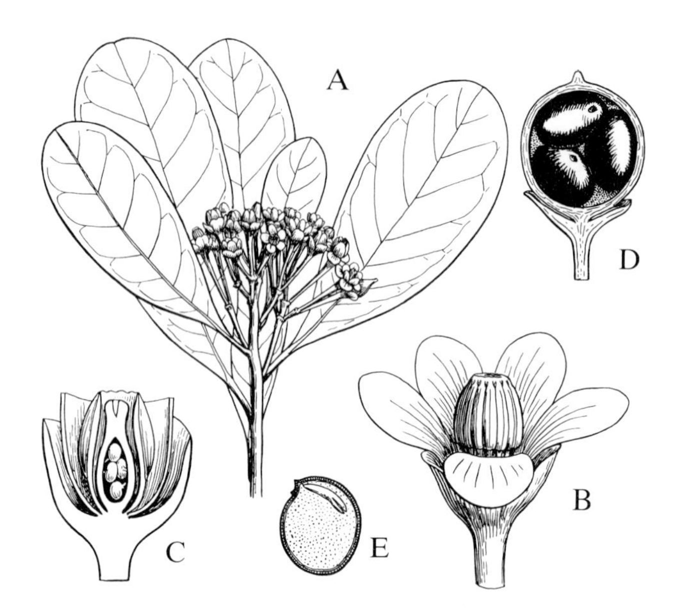

FIG. 77 **Canella winterana.** A, branch with leaves and flowers, x$^2/_3$. B, flower, x4. C, flower cut lengthwise, x4. D, fruit cut vertically, x2. E, seed cut lengthwise, x2. (F. & R.)

Family 19

LAURACEAE

Trees, shrubs, or leafless parasitic herbs; leaves alternate and simple; stipules lacking. Flowers perfect or dioecious, the axillary panicles or (in *Cassytha*) in spikes; perianth segments[1] usually 6, in two whorls, fused below into a short tube which develops into a cupule at the base of the fruit. Stamens in a double ring, an outer ring of 6 perfect stamens, and an inner ring of which 3 are perfect, all opening by valves, alternating with 3 staminodes which are often very small or apparently lacking. Ovary superior, free from the perianth-tube, 1-celled with 1 pendulous ovule. Fruit a 1-seeded berry (drupe), with the persistent perianth-tube usually forming a cupule at the base, or else enveloped by the fleshy receptacle; seed relatively large, without endosperm, the cotyledons thick and fleshy.

---

[1] Sepals and petals not distinguished.

A chiefly pantropical family of about 32 genera and more than 1,100 species, especially common in South America and southeast Asia. Many of them are important as timber trees, while others contain aromatic spices of commercial value. Examples of the latter are those producing cinnamon *(Cinnamomum zeylanicum)* and camphor *(Cinnamomum camphora)*.

KEY TO GENERA

1. Trees with green leaves:

   2. Cultivated tree; staminodes large, sagittate; perianth-segments persistent in fruit, a cupule lacking; fruit large and edible: **[1. Persea]**

   2. Wild tree; staminodes minute or lacking; perianth-segments deciduous in fruit, a cupule present; fruit small and not edible: **2. Ocotea**

1. Twining parasitic herb without green leaves; fruit enclosed by the enlarged and persistent perianth: **3. Cassytha**

[Genus 1. **PERSEA** Mill.

The "Avocado" or "Pear", *Persea americana* Mill., is often cultivated.

GRAND CAYMAN: *Millspaugh 1317.*
– Apparently indigenous to the region of Honduras, Guatemala, and southern Mexico, this species is a highly variable complex of cultivars whose genetic relationships are not well understood.]

Genus 2. **OCOTEA** Aubl.

Evergreen trees or rarely shrubs with coriaceous leaves, glabrous or variously pubescent. Flowers small, perfect or polygamo-monoecious, in axillary or terminal panicles; perianth 6-parted, its lobes nearly equal. Stamens 9, shorter than the perianth; anthers 4-celled, on very short filaments; staminodes none or small. Berry globose or ellipsoid, seated in a cupule (enlarged calyx-base).

A chiefly tropical American genus of more than 300 species. The genus *Nectandra* is here not considered separable from *Ocotea*.

1. Ocotea coriacea (Sw.) Britton in Britton & Millsp., Fl. Baham. 143. 1920.

Nectandra coriacea (Sw.) Griseb., 1860.

"Sweetwood".

A small tree to 10 m tall or more; leaves elliptic, lance-elliptic, or lance-oblong, chiefly 6-13 cm long, 2-4.5 cm broad, acuminate, glabrous or nearly so, glossy dark green and of hard texture, with finely reticulate veins. Panicles axillary, shorter than the leaves; flowers creamy-white, sweet-scented, on red pedicels. Berry black, ovoid or subglobose, 10-13 mm long, each seated in a 1-margined, bright red cupule.

GRAND CAYMAN: *Brunt 2160; Lewis 3862.* CAYMAN BRAC: *Proctor 29015, 29341.*

– Florida, West Indies, and the Yucatan Peninsula, in woodlands at low elevations. This tree has some use in carpentry; it is reported that the flowers yield good honey, and occasionally cattle eat the fruits. It is quite attractive when planted as a shade tree.

Genus 3. **CASSYTHA** L.

Twining parasitic herbs, adhering to the host by means of suckers (haustoria); leaves reduced to small scales. Flowers perfect, in spikes or racemes; perianth-segments 6, fused at base, arranged in two series, the 3 segments of the outer series much the smaller. Stamens 9, the staminodes 3. Ovary 1-celled with 1 ovule. Fruit globose and fleshy, enclosed by the succulent perianth-tube and crowned by the persistent perianth-segments.

A pantropical genus of about 20 species, the majority Australian.

1. **Cassytha filiformis** L., Sp. Pl. 1:35. 1753. **FIG. 78.**

"Old Man Berry".

Twining parasitic herb with slender, elongate yellow or greenish stems, freely branched, up to 4 m long or more, sometimes matted. Scales (reduced leaves) few, 1-2 mm long. Spikes few-several-flowered, 1-2 cm long; flowers white, c. 2 mm long, the inner perianth-segments ovate, larger than the outer. Fruit globose, white, 6 mm in diameter.

GRAND CAYMAN: *Brunt 1909; Hitchcock; Kings GC 165, GC 265; Proctor 15226.* LITTLE CAYMAN: *Kings LC 29, LC 82, LC 85; Proctor 28080.* CAYMAN BRAC: *Kings CB 94; Millspaugh 1168; Proctor 29096.*

– Pantropical, parasitic indiscriminately on other vegetation.

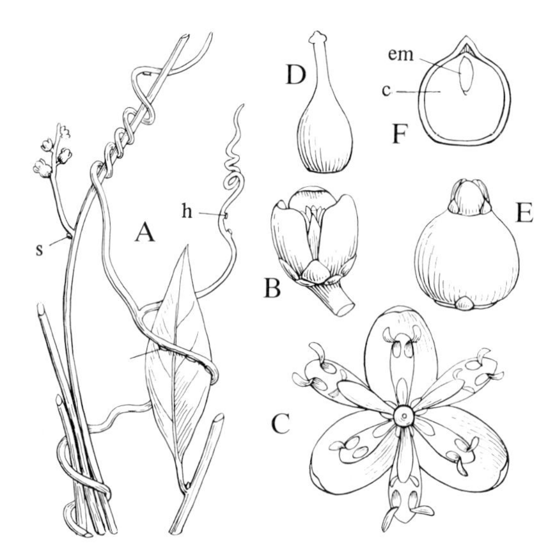

FIG. 78 **Cassytha filiformis.** A, habit, showing haustoria (h) and rudimentary leaf (s), x$^2/_3$. B, flower as normally seen, x6. C, flower flattened out, x7. D, pistil, x18. E, fruit, x2. F, long. section of fruit, showing inner face of cotyledon (c) and axis of embryo (em). (F. & R.)

Family 20

PIPERACEAE

Herbs, shrubs, or rarely small trees; leaves alternate, opposite, or whorled, nearly always simple and entire, palmately or pinnately veined, often succulent. Flowers minute, perfect, without a perianth, whorled or spirally arranged in tail-like spikes which may be terminal, opposite the leaves, or rarely axillary, sometimes several together on a common peduncle. Stamens 2–6, rarely more; ovary superior, sessile or rarely stalked, 1-celled, with 1 basal ovule. Fruit a small 1-seeded berry (drupe); seed with both perisperm and endosperm, usually pungently oily.

A pantropical family of about 9 genera and perhaps 2,000 species, or more, the majority in the Western Hemisphere.

KEY TO GENERA

1. Plants herbaceous; stigma 1: **1. Peperomia**

1. Plants more or less woody shrubs; stigmas 2–4: **2. Piper**

Genus 1. **PEPEROMIA** Ruiz & Pav.

Terrestrial or epiphytic herbs, creeping to erect, with more or less succulent stems. Leaves simple, variously arranged (alternate in Cayman species). Spikes terminal or axillary, simple or in branched inflorescences. Flowers numerous, minute, borne in the axils of round, peltate bracts; stamens 2; ovary sessile (in Cayman species), the apex obtuse, acute or beaked; stigma terminal, or lateral below the beak. Fruit very small, seed-like.

A large genus of nearly 1,000 species, of no economic importance except for a few used as ornamental pot-plants.

KEY TO SPECIES

1. Leaves acute to acuminate at apex, less than 4 cm long, minutely black-dotted; fruit not beaked: **1. P. glabella**

1. Leaves obtuse or retuse at apex, 5–13 cm long, not black-dotted; fruit beaked: **2. P. magnoliifolia**

1. Peperomia glabella (Sw.) A.Dietr. in L., Sp. Pl. ed. 6, 1:156. 1831.

A creeping or sometimes pendent herb, often rooting at the lower nodes, with ascending flowering branches. Leaves light green, petiolate, densely black-dotted, elliptic or lance-elliptic, chiefly 2–4 cm long, more or less acuminate at apex and narrowed at base, and palmately 3-nerved from near the base. Spikes slender, 1 mm thick, black-dotted, not densely-flowered, 6–12 cm long, often a cluster of several at the apex of the stem and solitary ones in the leaf-axils. Fruit ovoid, 0.3–1 mm long, the apex oblique with stigma below the apex.

GRAND CAYMAN: *Hitchcock; Kings GC 183.*
– West Indies and continental tropical America, epiphytic or on rocks. The Cayman plants were collected on shaded limestone rocks near Jackson Point.

2. Peperomia magnoliifolia (Jacq.) A.Dietr. In L., Sp. Pl. ed. 6, 1:153. 1831. **FIG. 79.**

"Wild Balsam", "Vine Balsam".

A creeping-ascending glabrous herb with succulent stems 3–5 mm thick, rooting at many of the nodes. Leaves thick and fleshy, rather dark green, mostly obovate on long, narrowly-winged petioles, 5–13 long, 3–6 cm broad above the middle, with 3-4 indistinct pinnate veins on each side. Spikes solitary in leaf-axils or up to 3

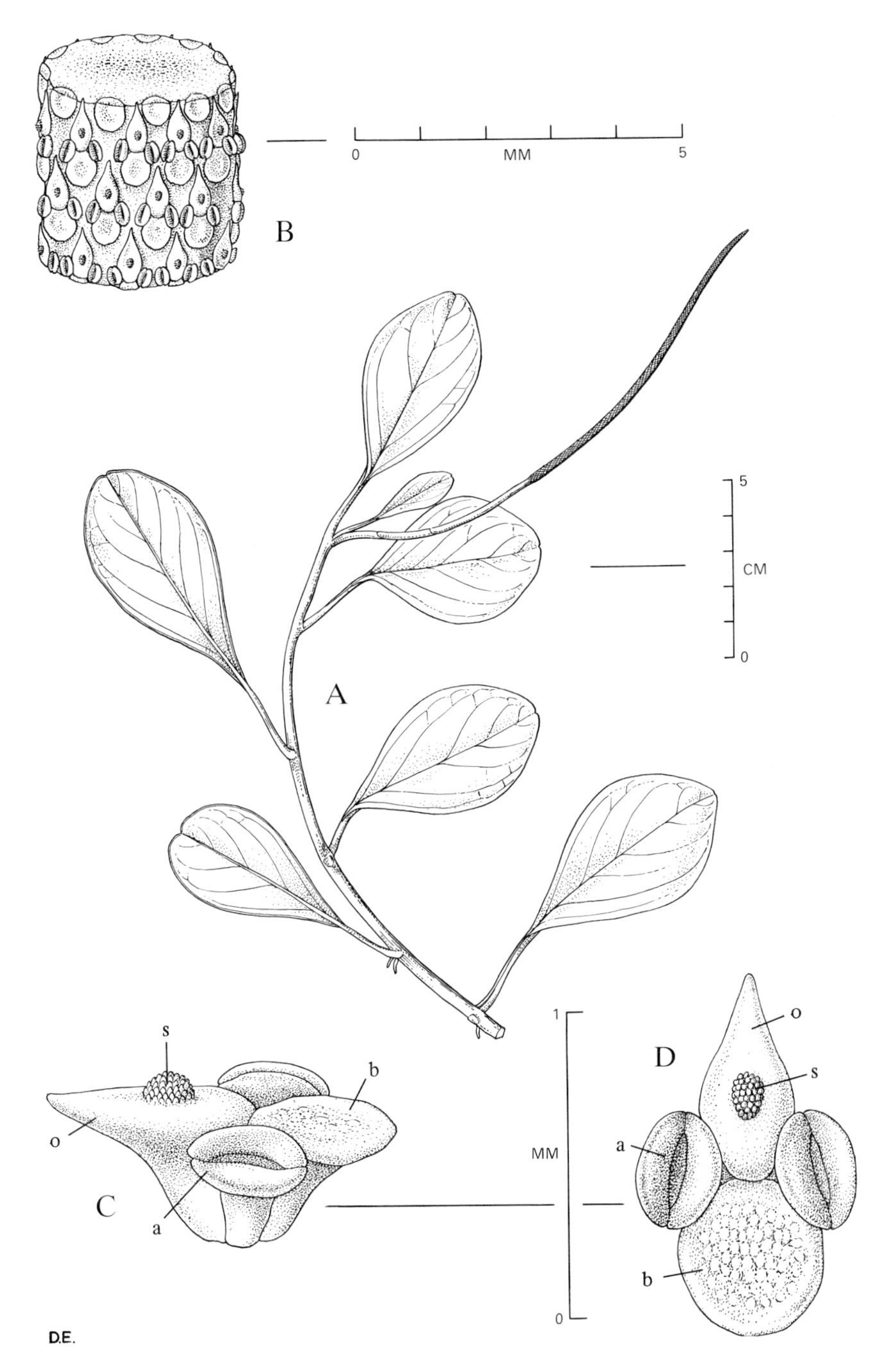

FIG. 79 **Peperomia magnoliifolia.** A, habit. B, portion of spike. C, D, two views of a flower showing bract (b), anther (a), ovary (o), and stigma (s). (D.E.)

terminal ones, the longest as much as 15 cm long, 3 mm thick, densely-flowered, on glabrous peduncles 2-3 cm long. Fruit ellipsoid, c. 1 mm long, with a curved beak not obviously hooked. (The closely related *P. obtusifolia* has peduncles puberulous and fruits with strongly hooked beaks)

GRAND CAYMAN: *Brunt 2157; Kings GC 368; Proctor 27986.*
– West Indies and continental tropical America. This species is apparently always terrestrial, rooting in accumulations of humus on rocks and boulders.

Genus 2. **PIPER** L.

Shrubs or small trees (rarely vines), with the branches often jointed at the nodes. Leaves alternate, often unequal-sided at the base, sometimes with pellucid dots, the venation pinnate or palmate. Spikes opposite the leaves, typically solitary and simple. Flowers usually perfect, numerous; stamens 2-6; ovary sessile; stigmas 2-4, sessile or on a short and thick style. Fruit variable, usually small.

A vast genus of more than 1,000 species, often numerically important in the understory of tropical vegetation, a few weedy species ubiquitous in clearings of many areas, chiefly where rainfall is high. The chief economic species is the Black Pepper, *Piper nigrum,* a subwoody vine.

1. Piper amalago L., Spl. Pl. 1:29. 1753.

"Jointer", "Pepper elder".

A shrub usually 2 or 3 m tall, nearly or quite glabrous, the slender branches with swollen joint-like nodes. Leaves chiefly ovate on distinct petioles and with long-acuminate apex, mostly 6-12 cm long and up to 5 cm broad below the middle, palmately 5-veined. Spikes 5-7 cm long and c. 1.5 mm thick, rather loosely flowered, minutely grayish-scurfy between the flowers. Fruit ovoid-conical, separated on the spike.

GRAND CAYMAN: *Brunt 2150; Hitchcock; Kings GC 233; Millspaugh 1299; Proctor 15010.*
– West Indies and continental tropical America, a common, variable, and widespread species.

Family 21

NYMPHAEACEAE

Aquatic herbs with thick, starchy, perennial rhizomes. Leaves usually floating, with very long petioles and peltate blades. Flowers solitary, large, floating on water surface or held a few inches above it; sepals 4, free; petals numerous, free, inserted in a close spiral on the receptacle. Stamens numerous, often with petaloid filaments. Carpels numerous, free or more or less united; ovules numerous, pendulous, parietal. Fruit a spongy berry, dehiscing by swelling of the mucilage surrounding the seeds; seeds with some endosperm and abundant perisperm.

A small, world-wide family of about 5 genera and 80 species.

Genus 1. **NYMPHAEA** L.

Rhizome fleshy, creeping, edible; leaves with roundish floating blades cleft nearly to the middle on one side. Flowers showy and often fragrant; inner stamens with narrow filaments and yellow anthers, passing into outer ones with broad petaloid filaments and small anthers. Carpels united into a many-celled, half-inferior ovary. Fruit ripening under water and breaking irregularly to free the seeds.

A widely-distributed genus of about 50 species, known generally as "water-lilies". The seeds of many species are both edible and nutritious, and are commonly eaten in several parts of the world.

1. **Nymphaea ampla** (Salisb.) DC., Syst. 2:54. 1821.

"Water Lily".

Leaves with blades 10–45 cm broad, the margin irregularly wavy-toothed, the upper surface green or yellowish-green, the lower surface purple and with conspicuous raised reticulate veins; petiole peltately attached almost centrally. Flowers 8–15 cm broad, opening during the day and closing at night, fragrant; sepals oblong-lanceolate, green with brown speckles on the outside; petals and petaloid filaments white; anthers of the outer stamens with connective prolonged into a appendage.

GRAND CAYMAN: *Hitchcock; Kings GC 190; Proctor 31025.*
– West Indies and Central America, in ponds and sluggish streams, apparently tolerant of brackish water. The Cayman plants were found in swampy ponds southeast of Georgetown.

Family 22

MENISPERMACEAE

Dioecious trees, shrubs, or vines with alternate entire leaves; stipules lacking. Flowers minute and inconspicuous, in axillary bracteate panicles. Sepals and petals various in number in dimerous or trimerous whorls, or sometimes solitary. Stamens of male flowers 4 or 6; staminodes in female flowers 6 or lacking; filaments free or united into a column. Carpels 3 or 1, free, each with a single ovule. Fruit a drupelet, often succulent; seeds horseshoe-shaped, with large embryo and scanty endosperm.

A chiefly tropical family with about 65 genera and 400 species. Members of this family often contain toxic alkaloids, and a number of South American species provide ingredients for arrow-poisons used by Amerindian tribes. These complex poisons, which vary in composition according to the species used, are collectively known as Curare.

Genus 1. **CISSAMPELOS** L.

Slender climbing shrubs or vines; leaves roundish, more or less cordate, sometimes peltate, palmately-veined. Panicles cymose, the staminate many-flowered with branches much exceeding the bracts, the pistillate few-flowered with branches enclosed by the bracts. Staminate flowers with 4 sepals and petals united into a short cup. Pistillate flowers with 1 obovate sepal and 1 smaller petal opposite to the sepal. Carpel 1, with 3-lobed style; drupelet globose, more or less hairy.

About 30 species, mostly tropical American.

1.Cissampelos pareira L. Sp. Pl. 2:1031. 1753. **FIG. 80.**

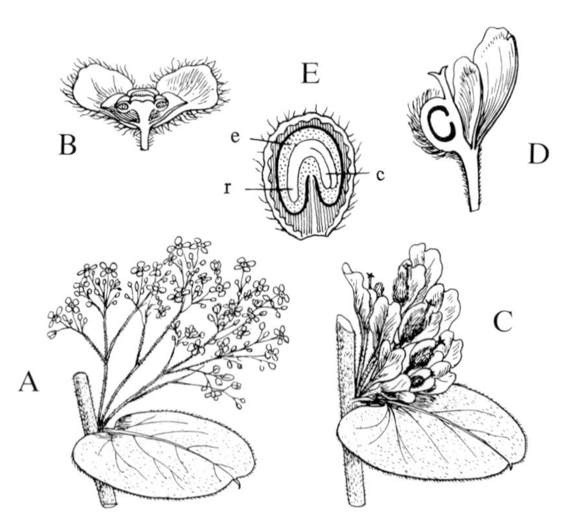

FIG. 80 **Cissampelos pareira.** A, staminate inflorescence, x$^2/_3$. B, male flower in section, x10. C, pistillate inflorescence, x6. D, female flower in section. E, drupe cut lengthwise, x4, showing endosperm (e), cotyledons (c), and radicle (r). (F. & R.)

"Quacori"

A slender, elongate vine, often high-climbing; leaves mostly 3-9 cm long and broad, varying from glabrous to softly velvety-hairy; petioles usually attached marginally but sometimes peltately. Flowers creamy-white, the staminate scarcely 1 mm across. Fruit scarlet, 3-5 mm in diameter.

GRAND CAYMAN: *Brunt 2041; Correll & Correll 51039. Kings GC 120.* – Pantropical, frequent in dryish thickets and woodlands. The roots contain alkaloids similar to those used in the preparation of curare. The only Cayman specimens seen are staminate.

Family 23

PAPAVERACEAE

Mostly herbs (rarely shrubs) with yellow or orange sap; leaves alternate, simple or variously lobed; stipules lacking. Flowers often rather large, solitary or rarely paniculate; sepals 2 or 3, deciduous; petals 4-6, free (rarely lacking). Stamens numerous, free. Ovary superior, 1-celled, with parietal placentas; ovules numerous; stigmas sessile or nearly so. Fruit a capsule, opening by valves or pores.

A chiefly Northern Hemisphere family most common in temperate regions, with about 26 genera and 200 species. Relatively few occur in the tropics or south of the equator. Many species contain characteristic alkaloids, the best-known being those of the Opium Poppy, *Papaver somniferum.*

Genus 1. **ARGEMONE** L.

Glaucous spiny herbs with yellow sap. Leaves sessile, pinnately lobed, with spiny teeth. Flowers solitary, terminal on branches, yellow or white; ovary with 4-7 placentas and numerous ovules. Capsule very spiny, opening by 4-7 short valves at the apex.

A genus of about 10 species, mostly of southwestern United States and Mexico.

1. **Argemone mexicana** L., Sp. Pl. 1:508. 1753. **FIG. 81.**

"Thistle", "Thorn Thistle".

Weedy, glabrous herb to 1 m tall, usually lower. Leaves very sharply spiny and noticeably glaucous, the lower ones 15 cm long or more, the upper ones much shorter. Flowers sessile at the apex of axillary or terminal branches, subtended by

1-3 leafy bracts; sepals c. 2 cm long, acuminate with spiny apex; petals yellow, 2-3 cm long, in 2 series. Capsule 3-4 cm long, ellipsoid; seeds black, numerous, c. 2 mm in diameter.

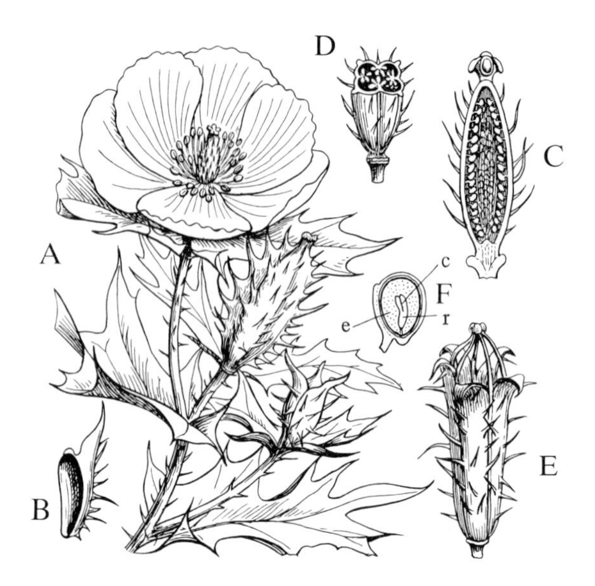

FIG. 81 **Argemone mexicana.** A, branch with bud, fruit and flower, x$^2/_3$. B, sepal, x$^2/_3$. C, pistil cut lengthwise, x$^2/_3$. D, x-section of ovary. E, ripe capsule, x$^2/_3$. F, seed cut lengthwise; e, endosperm; c, cotyledons; r, radicle. (F. & R.)

GRAND CAYMAN: *Brunt 1958; Hitchcock; Kings GC 129, GC 161; Millspaugh 1400; Proctor 27993.* CAYMAN BRAC: *Kings CB 40; Proctor 29100.*

– West Indies and continental tropical America; widely naturalized throughout the tropics as a weed of open waste ground. The yellow sap is said to be slightly caustic and useful for removing warts.

Family 24

ULMACEAE

Trees or shrubs, unarmed or sometimes spiny, with alternate, simple leaves 3-nerved and often asymmetric at base; stipules small, soon falling. Flowers usually in cymose clusters, unisexual or rarely perfect, regular, small; if unisexual, the plants monoecious; perianth sepaloid, the 4 or 5 segments free or united. Staminate flowers with erect stamens equalling the perianth-lobes in number and opposite them; rudimentary ovary (pistillode) present. Pistillate flowers with superior, 1-celled ovary containing 1 pendulous ovule; styles 2, simple or forked, stigmatic on the inner sides. Fruit (of Cayman species) a small hard drupe; seeds with straight embryo; endosperm none or scanty.

A world-wide family of about 15 genera and 200 species.

KEY TO GENERA

1. Leaves nearly smooth on upper (adaxial) surface; staminate perianth segments imbricate; pistillate perianth deciduous: **1. Celtis**

1. Leaves very rough on upper surface (like sandpaper); staminate perianth segments valvate; pistillate perianth persistent: **2. Trema**

Genus 1. **CELTIS** L.

Small trees, shrubs, or scramblers, unarmed or spiny; leaves entire or serrate. Staminate and pistillate flowers mixed in small axillary cymose clusters appearing with young foliage, the staminate several or numerous, the pistillate solitary or few. Perianth usually 5-partite. Drupe subglobose, with a somewhat flattened, reticulate or tuberculate stone-like seed.

A chiefly Northern Hemisphere genus of about 80 species.

KEY TO SPECIES

1. Leaves serrate, very oblique at base; erect unarmed tree; styles simple; fruit pedicel longer than petioles: **1. C. trinervia**

1. Leaves entire, not oblique at base; scrambling shrub with spines; styles forked; fruit pedicel shorter than petioles: **2. C. iguanaea**

1. Celtis trinervia Lam., Encycl. Méth. Bot. 4:140. 1797. **FIG. 82.**

Small unarmed tree to 10 m tall; leaves obliquely ovate, mostly 4–10 (–13) cm long, long-acuminate, the margins coarsely serrate except toward the base and apex, sparsely pubescent on both sides and finely black-dotted beneath. Staminate flowers greenish, in fascicles of 3 to 5; pistillate flowers usually solitary. Drupe globose-ovoid, 7–10 mm long, purple-black.

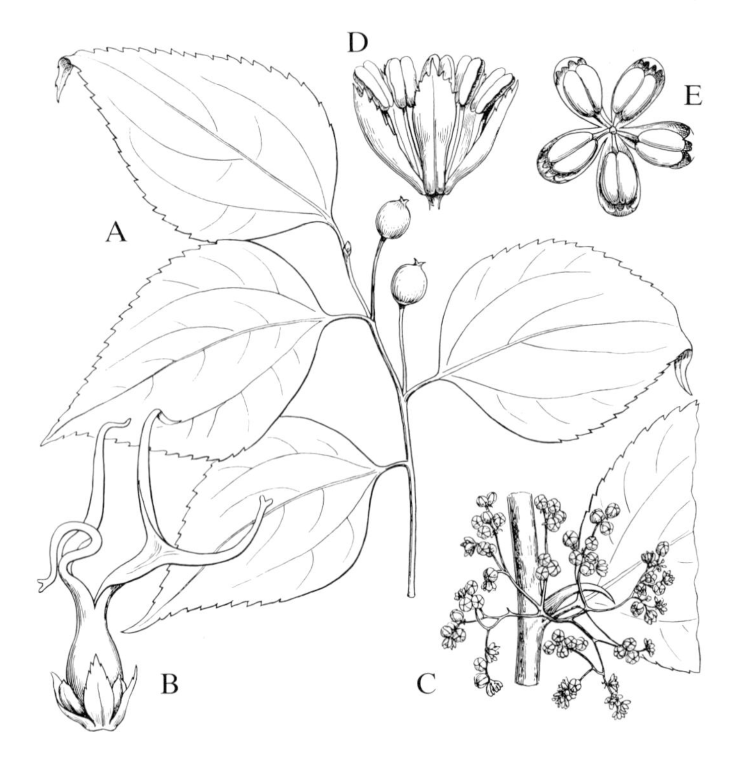

FIG. 82 **Celtis trinervia.** A, fruiting branch, x½. B, **C. iguanaea,** pistillate flower x4; C, staminate inflorescence, x½; D, E, staminate flower, x8. (F. & R.)

GRAND CAYMAN: *Proctor 15127.*
– Greater Antilles, in thickets and forests at low to medium elevations.

2. **Celtis iguanaea** (Jacq.) Sarg., Silv. N. Amer. 7:64. 1895. **FIG. 82.**

A scrambling or often high-climbing shrub armed with sharp recurved spines; leaves equilaterally ovate, elliptic or oblong (rarely slightly oblique at base), 5–12 cm long, blunt to subacuminate, the margins nearly entire, almost glabrous on both sides except for domatia in the main nerve-axils beneath. Flowers in short axillary cymes, the staminate in fascicles of 2–4, the pistillate solitary. Drupe globose-ellipsoid, 12–14 mm long, yellow.

LITTLE CAYMAN: *Proctor 28027.* CAYMAN BRAC: *Proctor 29344.*
– West Indies and continental tropical America, often growing in thickets on rocky limestone escarpments.

Genus 2. **TREMA** Lour.

Unarmed trees or shrubs; leaves serrate and often with scabrid upper surface. Staminate and pistillate flowers mixed in small axillary cymose clusters, the perianth mostly 5-partite; pistillate flowers sometimes with functional stamens. Drupe surrounded by the persistent perianth; seed with scant but fleshy endosperm.

A pantropical genus of about 30 species, characteristic of clearings, thickets, and forest-margins.

1. **Trema lamarckianum** (R. & S.) Blume in Mus. Bot. Lugd.-Bat. 2:58. 1856.

A shrub or small tree to 7 m tall or more, with slender, rough-pubescent twigs. Leaves more or less lance-oblong, 2–6 cm long, acute at apex, very rough on upper side, reticulate-veined and finely tomentose beneath. Flowers greenish, c. 2 mm across when open. Drupe ovoid, 2.8–3 mm long, red, glabrous.

GRAND CAYMAN: *Brunt 1911, 2040; Correll & Correll 51018; Kings GC 134; Proctor 27930; Sachet 421.* LITTLE CAYMAN: *Kings LC 79.*
– Florida and the West Indies south to St. Vincent, often common in thickets and on the borders of woodlands. The Cayman plants grow in sandy soils or in fragmented coral limestone.

Family 25

MORACEAE

Trees or shrubs, rarely herbs, often with milky sap (latex); leaves alternate, entire, toothed or lobed either pinnately or palmately; stipules present. Flowers unisexual, in spikes or heads, or on a flat entire or lobed receptacle, or on the inside of a closed receptacle; perianth sepaloid, of 2–5 (usually 4) free or fused segments. Male flowers with stamens generally as many as the perianth-lobes and opposite them, or only 1. Female flowers with 1-celled ovary, superior to inferior, bearing a simple, 2-toothed, or 2-partite style; ovule 1. Fruit 1-seeded, free or more commonly united with fruits of other flowers to form a more or less fleshy pseudocarp, in whose formation the receptacle may also be involved.

A mostly tropical family of about 53 genera and more than 1,000 species.

KEY TO GENERA

1. Flowers borne on the inner surface of a more or less globose, hollow receptacle, this having at the apex a small opening concealed by scales: **1. Ficus**

1. Flowers variously arranged but never inside a closed receptacle:

 2. Fruit a small globose head less than 1.5 cm in diameter; wild dioecious tree with spines: **2. Maclura**

 2. Fruit a large, head-like syncarp more than 10 cm in diameter; cultivated monoecious trees without spines: **[3. Artocarpus]**

Genus 1. **FICUS** L.

Trees or shrubs, rarely vine-like, often epiphytic at least when young, with milky or rarely watery sap. Leaves entire (in American species); stipules small, usually soon falling. Staminate and pistillate flowers mingled on the inner surface of a hollow, usually globose, fleshy receptacle, this with a small opening (ostiole) at the apex, covered or concealed by several small scales; receptacles axillary, each subtended by a lobate involucre of bracts. Staminate perianth of usually 3 small segments; stamens usually 1 or 2. Pistillate perianth of 4–6 small segments; ovary with 1 pendulous ovule. Fruit a small, seed-like achene, these numerous within a single receptacle, which as a whole is known as a "fig". All figs are edible, but most are rather insipid.

A pantropical genus of more than 600 species. The true edible fig, *Ficus carica* L. of the Mediterranean region, is widely cultivated.

KEY TO SPECIES

1. Figs distinctly stalked: **1. F. citrifolia**

1. Figs sessile in leaf-axils:

 2. Leaves obtuse; wild shrub or tree: **2. F. aurea**

 2. Leaves sharply acuminate; cultivated tree: **[3. F. benjamina]**

1. Ficus citrifolia Mill., Gard. Dict., ed. 8. 1768.

Ficus laevigata Vahl, 1805.
Ficus populnea Willd., 1806.
Ficus brevifolia Nutt., 1846.
"Barren Fig", "Wild Fig".

A much-branched glabrous tree up to 15 m tall, or shrub-like in exposed situations, often with dense masses of aerial roots hanging from the branches; when these roots reach the ground they thicken to form supplementary trunks or rib-like extensions of the main trunk. Leaves ovate to elliptic, long-petiolate, 5–10 cm long, 2–5 cm broad, more or less acute at apex, truncate or subcordate (rarely acutish) at base of blade, the upper surface dotted with numerous minute elevated papillae (cystoliths). Ripe figs yellow, sometimes red-spotted, globose, 7–12 mm in diameter, on slender peduncles usually 5–10 mm long, rarely longer.

GRAND CAYMAN: *Hitchcock; Kings GC 109, GC 130; Proctor 15124.* LITTLE CAYMAN: *Kings LC 88A.* CAYMAN BRAC: *Proctor 28982, 29085.*
– Florida, the West Indies, and parts of Central America, common, most frequent in rocky woodlands, coastal thickets, and along the base of cliffs; occasionally cultivated. The island of Barbados is supposed to have been named for this tree, whose abundant aerial roots often give it a "bearded" appearance.

2. Ficus aurea Nutt., Sylva 2:4. 1846. **FIG. 83.**

Ficus dimidiata Griseb., 1859.

"Wild Fig".

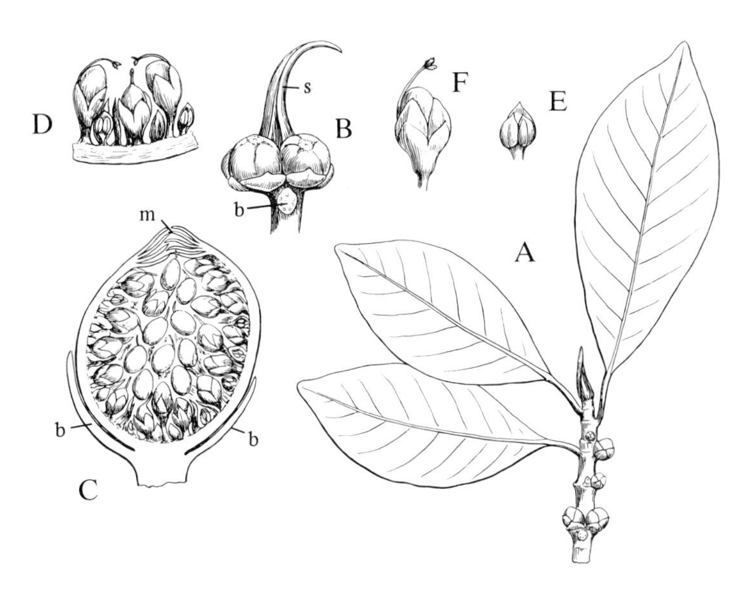

FIG. 83 **Ficus aurea.** A, flowering branch, x½. B, apex of shoot with two figs; s, stipule; b, leaf-scar. C, vertical section of fig, x4; b, basal bracts; m, ostiole. D, individual flowers within a fig, x5. E, staminate flower, x8. F, pistillate flower, x8. (F. & R.)

A much-branched glabrous tree up to 20 m tall, or diffuse and shrub-like in exposed situations, in habit like the preceding. Leaves elliptic to obovate, long-petiolate, 6–16 cm long, 2–7 cm broad, rounded or subacute at apex, narrowed or truncate at base of blade; minute punctate cystoliths present or absent on upper surface. Figs sessile and subtended by broad bracts, when ripe rose-red to crimson, depressed-globose, 4–7 mm in diameter.

GRAND CAYMAN: *Brunt 1929, 1992a; Hitchcock; Proctor 15029.* LITTLE CAYMAN: *Kings LC 87, LC 88; Proctor 28042.* CAYMAN BRAC: *Kings CB 96; Proctor 28937.*

– Florida, the Bahamas, and the Greater Antilles except Puerto Rico, also in the Swan Islands, frequent in coastal thickets and inland at low to medium elevations; in favourable situations it can become an enormous tree with a trunk as much as 2 m in diameter.

[3. **Ficus benjamina** L., a large Malayan species with glossy, ovate, abruptly acuminate leaves 4–10 cm long, is sometimes planted as a shade tree. GRAND CAYMAN: *Proctor 15149.*]

Genus 2. **MACLURA** Nutt.

Dioecious trees with yellowish latex, often armed with spines; leaves entire or toothed, pinnately-veined; stipules small, soon falling. Staminate flowers in axillary catkin-like spikes; perianth 4-parted; stamens 4, inflexed in bud; ovary rudimentary. Pistillate flowers in dense axillary heads, mingled with bracts similar to the perianth-segments; perianth 4-parted, the oblique ovary included, the style filiform, exserted; ovule laterally attached. In fruit the perianths become somewhat fleshy and are tightly crowded together to form a globose syncarp; achenes compressed, oblique at apex.

A genus of perhaps a dozen species, one temperate American, one Mexican, one tropical American, and the rest occurring in Africa and Madagascar.

1. **Maclura tinctoria** (L.) D.Don ex Steud., Nom. ed. 2, 2:87. 1841. **FIG. 84.**

Chlorophora tinctoria (L.) Gaud. ex Benth. in Benth. & Hook., 1880.

"Fustic".

A deciduous tree up to 10 m tall or more, with horizontal, often wide-spreading branches frequently armed with spines. Leaves lanceolate to oblong-elliptic or elliptic, short-petiolate, 3–8 cm long or more, mostly 1.5–3 cm broad, blunt to acuminate at apex, the margins entire or toothed, and chiefly glabrous except for minute hairs along the midrib beneath. Staminate catkins pendulous, 3-5 cm long or more, on short peduncles. Pistillate heads green, globose, 5-8 mm in diameter when flowering, increasing to 12 mm or more in fruit.

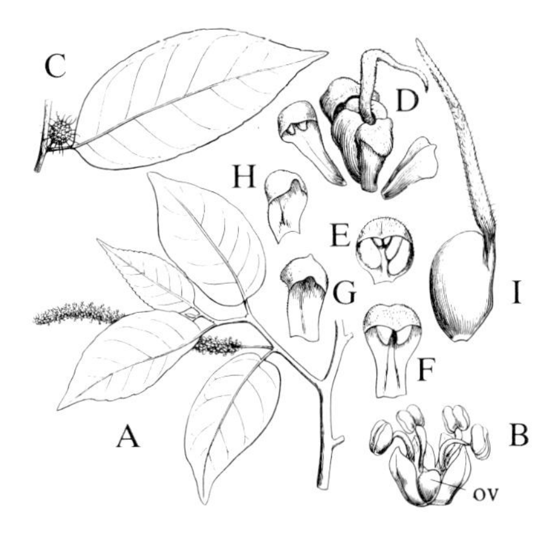

FIG. 84 **Maclura tinctoria.** A, branch with staminate inflorescence, x¼. B, staminate flower, x4; ov, rudimentary ovary. C, leaf with pistillate inflorescence, x¼. D, portion of pistillate inflorescence dissected, showing flower with enveloping bracts, x4. E, F, gland-bearing bracts, x4. G, H, outer and inner perianth segments, x4. I, pistil, x4. (F. & R.)

GRAND CAYMAN: *Brunt 2014; Kings GC 408; Proctor 27982.*
– West Indies and continental tropical America, common in woodlands at low to medium elevations. The fruits are edible (though scarcely palatable!); the wood is very tough and close-grained, and has many uses; also, a yellowish dye known as fustic or khaki is extracted from the wood, and is commonly used for dyeing the cloth used for military uniforms and schoolboys' clothing.

[Genus 3. **ARTOCARPUS** J.R. & G. Forst.

Monoecious trees with white latex; leaves medium-sized to large, entire or lobed, pinnately-veined; stipules paired, often large and papery. Inflorescences unisexual, axillary or cauliflorous on old wood. Staminate flowers in stout, cylindric spikes; perianth 2-4-lobed, with 1 exserted stamen. Pistillate flowers in fleshy more or less globose heads in which they are mixed with peltate bracts, all being fused together; style entire or 2-lobed. Mature syncarp (compound fruit) formed by enlargement of the entire pistillate head, when fertile thus enclosing numerous achenes each representing an individual fruit; seeds large, lacking endosperm but with fleshy cotyledons.

A genus of 47 species indigenous to the Indo-Malayan region eastward to the Solomon Islands. Two species, *Artocarpus altilis* (S. Parkinson) Fosberg, the Breadfruit, and *Artocarpus heterophyllus* Lam., the Jackfruit, are cultivated throughout the tropics for their edible fruits; both of these occur in the Cayman Islands, but no voucher specimens have been noted, except one for *A. altilis,* GRAND CAYMAN: *Sachet 446.* The ordinary Breadfruit is a seedless cultivar propagated by suckers; seeded Breadfruit trees are also grown in many West Indian islands.]

Family 26

URTICACEAE

Herbs, shrubs, or rarely small trees, without latex, sometimes armed with stinging hairs, and often with epidermal inclusions (cystoliths). Leaves opposite or alternate; stipules present, sometimes conspicuous. Flowers small and usually wind-pollinated, regular, unisexual, monoecious or dioecious, usually in cymose clusters or panicles; perianth segments 2–5, free or united, sepaloid. Staminate flowers with stamens as many as and opposite to the perianth segments, incurved in bud, straightening to an exserted position and dehiscing explosively when mature. Pistillate flowers with superior ovary, 1-celled with 1 basal ovule. Fruit an achene, sometimes enclosed in the persistent perianth.

A family of about 45 genera and 900 species, widely distributed in tropical and temperate regions.

Genus 1. **PILEA** Lindl.

Annual or perennial herbs, sometimes shrubby at base; leaves opposite, those of a pair usually unequal in size, usually 3-veined, and with conspicuous epidermal cystoliths; stipules connate, intrapetiolar. Flowers minute, monoecious or dioecious, the clusters axillary. Staminate perianth usually 4-partite; pistillate perianth 3-partite, the segments usually unequal, each subtending a scale-like staminode; stigma sessile, shortly penicillate. Achene compressed, more or less enclosed by the persistent perianth, the median segment of which is usually larger than the lateral ones. The achenes in at least some species are expelled by the catapult action of the enlarged inflexed staminodes, similar in mechanism to the explosive ejection of the pollen from staminate flowers.

A very large pantropical genus of more than 500 species, notable for their local endemism. More than two-thirds of the species occur in the American tropics, about 180 of them in the West Indies. A few species are commonly grown as ornamental pot-plants or as foliage-plants in gardens.

KEY TO SPECIES

1. Stems erect, spreading or tufted; leaf-blades oblanceolate to obovate, contracted gradually into the petiole, in strongly unequal pairs, the larger leaf of a pair 3–12 mm long: **1. P. microphylla**

1. Stems mostly prostrate, forming a delicate mat with short erect flowering branches; leaf-blades suborbicular, abruptly contracted into the petiole, in subequal or moderately unequal pairs, the larger leaf of a pair up to c. 4 mm long including the filiform petiole: **2. P. herniarioides**

1. Pilea microphylla (L.) Liebm. in Danske Vid. Selsk. Skr. V, 2:296. 1851.

A glabrous, pale green herb, annual or perennial in various forms, with succulent, freely-branched stems, these usually more or less erect, rarely prostrate or pendent. Leaves small, entire, 1-nerved, varying in shape from broadly obovate to oblanceolate or elliptic, the larger 2–12 mm long, the upper surface beset with linear cystoliths transversely arranged. Flowers in very small sessile or stalked cymules, these monoecious or unisexual. Achenes 0.5 mm long, slightly exceeding the persistent perianth-segment.

This taxon is a pantropical complex of differing forms, some of them weeds of disturbed soil, others occurring in various natural habitats. Several horticultural variants are known, often called "Lace-plant" or "Artillery-plant". One of the best-known is var. *trianthemoides* (Sw.) Griseb., a low shrubby herb with arching branches, commonly cultivated as an edging for flower-beds. Two varieties occur as wild plants in the Cayman Islands; these can be characterized as follows:

1. Annual; stems 0.5–1 mm thick; leaves thin; plants of damp shady ground near paths or habitations: **1a.** var. **microphylla**

1. Perennial; stems to 2.5 mm thick; leaves thick, succulent; plants of exposed limestone crevices: **1b.** var. **succulenta**

1a. Pilea microphylla var. **microphylla**

Characters as given in the key; upper leaf-surface with conspicuous parallel elongate-linear cystoliths.

GRAND CAYMAN: *Kings GC 352.*
– Pantropical.

1b. Pilea microphylla var. **succulenta** Griseb., Fl. Br. W.I. 155. 1860.

Characters as given in the key; cystoliths much shorter, less conspicuous, and less regularly arranged. The Cayman plants have much more narrow leaves than the typical form occurring in Jamaica, with the underside lacking the minutely foveolate (pitted) texture found in the latter.

GRAND CAYMAN: *Kings GC 393; Proctor 15209.* CAYMAN BRAC: *Kings CB 68, CB 100; Proctor 29043, 29118.*
– Jamaica and Hispaniola.

2. Pilea herniarioides (Sw.) Wedd. in Ann. Sci. Nat. sér. 3, Bot. 18: 207. 1852.

Stems filiform, 0.2–0.5 mm thick, few-branched, often rooting at the nodes. Leaves thin and delicate, with blades 1–3 mm long and broad, usually glabrous; cystoliths short-linear, more or less transversely scattered on the upper surface; petioles filiform, 1–2 mm long. Cymules in upper or terminal axils, sessile or stalked; perianth of staminate flowers 1.2–1.5 mm long, the pistillate flower smaller. Achenes ovoid-ellipsoid, 0.6–0.7 mm long.

GRAND CAYMAN: *Hitchcock, Jan. 18 1891.*
– Widely distributed in the Caribbean region from Florida to Curacao; also in Costa Rica. This species is quite rare throughout its range, usually occurring in hollows of honeycombed limestone near the sea. Some authors have treated it as a variety of *P. microphylla.*

Family 27

MYRICACEAE

Aromatic shrubs or small trees; leaves alternate, simple, entire or toothed (rarely lobed) resin-dotted; stipules usually absent. Flowers small, unisexual, monoecious or dioecious, sessile under scale-like bracts and grouped in short axillary spikes; bracteoles often present within the bract. Perianth lacking; staminate flower with usually 4–8 stamens inserted on a receptacle, the filaments short, distinct or slightly united; anthers 2-celled, dehiscing longitudinally. Pistillate flower a solitary 1-celled ovary subtended by 2–8 bracteoles; ovule 1, erect; style short and bearing 2 linear stigmas. Fruit a small oblong or globose drupe covered with resinous or waxy papillae; seed without endosperm, the embryo straight with flat cotyledons.

The family consists of a single genus with about 40 species, widely distributed in tropical, temperate and arctic regions.

Genus 1. **MYRICA** L.

With the characters of the family.

1. **Myrica cerifera** L., Sp. Pl. 2:1024. 1753. **FIG. 85.**

"Bayberry".

A shrub up to 4 m tall or more, the twigs glabrate or sparsely pubescent; leaves short-petiolate, narrowly oblanceolate, 5-10 cm long, with acute apex and attenuate base, the margins entire or with a few teeth, the surfaces with numerous minute golden resin-dots, those of the upper side sunken in small pits. Staminate spikes sessile, mostly 1 cm long or less; pistillate spikes longer. Drupes globose, 2-3 mm in diameter, densely covered with whitish wax.

GRAND CAYMAN: *Brunt 1950; Proctor 31042.*
– Eastern United States, the West Indies, and Mexico to Costa Rica, in widely various habitats. The Cayman plants grow in swampy woodlands toward the interior of the island. The wax of the fruits is often extracted in the United States and Central America for the making of candles; these burn with a pleasing aroma.

[Family 28

CASUARINACEAE

Trees of pine-like aspect, with green, jointed, angular, striate branchlets that perform the function of leaves; leaves reduced to minute scales, these borne in whorls at the nodes and often short-connate into a sheath. Flowers unisexual, the plants monoecious; staminate flowers borne in narrow cylindric spikes on the ends of short lateral branchlets, several bracts combining at each node of the spike to form a serrate cup over the edge of which hang several stamens; each stamen represents a male flowers with a concealed 1- or 2-segmented perianth and 2 bracteoles. Pistillate flowers in lateral, dense spherical heads, each female flower in the axil of a bract, without a perianth but subtended by 2 bracteoles; ovary small, 1-celled, with 2 long-exserted stigmas; ovules 1 or 2. After fertilization the pistillate flower-head becomes hard and cone-like, the woody bracts and bracteoles subtending winged achenes. Seed solitary, without endosperm.

A taxonomically isolated family, with a single genus of more than 40 species, these mostly Australian.

Genus 1. **CASUARINA** Adans.

With the characters of the family.

1. Casuarina equisetifolia L., Amoen. Acad. 4:123, 143. 1759; J.R. & G. Forst., 1776. **FIG. 86.**

Casuarina litorea L. in Stickman, 1754 (illegit.).

"Australian Pine", "Weeping Willow", "Beefwood".

A fast-growing tree up to 20 m tall or more, assuming a tall, narrow but open shape resembling a conifer; trunk to 1 m in diameter; branchlets pale or rather dark green, slender (resembling pine-needles), 0.6–0.8 mm thick; scales in whorls of 6–8, 1–3 mm long, acute, ciliate. Staminate spikes 1–4 cm long. Fruiting heads globose, c. 1.5 cm in diameter.

GRAND CAYMAN: *Kings GC 83; Proctor 15144, 31051.* CAYMAN BRAC: *Proctor (sight record).*
– Native of Australia, but widely planted and naturalized in most warm countries; common in the West Indies, especially in sandy or gravelly soils near the sea, but also flourishing inland. It is not definitely known when this species was introduced to the Cayman Islands, but probably during the second half of the nineteenth century. It is now well-established and naturalized in many localities. It is often planted and trimmed as a hedge, and is very useful as a windbreak near seashores because of its high tolerance to salt. The wood is very hard, heavy, and fine-grained, but unfortunately very susceptible to attack by termites.]

Family 29

PHYTOLACCACEAE

Herbs, shrubs or trees, mostly glabrous; leaves alternate, simple, entire; stipules minute or lacking. Flowers regular, perfect, in terminal and axillary racemes or spikes; perianth of 4 or 5 segments, persistent in fruit. Stamens 4, 5, or more, variously inserted. Ovary superior, of 1-several 1-ovuled carpels, the ovules basal; style short or lacking. Fruit an achene or berry of 1 to several carpels; seeds with mealy or fleshy endosperm.

A widespread, chiefly tropical family of up to 22 genera and about 120 species. Species of *Phytolacca* provide an edible pot-herb used in some countries.

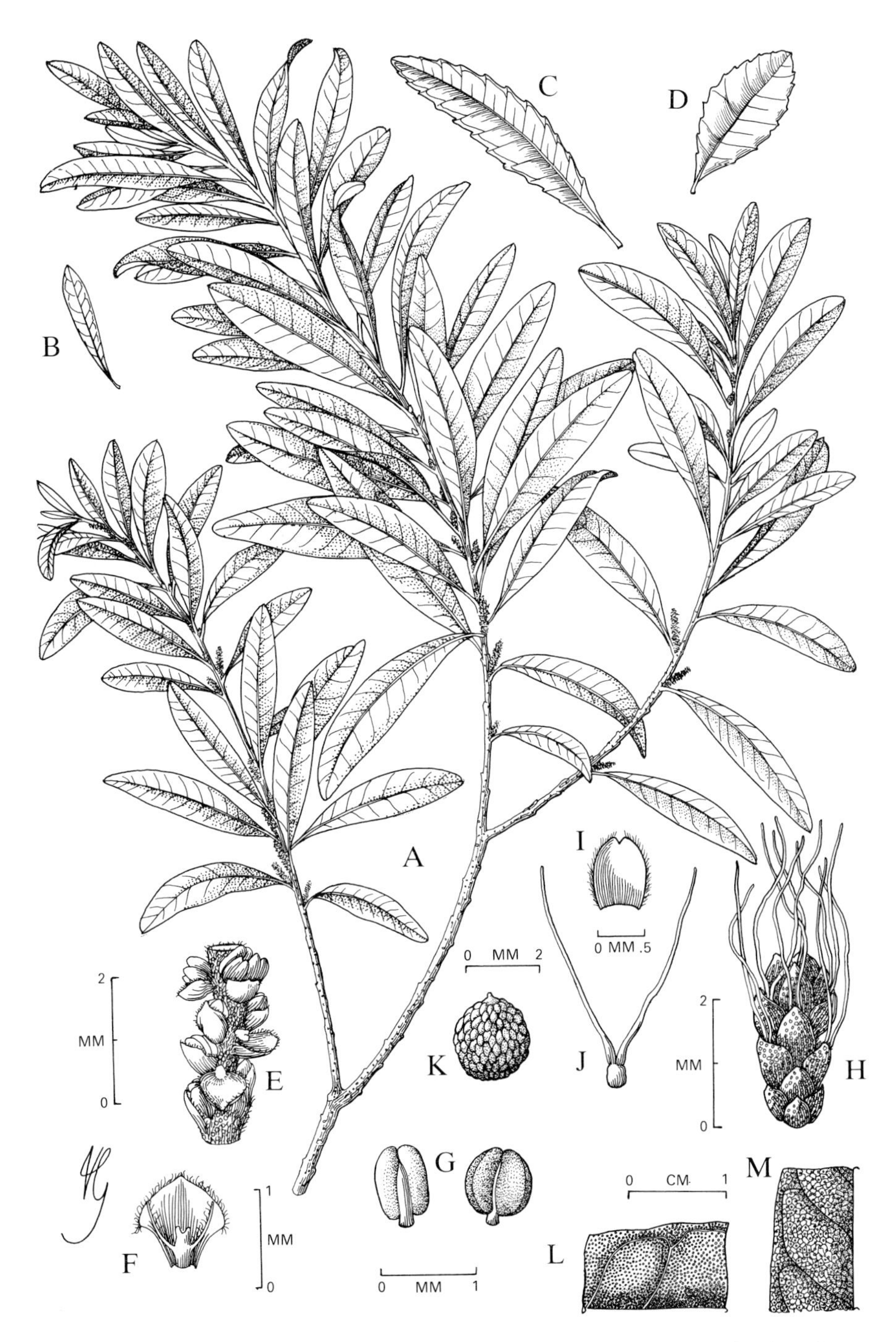

FIG. 85 **Myrica cerifera.** A, branch with staminate spikes. B, C, D, leaves of various form. E, portion of staminate inflorescence. F. bract of staminate flower with filaments. G, two views of anther. H, pistillate inflorescence. I, bracteole of pistillate flower. J, pistillate flower. K, fruit. L, upper surface of leaf. M, lower surface of leaf. (G.)

KEY TO GENERA

1. Erect herb; stamens 4; stigma capitate: **1. Rivina**

1. Climbing or scrambling shrub; stamens 8–16; stigma penicillate: **2. Trichostigma**

Genus 1. **RIVINA** L.

Erect herbs, somewhat shrubby at base; leaves alternate or subopposite. Flowers in slender terminal and axillary racemes, the bracts and bracteoles minute and deciduous. Perianth 4-partite, the segments not enlarging in fruit; stamens 4; ovary of 1 carpel, the stigma capitate on a very short style. Fruit a small globose berry with fleshy pericarp; seed with annular embryo and mealy endosperm.

A small tropical American genus of about 3 species.

1. Rivina humilis L., Sp. Pl. 1:121. 1753.

Rivina humilis var. *glabra* L., 1753.

Rivina humilis var. *laevis* (L.) Millsp., 1900.

"Fowl berry", "Blood berry".

A perennial herb up to 1.5 m tall (often much smaller), glabrous varying to finely pubescent. Leaves thin, long-petiolate, 2–14 cm long, the blades ovate to lanceolate, acute to acuminate at apex, obtuse or rounded at base. Racemes lax, 3–10 cm long; perianth white or creamy, 2-3 mm long, the stamens shorter. Fruit a crimson berry, at length becoming dry, c. 3 mm in diameter when dry.

GRAND CAYMAN: *Brunt 1891, 2030; Hitchcock; Kings GC 71; Lewis GC 53, 3832; Millspaugh 1343, 1352; Proctor 15102.* LITTLE CAYMAN: *Proctor 28036.* CAYMAN BRAC: *Kings CB 47, CB 75, CB 101; Proctor 29075.*
– Florida, Texas, West Indies and continental tropical America, frequent in shady thickets and glades, also sometimes a weed of gravelly waste ground.

Genus 2. **TRICHOSTIGMA** A. Rich.

Scrambling or vine-like shrubs; leaves alternate or subopposite, with minute, deciduous stipules. Flowers in terminal and axillary racemes, with deciduous bracts and persistent bracteoles. Perianth 4-partite, the segments somewhat enlarging and

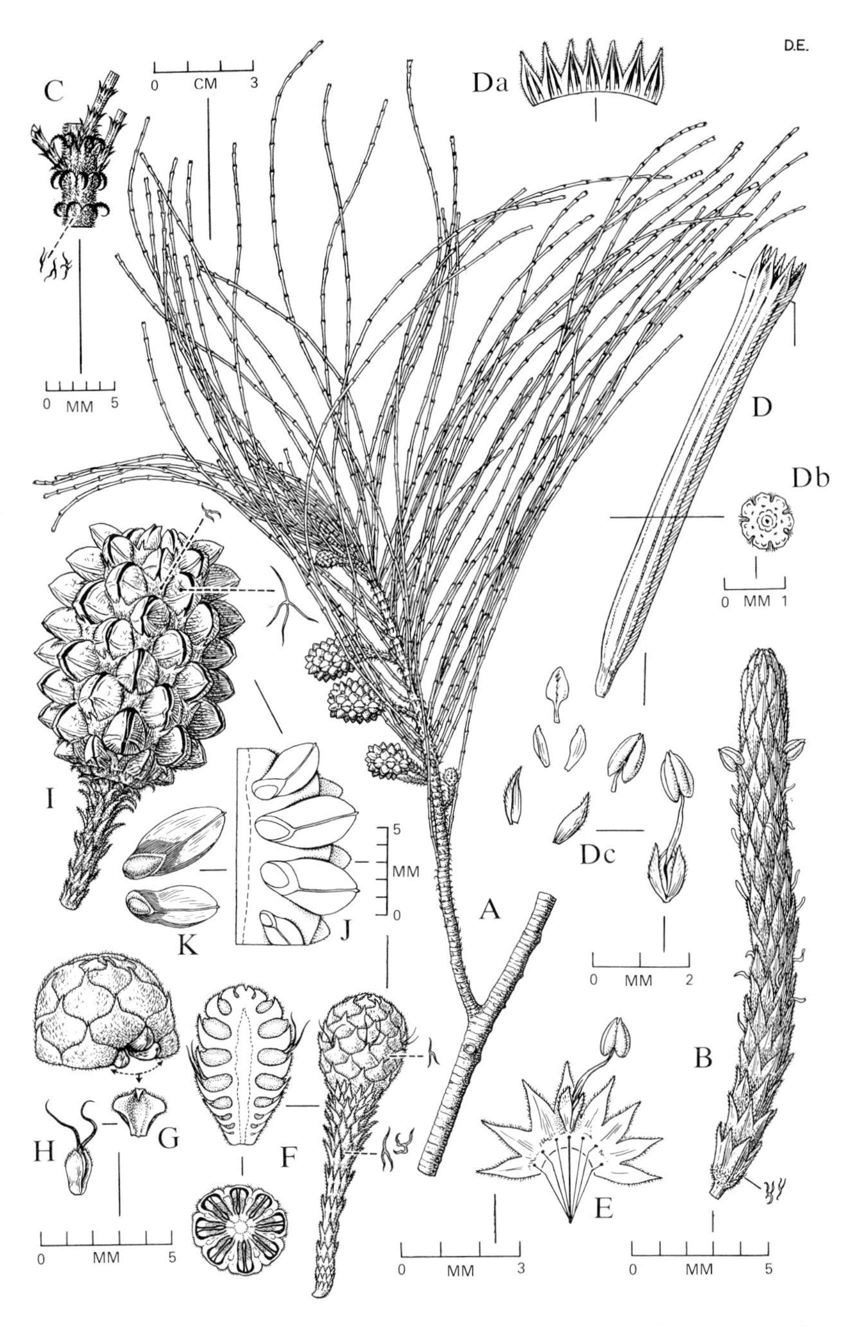

FIG. 86 **Casuarina equisetifolia.** A, branch bearing pistillate flowers and fruits. B, staminate inflorescence. C, portion of staminate spike with scales removed. D, perianth-like cup of united bracts; a, apex spread out; b, x-section; c, single staminate flower and separated parts. E, staminate flower in position on spread-out apex of bract cup. F, pistillate inflorescence in various aspects. G, pistillate bract. H, pistillate flower. I, fruiting inflorescence. J, long. section through portion of fruiting inflorescence. K, achenes. (D.E.)

becoming reflexed in fruit; stamens 8–16; ovary of 1 carpel, the stigma sessile, brush-like. Fruit a globose, 1-seeded berry, the leathery pericarp adhering to the seed.

A tropical American genus of 3 species.

1. Trichostigma octandrum (L.) H.Walt. in Engler. Pflanzenr. 4(83):109. 1909. **FIG. 87.**

A glabrous climbing shrub, the stems scrambling over other vegetation often to a height of 6 or 7 m with long, slender branches. Leaves membranous, long-petiolate, blades elliptic or lanceolate, in all 3–10 cm long or more, 1.5–4 cm broad, acuminate at apex, mostly cuneate at base. Racemes equalling the leaves or longer; perianth white or greenish, turning red or purple in fruit. Berry purple to nearly black, c. 5 mm in diameter.

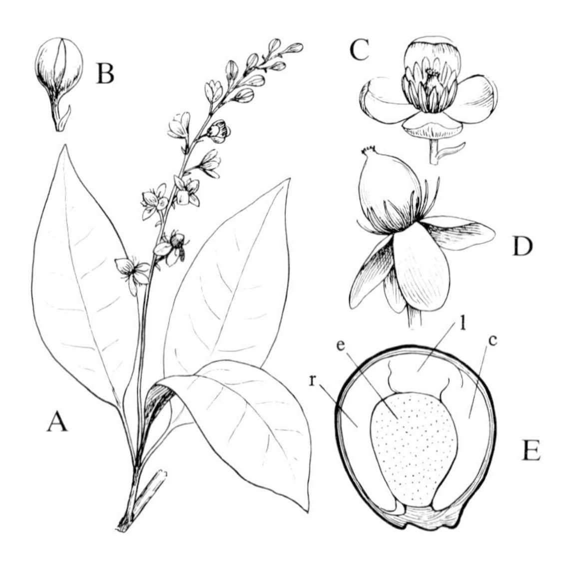

FIG. 87 **Trichostigma octandrum.** A, portion of branch with inflorescence, x$^2/_3$. B, flower-bud, x2. C, the same, opened, x2. D, fruit, x3. E, section of seed, x5; e, endosperm; c, cotyledons; l, lobed base of cotyledons; r, radicle. (F. & R.)

CAYMAN BRAC: *Proctor 15323.*
– Florida, West Indies, and continental tropical America, in rocky thickets and woodlands. The berries are often eaten by birds.

Family 30

NYCTAGINACEAE

Herbs, shrubs or trees; leaves opposite or sometimes partly alternate, simple, entire or nearly so; stipules lacking. Flowers regular, unisexual or perfect, usually in terminal or axillary cymes, the bracts minute or (in *Mirabilis*) forming a calyx-like involucre. Perianth segments usually 5, fused to form a tube, the base of which persists on the ripe fruit. Stamens 2–10, typically 5, the filaments often unequal in length. Ovary superior, enclosed by base of perianth-tube, 1-celled; style elongate, with capitate stigma. Fruit (called an anthocarp) consisting of an indehiscent utricle enclosed by the enlarged adhering base of the perianth; seed with large cotyledons enclosing the endosperm.

A pantropical family most abundant in the Western Hemishpere, in all with about 25 genera and more than 250 species. The ornamental woody vines *Bougainvillea spectabilis* Willd., *B. glabra* Choisy, and various hybrids of these, are widely cultivated.

KEY TO GENERA

1. Plants herbaceous:

  2. Perianth very small (less than 3 mm long), subtended by minute bracts: **1. Boerhavia**

  2. Perianth large and showy (up to 5 cm long), subtended by a calyx-like involucre: **2. Mirabilis**

1. Plants woody:

  3. Fruit dry, with 5 rows of glands; scrambling shrub armed with recurved spines: **3. Pisonia**

  3. Fruit drupe-like, fleshy, without glands; erect unarmed tree or shrub: **4. Guapira**

Genus 1. **BOERHAVIA** L.

Annual or perennial herbs often growing from a woody taproot, the main stem bifurcate near the base, otherwise erect, prostrate, or diffuse. Leaves opposite. Flowers very small, perfect, sessile or subsessile in small clusters on panicles; bracts minute. Perianth 5-lobed, bell- or funnel-shaped, with tubular base surrounding the ovary. Stamens 2 (rarely 3), the filaments fused at base. Ovary narrowed toward base; style with peltate stigma. Anthocarp small, angled or ribbed.

A pantropical genus of about 40 species, often weeds of open waste ground.

KEY TO SPECIES

1. Anthocarps glabrous, lacking glandular hairs; flowers white or pale rose: **1. B. erecta**

1. Anthocarps with glandular hairs; flowers crimson:

2. Anthocarps oblong-clavate, 4–6 mm long; leaves usually roundish at apex: **2. B. diffusa**

2. Anthocarps obovoid-clavate, 2.5–4 mm long; leaves usually acutish at apex: **3. B. coccinea**

1. Boerhavia erecta L., Sp. Pl. 1:3. 1753.

"Broomweed".

Annual herb, erect or ascending, 15–100 cm tall, the stems glabrate or sparsely puberulous. Leaves deltate-ovate or lanceolate, petiolate, 2–6 cm long, the apex sharply acute, the margins sinuate-repand, the lower surface whitish. Flowers white or pale rose, 2–6 in a cluster, many of these on a diffuse panicle; perianth-limb 1.5–2 mm long. Anthocarp club-shaped, 5-angled, 3–4 mm long, glabrous.

GRAND CAYMAN: *Brunt 1711; Correll & Correll 51025A. Kings GC 301; Millspaugh 1277.* LITTLE CAYMAN: *Proctor 28167.* CAYMAN BRAC: *Proctor 29114.*
– Southern United States, West Indies, and continental tropical America, a frequent weed of sandy paths and clearings at low elevations.

2. Boerhavia diffusa L., Sp. Pl. 1:3. 1753.

Boerhavia paniculata L.C. Rich., 1792.

"Chickweed".

Perennial herb with woody taproot, the main branches prostrate or decumbent, the flowering branches ascending; glabrate or minutely puberulous, often glutinous.

Leaves elliptic to roundish, petiolate, 2-6 cm long, the apex obtuse or roundish, the margins somewhat irregular and minutely ciliate, the lower surface pale or whitish. Flowers crimson, 2-6 in a cluster, the perianth-limb c. 1 mm long. Anthocarp oblong-clavate with truncate apex, glandular-hairy, 4-6 mm long.

LITTLE CAYMAN: *Kings LC 59; Proctor 28166.* CAYMAN BRAC: *Millspaugh 1189.*
– West Indies and continental tropical America, a weed of sandy clearings.

3. **Boerhavia coccinea** Mill., Gard. Dict. ed. 8. 1768. **FIG. 88.**

Boerhavia hirsuta Willd., 1794.

"Chick Weed".

A loosely-branched herb with long, somewhat sprawling stems, these dark and minutely puberulous. Leaves broadly ovate or subrhombic, petiolate, mostly 1-5 cm long, the apex obtuse to acute, sometimes apiculate, the margins somewhat irregular and minutely ciliate. Flowers crimson, 2-5 in a cluster, the perianth-limb 1.5-2 mm long. Anthocarp narrowly obovoid-clavate with rounded apex, glandular-hairy, 2.5-4 mm long.

GRAND CAYMAN: *Correll & Correll 51025B. Proctor 15056.* CAYMAN BRAC: *Kings CB 21; Proctor 29113.*
– West Indies and continental tropical America, a weed of sandy waste ground.

Genus 2. **MIRABILIS** L.

Perennial herbs with opposite leaves. Flowers perfect, in cymes; bracts forming a 5-lobed, calyx-like involucre subtending the perianth. Perianth-segments 5, fused to form a long, funnel-shaped tube, deciduous after flowering from a constriction level with the top of the ovary. Stamens 5 or 6, unequal in length, exserted; filaments fused at base to form a fleshy cup. Stigma capitate, hairy, at the end of a long-exserted style. Anthocarp ribbed; testa of seed adhering to the pericarp.

A tropical American genus of about 24 species.

1. **Mirabilis jalapa** L., Sp. Pl. 1:177. 1753. **FIG. 89.**

"Four-o'clock".

An erect herb up to 60 cm tall or more, freely branched. Leaves long-petiolate, 2-16 cm long, the blades ovate and long-acuminate, sometimes unequal-sided at

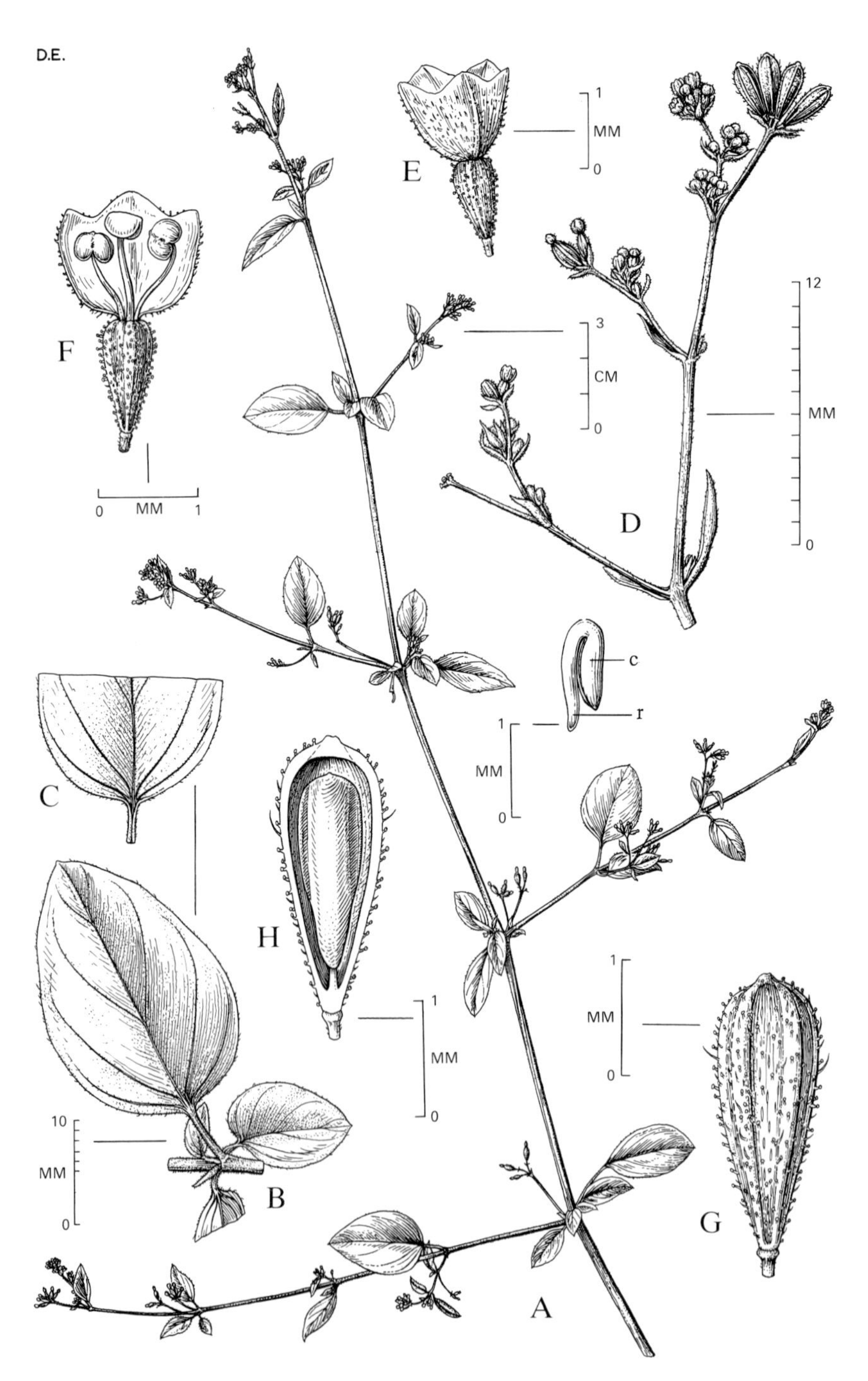

FIG. 88 **Boerhavia coccinea.** A, habit. B, node with leaves. C, base of leaf. D, branch with flowers and fruits. E, single flower. F, same with portion of perianth removed. G, anthocarp. H, same in long. section. (D.E.)

base, glabrous or minutely pubescent. Flowers in small terminal clusters, with calyx-like involucres 7-8 mm long; perianth opening about 4 p.m. (hence the common name), usually red, purple or white, c. 5 cm long. Anthocarp black, ovoid, c. 1 cm long, ribbed and tuberculate.

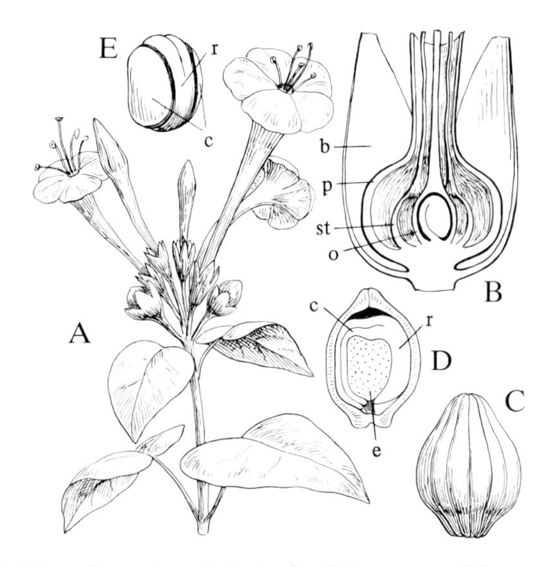

FIG. 89 **Mirabilis jalapa.** A, portion of plant, x$^2/_3$. B, lower part of flower cut lengthwise, x4; b, bract; p, perianth; st, stamen; o, ovary. C, anthocarp, x2. D, same cut lengthwise, x2; c, cotyledons; r, radicle; e, endosperm. E, embryo, x2; c, cotyledons; r, radicle. (F. & R.)

GRAND CAYMAN: *Millspaugh 1402.* CAYMAN BRAC: *Proctor (sight record).* – Florida and Texas, West Indies and continental tropical America, often cultivated and naturalized so that its true natural range is uncertain. It also occurs as an escape from cultivation in the Old World tropics.

Genus 3. **PISONIA** L.

Trees, shrubs or woody vines, some species armed with spines. Leaves mostly opposite. Flowers small, unisexual, in cymose panicles; plants dioecious. Staminate perianth funnel-shaped, 5-toothed, with 6-10 exserted stamens, the filaments slightly fused at the base. Pistillate perianth tubular, 5-toothed, with short usually sterile stamens; ovary with exserted style and capitate stigma. Anthocarps dry and rather hard, with 5 rows of extremely viscid stalked glands (rarely the glands restricted to the distal end).

A pantropical genus of about 25 species. The viscid-glandular anthocarps are capable of ensnaring and disabling birds.

1. Pisonia aculeata L., Sp. Pl. 2:1028. 1753.

A deciduous scrambling shrub often climbing high in trees, armed with stout recurved spines; young twigs minutely puberulous. Leaves rhombic-elliptic to ovate, long-petiolate, 3-15 cm long, the blades acute to acuminate at apex (rarely blunt or rounded), attenuate at base, glabrate or sparsely puberulous beneath. Cymes axillary and terminal, compact, puberulous; flowers yellowish-green, fragrant. Staminate perianth c. 3.5 mm long, the pistillate shorter. Anthocarps ellipsoid or clavate, 8-15 mm long, 5-angled.

LITTLE CAYMAN: *Proctor 35078.* CAYMAN BRAC: *Proctor 28960.*
– Pantropical, variable, uncommon in the Cayman Islands. In Little Cayman it occurs in dense rocky virgin forest near the center of the island; in Cayman Brac it grows in dense thickets along the base of cliffs.

Genus 4. **GUAPIRA** Aubl.

Unarmed trees or shrubs with opposite, often somewhat fleshy leaves. Flowers small, unisexual, in cymose panicles similar to those of *Pisonia;* plants dioecious. Stamens usually 10, exserted; stigma multifid. Anthocarp more or less fleshy, drupe-like, obovoid to ellipsoid or subglobose, lacking viscid glands.

A tropical American genus of 65 species or more. There is considerable doubt that this taxon can be maintained as sufficiently distinct from *Pisionia*, and some authors combine them under the latter name, which has priority.

1. Guapira discolor (Sprengel) Little in Phytologia 17:368. 1968.

Pisonia discolor Sprengel, 1825.

Torrubia discolor (Sprengel) Britton, 1904.

"Cabbage tree".

A shrub or small tree to about 6 m tall, with drooping branches; leaves thin, glabrous, long-petiolate, variable in size and shape, 2-8 cm long, blades oblong to narrowly or broadly elliptic, rarely obovate, obtuse at both ends or cuneate at base. Cymes lax, minutely puberulous, rather few-flowered, the flowers sessile or nearly so; staminate perianth c. 3.5 mm long, the pistillate a little shorter. Fruit crimson, ellipsoid, 5-8 mm long.

GRAND CAYMAN: *Kings GC 106; Proctor 15082.* LITTLE CAYMAN: *Kings LC 37; Proctor 28037, 28061.* CAYMAN BRAC: *Proctor 28958.*
– Bahamas and Greater Antilles, frequent in rather dry, rocky coastal woodlands. This species appears to be the chief Cayman host of the parasitic mistletoe *Phoradendron rubrum*, known locally as "Scorn-the-ground".

Family 31

CACTACEAE

Succulent perennials, sometimes subwoody, of various habit and usually spiny, the stems rounded, ribbed, angular or tuberculate, the surface dotted with small hairy cushions (areoles) from which variously arise branches, spines, flowers, hairs, glands, and leaves (if any). Leaves absent, or small and scale-like if present (foliaceous in *Pereskia*), soon falling; stipules absent. Spines of various types, usually clustered, sometimes lacking. Flowers regular, bisexual, usually solitary or sometimes in clusters; receptacle united with the inferior ovary, sometimes prolonged above it as a perianth tube. Perianth of few to numerous segments, often intergrading from sepaloid to petaloid or else sharply differentiated. Stamens usually numerous, the filaments usually borne on the throat of the perianth. Ovary 1-celled, sessile or partly immersed in a branch; ovules numerous on several parietal placentas; style long and simple, terminated by a several lobed stigma. Fruit a many-seeded berry, usually fleshy.

A large family of perhaps 125 ill-defined genera and about 1,800 species, confined in the natural state to the Western Hemisphere except for the genus *Rhipsalis*. However, a number of species of other genera (especially *Opuntia*) are now naturalized in the Old World and Australia. The curious spiny stems of cacti, which are adapted to endure extreme drought, are valued for their ornamental or bizarre appearance; in addition, many species bear beautiful flowers. Some species are planted to form "living fences". The large nocturnal flowers of *Epiphyllum* species, called "Night-blooming Cereus", are prized in horticulture. The fruits of many species are edible.

KEY TO GENERA

1. Ultimate divisions of the stems consisting of flattened, more or less elliptic, unribbed joints; areoles beset with minute barbed hairs, and when young bearing small terete leaves (these soon falling): **1. Opuntia**

1. All parts of stems cylindric and ribbed; areoles without minute barbed hairs, and leaves lacking at all stages:

  2. Plants erect or arching, without aerial roots:

    3. Stems relatively massive; flowering areoles bearing conspicuous tufts of white hairs; ovary naked: **2. Cephalocereus**

    3. Stems relatively slender; flowering areoles without conspicuous tufts of white hairs; ovary scaly, woolly or spiny:

4. Stems with thin, high, crenate ribs; perianth bell-shaped, pink; ovary with tufted spines: **3. Leptocereus**

4. Stems with relatively low, rounded ribs; perianth funnel-shaped with long tube, white; ovary bearing solitary rigid scales, each scale surrounded by a tuft of woolly hairs: **4. Harrisia**

2. Plants elongate and vine-like climbing on rocks and trees, the stems bearing aerial roots: **5. Selenicereus**

Genus 1. **OPUNTIA** Mill.

Succulent spiny perennials, branched from the base or developing a trunk, the branches conspicuously jointed, usually more or less flattened; leaves very small, terete, soon falling, the areoles (nodes) otherwise bearing spines, short barbed hairs (glochids), simple hairs, glands, and sometimes flowers. Flowers sessile, solitary; perianth bell-shaped, with numerous lobes. Ovary glabrous but with glochidiate areoles; style scarcely longer than the stamens, terminating in 2–7 erect stigmatic rays. Berry pear-shaped, depressed at the apex, with or without spines.

A widespread American genus of more than 250 species, occurring in both tropical and temperate areas. The fruits are edible.

KEY TO SPECIES

1. Plants noticeably spiny; stamens shorter than the perianth; wild species:

2. Stems spreading or bushy, branching at or near the base and not developing a trunk; flowers yellow, the perianth 3 cm long or more: **1. O. dillenii**

2. Stems developing a distinct trunk, becoming tree-like in habit; flowers orange, the perianth 1 cm long or less: **2. O. spinosissima**

1. Plants mostly lacking spines (except as seedlings); stamens much longer than the perianth; cultivated species: **[3. O. cochenillifer]**

1. Opuntia dillenii (Ker-Gawl.) Haw., Suppl. Pl. Succ. 79. 1819. **FIG. 90.**

Opuntia tuna of some authors, not Mill., 1768.

"Prickly Pear".

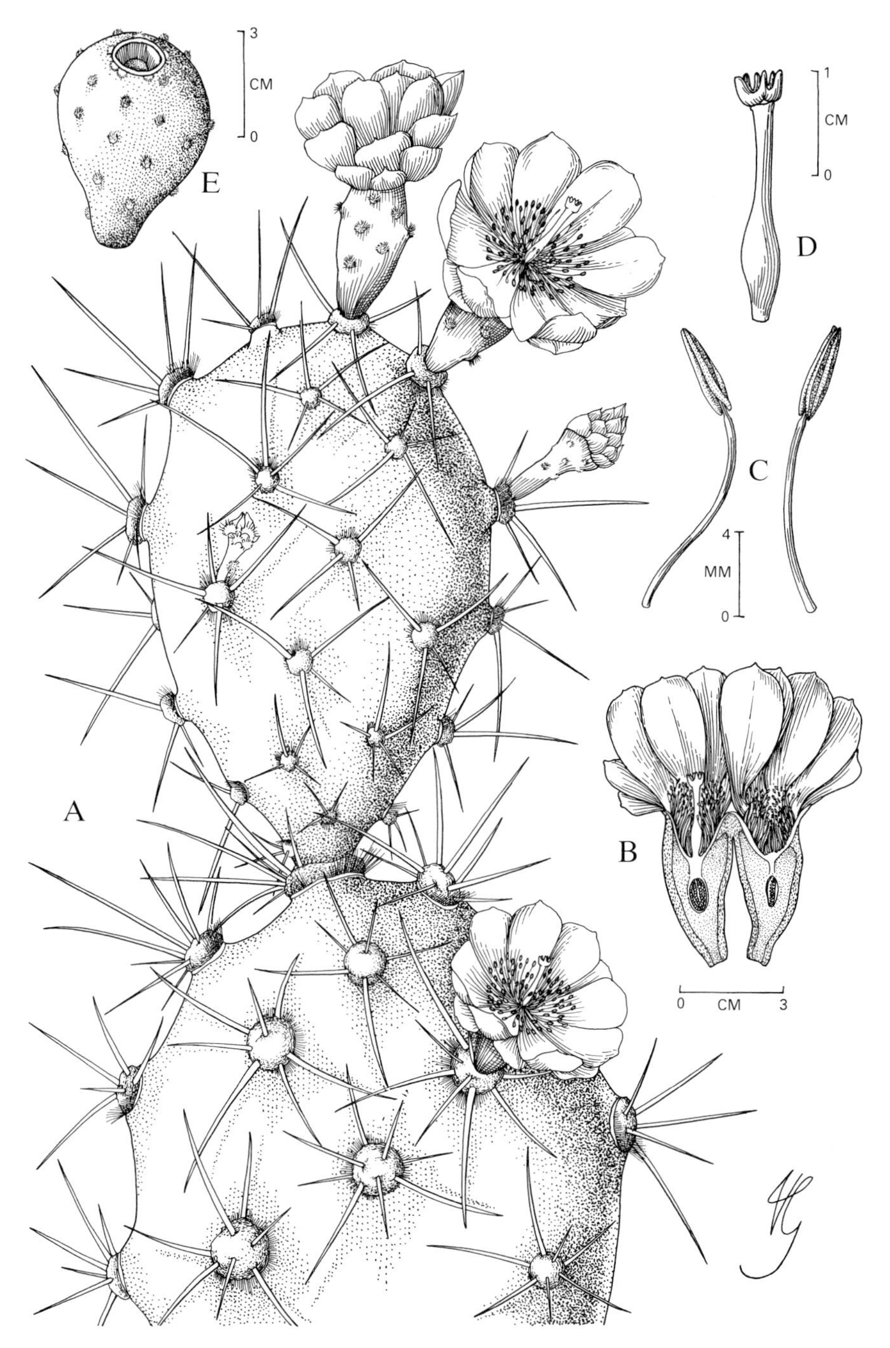

FIG. 90 **Opuntia dillenii.** A, habit. B, flower split lengthwise. C, two stamens. D, style terminated by stigmatic rays. E, fruit. (G.)

Plant usually 1-2 m tall, branched, sprawling and bushy; joints obovate or elliptic, 10–20 cm long; leaves 4-6 mm long; areoles large, with numerous glochids and 2-6 stout, slightly flattened, yellowish spines up to 4 cm long (but often much shorter), these straight or slightly curved-spreading. Perianth pale yellow, concolorous or sometimes shading to reddish at the base within, 3-5 cm long. Fruit obovoid, purplish-red, 5–7 cm long.

GRAND CAYMAN: *Kings GC 226; Proctor 15197.* LITTLE CAYMAN: *Proctor 35097.* CAYMAN BRAC: *Kings CB 29, CB 78; Proctor 29069.*
– Florida, the West Indies, and northern South America, chiefly coastal but also occurring inland in clearings and on dry stony plains. Cayman plants vary widely in length of spines, and also somewhat in size and colour of the flowers. In some West Indian islands the fleshy joints are peeled, diced, and cooked with rice as a vegetable.

2. **Opuntia spinosissima** Mill., Gard. Dict. ed. 8. 1768.

Extremely spiny plant of tree-like habit, up to 5 m tall, with an erect, cylindric, unjointed, densely long-spiny trunk; branches numerous, somewhat pendulous, the joints flat, more or less lance-oblong and often slightly curved or asymmetric, 10–25 cm long, 5-9 cm broad; areoles small, numerous, variably spiny, the spines needle-like, gray, 0.5–7 cm long. Ovary curved-cylindric, up to 5 cm long, with spiny and densely glochidiate areoles. Perianth orange to nearly scarlet, c. 1 cm long. Fruit obovoid, red, abruptly narrowed-acuminate and stalk-like at the base.

CAYMAN BRAC: *Kings CB 27; Proctor 29134.*
– Otherwise known only from Jamaica, where it is frequent in dry rocky thickets along the south side of the island.

[3. **Opuntia cochenillifer** (L.) Mill., Gard. Dict. ed. 8. 1768. **FIG. 91.**

"Cochineal".

Erect plant of somewhat tree-like habit at maturity, then up to 4 m tall or more with trunk 10–20 cm in diameter at the base, usually unarmed or nearly so; joints bright green, oblong or oblanceolate, 15-30 cm long or more; areoles with deciduous glochids. Flowers red or scarlet with pink, long-exserted stamens; perianth 1.5-2 cm long; stigma 6–7-lobed. Fruit red, c. 5 cm long.

GRAND CAYMAN: *Hitchcock; Kings GC 340.* LITTLE CAYMAN: *Proctor 35195.* CAYMAN BRAC: *Kings CB 33.*

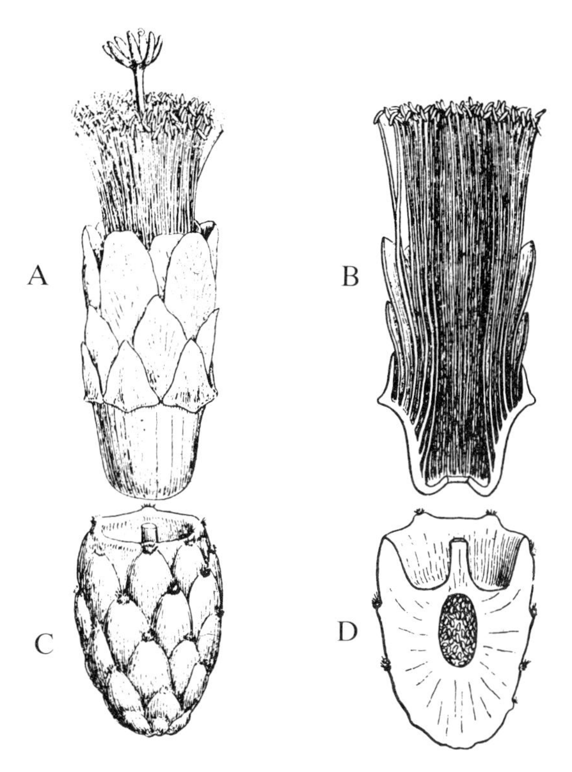

FIG. 91 **Opuntia cochenillifer.** A, flower removed from ovary, x1. B, same cut lengthwise and style removed, x1. C, ovary with base of style, x1. D, same cut lengthwise, x1 (F. & R.)

– Widely planted and locally spontaneous throughout the tropics, its natural origin probably Mexico or Central America. This is the chief species of cactus on which the cochineal insect was formerly grown for the production of a highly-prized red dye, an industry the Spanish invaders found already established when they conquered Mexico in 1518. The fruits are edible.]

Genus 2. **CEPHALOCEREUS** Pfeiffer

Erect columnar cacti, the stems simple or branched, the elongate joints leafless, ribbed and grooved, the upper areoles often densely woolly or long-bristly. Flowers solitary at the upper areoles, opening at night; perianth short-funnelform or campanulate; ovary subglobose, naked and spineless, rarely bearing a few small scales; style usually short-exserted. Fruit a more or less globose smooth berry with numerous small black or brown seeds.

A tropical and subtropical American genus of 40 or more species.

1. Cephalocereus swartzii (Griseb.) Britton & Rose in Contr. U.S. Nat. Herb. 12:425. 1909.

"Dildo".

Stems simple or branched, to 3 m tall or more, slightly glaucous; ribs about 10, obtuse, indented between the areoles; spines 8–20 from an areole, variable in length from 0.5 to 3 cm or more. Flowers greenish or pinkish, subtended by masses of woolly hairs; perianth 3–5 cm long. Fruit depressed-globose, 2–3.5 cm in diameter.

GRAND CAYMAN: *Proctor 15178.* LITTLE CAYMAN: *Proctor 35074.* CAYMAN BRAC: *Kings CB 30; Proctor 15312.*
– Otherwise found only in Jamaica and the Swan Islands, where it is widespread in rocky limestone habitats.

Genus 3. **LEPTOCEREUS** (Berg.) Britton & Rose

Arborescent or bush-like cacti, sometimes with diffuse or vine-like branches; joints elongate, with 3–8 prominent, thin, high, crenate ribs; aerial roots lacking; spines slender, needle-like. Flowers diurnal, small; ovary spiny; perianth short, campanulate; stamens numerous, borne at the base of the perianth-throat, scarcely exserted; stigma lobes a little exceeding the stamens. Fruit globose to oblong, more or less spiny; seeds numerous, black.

A small, Greater Antillean genus of 8 species, best represented in Cuba and lacking from Jamaica.

1. Leptocereus leonii Britton & Rose in Torreya 12:15. 1912.

Plants with slender rounded trunk 3 cm thick at the base, repeatedly branched upwardly, the ultimate branches elongate, 1.5 cm thick, 6–8-ribbed; areoles borne in depressions of the crenate ribs, each with 6–8 yellowish to gray spines 2–9 cm long. Flowers with ovary c. 3.5 cm long, the inner perianth-segments pink. Fruit globose-ovoid, 2 cm in diameter, with a few scattered spiny areoles.

CAYMAN BRAC: *Kings CB 28.*
– Otherwise known only from western Cuba. It apparently is always an inhabitant of limestone cliffs.

Genus 4. **HARRISIA** Britton

Arborescent, arching or rarely prostrate cacti with slender, branched stems, the elongate branches fluted or angled; areoles borne on ribs or prominences, spiny.

Flowers nocturnal, solitary at areoles near branch ends; perianth funnel-form, large, with a cylindric scaly tube as long as the limb or longer, the scales subtending areoles which bear tufts of woolly hairs. Stamens numerous, shorter than the perianth. Ovary and young fruit tuberculate. Fruit more or less globose, usually bearing sharp deciduous scales and small tufts of woolly hairs; seeds numerous, black.

A genus of about 17 species, widely distributed from Florida to Argentina, the majority West Indian.

1. **Harrisia gracilis** (Mill.) Britton in Bull. Torr. Bot. Club 35:563. 1908.

Erect cactus of columnar or tree-like habit, up to 7 m tall, with a short trunk and few to numerous elongate arching branches, these 9–11-ribbed and 1.5 -2.5 cm thick; ribs rounded, the grooves shallow; areoles with 6–10 or more needle-like spines of varying length from 0.5 to 3 cm. Flower-buds enveloped in tawny wool. Perianth 15–20 cm long, the inner segments white. Fruit globose, yellow, 4.5-6 cm in diameter, the white edible pulp containing numerous black seeds.

GRAND CAYMAN: *Kings GC 56.* LITTLE CAYMAN: *Proctor 35118.* CAYMAN BRAC: *Proctor 29005, 29068.*
– Otherwise known only from the southern side of Jamaica, where it is frequent in dry thickets, a similar plant, perhaps conspecific, occurs in the Swan Islands. The Cayman plants occur on exposed limestone rocks and cliffs.

Genus 5. **SELENICEREUS** (Berg.) Britton & Rose

Plants slender and elongate, the ribbed or angled stems trailing, clambering or climbing over rocks and trees, producing aerial roots at irregular intervals; areoles small, sometimes elevated on small knobs, usually spiny. Flowers large, nocturnal; perianth funnel-shaped with a slender elongate tube and flaring limb, the outer segments brownish or reddish, the inner ones white; ovary and perianth-tube furnished with scales, bristles, and hairs at the areoles. Stamens numerous, the weak filaments in two distinctly separate series, one from a circle at the top of the perianth-tube, the other from scattered points within the tube. Style exceeding the stamens; stigma-lobes slender, numerous. Fruit rather large, red or reddish, covered with tufts of deciduous spines, bristles, and hairs.

A genus of 16 or more species, ranging from Texas and the West Indies to South America. The flowers resemble those of the true "Night-blooming Cereus" species of the genus *Epiphyllum*, and are fully as ornamental, though apparently not as often or as prolifically produced.

KEY TO SPECIES

1. Spines acicular, to 9 mm long, the mature vegetative areoles usually lacking hairs; flowers not over 21 cm long, the inner perianth-segments gradually narrowed to a pointed apex: **1. S. grandiflorus**

1. Spines conical, less than 2 mm long, usually accompanied by numerous much longer rather stiff whitish hairs; flowers 24–39 cm long, the inner perianth-segments widest near the apex, abruptly terminated in a short-acuminate point: **2. S. boeckmannii**

1. Selenicereus grandiflorus (L.) Brit. & Rose in Contr. U.S. Nat. Herb. 12:430. 1909.

"Vine Pear"

Stems trailing or high-climbing, up to 10 m long or more, sometimes forming tangles, in Cayman plants seldom more than 1 cm thick, 8-ribbed on the average; spines 5-10 per areole, needle-like, rarely accompanied by a few shorter brown hairs; juvenile plants have the spines both fewer and shorter. Flower-buds and perianth-tube densely covered with brown hairs and sharp bristles; perianth 19-21 cm long, the outer segments light brown, the inner white and less than 1 cm broad. Fruit ovoid, c. 8 cm long.

GRAND CAYMAN: *Brunt 1697, 1919; Kings GC 56, GC 237; Proctor 15179, 31032.*

– Cuba and Jamaica; also said to be widely cultivated and escaping in various parts of tropical America. The indigenous Grand Cayman population differs from most plants of other areas particularly in its more slender stems. However, the extreme variability of this species, especially in Jamaica, mitigates against recognizing the Cayman plants as a distinct variety.

2. Selenicereus boeckmannii (Otto) Brit. & Rose in Contr. U.S. Nat. Herb. 12:429. 1909.

Stems light green, extensively trailing or climbing, often 10 m long or more, sometimes forming tangles, c. 1 cm thick and with up to 8 ribs; spines 3-6 per areole, most of them scarcely over 1 mm long, accompanied by 10-20 setaceous hairs, these up to 5 mm long or more. Perianth similar in coloration to that of *S. grandiflorus* but larger, the tube densely clothed with pale tawny hairs, the oblanceolate inner segments up to 3 cm broad near the apex. Fruit globular, 5-6 cm in diameter.

CAYMAN BRAC: *Proctor 35147.*

– Cuba, Hispaniola, and eastern Mexico; introduced into Florida and the Bahamas. Both this and the preceding species occur naturally in dense rocky woodlands.

Family 32

AIZOACEAE

Succulent unarmed herbs, the stems usually prostrate or semi-prostrate, the nodes often jointed. Leaves alternate, opposite or whorled, simple, entire. Flowers regular, perfect, terminal or axillary, solitary to several together or in branched inflorescences; perianth with 5 segments; stamens 5 to many. Ovary free, 1-many-celled; styles as many as the cells. Fruit a membranous dehiscent capsule, included in the persistent perianth; seeds usually many, rarely only 1.

A chiefly tropical and subtropical family of about 22 genera and 600 species, most abundant in Africa.

Genus 1. **SESUVIUM** L.

Perennial, mostly creeping herbs with branched fleshy stems; leaves opposite, the bases often dilated and connate; stipules absent. Flowers axillary, mostly solitary; perianth lobes united toward base, coloured and petaloid within; stamens 5 or numerous, inserted on the perianth-tube. Ovary 3-5-celled, with 3-5 styles. Capsule 3-5-celled, splitting transversely ("circumscissile"); seeds numerous, with annular embryo.

A pantropical genus of about 5 species.

KEY TO SPECIES

1. Flowers solitary, with numerous stamens; perianth more than 4 mm long:

   2. Leaves 2-6 cm long; capsule 9-11 mm long: **1. S. portulacastrum**

   2. Leaves 1-2 cm long; capsule 4-5 mm long: **2. S. microphyllum**

1. Flowers 1 or several together, each with 5 stamens; perianth less than 4 mm long: **3. S. maritimum**

1. Sesuvium portulacastrum L., Syst. Nat. ed. 10, 2:1058. 1759. **FIG. 92.**

"Sea-pusley"

Prostrate succulent herb, often widely and extensively creeping, occasionally with short suberect branches. Leaves linear-oblong to oblanceolate, mostly 2-6 cm long (rarely less or more), 2–10 mm broad, the bases clasping. Flowers solitary on peduncles 5–14 mm long; perianth cup c. 4 mm long, the lobes 5 mm long, usually pink within. Capsule oblong-conic, 9–11 mm long.

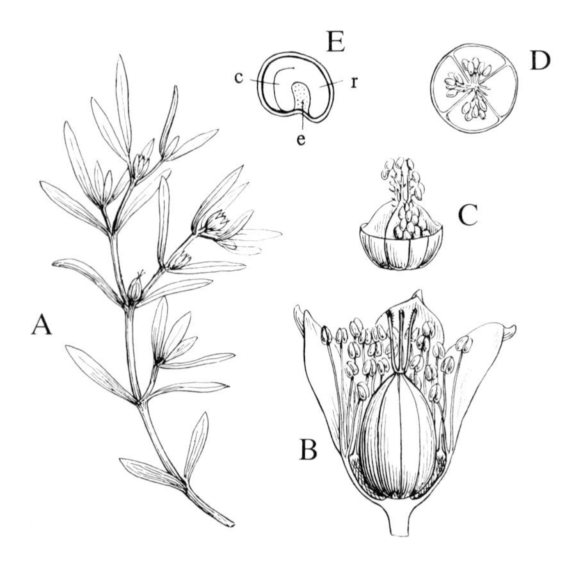

FIG. 92 **Sesuvium portulacastrum.** A, portion of plant, x$^2/_3$, B, flower with portion of perianth and some stamens removed, x3. C, capsule with upper part fallen, x3. D, lower part of C cut across, x3. E, section of seed, x10; e, endosperm; c, cotyledons; r, radicle. (F. & R.)

GRAND CAYMAN: *Brunt 2104; Hitchcock; Kings GC 34, GC 219, GC 282; Proctor 15201; Sachet 394.* LITTLE CAYMAN: *Proctor 28066.* CAYMAN BRAC: *Kings CB 55; Millspaugh 1221; Proctor 29128.*

– Pantropical, chiefly on sandy or rocky sea-shores or on saline flats near mangroves.

2. Sesuvium microphyllum Willd., Enum. Hort. Berol. 521. 1809.

Creeping succulent herb, forming mats, with numerous short tangled branches. Leaves oblanceolate or spatulate, chiefly 1–2 cm long. Flowers solitary on peduncles 3–7 mm long; perianth cup 1.5–2 mm long, the lobes 2–3.5 mm long, white within. Capsule ovoid-conic, 4–5 mm long.

LITTLE CAYMAN: *Proctor 28056.*

– Cuba and a few other West Indian islands, not common; often grows in sandy hollows among seaside rocks.

3. Sesuvium maritimum (Walt.) B.S.P., Prem. Cat. N.Y. 20. 1888.

Creeping succulent herb, forming mats or rosettes, with numerous short branches. Leaves oblanceolate or spatulate, chiefly 0.5–1.5 cm long. Flowers solitary or 2-3 together in leaf-axils, nearly sessile; perianth cup c. 1 mm long, the lobes 1.5–2 mm long; stamens 5. Capsule ovoid, c. 3 mm long.

GRAND CAYMAN: *Proctor 15115.*
– Atlantic coast of North America from Long Island to Florida; also in the Bahamas, Cuba, and Puerto Rico, often occurring on salinas and along the edge of mangrove swamps.

Family 33

PORTULACACEAE

Usually succulent unarmed herbs, prostrate or erect, rarely somewhat shrubby at the base. Leaves alternate, subopposite or opposite, simple, entire. Flowers regular, perfect, solitary, clustered at the tips of branches, or else in terminal racemes or panicles; sepals 2; petals usually 4–6, free. Stamens equal in number to the petals, or sometimes less or more. Ovary superior or half-inferior, 1-celled, with free-central placentation; style with 3–7 stigmatic branches; ovules 2 to many. Fruit a capsule, circumscissile or splitting by 3 valves; seeds 2 to numerous, with curved embryo.

A family of about 20 genera and 500 species, mostly American.

KEY TO GENERA

1. Flowers solitary or cymose-capitate; ovary half-inferior; capsule circumscissile; stipules scarious or represented by hairs: **1. Portulaca**

1. Flowers in racemes or panicles; ovary superior; capsule 3-valved; stipules absent: **2. Talinum**

Genus 1. **PORTULACA** L.

Fleshy herbs prostrate or erect; leaves alternate, subopposite or opposite, usually whorled around the flowers; stipules minute or reduced to hairs. Flowers solitary or clustered at the tips of branches; sepals united at base; stamens 8–20. Ovary with numerous ovules; style 3–9-cleft or -parted. Capsule many-seeded.

A genus of about 20 species, the majority American.

KEY TO SPECIES

1. Plants glabrous; leaves flat:

2. Leaves 6-30 mm long; sepals 3-5 mm long; petals 4-6 mm long; capsule c. 6 mm long: **1. P. oleracea**

2. Leaves 1-3.5 mm long sepals 2-3 mm long; petals c. 3.5 mm long; capsule 2-3.5 mm long: **2. P. tuberculata**

1. Plants hairy; leaves more or less cylindric:

3. Flowers yellow:

4. Annual; petals c. 3 mm long; capsule circumscissile below the middle; seeds black, 0.4 mm broad: **3. P. halimoides**

4. Perennial; petals more than 6 mm long; capsule circumscissile above the middle; seeds brown, 0.5-0.6 mm broad: **4. P. rubricaulis**

3. Flowers purple or crimson: **5. P. pilosa**

1. Portulaca oleracea L., Sp. Pl. 1:445. 1753. **FIG. 93.**

"Pusley", "Wild Parsley"

Annual fleshy glabrous herb, prostrate and radiating from a taproot, or sometimes ascending or erect, the stems often reddish. Leaves subopposite or sometimes alternate, subsessile, obovate or spatulate, up to about 30 mm long but usually less, the apex rounded or retuse. Flowers sessile; sepals broadly ovate, keeled, acute; petals 5, yellow, deeply notched at the apex, soon withering; stamens 6-12; stigmas 3-6. Capsule circumscissile at about the middle; seeds black, 0.7-0.8 mm in diameter, granulate.

GRAND CAYMAN: *Brunt 1859, 1959; Hitchcock; Kings GC 383; Proctor 15048.* LITTLE CAYMAN: *Kings LC 24; Proctor 28169, 28170.* CAYMAN BRAC: *Kings CB 23, CB 46; Proctor 29078.*

– Worldwide in distribution, occurring commonly in tropical and temperate regions of both hemispheres, a weed of sandy soil and open waste ground. This species is widely eaten as a vegetable, and is also used in salads and as an ingredient of soup.

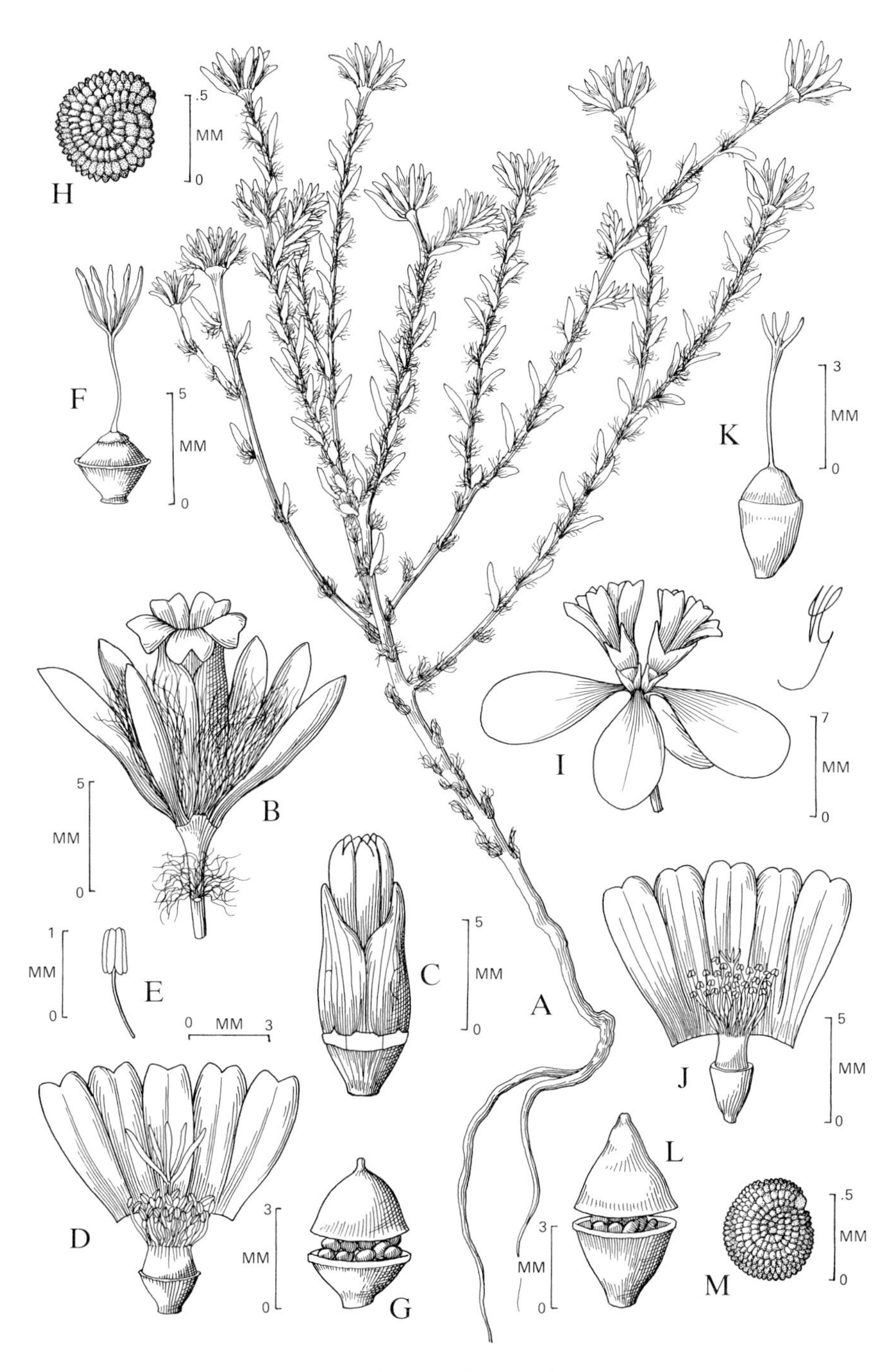

FIG. 93 **Portulaca pilosa.** A, habit. B, flower with subtending leaves and hairs. C, flower-bud. D, flower opened out to show stamens and pistil. E, stamen. F, ovary, style and stigmas. G, fruit showing dehiscence. H, seed. **Portulaca oleracea.** I, flowers with subtending leaves. J, flower opened out to show stamens & pistil. K, ovary, style and stigmas. L, fruit showing dehiscence. M, seed. (G.)

2. Portulaca tuberculata Leon in Contr. Ocas. Mus. Hist. Nat. Col. "De La Salle", no. 9:3. 1950.

Small fleshy herb of flaccid texture, with prostrate branches 2-10 cm long, forming rosettes from a taproot. Leaves opposite, flat, obovate to orbicular, 1-3.5 mm long. Flowers sessile, solitary or 2 together; sepals deltate-ovate; petals 4 or 5, yellow, scarcely exceeding the sepals; stamens 5-8; stigmas 4-5. Capsule terminated by a rounded-conic tubercle, and circumscissile at or slightly below the middle; seeds black or nearly so, 0.5-0.6 mm in diameter, granulate.

LITTLE CAYMAN: *Proctor 28191, 35089.* CAYMAN BRAC: *Proctor 29012, 29041, 29351.*
– Otherwise known only from eastern Cuba. The Cayman plants grow in small soil-filled pockets of limestone rocks and cliffs, or rarely in sandy clearings.

3. Portulaca halimoides L., Sp. Pl. ed. 2, 1:639. 1762.

An erect or diffuse annual herb usually less than 10 cm tall, the stems bearing tufts of white hairs in the leaf-axils; leaves subcylindric, 6-10 mm long, deciduous. Flowers sessile, embedded in a tuft of white hairs; sepals 2-2.5 mm long; petals yellow or white with yellow centre, ovate, 3 mm long; stamens 8-20. Capsule usually hidden in white hairs, circumscissile below the middle; seeds black, 0.4 mm in diameter, minutely tuberculate.

GRAND CAYMAN: *Kings GC 224.*
– West Indies and Mexico, in dry or occasionally moist clearings near the sea.

4. Portulaca rubricaulis Kunth in H.B.K., Nov. Gen. & Sp. 6:73. 1820.

Portulaca phaeosperma Urban, 1905.

A fleshy prostrate or ascending herb usually less than 15 cm high, often with red stems, and bearing small tufts of white axillary hairs. Leaves subcylindric, linear or lance-linear, up to 10 mm long or a little more. Flowers sessile, often c. 1 cm across when open; sepals 4-5 mm long; petals yellow, usually 6-7 mm long, occasionally less; stamens 12-16; stigmas 5-7. Capsule subglobose, 2.5-3 mm in diameter, circumscissile above the middle; seeds brown, 0.5-0.6 mm in diameter, rugulose.

GRAND CAYMAN: *Brunt 2006, 2122, 2149; Kings GC 329; Proctor 15221, 27943.* LITTLE CAYMAN: *Proctor 28176, 35088, 35140.* CAYMAN BRAC: *Proctor 28978, 28979.*
– Florida and the West Indies to northern South America, often in pockets of limestone rocks, occasionally in seasonally wet rocky pastures.

5. Portulaca pilosa L., Sp. Pl. 1:445. 1753. **FIG. 93.**

"Ten-o'clock"

Annual or semiperennial herb, the stems prostrate or ascending, less than 15 cm long, densely white- or brownish-pilose from the leaf-axils. Leaves linear-subcylindric, 5–15 mm long. Flowers sessile, surrounded by a tuft of whitish or pale brownish hairs and a whorl of 6–10 leaves; sepals ovate, 2–3 mm long, not keeled; petals purple or crimson, obovate-retuse, 3–5.5 mm long; stamens 15–22, with crimson filaments; stigmas 4–6. Capsule subglobose, 3–4 mm in diameter, circumscissile at about the middle; seeds black, 0.5 mm in diameter, minutely tuberculate.

GRAND CAYMAN: *Proctor 15204.* CAYMAN BRAC: *Proctor 29070.*
– Widespread from southern United States to South America, growing on open sandy ground or in soil-filled pockets of limestone rock.

The large-flowered *Portulaca grandiflora* Hook., native of Argentina, is sometimes cultivated. Its pink, red, yellow, orange or white flowers, which open only in the morning, have petals 15-25 mm long; "double" forms are also grown.

Genus 2. **TALINUM** Adans.

Fleshy glabrous erect or ascending herbs. Leaves alternate, without stipules. Petals 5; stamens 10–30, adherent to the base of the petals. Ovary free, with many ovules; style 3-lobed or 3-cleft. Capsule 3-valved; seeds numerous, borne on a central globose placenta.

A genus of about 15 species, chiefly American.

1. Talinum triangulare (Jacq.) Willd. in L., Sp. Pl. 2:862. 1800.

Annual herb with taproot, the erect stems to 50 cm tall or more; leaves oblanceolate to obovate, 2–6 cm long, 0.5–2 cm broad or more, rounded or acute at the apex. Racemes simple or branched, few-to many-flowered; pedicels 3-angled. Sepals persistent, c. 5 mm long; petals pink in Cayman plants, elsewhere often yellow or sometimes white, 6–9 mm long and 6 mm broad; stamens numerous. Capsule c. 5 mm long; seeds black, 0.8 mm long.

LITTLE CAYMAN: *Proctor 35142.*
– Widely distributed from the Bahamas and Greater Antilles to South America and in West Africa. The Little Cayman plants were found along sandy roadsides and may have escaped from cultivation.

Family 34

BASELLACEAE

Herbaceous glabrous vines, often rather succulent, with tuberous roots. Leaves alternate, simple, entire; stipules absent. Flowers small, regular, perfect, in axillary and terminal racemes or panicles; each pedicel with 1 bract at base and 2 bracteoles at top subtending the flower. Sepals 2, sometimes winged in fruit; petals 5; stamens 5, opposite the petals and inserted on a hypogynous disk adnate to the base of the corolla. Ovary superior, 1-celled, with 1 basal ovule; styles usually 3, free or united; stigmas entire or cleft. Fruit an indehiscent 1-seeded utricle, included in the perianth; seeds with endosperm.

A small family of 5 genera and about 22 species, mostly in tropical America.

Genus 1. **ANREDERA** Juss.

Climbing herbs with perennial tuberous roots and somewhat fleshy leaves. Flowers in axillary and terminal spike-like racemes, these simple or branched. bracteoles usually adnate to the base of the perianth; sepals petaloid, shorter than the petals; petals white or greenish-white, sometimes changing to purple with age, the stamens inserted at their base. Ovary ovoid, the styles separate or somewhat united. Seed with semiannular embryo.

A tropical American genus of 10 species.

1. **Anredera leptostachys** (Moq.) van Steenis in Fl. Males. (ser. 1) 5 (3):302. 1957. **FIG. 94.**

Boussingaultia leptostachys Moq., 1849.

Slender twining vine sometimes forming tangles; leaves short-petiolate, the blades ovate to elliptic, 2-6 cm long, 1.5-4 cm broad, acute to acuminate at both apex and base. Racemes lax, simple, slender, longer than the leaves; flowers pale greenish-yellow to white, on pedicels 1 mm long; petals c. 2 mm long. Fruits not seen.

GRAND CAYMAN: *Proctor 15190.* LITTLE CAYMAN: *Proctor 28147.*
– Florida, West Indies, and continental tropical America at low elevations, the Cayman plants found growing in sandy coastal thickets.

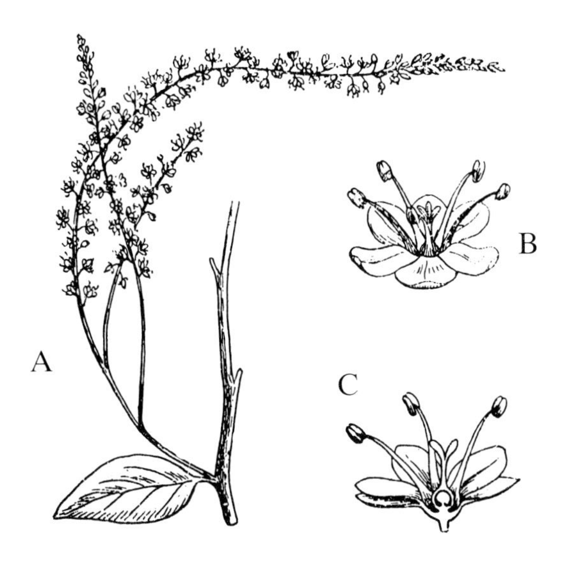

FIG. 94 **Anredera leptostachys.** A, portion of flowering branch, x$^{2}/_{3}$. B, flower of a related species (*Boussingaultia baselloides*), x3. C, the same cut lengthwise, x3 (F. & R.)

Family 35

CHENOPODIACEAE

Annual or perennial herbs, rarely shrubs; leaves alternate or rarely opposite, simple, sometimes reduced to scales; stipules lacking. Flowers small, regular, perfect or unisexual, variously clustered, usually greenish and wind-pollinated. Perianth absent or of 2–5 segments, these more or less united at the base, persistent after flowering; stamens as many as or fewer than the perianth-segments and opposite them, the anthers 2–4-celled. Ovary superior, 1-celled, with 1 basal ovule; styles 1–3, the stigmas capitate or 2–3-lobed and elongate. Fruit a 1-seeded, usually indehiscent utricle with a thin or hard pericarp, usually included in the persistent perianth; seed with annular embryo enclosing the endosperm, or endosperm sometimes lacking.

A large family of wide distribution, with over 100 genera and 1,400 species, best represented in Asia, and especially characteristic of deserts and saline soils. *Beta vulgaris* L., the European beet and its variants, and *Spinacia oleracea* L., spinach, are well-known vegetables of this family.

KEY TO GENERA

1. Leaves evident, alternate, more or less toothed or lobed; stem not jointed; stamens 3-5: **1. Atriplex**

1. Leaves reduced to opposite pairs of scales; stem jointed; stamens 2: **2. Salicornia**

Genus 1. **ATRIPLEX** L.

Herbs or low shrubs, more or less scurfy-canescent or silvery; leaves mostly alternate. Flowers dioecious or monoecious, small, green, in axillary capitate clusters or panicled spikes. Staminate flowers bractless, consisting of a 3-5-parted perianth and an equal number of stamens; filaments free or united by their bases; a pistillode sometimes present. Pistillate flowers subtended by 2 bracts which enlarge in fruit and are more or less united; perianth none; stigmas 2. Utricle completely or partly enclosed by the fruiting bracts; seeds with mealy endosperm.

A widely-distributed genus of about 150 species, often found in saline or arid habitats.

1. **Atriplex pentandra** (Jacq.) Standley in N. Amer. Fl. 21:54. 1916.

Annual bushy or suffrutescent herb, the branches procumbent or ascending, up to 8 cm long. Leaves silvery-scurfy especially on the under side, sessile or nearly so, oblong to rhombic or obovate, 1-2 cm long, with more or less repand-dentate margins. Flowers monoecious, the staminate in dense short terminal spikes, the pistillate clustered in the axils of leaves. Fruiting bracts united at base only, sharply toothed on the margins and crested or tuberculate on the sides.

LITTLE CAYMAN: *Sauer 3315.*
– Florida to Texas, Bermuda, the West Indies, and northern South America, on sandy seashores.

Genus 2. **SALICORNIA** L.

Succulent glabrous herbs with opposite jointed branches, the leaves reduced to mere opposite scales. Flowers perfect or the lateral ones staminate, sunken 3-7 together in the axils of the upper scales, thus forming narrow terminal spikes. Perianth obpyramidal or rhomboid, fleshy, 3-4-toothed or truncate, becoming spongy in fruit and falling with it. Stamens 2 or rarely solitary, exserted. Ovary ovoid, with 2 stigmas. Fruit enclosed by the adherent perianth; seed often hairy, without endosperm.

A cosmopolitan halophytic genus of about 35 species.

KEY TO SPECIES

1. Perennial; main stems prostrate and subwoody, with several or many erect branches: **1. S. virginica**

1. Annual; main stem erect with lateral branches: **2. S. bigelovii**

1. Salicornia virginica L., Sp. Pl. 1:4. 1753. **FIG. 95.**

Salicornia perennis of authors, not Mill., 1768.

Salicornia ambigua Michx., 1803.

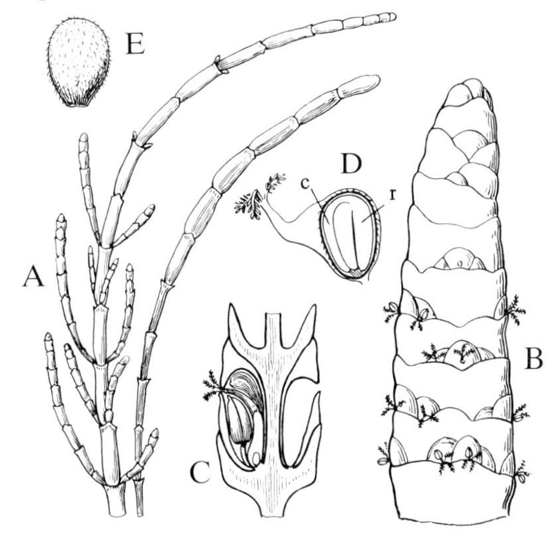

FIG. 95 **Salicornia virginica.** A, portion of stem and branch, x$^2/_3$. B, portion of stem in flower, x5. C, small portion of B cut lengthwise, showing a flower enclosed in the perianth and another perianth empty, x7. D, fruit cut lengthwise, x11; c, cotyledons; r, radicle. E, seed, x11. (F. & R.)

Perennial by a woody rootstock; main stems trailing or decumbent, up to 60 cm long or more, rooting at nodes, the branches ascending or erect; ultimate branchlets terete, c. 1 mm thick between the nodes. Scales connate to form a shallow cup surrounding each node of the branchlets. Fertile spikes 1–4 cm long, scarcely distinguishable from the sterile branchlets, their joints about as long as thick, the flowers about equalling the joints in length.

GRAND CAYMAN: *Brunt 1736; Kings GC 175, GC 176B; Proctor 27971.* – North American coasts of the Atlantic Ocean, also occurring in the north-western half of the West Indies and along the Gulf of Mexico, chiefly on saline shores and salt flats.

2. Salicornia bigelovii Torr., Bot Mex. Bound. Surv. 184. 1859.

Annual with an erect fleshy stem, few-many-branched upwardly, the nearly terete ascending branches 1.5–2 mm thick between the nodes. Scales triangular-ovate, sharply pointed, 2–3 mm long, connate at the sides. Fertile spikes 1–12 cm long, obviously thicker than the sterile branches, their joints about as long as thick, the flowers a little shorter.

LITTLE CAYMAN: *Kings LC 117; Proctor 35072.*
– Atlantic and Gulf coasts of North America and the northwestern half of the West Indies (but not reaching Jamaica), also in California; not common, occurring on sand-spits and salinas.

Chenopodium ambrosioides L., sometimes called "Mexican Tea", is an aromatic herb often cultivated in the West Indies as a vermifuge. There is no record of it from the Cayman Islands, but it may occur in a few old-time gardens. It tends to persist after cultivation, and sometimes becomes naturalized in sandy or gravelly waste places.

Family 36

AMARANTHACEAE

Herbs or rarely small shrubs, occasionally succulent or spiny; leaves opposite or alternate, simple, usually entire; stipules lacking. Flowers small, perfect or unisexual, bracteolate, variously clustered, usually in terminal or axillary inflorescences; perianth of 2–5 segments, scarious, persistent after flowering. Stamens 2–5, opposite the perianth segments; filaments free or commonly more or less fused, sometimes forming a corolla-like tube. Ovary superior, 1-celled; ovule 1 in Cayman genera, basal. Fruit a membranous or fleshy utricle, indehiscent, irregularly rupturing, or circumscissile, enclosed in or resting on the persistent perianth. Seed naked or arillate, usually lustrous and smooth or nearly so, with copious endosperm and annular or hippocrepiform embryo.

A world-wide family of about 65 genera and 850 species, of little or no economic importance. A few are edible ("Calalu"), and one species of *Amaranthus* produces seeds used as a cereal in Asia.

KEY TO GENERA

1. Leaves alternate: **1. Amaranthus**

1. Leaves opposite:

 2. Flowers in very elongate unbranched spikes: **2. Achyranthes**

 2. Flowers not in elongate spikes:

  3. Flowers in loose panicles; anthers 2-celled: **3. Iresine**

  3. Flowers in dense heads:

   4. Stems erect; heads 20–25 mm thick; anthers 2-celled: **[4. Gomphrena]**

   4. Stems prostrate; heads 5–10 mm thick; anthers 1-celled:

    5. Matlike herb with basal leaf-rosette; stamens 2, staminodes present: **5. Lithophila**

    5. Trailing succulent herb without basal leaf-rosette; stamens 5, staminodes absent: **6. Caraxeron**

Genus 1. **AMARANTHUS** L.

Annual erect or prostrate herbs, glabrous or pubescent; leaves entire, long-petiolate. Flowers unisexual, monoecious, in axillary clusters or terminal panicles; perianths of 3–5 equal segments. Stamens 2–5; filaments free; staminodes none; anthers 2-celled. Ovary with 2–3 stigmas. Fruit compressed, indehiscent or circumscissile, often with 2 or 3 beaks. Seed compressed, with annular embryo.

A world-wide genus of about 60 tropical and temperate species.

KEY TO SPECIES

1. Flowers in axillary clusters or very short spikes; leaf-blades oblong or obovate and less than 3.5 cm long: **1. A. crassipes**

1. Flowers in terminal panicles as well as axillary spikes; leaf-blades ovate or rhombic-ovate, mucronate, up to 10 cm long:

 2. Flowers with 3 stamens and perianth segments; utricle indehiscent: **2. A. viridis**

 2. Flowers with 5 stamens and perianth segments; utricle circumscissile: **3. A. dubius**

1. Amaranthus crassipes Schlecht. in Linnaea 6:757. 1831.

Glabrous herb with somewhat fleshy decumbent to ascending stems mostly less than 20 cm long or sometimes longer; leaves broadly oblong to obovate, 1-3.5 cm long, the blades with apex rounded and notched, the base attenuate, the tissue prominently whitish-veined beneath. Flowers straw-coloured, in small dense axillary clusters, the pistillate flowers with short, much-thickened peduncles that detach with the fruit, the staminate flowers in separate clusters in axils of the upper leaves. Perianth of the pistillate flowers with 4 or 5 segments, 1–1.5 mm long. Utricle finely wrinkled, indehiscent.

CAYMAN BRAC: *Proctor 29071.*

– Florida, West Indies, and on the continent from Mexico to Peru, usually a weed of dryish waste places. The Cayman Brac plants were growing in pockets of limestone rock near a path on The Bluff overlooking Spot Bay.

2. Amaranthus viridis L., Sp. Pl. ed. 2, 2:1405. 1763. **FIG. 96.**

"Calalu"

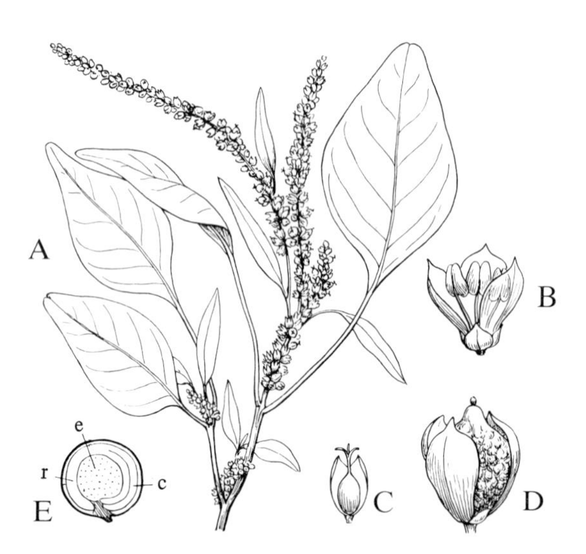

FIG. 96 **Amaranthus viridis.** A, portion of plant in flower, x$^{2}/_{3}$. B, staminate flower, x10. C, pistillate flower with one perianth-segment removed, x10. D, fruit with persistent perianth, x10. E, seed cut lengthwise, x10; c, cotyledons; r, radicle; e, endosperm. (F. & R.)

Erect herb up to 1 m tall, sparingly branched, nearly glabrous; leaves ovate or rhombic-ovate on long petioles, up to 8 cm long or sometimes much more, the apex blunt and notched, with a small mucro in the notch. Flowers green, in terminal panicles and axillary spikes or clusters; perianth of 3 segments; stamens 3. Utricle globose, wrinkled, indehiscent, about as long as the perianth.

GRAND CAYMAN: *Brunt 2097; Hitchcock; Kings GC 367; Millspaugh 1347; Proctor 15101, 27946.*
– Pantropical, often cultivated as a vegetable.

3. **Amaranthus dubius** Mart. ex Thell. in Mem. Soc. Sci. Nat. Cherbourg 38:203. 1912.

Amaranthus tristis of Griseb. not L., 1753.

Erect herb up to 1 m tall, more or less branched, glabrous or pubescent above. Leaves rhombic-ovate on long, often red petioles, mostly 2–10 cm long, the apex usually acute and mucronate. Flowers greenish, in terminal panicles and dense axillary clusters; perianth of 5 segments; stamens 5. Utricle with circumscissile dehiscence.

GRAND CAYMAN: *Hitchcock; Millspaugh 1390.* CAYMAN BRAC: *Kings CB 66; Millspaugh 1191.*
– West Indies and continental tropical America, also in tropical Africa, a weed of waste ground.

The weedy, pantropical "Spiny Calalu" or "Macca Calalu", *Amaranthus spinosus* L., has not been recorded from the Cayman Islands, but is to be expected.

Genus 2. **ACHYRANTHES** L.

Annual or perennial herbs, glabrous or pubescent, sometimes shrubby at base; leaves opposite, petiolate, entire. Flowers perfect, in long slender spikes, deflexed in fruit, the bracts spine-tipped. Perianth of 5 subequal segments (rarely 2 or 4), glabrous or pubescent, becoming hard and ribbed. Stamens 5 (rarely 2 or 4), alternating with laciniate staminodes; filaments united into a short cup at the base; anthers 4-celled. Ovary with filiform style and capitate stigma; ovule 1, suspended from an elongate funicle. Utricle indehiscent, included in the persistent perianth; seed oblong, with annular embryo.

A pantropical genus of about 10 species.

1. **Achyranthes indica** (L.) Mill., Gard. Dict. ed. 8. 1768. **FIG. 97.**

Achyranthes aspera var. *obtusifolia* (Lam.) Griseb., 1859.

"Devil's Horsewhip".

An erect herb to 1 m tall, the stems whitish-pubescent, with a few low branches; leaves obovate or rotund, 2–7 cm long and wide, rounded at the apex and sometimes with a small abrupt point, the surfaces pilose-sericeous especially beneath. Flowers green, in terminal whip-like spikes up to 40 cm long, more densely-flowered toward the end; bracts broadly ovate, 3 mm long, tipped by a rigid spine; bracteoles aristate, shorter than the perianth. Perianth segments 4 mm long, acuminate; stamens 1 mm long, twice as long as the staminodes. Utricle enclosed in the perianth and by the spiny bracteoles, readily detached and adhering to animals and human passers-by.

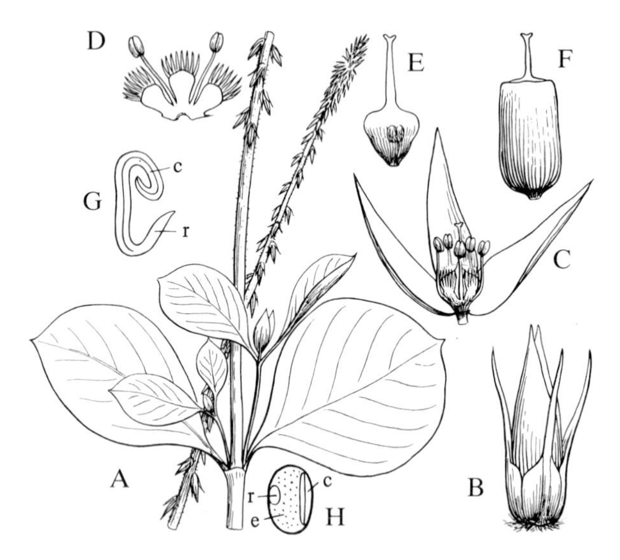

FIG. 97 **Achyranthes indica.** A, portion of plant, x$^2/_3$. B, flower with bract and bracteoles; x7. C, flower with two perianth-segments removed, x7. D, portion of staminal tube spread out, with 2 stamens and 3 staminodes, x7. E, ovary with ovule, x7. F, utricle, x7. G, embryo, x7; c, cotyledons; r, radicle. H, section through seed, x7; e, endosperm; r, radicle; c, cotyledons. (F. & R.)

GRAND CAYMAN: *Brunt 2111a; Hitchcock; Millspaugh 1272; Proctor 15247.* CAYMAN BRAC: *Proctor (sight record).*
– A pantropical weed of pastures and open waste ground.

Genus 3. **IRESINE** P. Browne

Erect to decumbent or scandent herbs, sometimes shrubs or small trees, glabrous or pubescent; leaves opposite, entire, petiolate. Flowers perfect or unisexual, monoecious or dioecious, with bracts and bracteoles, mostly in large panicles or panicled spikes; perianth of 5 segments, often hairy. Stamens 5, the subulate filaments united at base to form a short tube; anthers 2-celled. Style short or none; stigmas 2 or 3, subulate or filiform, or in the staminate flowers sometimes capitate but non-functional. Utricle more or less compressed, indehiscent; seed with annular embryo.

A chiefly tropical American genus of about 45 species, a few also in tropical Africa.

1. **Iresine diffusa** Humb. & Bonpl. ex Willd. in L., Sp. Pl. ed. 4, 4:765. 1806.

Iresine celosia L., 1759, illegit.

Annual or sometimes perennial, the stems erect, procumbent, or clambering, sometimes elongate and up to 3 m long, glabrous or nearly so. Leaves thin, broadly ovate to lanceolate, 2–14 cm long, acute to acuminate at apex. Flowers very numerous in panicled spikes, the pistillate with copious long wool at the base; bracts and perianth silvery or pale greenish, the perianth segments 1–1.5 mm long. Seeds shining dark red, 0.5 mm in diameter.

GRAND CAYMAN: *Hitchcock; Kings GC 70.*
– Southeastern United States, West Indies, and Central and South America, often in thickets or else a weed of waste places, common except at high elevations. The Cayman specimens are rather small for the species.

[Genus 4. **GOMPHRENA** L.

Pubescent annual or perennial herbs. Flowers perfect, in dense heads or spikes, white, yellow, or red; bracteoles keeled, the keel often crested. Perianth 5-lobed or 5-parted. Stamens 5, the filaments united into a lobed tube. Stigma 2-lobed, the lobes recurved. Utricle flattened; seed smooth.

A pantropical genus of about 100 species.

1. **Gomphrena globosa** L., Sp. Pl. 1:224. 1753.

Annual, usually branched or bushy herb 30–80 cm tall, the stems appressed-pilose, the nodes swollen. Leaves oblong-elliptic, 2–10 cm long, up to 4 cm broad, acute at the apex; petioles 5–20 mm long. Flower-heads subglobose, magenta, yellowish or white, usually subtended by 2 or 3 small leaves; perianth woolly, shorter than the acute to acuminate bracts and bracteoles.

GRAND CAYMAN: *Correll & Correll 51019.*
– Although originally described from India, this species probably originated in tropical America. It is widely cultivated in warm countries, and often escapes along roadsides and in open waste land.]

Genus 5. **LITHOPHILA** Sw.

Perennial herbs with stout woody taproot; leaves of two types: tufted elongate ones forming a basal rosette, and much smaller opposite ones along the flowering scapes. Flowers bracteolate, perfect, in small heads or short, dense spikes; perianth of 5 segments. Stamens 2, the filaments united at base into a short tube; anthers 1-celled; staminodes 3. Ovary with short style and 2 slender stigmas; ovule 1. Utricle compressed; seed lenticular.

A small genus of about 6 species, occurring in the West Indies, Venezuela, and the Galapagos Islands.

1. Lithophila muscoides Sw., Nov. Gen. & Sp. Pl. 14. 1788.

A prostrate, matlike herb, the branched stems usually less than 15 cm long; leaves whitish-villous near the base but otherwise glabrous, those of the basal rosette linear-oblanceolate and up to 5 cm long, those of the stems oblong or oblanceolate and mostly less than 1 cm long. Flower-heads sessile, more or less globose to subcylindric, 3-13 mm long; bracts whitish-membranous, ovate; perianth segments whitish with a median black spot, 1.5-2.5 mm long. Seed shining brown, 0.5 mm in diameter.

GRAND CAYMAN: *Brunt 1797; Kings GC 258, GC 331; Proctor 27972.* – Widespread in the West Indies but apparently lacking from Jamaica, rather variable; several varieties have been described. The Cayman plants grow in black humus in hollows of limestone pavements, in sheltered situations toward the interior of the island.

Genus 6. **CARAXERON** Raf.

Creeping perennial herbs, branched and rather succulent, glabrous or pubescent. Leaves opposite, narrow, entire. Flowers perfect, with bracts and bracteoles, in small dense heads or spikes, these axillary and terminal. Perianth of 5 subequal segments, thickened at the base and short-stipitate. Stamens 5, fused at the base; anthers 1- or 2-celled; staminodes absent. Ovary with a short style and 2 stigmas. Utricle compressed, indehiscent; seed lenticular, with annular embryo.

A small genus of 3 or 4 species, chiefly on seashores of tropical America and west Africa.

1. Caraxeron vermicularis (L.) Raf., Fl. Tellur. 3:38. 1836; Mears in Taxon 29(1): 88-89. 1980. **FIG. 98.**

Philoxerus vermicularis (L.) Beauv., 1818; Adams, 1972.

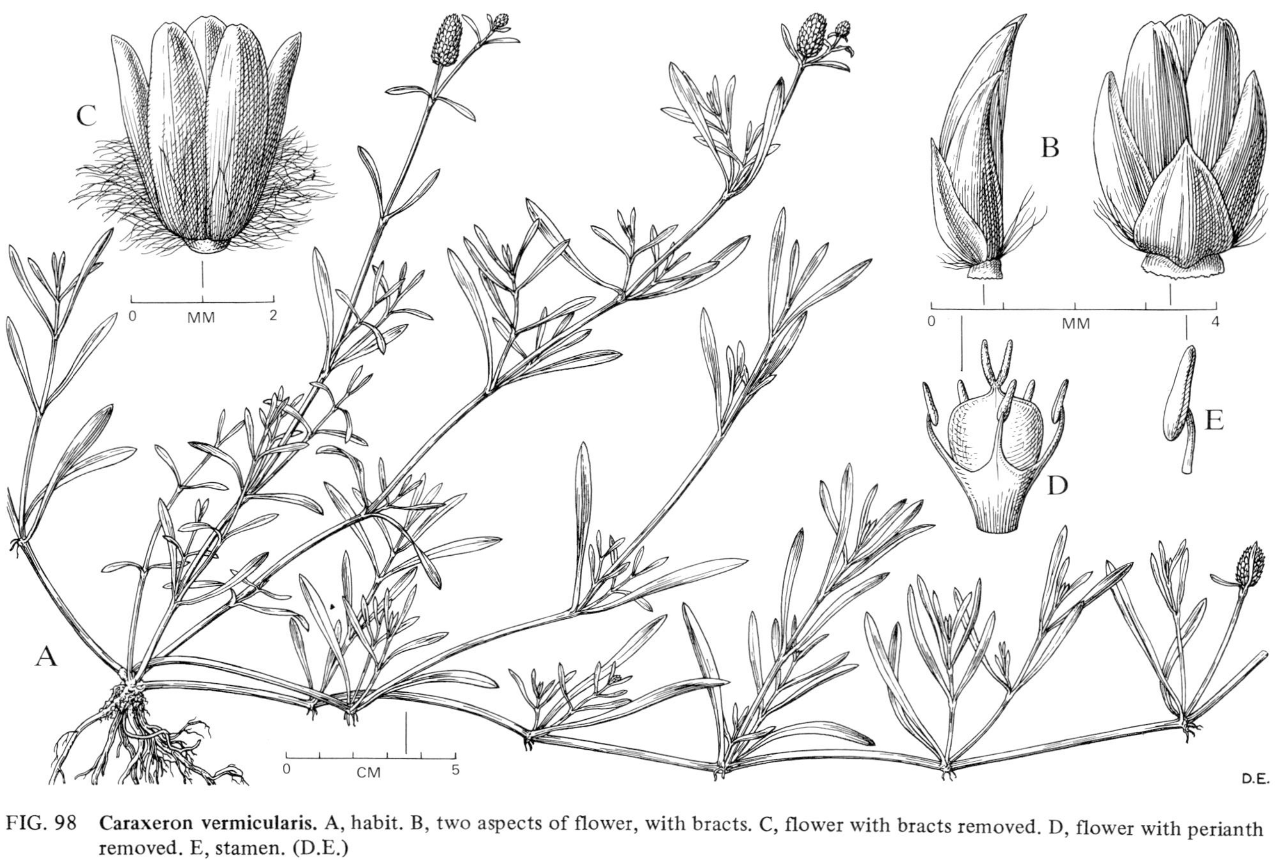

FIG. 98 **Caraxeron vermicularis.** A, habit. B, two aspects of flower, with bracts. C, flower with bracts removed. D, flower with perianth removed. E, stamen. (D.E.)

Fleshy creeping herb with main stems up to 60 cm long or more, rooting at the nodes, with short ascending flowering branches. Leaves sessile, linear to oblanceolate, mostly 1-3 cm long. Flower-heads sessile, globose to cylindric, 5-20 mm long; bracts and perianth whitish, the perianth segments 3-5 mm long, the inner ones woolly near the base. Seed orbicular, shining dark brown, 1 mm in diameter.

GRAND CAYMAN: *Brunt 1626, 1771, 1860; Hitchcock; Proctor 15298; Sachet 456.* LITTLE CAYMAN: *Kings LC 27; Proctor 28181.* CAYMAN BRAC: *Millspaugh 1220; Proctor 29362.*
– Florida, West Indies, continental tropical America, and the west coast of tropical Africa, frequent on seashores and in brackish situations.

Family 37

BATACEAE

Decumbent shrub with numerous opposite ascending branches; leaves opposite, sessile, succulent and semiterete, entire; stipules lacking. Flowers unisexual, dioecious, in fleshy axillary spikes. Staminate spikes sessile, with many persistent imbricated scales, each subtending a flower, the perianth cup-shaped and transversely 2-lobed above the middle. Stamens usually 4, inserted at base of the perianth, alternating with 4 staminodes; anthers 2-celled, opening inwardly. Pistillate spikes stalked, 4-12-flowered, with small roundish deciduous scales in alternating pairs; perianth lacking. Ovary sessile, 4-celled, those of a spike united to form eventually a fleshy compound fruit; ovule 1 in each cavity, erect; stigma sessile, 2-lobed. Seeds with no endosperm and large cotyledons.

A single genus with one species.

Genus 1. **BATIS** L.

The genus has the characters of the family.

1. Batis maritima L., Syst. Nat. ed. 10, 2:1289. 1759. **FIG. 99.**

A glabrous shrub with characteristic sweetish odour, less than 1 m high with stout spreading or prostrate main stems; branches angular. Leaves mostly 1-1.5 cm long, acutish or blunt. Spikes 5-14 mm long; stamens exserted. Fruit 1-2 cm long, yellowish-green.

GRAND CAYMAN: *Brunt 1634; Kings GC 280, GC 281A; Proctor 15136.* LITTLE CAYMAN: *Kings LC 118, LC 119.*
– Southeastern United States, West Indies, continental tropical America, California, and the Galapagos and Hawaiian Islands, frequent or common on salinas and coastal marshes.

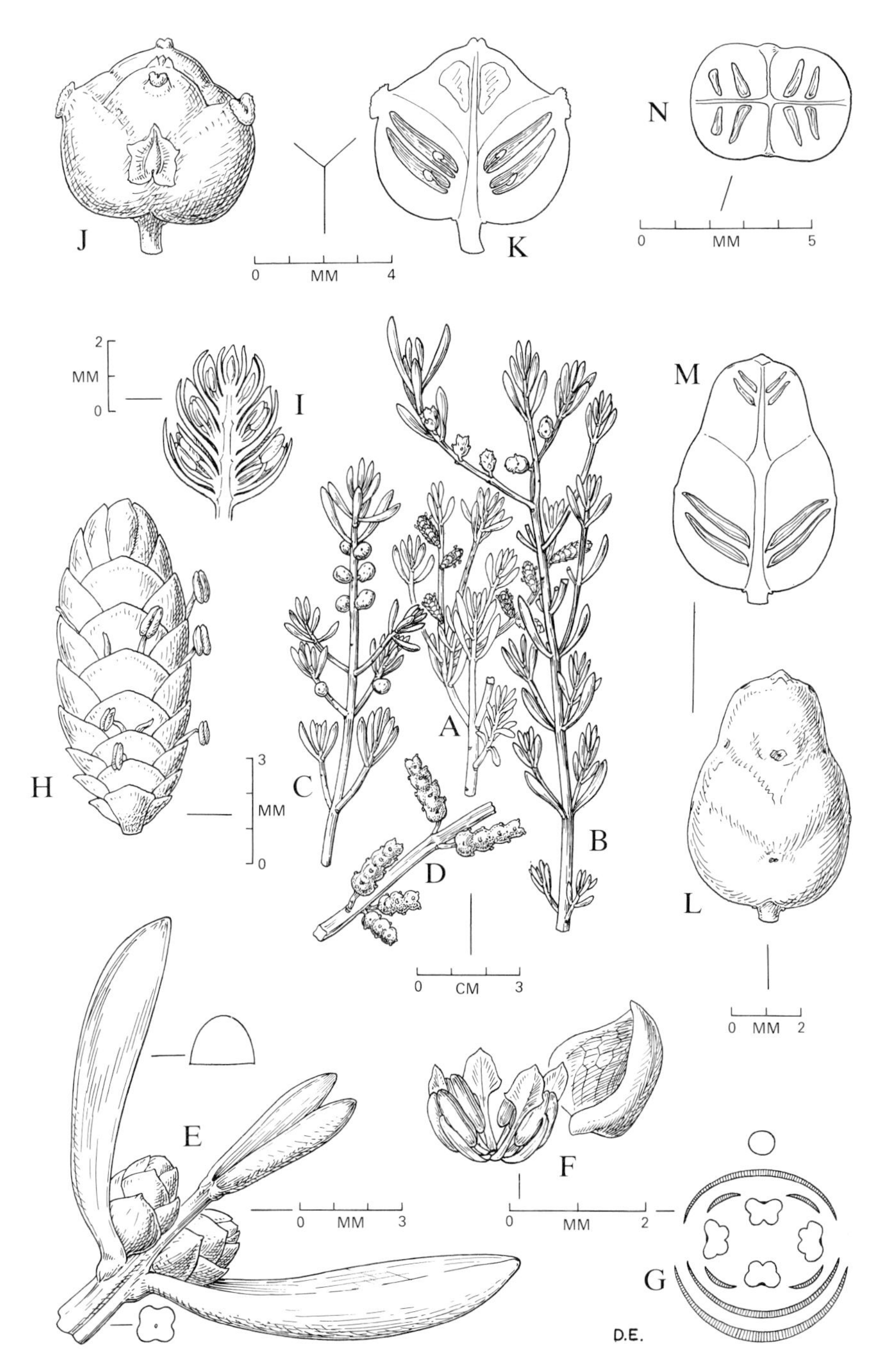

FIG. 99 **Batis maritima.** A, portion of staminate plant. B, C, portions of pistillate plant. D, fruiting branch. E, node with two staminate spikes (immature). F, staminate flower. G, floral diagram, staminate flower. H, staminate spike anthesis. I, long. section through portion of staminate spike. J, pistillate spike. K, long. section of same. L, fruiting spike. M, N, sections through same. (D.E.)

Family 38

POLYGONACEAE

Herbs, shrubs or trees of various habit; leaves alternate, simple, mostly entire, the petiole often dilated and clasping; stipules of characteristic tubular form (ocreae) ensheathing the stem above each leaf-base (except in *Antigonon*). Flowers small, regular, perfect or unisexual, in terminal and axillary racemose inflorescences; perianth of 3-6 segments, persistent after flowering, sometimes becoming fleshy. Stamens 3-9 from a central disk, the filaments free or fused at the base; anthers 2-celled, dehiscent by longitudinal slits. Ovary superior, 1-celled, with usually 3 styles; ovule 1, basal, sessile or erect at the apex of an elongate funicle. Fruit an achene, trigonous or compressed, usually surrounded by the persistent perianth; seed often grooved or lobate, with abundant endosperm, the embryo usually lateral and either curved or straight.

A family of about 30 genera and 800 species, of wide geographic distribution but the majority in temperate regions. Few of them are economically important, but "Buckwheat" (*Fagopyrum sagittatum* Gilib.) is a widely-grown Asiatic species whose seeds produce a kind of flour.

KEY TO GENERA

1. Plants woody, often tree-like: **1. Coccoloba**

1. Plants herbaceous:

 2. Plants vine-like, climbing by means of tendrils terminating the inflorescences; ocreae absent: **[2. Antigonon]**

 2. Plants not vine-like and without tendrils; ocreae present: **3. Polygonum**

Genus 1. **COCCOLOBA** L.

Trees or shrubs, glabrous or sometimes pubescent, with alternate, simple, entire leaves; leaves deciduous or persistent, the petioles not sheathing; ocreae deciduous or persistent, cylindric, truncate and not ciliate. Flowers small, unisexual or functionally so, dioecious, in spike-like, axillary or subterminal, simple or rarely branched racemes, and subtended by minute bracts (ocreolae); perianth of 5 subequal segments united at the base. Staminate flowers in small clusters along the raceme; stamens 8, the filaments fused at the base; ovary usually abortive. Pistillate flowers solitary (not in clusters) along the raceme, with abortive or rarely functional stamens; ovary 3-angled, with 3 styles. Fruit more or less 3-angled, enclosed by the thickened and succulent perianth, thus appearing drupe-like.

A chiefly tropical American genus of more than 150 species.

1. **Coccoloba uvifera** L., Syst. Nat. ed. 10, 2:1007. 1759. **FIG. 100.**

"Sea-grape".

Diffuse shrub or tree to 15 m tall, the branchlets finely pubescent when young, soon becoming glabrous. Leaves orbicular to reniform, mostly 8–15 cm long, sometimes wider than long, glabrous and minutely punctate on both sides, commonly bearded in the axils of the basal veins beneath; ocreae rigid, deciduous, 3–8 mm long, puberulous to pilose. Flowers creamy-white, the staminate in clusters of up to 7, with pedicels 1–2 mm long, the pistillate on pedicels 3–4 mm long. Fruit obpyriform, 1.2–2 cm long, in drooping clusters resembling bunches of grapes, edible; mature fruiting perianth rose-purple, the concealed achene black.

GRAND CAYMAN: *Brunt 1760; Hitchcock; Kings GC 298; Proctor 15119.* LITTLE CAYMAN: *Proctor 28051, 28099.* CAYMAN BRAC: *Kings CB 99; Millspaugh 1225.*
– From Florida throughout the Caribbean area to northern South America, a common plant of sandy coastal thickets, sometimes occurring inland. The fruits have an acidulous flavour, but are sometimes used to make jellies and preserves; they are also occasionally fermented with sugar to make an alcoholic beverage. The wood is hard, heavy, compact, and of fine texture; it takes a high polish and is sometimes used in cabinet-work. When cut, the bark yields an astringent red gum known as West Indian Kino; this was formerly used in medicine. The trees are quite ornamental in appearance and have considerable horticultural value in plantings near the sea, as they are quite tolerant of salt spray.

It seems quite remarkable that no other species of Coccoloba is known to occur in the Cayman Islands, as members of this genus are otherwise common throughout the West Indies, and several species are found very widely in habitats like the Cayman woodlands.

[Genus 2. **ANTIGONON** Endl.

Herbaceous vines, sometimes suffrutescent below; leaves petiolate, the blades cordate or deltate, entire; ocreae lacking or represented by a transverse line. Flowers usually pink, in racemes opposite the leaves or terminal, each raceme usually ending in a branched tendril. Perianth of 5 segments, the 3 outer ones larger and broadly cordate, the 2 inner ones narrower and oblong. Stamens 7–8, the filaments fused at the base. Ovary 3-angled, with 3 short styles, the stigmas capitate or peltate; ovule attached to a long funicle and at first pendulous. Achene 3-angled, hidden by the enlarged perianth; seed subglobose, 3–6-lobed, with ruminate endosperm.

A Mexican and Central American genus of 5 species, often cultivated for their showy pink flowers.

1. **Antigonon leptopus** Hook. & Arn., Bot. Capt. Beechey Voy. 308, t. 69. 1839.

"Coralilla".

Stems often 4 m long or more, climbing by means of axillary tendrils, finely pubescent. Leaves with deltate-ovate blades up to 9 cm long, the apex acute to acuminate, the base cordate and non-decurrent, densely puberulous on both sides in Cayman specimens; petioles 1–4 cm long, somewhat clasping at the base. Racemes 3–8 cm long; pedicels slender, up to 8 mm long; larger perianth segments c. 1 cm long at anthesis, bright pink, rarely white; filaments with glandular hairs.

GRAND CAYMAN: *Kings GC 125; Proctor 15150.* CAYMAN BRAC: *Proctor (sight record).*
– A native of Mexico, widely planted elsewhere as an ornamental, and frequently becoming naturalized in roadside thickets and waste places. The roots bear tubers said to be edible.]

Genus 3. **POLYGONUM** L.

Annual or perennial herbs, glabrous or pubescent, often glandular, the stems usually enlarged at the nodes. Leaves entire and rather thin, often glandular-punctate; petioles enlarged and sheathing at base; ocreae cylindric or funnel-shaped, usually membranous or hyaline, often ciliate or fringed with bristles. Flowers perfect, in spikes, racemes or narrow panicles, rarely capitate in corymbs; perianth of 4–6 lobes, these subequal or the outer ones larger, in fruit closely adhering to the achene. Stamens 3–9, inserted on the perianth. Ovary with 2 or 3 styles united below, with capitate stigmas; ovule usually stalked. Achene flattened or 3-angled; seed with embryo excentric.

A cosmopolitan, chiefly temperate genus of about 150 species, commonly occurring in marshy or wet places.

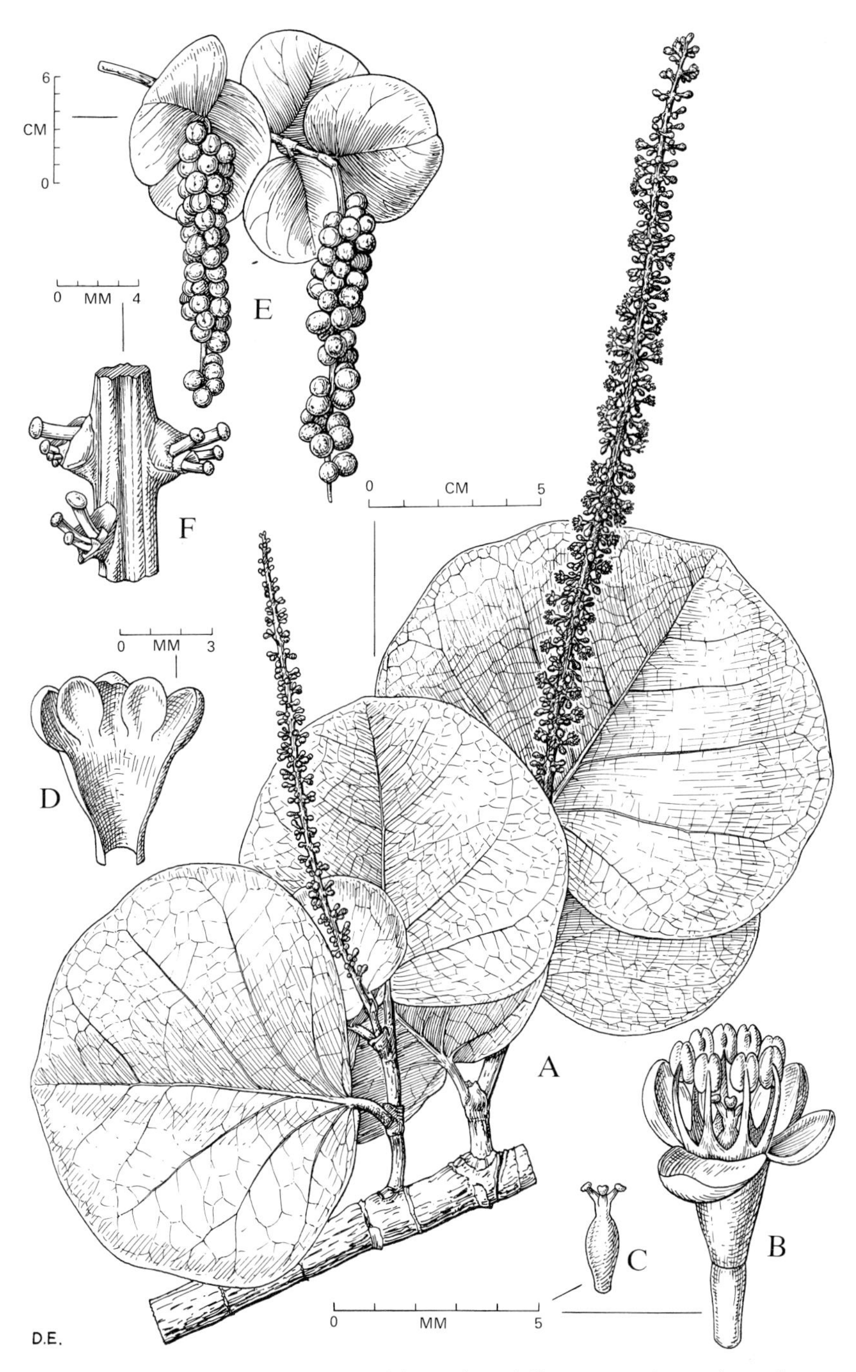

FIG. 100 **Coccoloba uvifera.** A, habit, with staminate inflorescences. B, staminate flower. C, sterile ovary. D, perianth. E, portion of staminate inflorescence after flower have fallen. F, portion of fruiting branch. (D.E.)

KEY TO SPECIES

1. Ocreae bristly-hairy: **1. P. punctatum**

1. Ocreae glabrous: **2. P. glabrum**

1. Polygonum punctatum Ell., Bot. S.C. & Ga. 1:455. 1817. **FIG. 101.**

"Smartweed".

A slender annual or perennial, often forming large colonies, the stems erect or the lower part creeping and rooting from nodes, simple or branched. Leaves lanceolate, mostly 3-8 cm long, shortly acuminate at each end, densely punctate; ocreae cylindric, 1-1.5 cm long, bristly-hairy and ciliate, persistent. Racemes linear, slender, often interrupted, 1-6 cm long; ocreolae ciliate, 2.5-3 mm long. Perianth white or greenish-white, 2 mm long, glandular-punctate. Achene usually 3-angled, 2 mm long, brown or black.

FIG. 101 **Polygonum punctatum.** A, leaf and flower-spikes, x$^{2}/_{3}$. B, portion of flower-spike, x5. C, perianth cut open, x5. D, stamen, x13. E, achene, x6. F, same in transverse section, x6; e, endosperm; c, embryo. (F. & R.)

GRAND CAYMAN: *Kings GC 178; Proctor 15291.*
– Common throughout temperate and tropical America, occurring in wet thickets, ditches and swales, along the border of streams and lakes, and in moist waste ground. In some countries (e.g., Guatemala), poultices of the leaves are applied to dogs suffering from mange.

2. Polygonum glabrum Willd. in L., Sp. Pl. 2:447. 1799.

A stout herb erect from a decumbent and rooting base, glabrous throughout. Leaves lanceolate, mostly 5-11 cm long, acuminate at both ends, densely glandular-punctate; ocreae loose, truncate, 1.3-2 cm long, the upper part deciduous from a cup-shaped base. Racemes continuous, 3-7 cm long, loosely but densely flowered, arranged in terminal panicles; perianth pink in bud, opening white, c. 3 mm long, glandular-punctate. Achene orbicular-biconvex, apiculate, c. 2 mm long, lustrous black.

GRAND CAYMAN: *Brunt 1798, 1823, 2100; Hitchcock; Kings GC 178, GC 179; Proctor 27934.*
– Pantropical, in moist situations, the Cayman plants often in *Typha* swamps and in seasonally flooded pastures.

Family 39

CLUSIACEAE

Trees, shrubs, or sometimes herbs, terrestrial or epiphytic, often with white or yellowish latex. Leaves mostly opposite and decussate, simple, entire, sometimes with black or transparent dots or lines; stipules lacking. Flowers regular, perfect, or unisexual and dioecious or monoecious, in terminal or axillary inflorescences or sometimes solitary, generally white or yellow, sometimes pink; sepals 2-6, persistent; petals 2-6. Staminate flowers with numerous stamens of indefinite number, these free or united at the base, an ovary lacking or rudimentary. Pistillate or perfect flowers with fewer stamens or staminodes, these often definite in number; ovary superior, with 1 to many cells; stigmas sessile or terminating separate styles; ovules 1 to numerous in each cell. Fruit a capsule, berry or drupe; seeds often enveloped in an aril, without endosperm.

A chiefly tropical family of about 45 genera and 1,000 species, related to the Theaceae but differing especially in the presence of oil glands or tubes in the tissues. Many species yield useful timber, and one of the most delicious tropical fruits, the Mangosteen (*Garcinia mangostana* L.), belongs to this family.

KEY TO GENERA

1. Flowers unisexual; fruit a resinous dehiscent capsule: **1. Clusia**

1. Flowers bisexual; fruit a large fleshy drupe: [**2. Mammea**]

Genus 1. **CLUSIA** L.

Glabrous trees or shrubs, rarely woody vines, with gummy latex, often epiphytic or growing on rocks or cliffs. Leaves usually thick and more or less leathery. Flowers unisexual or rarely perfect, mostly dioecious; inflorescences terminal, usually several-flowered and alternately branched, rarely 1-flowered. Bracteoles subtending the flowers in 1 to many decussate pairs, often resembling sepals. Sepals 4–6, roundish. Petals 4–10, free or somewhat connate at the base. Pistillate flowers with 5 to numerous staminodes, free or united, with or without anthers. Ovary 4–15-celled, with an equal number of sessile, radiating stigmas; ovules numerous in each cell, arillate. Fruit a leathery or fleshy resinous capsule, septicidally dehiscent.

A mostly tropical American genus of about 145 species, with 2 isolated species in New Caledonia and 1 in Madagascar. The aboriginal inhabitants of the West Indies made use of the waterproof latex of *Clusia* species to seal cracks in their water vessels and canoes, and even today similar use is occasionally made of this substance, by people who are far from being aboriginal.

KEY TO SPECIES

1. Leaves thick and rigid; bracteoles 2; petals 6, rosy-white; capsule 5–8 cm in diameter: **1. C. rosea**

1. Leaves thinner and more flexible; bracteoles 6–14; petals 4, thick and pale yellow; capsule 2–3 cm in diameter: **2. C. flava**

1. Clusia rosea Jacq., Enum. Pl. Carib. 34. 1760. **FIG. 102.**

"Balsam".

A shrub or more commonly a tree up to 10 m tall or more; leaves extremely thick and leathery, on very short, broad, winged petioles, rounded-obovate, mostly 12–16 cm long or sometimes more, with numerous lateral prominulous veins ascending at an angle of 45°. Inflorescences few-flowered; staminate flowers with stamens in several series, united at base to form a cup or ring. Pistillate flowers with staminodes wholly fused to form a cup surrounding the ovary; sepals 4, the inner ones up to 2 cm long; petals 3–4 cm long, of waxy texture. Capsule 6–8-celled, subglobose.

GRAND CAYMAN: *Proctor 32486.* CAYMAN BRAC: *Proctor 15311, 29050.* – West Indies and southern Mexico south to northern South America. The Cayman Brac trees grow in rocky woodlands on the central plateau.

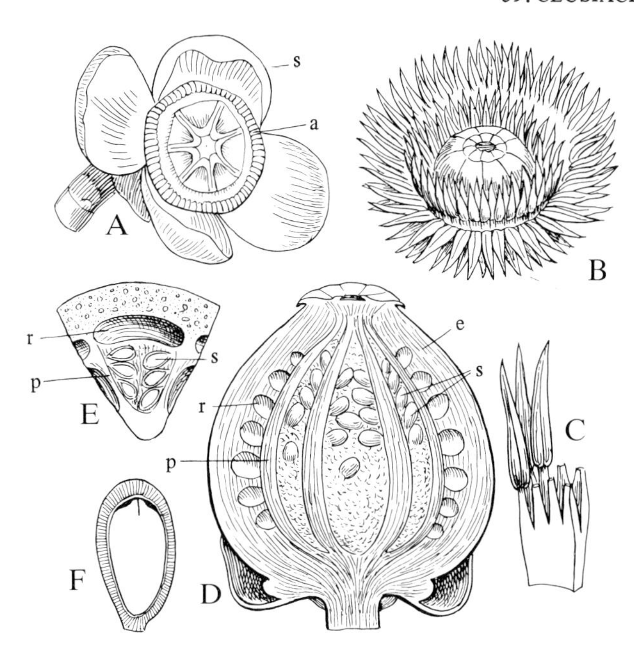

FIG. 102 **Clusia rosea.** A, pistillate flower with petals removed, x1; s, sepal; a, staminodes. B, double ring of stamens surrounding pistil of perfect flower, x1. C, portion of staminal ring, enlarged. D, fruit cut lengthwise, x$^{2}/_{3}$; e, exocarp; r, resin duct; p, placenta; s, seeds. E, portion of same cut across. F, seed cut lengthwise, x3. (F. & R.)

2. Clusia flava Jacq., Enum. Pl. Carib. 34. 1760.

"Balsam".

A diffuse shrub or small tree up to 9 m tall, terrestrial or epiphytic. Leaves leathery, on short petioles, cuneate-obovate, 6–16 cm long, with numerous lateral veins ascending at an angle of 45° or narrower. Staminate inflorescence recurved, with usually 2–7 flowers; stamens inserted on a receptacle, crowded, free, the sterile ovary with 4 3-rayed stigmas. Pistillate flowers usually solitary on a recurved pedicel; staminodes 8–12 in 4 bundles, free, bearing anthers; stigmas 12. Bracteoles sepal-like, decreasing in size gradually downward. Sepals 4, 9–11 mm long; petals 4, opposite the sepals, pale yellow and very thick, 2–2.5 cm long, one pair larger than the other. Capsule about 12-celled, subglobose.

GRAND CAYMAN: *Brunt 1644; Hitchcock; Kings GC 229; Millspaugh 1379; Proctor 15141, 31047; Sachet 431.* CAYMAN BRAC: *Proctor 29009.*
– Jamaica and Central America, in rocky woodlands. The Cayman Brac trees have leaves noticeably smaller than those of Grand Cayman.

[Genus 2. **MAMMEA** L.

Trees with resinous sap; leaves hard and leathery, dark green and with numerous pellucid dots, the lateral veins prominulous and reticulate with cross-veins. Inflorescences axillary, sessile, 1-3-flowered, or sometimes clustered along the branches below the leaves. Flowers perfect; calyx closed in bud, splitting into 2 segments when the flower opens; petals 4-6. Stamens numerous, free or united at base, the anthers linear and longitudinally dehiscent; ovary 2-celled, with 2 ovules in each cell, or 4-celled, the ovules solitary; ovules basal, erect. Style thick, capped by a large, shield-like, 2-lobed stigma. Fruit a large, fleshy drupe with 1-4 seeds; seeds large, rough-surfaced, the embryo with thick, fleshy cotyledons.

A genus of 4 species, one tropical American, the other 3 in tropical Africa.

1. **Mammea americana** L., Sp. Pl. 1:512. 1753.

"Mammee".

A glabrous tree up to 15 m tall or more; leaves oblong-elliptic or narrowly obovate, 10-25 cm long, short-petiolate, densely pellucid-dotted. Flowers white, fragrant; sepals 1-1.7 cm long; petals 1.5-2 cm long or more. Fruit subglobose, apiculate, 10-15 cm long, with rough, russet-coloured skin and yellowish edible flesh, rather firm and juicy but sometimes bitter; seeds about two-thirds as long as the fruit, reddish and rough.

GRAND CAYMAN: *Brunt 1992, 2112; Kings GC 198.*
– West Indies, apparently native on some islands, but introduced in Grand Cayman. The tree is widely planted in tropical countries for its handsome evergreen foliage, fragrant flowers, and edible fruits.]

Family 40.

TILIACEAE

Trees, shrubs, or sometimes herbs, the pubescence often of branched hairs; leaves alternate, rarely opposite, simple; stipules usually present. Flowers perfect, solitary or in few-flowered cymose inflorescences, axillary or terminal; sepals usually 5, free or united; petals usually 5 or else none. Stamens rather numerous, free or united at the base in fascicles; anthers 2-celled, opening by longitudinal slits or apical pores. Ovary superior, sessile, 2-10-celled; style usually simple, with as many stigmas as ovary-cells; ovules 1 to many in each cell, on axial placentae. Fruit a capsule, drupe or berry; seeds 1 to many, with endosperm and straight embryo.

A widely distributed family of about 50 genera and 400 species.

KEY TO GENERA

1. Fruit a more or less elongate capsule, lacking spines or bristles: **1. Corchorus**

1. Fruit globose, indehiscent, covered all over with stiff, often hooked, spines: **2. Triumfetta**

Genus 1. **CORCHORUS** L.

Herbs or low shrubs with pubescence of simple or stellate hairs; leaves mostly thin, with serrate or crenate margins; stipules present. Flowers yellow, small, solitary or in few-flowered clusters, axillary or opposite the leaves; sepals and petals 5, rarely 4. Stamens numerous or else twice as many as the petals, free, inserted on a flat torus. Ovary 2-5-celled; ovules numerous in each cell. Fruit a silique-like capsule, splitting lengthwise into 2-5 valves, many-seeded, sometimes with transverse partitions between the seeds.

A pantropical genus of about 30 species. Two Asiatic species (*C. olitorius* L. and *C. capsularis* L.) are the source of jute, a fibre of great economic importance. In Jamaica and the Cayman Islands, bags made of jute are called "crocus" bags, an obvious corruption of "Corchorus".

KEY TO SPECIES

1. Leaves and capsules nearly or quite glabrous:

   2. Calyx 6-7 mm long; capsule 2-celled, not winged, 3-7 cm long, with 4 minute teeth at the apex: **1. C. siliquosus**

   2. Calyx 3-4 mm long; capsule 3-celled, wing-angled, 1.5-2.5 cm long, with 3 horizontal beaks at the apex: **2. C. aestuans**

1. Leaves densely covered with stellate pubescence; capsules woolly, 4-celled: **3. C. hirsutus**

1. Corchorus siliquosus L., Sp. Pl. 1:529. 1753. **FIG. 103.**

A small shrub or shrubby herb up to 1 m tall or more, the tough stems usually bearing a line of short simple hairs. Leaves small, glabrous except for simple hairs on the petiole, lance-oblong to ovate, 0.4-2 cm long, the apex acute, the margins finely serrate; stipules hair-like. Flowers solitary or in pairs on short axillary peduncles; sepals linear, 6–7 mm long; petals obovate, shorter than the sepals. Capsule 3–7 cm long, linear, compressed, 2–3 mm broad, minutely puberulous along the join between the valves; seeds 3-angled, c. 1 mm long.

GRAND CAYMAN: *Brunt 1648; Hitchcock; Millspaugh 1344; Proctor 15095.* LITTLE CAYMAN: *Proctor 28160.* CAYMAN BRAC: *Proctor 28973.*
– West Indies and on the continent from Florida and Texas south to northern South America, a weedy persistent plant of open waste places, sandy roadsides, and pastures.

2. **Corchorus aestuans** L., Syst. Nat. ed. 10, 2:1079. 1759. **FIG. 103.**

An annual subwoody herb, the stems low-spreading or ascending, puberulous or short-pilose with simple hairs. Leaves glabrous or with scattered hairs, slender-petiolate, 1–6 cm long, the blades ovate to orbicular with obtuse or subacute apex, the margins finely serrate with the lowermost serration on one or both sides often elongated into a hair-like bristle; stipules subulate, 5–7 mm long. Flowers solitary or paired in the leaf-axils, almost sessile; sepals hooded, c. 4 mm long; petals obovate, 3–4 mm long. Capsule 1.5–2.5 cm long, narrowly oblong, 4–5 mm thick, triangular in cross-section, narrowly winged on the angles, glabrous; seeds discoid, dark brown, less than 1 mm broad.

CAYMAN BRAC: *Proctor 28938.*
– Pantropical but not usually very common. The Cayman Brac plants grow in moist hollows, thickets, and along the borders of pastures.

3. **Corchorus hirsutus** L., Sp. Pl. 1:530. 1753.

Shrub up to 2 m tall, clothed throughout with dense, soft, stellate pubescence. Leaves petiolate, the blades lance-oblong to ovate or elliptic, 1.5–6 cm long, the apex blunt to acute, the margins irregularly serrate or crenulate-serrate; stipules subulate, 4–5 mm long, soon deciduous. Flowers pedicellate, in axillary umbels of 2 to 8; sepals 5–6 mm long; petals obovate, about equalling the sepals. Capsule oblong or nearly globose, 7–11 mm long, densely woolly, with a short erect terminal beak; seeds irregularly ellipsoidal, black, 1.5–2 mm long.

GRAND CAYMAN: *Correll & Correll 51024.*
– West Indies and tropical Africa, mostly in rather dry thickets.

Genus 2. **TRIUMFETTA** L.

Herbs or shrubs, usually bearing stellate hairs. Leaves thin, variable, often 3–5-angled or -lobed, irregularly serrate. Flowers yellow or red, few or densely clustered

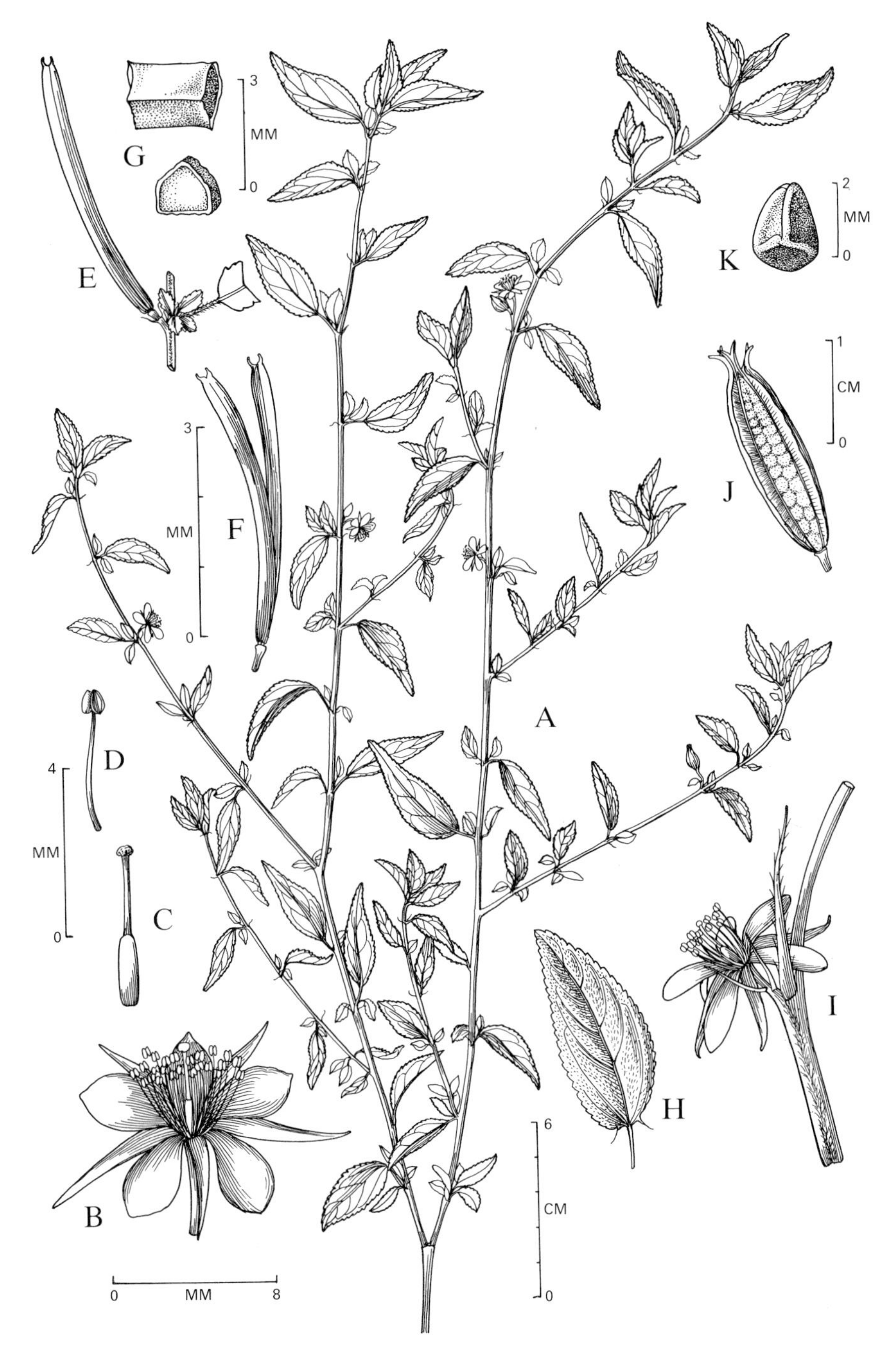

FIG. 103. **Corchorus siliquosus.** A, habit. B, flower. C, pistil. D, stamen. E, fruit. F, fruit after dehiscence. G, seed, 2 views. **Corchorus aestuans.** H, leaf. I, flower habit. J, fruit. K, seed. (G.)

in cymes or panicles, axillary or opposite the leaves, or sometimes in terminal racemes; sepals 5, somewhat hooded at the apex; petals 5 or sometimes lacking. Stamens numerous or sometimes 10, free, borne on the elevated 5-glandular receptacle. Ovary 2–5-celled with 2 ovules in each cell; stigma 2–5-toothed. Fruit a subglobose prickly capsule, indehiscent or separating into cocci; prickles hooked at the apex, by this means often clinging to animals or to clothing; seeds 1 or 2 in each cell, with endosperm.

A pantropical weedy genus of about 50 species.

KEY TO SPECIES

1. Flowers with petals; body of the fruit glabrous: **1. T. semitriloba**

1. Flowers lacking petals; body of the fruit stellate-pubescent: **2. T. lappula**

1. Triumfetta semitriloba Jacq., Enum. Pl. Carib. 22. 1760.

"Bur-weed".

A short-lived shrub or woody herb up to 1.5 m tall or more; leaves long-petiolate, the petioles 1–9 cm long; blades ovate to broadly ovate below, 2–10 cm long, or the upper lance-oblong and much smaller, acute at apex and rounded or truncate to cordate at base, the margins unequally dentate and often angled or shallowly 3-lobed, stellate-pubescent on both sides; stipules linear, 5–6 mm long. Inflorescences rather few-flowered; sepals 5–8 mm long, green; petals slightly shorter than the sepals. Ovary 3-celled. Fruit 3–5 mm in diameter (excluding prickles), the body glabrous or nearly so, the prickles with scattered retrorse hairs.

GRAND CAYMAN: *Hitchcock; Millspaugh 1297; Proctor 15019.*
– West Indies and continental tropical America, a common weed of waste ground, roadside thickets, and old fields.

2. Triumfetta lappula L., Sp. Pl. 1:444. 1753.

"Bur-weed".

A shrub or woody herb 1–2 m tall; leaves long-petiolate, the petioles 1–8 cm long; blades roundish-angulate or very broadly ovate below, up to 13 cm long, the upper ones ovate or elliptic and much smaller, acuminate at apex and usually truncate at base, the margins very unequally dentate and often angulate or 3-5-lobed, densely stellate-pubescent on both sides. Inflorescences many-flowered, dense; sepals 3–4 mm long, densely pubescent; petals absent. Ovary 2-celled (but in fruit often apparently 3- or 4-celled because of the development of both ovules in one or both cells). Fruit 2–3.5 mm in diameter (excluding the prickles), the body densely stellate-pubescent, the prickles with scattered or numerous retrorse hairs.

GRAND CAYMAN: *Hitchcock.*
– West Indies and continental tropical America; adventive at scattered localities in the Old World tropics, a weed of secondary thickets, waste ground and roadsides. The bark is said to contain a tough fibre suitable for making rope.

Family 41

STERCULIACEAE

Herbs, shrubs or trees, the pubescence often at least partly of stellate hairs. Leaves alternate, simple, and entire, dentate or lobate, rarely digitately compound; petioles often pulvinate at the apex; stipules usually present, but soon deciduous. Flowers perfect or rarely unisexual, in axillary or terminal racemes or cymose panicles, rarely solitary. Calyx 5-lobed, persistent. Petals 5, often persistent after withering, or sometimes lacking. Stamens 5 or more, united at the base or beyond the middle, then forming a tubular column on which the anthers often alternate with staminodes. Ovary superior, sometimes stalked, 2–5-celled (reduced to 1 carpel in *Waltheria*); styles as many as the carpels, free or more or less united; ovules 2 to many in each cell. Fruit with the carpels partly free, or else united to form a capsule, dehiscent or indehiscent; seeds with or without endosperm, the pericarp never woolly.

A pantropical genus of about 50 genera and 750 species. Chocolate (from *Theobroma cacao* L.) and cola (from *Cola* spp.) are economically important products of this family, widely used in the foods and beverage industries. They both contain a stimulating drug known as caffeine.

KEY TO GENERA

1. Ovary long-stalked, in age longer than the fruit; carpels spirally twisted in fruit: **1. Helicteres**

1. Ovary nearly or quite sessile; carpels not twisted in fruit:

 2. Herbs or low shrubs; petals flat, withering and persisting in fruit; capsules thin, splitting lengthwise:

  3. Ovary 1-celled; petals yellow: **2. Waltheria**

  3. Ovary 5-celled; petals pink or purple: **3. Melochia**

2. Tree; petals concave or hooded, deciduous; capsules hard and virtually indehiscent, covered with woody blackish tubercles: **4. Guazuma**

Genus 1. **HELICTERES** L.

Shrubs or small trees with stellate pubescence; leaves serrate or entire, petiolate. Flowers axillary, solitary or in clusters; calyx 2-lobed or else tubular with a 5-lobed apex; petals 5, flat, equal or unequal, clawed, with auriculate appendages on the claws. Stamens 6, 8, 10, or indefinite, the filaments slightly united in pairs at the base; anthers 2-celled, the cells sometimes confluent; staminodes present. Ovary 5-lobate and 5-celled, borne on a very long, pedicel-like gynophore, this curved and noose-like at first, finally straightening and long-exserted beyond the calyx and petals; ovules many. Fruit composed of 5 hard, cohering, tube-like carpels, these spirally twisted or sometimes straight, splitting open along the inner seams; seeds small, flattened-ovoid, with scanty endosperm and straight embryo.

A tropical American genus of about 30 species; the hard, screw-like fruits are unique.

1. **Helicteres jamaicensis** Jacq., Enum. Pl. Carib. 30. 1760. **FIG. 104.**

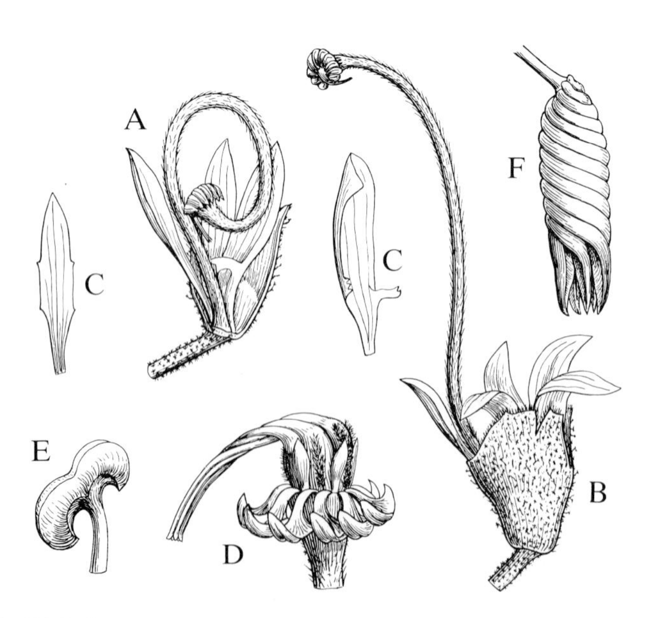

FIG. 104 **Helicteres jamaicensis.** A, bud of flower just opening cut lengthwise, x1. B, flower, x1. C, petals, x1. D, apex of gonophore with stamens and pistil, x4. E, stamen, x8. F, fruit, x$^{2}/_{3}$. (F. & R.)

"Wild Cow Itch", "Screw-bush".

Shrub, sometimes arborescent and up to 5 m tall, the twigs densely stellate-pubescent. Leaves long-petiolate, mostly 4–15 cm long, the blade ovate, acuminate at apex and cordate at base, the margins irregularly crenate-toothed, and both sides densely stellate-pubescent. Peduncles opposite the leaves, 1.5–3 cm long, 1-3-flowered. Calyx bell-shaped, 1.5–2 cm long, stellate-pubescent, 2-cleft and unequally 5-toothed. Petals oblong, white or creamy, longer than the calyx. Stamens 10, with short filaments; staminodes 5. Gonophore 5–8 cm long. Fruit more or less cylindric or ellipsoid, 2–3.5 cm long, woolly-pubescent with stellate hairs, the carpels twisted around about twice, free at the apex.

GRAND CAYMAN: *Brunt 1993; Correll & Correll 51013; Fawcett; Hitchcock; Kings GC 371; Millspaugh 1370; Proctor 15015.* LITTLE CAYMAN: *Kings LC 32; Proctor 28034.* CAYMAN BRAC: *Millspaugh 1183; Proctor 29026.*
– West Indies and Central America, usually in rocky thickets and woodlands.

Genus 2. **WALTHERIA** L.

Herbs or shrubs with stellate hairs; leaves petiolate, irregularly serrate; stipules small and narrow. Flowers yellow, small, in dense axillary or terminal cymose heads, each flower subtended by 3 linear bracteoles. Calyx 5-lobed. Petals 5, spatulate, withering-persistent. Stamens 5, united at base; staminodia absent. Ovary sessile, 1-celled; ovules 2; style slightly lateral and the stigma club-shaped or fringed. Capsule 2-valved, 1-seeded; seed with endosperm and straight embryo.

A chiefly tropical American genus of more than 30 species.

1. **Waltheria indica** L., Sp. Pl. 2:673. 1753. **FIG. 105.**

Waltheria americana L., 1753.

Erect or decumbent shrub or shrubby herb up to 1 m tall or a little more, densely stellate-pubescent throughout, or rarely glabrate. Leaves oblong to ovate or oblong-lanceolate, mostly 2–6 cm long, with apex obtuse, the base obtuse to subcordate, the petioles up to 2 cm long. Inflorescence with bracteoles 3–4 mm long; calyx 3.5–5 mm long, with linear-lanceolate lobes; petals 5–6 mm long. Staminal tube 2 mm long. Capsule 2–3 mm long.

GRAND CAYMAN: *Brunt 1637, 1898a; Hitchcock; Millspaugh 1325, 1327, 1336, 1382; Proctor 15084; Sachet 415.* LITTLE CAYMAN: *Proctor 28084.*
– West Indies and continental tropical America; naturalized in the Old World tropics. In the Cayman Islands, this species grows in dry sandy thickets, rough pastures, and in rocky scrublands.

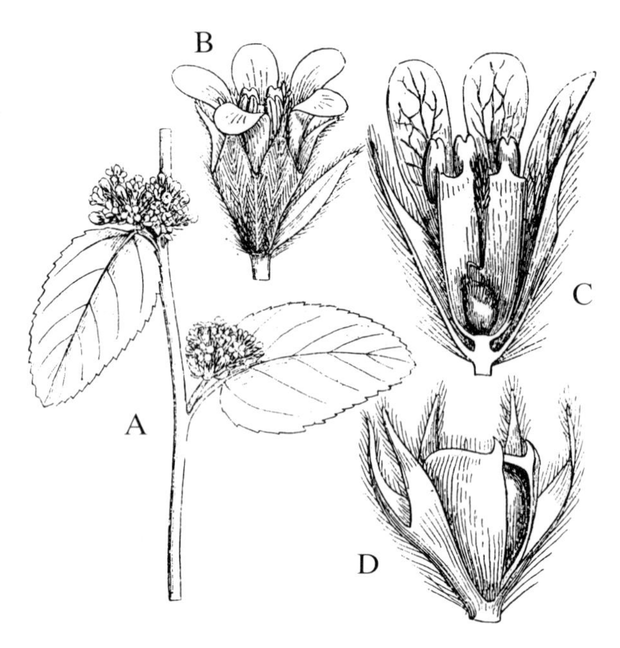

FIG. 105 **Waltheria indica.** A, portion of branch with leaves and flowers, x$^2/_3$. B, flower with bracteole, x4. C, flower cut lengthwise, x7. D, fruit with persistent calyx, showing seed, x7. (F. & R.)

Genus 3. **MELOCHIA** L.

Herbs or shrubs, rarely trees, with pubescence chiefly of stellate hairs; leaves petiolate, narrow to broad, with serrate margins. Flowers mostly small, perfect but heterostylous (i.e., with either long or short styles), more or less densely clustered in the leaf-axils or opposite the leaves, or else in terminal spikes, cymes, or panicles. Calyx 5-lobed or 5-toothed, bell-shaped; petals 5, usually pink, purple or violet, sometimes white, spatulate to oblong, withering-persistent. Stamens 5, opposite the petals, more or less united into a tube at the base, rarely alternating with 5 tooth-like staminodes. Ovary 5-celled, sessile or short-stalked; ovules 2 in each cell; styles 5, free or more or less united. Fruit a 5-valved capsule, longitudinally dehiscent, 5–10-seeded; seeds with endosperm and straight embryo.

A pantropical genus of about 60 species, the majority in tropical America.

KEY TO SPECIES

1. Leaves and stems densely whitish-pubescent with stellate hairs: **1. M. tomentosa**

1. Leaves and stems glabrous or minutely puberulous with mostly simple hairs:

2. Inflorescence stalked, opposite the leaves; capsule pyramidal: **2. M. pyramidata**

2. Inflorescence sessile, axillary; capsule depressed-globose: **3. M. nodiflora**

1. Melochia tomentosa L., Syst. Nat., ed. 10, 2:1140. 1759.

"Velvet-leaf".

Shrub up to 2 m tall or sometimes more, densely whitish stellate-pubescent; leaves lance-oblong to ovate 1–7 cm long, rounded to acute at apex, the base rounded or subcordate, the veins channelled on the upper side and prominently raised beneath. Flowers in small axillary and terminal pedunculate cymes; calyx 6–8 mm long, deeply and narrowly lobed, and densely stellate-pubescent on the outside; petals pink, mauve, or rosy-purple, 10–12 mm long or more. Stamens 6–8 mm long, united for about half their length. Ovary short-stalked; styles united for about half their length. Capsule broadly pyramidal, long-beaked, c. 9 mm long; seeds 1 or 2 in each cell.

GRAND CAYMAN: *Brunt 1810, 1999; Fawcett; Kings GC 423; Millspaugh 1288, 1326; Proctor 15054; Sachet 405.* LITTLE CAYMAN: *Kings LC 95; Proctor 28141.* CAYMAN BRAC: *Kings CB 15; Millspaugh 1184; Proctor 28961.*
– West Indies and continental tropical and subtropical America, in sandy or rocky thickets and scrublands.

2. Melochia pyramidata L., Sp. Pl. 2:674. 1753.

A decumbent to erect shrubby herb to 1 m tall but often lower, the stems usually minutely puberulous on one side or nearly glabrous. Leaves thin, lanceolate to ovate, 1–6 cm long, acute or obtuse at the apex, rounded at the base, glabrous or minutely puberulous with simple hairs mixed with a few branched or stellate ones. Flowers in pedunculate umbels of 2–5, opposite the leaves; calyx 3.5–4 mm long, with linear lobes; petals pink or purplish, obovate, 6–7 mm long, clawed. Capsule pyramidal, 5-angled, 5–8 mm long, bearing a few minute stellate hairs.

GRAND CAYMAN: *Kings GC 324; Millspaugh 1345; Proctor 15252, 27995.*
– West Indies and continental tropical America, chiefly along roadsides and in weedy fields.

3. Melochia nodiflora Sw., Nov. Gen. & Sp. Pl. 97. 1788.

Slender low shrub up to 1.5 m tall, often suffrutescent, the stems dark red, the younger parts usually puberulous with mostly simple hairs. Leaves on slender petioles, the blades oblong to very broadly ovate, mostly 1.5–7 cm long and up to

5.5 cm broad, the apex acute to acuminate, the base rounded or subcordate, conspicuously veined beneath. Flowers mingled with thin brown bracteoles in dense sessile axillary clusters. Calyx 3-4 mm long, with lanceolate lobes; petals pink, often striped, c. 5 mm long, short-clawed. Stamens completely united, forming a tube c. 2 mm long. Ovary sessile, 5-lobed; styles free. Capsule 5-lobed, pubescent, the carpels separating when ripe.

GRAND CAYMAN: *Correll & Correll 51011; Hitchcock; Lewis 3834; Proctor 15106.* CAYMAN BRAC: *Proctor 28948.*
– West Indies and continental tropical America, a weed of roadsides, pastures and old fields.

Genus 4. **GUAZUMA** Adans.

Trees with stellate pubescence; leaves short-petiolate, sometimes inequilateral at the base, and irregularly toothed; stipules present. Flower small, in axillary short-pedunculate cymes; calyx 2-3-parted; petals 5, clawed and hooded-concave, the apex 2-cleft and bearing a terminal linear 2-cleft appendage. Stamens united to form a 5-lobed tube, the acuminate lobes (equivalent to staminodes) alternate with the petals, the short-stalked anthers in groups of 2 or 3 in the sinuses, 2-celled. Ovary sessile or short-stalked, 5-lobed, 5-celled; ovules few to numerous in each cell, on axile placentae; styles more or less united. Fruit a subglobose, woody capsule, covered with short, hard tubercles or (in one species) plumose-setose, indehiscent or imperfectly 5-valvate at the apex; seeds with endosperm and slightly curved embryo, the cotyledons leaf-like and inflexed-folded.

A tropical American genus of 4 species (ref. Freytag in Ceiba 1(4):193-225. 1951).

1. Guazuma tomentosa Kunth in H.B.K., Nov. Gen. 5:320. 1823. **FIG. 106.**

"Ba'cedar".

A small tree up to 10 m tall or more; leaves ovate to ovate-oblong, 3-15 cm long, acute to acuminate at the apex, usually cordate or subcordate and inequilateral at base, pubescent with stellate hairs on both sides. Inflorescences 2-3 cm long; sepals 3-4 mm long, reflexed; petals yellowish or cream, 3-4 mm long, minutely stellate-pubescent on the outside, the appendage 4-6 mm long and 0.2-0.5 mm wide. Fruit globose or oblongoid, 17-37 mm long, indehiscent, the blackish tubercles separating deeply and irregularly at maturity; seeds numerous, obovoid, 2-3.8 mm long.

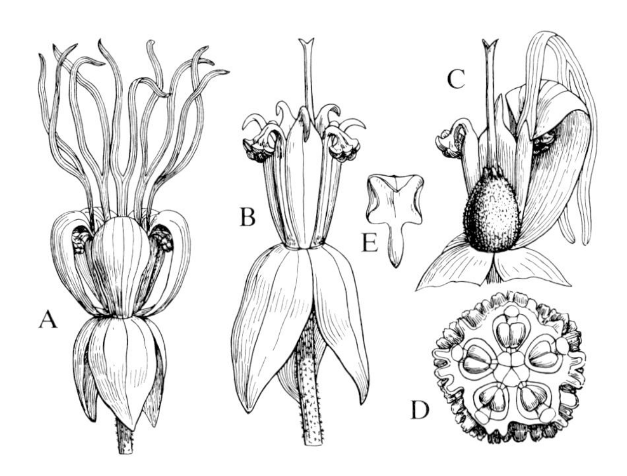

FIG. 106 **Guazuma tomentosa.** A, flower, x4. B, same with petals removed, showing staminal tube with stamens and staminodes, x7. C, portion of flower showing the apex of a petal with appendage lying on a stamen, also ovary and style, x7. D, x-section of fruit, x1. E, embryo, x11. (F. & R.)

GRAND CAYMAN: *Hitchcock.* Apparently very rare in the Cayman Islands. – West Indies and continental tropical America, usually in dry open woodlands or on the borders of pastures. It is a rather useful tree, because in addition to being resistant to drought, rope and twine can be made from the tough fibrous bark, the flowers are a good source of honey, and both the foliage and the immature fruits make nutritious fodder for cattle and other stock. A closely related species is *G. ulmifolia* Lam., differing in having narrower glabrate leaves, slightly larger flowers, and dehiscent fruits. The constancy of these differences, however, is open to question, and many authors combine both forms under the name *G. ulmifolia.*

Family 42

MALVACEAE

Herbs, shrubs or trees, often with stellate hairs; leaves alternate, simple but often lobed, usually palmately-veined, at least at the base; stipules present. Flowers regular, usually perfect, axillary and solitary, or else in axillary or terminal racemes, fascicles, or panicles; each flowers subtended by 3 or more bracteoles, these often large and forming an "epicalyx" outside the true calyx. Sepals usually 5, more or less united, the lobes valvate; petals 5, free but adnate to the base of the staminal column (thus appearing gamopetalous), twisted and overlapping in bud. Stamens numerous, rarely 5 or 10, the filaments united below into a tube (column); anthers 1-celled. Ovary superior, 2- to many-celled; style single at the base, dividing above into as many branches as there are ovary-cells; ovules 1 or more in each cell, attached along its inner angle. Fruit usually a dry schizocarp or capsule; seeds with scant endosperm and curved embryo, the cotyledons folded and foliaceous.

A nearly world-wide family of more than 45 genera (up to 75 recognized by some authorities) and about 1,500 species. Many are of economic value for food (e.g., Okra, *Hibiscus esculentus* L., and Sorrel, *Hibiscus sabdariffa* L.); textile fibres (e.g., Cotton, *Gossypium* spp.); timber (e.g., Blue mahoe, *Hibiscus elatus* Sw.); and ornament (e.g., *Hibiscus* spp., *Malvaviscus* spp., etc.)

KEY TO GENERA

1. Petals convolute, never opening; style-branches 10; carpels fleshy, united into a berry: **9. Malvaviscus**

1. Petals spreading; style-branches 5, separate or united; carpels dry, not berry-like:

2. Fruit a schizocarp, the carpels more or less separating at maturity into separate cocci:

3. Flower subtended by large, cordate-ovate, leaf-like bracts and almost concealed by them; staminal column bearing anthers on the side; plants with harsh stinging hairs: **7. Malachra**

3. Flowers not subtended by leaf-like bracts (but with or without an epicalyx); staminal column bearing anthers at the apex:

4. Epicalyx absent:

5. Carpels 1-seeded:

6. Carpels with a transverse partition or ring inside; leaves whitish-tomentose beneath: **1. Wissadula**

6. Carpels not partitioned or ringed inside; leaves glabrous or pubescent, but not whitish-tomentose beneath: **5. Sida**

5. Carpels 2- or 3-seeded:

7. Carpels leathery, beaked: **2. Abutilon**

7. Carpels membranous, inflated and bladderlike, rounded at apex and not beaked: **3. Herissantia**

4. Epicalyx present:

8. Carpels of fruit covered with numerous short, barbed spines: **8. Urena**

8. Carpels of fruit unarmed or with 1-3 long, simple spines: **4. Malvastrum**

2. Fruit a capsule splitting at maturity, the carpels not separating:

9. Epicalyx absent: **6. Bastardia**

9. Epicalyx present:

10. Epicalyx of 3 bracteoles larger than the sepals: **13. Gossypium**

10. Epicalyx of 5 or more bracteoles smaller than the sepals:

11. Cells of the capsule 1-seeded; herbs with hastate leaves and stinging hairs: **11. Kosteletzkya**

11. Cells of the capsule with 2 or more seeds; shrubs or trees:

12. Styles distinctly branched; calyx deeply cleft or lobed: **10. Hibiscus**

12. Styles united to the apex or nearly so; calyx truncate with 5 minute teeth: **12. Thespesia**

Genus 1. **WISSADULA** Medic.

Shrubs or shrubby herbs, usually densely clothed in whitish or yellowish tomentum composed of minute stellate hairs; leaves often long-petiolate, cordate, and long-acuminate. Inflorescences axillary and terminal, loosely paniculate; flowers on long slender pedicels, lacking an epicalyx; calyx broadly 5-lobed; petals yellow or orange. Staminal column divided at apex into an indefinite number of filaments. Ovary 5-celled; ovules 1 or 3 in each cell; style-branches with capitate stigmas. Fruits somewhat top-shaped; ripe carpels beaked, the beaks erect or pointing outward, partially divided inside by an incomplete transverse partition sometimes represented by a ring, and opening by two valves; seeds 1-3 in each carpel.

A genus of about 40 species, all but a single pantropical one occurring in tropical America.

1. Wissadula fadyenii Planch. ex. R.E. Fries in Svensk. Vet. Akad. Handl. 43(4):30, t.1, f. 1-2; t. 6, f. 2-4. 1908.

Wissadula divergens Griseb., 1859, not Benth.

A shrubby herb to 1 m tall or more; leaves 3–17 cm long, the blades triangular-ovate and long-acuminate, cordate or subcordate at base, the margins entire, densely soft stellate-pubescent beneath, and 5–7-nerved from the base; petioles varying from 0.5 cm on the upper leaves to 6 cm on the lower ones. Pedicels to 5 cm long. Calyx 3–3.5 mm long, divided about halfway into triangular lobes; petals 4–5 mm long. Fruit c. 6 mm broad at maturity, the dark brown carpels minutely puberulous, their beaks c. 0.5 mm long; seed one in each carpel, c. 2 mm long, minutely hairy.

GRAND CAYMAN: *Hitchcock*, collected January 19, 1891. No subsequent collector has found this species in the Cayman Islands.
– Jamaica, Trinidad, and northern South America, a weedy plant of clearings and old fields.

Genus 2. **ABUTILON** Mill.

Herbs or shrubs, rarely trees, usually velvety-pubescent with stellate hairs; leaves mostly cordate, often angulate or lobed. Flowers small or rather large, axillary and solitary or in small cymes, or else terminal and paniculate. Calyx 5-lobed; petals commonly white, yellow or red. Staminal column divided at the apex into numerous slender filaments. Ovary 5- to many-celled; ovules 3–9 in each cell; style-branches as many as the cells, filiform or club-shaped. Mature carpels coalescent at the base or completely separating, 2-valved; seeds subreniform.

A genus of more than 100 species, occurring widely in tropical and subtropical regions.

1. **Abutilon permolle** (Willd.) Sweet, Hort. Brit. 1:53. 1826.

A shrubby herb with an erect stem up to 1.5 m tall, all parts softly stellate-pubescent; leaves mostly 4–10 cm long, the blades broadly ovate and deeply cordate, the apex acuminate, the margins irregularly crenate; petioles up to 5 cm long. Flowers axillary, solitary, long-pedunculate; calyx 8–10 mm long, densely stellate-pubescent; petals 1.2–1.7 cm long, yellow. Carpels 7–10, when ripe 9–10 mm long, dark brown, villous, 3-seeded, beaked at the apex; seeds 2 mm in diameter.

GRAND CAYMAN: *Hitchcock; Millspaugh 1799.*
– Florida, West Indies, southern Mexico and northern Guatemala, often in coastal thickets and dry fields.

Genus 3. **HERISSANTIA** Medic.

Annual or sometimes short-lived perennial herb, the slender branches trailing, spreading, or weakly erect, more or less finely stellate-tomentose; leaves cordate. Flowers mostly solitary on axillary peduncles, the peduncles filiform. Calyx deeply 5-cleft; petals 5, whitish. Carpels about 12, the styles slender with terminal stigmas; ovules 2-6 in each carpel. Seeds glabrous or thinly setulose.

A monotypic genus, native of tropical and subtropical America, now pantropical in distribution.

1. Herissantia crispa (L.) Briz. in Jour. Arnold Arb. 49(2): 278-9. 1968. **FIG. 107.**

Abutilon crispum (L.) Medic., 1787.

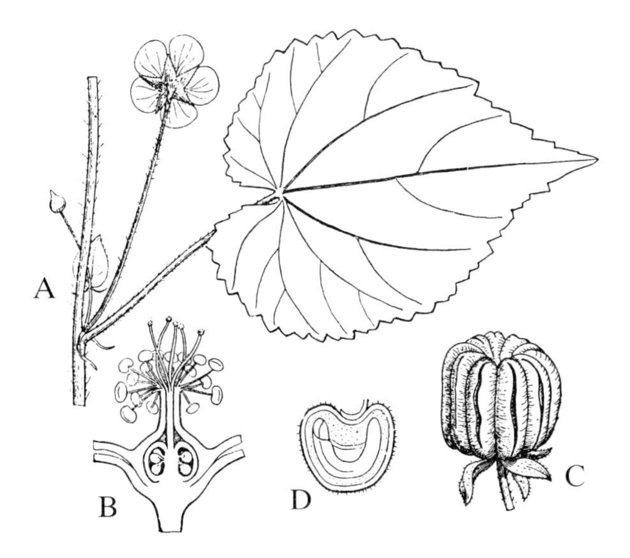

FIG. 107 **Herissantia crispa.** A, leaf and flower, x$^2/_3$. B, flower cut lengthwise, with calyx and petals removed, enlarged. C, capsule, x1. D, section of seed, x6. (F. & R.)

Stems up to 1.2 m long, frequently much shorter in arid situations; leaves 2-8 cm long, the petioles equalling the blades; blades acute to shortly acuminate, the margins crenate-dentate, stellate-tomentose on both sides. Peduncles 1.5–5 cm long. Calyx 4–8 mm long; petals as long as to twice as long as the calyx, obovate.

Carpels when ripe aggregated in a head 10–15 mm in diameter, thinly pubescent; seeds black, c. 2 mm in diameter.

GRAND CAYMAN: *Hitchcock; Proctor 15220.* CAYMAN BRAC: *Proctor 29059.*

– Florida and Texas southward throughout the West Indies and continental tropical America; also said to occur in southeast Asia. This species chiefly grows in sandy waste ground, dry fields, and gravelly clearings in woodlands.

Genus 4. **MALVASTRUM** A. Gray

Annual or perennial herbs, sometimes subwoody, with pubescence of simple, branched and stellate hairs. Leaves entire, serrate, lobed or cleft; linear stipules present. Flowers axillary or in terminal clusters or spikes, stalked or sessile; epicalyx of 3 linear to lanceolate bracteoles about as long as the calyx, rarely lacking. Calyx 5-lobed; petals 5, usually yellow or white. Ovary of 5 to many carpels; ovules 1 in each cell; style-branches as many as the carpels; stigmas linear, club-shaped or capitate. Ripe carpels separating, somewhat bivalvate or else indehiscent; seeds reniform.

A genus of 50 or more species, the majority in tropical America and a few in South Africa.

KEY TO SPECIES

1. Flowers in dense terminal and axillary spikes; stellate hairs with 5 or more radiating branches: **1. M. americanum**

1. Flowers mostly solitary in the axils and in small terminal heads; hairs 4-branched, with paired branches directed forward and backward:

  2. Carpels without spines: **2. M. corchorifolium**

  2. Carpels with 3 spines on the back: **3. M. coromandelianum**

1. Malvastrum americanum (L.) Torr. in Emory, Rep. U.S. & Mex. Bound. Surv. 2(1): 38. 1859.

Malvastrum spicatum (L.) A. Gray, 1849.

Perennial shrubby herb up to 2 m tall but usually lower, stellate-pubescent throughout; leaves 3–20 cm long, the blades ovate or triangular-ovate, acute or subacuminate at apex, truncate at base, the margins unequally serrate; petioles

1-9 cm long. Flowers sessile in dense terminal and axillary spikes intermingled with leaf-like bracts, the axillary spikes sometimes of only 2 or 3 flowers; bracteoles 5-7 mm long. Calyx c. 5 mm long; petals yellow, 5-6 mm long. Carpels up to 15, beaklike at the apex but without spines, hispid on the upper side.

CAYMAN BRAC: *Millspaugh 1167-bis, 1188.*
– Pantropical, often a weed of roadsides and waste places. The stems contain a strong, hemp-like fibre.

2. **Malvastrum corchorifolium** (Desr.) Britton in Small, Fl. Miami 119. 1913.

A woody herb up to 2 m tall but usually much lower, the stems and leaves clothed with stiff adpressed 4-branched hairs; leaves 1-8 cm long, the blades oblong to ovate, acutish, with sharply serrate margins; petioles 0.5-3.5 cm long. Flowers subsessile, solitary in leaf-axils and clustered in small terminal heads; bracteoles about equalling the calyx. 4-5 mm long, clothed with long simple and minute stellate hairs; petals orange or yellow, c. 5 mm long. Carpels up to 15, without spines, but bearing long simple hairs on the upper part of the back.

GRAND CAYMAN: *Brunt 1706; Proctor 15100.* CAYMAN BRAC: *Proctor 28934.*
– Florida and the West Indies, often in dry sandy fields and along roadsides.

3. **Malvastrum coromandelianum** (L.) Garcke in Bonplandia 5:295. 1857. **FIG. 108.**

Malvastrum tricuspidatum (Ait.) A. Gray, 1852.

M. americanum of some authors, not Torr., 1859.

A woody annual herb, decumbent or up to 1 m tall, the stems clothed with stiff adpressed 4-branched hairs, the leaves with similar hairs and also simple ones; leaves 1-10 cm long, the blades oblong to ovate, obtuse, with coarsely serrate margins; petioles 0.5-3 cm long. Flowers short-stalked, solitary in the leaf-axils or occasionally several together at apex of the stem; bracteoles about equalling the calyx and inserted on its base. Calyx 5-8 mm long, hairy; petals pale yellow or dull orange, 7-9 mm long. Carpels 10 or more, with 2 short spines at the middle of the back and a longer one at the top, surrounded by hairs.

GRAND CAYMAN: *Hitchcock.*
– Pantropical, said to be introduced from America to the Old World, a weed of roadsides and open waste ground.

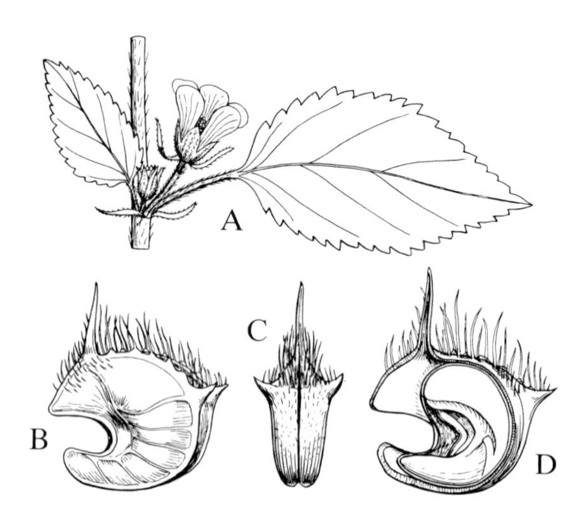

FIG. 108 **Malvastrum coromandelianum.** A, portion of flowering branch, x1. B, ripe carpel, x7. C, back of same beginning to split open, x7. D, ripe carpel with seed, cut through, x7. (F. & R.)

Genus 5. **SIDA** L.

Herbs or shrubs with stellate or simple hairs; leaves usually more or less serrate; narrow stipules present. Flowers axillary or terminal, solitary or in heads, spikes, racemes, cymes or panicles; epicalyx lacking. Calyx 5-lobed or 5-toothed; petals mostly yellow, orange, or white. Staminal column bearing anthers at the apex. Carpels 5 or more; ovules 1 in each cell. Ripe carpels separating, 2-valvate above and often bearing spines at the apex; seeds solitary in each carpel.

A pantropical genus of about 125 species.

KEY TO SPECIES

1. Peduncles adnate to the petioles of leaf-like bracts; flowers orange; stems prostrate from a woody taproot: **4. S. ciliaris**

1. Peduncles free; flowers yellow or whitish; stems erect or ascending:

 2. Carpels 5:

  3. Flowers mostly solitary in the leaf-axils; leaves lance-oblong, minutely stellate-pubescent, not over 3 cm long: **2. S. spinosa**

3. Flowers in clusters, axillary and terminal; leaves ovate-cordate; up to 10 cm long:

 4. Stems and leaves with long simple hairs and not viscid-glandular; flowers in small dense axillary and terminal heads: **1. S. urens**

 4. Stems viscid-glandular and leaves with small soft stellate hairs; flowers in loose cymes or cymose panicles: **3. S. glutinosa**

2. Carpels more than 5:

 5. Leaves densely and softly stellate-pubescent, the blades ovate; calyx densely pubescent; carpel-spines with simple hairs: **5. S. cordifolia**

 5. Leaves nearly or quite glabrous, the blades lance-oblong; calyx glabrate; carpel-spines with minute stellate hairs: **6. S. stipularis**

1. **Sida urens** L., Syst. Nat. ed. 10, 2:1145. 1759.

Annual or sometimes persisting herb, much-branched, the stems trailing or ascending, clothed throughout with long, stiff, simple hairs; leaves long-petiolate, 3–10 cm long, the blades ovate or lance-ovate, long-acuminate, the base cordate, the margins unequally serrate. Flowers several together in dense globose clusters, these axillary and terminal, subsessile or pedunculate. Calyx 6–8 mm long, 5-angulate, the lobes long-acuminate; petals yellow with red spot at the base, c. 7 mm long. Carpels 5, glabrous, spineless or with 2 short teeth at the apex.

GRAND CAYMAN: *Millspaugh 1284.*
– West Indies, continental tropical America and tropical Africa, a weed of fields, roadsides, pastures and secondary thickets.

2. **Sida spinosa** L., Sp. Pl. 2:683. 1753.

Annual or perennial herb with erect or spreading stems, all the younger parts clothed with very minute stellate pubescence, the stems often with 1 or 2 small spine-like tubercles just below the attachment of each leaf. Leaves petiolate, 1–3 cm long, the blades linear or narrowly lance-oblong to ovate, obtuse or acute at the apex, the base truncate or subcordate, the margins serrate. Flowers mostly solitary in the leaf-axils on short peduncles. Calyx densely stellate-pubescent, 5–7 mm long, with 5 triangular lobes; petals yellow, equalling the calyx. Carpels 5, each bearing 2 short spines at the apex.

GRAND CAYMAN: *Brunt 2114; Kings GC 382; Proctor 27935.*
– Pantropical, also adventive in temperate North America, a weed of roadsides and pastures.

3. Sida glutinosa Commers. ex Cav., Monad. Diss. 1:16, t. 2, f. 8. 1785.

Wissadula divergens of Millspaugh, 1900, non Benth.

A somewhat shrubby erect herb up to 1 m tall or more, the stems densely puberulous with simple, often glandular-viscid hairs; leaves long-petiolate, 3–12 cm long, the blades ovate, acuminate, cordate at the base, and with crenate-serrate margins, both surfaces clothed with fine stellate pubescence. Flowers initially solitary from the leaf-axils, but almost immediately becoming accompanied by elongate pedunculate cymes of several flowers from the same axils; these, together with a terminal cyme, form a large paniculate inflorescence with numerous leaf-like bracts. Calyx angulate with acuminate lobes, 5–6 mm long, clothed with numerous simple hairs; petals salmon-yellow, pale orange, or buff, c. 8 mm long. Carpels 5, puberulous on the inner part and with 2 short beaks at the apex.

GRAND CAYMAN: *Millspaugh 1346, 1350.*
– West Indies and continental tropical America, in clearings, along roadsides, and in dry waste ground.

4. Sida ciliaris L., Syst. Nat. ed. 10, 2:1145. 1759.

Prostrate trailing herb with perennial woody taproot, the stems to 30 cm long but usually shorter, clothed with adpressed stellate hairs. Leaves petiolate, mostly 6–15 mm long, the blades oblong or oblanceolate, the margins serrate toward the apex, glabrous on the upper side and stellate-pubescent beneath. Flowers solitary or several together, terminal, the short peduncles adherent to the petiole of a leaf-like bract. Calyx 4–5 mm long, with long simple hairs and very minute stellate ones; petals orange or salmon, often purplish-red at the base, c. 8 mm long. Carpels 7–8, covered with short tubercles toward the apex.

GRAND CAYMAN: *Brunt 2119, Kings GC 381-A; Proctor 15099.* CAYMAN BRAC: *Proctor 28993.*
– Florida and Texas, West Indies, and continental tropical America, in sandy clearings, open fields, and in pockets of limestone rocks.

5. Sida cordifolia L., Sp. Pl. 2:684. 1753.

Annual or persisting erect herb up to 1.5 m tall but usually less, all parts densely and softly stellate-pubescent. Leaves petiolate, 3–8 cm long or more, the blades ovate or lance-oblong, obtuse at the apex, truncate or slightly cordate at the base, the margins serrate. Flowers pedunculate, axillary and terminal in dense racemes or corymbs. Calyx angulate, 6–7 mm long; petals orange or salmon, or yellow with an orange spot at the base, c. 10 mm long. Carpels 7–12, 2-beaked at the apex, the beaks divergent, hairy.

CAYMAN BRAC: *Proctor 29058.*
– Pantropical, in clearings, sandy fields, and along paths.

6. **Sida stipularis** Cav., Monad. Diss. 1:22, t. 3, f. 10. 1785.

Sida acuta of many authors, non Burm. f., 1768.

Sida carpinifolia var. *antillana* Millsp., 1900.

"Broom-weed".

A shrubby annual or persisting herb with erect (rarely decumbent), tough stems up to 1 m tall, the young parts very minutely stellate-pubescent, becoming mostly glabrate with age. Leaves short-petiolate, mostly 1.5–6 cm long, the blades lance-oblong to elliptic, blunt or acute at both ends, the margins finely to coarsely serrate, essentially glabrous and often reddish on the upper side, but with scattered very minute stellate hairs beneath. Flowers axillary, mostly solitary but sometimes 2 or 3 together, short-stalked or subsessile. Calyx 6–8 mm long, glabrate or with a few very minute stellate hairs on the outside and a few simple hairs on the margins of the lobes; petals pale yellow, buff, or whitish, c. 10 mm long. Carpels 7–12, with 2 spines at the apex, these bearing very minute stellate hairs.

GRAND CAYMAN: *Brunt 2078; Kings GC 361; Millspaugh 1303; Proctor 15099; Sachet 448.* LITTLE CAYMAN: *Proctor 28164.* CAYMAN BRAC: *Proctor 28974.*
– Pantropical, also in warm-temperate North America, an often abundant weed of pastures, roadside banks, and waste ground.

Genus 6. **BASTARDIA** Kunth

Herbs or shrubs with stellate pubescence and often also glandular-viscid; leaves petiolate, cordate, and entire, crenate or dentate; stipules filiform, deciduous. Flowers axillary, solitary or 2 to 3 together; epicalyx lacking. Calyx 5-lobed; petals yellow. Anthers borne at the apex of the staminal column. Ovary 5–8-celled; ovules 1 in each cell, pendent, attached above at the inner angle; style-branches as many as the carpels. Fruit a dehiscent 5–8-valved capsule, the more or less hairy seeds solitary in the cells.

A tropical American genus of 6 species, differing from *Sida* in the capsular fruit.

1. **Bastardia viscosa** (L.) Kunth in H.B.K. Nov. Gen. 5:256. 1822.

An erect shrubby herb up to 1 m tall or more, viscid-glandular throughout and with a strong unpleasant odour. Leaves 2–5 cm long, acuminate, the margins glandular-dentate, and both surfaces densely pubescent with minute stellate hairs. Peduncles slender, 1–3 cm long; calyx 3.5–4 mm long, deeply lobed; petals yellowish, c. 5 mm long. Capsule 5–8-celled, 3–4 mm long, minutely stellate-pubescent, not beaked; seeds black, puberulous with white hairs.

GRAND CAYMAN: *Brunt 1964, 2145; Proctor 15264.* The Cayman specimens represent the small-leafed variety known as var. *parvifolia* (Kunth) Griseb.
– West Indies and continental tropical America, a foetid weed of waste places.

Genus 7. **MALACHRA** L.

Annual to somewhat persisting herbs, often bristly with stiff, harsh, simple or branched hairs; leaves long-petiolate, frequently palmately-angled or -lobed; stipules linear. Flowers in dense axillary or terminal heads subtended by large foliaceous bracts; bracts (rarely more), prominently veined, folded down the middle, shortly stalked or sessile, each bract with 2 or 4 stipule-like outgrowths at the base; epicalyx lacking. Calyx 5-lobed; petals 5, yellow or white (rarely purple). Staminal column short, truncate or 5-toothed, with anthers borne on the outside. Ovary 5-celled, each cell with 1 ovule; style 10-branched. Fruit a schizocarp, the mature obovoid carpels separating from the central axis; seeds reniform.

A small genus of 9 species indigenous to tropical and subtropical America, but some of them widely naturalized as weeds in the Old World tropics.

1. **Malachra alceifolia** Jacq., Collect. 2:350. 1789.

"Wild Okra".

A coarse weedy herb up to 1 m tall or more, the stems erect and often unbranched, clothed with numerous small soft stellate hairs mixed with much longer, stiff, yellowish ones, these simple or stellate. Leaves 5–18 cm long, the blades roundish or broadly ovate, somewhat angled to shallowly 3–5-lobed, the margins dentate, the surfaces stellate-pubescent, especially beneath. Flower-heads axillary, sessile or stalked; bracts triangular-ovate, acute to acuminate at the apex, 1.5–2.5 cm long, often white-spotted between the veins; calyx 4–6 mm long, the lobes lanceolate; petals yellow, 1.5 cm long. Carpels of the ripe fruit 3–3.5 mm long, puberulous with simple hairs.

GRAND CAYMAN: *Brunt 2095, 2141; Correll & Correll 51048; Hitchcock; Proctor 27952.*
– West Indies and continental tropical America. The Cayman plants often occur in seasonally wet hollows in pastures.

Genus 8. **URENA** L.

Herbs or shrubs, the pubescence all or chiefly of stellate hairs; leaves long-petiolate, usually palmately angled or lobed, the 3 central nerves each bearing a slit-like gland near its base; stipules linear. Flowers axillary, solitary or in small clusters, sometimes forming long terminal interrupted spikes; epicalyx of 5 united bracteoles, adherent to the calyx. Calyx 5-lobed, the lobes alternating with the lobes of the epicalyx; petals 5, pink, obovate or obcordate. Staminal tube about as long as the petals, bearing anthers on the outside. Ovary 5-celled, each cell with 1 ovule; style 10-branched. Fruit a schizocarp, each carpel covered with barbed spines.

A pantropical genus of 6 species.

1. **Urena sinuata** L., Sp. Pl. 2:692. 1753.

An erect, branched herb up to 1 m tall; leaves 2–9 cm long, the blade rounded- or ovate-angulate in outline, more or less deeply lobed, in larger leaves the lobes narrowed downward forming rounded bays between them. Epicalyx 4–4.5 mm long in flower, slightly lengthening in fruit, stellate-pubescent; calyx nearly as long as the epicalyx; petals c. 1.5 cm long. Ripe carpels c. 5 mm long, puberulous and covered with spines, the spines each bearing 2–4 retrorse barbs at the tip.

GRAND CAYMAN: *Hitchcock; Millspaugh 1321.*
– A pantropical weed, readily spread by its adherent spiny carpels. This species intergrades in many regions, presumably by hybridization, with *Urena lobata* L., but the latter has not been found in the Cayman Islands. Both species yield a strong fibre which can be used as a substitute for flax.

Genus 9. **MALVAVISCUS** Adans.

Shrubs, sometimes arborescent, the pubescence chiefly of stellate hairs; leaves long-petiolate, the blades usually dentate and often angled or lobed, the tissue with numerous minute pellucid dots; stipules linear or subulate. Flowers mostly solitary in the upper leaf-axils, or sometimes forming short terminal corymbs or racemes; epicalyx of 5 to many, more or less linear bracteoles. Calyx 5-lobed, bell-shaped, somestimes with 2 or 3 of the lobes united; petals 5, usually bright red or pink, convolute into a loose tube and not opening. Staminal column spirally grooved, long-exserted beyond the petals, the anthers borne on the outside toward the apex. Ovary 5-celled, each cell with 1 ovule; style 10-branched, with capitate stigmas. Fruit fleshy, berry-like at first, the carpels ultimately separating, indehiscent.

A tropical American genus of 3 or more species, two of them rare, the other extremely variable and difficult to classify.

1. **Malvaviscus arboreus** Cav., var. **cubensis** Schlecht. in Linnaea 11:360. 1837.

Malvaviscus jordan-mottii Millspaugh, 1900.

"Mahoe".

A somewhat straggling shrub, sometimes arborescent and up to 5 m tall, the upper branches, petioles and pedicels subglabrous or pubescent with fine stellate hairs. Leaves 5–17 cm long, the blades lanceolate to very broadly ovate, the apex acutish to short-acuminate, the base cordate, the margins sinuate-dentate, the surfaces glabrate to densely stellate-pubescent, especially beneath. Epicalyx and calyx c. 1 cm long, stellate-pubescent; petals about twice as long as the calyx. Staminal column projecting about 1 cm beyond the petals.

GRAND CAYMAN: *Brunt 2159; Correll & Correll 51017; Hitchcock; Kings GC 404; Millspaugh 1313; Proctor 11980, 15211; Rothrock 180, 237.* LITTLE CAYMAN: *Proctor 35103.* CAYMAN BRAC: *Millspaugh 1166 (type of M. jordan-mottii); Proctor 29008.*
– Bahamas, Cuba, and perhaps the Yucatan area of Central America, usually in rocky limestone woodlands at low elevations. Several other varieties occur in various regions. The Grand Cayman and Cayman Brac populations differ markedly in pubescence. In the former island, the leaves are covered beneath with a fine, soft stellate pubescence in which also occur numerous much larger, stiff and harsh, 3-branched hairs. Cayman Brac specimens, on the other hand, lack the fine pubescence, but the larger hairs are usually even more abundant, making the foliage very unpleasant to handle. Sometimes, however, almost glabrous plants occur (including those of Little Cayman), and one such was the basis of the name *M. jordan-mottii*. Millspaugh (1900) reported that the nettle-like quality of "Mahoe" has caused it to be used in the Cayman Islands as a flagellant for rheumatic patients, but the writer has heard no recent report of such a drastic treatment.

Malvaviscus arboreus var. *penduliflorus* (DC.) Schery, sometimes called "Sleeping Hibiscus" or "Pepper Hibiscus", is often cultivated in the West Indies as an ornamental; it differs from var. *cubensis* in having very much larger flowers (more than 4.2 cm long) and in always lacking nettle-like stinging hairs.

Genus 10. **HIBISCUS** L.

Herbs, shrubs or trees, variously pubescent or almost glabrous; leaves various, often lobed or toothed, sometimes simple and entire. Flowers usually solitary in the axils of the upper leaves, often large and showy; epicalyx of usually many bracteoles, rarely as few as 5, free or united, sometimes more or less adherent to the calyx. Calyx 5-lobed; petals 5, variously coloured, often large and showy. Ovary

5-celled, each cell with 3 to many ovules; style 5-branched, the stigmas more or less capitate; seeds reniform or subglobose, glabrous or pubescent, often numerous.

A pantropical genus of about 200 species, a few also occurring in temperate regions.

KEY TO SPECIES

1. Leaves sharply toothed or lobed; stipules linear-subulate; bracteoles free:

   2. Wild shrub; leaves 3-5-angled or lobed, softly pubescent; calyx-lobes ovate, more than half as long as the petals: **1. H. clypeatus**

   2. Cultivated shrub; leaves coarsely toothed but not lobed, glabrate; calyx-lobes lance-acuminate, about one-fourth as long as the petals or less: **[2. H. rosa-sinensis]**

1. Leaves entire or crenulate; stipules oblong, large; bracteoles united to form a cup, free from the calyx: **3. H. tiliaceus**

1. Hibiscus clypeatus L., Syst. Nat. ed. 10, 2:1144. 1759.

A shrub up to 4 m tall or more, softly stellate-pubescent throughout. Leaves long-petiolate, 7-25 cm long or more, the blades very broadly angular-ovate, usually shallowly but rather sharply 3-lobed toward the apex, cordate at base, the margins glandular-dentate; stipules linear-subulate, c. 1.5 cm long. Flowers on stout peduncles 5-11 cm long; epicalyx of 9-11 linear bracteoles, these unequal in length and much shorter than the calyx. Calyx 3.5-4 cm long, with foliaceous 5-nerved lobes. Petals 4.5-6 cm long, pale yellow, velvety pubescent on the outside. Capsule shorter than the calyx, yellowish-hairy; seeds glabrous, dark brown, c. 4 mm long.

CAYMAN BRAC: *Fawcett 10.* Not collected in the Cayman Islands since 1888. – Greater Antilles, Yucatan and Peten, in thickets and dry rocky woodlands.

[2. **Hibiscus rosa-sinensis** L., Sp. Pl. 2:694. 1753.

The commonly cultivated ornamental Hibiscus or "Shoe-black" occurs in many color-forms. It is a native of tropical Asia.]

3. **Hibiscus tiliaceus** L., Sp. Pl. 2:694. 1753.

Pariti tiliaceum (L.) Britton, 1918.

"Seaside Mahoe".

An arborescent shrub or small diffuse tree to 6 m tall, all parts finely stellate-puberulous except the upper side of the leaves; leaves long-petiolate, 5–28 cm long, the blades roundish-reniform or very broadly ovate, abruptly acuminate, the base cordate, the margins entire or finely crenulate, the upper surface green and glabrous or nearly so, beneath finely whitish-tomentose, the median 1 or 3 nerves bearing a slit-like gland near the base; stipules oblong, 2–4 cm long, deciduous. Flowers solitary or occasionally 2 or 3 together, axillary and terminal, on short peduncles mostly 0.5–1.5 cm long; epicalyx cup-shaped, 8–11-lobed, shorter than the calyx. Calyx 2-3 cm long, 5-lobed to the middle; petals yellow or orange, 5–7 cm long. Staminal column nearly as long as the petals, bearing anthers along its entire length. Capsule ovoid, hairy, 1.5–2 cm long; seeds numerous, 4 mm long, glabrous and papillose.

GRAND CAYMAN: *Brunt 1625, 1791; Fawcett 9.* LITTLE CAYMAN: *Kings LC 71; Proctor 35110.*
– Pantropical, on protected seashores and borders of mangrove swamps. The wood is very light and can be used as a substitute for cork. The bark contains strong fibres which have the quality of becoming stronger when wet; in many tropical regions this material is used for making rope, mats, and coarse cloth.

Genus 11. **KOSTELETZKYA** Presl

Herbs or shrubs, usually with harsh simple and stellate pubescence; leaves mostly sagittate or often angular-lobed; stipules linear or subulate. Flowers axillary and terminal, solitary or in few-flowered open cymes; epicalyx of 5–10 bracteoles, these sometimes minute or lacking. Calyx 5-lobed or 5-toothed; petals white, pink, purple or yellowish, either spreading or erect and convolute. Staminal column with anthers on the outside. Ovary 5-celled, each cell with 1 ovule; style 5-branched, with capitate or dilated stigmas. Fruit a depressed, 5-angled, dehiscent capsule; seeds reniform, glabrous or puberulous.

A genus of 12 species occurring in America, Africa, and the Mediterranean region.

1. **Kosteletzkya pentasperma** (Bert.) Griseb., Fl. Brit. W. Ind. 83. 1859. **FIG. 109.**

A slender, branched herb up to 1 m tall or more, the stems green, finely puberulous in a line on one side, otherwise plentifully clothed with long, stiff, nettle-like simple hairs each from a pustulate base; leaves petiolate, 2–9 cm long, the blades varying from narrowly lanceolate or hastate to ovate, the apex narrowly acute, the base cordate, the margins crenate-serrate, clothed on both sides with harsh appressed hairs, those on the upper side mostly simple, those beneath 3-branched. Flowers solitary on long slender peduncles from the leaf-axils, often accompanied by a secondary cymose branchlet bearing several flowers; bracteoles linear or subulate, ciliate, shorter than the calyx. Calyx 4.5–5 mm long, bearing 3-branched hairs; petals white, c. 10 mm long. Capsule c. 9 mm broad and 5 mm high, deeply 5-lobed, recurved-hairy on the angles; seeds minutely puberulous.

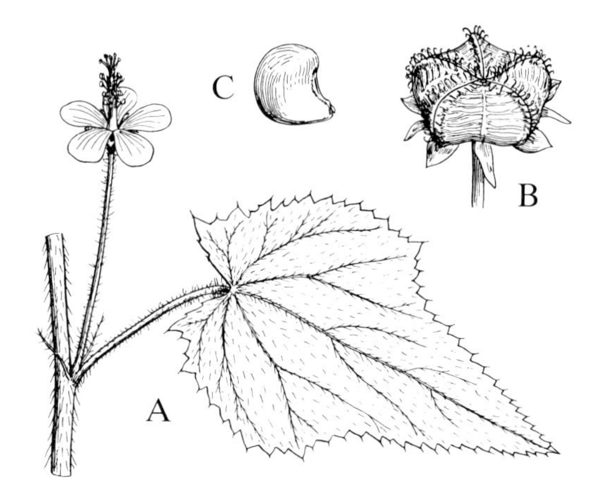

FIG. 109 **Kosteletzkya pentasperma.** A, leaf and flower, x$^2/_3$. B, fruit, x2. C, seed, x4. (F. & R.)

GRAND CAYMAN: *Brunt 1964a, 2091, 2108; Kings GC 287, GC 327; Proctor 27951.*

– West Indies and continental tropical America, often occurring in swamps, wet ditches and swales. The harsh, nettle-like hairs covering this plant make it very unpleasant to handle.

Genus 12. **THESPESIA** Soland. ex. Correa

Trees or sometimes shrubs; leaves long-petiolate, entire or sometimes lobed; stipules linear, deciduous. Flowers solitary in the leaf-axils, usually large; epicalyx of 3–5 narrow deciduous bracteoles. Calyx truncate and 5-toothed or rarely 5-lobed; petals yellow. Ovary 5-celled, each cell with several ovules; style club-shaped at the apex, with 5 grooves or short branches. Capsule hard or leathery, indehiscent or sometimes splitting by 5 valves; seeds usually 2 or 3 in each cell, glabrous or hairy, the cotyledons much-folded and black-dotted.

A small pantropical genus of about 17 species.

1. Thespesia populnea (L.) Soland. ex. Correa in Ann. Mus. Hist. Nat. Paris 9: 290, t. 25, f. 1. 1807. **FIG. 110.**

"Popnut" or "Plopnut".

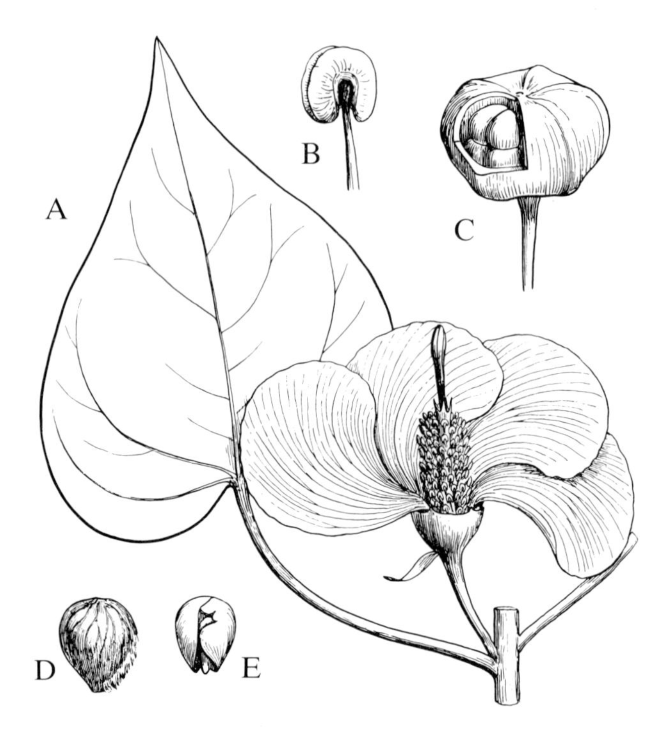

FIG. 110 **Thespesia populnea.** A, leaf and flower with a petal removed, x$^2/_3$. B, stamen, x7. C, fruit partly cut open, x$^2/_3$. D, seed, x1. E, embryo, x1. (F. & R.)

A shrub or small tree up to 10 m tall, the younger parts clothed with numerous very minute peltate scales, the whole plant otherwise glabrous. Leaves mostly 7–20 cm long, the petioles often as long as the blades; blades broadly ovate, acuminate at apex, cordate at base, the margins entire. Flowers on stout peduncles shorter than the petioles; bracteoles 1–2.5 cm long. Calyx cup-shaped, 7–9 mm long; petals 5–6 cm long, pale yellow with red spot at the base, the whole turning reddish with age, the margins ciliate toward the base. Staminal column much shorter than the petals. Capsule subglobose, c. 3 cm in diameter; seeds 8–10 mm long, finely hairy on the angles.

GRAND CAYMAN: *Brunt 1664, 1673; Hitchcock; Kings GC 62; Millspaugh 1238; Proctor 15098.* LITTLE CAYMAN: *Kings LC 70; Proctor 28076; Sauer 4174.* CAYMAN BRAC: *Proctor 29104.*

– Pantropical, chiefly near seashores and along the borders of mangrove swamps. Sometimes planted for ornament.

Genus 13. **GOSSYPIUM** L.

Subwoody herbs or shrubs, sometimes arborescent, generally marked all over with numerous minute black dots, often pubescent with simple or stellate hairs. Leaves usually long-petiolate and 3-7-lobed, or sometimes entire. Flowers solitary in the leaf-axils, large, on stout peduncles; epicalyx of 3 large cordate bracteoles, these more or less deeply incised or rarely entire. Calyx truncate or 5-toothed; petals yellow or red. Ovary 5-celled, the cells with many ovules; style club-shaped at the apex, grooved and bearing 3–5 stigmas. Fruit a dehiscent capsule; seeds densely woolly with very long or short hairs, or both, or sometimes nearly glabrous; endosperm scant or none.

A widespread tropical and subtropical genus of more than 20 species (or twice as many recognized by some authors). Several species have been cultivated (as "Cotton") for a very long time, and as they hybridize freely their classification is often very difficult.

KEY TO SPECIES

1. Seeds with long, loose, easily detachable hairs and without a covering of "fuzz" or short hairs; staminal column relatively long, with anthers compactly arranged on short filaments which are all about the same length: **1. G. barbadense**

1. Seeds with a double coat, consisting partly of long, firmly adherent hairs and also a dense covering of short "fuzz"; staminal column short, with anthers loosely arranged on filaments longer above than below: **2. G. hirsutum**

1. Gossypium barbadense L., Sp. Pl. 2:693. 1753. **FIG. 111.**

"Sea-island Cotton", "Long-staple Cotton".

A coarse herb or a shrub up to 4 m tall; leaves 5–15 cm long, the blades 3-5-lobed or nearly entire, cordate at base, usually glabrous or nearly so. Bracteoles broadly cordate, 4–7 cm long, with 3 to 7 lacerations. Petals 4.5–8 cm long, pale yellow turning dull reddish. Capsule ovoid, acuminate, mostly 3-celled; seeds with long, white, easily detachable lint but no short fuzz.

GRAND CAYMAN: *Brunt 1945; Hitchcock; Millspaugh 1367; Proctor 15053.* – West Indies and tropical South America; widely cultivated and spontaneous in tropical and subtropical regions.

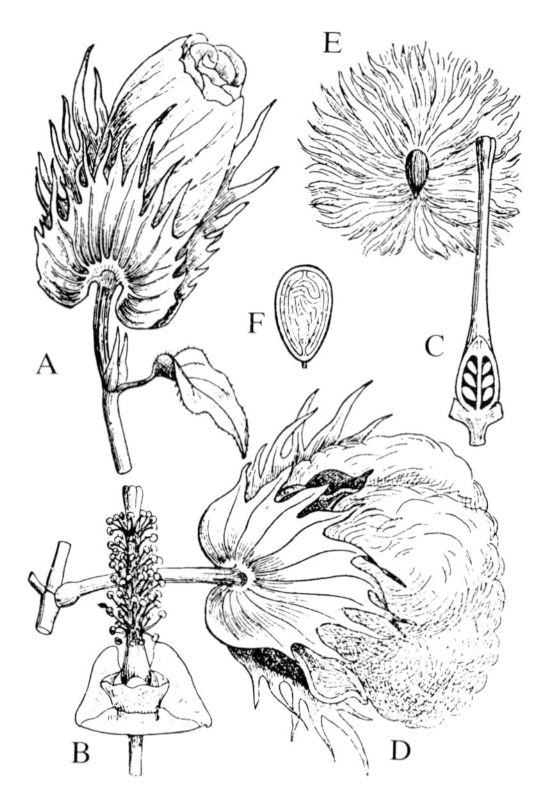

FIG. 111 **Gossypium barbadense** A, flower about to open, x²/₃. B, flower with calyx and corolla cut away, showing staminal tube enclosing pistil, x²/₃. C, pistil with ovary cut lengthwise, x1. D, capsule open, showing mass of cotton, x²/₃. E, seed with cotton attached, x²/₃. F, seed cut lengthwise, showing twisted embryo, x1½ (F. & R.)

2. **Gossypium hirsutum** L., Sp. Pl. ed. 2, 2:975. 1763.
var. **punctatum** (Schum.) J.B. Hutch., Sil & Steph. Evol. Gossyp. 40. 1947.

"Short-staple Cotton".

A coarse herb or a shrub up to 4 m tall, often clothed with chiefly simple hairs; leaves 5-15 cm long, the blades mostly 3-lobed and cordate. Bracteoles broadly cordate, 3-6 cm long, with 9 to 13 lacerations. Petals 3.5-5 cm long, pale lemon-yellow turning pink. Capsule ovoid-ellipsoid, acuminate, rough, mostly 3-celled; seeds with short dense fuzz and long white adherent lint.

GRAND CAYMAN: *Sauer 4102.* LITTLE CAYMAN: *Kings LC 10, LC 11; Proctor 28150.* CAYMAN BRAC: *Kings CB 43, CB 85; Proctor 28980; Sauer 4140.* – Coasts and islands of the Gulf of Mexico, Florida, Bahamas, and Hispaniola; in various forms cultivated in many regions and frequently becoming naturalized. Cultivars of this species are commonly and abundantly grown as "Cotton" in the United States, and are among the world's most important economic plants.

[LECYTHIDACEAE

Barringtonia asiatica (L.) Kurz, a large tree with evergreen foliage, large flowers with innumerable long, pink stamens, and curious 4-cornered green fruits up to 15 cm in diameter or more, is sometimes planted for shade or ornament.

GRAND CAYMAN: *Kings GC 232.*]

Family 43

FLACOURTIACEAE

Trees or shrubs, sometimes armed with spines or thorns, these simple or branched; leaves alternate, simple, entire or toothed, sometimes pellucid-dotted; stipules small and deciduous. Flowers small, regular, perfect or unisexual, clustered or in racemes or panicles, with jointed, bracteolate pedicels. Calyx persistent, of 5 to 7 sepals, these free or commonly united below, imbricate or valvate. Petals usually as many as the calyx-lobes, or else none. Stamens few or indefinite, often alternating with staminodes, arising from a small disc or torus, or else inserted on the calyx near its base. Ovary superior and free or more or less inferior, sessile, 1-celled or rarely 2–5-celled; ovules usually numerous, on 3 to 5 parietal placentae; style short, simple or divided. Fruit a leathery or fleshy capsule, dehiscent by 3 to 5 valves; seeds usually few, often arillate; endosperm fleshy, and the embryo with broad, often cordate cotyledons.

A pantropical family of about 85 genera and more than 800 species.

KEY TO GENERA

1. Flowers perfect; staminodes present, alternating with the stamens; spines absent, or if present simple and spur-like:

2. Style present; stamens 6–15: **1. Casearia**

2. Style absent; stamens 20–40: **2. Zuelania**

1. Flowers unisexual, the plants dioecious; staminodes absent; at least the trunk armed with branched spines: **3. Xylosma**

Genus 1. **CASEARIA** Jacq.

Trees or shrubs, usually unarmed but sometimes the trunk spiny or the branches with spine-like spurs. Leaves petiolate, entire or toothed, often with pellucid dots and lines in the tissue; stipules small and soon falling. Flowers perfect, clustered or umbellate in the leaf-axils, rarely solitary or racemose, with jointed pedicels bracteate at the base. Calyx of 4 to 6 overlapping lobes; petals none. Stamens 6–15, inserted on the calyx near its base, alternating with an equal number of staminodes; filaments free or united with the staminodes to form a ring. Ovary superior, free, ovoid or oblong, narrowed above into a short style, the stigma capitate or divided; ovules numerous, produced on 3 parietal placentas. Fruit a fleshy or dry capsule, 3-4-valvate, with numerous seeds; seeds with fleshy aril.

A pantropical genus alleged to have about 200 species.

KEY TO SPECIES

1. Leaves velvety-pubescent beneath: **3. C. hirsuta**

1. Leaves glabrous or nearly so:

  2. Stigma 2-3-lobed; calyx 1.5–2.5 mm long: **1. C. sylvestris**

  2. Stigma entire, capitate; calyx 4–5 mm long:

    3. Leaves mostly 9–18 cm long; pedicels jointed close to the base: **2. C. guianensis**

    3. Leaves 2–7 cm long; pedicels jointed below the middle:

      4. Sepals united for 1/4 to 1/3 their length; branches often with stout, naked or leafy, spinelike spurs, these bearing flower-clusters: **4. C. aculeata**

      4. Sepals essentially free, united only at the base; branches always unarmed: **5. C. odorata**

1. Casearia sylvestris Sw., Fl. Ind. Occ. 2:752. 1798.

An unarmed shrub or small tree, often with long, slender, puberulous branches; leaves lance-oblong, 5–10 cm long, long-acuminate at apex, glossy green and densely pellucid-dotted; stipules roundish, 1–1.5 mm long. Flowers in dense, sessile, axillary clusters, yellow-green or whitish; pedicels mostly less than 3 mm long or often nearly obsolete. Calyx minutely puberulous and ciliolate. Stamens 10, free; staminodes hairy. Fruit subglobose, 3–5 mm in diameter, red, 3-valved, 2–6-seeded; seeds flattened-ellipsoidal, 2 mm long.

GRAND CAYMAN: *Proctor 15011.*
– West Indies and continental tropical America. The Grand Cayman plants were found to be rare in dry rocky woodland between North Side and Forest Glen.

2. Casearia guianensis (Aubl.) Urban, Symb. Ant. 3:322. 1902.

An unarmed shrub, sometimes arborescent; leaves obovate or elliptic, up to 18 cm long, short-acuminate at the apex, the margins distantly and minutely glandular-denticulate, the tissue with numerous pellucid dots and lines; stipules linear-subulate, 2–5 mm long. Flowers in loose or rather dense axillary clusters, greenish-white or cream; pedicels 5–7 mm long, finely appressed brownish-hairy; calyx 4–5 mm long, hairy like the pedicel. Stamens 8; staminodes hairy. Stigma capitate. Fruit subglobose or ellipsoid, 6–12 mm in diameter, white or greenish, sometimes flushed with red, obtusely 6-angled, 3–10-seeded; seeds subovoid, 3–3.5 mm long.

CAYMAN BRAC: *Proctor 29116.*
– West Indies and continental tropical America; the Cayman Brac specimen was collected in rocky woodland near the base of cliffs at Spot Bay.

3. Casearia hirsuta Sw., Fl. Ind. Occ. 2:755. 1798.

An unarmed shrub, the young branches velvety brownish-pubescent; leaves narrowly obovate to oblong-elliptic or elliptic, 5–12 cm long, the apex blunt to very shortly acuminate, the margins distantly and minutely glandular-denticulate, the tissue with numerous pellucid dots but scarcely any lines, finely puberulous on the upper side, beneath velvety brownish-pubescent; stipules linear-lanceolate, 3 mm long. Flowers in axillary clusters or at leafless nodes, greenish-white, fragrant; pedicels 4–6 mm long, spreading or bristly-puberulous; calyx 4–5 mm long, hairy like the pedicel. Stamens 8 or 10; staminodes hairy. Fruit ovoid, 10–15 in diameter, green or yellow-green, 3-angled; seeds oblong-conic, 2.5–3 mm long, minutely pitted.

GRAND CAYMAN: *Proctor 15012, 15083.* CAYMAN BRAC: *Proctor 15318, 28966.*
– Cuba, Hispaniola,* Jamaica, and northern South America, in rocky thickets and woodlands. This and the preceding species are sometimes called "Wild Coffee".

---

* Haiti and Dominican Republic.

4. Casearia aculeata Jacq., Enum. Pl. Carib. 21. 1760.

"Thom Prickle"

A shrub up to 3 m tall or more, or frequently wand-like and few-branched, or semi-scrambling, often beset with sharp, spine-like, woody spurs 1.5-3 cm long, these representing modified branchlets and usually bearing 1 or 2 leaves and a cluster of flowers, or even developing tiny secondary branchlets, the youngest parts very minutely puberulous. Leaves narrowly or broadly elliptic or obovate, mostly 1.5-5 cm long, blunt to acute at the apex, the tissue with numerous pellucid dots and lines; stipules triangular-acuminate, 1 mm long. Flowers cream, in clusters often on spur-like branchlets; pedicels puberulous, jointed about the middle or below; calyx 4-5 mm long. Stamens 8; staminodes hairy. Fruit subglobose, obtusely 3-angled, 6-12 mm long.

GRAND CAYMAN: *Brunt 1898, 2087, 2163; Proctor 27990.*
– West Indies and continental tropical America, in dry thickets and along fencerows.

5. Casearia odorata Macf., Fl. Jam. 1:215. 1837.

An unarmed shrub up to 4 m tall, sometimes with straggling branches, rarely tree-like; leaves elliptic or obovate, 2-9 cm long, the apex obtuse or acutish, the tissue with numerous pellucid dots but scarcely any lines; stipules lance-subulate, 0.5-1 mm long. Flowers cream, very fragrant, in axillary clusters; pedicels appressed-puberulous, jointed about the middle or below; calyx 4-5 mm long. Stamens 8; staminodes hairy. Fruit globose, obscurely angled, 8-10 mm long.

GRAND CAYMAN: *Kings GC 359; Proctor 27931.*
– Jamaica. It is possible that this and the preceding are really only forms of a single variable species.

Genus 2. **ZUELANIA** A. Rich.

Unarmed deciduous trees or shrubs, usually flowering when leafless or just as the new leaves appear; young parts softly brownish-puberulous. Leaves petiolate, usually inequilateral at the base, with crenulate or crenate margins and numerous pellucid dots. Flowers perfect, in dense terminal clusters, the flowering shoots prolonged by new growth after flowering so that the fruits are lateral on the stems below the new leaves; pedicels jointed, with bracts at the base; calyx of 4-5 overlapping lobes; petals lacking. Stamens 20-40, alternating with an equal number of staminodes. Ovary superior, free; stigma sessile or subsessile, peltate; ovules numerous on 3 placentas. Fruit a large fleshy berry-like capsule, at length splitting open by 3 valves; seeds numerous, arillate.

A chiefly West Indian genus of 2 or 3 species, closely related to *Casearia.*

1. Zuelania guidonia (Sw.) Britton & Millsp., Bahama Fl. 285. 1920. **FIG. 112**

"Jeremiah-bush"

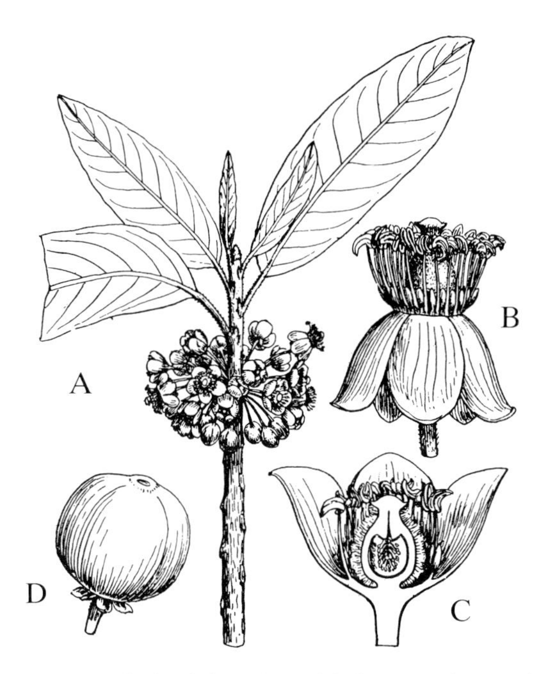

FIG. 112 **Zuelania guidonia.** A, branchlet with flowers and young leaves, x⅔. B, flower, x3. C, opening flower cut lengthwise, x3. D, fruit, x½. (F. & R.)

A shrub or small tree to 10 m tall or more; leaves lance-oblong or elliptic, 5-20 cm long, subacuminate at the apex, obtuse to subcordate at base, the margins subentire to crenate, mostly glabrescent on the upper surface, densely soft-puberulous beneath. Flowers with puberulous pedicels 6-10 mm long; calyx whitish or greenish, 6-7 mm long. Ovary tomentose. Fruit subglobose, green and juicy, 3-5 cm in diameter; seeds 5 mm long.

GRAND CAYMAN: *Kings GC 143; Proctor 15251.* LITTLE CAYMAN: *Proctor 35213.* CAYMAN BRAC: *Proctor 29342.*
– Bahamas, Cuba, Hispaniola, Jamaica and Central America, in rocky woodlands. The wood is sometimes used for construction.

Genus 3. **XYLOSMA** Forster

Shrubs or trees, often armed with simple, straight spines in the leaf-axils or branched ones on the trunk. Leaves toothed or sometimes entire; stipules lacking. Flowers unisexual and dioecious, in small axillary clusters or short racemes. Sepals 4 or 5, usually small, scale-like, and ciliate; petals lacking. Male flowers with several to many stamens, these often surrounded by a glandular disk or mixed with glandular staminodes. Female flowers with ovary inserted on a glandular disk; ovules 2 on each of 2 (rarely 3–6) parietal placentas; style short, entire or divided, or rarely the stigmas subsessile and peltate-lobed. Fruit a small berry with 2–8 seeds obovoid, smooth and hard.

A pantropical genus of perhaps 65 species, many of them variable and difficult to distinguish.

1. **Xylosma bahamense** (Britton) Standl. in Tropical Woods 34:41. 1933.

"Shake Hand"

An intricately-branched shrub or small tree to 6 m tall, the trunk and often the larger limbs densely armed with branched spines up to 5 cm long, the leafy twigs with or without simple spines. Leaves mostly ovate or broadly elliptic, 6–15 (-30) mm long, entire or with 1-4 blunt teeth, the apex acute, sometimes mucronate; petiole c. 1 mm long. Fruit obovoid-oblong, 6 mm long.

GRAND CAYMAN: *Kings GC 332; Proctor 15138, 15183.*
– Otherwise known only from the northern Bahamas.

Family 44

TURNERACEAE

Herbs or shrubs, rarely trees, glabrous or pubescent, the hairs simple or branched; leaves alternate, simple, usually serrate, often 2-glandular at the base; stipules small or lacking. Flowers perfect, regular, axillary, solitary or few, sessile or stalked, rarely in racemes; peduncles free or united with the petiole, often jointed and 2-bracteolate. Calyx tubular, 5-lobed, soon deciduous. Petals 5, often yellow, clawed and inserted in the throat of the calyx, sometimes bearing a fimbriate scale at the apex of the claw. Stamens 5, inserted at the base, middle, or throat of the calyx tube, rarely hypogynous; filaments free; anthers oblong, opening inwardly. Ovary superior, free, 1-celled; styles 3, terminal, thread-like, simple or divided; stigmas brush-like or rarely merely dilated; ovules numerous, 2-seriate on 3 parietal placentas. Fruit a capsule splitting partially or completely by 3 valves; seeds oblong-cylindric, slightly curved, arillate, with hard pitted coat; endosperm abundant; cotyledons plano-convex.

A pantropical family of 8 genera and about 150 species, the majority American.

Genus 1. **TURNERA** L.

Herbs or low shrubs, glabrous or pubescent with simple hairs; leaves serrate or rarely entire, often 2-glandular at the base. Flowers solitary in the axils, small or large, the peduncles usually adnate to the petioles, each with 2 bracteoles. Calyx with short tube and oblong, linear or lanceolate lobes; petals usually yellow, obovate or spatulate. Stamens inserted below the petals, sometimes hypogynous. Ovary sessile; styles simple, the stigmas brush-like or fanlike. Capsule splitting to the base, usually many-seeded; aril unilateral.

A chiefly tropical American genus of more than 60 species.

KEY TO SPECIES

1. Leaves more than 2 cm long (up to 8 cm or more), with 2 protuberant glands at the base of the blade; calyx 1 cm long or more: **1. T. ulmifolia**

1. Leaves less than 1.5 cm long, without protuberant glands at the base of the blade; calyx 0.4–0.7 cm long: **2. T. diffusa**

1. Turnera ulmifolia L., Sp. Pl. 1:271. 1753. **FIG. 113.**

Turnera ulmifolia var. *angustifolia* (Mill.) Willd., 1979.

Turnera triglandulosa Millsp. in Field Mus. Bot. 2:77. 1900.

"Cat-bush"

An erect bushy herb or short-lived shrub up to c. 1 m tall, the younger parts and leaves finely appressed-puberulous. Leaves narrowly to broadly lanceolate, mostly 2–8 cm long, acute or acuminate at the apex, the margins irregularly crenate-serrate; basal glands 0.7–0.8 mm in diameter. Peduncles almost completely united to the petioles; bracteoles leaf-like, 1–3 cm long. Calyx-tube c. 1 cm long; petals bright yellow, 2–3 cm long or sometimes more. Capsule 6–9 mm long; seeds c. 2.5 mm long, densely pitted.

GRAND CAYMAN: *Brunt 1849, 2079; Hitchcock; Kings GC 391; Lewis 3847; Millspaugh 1260; Proctor 15086.* LITTLE CAYMAN: *Kings LC 102; Proctor 28049.* CAYMAN BRAC: *Kings CB 25; Millspaugh 1152* (type of *T. triglandulosa*), *1195, 1209; Proctor 29031.*

– West Indies and continental tropical America; introduced in the Old World tropics. A somewhat weedy, variable species, but quite handsome in flower.

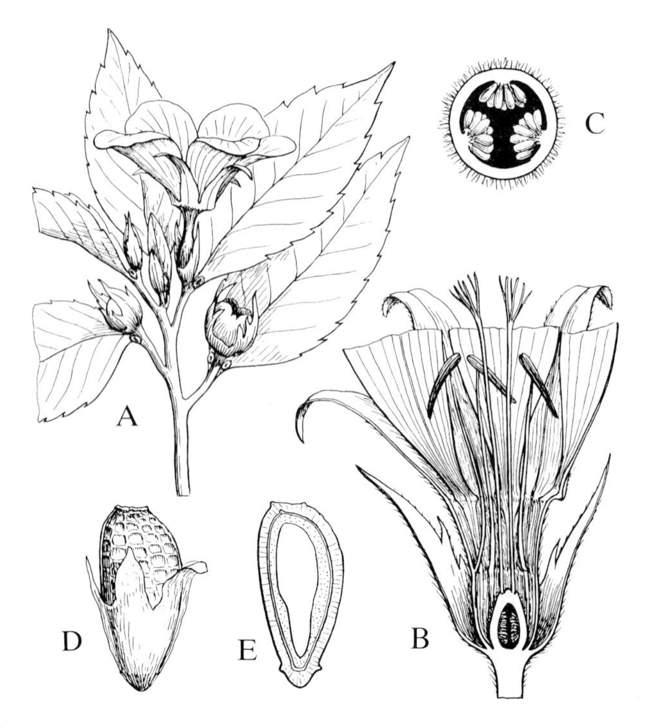

FIG. 113 **Turnera ulmifolia.** A, branchlet with flower and fruit, x$^{2}/_{3}$. B, flower cut lengthwise (petals cut), x2. C, x-section of ovary, x6. D, seed with aril, x10. E, seed cut lengthwise, x10. (F. & R.)

2. **Turnera diffusa** Willd. ex Schult. in L., Syst. Veg. 6:679. 1820.

A low shrub, often densely branched, less than 1 m tall; younger parts and underside of leaves densely puberulous. Leaves oblanceolate or obovate, 6–15 mm long, coarsely crenate-dentate, the nerves grooved on the upper side and prominant beneath, the underside minutely and densely glandular all over the surface as well as puberulous. Peduncles very short or obsolete; bracteoles linear-subulate, 2–4 mm long. Calyx 5-toothed, 3–4 mm long; petals bright yellow, up to 8 mm long. Capsule 3–4 mm long; seeds 1.5–2 mm long, reticulate-striate.

LITTLE CAYMAN: *Proctor 28113, 35127.*
– West Indies and continental tropical America, usually in arid rocky scrublands. In some countries the leaves are used to make an aromatic tea, which is reputed to be an aphrodisiac.

Family 45

PASSIFLORACEAE

Vines, trailing or climbing by means of axillary unbranched tendrils, herbaceous or woody, or rarely erect shrubs or trees. Leaves alternate, simple, entire or lobed, or very rarely compound, the petiole usually bearing glands; stipules 2, thread-like or large and ovate. Flowers axillary, perfect or rarely unisexual, solitary or variously cymose-racemose or paniculate; peduncles usually jointed; bract and bracteoles 3, small and distant from the flower, or large, foliaceous and close to the flower, forming an involucre. Calyx-tube (receptacle) almost flat to saucer-like or bell-shaped, giving rise in the middle to a gynophore (ovary-stalk), and nearly always bearing toward the margin or within its throat a ring of one or several series of erect or radiate filamentous or membranous outgrowths (called the corona); sepals usually 5, inserted in the throat of the receptacle, overlapping, often with a small horn-like process on the back near the apex, and usually coloured inside. Petals either lacking or usually 5, alternate with the sepals, free, overlapping, and withering-persistent. Stamens usually 5, attached below the ovary on the gynophore. Ovary superior and free, located at the apex of the gynophore, 1-celled; style simple or 3-4-branched, or else 3 separate styles present; ovules numerous, attached to 3 or 5 parietal placentas. Fruit berry-like or sometimes a capsule opening by 3 apical valves; seeds numerous, usually compressed-ovoid, covered with a fleshy aril or surrounded by pulp; endosperm fleshy; cotyledons often foliaceous.

A widespread tropical and subtropical family of about 12 genera and 600 species, especially concentrated in South America. Most of the species belong to the single genus *Passiflora;* several of these produce edible fruits known by such names as "Passion-fruit", "Granadilla", "Sweet-cup", and "Golden Apple".

Genus 1. **PASSIFLORA** L.

Vines with characters as given for the family; receptacle shorter than the rest of the flower; corona prominent and often brightly coloured, its filaments distinct or more or less united. Sepals and petals 4 or 5, or petals sometimes lacking. Stamens 4 or 5, the filaments joined to the gynophore below, free above. Styles 3; stigmas capitate. Fruit a dry or pulpy berry.

A largely American genus of more than 500 species. These are easily recognized by their complicated and rather curious flowers, which to the early Spanish explorers suggested some of the emblems associated with the Crucifixion, hence the Spanish name "Pasionaria", the English name "Passion-flower", and the Latin term *Passiflora.*

FIG. 114 **Passiflora suberosa.** A, habit. B, C, D, E, F, various forms of leaf. G, flower. H, stamen. I, pistil. J, fruit. **Passiflora cupraea.** K, flower. L, two stamens. M, fruit. (G.)

KEY TO SPECIES

1. Flowers large (6–10 cm across) and intensely fragrant, subtended by an involucre of 3 large segments; fruit ellipsoidal, 7–8 cm long: **[1. P. laurifolia]**

1. Flowers smaller, not noticeably fragrant, and lacking an involucre; fruit globose, less than 2 cm in diameter:

  2. Leaves entire or commonly 3-lobed, the apex more or less acute; flowers greenish or cream; petals lacking: **2. P. suberosa**

  2. Leaves always entire, the apex usually notched and bearing a small bristle in the notch; flowers deep red-purple; petals present: **3. P. cupraea**

[**1. Passiflora laurifolia** L., Sp. Pl. 2:956. 1753.

"Golden Apple", "Water Lemon"

High-climbing vine, the stems grooved-striate; leaves glabrous, elliptic or broadly oblong, 6–14 cm long, 1-nerved, the apex abruptly pointed, the base rounded or subcordate; petioles with 2 glands near the apex; stipules linear, 6–9 mm long. Flowers solitary on peduncles longer than the petioles; segments of the involucre 3.5–4 cm long, leafy, not united, with margins usually glandular-dentate. Sepals pale green, horned below the apex; petals shorter and narrower than the sepals, cream with minute red speckles; corona in several unlike series of various lengths, bright purple with white transverse bands toward the base. Fruit narrowly ellipsoid, 7–8 cm long, orange-yellow, edible.

GRAND CAYMAN: *Kings GC 293.* This was probably gathered from a cultivated plant.
– West Indies and northern South America, indigenous and cultivated.]

2. Passiflora suberosa L., Sp. Pl. 2:958. 1753. **FIG. 114.**

Passiflora pallida L., 1753.

Passiflora minima L., 1753.

Passiflora angustifolia Sw., 1788.

"Wild Pumpkin"

A small, chiefly herbaceous vine, creeping or climbing, glabrous nearly throughout or sometimes pubescent (at least on the petioles); leaves usually membranous, extremely variable in outline even on the same plant, frequently deeply 3-lobed, but also often entire and narrowly to broadly elliptic, 2–13 cm long, the apex acutish and bristle-tipped, and with 2 short-stalked glands near apex of the petiole. Flowers 1.5–3 cm across, 1 or 2 in the leaf-axils on filamentous peduncles longer than the petioles; involucre lacking. Sepals greenish or cream, 5–8 mm long; petals absent; corona half as long as the sepals, 2-seriate, the outer ring with filaments recurved, yellow at apex, white in the middle, and purple or violet toward base. Berry globose, usually black, 6–15 mm in diameter; seeds 3–4 mm long, abruptly acuminate, coarsely reticulate.

GRAND CAYMAN: *Brunt 2144, 2164; Hitchcock; Kings GC 238; J. Popenoe in 1968 (MO); Proctor 1194.* LITTLE CAYMAN: *Proctor 28057.* CAYMAN BRAC: *Proctor 29010.*

– Widespread and common in the warmer parts of the Western Hemisphere, often in thickets, rocky woodlands, and pastures.

3. **Passiflora cupraea** L., Sp. Pl. 2:955. 1753. **FIG. 114.**

A subwoody, glabrous vine, trailing or high-climbing; leaves entire, ovate to very broadly elliptic or rotund, 3–7 cm long, shallowly notched and bristle-tipped at apex, subcordate at base, reticulate-veined on both sides, and with a few distant, flat, circular glands beneath. Flowers solitary in the axils, usually red-purple, on peduncles longer than the petioles; involucre lacking. Sepals narrowly oblong, c. 2.5 cm long; petals narrower and shorter; corona 1-seriate, yellowish, the filaments linear-oblong, 2–4 mm long. Gynophore 2–3 cm long. Berry globose, dark blue or purple, 1.5–2 cm in diameter; seeds c. 3 mm long, blunt or acute, transversely rugose.

GRAND CAYMAN: *Correll & Correll 51026.* LITTLE CAYMAN: *Proctor 35113.* CAYMAN BRAC: *Proctor 29053.*

– Bahamas, eastern Cuba, and Tortue Island (Haiti), in rocky woodlands, scrublands, and coastal thickets.

[Family 46

CARICACEAE

Coarse or giant unbranched or few-branched herbs, often tree-like or shrubby, with terminal crowns of leaves and milky sap; leaves alternate, usually very large and long-petiolate, palmately-lobed or -foliolate; stipules lacking. Flowers yellow, whitish or greenish, in axillary inflorescences, unisexual or perfect, monoecious, dioecious or polygamous; staminate and polygamous inflorescences pendulous and paniculate, cymose or racemose; pistillate inflorescences short, few-flowered and cymose. Calyx 5-lobed; corolla of 5 petals, in staminate and perfect flowers united below into a long tube, in pistillate flowers almost free. Stamens usually 10, inserted at the mouth of the corolla-tube; filaments with adnate 2-celled anthers. Pistillate flowers without staminodes; ovary free, sessile, 1-celled or imperfectly 5-celled; stigmas sessile or nearly so, 3-5-lobed; ovules numerous, in 2 or more series on 5 placentas. Fruit a large, fleshy berry; seeds numerous, compressed or globose, with fleshy endosperm.

A small family of about 4 genera and 40 species; one genus occurs in tropical Africa, the others all being tropical American.

Genus 1. **CARICA** L.

With the characters of the family, the stems unarmed. Flowers dioecious; stamens inserted in 2 series, free, the outer ones with elongate filaments, the inner ones short, the connective often produced into a ligule beyond the anther-cells; stigmas 5, linear or variously cleft. Fruit filled with pulp, or a large cavity often present; seeds rugose-tuberculate, covered with a succulent membrane.

About 30 species, the majority in South America.

1. **Carica papaya** L., Sp. Pl. 2:1036. 1753. **FIG. 115.**

"Pawpaw"

A giant tree-like herb up to 6 m tall or more, the mostly simple, columnar trunk soft and hollow, and marked with the scars of fallen leaves; leaves up to 1 m long or more, with long hollow petioles; blade mostly 30-60 cm broad, palmately 5-7-lobed, the segments pinnately lobed, more or less glaucous beneath. Flowers with corolla twisted in the bud; staminate flowers 2-3.5 cm long, the pistillate ones 4-5 cm long, creamy or pale yellow. Fruit pendulous, varying greatly in size, shape and colour, up to 50 cm long, edible; seeds black, 6-7 mm long.

GRAND CAYMAN: *Crosby, Hespenheide & Anderson 28 (GH); Proctor (sight record).* LITTLE CAYMAN: *Kings LC 80.* CAYMAN BRAC: *Proctor 29340.* – Native of tropical America, but the precise extent of its original range is unknown; cultivated and becoming naturalized in almost all tropical countries. The milky sap of this and related species contains a substance called papain, which is often used to tenderize meat. The ripe Pawpaw or Papaya is one of the most delicious of all tropical fruits.]

Family 47

CUCURBITACEAE

Climbing or creeping herbs of rapid growth and with abundant watery sap, annual or perennial, usually with lateral tendrils which are interpreted as modified stipules; leaves alternate, simple or palmately lobed or divided, often cordate and of

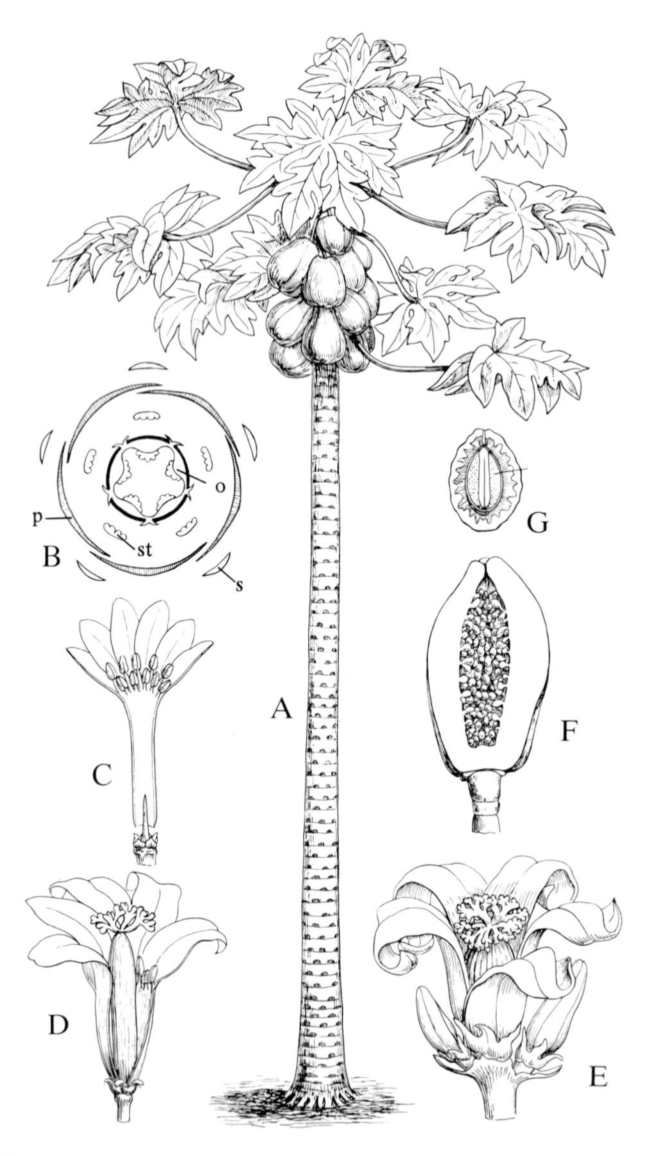

FIG. 115 **Carica papaya.** A, pistillate (fruiting) plant, much reduced. B, floral diagram, perfect flower; s, sepal; p, petal; st, stamen; o, ovary. C, staminate flower cut open, x1½. D, perfect flower cut open, x1. E, pistillate flower cut open, x1. F, young fruit cut lengthwise, much reduced. G, seed cut lengthwise, x2; e, endosperm. (F. & R.)

membranous texture, and usually with long petioles. Flowers regular, axillary, solitary or in racemes, cymes or panicles, unisexual, monoecious or rarely dioecious. Calyx of 5 (rarely of 3 or 6) sepals united below into a bell-shaped tube. Petals 5 (rarely 3 or 6), free or united. Stamens 3 (rarely 5) of which one has a 1-celled anther, the others 2-celled; filaments short; anthers free or cohering into a head, often contorted. Ovary inferior (or in *Sechium* with the apex free), 3-celled, each cell representing a carpel, the placentas meeting in the axis; style terminal, simple or lobed; ovules few to numerous. Fruit usually a fleshy or corky berry (rarely a capsule), usually indehiscent (but in *Momordica* opening by valves); seeds 1 or few to many, compressed, without endosperm, the cotyledons foliaceous.

A largely pantropical family of more than 100 genera and perhaps 850 species. Many produce edible fruits, but a few are poisonous.

KEY TO GENERA

1. Leaves nearly entire, or else angled or shallowly 3–5-lobed:

  2. Tendrils simple; fruit a small berry less than 1.5 cm long: **5. Melothria**

  2. Tendrils branched; fruit much larger:

    3. Perennial with high-climbing stems; flowers clustered, greenish or white; fruits 1-seeded: **[3. Sechium]**

    3. Annual with trailing stems; flowers solitary, yellow; fruits many-seeded: **[4. Cucurbita]**

1. Leaves deeply 5 to many-lobed:

  4. Tendrils simple; fruit prickly or tuberculate:

    5. Fruits indehiscent, prickly, with greenish or yellowish pulp, not bitter: **[6. Cucumis]**

    5. Fruits splitting open by 3 valves, tuberculate, with crimson pulp, very bitter: **[1. Momordica]**

  4. Tendrils branched; fruits smooth: **[2. Citrullus]**

[Genus 1. **MOMORDICA** L.

Slender herbaceous vines, annual or perennial, with simple tendrils; leaves lobed or in some species entire. Peduncles often with a small leafy bract. Flowers yellow, monoecious or dioecious. Staminate flowers solitary or in panicles; calyx-tube short, closed by 2 or 3 oblong, incurved scales; petals 5, free or nearly so; stamens 3, the anthers at first coherent, at length free, contorted. Pistillate flowers solitary, with calyx and corolla like those of the staminate flowers; staminodes none or represented by 3 glands at the base of the style; stigmas 3; ovules numerous. Berry opening by 3 valves or indehiscent.

A genus of 45 species indigenous to the Old World tropics; one or two are naturalized in nearly all warm countries.

1. **Momordica charantia** L., Sp. Pl. 2:1009. 1753.

"Serasee", sometimes spelled "Cerasee"

Tender climber with stems 1-3 mm thick; leaves 2-7 cm long and wide, deeply palmately lobed into 5-7 obovate, irregularly toothed lobes, glabrous or nearly so except on the veins. Staminate flowers solitary on long filiform peduncles 5-15 cm long, with a roundish green bract near the middle. Pistillate flowers on short peduncles 2-4 cm long, with a bract near the base. Fruit bright orange, 4-6 cm long (var. *charantia*) or 1.5-3 cm long (var. *abbreviata* Ser. ex DC., 1828)[1], ovoid or spindle-shaped, tuberculate, when ripe splitting by 3 valves from the apex; seeds embedded in crimson pulp.

GRAND CAYMAN: *Brunt 1717, 1897a; Hitchcock; Kings GC 223; Maggs II 63; Millspaugh 1329; Proctor 15078; Sachet 381.* CAYMAN BRAC: *Proctor 29122.* All the Cayman specimens appear to represent var. *abbreviata.*

– Now occurring in all warm countries, cultivated and naturalized. Both the leaves and the fruits are believed by many people to be of medicinal value, perhaps because they are so bitter.]

[Genus 2. **CITRULLUS** Schrad.

Citrullus lanatus (Thumb.) Matsumura & Nakai *(C. vulgaris* Schrad.) the Watermelon, is cultivated and often found growing as an escape in open waste ground.

GRAND CAYMAN: *Brunt 2129; Proctor 27960.*

– Native of tropical and southern Africa.]

---

[1] *M. charantia* var. *zeylanica* (Mill.) Hitchc., 1893.

[Genus 3. **SECHIUM** Juss.

Sechium edule (Jacq.) Sw., the Chocho, is sometimes cultivated. It is native to continental tropical America, where it is usually called "Chayote", and is cultivated and seminaturalized in many tropical countries. Both the fruits and the root are edible, the latter resembling a yam when cooked.]

[Genus 4. **CUCURBITA** L.

Cucurbita pepo L., the Pumpkin, is cultivated and has been observed as an escape, but scarcely persisting.

GRAND CAYMAN: *Proctor (sight record).* CAYMAN BRAC: *Proctor 29347.* – Native of continental tropical America, and almost worldwide in cultivation. Many varieties occur.]

Genus 5. **MELOTHRIA** L.

Slender herbaceous vines with simple tendrils; leaves lobed, angular, or nearly entire. Plants monoecious in the Cayman species; flowers small, yellow or white. Staminate flowers racemose (rarely solitary); calyx 5-toothed; corolla deeply 5-parted; stamens 3. Pistillate flowers solitary or clustered, on long slender peduncles, the calyx and corolla as in the staminate flowers; staminodes 3 or none; ovary constricted below the calyx-tube; placentas 3; style short, surrounded at the base by a ring-like disk; stigmas 3. Berry small, globose or ovoid.

A genus of 10 tropical American species.

1. Melothria pendula L., Sp. Pl. 1:35. 1753. **FIG. 116.**

Melothria guadalupensis (Spreng.) Cogn. in DC., 1881; Adams, 1972.

A tender glabrous climber; leaves broadly ovate-cordate, 2–5 cm long or more, 3–5-angled or shallowly 3–5-lobed, the margins distantly denticulate, the upper (adaxial) surface very rough, the lower less so. Staminate flowers several toward the end of a peduncle, on pedicels 1–2 mm long; calyx-tube 2 mm long; corolla yellow, 2 mm long. Pistillate flowers solitary, sometimes in the same axil as the staminate ones but on much longer peduncles; calyx and corolla similar to the staminate ones; ovary 4–5 mm long. Fruit purplish-black, ovoid, 1–1.5 cm long; seeds flat, 4 mm long.

GRAND CAYMAN: *Proctor 27984.*
– West Indies and continental tropical America, variable, occurring in rocky thickets and woodlands.

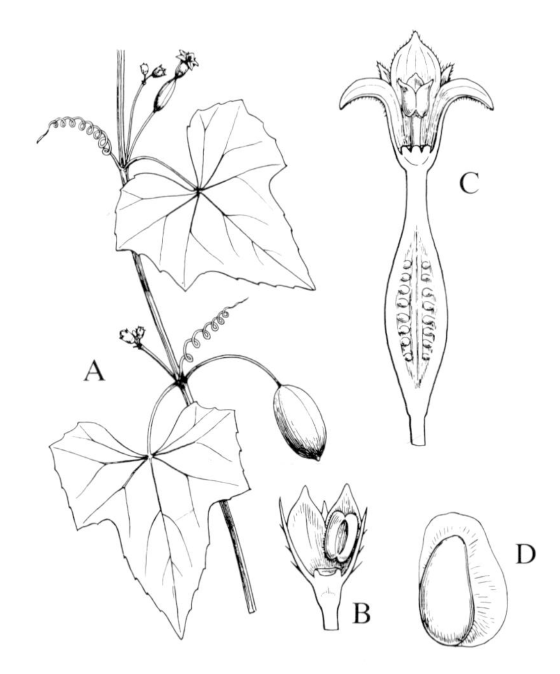

FIG. 116 **Melothria pendula.** A, portion of branch with leaves, flowers and fruit, x$^{2}/_{3}$. B, staminate flower of a related species *(M. cucumis)* cut lengthwise, one stamen removed, x5. C, pistillate flower of same, cut lengthwise, x3. D, seed, x4. (F. & R.)

[Genus 6. **CUCUMIS** L.

Annual or perennial herbs, usually trailing, or sometimes climbing by means of simple tendrils; leaves deeply lobed. Plants usually monoecious, with yellow flowers. Staminate flowers clustered or rarely solitary, the calyx 5-lobed, the corolla more or less campanulate, 5-parted; stamens 3, with linear or curved anther-cells, the connective prolonged above; ovary represented by a gland. Pistillate flowers solitary, the calyx and corolla as in the staminate flowers; staminodes 3; ovary with 3-5 placentas; stigmas 3-5; ovules numerous. Fruit fleshy and often juicy, of various shapes, indehiscent, many-seeded.

A chiefly African genus of 25 species, the following usually considered native to tropical America, but actually probably introduced at an early date.

1. Cucumis anguria L., Sp. Pl. 2:1011. 1753. **FIG. 117.**

"Wild Cucumber"

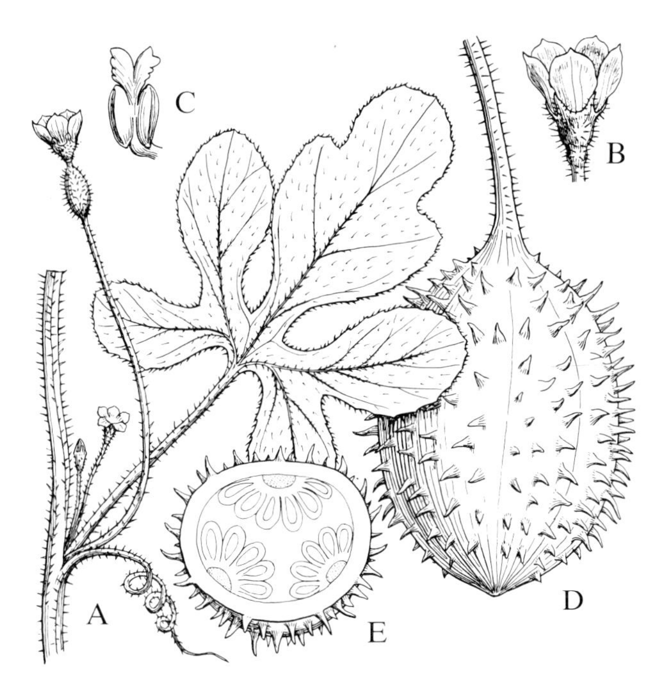

FIG. 117 **Cucumis anguria.** A, portion of branch with leaf, tendril, and staminate and pistillate flowers, x$^{2}/_{3}$. B, staminate flower, x3. C, stamen much enlarged. D, fruit, x$^{2}/_{3}$. E, x-section of same, x$^{2}/_{3}$ (F. & R.)

An annual trailing herb, the angled stems and leaves rough with stiff white hairs; leaves 5-9 cm long, cordate and deeply 5-lobed, with toothed and somewhat wavy margins; petioles equalling the blades or longer. Flowers small, the staminate on peduncles 1-2 cm long and with calyx c. 6 mm long; anthers with a 2-lobed appendage. Pistillate flowers solitary on peduncles 5-10 cm long. Fruit pale yellow, ovoid, 4-7 cm long, more or less prickly, edible; seeds 4-4.5 mm long.

GRAND CAYMAN: *Brunt 2115; Kings GC 417; Proctor 27959.*
– Tropical Africa; West Indies and continental tropical and subtropical America, often a weed of open fields and waste ground. The Cucumber *(Cucumis sativa L.)* and Muskmelon or Cantaloupe *(Cucumis melo* L.) are related species of African origin].

Family 48

CAPPARACEAE

Herbs, shrubs or trees, glabrous or pubescent, sometimes glandular or beset with closely adhering scales; leaves usually alternate, simple or palmately compound; stipules present or absent, sometimes represented by spines. Flowers regular or irregular, perfect, solitary in the leaf-axils or in axillary or terminal racemes or corymbs; sepals 4, free or partly united; petals 4 or rarely none, sessile or clawed. Stamens 4 to numerous, the filaments usually free or sometimes united to the stalk of the ovary. Ovary borne on a stalk (gynophore) or rarely sessile, 1-celled or sometimes apparently several-celled by false septa; style usually short or none, the stigma orbicular; ovules numerous on parietal placentas in 1 to many series, rarely solitary. Fruit a silique-like capsule or a berry, rarely a drupe; seeds with little or no endosperm and curved embryo.

A widely distributed, chiefly tropical family of about 40 genera and 700 species. Few are of any economic value, but Capers, the pickled flower-buds of *Capparis spinosa* L. of the Mediterranean region, are used as a condiment.

KEY TO GENERA

1. Shrubs or trees; fruit a globose to elongate berry, the seeds embedded in pulp: **1. Capparis**

1. Herbs; fruit an elongate capsule, the seeds not embedded in pulp: **2. Cleome**

Genus 1. **CAPPARIS** L.

Shrubs or trees, unarmed or sometimes spiny, glabrous, pubescent, or lepidote. Leaves simple, often of hard or leathery texture; stipules present or absent. Flowers usually several together in axillary and terminal racemes or corymbs; sepals free or partly united, or rarely united in bud and irregularly rupturing, and often with a gland at the base. Petals imbricate, usually white. Stamens numerous, with filiform free filaments. Ovary on a long gynophore, 1-4-celled, with 2 to 6 placentas; ovules numerous; stigmas sessile. Fruit a stalked globose to cylindric berry, indehiscent or rarely irregularly dehiscent; seeds numerous, with convolute embryo.

A widespread, chiefly tropical genus of about 150 species.

KEY TO SPECIES

1. Fruits subglobose; younger parts and underside of leaves covered with minute stellate hairs: **1. C. ferruginea**

1. Fruits elongate; stellate hairs absent:

 2. Younger parts and underside of leaves covered with minute peltate scales; mature fruit dry within, rupturing irregularly: **2. C. cynophallophora**

 2. All parts glabrous and devoid of scales; mature fruits with seeds embedded in red pulp, and rupturing by two valves (like a legume pod): **3. C. flexuosa**

1. **Capparis ferruginea** L., Syst. Nat. ed. 10, 2:1071. 1759.

"Devil Head"

Shrub or small tree to 5 m tall or more, the young branches densely clothed with minute rusty stellate hairs; leaves oblanceolate or narrowly elliptic, mostly 3-8 cm long, sharply acute or subacuminate at the apex, glabrous above, beneath densely clothed with minute whitish stellate hairs. Flowers in pedunculate axillary corymbs; calyx 2-2.5 mm long, deeply cleft into narrow segments; petals c. 5 mm long; stamens usually 8, about as long as the petals. Fruit subglobose, c. 1.5 cm long, covered with minute rusty stellate hairs, and rupturing irregularly at maturity.

GRAND CAYMAN: *Brunt 1786, 2134; Kings GC 240; Proctor 15049.*
– Cuba, Hispaniola and Jamaica, in dry rocky woodlands and coastal thickets.

2. **Capparis cynophallophora** L., Sp. Pl. 1:504. 1753. **FIG. 118.**

Capparis longifolia Sw., 1788 (juvenile form).

"Headache Bush"

Shrub or small tree to 6 m tall or more, the young branches covered with minute, closely adhering peltate scales. Leaves oblong, lance-oblong, or elliptic (or in some juvenile forms narrowly linear), mostly 3-8 (-12) cm long, the apex shallowly notched, rounded, or sometimes acutish, the upper surface naked and dark shining green, the underside pale and densely covered with minute, closely adhering, peltate scales. Flowers in loose, few-flowered subterminal corymbs, the peduncles arising from axils of the upper leaves; sepals free, 8-11 mm long; petals 10-14 mm long, scaly outside, at first white, turning purplish. Stamens numbering 20-30, two or three times as long as the petals, the filaments purple. Gynophore as long as the filaments. Fruit torulose, 5-30 cm long.

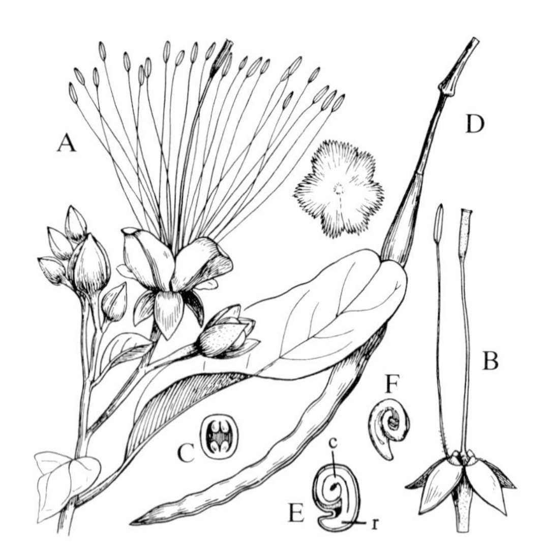

FIG. 118 **Capparis cynophallophora.** A, portion of flowering branch, x²/₃. B, flower with petals and all but one stamen removed, x²/₃. C, x-section of ovary, x5. D, fruit, x²/₃. E, seed cut lengthwise, slightly enlarged; c, cotyledon; r, radicle. F, embryo slightly enlarged. G, scale of leaf, very much enlarged. (F. & R.)

GRAND CAYMAN: *Brunt 1757, 1784, 2002; Kings GC 142; Proctor 15125.* LITTLE CAYMAN: *Proctor 35214, 35218.* CAYMAN BRAC: *Proctor 29138.* – Florida, West Indies, and continental tropical America, in rocky thickets and woodlands.

3. **Capparis flexuosa** (L.) L., Sp. Pl. ed. 2, 1:722. 1762.

"Raw Bones", "Raw Head", "Bloody Head"

A shrub or small tree, sometimes scrambling or with arching or flexuous branches, entirely glabrous and devoid of scales. Leaves oblong, lance-oblong, or elliptic, 4–9 cm long, blunt or often notched at the apex, an oblong gland in each axil. Flowers nocturnal, large; sepals 6–10 mm long; petals white or pale greenish, 1.5–3 cm long; stamens numerous, the white filaments 4–6 cm long. Ovary 0.6–1 cm long, on an elongate gynophore. Fruit red, 10–20 cm long, cylindric or torulose, with scarlet pulp and several to numerous large white seeds.

GRAND CAYMAN: *Brunt 1667, 1700, 2044; Correll & Correll 51004; Kings GC 289; Proctor 15126.* LITTLE CAYMAN: *Kings LC 109; Proctor 28044.* CAYMAN BRAC: *Kings CB 16; Proctor 29006.*
– Florida, West Indies and continental tropical America, in rocky thickets and woodlands.

Genus 2. CLEOME L.

Herbs, sometimes somewhat shrubby, rarely scandent, glabrous or pubescent; leaves simple or palmately 3-7-foliolate, the leaflets entire or serrulate. Flowers solitary in the upper axils, forming a more or less leafy raceme; calyx 4-toothed or 4-parted, persistent or deciduous; petals subequal, sessile or clawed. Stamens usually 6, rarely 4 or 12-20, all or only 2 bearing anthers; filaments usually unequal and declinate. Ovary sessile or at the apex of a gynophore; style very short or the stigma sessile; ovules numerous. Fruit a more or less elongate 1-celled capsule with membranous valves; seeds more or less reniform, usually rough or pubescent.

A pantropical genus of about 75 species.

KEY TO SPECIES

1. Leaves simple, linear-lanceolate; plant perennial, decumbent from the apex of a woody taproot: **1. C. procumbens**

1. Leaves palmately compound; plants annual, erect:

  2. Stems armed with stipular spines; flowers white or pale rose; stamens 4-6: **2. C. spinosa**

  2. Stems unarmed; flowers yellow; stamens 12-20: **3. C. viscosa**

1. Cleome procumbens Jacq., Sel. Stirp. Amer. 189, t. 120. 1763.

A low, more or less decumbent herb arising from a woody perennial taproot, glabrous throughout; leaves simple, linear-lanceolate, 0.5-2 cm long. Flowers on slender pedicels less than 1 cm long; sepals free, unequal, up to 3 mm long; petals yellow, c. 5 mm long. Capsule narrowly cylindric, usually 15-20 mm long, the apex apiculate with the persistent 1.5-2 mm style.

GRAND CAYMAN: *Brunt 2120; Correll & Correll 51036; Proctor 27937; Sachet 411.*
– Cuba, Hispaniola, and Jamaica, in seasonally wet savannas and pastures.

2. Cleome spinosa Jacq., Enum. Pl. Carib. 26. 1760. **FIG. 119.**

"Cat's Whiskers"

A coarse, erect herb up to 1 m tall or more, usually glandular-puberulous throughout, the stems armed with pairs of sharp, yellowish stipular spines. Leaves

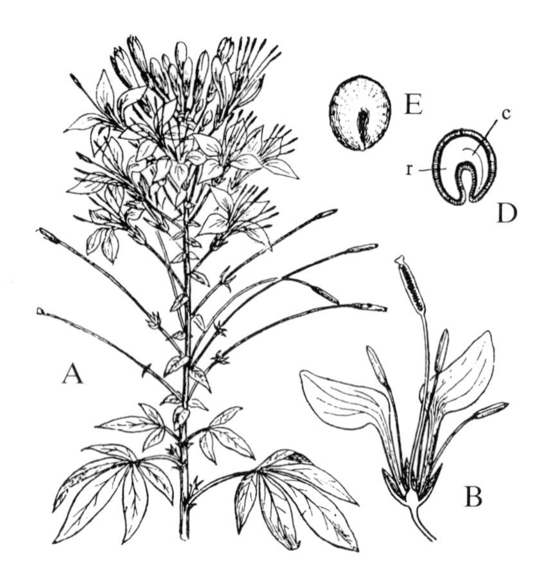

FIG. 119 **Cleome spinosa.** A, portion of flowering branch. x½. B, flower cut lengthwise, slightly enlarged. C, seed x5. D, seed cut lengthwise, x5; c, cotyledons; r, radicle. (F. & R.)

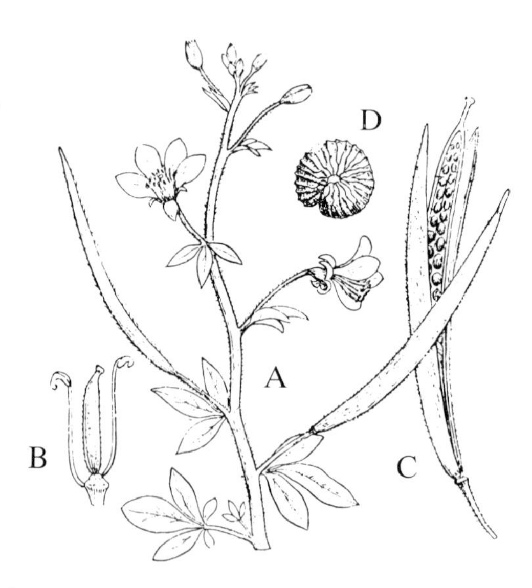

FIG. 120 **Cleome viscosa.** A, portion of flowering branch, x1. B, flower with sepals, petals, and all but two stamens removed, x2. C, ripe fruit, x1. D, seed, much enlarged. (F. & R.)

palmately (3-) 5-7-foliolate, the leaflets 2-9 cm long; petioles 2-11 cm long, usually armed with prickles. Flowers in terminal racemes, each pedicel subtended by a large sessile oval bract; sepals linear, 4-5 mm long; petals white or light purple, long-clawed. Stamens 6, crimson, long-exserted. Gynophore much longer than the pedicel, the ovary glandular, 7-8 mm long, with sessile stigma. Capsule linear-cylindric, 5-12 cm long, glabrous or puberulous; seeds nearly 2 mm in diameter, almost smooth.

CAYMAN BRAC: *Proctor 29093.*
– West Indies and continental tropical and subtropical America, usually a roadside or pasture weed, sometimes cultivated for ornament.

3. Cleome viscosa L., Sp. Pl. 2:672. 1753. **FIG. 120.**

Cleome icosandra L., 1753.

Polanisia viscosa (L.) DC., 1824.

A coarse erect viscid herb to about 1 m tall, glandular-pubescent throughout, the stems unarmed. Leaves 3-5-foliolate, the obovate or elliptic leaflets 1-4 cm long. Flowers solitary in the upper leaf-axils; sepals narrowly oblong, 5-7 mm long, deciduous; petals yellow, c. 10 mm long. Stamens 12-20, usually shorter than the petals. Ovary sessile; style short, elongating in fruit, with capitate stigma. Capsule linear-cylindric, 6-8 cm long, densely glandular-pubescent, the apex terminated by the persistant style; seeds transversely crested.

CAYMAN BRAC: *Proctor 29040.*
– A pantropical weed of open waste ground.

Family 49

CRUCIFERAE

Annual or perennial herbs with watery acrid sap, the pubescence of simple or branched hairs; leaves alternate, simple or compound, the basal ones often forming a rosette; stipules absent. Flowers regular, perfect, in terminal or axillary racemes; sepals 4, free, the inner two sometimes pouch-like at the base; petals 4, rarely absent. Stamens usually 6, with 4 long and 2 short; anthers mostly 2-celled, longitudinally dehiscent. Ovary superior, sessile or rarely stalked, 2-carpellate and usually 2-celled; style simple with 2 stigmas; ovules usually numerous. Fruit a silique (a 2-valved capsule whose 2 parietal placentas persist as a "replum" from which the valves separate) or (in *Cakile*) an indehiscent capsule that is transversely 2-jointed; seeds small, often mucilaginous when wet, frequently winged or marginate, mostly lacking endosperm, or sometimes an oily endosperm present.

A large, worldwide, but chiefly temperate-climate family of about 225 genera and more than 2,500 species. There are no poisonous plants in this family, and a number have found wide use as food. Among these are the turnip (*Brassica campestris* L.), the cabbage and its several races, such as cauliflower, broccoli, and brussels sprouts (all forms of *Brassica oleracea* L.), and mustard (*Brassica nigra* (L.) Koch).

Genus 1. **CAKILE** Mill.

Succulent, glabrous, annual or biennial herbs, the branched stems often decumbent; leaves entire to pinnatifid. Flowers in rather long racemes, the pedicels short, becoming thickened in fruit; sepals erect, the outer ones obtuse and somewhat hooded at the apex; petals clawed and white, pink or purple. Ovary cylindric, sessile, articulate near the middle, each joint containing 1 ovule; stigma depressed-capitate, narrower than the style. Capsule indehiscent, consisting of two 1-seeded joints, the upper terminating in a beak and easily separating from the lower; seeds rather large, somewhat rugulose.

A widely-distributed genus of perhaps 4 species, chiefly on seashores.

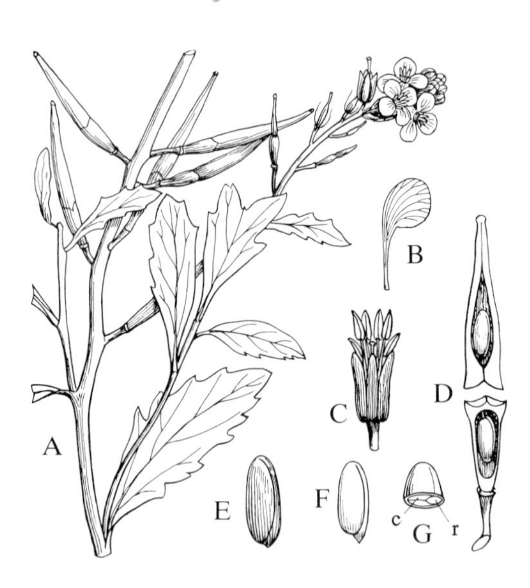

FIG. 121 **Cakile lanceolata.** A, portion of plant, x²/₃ B, petal, x2. C, flower without petals, x2. D, fruit with the joints separated, slightly enlarged. E, seed, x2. F, embryo, x2. G, seed cut across, x2; c, cotyledons; r, radicle. (F. & R.)

1. **Cakile lanceolata** (Willd.) O.E. Schulz in Urban, Symb. Ant. 3:504. 1903. **FIG. 121.**

Cakile maritima of some authors, not Scop.

With characters of the genus; leaves petiolate, linear-oblanceolate to oblong-elliptic, 3-7 cm long, entire, variably toothed or undulate. Petals white, 6-8 mm long. Fruit narrowly spindle-shaped, mostly 1.8-2 (-3) cm long, 2-4 mm thick, the lower joint cylindric, the upper much longer and narrowly conic-subacuminate.

GRAND CAYMAN: *Kings GC 271; Millspaugh 1308; Proctor 15052.* LITTLE CAYMAN: *Kings LC 91.* CAYMAN BRAC: *Kings CB 56; Millspaugh 1159; Proctor 28922.*
– Florida, West Indies, and the Caribbean coasts of Central and South America, on sandy seashores.

[Family 50

MORINGACEAE

Trees with pungent roots, the bark exuding gum; leaves alternate, 3-4-pinnately compound, the divisions opposite, the ultimate leaflets entire; stipules none or reduced to glands. Flowers perfect, irregular, in axillary panicles; sepals 5, unequal, united at base into a short, cup-like tube; petals 5, the upper two smaller. Stamens 10, 5 perfect ones alternating with 5 sterile, all inserted on the edge of a disk lining the calyx-tube; filaments free; anthers 1-celled. Ovary superior, stalked, 1-celled with 3 parietal placentas; style tubular, open at the apex; ovules numerous. Fruit a many-seeded elongate capsule opening by 3 valves; seeds large, 3-winged or wingless; embryo without endosperm.

A small family of 1 genus and 10 species, indigenous to Africa, Arabia and India.

Genus 1. **MORINGA** Adans.

With characters of the family.

1. **Moringa oleifera** Lam., Encycl. Meth. Bot. 1:398. 1785.

"Maronga", "Horseradish Tree"

A small tree to 6 m tall or more, with whitish bark, the branchlets and leaves usually puberulous. Leaves to 40 cm long, with numerous leaflets, these 1-2 cm long.

Flowers numerous, fragrant, the petaloid sepals 9–13 mm long, white tinged with crimson; petals slightly larger and white to yellowish, tinged crimson near the base outside. Capsule obtusely 3-angled, pendent, 20–45 cm long, 1–2 cm thick; seeds broadly 3-winged, 2.5–3 cm long.

GRAND CAYMAN: *Hitchcock; Kings GC 231.*

– Native of eastern Africa but widely planted and naturalized in tropical America. The roots have the flavour of horseradish, for which they can be used as a substitute. The young leaves, pods, and flowers can be cooked as an excellent vegetable. The seeds yield "ben oil", used for lubricating watches and other delicate machinery.]

Family 51

SAPOTACEAE

Trees or shrubs, often with milky sap (latex); pubescence often of 2-branched hairs; leaves mostly alternate, simple, entire and without stipules. Flowers rather small, usually perfect, solitary or clustered in the leaf-axils; sepals 4–12, imbricate; corolla 4–8-lobed but usually the lobes 5, often externally appendaged between the lobes. Stamens inserted on the corolla-tube opposite the lobes and equal to them in number, frequently alternating with more or less petaloid staminodes; filaments distinct; anthers 2-celled. Ovary superior, 4–14-celled; style simple; ovules solitary in each cell. Fruit usually a berry or drupe, sometimes of large size; seeds large, smooth and shining, with a rather conspicuous basal or lateral scar (hilum); endosperm present or lacking, usually scanty; embryo large, with broad foliaceous cotyledons.

A chiefly pantropical family of between 35 and 75 ill-defined genera and about 800 species. Many are important timber trees, while several are valuable for their edible fruits. The latex of several Central American species is an important ingredient of chewing-gum.

KEY TO GENERA

1. Sepals imbricate or spiralled in 1 series; appendages of the corolla-lobes lateral or absent; fruits smooth, not mealy-roughened:

 2. Staminodes present; leaves not persistently silky-hairy beneath:

  3. Fruit less than 3 cm long; seed-scar basal or nearly so:

   4. Leaves minutely inrolled at base on upper side, forming a small pouch; corolla-lobes without lateral appendages: **1. Mastichodendron**

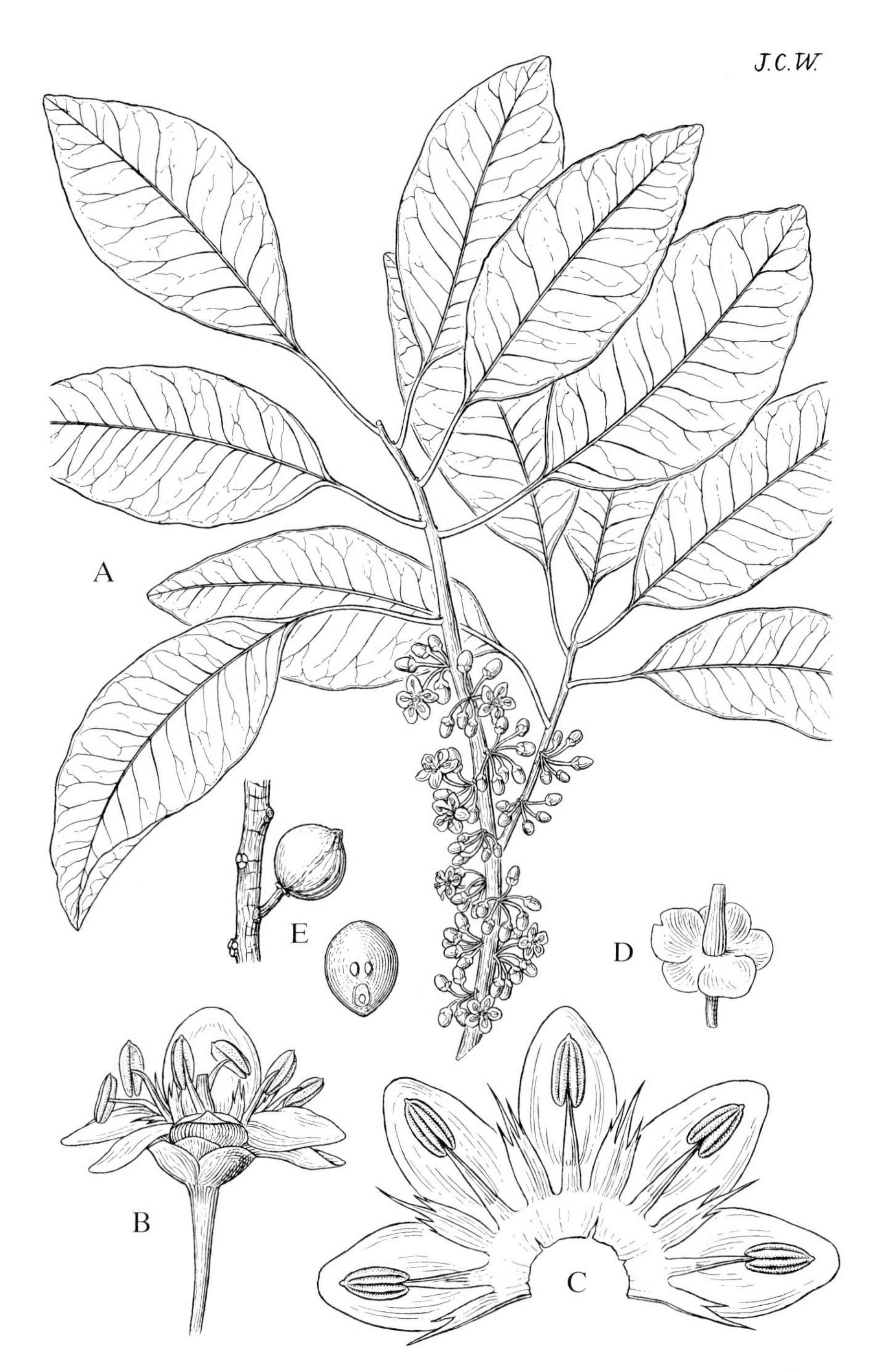

FIG. 122 **Mastichodendron foetidissimum.** A, branch with leaves and flowers, x⅓. B, flower, x4. C, corolla spread out, x6, showing anthers and staminodes. D, calyx and pistil, x4. E, two views of fruit, x⅔, x1. (J.C.W., ex St.)

4. Leaves flat at base on upper side; corolla-lobes with lateral appendages: **2. Bumelia**

3. Fruit over 4 cm long; seed-scar lateral, extending nearly the whole length of the seed: **[3. Pouteria]**

2. Staminodes absent; leaves persistently silky-hairy beneath: **[4. Chrysophyllum]**

1. Sepals in two series (usually 3 + 3); corolla-lobes often with paired dorsal appendages; fruit mealy-roughened, the seeds with lateral scar: **5. Manilkara**

Genus 1. **MASTICHODENDRON** (Engl.) H.J. Lam

Small to rather large trees; leaves alternate or subopposite, the midrib channeled on the upper side and (in the single Cayman species) arising from a small pouch at the inrolled base of the blade. Flowers in dense clusters arising from leaf-axils or at defoliated nodes below the foliage; sepals 5; corolla 5-lobed; stamens and staminodes each 5, the latter not petaloid. Ovary glabrous; ovules attached basilaterally, usually only 1 maturing in each fruit. Fruit somewhat fleshy; seed with nearly basal scar; endosperm copious; embryo vertical.

A tropical American genus of about 8 species; similar Old World species are probably not congeneric.

1. Mastichodendron foetidissimum (Jacq.) H.J. Lam in Rec. Trav. Bot. Neerl. 36:521. 1939. **FIG. 122.**

"Mastic"

A glabrous, medium-sized tree often 10–15 m tall or more; leaves mostly 6–11 cm long, long-petiolate, the blades broadly elliptic or rotund. Flowers strongly scented, cream or yellowish, the sepals 1.3–2.1 mm long, broadly rounded; corolla 3.5–5 mm long, the filaments and lacerate or toothed staminodes somewhat shorter. Fruit yellow, ellipsoid or subglobose, usually 1.5–2.5 cm long (rarely more), with thin but fleshy pulp covering the single large seed.

GRAND CAYMAN: *Proctor 15002.*
– Florida and the West Indies as to typical variety; a related form occurs in Mexico and Central America. This is a species of dry rocky woodlands; its wood is hard, heavy, strong and durable, and is used for construction, boats and furniture. It has become uncommon or rare through much of its range because of over-cutting and the total neglect of replanting.

Genus 2. **BUMELIA** Sw.

Shrubs or trees, unarmed or bearing spines; leaves alternate or sometimes casually opposite. Flowers clustered in the leaf-axils, mostly with 5 sepals and 5 corolla-lobes (but sometimes less or more); corolla-lobes each with a pair of lateral lobes or appendages at the base, or these sometimes lacking. Staminodes petaloid, entire to erose or lacerate. Ovary glabrous or hairy; ovules 5, attached basilaterally. Fruit fleshy, mostly 1-seeded, usually small; seeds with nearly basal scar; endosperm present or absent.

An American genus of about 40 species, occurring in both tropical and warm-temperate areas.

1. Plant an unarmed tree with leaves usually 5-11 cm long: **1. B. salicifolia**

1. Plant a spiny shrub with leaves mostly less than 1.5 cm long: **2. B. glomerata**

1. Bumelia salicifolia (L.) Sw., Nov. Gen. & Sp. Pl. 50. 1788; Fl. Ind. Occ. 1:491. 1797. **FIG. 123.**

Dipholis salicifolia (L.) A.DC., 1844.

"Wild Sapodilla", "White Bullet"

A small unarmed tree to 8 m tall or more, the youngest parts clothed with fine reddish-brown hairs; leaves oblanceolate to narrowly or rather broadly elliptic, 5-11 cm long, with petioles mostly 1 cm long or less. Flowers numerous in clusters at defoliated nodes or sometimes in leaf-axils, on pedicels 1-4 mm long; sepals 1.4-3 mm long, finely hairy; corolla 3.3-4.5 mm long, the lobes narrowed to a claw-like base. Staminodes 1.5-2 mm long, erose-lacerate. Ovary glabrous or slightly hairy. Fruit black, subglobose, 6-10 mm in diameter, 1-seeded or rarely with 2 or 3 seeds.

GRAND CAYMAN: *Brunt 2161; Proctor 27988.* LITTLE CAYMAN: *Proctor 28115, 35173, 35217.* CAYMAN BRAC: *Proctor 15322.*
– Florida, West Indies and Central America, frequent in rocky woodlands. The wood is said to be hard, very heavy, strong, tough, and moderately durable.

2. Bumelia glomerata Griseb. in Mem. Amer. Aca. II, 8:518. 1862. **FIG. 124.**

"Shake Hand", "White Thorn"

A spiny shrub, nearly prostrate or erect and up to 3 m tall or more, the youngest parts minutely reddish-strigillose, otherwise glabrous. Leaves oblanceolate to

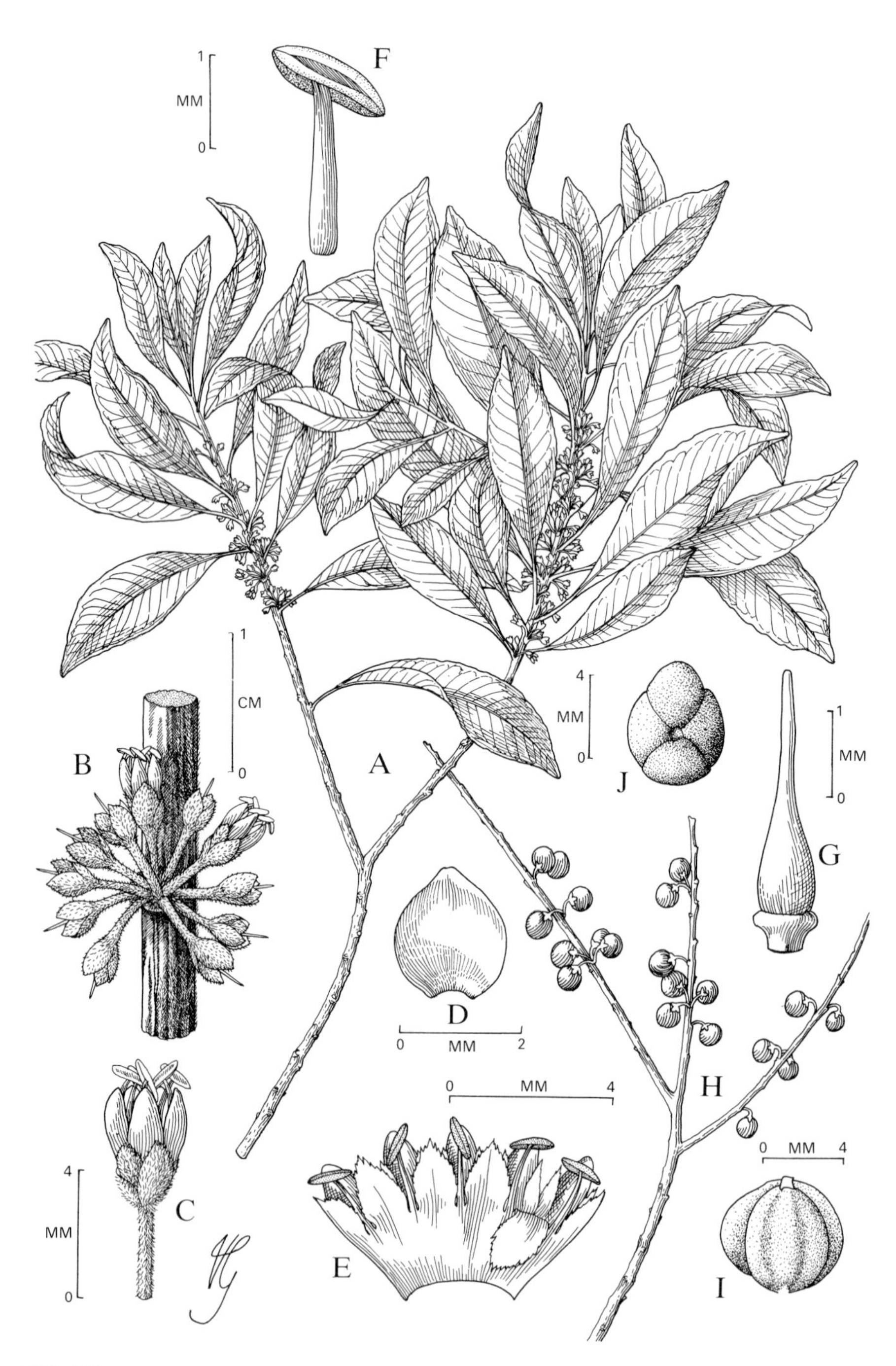

FIG. 123 **Bumelia salicifolia.** A, branch with leaves and flowers. B, single cluster of flowers. C, flower. D, sepal (inner side). E, corolla spread out to show stamens and staminodes; one staminode bent down. F, stamen. G, pistil. H, fruiting branch I, lateral view of fruit. J, basal view of fruit. (G.)

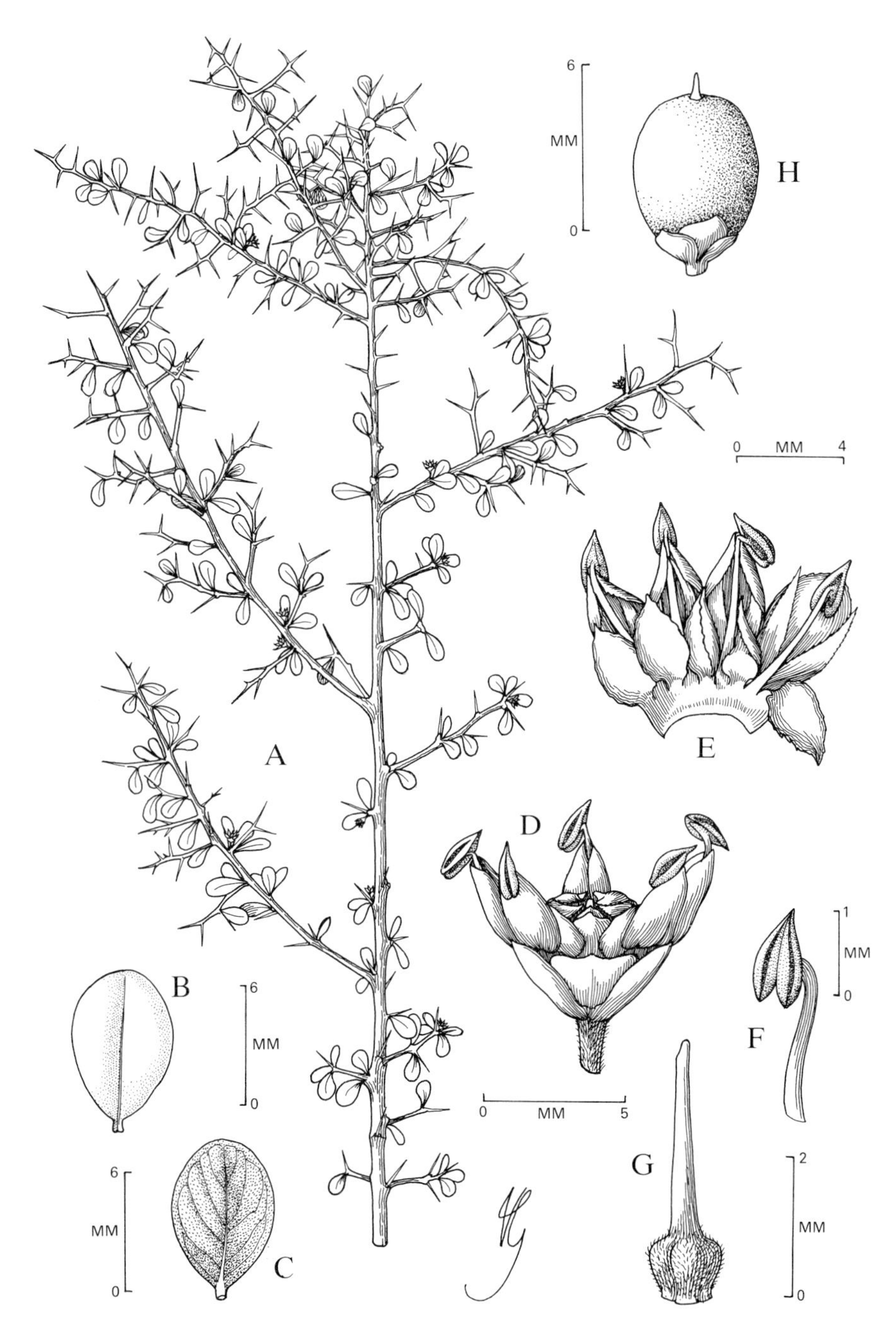

FIG. 124 **Bumelia glomerata.** A, branch with leaves and flowers. B, leaf, dorsal (upper) surface. C, leaf, ventral (lower) surface. D, flower. E, corolla spread out to show stamens and staminodes; one staminode bent down. F, stamen. G, pistil. H, fruit. (G.)

broadly obovate or suborbicular, rounded at the apex, the veins (except the midrib) generally not visible. Flowers 1 to 4 together, subsessile; sepals 1.1-2.2 mm long; corolla 3.1-3.8 mm long; stamens included, with filaments c. 1 mm long. Staminodes ovate to reniform, 1-1.3 mm long. Fruit subglobose, green or dark red, mostly 5-9 mm in diameter.

GRAND CAYMAN: *Brunt 1781, 1782; Kings GC 326.* LITTLE CAYMAN: *Proctor 28089.* CAYMAN BRAC: *Kings GC CB 38.*

– Cuba and Haiti, in dry rocky thickets and woodlands. This is perhaps the most obtrusive of several unrelated spiny shrubs known as "Shake Hand" in the Cayman Islands, the others being *Xylosma bahamense* (Flacourtiaceae) and *Zanthoxylum coriaceum* (Rutaceae). The three can usually be distinguished easily by differences in the leaves, those of *Bumelia* being simple and entire, and minutely reddish-hairy beneath when young; those of the *Xylosma* being simple and often with 1-4 blunt teeth, and always glabrous; while those of *Zanthoxylum* are pinnately compound, the leaflets being very shiny and with prominulous venation.

[Genus 3. **POUTERIA** Aubl.

Trees with alternate or occasionally subopposite leaves, these with primary lateral veins widely apart, strongly developed, and distinctly upcurved toward the margins. Flowers solitary or few together in the leaf-axils, or sometimes at defoliated nodes; sepals 4-12, paired when 4, otherwise imbricate or spiralled; corolla usually 4-6-lobed. Stamens attached at base of corolla or else to near the level of the corolla-tube sinuses; staminodes petaloid or not, rarely absent. Ovary more or less hairy, 1-10-celled; ovules laterally attached. Fruit usually fleshy, rarely hard, containing 1-6 or more seeds, these with a long lateral scar; endosperm absent.

A tropical American genus of 50 or more species.

1. **Pouteria campechiana** (Kunth) Baehni in Candollea 9:398. 1942.

"Egg Fruit"

Tree to 15 m tall or more (often smaller), nearly glabrous except for minute reddish hairs on the growing tips; leaves oblanceolate to narrowly obovate, mostly 5-18 cm long, acuminate at apex. Flowers solitary or several together on pedicels averaging 10 mm long; sepals 4-6, minutely pubescent outside, the margins ciliolate; corolla greenish-cream, 7-12 mm long, with 4 to 7 lobes. Stamens attached near level of sinuses, the filaments 1.3-2 mm long; staminodes linear or lanceolate, 2.4-2.9 mm long. ovary 4- to 10-celled. Fruit more or less pear-shaped, 4-7 cm long, with pointed apex and yellow edible flesh; seeds 4-7 (rarely less), 2-4 cm long, lustrous brown with elongate lateral scar.

GRAND CAYMAN: *Brunt 2047, 2048; Lewis 2868; Proctor 15114, 15235.*
– Central America; occurs as an escape from cultivation on the Florida Keys, the Bahamas, and Cuba. There is a tradition on Grand Cayman that this tree was introduced from Central America and cultivated for its edible fruits. However, it now occurs as a component of some of the natural coastal woodlands, and in such areas cannot be distinguished from the indigenous flora.]

[Genus 4. **CHRYSOPHYLLUM** L.

Shrubs or small to medium-sized trees; leaves alternate, with secondary lateral veins nearly parallel to the primary ones. Flowers solitary to numerous in the leaf-axils; sepals usually 5, nearly free; corolla usually 5-lobed. Stamens arising from near the level of the sinuses, often connected by a slightly thickened ring in the corolla-throat; staminodia nearly always absent. Ovary 4–12-celled; ovules attached laterally or basilaterally. Fruit fleshy, with 1 to 5 or more seeds; seeds with large lateral or basilateral scar; endosperm copious.

A tropical, chiefly American genus of about 150 species.

1. **Chrysophyllum cainito** L., Sp. Pl. 1:192. 1753.

"Star Apple"

A small to rather large tree, the leaves densely reddish-sericeous beneath, glabrous and shining green on the upper side. Fruit green or purple, commonly globose or nearly so, 3–10 cm in diameter, several-seeded, edible; seeds with lateral scar.

GRAND CAYMAN: *Kings GC 358,* collected from a cultivated plant in George-town. The species does not occur wild in the Cayman Islands.
– West Indies; widely cultivated and naturalized both in the West Indies and continental tropical America, so that its original natural range would be difficult to determine.]

Genus 5. **MANILKARA** Adans.

Trees with alternate leaves often of hard, leathery texture, the primary lateral veins parallel, straight, and often obscure. Flowers perfect or rarely unisexual, solitary or few together in the leaf-axils; sepals biseriate, commonly 6 (3, 3), occasionally 8 or 4; corolla with lobes as many as the sepals each with a pair of dorsal more or less petaloid appendages at the base, sometimes partly or wholly fused to the lobes, or rarely vestigial. Stamens as many as the corolla-lobes and opposite them, usually alternating with an equal number of staminodes. Ovary 6–14-celled; ovules attached laterally. Fruit often mealy-roughened, several-seeded; seeds with long lateral scar; endosperm copious.

A pantropical genus of about 35 species.

1. **Manilkara zapota** (L.) van Royen, Blumea 7:410. 1953.

Achras zapota L., 1759, not L., 1753.

Sapota achras Mill., 1768.

Manilkara zapotilla (Jacq.) Gilly. 1943.

"Naseberry", a corruption of the Spanish "Nispero", a name originally applied to the European Medlar-tree (*Mespilus germanica* L., Rosaceae), whose fruits those of *Manilkara* resemble.

A medium-sized to large tree with copious white latex; leaves clustered toward the ends of the branchlets, elliptic or nearly so, mostly 5–12 cm long, loosely reddish-tomentose beneath when young, soon glabrate; petioles mostly 1–2 cm long. Flowers solitary in the leaf-axils, the pedicels about equalling the petioles; sepals 6–10 mm long, minutely pubescent; corolla 5–13 mm long, the lobes entire or toothed at the apex. Staminodes petaloid and similar to the corolla-lobes. Ovary 9–12-celled, densely short-hairy. Fruit brown, mealy-roughened, and varying from ellipsoid to ovoid or subglobose, up to 10 cm in diameter, edible; seeds strongly flattened, 16–23 mm long.

GRAND CAYMAN: *Kings GC 347; Proctor 15088.* LITTLE CAYMAN: *Proctor 28148.* CAYMAN BRAC: *Proctor 28967.*
– Florida, West Indies and Central America, often planted and naturalized, so that its true natural range would be difficult or impossible to ascertain. It is doubtfully native in the Cayman Islands, although it can be found growing in woodlands remote from habitations. The fruit is one of the most important in regions where it grows. The very heavy light red wood is hard and durable, and is sometimes used to make furniture. This is one of the species used as a source of "chiclé" in Central America.

Family 52

THEOPHRASTACEAE

Trees or shrubs; leaves evergreen and mostly of stiff, hard texture, alternate, opposite or whorled, and often gland-dotted; stipules absent. Flowers regular, perfect or unisexual, solitary or in axillary or terminal clusters or racemes. Calyx 5-parted, with obtuse, overlapping lobes. Corolla 5-lobed, the tube bell-shaped or cylindric. Stamens 5, borne near the base of the corolla-tube and opposite the lobes, alternating with 5 staminodes. Ovary superior, 1-celled, consisting of 5 fused carpels; ovules numerous on a basal placenta; style rather short, with a capitate or discoid stigma. Fruit a hard or fleshy indehiscent berry, few- or several-seeded; seeds more or less flattened; endosperm cartilaginous.

A tropical American family of 5 genera and about 60 species. Some of them are extremely ornamental, but they are not often cultivated.

Genus 1. **JACQUINIA** L.

Shrubs or small trees, often with pale or whitish bark; leaves scattered or whorled, short-petioled, usually gland-dotted. Flowers perfect, mostly in terminal or axillary racemes, rarely solitary; pedicels often thickened upwardly; sepals united at the base; corolla with spreading lobes alternating with petaloid staminodes. Stamens attached to a fleshy ring at base of the corolla-tube; staminodes inserted near top of the corolla-tube. Fruit ovoid or globose, of rather hard texture and relatively few-seeded.

A genus of about 30 species, widely distributed in the West Indies and continental tropical America.

KEY TO SPECIES

1. Petioles glabrous:

   2. Leaves less than 4 cm long; calyx and corolla-lobes c. 2 mm long; corolla yellow: **1. J. berterii**

   2. Leaves more than 4 cm long; calyx and corolla-lobes 3–4 mm long; corolla white: **2. J. arborea**

1. Petioles glandular-puberulous: **3. J. keyensis**

1. Jacquinia berterii Spreng. in L., Syst. Veg. 1:668. 1825.

"Wash-wood"

A much-branched glabrous shrub 1.5–3 m tall, the young twigs covered with minute granulate scales. Leaves narrowly to broadly obovate 1.5–4 cm long, flat but the margins narrowly revolute, the apex acutish or retuse, sometimes mucronate. Inflorescence usually shorter than the leaves, the axis more or less reflexed and 0.5–2 cm long; bracteoles appressed to bases of the pedicels, c. 1 mm long, brown-tipped; pedicels 3–6 mm long. Flowers fragrant; calyx c. 2 mm long, with orbicular lobes; corolla yellow, with lobes reflexed and c. 2 mm long. Fruit yellowish-green, 4–5 mm in diameter.

LITTLE CAYMAN: *Proctor 28105, 28184, 28189, 35077.* CAYMAN BRAC: *Proctor 15328.*
– Bahamas and Cuba to Guadeloupe, in several variable races. The Cayman population appears to represent var. *portoricensis* Urban.

2. **Jacquinia arborea** Vahl, Eclog. Amer. 1:26. 1796.

Jacquinia barbasco (Loefl.) Mez, 1903, in part.

"Wash-wood"

A shrub or small tree to 5 m tall or more, the young branchlets minutely whitish-scurfy; leaves mostly whorled at the upper nodes, narrowly to broadly obovate, 4–10 (–13) cm long, the very short glabrous petioles usually orange. Flowers fragrant, in terminal racemes usually a little shorter than the leaves, the axils 2.5–4 cm long, minutely glandular-puberulous in the axils of the pedicels; bracteoles whitish, c. 2 mm long; pedicels 7–12 mm long. Calyx 3–4 mm long; corolla white, the lobes c. 3.5 mm long. Fruit bright red, 9–10 mm in diameter; seeds 4.5 mm long.

GRAND CAYMAN: *Kings GC 334.*
– West Indies, frequent in coastal thickets.

3. **Jacquinia keyensis** Mez in Urban, Symb. Ant. 2:444. 1901. **FIG. 125.**

"Wash-wood"

A shrub or small tree up to 5 m tall or more, with smooth gray bark, the young branchlets minutely whitish-scurfy. Leaves alternate or crowded, oblong to obovate, mostly 2–5 cm long, often bullate-revolute, the apex obtuse or retuse, the very short petioles densely glandular-puberulous. Inflorescence equalling or usually longer than the leaves, the axis 2–8 cm long; bracteoles 0.5 mm long, blackish and bullate with a pale triangular apex; pedicels 5–8 mm long. Flowers fragrant; calyx c. 2 mm long; corolla cream, the lobes 3–4 mm long. Fruit orange, 7–9 mm in diameter; seeds 2.5 mm long.

GRAND CAYMAN: *Brunt 1885, 2102; Kings GC 172; Proctor 27965.*
– Florida, the Bahamas, Cuba, and Jamaica, in coastal scrub-lands and thickets.

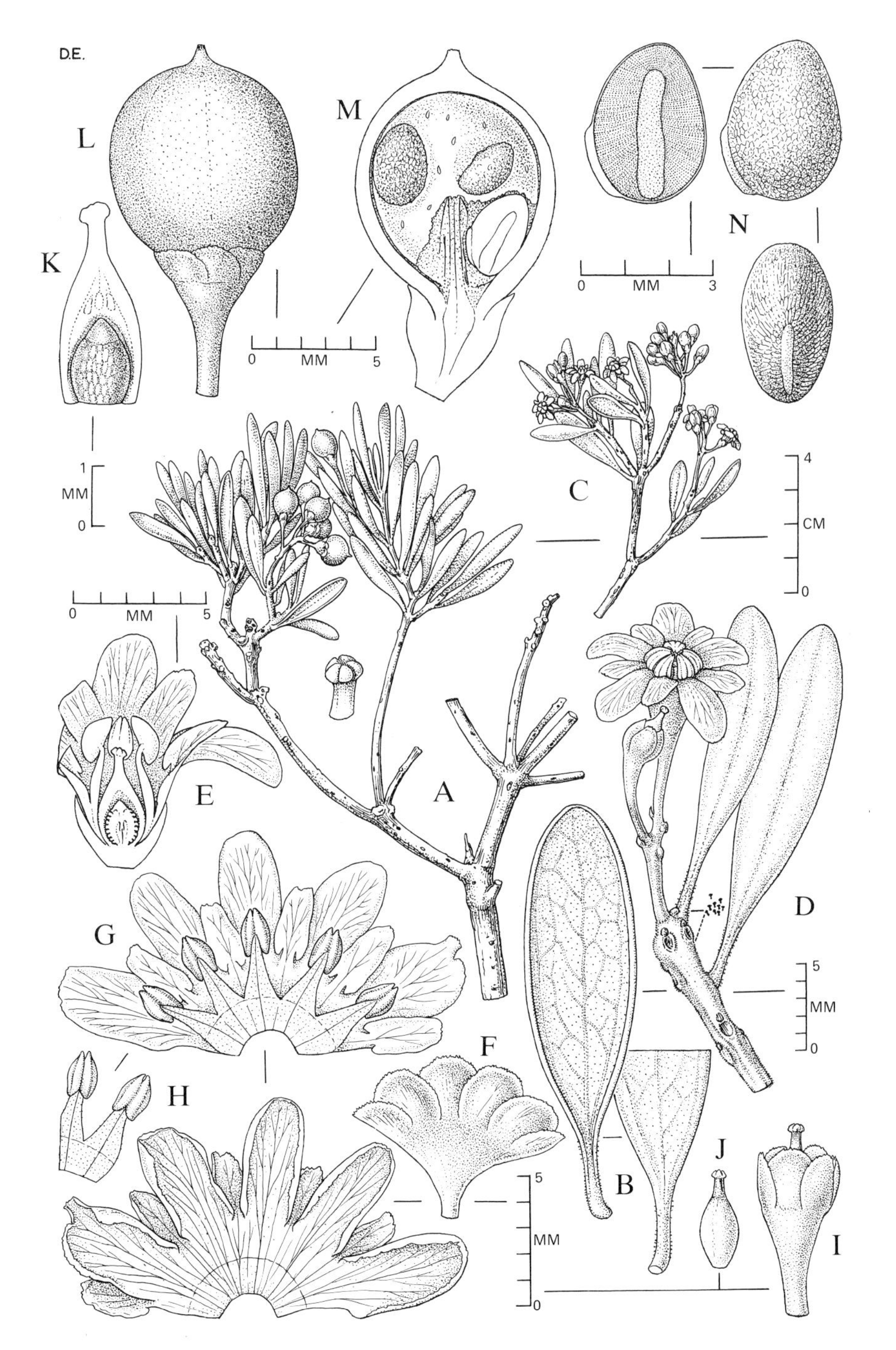

FIG. 125 **Jacquinia keyensis.** A, fruiting branch. B, two views of leaves. C, flowering branch. D, portion of same, enlarged. E, long. section of flower. F, calyx spread out. G, two views of corolla spread out. H, two stamens. I, calyx with pistil enclosed. J, pistil K, long. section of pistil. L, fruit. M, same, long. section. N, three views of seed. (D.E.)

Family 53

MYRSINACEAE

Trees or shrubs, glabrous or pubescent; leaves alternate, simple, entire or with crenulate or serrate margins, the surfaces often gland-dotted; stipules none. Flowers perfect, or unisexual and dioecious, regular, with parts in 4's or 5's in terminal or axillary clusters, racemes, or panicles. Calyx segments free or somewhat connate, often ciliate and gland-dotted; corolla with segments united at least toward the base. Stamens as many as the corolla-lobes, shorter than these and opposite them, almost free or inserted on and somewhat connate with the corolla-tube; anthers opening by slits or apical pores; staminodes none. Ovary superior, sessile, 1-celled, with a free central placenta; ovules numerous or few; style short or long; stigma simple and variously capitate, lobed, or fimbriate. Fruit a 1-seeded drupe; seed with copious endosperm.

A pantropical family of about 30 or 40 rather weakly-differentiated genera and 900 species, few if any of economic value.

Genus 1. **MYRSINE** L.

Shrubs or trees, glabrous or pubescent; leaves petiolate, entire or nearly so. Flowers small, dioecious, 4–5-parted, densely clustered on short, spur-like racemes or glomerules in the leaf-axils or at defoliated nodes. Sepals connate at base; corolla-lobes usually united below, rarely free, usually dark-lined, often papillose on the margins. Stamens inserted on throat of the corolla, the filaments obsolete, the anthers sessile, opening by slits, producing no pollen in pistillate flowers. Ovary globose or ellipsoid, in pistillate flowers the sessile stigma relatively large and subcapitate or variously lobed; ovules few, uniseriate. Fruit dry or fleshy, with a hard endocarp; seed globose and smooth; embryo curved, transverse.

A pantropical genus of more than 200 species.

1. Myrsine acrantha Krug & Urban in Notizbl. Bot. Gart. Berlin 1:79. 1895. **FIG. 126.**

Rapanea acrantha (Krug & Urban) Mez, 1901.

A glabrous shrub or small tree to 7 m tall or more; leaves oblong or narrowly obovate to obovate, mostly 5–8 (–12) cm long and usually over 2.5 cm broad, blunt or minutely notched at the apex, the lower surface minutely gland-dotted. Flowers

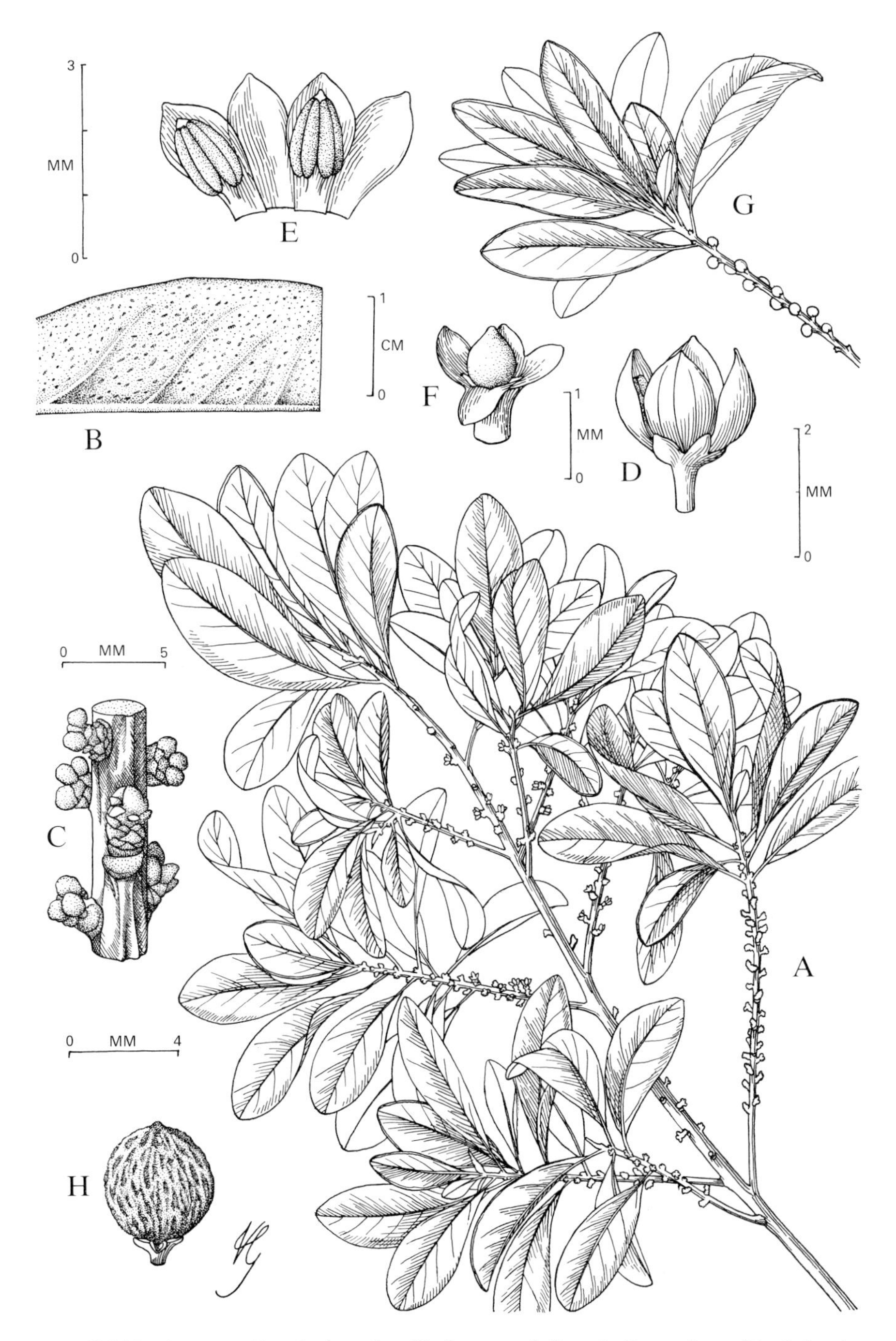

FIG. 126 **Myrsine acrantha.** A, branch with leaves and flowers. B, portion of leaf showing lower (ventral) surface. C, portion of stem showing clusters of buds in glomerules. D, staminate flower. E, corolla of staminate flower, spread out, showing stamens. F, pistillate flower. G, fruiting branch. H, fruit. (G.)

4-parted, in dense clusters of 3 to 7 on lateral spurs shorter than the petioles, mostly along naked branches below the foliage; corolla-lobes elliptic, free to the base. Fruits c. 5 mm in diameter, greenish with longitudinal glandular markings.

GRAND CAYMAN: *Brunt 1977; Proctor 11997.*
– Jamaica and perhaps Cuba, at medium to rather high elevations. The Cayman plants grow at or near sea-level in dry, rocky woodlands.

[Family 54

CRASSULACEAE

Succulent annual or perennial herbs, sometimes suffrutescent; leaves opposite, simple or sometimes compound, usually very thick; stipules absent. Flowers regular, perfect, usually in cymes or panicles; sepals 4 or 5, united below; corolla 4–5-lobed, the lobes spreading. Stamens as many as or twice as many as the corolla-lobes, the anthers dehiscent by longitudinal slits. Ovary superior, with as many carpels as sepals, distinct or partly united, with a scale at the base of each; ovules numerous, in 2 series along the ventral suture of the carpel. Fruit consisting of 4 or 5 1-celled follicles containing numerous small or minute seeds; seeds with fleshy endosperm; embryo terete, with short, obtuse cotyledons.

A widespread family of at least 33 genera and more than 1,000 species, in this hemisphere best represented in Mexico.

Genus 1. **BRYOPHYLLUM** Salisb.

Erect, often large perennial herbs; leaves simple or pinnate with a terminal leaflet. Flowers nodding, in paniculate, many-flowered cymes; calyx bell-shaped, inflated, 4-toothed; corolla more or less urn-shaped, with 4 short lobes. Stamens 8, inserted on the corolla-tube at or below the middle; anthers borne at or near the mouth of the corolla-tube. Carpels 4, free or partly united; fruit of 4 follicles.

A genus of about 20 species, all native to Madagascar, one of them (*B. pinnatum*) often cultivated and widely naturalized in tropical countries.

1. Bryophyllum pinnatum (Lam.) Oken in Allgem. Naturgesch. 3(3):1966. 1841. **FIG. 127.**

"Leaf-of-life", "Curiosity Plant"

Fleshy, glabrous herb to 1 m tall; lower leaves simple, the upper pinnate, the leaves or leaflets oblong to elliptic and up to 14 cm long or more, with coarsely crenate margins, often bearing plantlets in the notches. Calyx greenish-yellow, the tube 2.5–3.5 cm long; corolla reddish, 2.5–4.5 cm long, the tube constricted below the middle, the lobes acute and 1–1.5 cm long. Stamens attached to the constriction of the corolla-tube. Carpels 1.2–1.5 cm long.

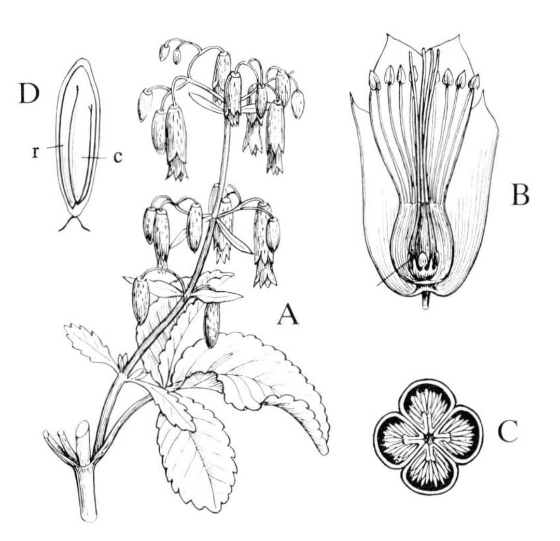

FIG. 127 **Bryophyllum pinnatum.** A, inflorescence and leaf. x⅙. B, flower cut lengthwise, x1; sc, scale. C, x-section of ovary, x3. D, seed, x40; c, cotyledon; r, radicle. (F. & R.)

GRAND CAYMAN: *Hitchcock; Kings GC 35a, GC 97a, GC 117, GC 146; Millspaugh 1311.* LITTLE CAYMAN: *Proctor 28178.* CAYMAN BRAC: *Kings CB 69; Proctor 29029.*

– Naturalized throughout the tropics. The leaves are sometimes juiced and with added salt taken as a cold remedy; it has a very unpleasant taste. An allegedly medicinal tea is also made from the leaves.]

Family 55

CHRYSOBALANACEAE

Trees or shrubs with alternate, simple, entire leaves; stipules present. Flowers perfect or rarely unisexual, usually more or less zygomorphic, in simple or compound racemes or cymes. Calyx 5-lobed from a turbinate or bell-shaped tube, more or less unequal or spurred at the base. Petals 5 or lacking, often unequal, short-clawed, inserted in the mouth of the calyx-tube. Stamens 2 to many, inserted with the petals, often with larger fertile ones opposite the larger calyx-lobes; filaments slender, exserted. Ovary superior, sessile or more often stalked, 1-celled with 2 basal erect ovules, or rarely 2-celled, each cell with 1 ovule; style simple, lateral or almost gynobasic. Fruit a sessile or stalked drupe, with bony almost 2-valved endocarp, or sometimes a crustaceous berry; seed lacking endosperm; embryo with thick, fleshy cotyledons.

A widespread tropical and subtropical family of 10 genera and about 400 species, the majority in tropical America. This family is sometimes included with the Rosaceae.

Genus 1. **CHRYSOBALANUS** L.

Shrubs or small trees; leaves glabrous, the stipules small and deciduous. Flowers small, white or cream, in terminal and axillary pubescent cymes; stamens 10 to numerous, a few of them sterile. Ovary of 1 carpel, sessile at base of the calyx-tube; style attached laterally near base of the ovary; ovules 2. Drupe fleshy, 1-seeded.

A genus of about 3 species, 2 of them American, the other in Africa.

1. **Chrysobalanus icaco** L., Sp. Pl. 1:513. 1753. **FIG. 128.**

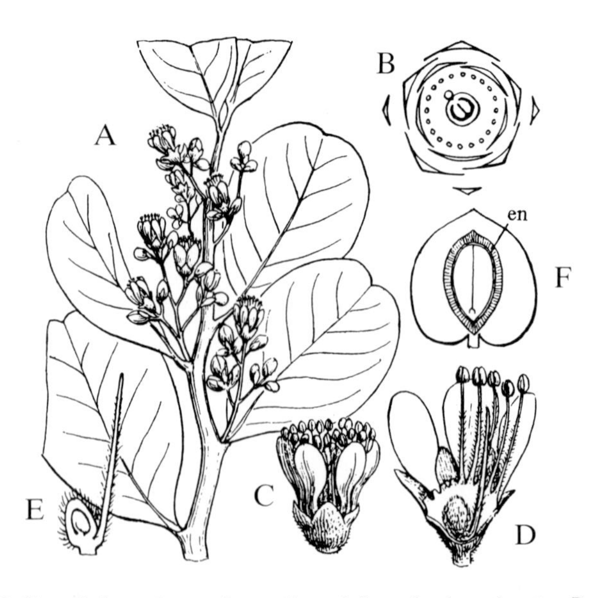

FIG. 128 **Chrysobalanus icaco.** A, portion of flowering branch, x$^2/_3$. B, floral diagram. C, flower, x2. D, flower cut lengthwise and four stamens removed, x3. E, pistil with ovary cut lengthwise, x4. F, fruit cut lengthwise, x$^2/_3$; en, endocarp. (F. & R.)

"Coco-plum"

Usually a shrub or sometimes a small tree with glabrous branches; leaves roundish-elliptic or very broadly obovate, mostly 3–7 cm long, the apex blunt, rounded or retuse. Cymes paniculate, mostly shorter than the leaves; flowers finely woolly, the calyx-lobes 2.5 mm long, the petals c. 5 mm long; stamens longer than the petals, with hairy filaments. Drupe purplish-red or blackish, more or less ellipsoid, 1.3–3 cm long, the pericarp fleshy and edible.

GRAND CAYMAN: *Brunt 2053; Hitchcock; Kings GC 355; Lewis 3824; Proctor 15118, 15223.* LITTLE CAYMAN: *Proctor 28144.* CAYMAN BRAC: *Proctor 29004.*

– Florida, West Indies, and on the continent from Mexico to northern South America, common especially in sandy thickets near the sea. The fruit is edible but rather insipid. The seeds are said to be rich in oil, and the Carib Indians are alleged to have strung them on sticks and burned them like candles.

Family 56

LEGUMINOSAE

Trees, shrubs or herbs of various habit, sometimes climbing, often armed with spines or prickles. Leaves alternate or rarely opposite, sometimes simple but mostly pinnately compound or often bipinnate, the leaflets few or numerous, small or large, entire, lobed, or rarely toothed; stipules always present, stipels sometimes so. Flowers mostly irregular and perfect, or regular and polygamous, the peduncles axillary or terminal, 1-many-flowered; pedicels solitary, paired, or clustered, usually in the axil of a bract; bracteoles usually 2 at the base of the calyx. Sepals free or more or less united, usually 5 or sometimes 4 in irregular flowers, 3–6 but most often 5 in regular flowers. Petals as many as the sepals, in irregular flowers one borne higher than the rest, either outside and enclosing the others in bud, or inside and enclosed by the others in bud. Stamens usually twice as many as the petals, rarely as many or fewer, or (in Mimosoideae) often more numerous, inserted on the margin of the receptacle, free or partly connate; anthers 2-celled, mostly opening by longitudinal slits. Ovary superior and usually elongate, sessile, 1-celled; ovules numerous or rarely 1, attached along the inner angle of the ovary in 1 or 2 series. Fruit a pod, usually dry but sometimes more or less pulpy within, rarely drupe-like, most often 2-valved but sometimes indehiscent, continuous within or septate; seeds 1 to many, with a hilum or scar on one edge, attached along the upper valve of the pod; endosperm usually none or scanty, rarely copious; cotyledons flat and leaf-like or thick and fleshy.

One of the largest families of flowering plants and of world-wide distribution, with about 500 genera and nearly 13,000 species. Included are many of the most valuable members of the plant kingdom, variously useful for food or other purposes. Especially important are those with edible seeds, such as the Peanut (*Arachis hypogaea* L.), Congo or Pigeon Pea (*Cajanus cajan* (L.) Millsp.), Soy Bean (*Glycine max* (L.) Merr.), Kidney Bean or "Red Peas" (*Phaseolus vulgaris* L.), and the Pea or "Green Peas" (*Pisum sativum*) L.); there are many others, too numerous to mention here. Numerous other species are useful for fodder, timber, fibre, dyes, gums and resins, and medicinal products.

The roots of most leguminous plants possess small nodules which contain minute bacterial organisms that are able to "fix", or chemically combine, atmospheric nitrogen with other elements. This process enriches soils where leguminous plants grow.

Some botanists divide the group into three families, but their close relationship is so evident that it seems more logical to treat these taxa as subfamilies.

KEY TO SUBFAMILIES

1. Flowers more or less irregular, often conspicuously so; stamens 10 or less; mostly shorter than the longest petals:

  2. Flowers pea-like, the uppermost petal outside the others and covering them in bud: *SUBFAM. 1. FABOIDEAE*

  2. Flowers not pea-like, the uppermost petal attached inside the others and covered by them in bud: *SUBFAM. 2. CAESALPINIOIDEAE*

1. Flowers regular, the petals all equal; stamens 10 or more numerous, usually much more conspicuous than the petals: *SUBFAM. 3. MIMOSOIDEAE*

SUBFAMILY 1. FABOIDEAE

Herbs, shrubs or trees; leaves mostly pinnately or rarely palmately compound, sometimes simple or trifoliate, the leaves and leaflets mostly entire. Flowers irregular; sepals usually 5, partly united below into a cup; petals 5, overlapping, the uppermost (the standard) outside and enclosing the others in bud, the two lateral (the wings) more or less parallel to each other, the two lower innermost, generally parallel and united by their lower margin to form a keel. Stamens usually 10, the filaments partly united and sheath-like below, sometimes 9 of them so united and the uppermost free or absent.

Among the Leguminosae, this subfamily is especially characteristic of temperate regions, though many species also occur in tropical countries.

KEY TO GENERA

1. Leaves simple or with one leaflet:

 2. Pods continuous, not jointed; plants erect or climbing, herbaceous or woody:

  3. Pod enveloped in a large membranous bract; shrub: **[22. Moghania]**

  3. Pod not enveloped in a bract:

   4. Erect herbs; pods inflated: **1. Crotalaria**

   4. Woody scrambler or an arching shrub; pods flat: **23. Dalbergia**

 2. Pods jointed, the joints separating; plant a more or less prostrate herb from a woody taproot: **[8. Alysicarpus]**

1. Leaves compound:

 5. Leaves abruptly pinnate, without a terminal leaflet:

  6. Slender twining woody vine; seeds red with a black spot: **9. Abrus**

  6. Erect shrubs or trees; seeds brown: **[5. Sesbania]**

 5. Leaves pinnate with a terminal leaflet:

  7. Leaflets 3:

   8. Pods jointed, the joints separating:

    9. Stipules united to the petiole; leaflets without stipels; flowers yellow: **6. Stylosanthes**

    9. Stipules free from the petiole; leaflets subtended by stipels; flowers not yellow: **7. Desmodium**

   8. Pods not jointed (but sometimes constricted between the seeds):

    10. Erect trees or shrubs:

     11. Tree with orange or scarlet flowers; seeds not edible: **[13. Erythrina]**

11. Shrub with yellow flowers (sometimes reddish-tinged); seeds commonly eaten: **[20. Cajanus]**

10. Twining vines or suberect herbs:

12. Standard more then 1 cm long:

13. Standard different in size from the other petals:

14. Standard flat, much larger than the other petals: **10. Centrosema**

14. Standard folded, shorter than the other petals: **14. Mucuna**

13. Standard about the same size as the other petals:

15. Suberect herb; flower deep red; keel incurved, forming a complete spiral: **17. Phaseolus**

15. Twining vines; flower not deep red; keel not forming a complete spiral:

16. Style hairy:

17. Flowers yellow; stigma oblique or lateral; pod smooth: **18. Vigna**

17. Flowers white, purplish or violet; stigma terminal; pod with warty projections along the margins: **[19. Lablab]**

16. Style glabrous:

18. Calyx 4-lobed, the lobes equal and acuminate; pod not ridged: **15. Galactia**

18. Calyx 2-lipped, the lips more or less rounded, the upper larger than the lower; pod ridged longitudinally on each side near the upper margin: **16. Canavalia**

12. Standard less than 1 cm long:

19. Flowers white; pod more than 3 cm long: **12. Teramnus**

19. Flowers yellow; pod less than 2 cm long: **21. Rhynchosia**

7. Leaflets 5 or more:

20. Trees:

21. Flowers white or nearly so; all 10 stamens united into a sheath; pod broadly 4-winged: **24. Piscidia**

21. Flowers pink; 9 stamens united, 1 free; pod smooth and unwinged: **[4. Gliricidia]**

20. Herbs or shrubs:

22. Flowers yellow; stamens all free; pod constricted between the seeds: **25. Sophora**

22. Flowers not yellow; stamens all united except the uppermost one; pod not constricted between the seeds:

23. Twining vine; flowers blue: **[11. Clitoria]**

23. Erect herbs or subshrubs; flowers not blue:

24. Leaflets subtended by minute stipels; racemes axillary; pods subcylindric: **2. Indigofera**

24. Leaflets without stipels; racemes terminal or opposite the leaves; pods flat: **3. Tephrosia**

Genus 1. **CROTALARIA** L.

Herbs or shrubs; leaves simple or digitately 3–5-foliolate; stipules free from the petiole, sometimes decurrent on the stem, often small or none. Flowers yellow or sometimes blue or purplish, in racemes terminal or opposite the leaves, rarely solitary and axillary. Calyx with free lobes or the calyx rarely 2-lipped. Standard roundish, short-clawed; wings obovate or oblong, shorter than the standard; keel incurved, beaked. Stamens 10, united below into a 2-cleft sheath; anthers alternately small, versatile, and large, basifixed. Ovary with 2 to many ovules; style incurved or abruptly inflexed above the ovary, minutely barbed above along the inner side. Pod globose or oblong, 2-valved, more or less inflated, continuous within.

A widespread, chiefly tropical genus of more than 200 species. *C. juncea* L. or "Sunn Hemp" is cultivated in some countries for its valuable fibre, usable as a substitute for flax and true hemp (*Cannabis*).

KEY TO SPECIES

1. Leaves simple:

2. Stipules roundish and leaf-like; flowers lavender-blue; pods hairy: **1. C. verrucosa**

2. Stipules minute or lacking; flowers yellow; pods glabrous: **2. C. retusa**

1. Leaves 3-foliolate: **3. C. incana**

1. Crotalaria verrucosa L., Sp. Pl. 2:715. 1753.

"Sweet Pea"

A low bushy annual less than 1 m tall, with 4-angled stems. Leaves simple, somewhat glaucous, ovate or roundish from an abruptly tapered base, 2–7 cm long, glabrous above, minutely pubescent beneath; stipules roundish-lunate and deflexed, up to 10 mm long. Racemes densely several- to many-flowered; calyx 2-lipped, 7–9 mm long, about half as long as the corolla, the lobes narrowly triangular and acuminate; standard whitish or pale lavender-blue; keel violet or purple. Pod obovoid-oblong, beaked, 3–3.5 cm long, finely appressed-puberulous.

GRAND CAYMAN: *Brunt 2017; Kings GC 307; Proctor 15080.* LITTLE CAYMAN: *Proctor 28064.* CAYMAN BRAC: *Kings CB 5, CB 20a; Proctor 28986.* – Pantropical, in pastures and along sandy roadsides.

2. Crotalaria retusa L., Sp. Pl. 2:715. 1753. **FIG. 129.**

Tough-stemmed annual, the stems striate and appressed-puberulous; leaves simple, oblanceolate, 3–10 cm long, the apex rounded or retuse, glabrous above, minutely puberulous beneath, the tissue with numerous minute pellucid dots; stipules awl-shaped, c. 1 mm long. Racemes elongate, many-flowered; calyx 2-lipped, c. 12 mm long, about half as long as the corolla, the lobes triangular-acuminate; petals yellow, the standard up to 2.4 cm broad, tinged reddish on the back and with fine purplish lines within. Pod oblong, blackish, 3–4 cm long, glabrous.

GRAND CAYMAN: *Brunt 1878; Kings GC 209.*
– Pantropical, a common weed of sandy roadsides, pastures, and open waste ground. The stems are said to contain a useful fibre.

3. Crotalaria incana L., Sp. Pl. 2:716. 1753.

Erect, shrubby annual up to 1 m tall, the stems, petioles and racemes densely brownish-hirsute; leaves long-petiolate, trifoliolate, the leaflets broadly obovate or

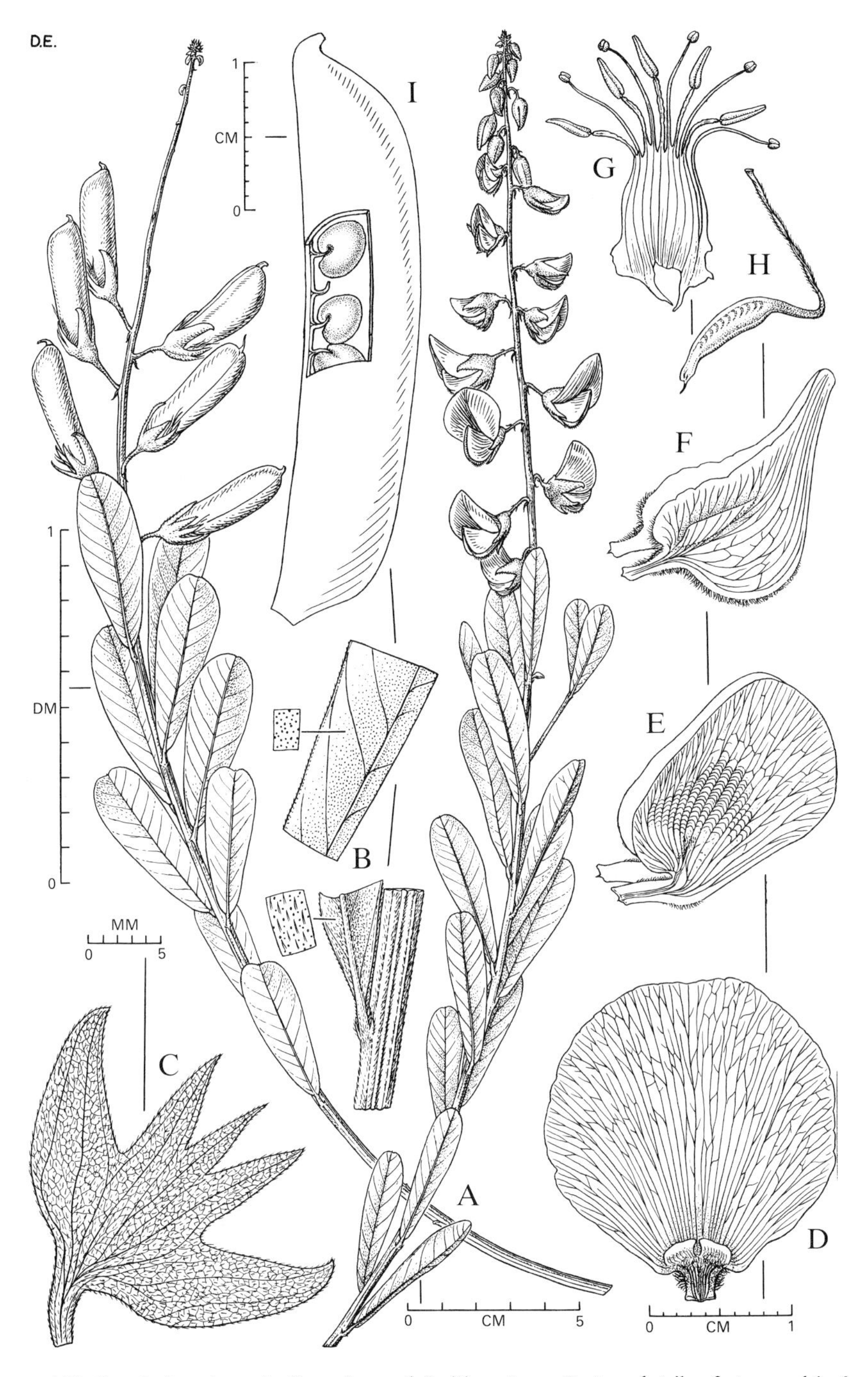

FIG. 129 **Crotalaria retusa.** A, flowering and fruiting stems. B, two details of stem and leaf. C, calyx spread out. D, standard. E, wings. F, keel. G, stamens, H, pistil. I, pod with portion cut away to show seeds. (D.E.)

roundish, 1–5 cm long, of thin texture, glabrous above, pale beneath and appressed-pilose especially along the midvein; stipules awl-shaped, small, deciduous. Racemes rather few-flowered; calyx 5-lobed, 7–9 mm long, with lance-acuminate lobes; corolla yellow, 10–13 mm long. Pod pendulous, oblong, 2.5–3 cm long, soft-hairy.

GRAND CAYMAN: *Hitchcock; Millspaugh 1387.*
– Pantropical, a weed of roadsides and open waste ground.

Genus 2. **INDIGOFERA** L.

Herbs or shrubs, often with strigose pubescence, the hairs attached by the middle; leaves pinnate with a terminal leaflet, rarely 3-foliolate or 1-foliolate; stipules small, subulate, slightly adnate to the petioles. Racemes axillary; calyx 5-lobed, the lobes sometimes unequal; standard broad or roundish, sessile or short-clawed, strigillose outside; keel usually spurred on each side near the base. Stamens 10, the uppermost free, the others united into a sheath. Ovary sessile; style bent upward, with capitate stigma; ovules 1-many. Pods narrow, subcylindric, 2-valved but usually opening along the upper suture only, partitioned inside.

A pantropical genus of more than 300 species, most abundant in Africa.

KEY TO SPECIES

1. Pod sickle-shaped, less than 2 cm long, with 3–6 seeds: **1. I. suffruticosa**

1. Pod straight or only slightly curved, 2–4 cm long, with 8–15 seeds: **2. I. tinctoria**

1. Indigofera suffruticosa Mill., Gard. Dict., ed. 8, no. 2. 1768.

Indigofera anil L., 1771.

"Wild Indigo"

A stiff, shrubby herb up to 1.5 m tall but usually lower, finely white-strigillose throughout; leaves with 9–15 leaflets, these elliptic or oval, 1.5–3 cm long, mucronate at the apex. Racemes shorter than the leaves; calyx c. 1.5 mm long; corolla salmon-red, 5–6 mm long. Pods numerous and densely crowded, 1.5–2 cm long, strongly curved, 2 mm thick.

GRAND CAYMAN: *Hitchcock; Kings GC 273.*
– West Indies and continental tropical America, often a weed of sandy waste ground or along roadsides; naturalized in the Old World tropics. This plant was formerly much cultivated, especially in Central America, as the chief source of the blue dye known as indigo, a colour supplanted in modern times by synthetic substitutes.

[2. **Indigofera tinctoria** L., Sp. Pl. 2:751. 1753. **FIG. 130.**

"Indigo"

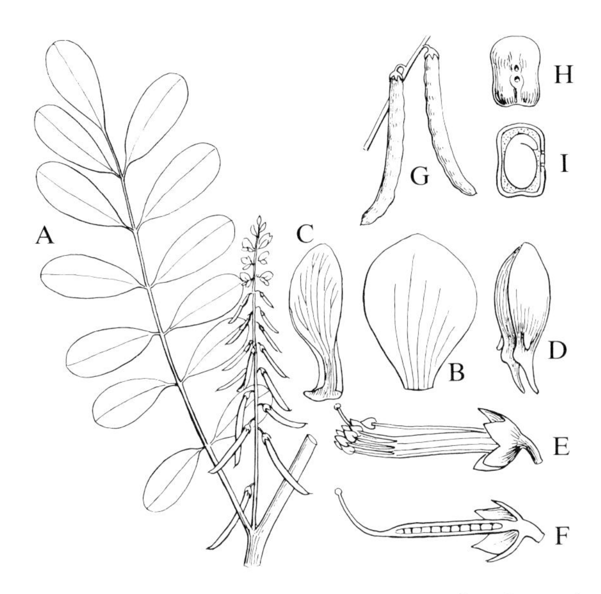

FIG. 130 **Indigofera tinctoria.** A, leaf and raceme, x$^2/_3$. B, standard, x6. C, wing, x6. D, keel, x6. E, flower with petals removed, x7. F, ovary and calyx cut lengthwise, x7. G, ripe pods, x$^2/_3$. H, seed, x3. I, seed cut lengthwise, x3. (F. & R.)

A shrubby herb up to 1 m tall, the stems sparsely white-strigillose, or densely so near the ends; leaves with 9-15 leaflets, these broadly obovate or oval, 1-2.5 cm long, mucronate at the apex, finely strigillose. Racemes shorter than the leaves; calyx c. 1.5 mm long; corolla salmon-red, 5-6 mm long. Pods numerous, 2-4 cm long, slightly curved or straight, 1.5-2 mm thick, beaked at the apex.

GRAND CAYMAN: *Proctor 27996.* LITTLE CAYMAN: *Proctor 35196.*
– Native of southern India, but now naturalized in most warm countries; formerly cultivated as a source of indigo.]

Genus 3. **TEPHROSIA** Pers.

Annual or perennial herbs or sometimes shrubs; leaves pinnate with a terminal leaflet; leaflets few or often numerous, with many conspicuous lateral nerves beneath, these parallel and very oblique; stipules setaceous or broader, stipels absent. Flowers in rather few-flowered racemes, these terminal or opposite the leaves, rarely axillary; bracteoles absent. Calyx 5-lobed, the lobes subequal or the lower one longer, the upper 2 usually more or less joined. Petals clawed, the standard roundish and more or less sericeous outside. Stamens 10, the uppermost more or less free. Ovary sessile; style inflexed or incurved, usually glabrous; ovules numerous. Pods flat, 2-valved, beaked, many-seeded.

A widely-distributed genus of more than 125 species, commonest in tropical regions. Several species have been used as fish poisons, especially in South and Central America.

KEY TO SPECIES

1. Leaflets 9-15, narrowly oblanceolate, obtuse or acute; racemes with short peduncles; corolla 10-15 mm long: **1. T. cinerea**

1. Leaflets 5-9, more or less obovate, rounded or retuse at the apex; racemes with long peduncles; corolla 7-10 mm long: **2. T. senna**

1. Tephrosia cinerea (L.) Pers., Syn. Pl. 2:328, 1807.

Perennial herb, trailing, decumbent or ascending from a woody taproot, often much-branched, finely strigose; leaves with 9-15 leaflets, these mostly linear-oblanceolate or oblong-oblanceolate, 1-3 cm long, 2.5-5 mm wide, usually acutish and mucronate, whitish-strigose beneath; stipules lance-acuminate, 3-8 mm long. Racemes opposite the leaves, on peduncles mostly 1-3 cm long. Calyx c. 5 mm long; corolla purple, 10-15 mm long. Pods 4-5 cm long, 3-5 mm broad, finely strigose, 6-12-seeded.

GRAND CAYMAN: *Hitchcock; Proctor 15238.*
– West Indies and continental tropical America, especially in sandy soils along roadsides and in open ground near the sea.

2. Tephrosia senna Kunth in H.B.K., Nov. Gen. 6:458. 1824.

Tephrosia cathartica (Sesse & Moç.) Urban, 1905.

An erect perennial herb, much-branched and finely strigose; leaves with 4-9 leaflets, these obovate or cuneate-oblong, 1-3.5 cm long, 4-16 mm wide, usually rounded or retuse with a small mucro in the notch, whitish-strigillose on both surfaces; stipules subulate, 7-10 mm long. Racemes opposite the leaves, on peduncles 3.5-7 cm long; flowers widely spaced. Calyx c. 5 mm long; corolla rose-red or pale mauve, 7-10 mm long. Pods 2.5-4 cm long, 3-4 mm broad, finely strigose, 6-8-seeded.

CAYMAN BRAC: *Kings CB 82; Millspaugh 1158; Proctor 29067.*
– West Indies except Jamaica, also in the Yucatan area and Colombia, in sandy or loamy clearings and along borders of thickets.

[Genus 4. **GLIRICIDIA** Kunth ex Endl.

Deciduous shrubs or small trees; leaves pinnate with a terminal leaflet; stipules small, stipels lacking. Flowers in axillary racemes, often appearing during the dry season when the plant is more or less leafless; bracteoles absent. Calyx shortly bell-shaped, truncate or 5-toothed. Corolla usually pink or whitish; standard roundish, reflexed, short-clawed, often with 2 callosities at the base of the blade within; wings oblong, free; keel curved, the petals clawed and free at the base, united at the apex. Stamens 10, 9 united and the uppermost free. Ovary stalked; style glabrous, bent at almost a right angle, the stigma capitate and papillose; ovules 7-12. Pod elongate, short-stalked, flat, 2-valved; seeds roundish, compressed.

A tropical American genus of about 5 species.

1. **Gliricidia sepium** (Jacq.) Kunth ex Griseb. in Abh. Ges. Wiss. Götting. 7, Phys. Cl.:52. 1857.

A small tree up to about 10 m tall, or sometimes shrub-like, branching from near the base; leaves with 7-17 leaflets, these lance-oblong, elliptic or ovate, 3-6 cm long, acutish at both ends, glabrate above and sparsely fine-hairy beneath; stipules ovate or lanceolate, 2 mm long. Racemes 5-10 cm long, often densely-flowered; pedicels 5 mm long; calyx 4-5 mm long; corolla 1.5-2 cm long, usually light pink. Pods 10-12 cm long, 1-1.5 cm broad, glabrous; seeds dark brown, c. 1 cm in diameter.

GRAND CAYMAN: *Proctor 15234.* CAYMAN BRAC: *Proctor 29139.*
– Central and northern South America, commonly planted and becoming naturalized in the West Indies. Uncommon in the Cayman Islands. The wood is very hard, heavy, strong and durable, and takes a high polish. The leaves are apparently safely eaten by cattle, but all parts of the plant are poisonous to rats and mice, and also to dogs. Poultices of the fresh crushed leaves are said to have a healing effect on skin ulcers and sores.]

[Genus 5. **SESBANIA** Scop.

Sesbania grandiflora (L.) Poir., a small tree probably native either to India or Australia, is sometimes cultivated for ornament; it is locally called "Picashia" or "Spanish Armada". It has short, 2-flowered racemes, curved buds, and large white to red flowers 7–8 cm long.

GRAND CAYMAN: *Kings GC 147, GC 288.*]

Genus 6. **STYLOSANTHES** Sw.

Mostly low herbs, annual or perennial, or sometimes small shrubs, often with viscid pubescence; leaves pinnately 3-foliolate, stipules united to the base of the petiole, stipels absent. Flowers small, yellow, in chiefly terminal spikes or heads, each flower nearly sessile in the axil of a 2-dentate or divided bract, the pedicel very short and adnate to the bract, sometimes accompanied by a bristle-like sterile flower. Calyx with a long, stalk-like tube and with the 4 upper lobes joined, the lower one narrower and distinct. Petals and stamens inserted at top of the calyx-tube; standard roundish; stamens 10, all united into a closed sheath, the anthers alternately long and short. Ovary subsessile; style hair-like; ovules 2 or 3. Pods compressed, 1–2-jointed, the joints reticulate or muricate, the apical one with a hooked beak.

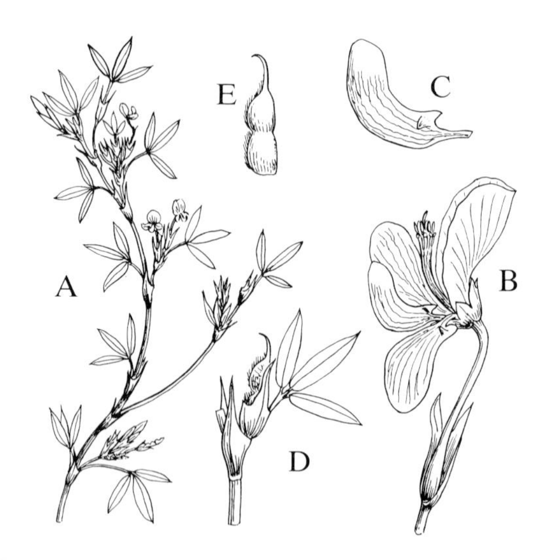

FIG. 131 **Stylosanthes hamata.** A, branch with leaves and flowers, x$^2/_3$. B, flower with one wing turned down, x5. C, wing, inner face, x5. D, portion of branch with pod, x2. E, pod, x2. (F. & R.)

A genus of about 30 species, chiefly in tropical America, a few in tropical Africa and Asia.

1. Stylosanthes hamata (L.) Taub. in Verh. Bot. Ver. Brandenb. 32:22. 1889. **FIG. 131.**

Stylosanthes procumbens Sw., 1788.

"Lucy Julia", "Pencil Flower"

Perennial subwoody herb with prostrate or ascending slender stems reaching about 45 cm in length; leaflets lanceolate to elliptic, 5-15 mm long, 2-5 mm broad, glabrate but with finely ciliate margins, the nerves prominent beneath; stipules 6-8 mm long. Flowers subtended by leaf-like bracts; calyx-tube 3-4 mm long; corolla 4-5 mm long. Pods 7-10 mm long, with an apical beak about as long as one joint.

GRAND CAYMAN: *Brunt 1875, 2086, 2121; Hitchcock; Kings GC 69; Millspaugh 1335; Proctor 15061.* LITTLE CAYMAN: *Kings LC 75; Proctor 28062.* CAYMAN BRAC: *Proctor 29095.*
– Florida, West Indies, and Central America south to Colombia, often a weed of pastures and lawns, and to be considered beneficial in such situations.

Genus 7. **DESMODIUM** Desv.

Annual or perennial herbs, procumbent to erect, sometimes scandent or vine-like, often somewhat woody; pubescence often of hooked hairs. Leaves trifoliolate or rarely 1-foliolate, stipulate; leaflets with stipels. Flowers in terminal or axillary racemes or panicles, rarely solitary or in small clusters, usually pink, purple, or sometimes white. Calyx with short tube, 2-lipped, the upper lip with 2 teeth, the lower with 3 acute or attenuate lobes. Standard obovate or roundish; wings obliquely oblong, adhering to the keel; keel nearly straight, obtuse. Stamens 10, the upper sometimes more or less free; anthers all alike. Ovules 2 to many. Pod of 2 to many flat or twisted 1-seeded joints, usually with hooked hairs, indehiscent and separating easily from each other.

A cosmopolitan genus of more than 500 species. The fruits adhere tenaciously to clothing, feathers, or fur by means of their minute hooked hairs, and by this means are often widely dispersed.

KEY TO SPECIES

1. Flowers solitary or in small clusters of 2-4; plants small, creeping, with leaflets less than 1 cm long: **1. D. triflorum**

1. Flowers in more or less elongate racemes; stems ascending or erect; leaflets more than 1 cm long:

  2. Upper margin of pod continuous, not notched: **2. D. incanum**

  2. Upper and lower margins of pod equally notched: **3. D. tortuosum**

1. Desmodium triflorum (L.) DC., Prodr. 2:334. 1825.

A prostrate creeping herb with pubescent stems, often rooting at the nodes. Leaves with obovate or obcordate leaflets mostly 3–8 mm long. Flowers axillary or opposite the leaves; calyx 2–3 mm long, with lanceolate lobes; corolla pink or purplish, 3–4 mm long. Pods flat, curved, with continuous upper margin; joints 2-6, each c. 3 mm long, the sides prominently net-veined.

GRAND CAYMAN: *Correll & Correll 51053; Kings GC 400; Millspaugh 1368, 1800.*
– Pantropical, a frequent weed of lawns and pastures.

2. Desmodium incanum DC., Prodr. 2:332. 1825; Nicolson in Taxon 27(4):365-370. 1978.

Desmodium supinum (Sw.) DC., 1825.

D. canum (J.F. Gmel.) Schinz & Thell., 1914; Adams, 1972.

"Chick Weed"

Decumbent woody herb with ascending branches, more or less pubescent throughout. Leaves petiolate, the leaflets chiefly oblong-elliptic or elliptic, 1–6 cm long or more, the terminal the largest, usually pale beneath. Racemes terminal; calyx red, c. 3 mm long, with triangular teeth; corolla pink, fading to bluish; standard c. 6 mm long and broad. Uppermost stamen free. Pods flat, deeply notched on the lower side, the upper side continuous; joints 3–8, each c. 4 mm long.

GRAND CAYMAN: *Brunt 1899, 2127; Hitchcock; Kings GC 249, GC 277; Millspaugh 1330, 1384; Proctor 15262.* CAYMAN BRAC: *Kings CB 22; Proctor 28970.*
– Florida, West Indies, continental tropical America, and tropical Africa, common weed of roadsides, pastures, clearings, and thickets.

3. **Desmodium tortuosum** (Sw.) DC., Prodr. 2:332. 1825.

An erect subwoody herb to 1 m tall or more, the stem finely pubescent with hooked hairs. Leaves petiolate, the leaflets lanceolate, mostly ovate or elliptic, 1-7 cm long. Racemes terminal and in the upper axils, together forming a large, open, paniculate inflorescence; flowers often 2 or 3 together, on slender pedicels 1-1.5 cm long. Calyx 2.5-3 mm long; corolla pink or purple, c. 4 mm long. Pods more or less twisted, equally indented on both margins; joints 3-7, each c. 4 mm long.

GRAND CAYMAN: *Brunt 2074; Kings GC 93, GC 360; Proctor 27926; Sachet 371.*
– West Indies, continental tropical America, and tropical Africa, a weed of open waste ground and roadside banks.

[Genus 8. **ALYSICARPUS** Desv.

Perennial herbs with 1-foliolate leaves; stipules dry, papery, enclosing 2 stipels. Flowers small, in short terminal or axillary racemes; calyx deeply 5-lobed, the lobes elongate and rigid; corolla equalling or shorter than the calyx and included in it; standard roundish or obovate, clawed; wings obliquely oblong, adhering to the keel; keel obtuse, incurved. Stamens 10, 9 of them united, the uppermost free. Pods cylindric with several or numerous joints, the joints separating, indehiscent.

A genus of about 16 species indigenous to the tropics of Africa, Asia and Australia.

1. **Alysicarpus vaginalis** (L.) DC., Prodr. 2:353. 1825.

An herb with prostrate or trailing stems growing from a woody taproot, nearly glabrous throughout; leaves with leaflet stalked and variously lanceolate, oblong, obovate or roundish, 0.4-4 cm long, the apex rounded but with a minute mucro, the base subcordate; stipules lanceolate, c. 7 mm long, whitish. Racemes mostly 1-3 cm long; calyx 4-5 mm long; corolla deep salmon-red or purple, about as long as the calyx. Pods up to 2 cm long, 2-8-jointed, minutely puberulous.

GRAND CAYMAN: *Correll & Correll 51052; Proctor 15279.* CAYMAN BRAC: *Proctor 28985, 29358.*
– Native of tropical Asia and Africa; widely naturalized in the West Indies, the Cayman plants occurring in second-growth thickets, damp pastures, and in soil-pockets of exposed limestone pavements.]

Genus 9. **ABRUS** Adans.

Slender woody vines; leaves even-pinnate, lacking a terminal leaflet, the rhachis terminating in a bristle; leaflets numerous; stipels absent. Flowers in short racemes, terminal or axillary on short branches, the pedicels clustered at nodes of the raceme. Calyx truncate or very shortly 5-toothed; corolla usually pinkish or whitish; standard ovate, the short claw adherent to the stamen-tube; wings oblong-falcate; keel curved, longer than the wings. Stamens 9, united into a sheath split above; anthers all alike. Ovary subsessile; style short, incurved, with capitate stigma; ovules numerous. Pod somewhat compressed, 2-valved, with partitions between the seeds.

A small pantropical genus of about 5 species.

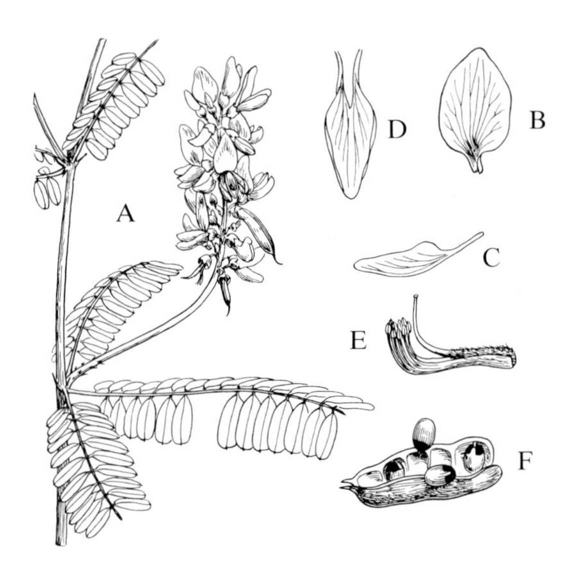

FIG. 132 **Abrus precatorius.** A, leaves and raceme, x$^2/_3$. B, standard, x1½. C, wing, x1½. D, keel, x 1½. E, stamens and pistil x1½. F, pod and seeds, x$^2/_3$. (F. & R.)

1. Abrus precatorius L., Syst. Nat. ed. 12, 2:472. 1767. **FIG. 132.**

"Licorice", "John Crow Bead"

An elongate, sometimes high-climbing vine, the younger green stems sparsely pubescent. Leaves 4–10 cm long; leaflets in 10–20 pairs, oblong or obovate-oblong, 5–18 mm long, mucronulate at the apex, glabrous above and sparsely appressed-pubescent beneath. Racemes 4–8 cm long, many-flowered; calyx 2-3 mm long;

corolla pinkish or rose with whitish keel, 8–11 mm long. Pods oblong, beaked, 2–2.5 cm long, finely pubescent, 3–5-seeded; seeds c. 5 mm long, bright scarlet with a black spot at the hilum.

GRAND CAYMAN: *Brunt 2172; Kings GC 182a; Proctor 15231; Sachet 460a.* LITTLE CAYMAN: *Proctor 28065.* CAYMAN BRAC: *Kings CB 87; Proctor 28951, 29357.*

– Pantropical, in thickets and woodlands, usually near sea-level. The roots allegedly contain glycerrhizin, the same substance as is found in commercial licorice. There are conflicting reports in the literature about the use of this material as a substitute for true licorice. The seeds, on the other hand, are known to be extremely poisonous if eaten; however, they are often used to make bracelets, necklaces, and other ornaments, and have been much used in some Asian countries as weights by jewel merchants. The unit of weight called the "carat" (one twenty-fourth of an ounce) is said to have been associated with the weight of *Abrus* seeds, 2 seeds allegedly weighing 1 carat.

Genus 10. **CENTROSEMA** (DC.) Benth.

Herbaceous twining vines, sometimes woody below; leaves pinnately 3-foliolate, with stipels; stipules persistent, striate. Flowers rather large, on 1- to several-flowered axillary peduncles, these solitary or paired; bracteoles relatively large, striate, appressed to the calyx. Calyx bell-shaped, unequally 5-lobed; standard large, roundish, flattened, spurred or slightly pouched at the base; wings and keel much shorter than the standard, the keel incurved. Stamens 10, all united or the uppermost free; anthers all alike. Ovary subsessile; style incurved, somewhat dilated at the apex; ovules numerous. Pod thin and flat, 2-valved, often with a longitudinal flange near each margin, and terminated by a long straight beak, partially septate within between the seeds; seeds transversely oblong.

A genus of about 30 species in tropical and warm-temperate America.

1. **Centrosema virginianum** (L.) Benth., Comm. Legum. Gen. 56. 1837.

A slender twining herbaceous vine, the stems nearly glabrous; leaflets glabrous, narrowly ovate or narrowly elliptic, 1.5–4 cm long or more. Flowers 1–3 on each peduncle; bracteoles ovate, c. 5 mm long, minutely puberulous; calyx-lobes lance-linear, 8–12 mm long; corolla lavender-pink sometimes bluish or nearly white; standard 2.5–3 cm broad, minutely puberulous on the back. Pods 7–10 cm long, 3–4 mm wide, the terminal beak up to 10 mm long.

GRAND CAYMAN: *Sachet 369.* CAYMAN BRAC: *Proctor 28950.*

– Widespread in warm-temperate and tropical America, consisting of many varying races. The Cayman plants grow in secondary woodland thickets and are quite rare.

[Genus 11. **CLITORIA** L.

Erect, trailing, or climbing herbs or shrubs, or sometimes large trees; leaves simple or more usually pinnately 3–7-foliolate, rarely 9–11-foliolate; stipules striate, persistent; stipels usually present. Flowers rather large, solitary or clustered in the leaf-axils, or in short racemes; bracts paired, persistent; bracteoles mostly larger than the bracts, striate. Calyx tubular, 5-lobed; standard large, rounded, retuse at the apex, narrowed and without appendages at the base; wings falcate-oblong, adherent to middle of keel; keel shorter than the wings, acute, incurved. Stamens 10, the uppermost free, the rest united; anthers all alike. Ovary stalked; style incurved, dilated at the apex, hairy on inner side. Pod stalked, elongate, compressed, 2-valved, interrupted or continuous within; seeds subglobose or compressed.

A pantropical genus of about 30 species.

1. Clitoria ternatea L., Sp. Pl. 2:753. 1753. **FIG. 133.**

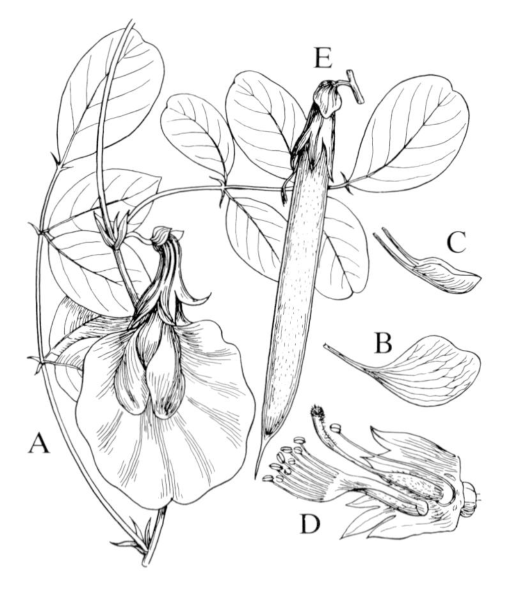

FIG. 133 **Clitoria ternatea.** A, portion of branch with leaves and flower, x$^2/_3$. B, wing, x$^2/_3$. C, keel, x$^2/_3$. D, calyx, stamens and pistil, x1. E, pod, x$^2/_3$. (F. & R.)

"Bluebell"

A twining herbaceous vine, more or less woody at the base, finely pubescent more or less throughout, but often sparsely so. Leaves petiolate, 5–7-foliolate; leaflets elliptic or ovate, 2–5 cm long; stipules linear, 5–7 mm long. Flowers solitary on axillary peduncles 3-8 mm long; bracteoles roundish, 6-7 mm long; calyx 1.5-2 cm long; standard intense blue with white blotch in the middle, rarely all white, 4-4.5 cm long; wings and keel much shorter. Pods 6-10 cm long, 8-9 mm broad, beaked at the apex; seeds 5-6 mm long, mottled dark brown.

GRAND CAYMAN: *Brunt 1645, 1727; Hitchcock; Kings GC 43, GC 186, GC 227; Maggs II 54; Millspaugh 1318, 1883; Proctor 15077, 15182; Sachet 377.* CAYMAN BRAC: *Kings CB 63; Proctor 28949, 29353* (flowers white). – Native of tropical Africa, but widely cultivated as an ornamental, and now somewhat naturalized in nearly all warm countries.]

Genus 12. **TERAMNUS** P. Br.

Slender twining herbaceous vines, more or less pubescent; leaves pinnately 3-foliolate, with stipels; stipules small. Flowers very small, in axillary racemes or clusters; bracteoles linear or lanceolate; calyx-lobes 5, or 4 with the upper one 2-toothed; standard obovate, narrowed below into a long claw, not appendaged; wings narrow, adherent to the keel; keel shorter than the wings, almost straight, obtuse. Stamens 10, all united into a closed sheath, the filaments alternately long and short, the anthers on short filaments sterile. Ovary sessile, the style thick and glabrous, the stigma capitate; ovules many. Pod linear-elongate, compressed, 2-valvate, straight or curved, septate within, at apex often beaked.

A small pantropical genus of about 4 species.

KEY TO SPECIES

1. Leaflets mostly 1.5-4 cm long; pods 2-4 cm long, sparsely pubescent: **1. T. labialis**

1. Leaflets mostly 5-8 cm long; pods 5 cm long or more, densely brown-pubescent: **2. T. uncinatus**

1. Teramnus labialis (L.f.) Spreng. in L., Syst. Veg. ed. 16, 3:235. 1826.

A slender twiner, finely pubescent throughout; leaflets ovate or elliptic, 1.5-4 cm long (rarely longer), acutish and mucronate at the apex; stipules lance-linear, 2-3 mm long. Racemes slender, about as long as the leaves, the flowers scattered; calyx 5-lobed, 3-4 mm long, the lobes mostly shorter than the tube;

corolla white, slightly longer than the calyx. Pods 2-4 cm long, rarely more, c. 3 mm wide, slightly curved, with an upturned beak 1.5-2 mm long; seeds dark brown, c. 2 mm long.

GRAND CAYMAN: *Correll & Correll 51008; Proctor 31037.* CAYMAN BRAC: *Proctor 29369.*

– Pantropical, a frequent weed of thickets and overgrown waste ground.

2. **Teramnus uncinatus** (L.) Sw., Nov. Gen. & Sp. Pl. 105. 1788.

An elongate slender twiner, the stems densely pilose with reflexed hairs; leaflets oblong to lance-linear, mostly 5-8 cm long, obtuse and mucronate at the apex, appressed hairy on both-sides, densely so beneath; stipules lanceolate, 3-5 mm long. Racemes usually much longer than the leaves, the flowers remotely scattered; occasionally some flowers solitary in the leaf-axils; calyx 5-lobed, pilose, 5-6 mm long, the lobes longer than the tube; corolla purplish or whitish, scarcely longer than the calyx. Pods mostly 5-7 cm long, 3-4 mm wide, densely brown-pilose, nearly straight, with right-angled beak 3-5 mm long; seeds lustrous orange-brown, c. 4 mm long.

GRAND CAYMAN: *Correll & Correll 51007; Hitchcock.*

– West Indies and continental tropical America, usually in rather moist second-growth thickets.

[Genus 13. **ERYTHRINA** L.

Erythrina velutina Willd. was collected by Hitchcock near Georgetown during January, 1891, and erroneously reported by him as *E. corallodendron.* This probably was from a cultivated tree.]

Genus 14. **MUCUNA** Adans.

Herbaceous or woody vines; leaves pinnately 3-foliolate, the lateral leaflets usually unequal-sided; stipules deciduous; stipels present. Flowers rather large, in axillary long-stalked racemes or clusters, the peduncles often greatly elongate and pendent; calyx shortly bell-shaped, 4-lobed; corolla mostly dark purple or yellowish; standard shorter than the wings, with inflexed auricles at the base; wings incurved, often adherent to the keel; keel equalling the wings or longer, incurved at the acute or beak-like apex. Stamens 10, 9 of them united, the uppermost free; anthers alternately long and short, the short ones hairy. Ovary sessile, hairy; style glabrous; ovules few. Pod more or less oblong, thick, leathery, hairy (the hairs often stinging), septate within; seeds few, large, rounded-oblong.

A pantropical genus of about 50 species. Several, such as the Velvet Bean (*Mucuna deeringiana* (Bort) Merrill), have been widely grown as cattle-fodder. The seeds of that species are also sometimes parched and ground as a substitute for coffee.

1. **Mucuna pruriens** (L.) DC., Prodr. 2:405. 1825.

"Cow-itch"

An elongate herbaceous vine, growing over shrubs and small trees; leaflets rhombic-ovate, the lateral very asymmetric, 7–14 cm long, of thin texture, appressed-pilose especially beneath. Racemes mostly short-pedunculate, few-flowered; calyx broader than long, the lowest lobe up to 10 mm long; corolla dull dark purple; standard 2 cm long, half as long as the wings; keel 4.5 cm long, yellowish-beaked. Pods curved-oblong, mostly 5–7.5 cm long, 1–1.5 cm thick, covered with abundant easily-detached irritant hairs; seeds 2–6, transverse-oblong, 1–1.5 cm long, black with a white hilum.

GRAND CAYMAN: *Hitchcock; Kings GC 228.*
– Pantropical. The stinging hairs of the pods can cause almost unbearable itching and burning of the skin, and are very dangerous to the eyes.

Genus 15. **GALACTIA** P. Br.

Slender herbaceous or woody vines, rarely erect shrubs; leaves pinnately 3-foliolate, with stipels; stipules small, often deciduous. Flowers paired or clustered at raised nodes on axillary racemes, rarely solitary in the leaf-axils; bracteoles minute. Calyx 4-lobed, the lobes acuminate, unequal; corolla pink or white; standard roundish, the margins slightly inflexed at the base; wings adherent to the keel and equalling it in length. Stamens 10, the lower 9 united, the uppermost free or partly so; anthers all alike. Ovary nearly sessile; style very slender, glabrous, with a minute stigma. Pod elongate, straight or slightly curved, usually compressed, 2-valved, partly septate within.

A pantropical genus of more than 80 species.

1. **Galactia striata** (Jacq.) Urban. Symb. Ant. 2:320. 1900.

A slender and often elongate herbaceous vine, the stems finely pubescent; leaflets elliptic or ovate-elliptic, 2.5–8 cm long, mucronate at the apex, the upper side glabrate and with finely prominulous venation, the underside rather densely appressed-pubescent. Inflorescence up to 15 cm long but often less, short- to long-pedunculate; flowers paired at the nodes; calyx appressed-pubescent, 5–7 mm long, with narrowly lance-acuminate lobes; corolla 8–9 mm long, light rose-violet, the standard, wings and keel of about equal length. Pods 4–5 cm long, c. 5 mm broad, pubescent.

GRAND CAYMAN: *Kings GC 42, GC 311; Proctor 31040.*
– Florida, West Indies and continental tropical America, in thickets and woodlands. Over its wide range this is an extremely variable species; the above description is based chiefly on Cayman material.

Genus 16. **CANAVALIA** Adans.

Climbing or trailing herbs; leaves pinnately 3-foliolate, with stipels; stipules small, deciduous. Flowers rather large, borne in clusters or solitary at swollen nodes of axillary racemes; calyx 2-lipped, the upper lip large and truncate or 2-lobed, the lower much smaller and entire or 3-cleft; standard roundish, reflexed; wings narrow, subfalcate or somewhat twisted; keel broader than the wings, incurved, obtuse or with an inflexed or spiral beak. Stamens 10, all united above, the uppermost one free at the base; anthers all alike. Ovary subsessile; style incurved or spiral with the keel, glabrous, the stigma small; ovules few. Pod oblong or broadly linear, compressed or convex, winged or ribbed longitudinally, 2-valved; seeds rounded or bean-like, with a linear hilum.

A pantropical genus of about 50 species, a few sometimes cultivated as forage or for the edible seeds and young pods.

KEY TO SPECIES

1. Nodes of inflorescence each with 3 or more flowers; pods with sutural ribs only; seeds concolorous, with hilum nearly as long as the seed: **1. C. nitida**

1. Nodes of inflorescence each with 1 or 2 flowers; pods with an additional rib on either side 3–5 mm from the ventral suture; seeds marbled orange and brown, with hilum less than half as long as the seed: **2. C. rosea**

1. Canavalia nitida (Cav.) Piper in Contr. U.S. Nat. Herb. 20:559, 562. 1925.

Canavalia ekmanii Urban, 1918.

"Horse Bean"

Herbaceous vine, climbing on rocks or over bushes, the stems glabrous or nearly so; leaflets glabrous, oblong-elliptic or elliptic, mostly 5–8 cm long, 2.5–3.5 cm broad, bluntly subacuminate and minutely retuse at the apex, the fine reticulate venation slightly raised and evident on both sides. Racemes longer than the subtending leaves; calyx with the upper lip deeply 2-cleft, the lower lip 3-lobed; corolla purple or dark violet with a pink keel, twice as long as the calyx, up to 1.6 cm long. Pods oblong, 10–15 cm long and c. 3 cm broad, strongly convex and of thick texture; seeds lustrous wine-red or brick-red, the linear hilum blackish.

GRAND CAYMAN: *Kings GC 104,* growing on "the cliff at East End".
– Originally thought to be endemic to Cuba, this species has more recently been ascribed a broader range including the Bahamas, the Greater Antilles (except Jamaica), the Virgin Islands and Mexico.

2. **Canavalia rosea** (Sw.) DC., Prodr. 2:404. 1825.

Canavalia maritima Thouars, 1813, illegit.

C. obtusifolia (Lam.) DC., 1825, in part.

"Sea Bean"

A long-trailing herb, the stems becoming rope-like; puberulous when young. Leaflets sparsely strigillose, roundish or very broadly obovate, mostly 5–12 cm long, 2.5–10 cm broad, rounded to broadly retuse at the apex, the fine venation scarcely evident. Racemes often much longer than the leaves; calyx helmet-like, the upper lip broad and marginate, the lower much shorter and 3-toothed; corolla pink or rose, fading bluish-purple, c. 2 cm long. Pods oblong, 7–13 cm long and 2.5–3 cm broad, somewhat compressed and with a distinct flange on either side parallel to the ribs of the ventral suture and 3–5 mm from them; seeds marbled orange and brown, 12–16 mm long (usually less than 15 mm), the whitish hilum less than half as long as the length of the seed.

GRAND CAYMAN: *Brunt 2052; Hitchcock; Kings GC 68; Millspaugh 1307; Proctor 15060; Sachet 404; Sauer 3320.* LITTLE CAYMAN: *Kings LC 23; Proctor 28085; Sauer 4164.* CAYMAN BRAC: *Kings CB 13; Millspaugh 1170; Proctor 29074.*
– Pantropical, especially common in sandy clearings near the sea.

Genus 17. **PHASEOLUS** L.

Annual or perennial herbs, twining or erect, rarely prostrate, sometimes with woody or tuberous roots; leaves pinnately 3-foliolate with stipules; stipules striate, persistent. Flowers in clusters at enlarged nodes along the axes of axillary racemes, the nodes bracteolate. Calyx 4- or 5-toothed or -lobed, the upper two teeth or lobes often more or less united; corolla white, yellow, red, or purple;

standard roundish, recurved-spreading or somewhat twisted; wings usually obovate, equalling or longer than the standard; keel linear to obovate, with a low spirally twisted or coiled beak. Stamens 10, the uppermost free, the rest united; anthers all alike. Ovary subsessile; style twisted with the keel, usually hairy on one side; stigma oblique; ovules few or many. Pod linear or falcate, compressed or subcylindric, 2-valved, with tissue between the seeds; seeds rounded-oblong or flattened, the hilum small or linear.

A world-wide genus of more than 100 species, some often cultivated. As a source of food, this genus is of major importance, as it includes the majority of the various kinds of beans. The only wild Cayman species is often placed in a segregate genus *Macroptilium* distinguished especially on the basis of having 5 (instead of usually 4) calyx lobes.

1. **Phaseolus lathyroides** L., Sp. Pl. ed. 2, 2:1018. 1763.

Phaseolus semierectus L., 1767.

Macroptilium lathyroides (L.) Urban, 1928; Adams, 1972.

An erect or straggling, annual or persistent herb with subwoody rhizomes; leaflets mostly lanceolate to elliptic, 1.5–5 cm long or more, acutish at the apex, glabrous or nearly so except for the densely pubescent petiolules. Racemes few-flowered, on peduncles much longer than the leaves; flowers in pairs remotely scattered on the puberulous rachis; bracts and bracteoles subulate. Calyx 4–6 mm long, 5-toothed; corolla deep red-purple; standard 1.5 cm long; keel forming a single spiral. Pods narrowly linear, 8–10 cm long, c. 2 mm broad, subcylindric, strigillose; seeds oval, c. 3 mm long, brownish-gray speckled with black.

GRAND CAYMAN: *Brunt 1902, 1968; Hitchcock; Kings GC 323; Millspaugh 1324; Proctor 15035; Sachet 454.* CAYMAN BRAC: *Kings PCB 2; Proctor 29089.* – West Indies and continental tropical America; naturalized in the East Indies and Polynesia. This is commonly a weed of open fields and roadsides.

Genus 18. **VIGNA** Savi

Twining herbaceous vines, sometimes prostrate or erect; leaves pinnately 3-foliolate, with stipels; stipules sessile or sometimes extended at the base below the point of attachment. Flowers in short, crowded racemes at the ends of long axillary peduncles; bracts and bracteoles small, soon falling. Calyx bell-shaped, 5-lobed or the upper 2 lobes united; corolla yellow or rarely pale purple; standard roundish, with inflexed basal auricles; wings falcate-obovate, shorter than the standard; keel equalling the wings, incurved but not forming a complete loop or spiral. Stamens 10, the uppermost free, the rest united; anthers all alike. Ovary sessile; style thread-like or thickened above, often bearded along the inner side, the stigma oblique or lateral; ovules many. Pod linear, straight or nearly so, subcylindric, 2-valved, interrupted within between the seeds; seeds reniform or subquadrate with a short lateral hilum.

A pantropical genus of about 40 species.

1. **Vigna luteola** (Jacq.) Benth. in Mart., Fl. Bras. 15(1):194. 1859.

Vigna repens (L.) Kuntze, 1891, not Baker, 1876.

A trailing or twining herbaceous vine, glabrous or somewhat pubescent, sometimes rampant and forming dense tangles; leaflets lanceolate to ovate, 3–9 cm long, blunt or acutish at the apex, glabrous or nearly so except for the pubescent petiolules. Peduncles longer than the leaves; calyx 4-lobed, 4–5 mm long, the upper lobe rounded with 2 small teeth; corolla yellow; standard c. 1.5 cm long, retuse at the apex; keel obtuse. Pods straight or slightly curved, 4–7 cm long, c. 5 mm wide, sparsely pilose; seeds 4–5 mm long, lustrous black with white hilum.

GRAND CAYMAN: *Brunt 1748, 1914, 2126; Hitchcock; Kings GC 79, GC 212; Lewis 3828; Maggs II 52; Millspaugh 1241.* LITTLE CAYMAN: *Kings LC 73.* CAYMAN BRAC: *Proctor 29372.*
– Pantropical, a weedy species of damp thickets and more or less open waste ground.

[Genus 19. **LABLAB** Adans.

Twining herbaceous vines, perennial from a subwoody root; leaves pinnately 3-foliolate, with stipels. stipules small, striate. Peduncles axillary, elongate; flowers clustered at nodes of a raceme; bracts and bracteoles small, soon falling. Calyx campanulate, 4-toothed; corolla white or purple; standard roundish, clawed, with 2 auricles at the base; wings falcate-obovate; keel incurved at a right-angle. Stamens 10, the uppermost free, the rest united; anthers all alike. Ovary sessile; style flattened toward apex and bearded along the inner side; stigma terminal. Ovules few. Pod falcate-oblong, slightly curved, strongly compressed, 2-valved with thickened margins, beaked at the apex, and 2-4-seeded; seeds somewhat compressed, with linear hilum.

A genus of 1 species, probably native of Africa but cultivated widely both for its edible seeds and young pods, and as an ornamental.

1. **Lablab purpureus** (L.) Sweet, Hort. Brit. 481. 1827.

Dolichos lablab L., 1753.

D. purpureus L., 1763

"Bonavist"

With the characters of the genus. Leaflets ovate-deltate or ovate-rhombic, 4–13 cm long, the lateral ones unequal-sided. Racemes many-flowered; standard 1.5 cm long and somewhat broader. Pods c. 7.5 cm long and 2.5 cm broad, the lower suture curved and finely warty-roughened; seeds c. 1 cm long.

GRAND CAYMAN: *Millspaugh 1396.*
– Now pantropical in distribution, cultivated or often spontaneous in roadside thickets and open waste ground.]

[Genus 20. **CAJANUS** DC.

Cajanus cajan (L.) Millsp., the Congo ("Gungo") or Pigeon Pea, is sometimes cultivated for its edible seeds. It shows little or no tendency to naturalize.

GRAND CAYMAN: *Hitchcock*.]

Genus 21. **RHYNCHOSIA** Lour.

Twining herbaceous or subwoody vines, rarely erect; leaves pinnately 3-foliolate, with or without stipels; leaflets minutely gland-dotted beneath; stipules lanceolate to ovate. Flowers borne singly or in pairs along the axillary racemes; calyx 4–5-lobed, the 2 upper lobes more or less united; corolla yellow, often tinged or striped with dark red; standard roundish, spreading or reflexed, with inflexed auricles at the base; wings narrow; keel falcate or incurved at the apex. Stamens 10, the uppermost free, the rest united. Ovary subsessile; style thread-like, incurved above, the stigma terminal; ovules 2 or rarely 1. Pod oblong or falcate, compressed, beaked at the apex, 2-valved, continuous within or rarely septate; seeds usually 2, roundish-compressed, often red, with short lateral hilum.

A large pantropical genus of about 150 species, a few extending into temperate North America.

1. **Rhynchosia minima** (L.) DC., Prodr. 2:385. 1825. **FIG. 134.**

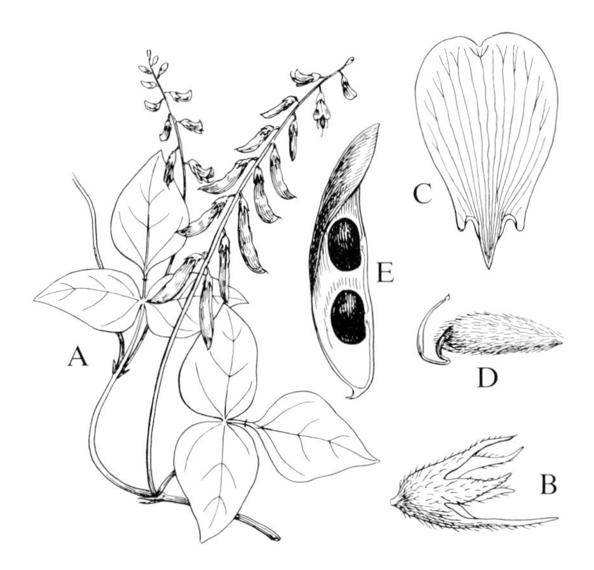

FIG. 134 **Rhynchosia minima.** A, leaves, inflorescence, and pods, x$^2/_3$. B, calyx, x5. C, standard, x5. D, pistil, x5. E, one valve of pod with seeds, x2. (F. & R.)

A small twining or trailing vine, the stems angled or grooved, puberulous when young; leaflets rhombic or subrhombic, 1-2.5 cm long, minutely puberulous on both sides; stipules subulate, c. 2 mm long. Racemes slender, 5-10 cm long, often exceeding the leaves; flowers remote, not paired, reflexed in age; calyx 3-4 mm long, puberulous; corolla pale yellow, 5-6 mm long; standard puberulous and gland-dotted. Pods 1-1.7 cm long, 3-5 mm broad, puberulous and gland-dotted, slightly constricted between the seeds; seeds 3-3.5 mm long, mottled with light and dark brown.

GRAND CAYMAN: *Brunt 1969; Hitchcock; Kings GC 74; Millspaugh 1283, 1349; Proctor 15113; Rothrock 167.* LITTLE CAYMAN: *Proctor 28161.* CAYMAN BRAC: *Kings CB 4; Proctor 28996.*
– A pantropical weed of open waste ground and roadsides, or sometimes in pastures and thickets.

[Genus 22. **MOGHANIA** J. St. Hil.

Erect shrubs or woody herbs; leaves 1-foliolate or digitately 3-foliolate; leaflets prominently veined and with numerous minute glandular dots beneath. Flowers in axillary and terminal spikes or racemes, or in small cymes along a raceme-like axis; bracts small and soon falling, or large, inflated and persistent. Calyx 5-lobed, the lowest lobe longer than the others; standard roundish, auricled at the base. Stamens 10, the uppermost free, the rest united. Ovary with 2 ovules; stigma small, terminal. Pod oblong, 2-valved, inflated; seeds 2 or sometimes 1.

A genus of about 35 species occurring in tropical Africa, Asia, and Australia.

1. Moghania strobilifera (L.) J. St. Hil. ex Ktze., Revis. Gen. Pl. 1:199. 1891.

Flemingia strobilifera (L.) Ait. f., 1812.

"Wild Hops"

A weedy shrub up to 1.5 m tall or more, the young stems and petioles appressed-pubescent; leaflet 1, elliptic to ovate, mostly 5–15 cm long, 3–10 cm broad, acutish at the apex. Inflorescences axillary and terminal, paniculate, consisting of 2 to 5 raceme-like branches, these densely beset with large, distichously arranged, inflated pale green bracts, each enclosing a small cyme of flowers; racemes 5-15 cm long; bracts cordate, 1-2.5 cm long, broader than long, pubescent, turning light brown with age. Calyx c. 5 mm long; corolla whitish, 5-6 mm long. Pods c. 1 cm long, concealed by the persistent bracts.

GRAND CAYMAN: *Kings GC 294.*
– Native of the East Indies, but long naturalized almost throughout the West Indies and common on many islands. It appears, however, to be rare in the Cayman Islands.]

Genus 23. DALBERGIA L.

Trees or shrubs, the branches often scandent or trailing; leaves pinnate with a terminal leaflet or (in Cayman species) 1-foliolate. Flowers white or purplish, in axillary or terminal racemes, cymes or panicles; bracts and bracteoles minute. Calyx 5-lobed, the lobes unequal, the upper two broader and the lowest one longer than the others; standard ovate or roundish; wings oblong; keel obtuse, its petals united dorsally at the apex. Stamens 10, the uppermost one more or less free, the sheath split on the upper side, sometimes also on the lower side; anthers small, erect, and paired. Ovary stalked; style short, incurved, with small terminal stigma; ovules few. Pod roundish to oblong or linear, very compressed and flat, indehiscent, 1–4-seeded; seeds kidney-shaped, compressed.

A pantropical genus of more than 100 species.

KEY TO SPECIES

1. Leaves glabrous on both sides and more or less cordate at the base; pods oblong, 3-4-seeded: **1. D. brownei**

1. Leaves puberulous beneath and truncate or rounded at the base; pods roundish, 1-seeded: **2. D. ecastaphyllum**

1. Dalbergia brownei (Jacq.) Urban, Symb. Ant. 4:295. 1905. **FIG. 135.**

"Cocoon"

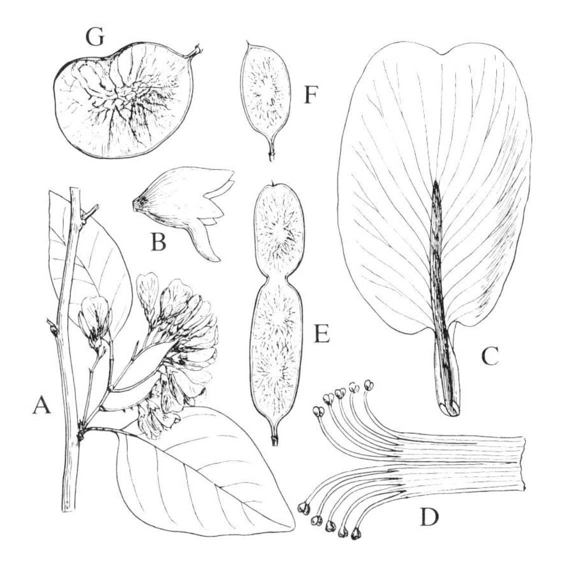

FIG. 135 **Dalbergia brownei.** A, portion of flowering branch, x$^2/_3$. B, calyx, x3. C, standard, x5. D, stamens, x5. E, F, pods, x$^2/_3$. G, pod of **D. ecastaphyllum** x$^2/_3$. (F. & R.)

A suberect, tangled or trailing shrub, the glabrous branches often ending in a woody tendril; leaflet 1, glabrous, ovate, 4–8.5 cm long, acutish at the apex, subcordate or cordate at the base, very lustrous on the upper side. Flowers white, fragrant, in small dense panicles scarcely exceeding the petioles; panicle-branches and petioles minutely puberulous. Calyx c. 4 mm long; standard c. 8 mm long; keel much shorter than the wings. Pods more or less oblong from a stalk-like base, mostly 2–5 cm long, 3-4-seeded or occasionally 1-seeded, glabrous and covered with fine raised reticulations.

GRAND CAYMAN: *Brunt 1629, 1912; Hitchcock; Proctor 15192.*
– Florida, the Greater Antilles, and continental tropical America, chiefly along the borders of mangrove swamps.

2. **Dalbergia ecastaphyllum** (L.) Taub. in Engl. & Prantl, Nat. Pflanzenfam. 3(3):335. 1894.

Ecastaphyllum brownei Pers., 1807.

A rambling shrub, sending up long vertical shoots from the more or less tangled main branches, reaching 3 m high or more, the young leafy branches finely pubescent; leaflet 1, oblong-ovate or elliptic, 5-12 cm long, shortly acuminate at the apex and usually rounded or truncate at the base, glabrous above, paler beneath and finely puberulous. Flowers white, faintly scanted, in small dense panicles about 2 cm long, about twice as long as the petioles. Calyx c. 3 mm long, pubescent; standard 7-8 mm long. Ovary with stalk longer than the calyx. Pods roundish or kidney-shaped, 2-3 cm across, finely pubescent, on a short stalk, always 1-seeded.

GRAND CAYMAN: *Brunt 2045.* LITTLE CAYMAN: *Kings LC 89, LC 90; Proctor 35208.*
– West Indies, continental tropical America, and west tropical Africa, along borders of mangrove swamps and borders of streams near the coast.

Genus 24. **PISCIDIA** L.

Deciduous trees or large shrubs, the dark twigs marked with numerous small whitish lenticels; leaves alternate, pinnate with a terminal leaflet; leaflets opposite, in 3-5 pairs; stipels absent. Flowers in small lateral panicles, appearing while the tree is leafless; calyx 5-lobed, the lobes short, triangular; corolla whitish or pink; standard rounded, retuse; wings and keel clawed at the base. Stamens 10, all united above, the uppermost one free toward the base. Ovary sessile, with many ovules; style thread-like, incurved, the stigma small, terminal. Pod linear, compressed, indehiscent, 3-7-seeded, with 4 broad longitudinal wings; seeds oval, compressed.

A small tropical American genus of about 6 species.

1. **Piscidia piscipula** (L.) Sarg. in Gard. & For. 4:436. 1891.

Ichthyomethia piscipula (L.) Hitchc., 1891.

"Dogwood"

A small tree usually 6-10 m tall; leaflets elliptic-oblong or obovate, 5-10 cm long, shortly acuminate at the apex, glabrous above, finely pubescent beneath. Panicles 8-20 cm long, somewhat dense with numerous flowers; pedicels 2-7 mm long; calyx 6–7 mm long, grayish-strigillose; corolla whitish or pale pink; standard c. 1.5 cm long, about equalled by the wings and keel. Pods 2–8 cm long, the body 4–5 mm wide, much exceeded in width by the wings; wings thin, glabrate, pale green, undulate or ruffled.

GRAND CAYMAN: *Kings GC 318.*
– Florida, West Indies, and continental tropical America, occasional or frequent in dry woodlands. The wood of this species is hard, heavy, strong and durable; it is difficult to work but takes a high polish. The bark, especially of the roots, is well-known for its narcotic and poisonous properties. In some places, people apply this material locally to relieve toothache, and in Jamaica it is said to be used occasionally for curing mange in dogs. If the bark and leaves are crushed and thrown into water, most nearby fish will soon become stupefied and will float on the surface.

Genus 25. **SOPHORA** L.

Herbs, shrubs or trees; leaves pinnate with a terminal leaflet; leaflets opposite, subopposite, or irregularly alternate; stipels absent. Flowers in terminal racemes or panicles; calyx bell-shaped, obliquely truncate or with 5 short teeth; corolla yellow; standard elliptic or rounded; wings and keel about the same length, clawed. Stamens 10, all free. Ovary short-stalked; style slightly incurved, with a minute terminal stigma; ovules rather few. Pod stalked, elongate, constricted between the seeds, indehiscent; seeds rounded-oblong or somewhat compressed.

A tropical and warm-temperate genus of about 50 species, some of the arboreous ones noted for their extremely hard wood.

1. **Sophora tomentosa** L., Sp. Pl. 1:373. 1753. **FIG. 136.**

"Micar"

A shrub usually 1–2.5 m tall, variably more or less tomentose throughout; leaves with 11–17 subopposite or irregularly alternate leaflets, these ovate or elliptic, 2–5 cm long, blunt or slightly retuse at the apex, densely pubescent beneath or sometimes glabrate. Racemes mostly 10–30 cm long, often many-flowered; calyx truncate or obscurely toothed, c. 7 mm long; standard elliptic, folded, 2–3 cm long. Pods 5–15 cm long, 5–9-seeded, tomentose; seeds rounded-oblong, 7–8 mm long, orange-brown.

GRAND CAYMAN: *Kings GC 387; Proctor 15147.*
– Pantropical in coastal thickets.

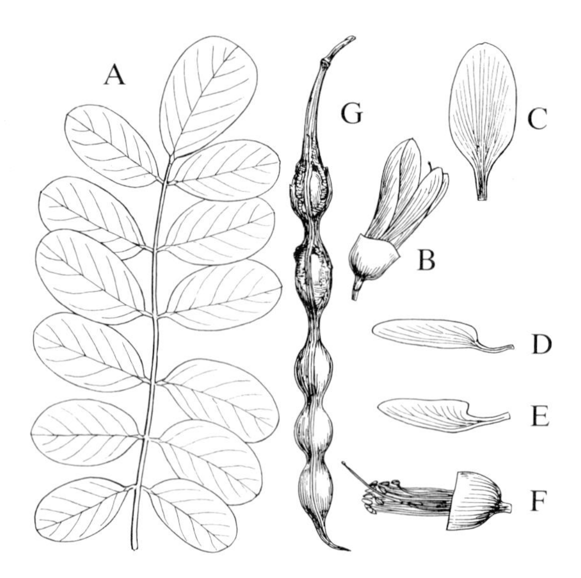

FIG. 136 **Sophora tomentosa.** A, leaf, x½. B, flower, x²/₃. C, standard, x²/₃. D, wing, x²/₃. E. keel, x²/₃. F, calyx, stamens and pistil, x²/₃. G, pod partially decayed above, showing seeds, x²/₃. (F. & R.)

SUBFAMILY 2. CAESALPINIOIDEAE

Herbs, shrubs or trees, with or without prickles, erect or sometimes trailing, scandent or climbing; leaves usually pinnate or bipinnate, or (in *Bauhinia*) with 2 leaflets partly united; stipules usually present. Flowers irregular or rarely regular, terminal or axillary, solitary or in racemes or panicles; sepals usually 5, imbricate or partly united; petals usually 5, imbricate, the uppermost surrounded by the rest in bud. Stamens 10, rarely fewer, sometimes some of them sterile; filaments usually free or sometimes partly united. Pods dehiscent or indehiscent.

The members of this subfamily are chiefly found in the tropics.

KEY TO GENERA

1. Leaves pinnate, or rarely bipinnate at the base only:

 2. Leaflets 2, more or less united: **30. Bauhinia**

 2. Leaflets more than 2:

3. Petals 3; stamens 3, united into a sheath; pods not strongly flattened, indehiscent, containing edible pulp: **[31. Tamarindus]**

3. Petals 5; stamens 6-10; pods strongly flattened:

4. Pods splitting at the middle of the valves, not at the margins; leaflets obovate; flowers numerous in racemes: **[27. Haematoxylum]**

4. Pods splitting at one or both margins, or not splitting; leaflets various, but if obovate, the flowers solitary: **29. Cassia**

1. Leaves amply bipinnate:

5. Sprawling or erect shrubs, with or without prickles; pods less than 12 cm long: **26. Caesalpinia**

5. Tree, without prickles; pods 20-50 cm long: **[28. Delonix]**

Genus 26. **CAESALPINIA** L.

Trees or shrubs, erect or sometimes sprawling or scandent, variously prickly or else unarmed; leaves bipinnate without a terminal pinna; pinnae with an odd number of leaflets, these small and numerous or few and relatively large; stipels sometimes present, or sometimes represented by spines; stipules large or minute, or rarely absent. Flowers usually yellow or red, often showy, in axillary racemes or terminal panicles; bracteoles absent. Calyx with short tube and 5 imbricate lobes; petals roundish to oblong, strongly imbricate, slightly unequal, sometimes clawed. Stamens all free, usually hairy or glandular at the base; anthers all alike. Ovary sessile, free from the calyx; style thread-like with small terminal stigma; ovules few. Pods compressed or somewhat turgid, 2-valved or sometimes indehiscent; seeds roundish or globose, sometimes flattened; endosperm lacking.

A pantropical genus of 70 or more species.

KEY TO SPECIES

1. Sprawling or scrambling wild shrubs, often prickly; seeds oblong to subglobose, not flattened:

2. Pods nearly or quite without prickles: **1. C. caymanensis**

2. Pods densely beset with prickles:

3. Stipules large and leaf-like; seeds gray: **2. C. bonduc**

3. Stipules minute or lacking; seeds yellow or olive:

   4. Leaflets mostly 2–3.5 cm long; pods less than 3.5 cm broad, often narrowed toward base: **3. C. wrightiana**

   4. Leaflets mostly 4–6 cm long; pods c. 4 cm broad, abruptly rounded or subtruncate at base: **4. C. intermedia**

1. Erect cultivated shrub; pods compressed, without prickles, the seeds strongly flattened: **[5. C. pulcherrima]**

1. Caesalpinia caymanensis Millsp. in Field Mus. Bot. Ser. 2:49, t. 60. 1900.

Sprawling shrub, the young branches white-ciliate and with few or no prickles, soon becoming glabrous; leaves golden-tomentose throughout, 20–25 cm long; pinnae about 6 pairs; leaflets about 15 per pinna, oblong-elliptic, 2–2.7 cm long, blunt and mucronate, each subtended on the rhachilla by a recurved thorn. Flowers undescribed. Pods dark brown, broadly oblong, 5.5–8 cm long, spineless or nearly so; seeds usually 2, gray or greenish-gray, subglobose, c. 2 cm long.

GRAND CAYMAN: *Kings GC 96, GC 97; Millspaugh 1263 (type).*
– Endemic. This species used to occur in extensive thickets just north of Georgetown, where in recent years it appears to have been exterminated by growth of the town and the construction of hotels.

2. Caesalpinia bonduc (L.) Roxb., Fl. Ind. 2:362. 1832. **FIG. 137.**

Caesalpinia bonducella (L.) Fleming, 1810.

Caesalpinia crista L., 1753, in part but not as to present interpretation.

"Cockspur", "Gray Nickel"

Sprawling or scrambling thorny shrub, the young branches finely pubescent and often bearing numerous prickles of variable length and stoutness; leaves finely golden-tomentose more or less throughout when young, 25–60 cm long, the rhachis and rhachillae with numerous recurved thorns. Pinnae in 5–8 pairs; leaflets about 15 per pinna, ovate or oblong-elliptic, 2–4 cm long, mucronate; stipules leafy, of 2 or 3 broadly-rounded segments to 3 cm long. Racemes simple or branched, 10–20 cm long, with numerous crowded flowers; calyx 6–7 mm long, reddish-woolly;

petals yellow, 9-12 mm long. Stamens shorter than the petals. Pods dark brown, broadly elliptic-oblong, 5-8 cm long, covered with numerous straight prickles; seeds usually 2, gray or greenish-gray, subglobose, c. 2 cm long.

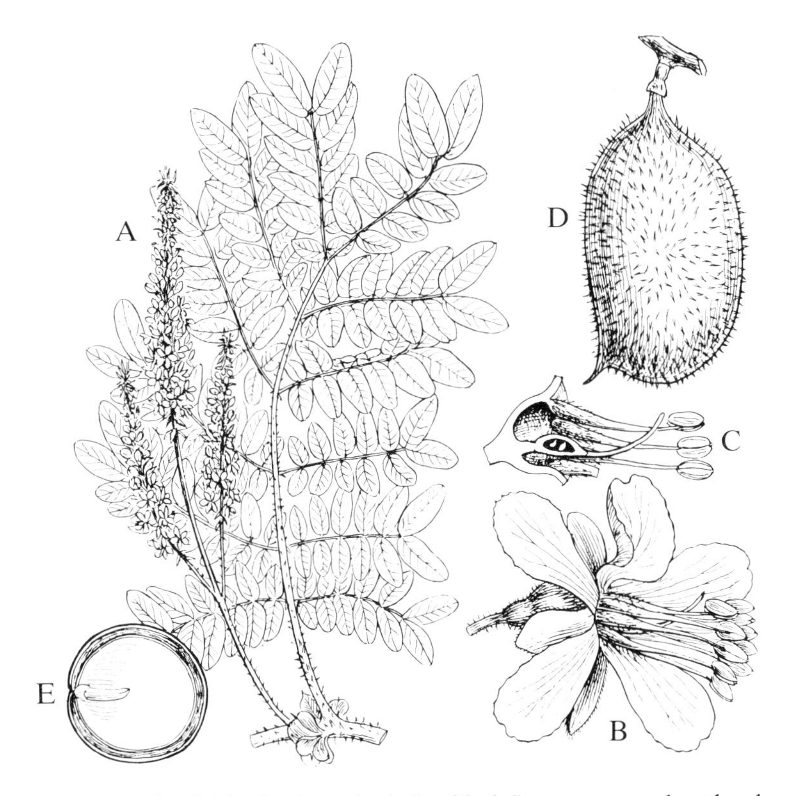

FIG. 137 **Caesalpinia bonduc.** A, leaf with inflorescence, much reduced. B, flower, x2. C, flower cut lengthwise, with sepals and petals removed, x2. D, pod, x½. E, seed cut to show cotyledon and plumule, x⅔. (F. & R.)

GRAND CAYMAN: *Brunt 1762, 1803; Hitchcock; Kings GC 98, GC 99; Milspaugh 1250; Proctor 15120, 31043; Sachet 425.* LITTLE CAYMAN: *Kings LC 46; Proctor 28032.* CAYMAN BRAC: *Kings CB 77; Proctor 29119.*

– Pantropical, on sandy seashores and in coastal thickets or open waste land. The seeds, when roasted, ground, and prepared as a beverage like coffee, are alleged to have medicinal value in controlling oedema, but unroasted are said to be poisonous.

3. Caesalpinia wrightiana Urban, Symb. Ant. 2:274. 1900.

"Cockspur", "Yellow Nickel"

Trailing or scrambling thorny shrub, the young branches minutely puberulous or glabrate, with rather few uniform prickles; leaves lustrous, puberulous when very young, soon becoming glabrate, 20–50 cm long or more, the rhachis and rhachillae armed with strongly recurved thorns. Pinnae mostly in 3–6 pairs; leaflets mostly 9–13 per pinna, ovate, rather coriaceous, short-acuminate, mucronate, glossy on the upper surface; stipules minute and subulate or usually lacking. Racemes simple or branched, mostly 8–12 cm long; pedicels slender, 2–3 (–6) mm long; calyx 7 mm long; petals yellow, c. 5 mm long. Stamens about equalling the petals. Pods light brown, often slightly glaucous, obovate-oblong, 5–6 cm long, covered with numerous prickles up to 1 cm long; seeds usually 2, yellow or olive, oblong, mostly 1.7–1.9 cm long.

LITTLE CAYMAN: *Proctor 35109, 35124.*
– Cuba, Jamaica, and perhaps the Swan Islands, occurring in coastal thickets and dry woodlands over limestone.

4. **Caesalpinia intermedia** Urban, Symb. Ant. 2:274. 1900.

Caesalpinia major of Adams in Fl. Pl. of Jamaica, 1972, not Dandy & Exell, 1938.

"Cockspur", "Yellow Nickel"

Scrambling shrub like the last, but with larger, often more elliptic leaflets of a thinner texture, the tips more strongly acuminate. Racemes simple or branched, 15–30 cm long; pedicels rather stout, 6–8 mm long; calyx 8–9 mm long; petals yellow, 10–11 mm long. Stamens shorter than the petals. Pods dark brown at maturity, ovate-oblong, 5–7 cm long, prickly more or less like the last. Seeds oblong-rounded to subglobose, c. 2 cm long or sometimes slightly less.

GRAND CAYMAN: *Kings GC 127; Proctor 15143.* LITTLE CAYMAN: *Proctor 28063.*
– Cuba and Jamaica. This species and the previous one (*C. wrightiana*) have long been confused and have usually been listed under one or another incorrect name. Type material of both species has been examined in order to establish their correct identity; for access to these authentic specimens the author is indebted to the late Dr. William T. Gillis of Michigan State University.

[5. **Caesalpinia pulcherrima** (L.) Sw., Obs. Bot. 166. 1791.

"Barbados Pride"

An erect, glabrous shrub to 3 m tall, unarmed or with a few scattered prickles; leaves 10–40 cm long; pinnae in 5–20 pairs; leaflets 17–25 per pinna, oblong,

1-2 cm long. Racemes usually terminal, pyramidal, the lower pedicels much longer than the upper. Sepals unequal, 10-16 mm long; petals yellow or red, long-clawed, 2-2.5 cm long. Stamens 6-8 cm long, with yellow or red filaments, these long-exserted. Pods flat, irregularly oblong, 8-10 cm long or sometimes a little longer, 6-8-seeded.

GRAND CAYMAN: *Hitchcock; Kings GC 397.* CAYMAN BRAC: *Proctor (sight record, cultivated).*
– Origin unknown; widely cultivated in most warm countries, and sometimes becoming naturalized.]

[Genus 27. **HAEMATOXYLUM** L.

Glabrous trees, often armed with spines; leaves even-pinnate or occasionally bipinnate by elongation of the lowest pair of pinnae; pinnae 2 to 4 pairs. Flowers small, yellow, in axillary racemes; bracteoles absent; calyx deeply 5-lobed, the lobes imbricate and slightly unequal; petals oblong, spreading, not clawed. Stamens free, hairy at the base; anthers all alike. Ovary short-stalked; style thread-like with small terminal stigma; ovules 2 or 3. Pods flat and wing-like, opening by a longitudinal slit along the middle of each valve; seeds oblong, compressed, without endosperm, the cotyledons 2-lobed.

A tropical American genus of 2 species.

1. **Haematoxylum campechianum** L., Sp. Pl. 1:384. 1753. **FIG. 138.**

"Logwood"

A shrub or small tree with deeply fluted trunk, usually less than 8 m tall, often armed with stout spines; pinnae cuneate-obovate, 0.7-2.5 cm long, rounded, truncate, or emarginate at the apex, finely many-nerved; stipules sometimes spine-like. Racemes mostly dense, 2-8 cm long; calyx 3-4 mm long; petals 5-6 mm long. Pods oblanceolate or narrowly oblong-elliptic, 3-5 cm long, 1-3-seeded.

GRAND CAYMAN: *Brunt 1800, 1808; Hitchcock; Kings GC 416; Millspaugh 1369, 1731; Sachet 409.*
– Native to Central America, chiefly in the Yucatan Peninsula; widely naturalized in the West Indies, where it was introduced early in the 18th century. The heartwood, which is bright red when freshly cut, is the source of a dye formerly much used for textiles, and which is still highly valued as a bacteriological and cytological stain. It is extracted by boiling the wood chips in water, and must be used with a chemical "mordant" in order to be permanent. The usual final dye-colour obtained is black. The wood itself is very hard and heavy, and normally will not float; although brittle, it is very durable and takes a high polish.]

[Genus 28. **DELONIX** Raf.

Delonix regia (Bojer) Raf., the Poinciana, is often planted for ornament. This tree, whose flowers occur in several colour-forms, is a native of Madagascar.

GRAND CAYMAN: *Brunt 2027; Kings GC 77.* CAYMAN BRAC: *Proctor (sight record).*]

Genus 29. **CASSIA** L.

Herbs, shrubs or trees of various habit; leaves even-pinnate, often with glands on the rhachis and petiole. Flowers solitary, or in axillary or terminal racemes, or in terminal panicles; calyx of 5 imbricate sepals, these deciduous; petals 5, subequal or the lower ones smaller. Stamens usually 10, all perfect, or with 6-7 perfect and the remainder staminodial, or else 3 lacking; anthers all alike or those of the lower stamens larger, all opening by an apical pore or slit. Ovary sessile, stalked, free from the calyx; ovules few or many. Pods cylindric or flattened, indehiscent or 2-valved, rarely with longitudinal wings, continuous or septate within, or often filled with pulp; seeds more or less compressed, with endosperm.

A very large and complex genus of more than 450 species, most plentiful in tropical regions, by some botanists split into a number of smaller genera. The laxative drug "senna" consists of the dried leaves of several Asiatic species of *Cassia;* otherwise, few are of any economic value. However, a number of species (e.g. *C. fistula, G. nodosa,* etc.) are often planted as ornamentals.

KEY TO SPECIES

1. Leaflets 2–4 pairs, pubescent all over:

 2. Weedy herbs with leaflets mostly 2–5 cm long:

  3. Petals 5–7 mm long; pods straight, 2.5–5 cm long, transversely grooved between the seeds: **1. C. uniflora**

  3. Petals c. 10 mm long; pods curved, 15–20 cm long, not transversely grooved: **2. C. obtusifolia**

 2. Small bushy shrub with leaflets less than 1.5 cm long: **7. C. clarensis**

1. Leaflets mostly 5 pairs or more, glabrous or nearly so:

 4. Petals 1.2 cm long or more; pods glabrous or puberulous, more than 5 cm long:

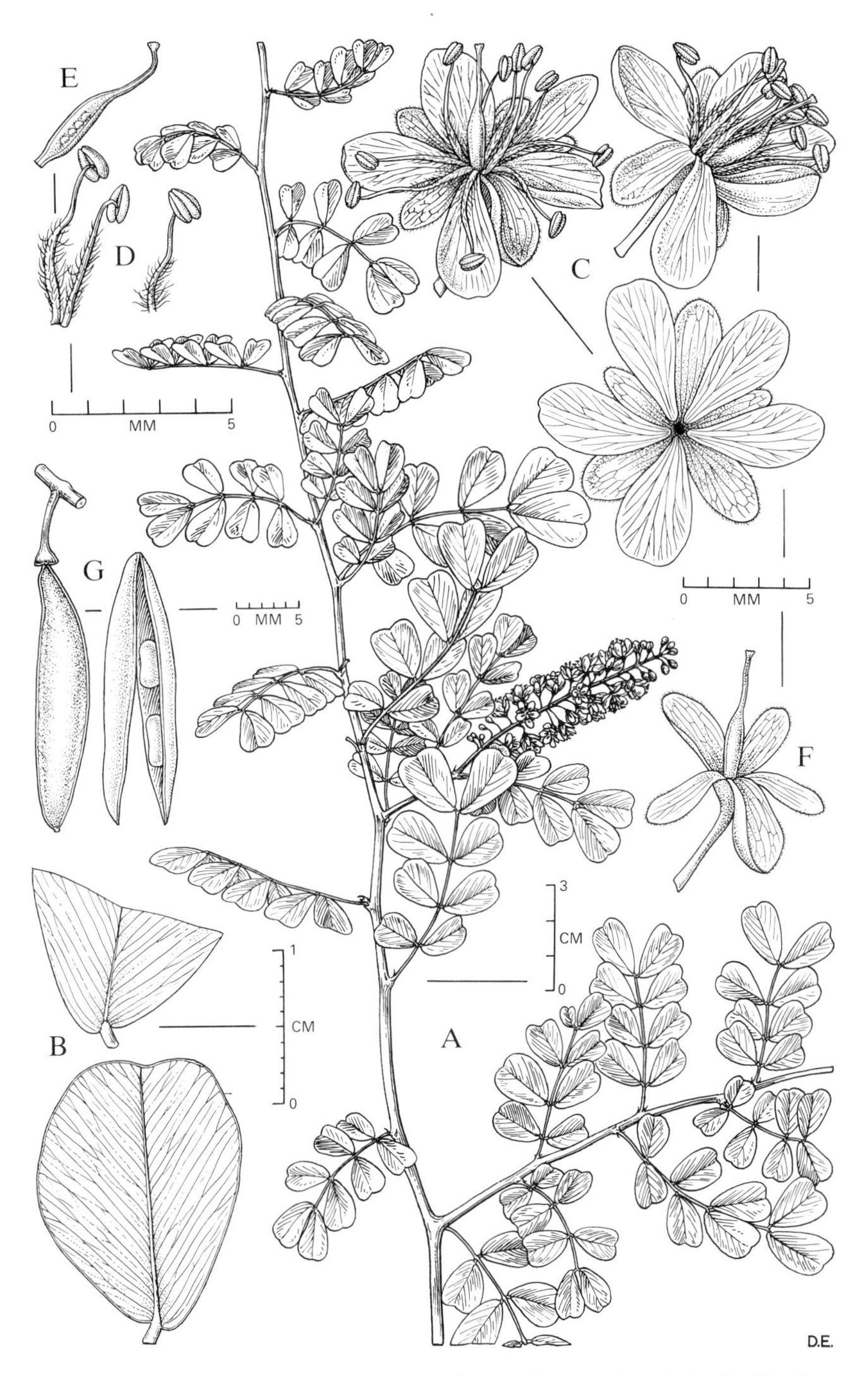

FIG. 138 **Haematoxylum campechianum.** A, portion of flowering branch. B, details of two leaflets. C, details of three flowers. D, stamens. E, pistil. F, pistil and calyx. G, two views of fruit. (D.E.)

5. Leaves glandular at the base of the petiole; leaflets acute or acuminate:

6. Petiolar glands globose-tuberculate; seeds dark olive: **3. C. occidentalis**

6. Petiolar glands cylindric-pointed; seeds black with gray margins: **4. C. ligustrina**

5. Leaves glandular between the lowest pair of leaflets; leaflets obtuse or rounded at the apex: **5. C. biflora**

4. Petals less than 1 cm long; pods hairy, 1.5-3 cm long: **6. C. nictitans** var. **aspera**

1. Cassia uniflora Miller, Gard. Dict. ed. 8. 1768. **FIG. 139.**

An erect annual herb to 1 m tall, clothed throughout with more or less appressed, brownish hairs. Leaves petiolate, the petiole not glandular; rhachis with 1 or more cylindric or club-shaped glands between the leaflets; leaflets 2-4 pairs, oblong to broadly obovate, 2-5 cm long, rounded and mucronate at the apex. Racemes short, axillary; sepals rounded, 4-6 mm long; petals yellow, about twice

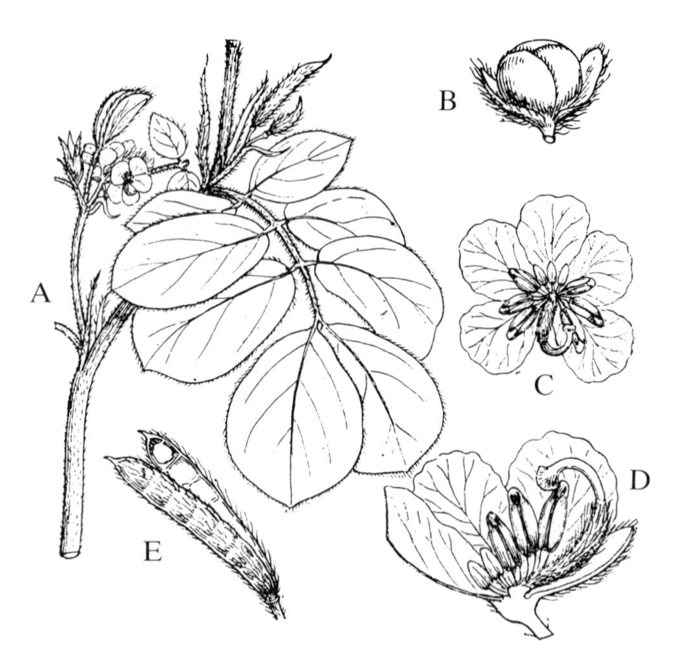

FIG. 139 **Cassia uniflora.** A, portion of branch showing leaf, flowers, and young pods, x½. B, calyx in bud, x2. C, flower from above, x2. D, vertical section through flower, x3. E, pod, x1. (F. & R.)

as long as the sepals. Perfect stamens 7. Pods linear, 2.5-5 cm long, 3-4 mm broad, deeply grooved between the seeds and ultimately separating into joints; seeds 5-10, trapezoidal, c. 4 mm long, dark brown.

GRAND CAYMAN: *Kings GC 48, GC 49, GC 90.*
– Bahamas, Greater Antilles, Mexico to northern South America, also in the Galapagos Islands, a weed of grassy roadsides, clearings, and thickets.

2. **Cassia obtusifolia** L., Sp. Pl. 1:377. 1753.

Annual shrubby herb to 1 m tall or more; leaflets 2-3 pairs, obovate or broadly oblong-obovate, puberulous on both sides, 1.5-4 cm long, 1-2.5 cm broad, broadly obtuse and shallowly emarginate, with a gland on the leaf-rhachis between the lowest pair of leaflets only. Flowers mostly paired on short peduncles in the leaf-axils; petals light yellow or orange-yellow. Pods narrowly linear, slightly 4-angled, smooth; seeds 20-24 per pod.

GRAND CAYMAN: *Sachet 383.*
– Pantropical except for Australia, a weed of open waste ground, roadsides and clearings.

3. **Cassia occidentalis** L., Sp. Pl. 1:377. 1753.

"Dandelion"

An erect annual herb up to about 1 m tall, often subwoody near the base, glabrous or nearly so. Leaves long-petioled, the petiole bearing a sessile globose gland near the base; leaflets mostly 4-6 pairs, ovate or lance-ovate, 3-7 cm long, acuminate at the apex, the margins finely ciliate. Racemes few-flowered, in the upper axils; sepals greenish, 6-9 mm long; petals yellow, twice as long as the sepals. Perfect stamens 6, two of them longer than the other four, and 4 staminodes. Ovary pubescent. Pods oblong-linear, slightly curved, 6-12 cm long, 6-9 mm broad, slightly grooved transversely between the seeds, minutely puberulous or glabrate; seeds flattened-obovoid, brown, 3-4 mm long.

GRAND CAYMAN: *Hitchcock; Kings GC 220; Millspaugh 1393; Proctor 15112.* CAYMAN BRAC: *Proctor 28971.*
– Southern United States, West Indies, and continental tropical America; naturalized in the Old World tropics. In some countries, the roasted and pulverized seeds are used as a substitute for coffee.

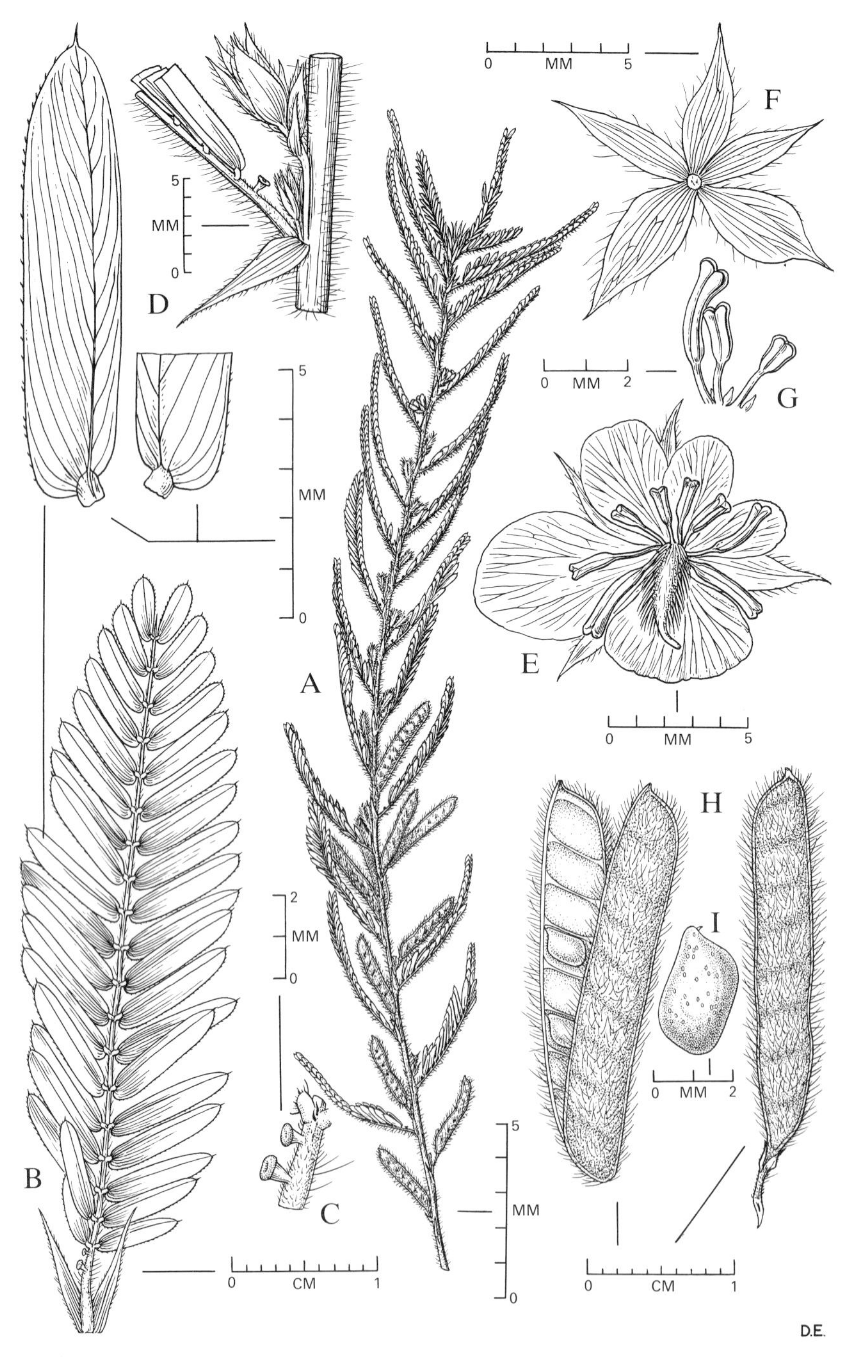

FIG. 140 **Cassia nictitans var. aspera**. A, branch with flowers and fruits. B, leaf. C, glands of petiole. D, details of stem and leaflets. E, complete flower. F, calyx. G, stamens. H, fruits. I, seed. (D.E.)

4. Cassia ligustrina L., Sp. Pl. 1:378. 1753.

A small erect shrub up to 1.5 m tall or more, the young branches puberulous. Leaves long-petioled, the petiole bearing a pointed-cylindric or conic gland below the middle; leaflets 5-7 pairs, lanceolate and unequal-sided, 1.5-4 cm long or more, narrowly acute, glabrous except for the finely ciliate margins. Racemes few-flowered, in the crowded upper axils, forming a panicle-like leafy inflorescence; sepals 6-8 mm long; petals yellow, dark-veined, 1.2-1.5 cm long. Stamens as in *C. occidentalis.* Pods oblong-linear, somewhat curved, 7-10 cm long, 6-8 mm broad, glabrous or sparsely puberulous; seeds numerous, oblong-ellipsoid, 4-4.5 mm long, black with a gray rim.

GRAND CAYMAN: *Fawcett; Proctor 15266.*
– Florida, Bermuda, Bahamas, and the Greater Antilles, in thickets and clearings.

5. Cassia biflora L., Sp. Pl. 1:378. 1753.

A shrub usually 2-4 m tall, the young twigs, petioles, and inflorescence glabrous or puberulous. Leaves petiolate, the petiole not glandular, but a cylindric and often acuminate gland occurs between the lowest pair of leaflets; leaflets 4-11 pairs, oblong-elliptic or narrowly obovate, 1-3.5 cm long, rounded and mucronulate at the apex. Inflorescence a subumbellate axillary raceme with 2-4 rather large flowers, the pedicels bearing cylindric glands at the base; calyx 5.5-8 mm long, the lobes unequal; petals yellow, very unequal, the largest 2-2.3 cm long, subsessile, the smaller ones clawed. Perfect stamens 7, three of them larger than the others and beaked. Ovary sessile. Pods linear, compressed, 2-valved, 7-15 cm long, 5-8 mm broad; seeds 14-20, oblong with notch at one end, brown, 5-6 mm long.

CAYMAN BRAC: *Proctor 28956.*
– West Indies, Central America and northern South America, in thickets and woodlands.

6. Cassia nictitans (L.) Greene.
var. **aspera** (Muhl. ex Elliott) Torr. & Gray, Fl. N. Am. 1:396. 1840. **FIG. 140.**

Chamaecrista riparia of Britton & Millsp., 1920, not *Cassia riparia* Kunth, 1824.

Chamaecrista confusa Britton, 1930, not *Cassia confusa* Phil., 1894.

Cassia caymanensis C. D. Adams, 1970.

"Wild Shame-face"

A small woody herb or sub-shrub, the villous-pubescent stems ascending or rigidly erect, branched or unbranched. Leaves much longer than wide, 2.5-6.5 cm long, short-petiolate, the pubescent petioles with 1 or 2 stalked glands; rhachis pubescent; leaflets glabrous, in 10-25 pairs, linear, unequal-sided, 6-12 mm long, aristate at the apex. Flowers solitary on axillary peduncles shorter than the petioles; calyx c. 5 mm long, hairy; petals yellow, the larger 7-8 mm long. Perfect stamens 10. Pods compressed, hairy, linear-oblong, 15-25 mm long, 3-4 mm wide, 2-valved and elastically dehiscent; seeds 4-7, flattened obovoid, dark brown, c. 3 mm long.

GRAND CAYMAN: *Brunt 1904; Kings GC 44, GC 92; Millspaugh 1305; Proctor 15072, 27929, 31048; Sachet 370.*
– Southeastern U.S.A., Bahamas and Jamaica, in pastures and sandy clearings.

7. **Cassia clarensis** (Britton) Howard in J. Arnold Arb. 28:126. 1947.

A bushy, much-branched shrub up to 1.5 m tall, densely appressed-puberulous throughout. Leaves often broader than long, the short petiole bearing a round sessile gland above the middle; leaflets 2-4 pairs, oblanceolate or oblong-obovate, unequal-sided, 8-15 mm long, rounded or truncate and mucronate at the apex, the lateral veins prominent. Flowers axillary, solitary or 2 together, on slender pedicels c. 1 cm long, much longer than the petioles; calyx c. 8 mm long, puberulous in a longitudinal line on the back; petals yellow, the larger 9-10 mm long. Perfect stamens 10. Pods compressed, appressed-puberulous, linear and slightly curved, 20-35 mm long, 3-4 mm wide, apiculate, 2-valved and elastically dehiscent; seeds 8-11, oblong-stipitate, light brown, c. 3 mm long.

GRAND CAYMAN: *Brunt 1907.*
– Cuba and perhaps Jamaica, in sand or pockets of exposed limestone, usually near the sea. This species is not very distinct from several others occurring at various West Indian localities, in particular *Cassia lineata* Sw.

Genus 30. **BAUHINIA** L.

Trees or shrubs, erect or climbing, with or without spines or tendrils; leaves simple, of 2 united leaflets more or less parted at the apex, or rarely completely 2-foliolate; stipules small and soon falling. Flowers solitary or in racemes, these simple and terminal or axillary, or paniculate; calyx with a short elongate tube, the limb more or less spathe-like, before anthesis either closed and entire, or else contracted at the apex and 5-toothed, after anthesis remaining entire or variously splitting. Petals 5, slightly unequal. Perfect stamens 10 or less, some or most often being reduced to staminodia or lacking; anthers attached at the middle and opening longitudinally. Ovary stalked or subsessile, with 2 to many ovules; style various, the stigma small or often dilated and peltate. Pods oblong or linear, compressed, 2-valved with elastic valves, or indehiscent; seeds roundish, compressed, and with endosperm; cotyledons flat, more or less fleshy.

A pantropical genus of more than 200 species, several often cultivated for ornament.

1. Bauhinia divaricata L., Sp. Pl. 1:374. 1753. **FIG. 141.**

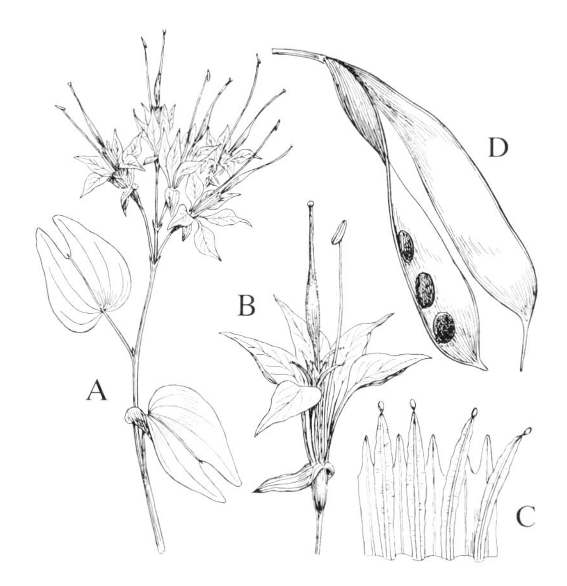

FIG. 141 **Bauhinia divaricata.** A, inflorescence and two leaves. x⅓. B, flower, x⅔. C, sterile stamens opened flat, x2. D, pod, x⅔. (F. & R.)

Bauhinia porrecta Sw., 1788.

"Bull-hoof"

An erect to arching shrub or small tree to 4 m tall or more, the branches glabrous or nearly so; leaves petioled, the bifurcate blades mostly 2–6 cm long, often wider than long, the lobes variously united from 1/4 to 7/8 of their length, the nerves puberulous beneath. Racemes short and dense, with up to about 10 flowers, all perfect or often several with abortive ovary and functionally staminate only; calyx 12–15 mm long, terminated by short, bristle-like teeth; petals at first white, then turning pink, 2–2.5 cm long, slender-clawed. Fertile stamen 1, twice as long as the petals, the 9 sterile ones much shorter and united into a tube for most of their length. Ovary long-stalked. Pods linear- or oblanceolate-falcate, 5–12 cm long, 9–15 mm wide, long-stalked, finely puberulous or glabrate; seeds 3–10, flattened-ellipsoidal, 6–8 mm long, dark brown.

GRAND CAYMAN: *Brunt 1787, 1887; Correll & Correll 50998; Hitchcock; Lewis GC 32; Millspaugh 1289; Proctor 15024.* LITTLE CAYMAN: *Kings LC 106; Proctor 28040.* CAYMAN BRAC: *Kings CB 14, CB 91a; Millspaugh 1160, 1207, 1208; Proctor 29047.*

– West Indies and on the continent from southwestern Texas south to Honduras, often frequent in rocky woodlands. Such ornamental species as *Bauhinia galpinii* N.E.Br. and *Bauhinia monandra* Kurz are sometimes cultivated.

[Genus 31. **TAMARINDUS** L.

Unarmed trees; leaves even-pinnate, with small, numerous leaflets, the petiole and rhachis not glandular; stipules minute, soon falling. Flowers in short terminal racemes; bracts and bracteoles ovate-oblong, soon falling. Calyx 4-lobed; petals 5, three of them subequal and evident, the other two minute and scale-like. Perfect stamens 3, united into a sheath; anthers oblong, opening longitudinally. Ovary stalked, the stalk adnate to the calyx-tube; ovules numerous; style elongate. Pods more or less oblong, thick and scarcely compressed, indehiscent, the outer covering (epicarp) crust-like and fragile, enclosing an edible pulp; seeds roundish, compressed, separated by hard partitions; endosperm lacking, the cotyledons thick.

The genus consists of a single species, native of tropical Asia, but now cultivated in most tropical countries and often becoming naturalized.

1. **Tamarindus indica** L., Sp. Pl. 1:34. 1753.

"Tamarind"

A handsome evergreen tree 10 m tall or more; leaves glabrous or nearly so; leaflets 10–18 pairs, oblong, 12–25 mm long, the venation finely prominulous. Racemes mostly shorter than the leaves; calyx 8–10 mm long; petals pale yellow with red veins, the larger petals slightly longer than the calyx. Pods 5–10 cm long or more, 2 cm thick, brown and finely scaly; seeds 1 cm in diameter, lustrous brown.

GRAND CAYMAN: *Brunt 2171; Kings GC 81.* CAYMAN BRAC: *Kings CB 6.*

– The juicy, acidulous pulp of the fruit is used to make a refreshing drink, and as an ingredient of candies and condiments. The plant itself makes a fine shade tree, resistant to drought, and there are also numerous uses for the leaves and wood.]

SUBFAMILY 3. MIMOSOIDEAE

Herbs, shrubs, or trees, with or without prickles; leaves usually bipinnate. Flowers regular, perfect or rarely polygamous, in axillary globose heads, cylindrical spikes or racemes, or in terminal panicles of globose heads; calyx usually 4–5-lobed or -toothed, the lobes valvate; petals 4 or 5, free or united below, valvate. Stamens as many or twice as many as the petals, or more numerous up to 100 (always 10 or more in Cayman genera except *Mimosa*); filaments thread-like, usually elongate, free or united below, the tube so formed often exserted and showy.

A subfamily chiefly occurring in the tropics.

KEY TO GENERA

1. Leaflets alternate; flowers in racemes; stamens scarcely longer than the petals; seeds bright red: **[32. Adenanthera]**

1. Leaflets opposite; flowers in small round heads; stamens long-exserted; seeds not red:

  2. Pods with continuous persistent margins, the valves separating from them and breaking into joints; flowers pink; stamens 4 per flower; leaves sensitive to the touch: **34. Mimosa**

  2. Pods with valves not separating from the margins and not breaking into joints; flowers white or yellow; stamens 10 or more per flower; leaves not sensitive to the touch:

    3. Flowers yellow; pods swollen, marked with lines: **36. Acacia**

    3. Flowers white; pods flat, not marked with lines:

      4. Pods more than 1 cm broad: **35. Leucaena**

      4. Pods less than 1 cm broad:

        5. Green-stemmed wiry herb; leaflets less than 1 cm long; stamens 10, free: **33. Desmanthus**

        5. Large shrub or small tree; leaflets more than 1 cm long; stamens more than 10, united at the base: **37. Calliandra**

[Genus 32. **ADENANTHERA** L.

Unarmed trees; leaves bipinnate, the pinnae subopposite, the leaflets alternate. Flowers very small, in long slender racemes, perfect or polygamous; calyx bell-shaped, 5-toothed; petals 5, united below the middle or nearly free. Stamens 10, free, the filaments about as long as the petals; anthers bearing a deciduous gland. Ovary sessile, with many ovules. Pods flat, linear, 2-valved, the valves becoming twisted and curled after dehiscence; seeds roundish, thick and hard.

A small genus of 3 or 4 species, native of Africa, Asia and Australia.

1. **Adenanthera pavonina** L., Sp. Pl. 1:384. 1753.

"Curly bean"

A tree to 10 m tall or more, nearly glabrous; leaves with 2 to 5 pairs of sub-opposite pinnae, these 10–20 cm long; leaflets alternate, elliptic or oblong-elliptic, 2–4 cm long, obtuse. Racemes 5–15 cm long, simple from the leaf-axils, panicled at the ends of branches. Flowers cream, turning yellow or pale orange, on slender pedicels; calyx c. 1 mm long; petals c. 3 mm long. Pods 15–25 cm long, 12–16 mm broad, swollen over the seeds; seeds somewhat compressed, c. 8 mm in diameter, bright red.

GRAND CAYMAN: *Hitchcock; Kings GC 76.*
– Native of tropical Asia, but now planted and becoming naturalized in most warm countries. The wood is hard, close-grained, strong and durable. The seeds are often used to make necklaces; in India, they are also used as a standard measure of weight, 1 seed weighing about 4 grains.]

Genus 33. **DESMANTHUS** Willd.

Unarmed woody herbs or slender shrubs with angulate-striate branches; leaves bipinnate, with a sessile gland on the petiole just below the lowest pair of pinnae; leaflets very small; stipules bristle-like, persistent. Flowers in small, few-flowered heads, all perfect or the lowest sterile or staminate only, on solitary axillary peduncles. Calyx shortly 5-toothed; petals 5, free or slightly coherent, valvate. Stamens usually 10, free, exserted; anthers without glands. Ovary subsessile, with many ovules. Pods linear, straight or falcate, flat, 2-valved, continuous or septate within; seeds flattened-ovoid.

A tropical American genus of perhaps more than 20 species, the exact number uncertain due to great variability within the group.

1. Desmanthus virgatus (L.) Willd. in L., Sp. Pl. 4:1047. 1806. **FIG. 142.**

Desmanthus depressus Humb. & Bonpl. ex Willd. in L., 1806.

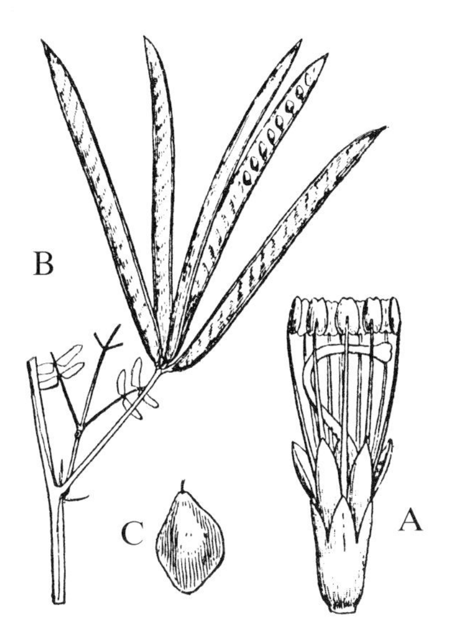

FIG. 142 **Desmanthus virgatus.** A, flower, x5. B, peduncle and ripe pods, x$^{2}/_{3}$. C, seed much enlarged. (F. & R.)

A somewhat shrubby herb or slender shrub, variable in habit from decumbent to erect, to 1 m tall or more, glabrous or nearly so; leaves with pinnae in mostly 3-5 pairs; leaflets 8-20 pairs, linear or linear-oblong, 3-9 mm long, of thin texture. Peduncles 1-5 cm long; flowers whitish, in small brush-like heads 3-5 mm across. Pods 3-8 cm long, 3-5 mm broad, acute or acuminate; seeds oblique, brown, 2 mm long.

CAYMAN BRAC: *Proctor 28972.*
– Florida and Texas southward to South America, and also naturalized in tropical Asia, a weed of secondary thickets and open waste ground. Some writers separate the smaller and larger forms into two species, but there appears to be complete intergradation between the extremes. The Cayman specimen combines the leaf and fruit characters of *D. "depressus"* with the erect, shrubby habit of typical *D. virgatus.*

Genus 34. **MIMOSA** L.

Herbs, shrubs or trees, or sometimes woody vines, usually armed with prickles; leaves bipinnate, often sensitive to the touch, and the petioles mostly lacking glands. Flowers perfect or polygamous, in stalked heads, mostly 4-5-parted, or sometimes the parts in 3's or 6's; peduncles axillary and solitary or clustered, or sometimes terminal and paniculate. Calyx usually minute; petals more or less united. Stamens as many or twice as many as the petals, free, exserted. Ovary sessile or rarely stalked, and with 2 to many ovules. Pods oblong or linear, usually flat, the valves jointed and separating from the persistent continuous margins, the joints 1-seeded; seeds compressed, roundish.

A pantropical genus of more than 400 species, best represented in the warmer parts of America.

1. **Mimosa pudica** L., Sp. Pl. 1:518. 1753. **FIG. 143.**

"Shame-face", "Shame-lady", "Sensitive Plant"

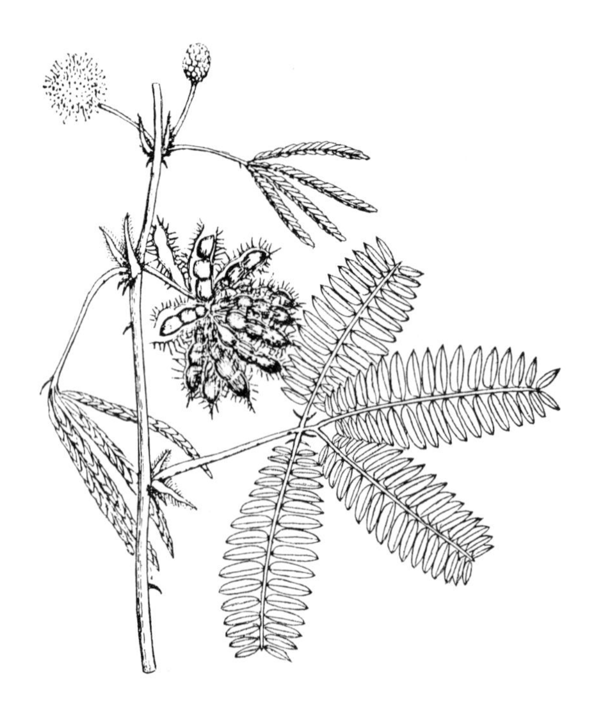

FIG. 143 **Mimosa pudica.** Portion of branch with flowers in bud and open, and cluster of ripe pods. (F. & R.)

A subwoody annual or persistent herb, prostrate or erect, the stems up to 60 cm long, bearing long spreading hairs or glabrate, also armed with sharp recurved prickles below the stipules and sometimes elsewhere. Leaves collapsing when touched; pinnae 2 pairs, 2.5-8 cm long; leaflets in 10-20 pairs, oblong-linear, 5-10 mm long, very inequilateral at the base, and sparsely hairy. Flowers pink, in globose heads 1-1.5 cm in diameter, the peduncles 1-3 in the leaf-axils; petals

c. 2 mm long; stamens 4, pink, 7-8 mm long. Pods linear-oblong, 1-1.5 cm long, 3-4 mm broad, with 2-5 joints, glabrous, the margins densely armed with bristles.

GRAND CAYMAN: *Kings GC 421.*
– West Indies and continental tropical America, also naturalized in the Old World tropics. This is chiefly a weed of pastures and open waste ground, often in moist places and along paths.

Genus 35. **LEUCAENA** Benth.

Unarmed shrubs or small trees; leaves bipinnate, usually with a gland on the petiole; stipules small. Flowers white, in globose, stalked, many-flowered heads, the peduncles solitary or clustered in the leaf-axils, or sometimes arranged in terminal naked racemes, each peduncle bearing 2 bracts at or below the apex. Calyx tubular, 5-toothed; petals 5, free, valvate. Stamens 10, free, exserted, the anthers often hairy. Ovary stalked, with filiform style; ovules many. Pods oblong-linear, flat, 2-valved, continuous within, short-stalked; seeds transverse, compressed-ovoid.

An American genus of perhaps as many as 50 species, but probably less.

1. **Leucaena leucocephala** (Lam.) DeWit in Taxon 10:54. 1961.

Leucaena glauca Benth., 1842, in part, not *Mimosa glauca* L., 1753.

A shrub or small tree to about 6 m tall (in our area), the younger parts and leaflet-margins puberulous; pinnae in 3-8 pairs, 6-9 cm long; leaflets in 8-14 pairs, linear-oblong and inequilateral, 7-16 mm long, 2-4 mm broad, often pale or glaucous beneath. Flower-heads c. 2 cm in diameter, on axillary peduncles 2-4 cm long; calyx 2-3 mm long; petals 4-5 mm long, puberulous. Stamens 8-10 mm long. Pods dark brown, pointed, 10-18 cm long, 1.3-1.8 cm broad, usually 3-10 in a cluster.

GRAND CAYMAN: *Brunt 1891, 2083; Hitchcock; Kings GC 78; Millspaugh 1391; Proctor 15205.* CAYMAN BRAC: *Kings CB 83; Proctor 28933.*
– Florida, West Indies, and continental Caribbean countries; naturalized in the Old World tropics; characteristic of second-growth thickets, old fields, and roadsides. It is alleged that if horses or mules eat any part of this plant, their hair will fall out, but that cattle and goats are not affected.

Genus 36. **ACACIA** L.

Shrubs or trees, armed with prickles or unarmed; leaves bipinnate, usually with a petiolar gland; leaflets mostly small and numerous; stipules often spine-like.

Flowers in globose heads or cylindric spikes, perfect or polygamous, usually yellow, on solitary or clustered peduncles from the leaf-axils, or in terminal panicles. Sepals 4–5-toothed or -lobed; petals 4 or 5, more or less united. Stamens numerous (sometimes 50 or more), free or nearly so, exserted. Ovary sessile or stalked, with hair-like style; ovules 2 to many. Pods various in form, cylindric or compressed, linear to ovate, straight, curved or contorted, membranous to woody, 2-valved or indehiscent; seeds transverse or longitudinal, usually compressed-ovoid.

A large, widespread genus of about 500 species, most plentiful in tropical America, Africa and Australia. It is sometimes subdivided into a number of smaller genera, but these are mostly not very distinct.

1. **Acacia farnesiana** (L.) Willd. in L., Sp. Pl. ed. 4, 4:1083. 1806.

A small bushy tree or thicket-forming shrub, the branches glabrous or nearly so, with prominent lenticels, and armed with stipular spines; leaves with 1-3 pairs of pinnae, rarely more, 1–4 cm long; petiole puberulous and with a small gland; leaflets in 10–20 pairs, oblong-linear, 2-6 mm long, c. 1 mm broad; stipular spines 4–30 mm long. Flowers yellow, fragrant, in globose heads 7–15 mm in diameter; peduncles solitary or clustered, 2-4 cm long, puberulous. Calyx 1-1.5 mm long; corolla 2-3 mm long. Pods dark brown and marked with fine longitudinal lines, glabrous, linear-oblong, curved, sub-cylindric, 4–7 cm long, 1–1.5 cm thick, filled with sweet pulp.

GRAND CAYMAN: *Brunt 1890.* CAYMAN BRAC: *Proctor 29087.*
– Southern United States, West Indies, and continental tropical America, often planted and naturalized, so that its true natural range is obscure; introduced in the Old World tropics and subtropics. Commonly occurs in thickets along the borders of pastures; the leaves and pods are much eaten by livestock. The wood is hard and close-grained, but used chiefly as fuel. The bark and pods are rich in tannin. The viscid juice of the pods can be used to mend china, while the gum exuding from the trunk is similar to gum arabic and is suitable for making mucilage. The flowers yield a delicious, high-priced perfume by petroleum ether extraction further purified by alcoholic extraction; commercial production at present is mostly confined to Lebanon.

Genus 37. **CALLIANDRA** Benth.

Mostly unarmed shrubs or small trees, rarely herbs; leaves bipinnate, the leaflets small and numerous or large and few; stipules usually persistent, often crowded at the base of young shoots, rarely spine-like. Peduncles solitary or clustered in the leaf-axils or in terminal racemes; flowers in globose heads, polygamous; calyx 5-toothed or -lobed; corolla funnel-shaped, 5-lobed to about the middle, the lobes valvate. Stamens numerous (10–100), united toward the base, long-exserted;

anthers usually minute and either glandular-pubescent or glabrous. Ovary sessile, with hair-like style; ovules many. Pods linear, straight or nearly so, narrowed at the base, flat with thickened margins, 2-valved, the valves elastically dehiscent, continuous within; seeds roundish, compressed.

A genus of more than 150 species, the majority in tropical America. Several with showy red stamens are commonly cultivated for ornament.

1. **Calliandra cubensis** (Macbr.) Léon in Contr. Ocas. Mus. Hist. Nat. Col. "De La Salle", no. 9:7. 1950.

Calliandra gracilis of Hitchc., 1893, not Griseb., 1861.

Calliandra formosa var. *cubensis* Macbr., 1919.

A small glabrous tree to 5 m tall or more; leaves with 2 or 3 pairs of pinnae; leaflets 6–12 pairs, oblong or oblong-obovate, mostly 10–24 mm long with rounded apex and very unequal-sided base. Peduncles 3–5 cm long; calyx 2.5–3 mm long; corolla 4–6 mm long. Stamens with white filaments 1–1.5 cm long. Pods 5–9 cm long, usually 7–8 mm broad, rounded-truncate and obliquely short-apiculate at the apex, long-attenuate to a stalk-like base, 8–10-seeded.

GRAND CAYMAN: *Hitchcock.* LITTLE CAYMAN: *Proctor 35189.* CAYMAN BRAC: *Proctor 29034, 29355.*
– Eastern Cuba and the Bahamas, in rocky limestone woodlands.

Family 57

LYTHRACEAE

Herbs, shrubs or trees, the young stems often 4-angled; leaves usually opposite, simple and entire, or rarely alternate; stipules absent or sometimes present. Flowers regular or zygomorphic, usually perfect, 3–16-parted, axillary or extra-axillary and solitary or in cymes or clusters, or rarely in terminal panicles. Calyx tubular, with valvate primary teeth or lobes, sometimes also with "epicalyx" of as many accessory teeth; petals as many as the calyx-lobes, or fewer, or none, inserted in the throat of the calyx between the lobes, often crumpled in bud. Stamens as many as, or fewer or more than, the calyx-lobes, inserted at different levels on the inside of the calyx-tube. Ovary superior, sessile or stalked, completely or incompletely 2–6-celled, the style simple or none, the stigma small and capitate or rarely 2-lobed; ovules numerous on axile placentas, or rarely only 2. Fruit a dry dehiscent or indehiscent capsule; seeds usually small and many, lacking endosperm.

A world-wide but chiefly tropical and subtropical family of about 22 genera and 475 species. Few are of any commercial value, but several, especially species of *Lagerstroemia* (with alternate leaves and 15 to many stamens) are widely cultivated for ornament.

KEY TO GENERA

1. Herbs; flowers sessile and solitary or few together in the leaf-axils: **1. Ammannia**

1. Shrubs; flowers stalked and numerous in terminal panicles: [**2. Lawsonia**]

Genus 1. **AMMANNIA** L.

Annual herbs, glabrous or nearly so, with stems more or less 4-angled; leaves narrow, opposite, sessile. Flowers small, sessile, solitary or clustered in the leaf-axils; calyx bell-shaped to globose or void, 4-toothed, with 4 accessory teeth in the sinuses; petals 4, deciduous, or else none. Stamens 4 or 8. Ovary sessile, 1-5-celled, with very short or longer and exserted style. Capsule subglobose, thinly membranous, more or less enclosed by the persistent calyx, rupturing irregularly.

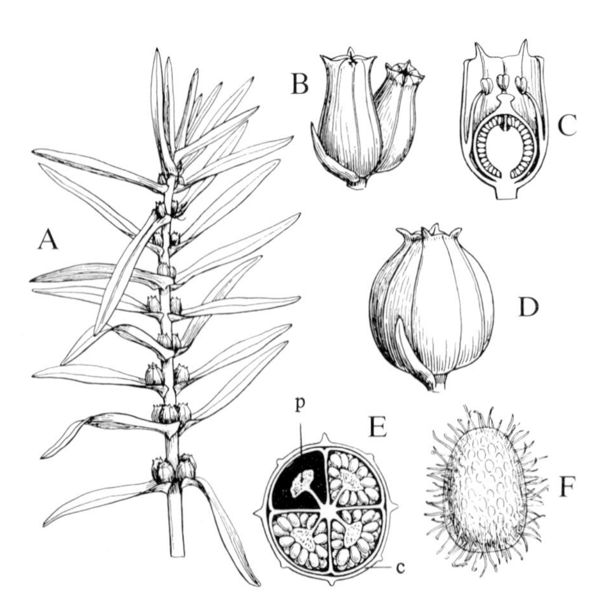

FIG. 144 **Ammannia latifolia**. A, portion of branch with flowers and fruit, x$^2/_3$. B, two flowers from leaf-axil, x4. C, flower cut lengthwise and spread out, x4. D, fruit enclosed in the globose calyx, x4. E, x-section of fruit, x4; c, calyx; p, placenta with seeds removed. F, seed, x30. (F. & R.)

A mostly tropical or subtropical genus of about 20 species, mostly occurring in swamps or moist places.

1. **Ammannia latifolia** L., Sp. Pl. 1:119. 1753. **FIG. 144.**

A sprawling or erect herb to 1 m tall or more, simple or much-branched; leaves linear or linear-lanceolate, 3–8 cm long, 3–10 mm broad, clasping at the base. Flowers solitary or 2–3 together in the leaf-axils; calyx c. 4 mm long, the teeth indistinct or nearly obsolete, the accessory teeth much larger and spreading; petals none. Stamens and style enclosed by the calyx. Capsule 4–5 mm in diameter.

GRAND CAYMAN: *Brunt 1826, 1879, 1880, 1881, 2106; Hitchcock; Kings GC 253; Proctor 15246, 15302, 27948.*
– West Indies and continental tropical America, commonly in swampy places.

[Genus 2. **LAWSONIA** L.

A glabrous shrub or small tree, the 4-angled branchlets sometimes with a spine-like tip; leaves rather small, opposite, with short petioles; stipules minute, conic, whitish. Flowers small, 4-parted, fragrant, in terminal panicles; calyx broadly top-shaped, with lobes slightly longer than the tube, lacking accessory teeth; petals short-clawed. Stamens 8, the filaments thick, exserted. Ovary sessile, 2–4-celled, with a stout style. Capsule globose, indehiscent or rupturing irregularly; seeds thick, trigonous-pyramidal, spongy at the apex.

The genus consists of a single species.

1. **Lawsonia inermis** L., Sp. Pl. 1:349. 1753.

"Henna"

A bushy shrub 2 m tall or more; leaves elliptic to obovate, 1–3.5 cm long, sharply acuminate. Panicles 5–20 cm long; calyx 3–5 mm long; petals dull creamy or tawny, 4–6 mm long. Capsules 4–7 mm in diameter.

GRAND CAYMAN: *Brunt 1856; Howard & Wagenknecht 15027-A.* LITTLE CAYMAN: *Proctor (seen in cultivation).* CAYMAN BRAC: *Proctor 28939.*
– Probably native of eastern Africa and tropical Asia, but now planted and often naturalized in most warm countries. In the Cayman Islands it now grows wild in thickets and along the borders of pastures. In some countries, chiefly in Asia, the leaves are used to make a yellow dye for staining the nails, hands and feet. If a paste of the leaves is applied to the hair, it soon produces a bright red colour. The plant also yields a dull red dye for cloth, and an excellent perfume can be extracted from the flowers. However, the yield of essential oil from the flowers is relatively so small as to be commercially unprofitable.]

Family 58

THYMELAEACEAE

Unarmed trees or shrubs, rarely herbs, the inner bark made up of very strong meshed fibres; leaves opposite or alternate, entire, mostly pinnate-nerved; stipules lacking. Flowers regular, perfect or by abortion polygamous or unisexual, rarely solitary, usually in heads, umbels, racemes or spikes, these stalked or sessile, terminal or axillary. Calyx tubular or urn-shaped, 4-5-lobed, the lobes equal or unequal, imbricate; petals as many or twice as many as the calyx-lobes, or often absent. Stamens as many or twice as many as the calyx-lobes, attached within the tube above the middle; anthers 2-celled, opening by longitudinal slits. Ovary superior, sessile or short-stalked, 1- or 2-celled, subtended by a ring-like or cup-like disk, sometimes represented by 4 or 5 scales, or lacking; stigma terminal, more or less capitate; ovules solitary in each cell, laterally attached near the apex of the cell, anatropous. Fruit nut-like or drupe-like, indehiscent; seeds with or without endosperm, the embryo straight and with fleshy cotyledons.

A widespread family of about 40 genera and more than 500 species, the majority in Australia and South Africa. The strong bark-fibres of many species have been used for making rope.

Genus 1. **DAPHNOPSIS** Mart. & Zucc.

Mostly dioecious shrubs or small trees; leaves alternate or sometimes apparently whorled. Flowers unisexual, axillary or terminal in heads, umbels or racemes, the latter simple or branched. Calyx 4-lobed; petals 8, 4, or absent. Stamens in staminate flowers 8, inserted in two series in the calyx-tube, the anthers sessile or nearly so. Pistillate flowers usually smaller than the staminate, with 8, 4, or no staminodes; ovary sessile, 1-celled. Fruit a small drupe; seeds without endosperm.

A tropical American genus of 46 species.

1. Daphnopsis occidentalis (Sw.) Krug & Urban in Engl. Bot. Jahrb. 15:349. 1892. **FIG. 145.**

A shrub or small tree to 7 m tall or more, all the younger parts lightly clothed with fine appressed hairs; leaves oblanceolate or narrowly obovate, mostly 4–11 cm long, acuminate or blunt to rounded at the apex, usually pale beneath. Flowers apparently monoecious in small heads on peduncles 1–5 cm long, greenish- or yellowish-white, the tube appressed-hairy; petals absent; staminate calyx-tube 8 mm long, the lobes 4 mm long; pistillate calyx-tube 4 mm long, the lobes 2–2.5 mm long. Drupe ellipsoidal, 1.4–1.8 cm long, whitish when ripe.

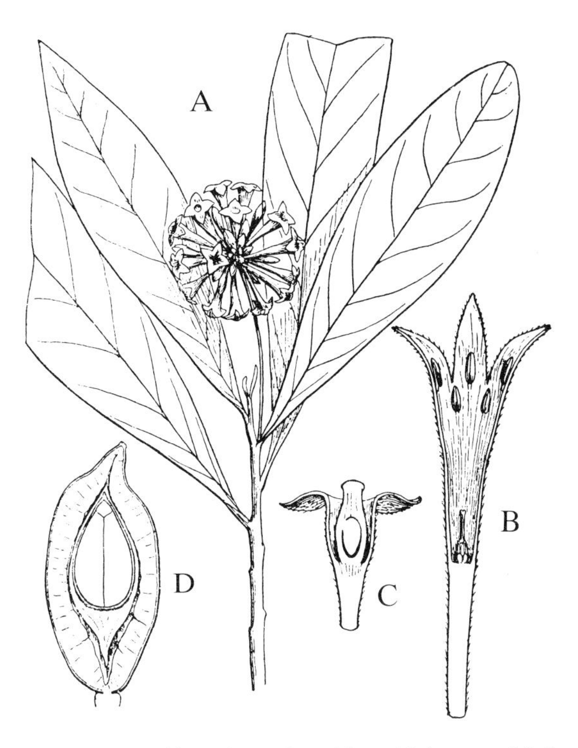

FIG. 145 **Daphnopsis occidentalis.** A, branchlet with leaves and inflorescence, x$^{2}/_{3}$. B, staminate flower cut lengthwise, x4. C, pistillate flower cut lengthwise, x4. D, fruit cut lengthwise, x3. (F. & R.)

GRAND CAYMAN: *Proctor 31041.* CAYMAN BRAC: *Proctor 29020.* – Jamaica; usually occurs in rocky limestone woodlands.

Family 59

MYRTACEAE

Unarmed shrubs or trees with simple, entire, opposite or alternate leaves, these pinnate-veined and usually punctate with resinous or pellucid glands; principal lateral veins usually united toward the margins by an often obscure submarginal vein extending the length of the blade; stipules absent. Flowers axillary or terminal, solitary or in bracteate inflorescences with opposite branching, usually modified in various ways, e.g. (a) elongation of the axis and reduction of the lateral branches to one flower each, forming a raceme; (b), suppression of the axis and reduction of the lateral branches to one flower, forming "glomerules" or umbel-like clusters; (c), reduction of the lateral branches to one pair, these arising just below the flower terminating the main axis, forming a "dichasium"; (d), elongation of both central axis and lateral branches, resulting in a panicle; and (e), transitional forms in which a panicle terminates in triads or dichasia. Flowers usually perfect, regular or essentially so. Ovary inferior, the calyx-tube ("hypanthium") adnate to the ovary its whole length or sometimes prolonged beyond it; calyx-lobes usually 4 or 5, either distinct and subequal, or the calyx more or less closed and rupturing irregularly at anthesis, or else the calyx closed and circumscissile, the top lifting off like a lid. Petals usually 4 or 5, or sometimes absent. Stamens indefinitely many, in one to many series originating around the margin of the thickened calyx-disk, usually inflexed in bud; anthers mostly 2-celled, usually opening by longitudinal slits. Ovary 2- to many-celled, bearing a simple, elongate style with a small capitate stigma; ovules 2 or more to each cell, borne on axial or parietal placentas. Fruit a berry, drupe or capsule; seeds usually without endosperm.

A family of perhaps 60 genera and nearly 3,000 species, chiefly in the tropics, but well represented in subtropical and temperate areas of the Southern Hemisphere.

The wood of this family is usually hard, tough and close-grained, and many species yield useful timber. Many also produce edible fruit, while the aromatic oils of numerous *Eucalyptus* species have found a wide use in the pharmaceutical, soap, and perfume industries. The Clove tree (*Syzygium aromaticum*) is well-known as a source of culinary spice and an aromatic oil used in various food-products, dental preparations, and perfumes. In the West Indies, the Pimento or Allspice tree (*Pimenta dioica*) and the Bay Rum tree (*Pimenta racemosa*) likewise produce commercially important spice or oil.

KEY TO GENERA

1. Calyx without lobes, closed in bud, the top falling off like a lid at anthesis; petals absent or minute and adherent to the inside of the calyx-lid; vegetative branching dichotomous: **1. Calyptranthes**

1. Calyx with 4 or 5 evident lobes, these usually persistent; petals distinct; vegetative branching not dichotomous:

 2. Calyx open in bud, the lobes 4, subequal; stigma not or scarcely thicker than the style; fruits with 1–4 seeds:

  3. Flowers mostly 3 or 7 in a stalked dichasium: **2. Myrcianthes**

  3. Flowers solitary, clustered, or in racemes: **3. Eugenia**

2. Calyx closed in bud or nearly so, at anthesis splitting to the disk in 4 or usually 5 irregular or unequal lobes; stigma capitate; fruits with numerous seeds: **4. Psidium**

Genus 1. **CALYPTRANTHES** Sw.

Shrubs or small trees with more or less dichotomous branching, the young branchlets often flattened, keeled or winged, glabrous or pubescent; if pubescent, the hairs usually attached near the middle (dibrachiate). Flowers usually in axillary or terminal cymose panicles, rarely solitary or few in a simple cluster. Calyx completely closed in bud, circumscissile at anthesis, the cap-like top falling away or sometimes remaining attached at one edge. Petals none, or 2-5 and very small and inconspicuous, often falling with the calyx-cap. Stamens numerous in several series. Ovary 2-celled, with 2 ovules in each cell. Fruit a 1-4-seeded berry, crowned by the basal part of the calyx; seeds sub-globose; cotyledons relatively large, thin and contorted.

A tropical American genus of more than 100 species.

KEY TO SPECIES

1. Branchlets, leaves and inflorescence entirely glabrous; individual flowers distinctly stalked: **1. C. zuzygium**

1. Branchlets, leaves and inflorescence more or less densely sericeous; flowers sessile in small clusters at the ends of panicle-branches: **2. C. pallens**

1. Calyptranthes zuzygium (L.) Sw., Nov. Gen. & Sp. Pl. 79. 1788.

Shrubs or small tree to 10 m tall, glabrous throughout; leaves stiff, short-petioled, elliptic or narrowly obovate, 2.5-7 cm long, with obtuse or bluntly acuminate apex, the midrib prominent on the upper side toward the base. Panicles trichotomous, rather few-flowered, the flowers fragrant; pedicels to 8 mm long; mature unopened buds c. 4 mm long, glabrous; hypanthium 2.5 mm long. Berries globose, 7-9 mm in diameter, red turning blackish-glaucous, edible.

LITTLE CAYMAN: *Kings LC 68?*
– Florida, Bahamas, the Greater Antilles, and the Swan Islands, chiefly in woodlands over limestone. The presence of this species in the Cayman Islands needs further confirmation.

2. **Calyptranthes pallens** Griseb. in Gött. Abh. 7:67. 1857.

"Bastard Strawberry", "Strawberry Tree", "Red Strawberry"

Shrub or small tree to 7 m tall or more, the young branchlets reddish-sericeous and somewhat 2-edged. Leaves stiff, petiolate, lanceolate to elliptic, 2.5–7 cm long, the apex often long-acuminate, the midrib grooved on the upper side toward the base, the tissue beneath (abaxial side) sericeous and often distinctly paler. Panicles many-flowered, the flowers in small sessile clusters; mature unopened buds 2–3 mm long, sericeous; hypanthium 1–2 mm long. Berries globose, 4–5 mm in diameter, dark red, edible.

GRAND CAYMAN: *Brunt 1806, 1894, 2050; Kings GC 284, GC 413; Proctor 15301.* LITTLE CAYMAN: *Kings LC 16; Proctor 28127.* CAYMAN BRAC: *Kings CB 58a; Proctor 15329. 28911, 29057.*
– Florida, West Indies, and Mexico, in rocky or sandy woodlands and in thickets bordering pastures.

Genus 2. **MYRCIANTHES** Berg

Small or medium-sized trees; leaves opposite or sometimes in 3's. Flowers solitary or usually in 3- or 7-flowered dichasia (rarely more), the peduncles 1 to several in the upper leaf-axils, or sometimes the inflorescence modified into a terminal panicle bearing the flowers in small dichasia at the tips; central flower of a dichasium usually sessile. Calyx mostly 4-lobed; petals 4 or in some species 5. Ovary usually 2-celled, with numerous (usually 8–20) ovules radiating from a centrally-attached placenta. Fruit a 1–4-seeded berry; seeds with fleshy, plano-convex cotyledons, short terete radicle, and evident plumule.

A widespread American genus of perhaps 25 species, the majority in South America.

1. **Myrcianthes fragrans** (Sw.) McVaugh in Fieldiana, Bot. 29:485. 1963.

Eugenia fragrans (Sw.) Willd., 1800.

Anamomis fragrans (Sw.) Griseb., 1860.

Anamomis lucayana Britton, 1920.

"Cherry"

Aromatic shrub or small tree to 7 m tall or more, the bark smooth and whitish, the young branchlets and petioles puberulous; leaves narrowly obovate or elliptic, mostly 1.5–4 cm long, blunt or slightly notched at the apex, the margins more or less revolute, densely gland-dotted on both sides, glabrous except for the short petiole. Peduncles usually solitary, 1.5–3.5 cm long, usually 3-flowered, but occasionally the flowers 1 or 7. Hypanthium puberulous, 2–2.5 mm long; calyx-lobes in two unequal pairs; petals white, 4–6 mm long, gland-dotted. Berry black, up to 1 cm in diameter, usually with a single bean-shaped seed maturing.

GRAND CAYMAN: *Brunt 1778, 2189, 2193; Kings GC 379; Proctor 15202, 27975.* LITTLE CAYMAN: *Kings LC 65; Proctor 28092, 35076.* CAYMAN BRAC: *Proctor 29056, 35157.*

– Florida, West Indies, Mexico and Central America, and northern Venezuela, varying considerably from one region to another. The Cayman population described above closely resembles that of Cuba and the Bahamas, and is quite unlike that typical of Jamaica. It grows in rocky thickets and woodlands.

Genus 3. **EUGENIA** L.

Shrubs or trees, glabrous or pubescent; leaves opposite. Flowers axillary, usually at leafy nodes, solitary or usually in racemes, the axis elongate or variously shortened, or sometimes lacking so that the flowers appear to be in umbels or glomerules. Calyx 4-lobed, the lobes distinct in bud and usually persistent on the fruit; hypanthium closely subtended by a pair of bracteoles, at the apex little or not at all prolonged beyond the summit of the ovary. Petals 4, usually white and conspicuous. Stamens borne on a flat disk surrounding the base of the style. Ovary usually 2-celled, with numerous ovules in each cell (rarely only 2). Fruit a small drupe or berry, usually with thin pulp over a single massive seed, or occasionally the fruit 2-seeded; embryo with short, thick cotyledons usually not distinguishable from each other.

A very large tropical and subtropical genus of at least 1,000 species.

KEY TO SPECIES

1. Leaves with midrib grooved on the upper side; flowers several in a very short but distinct raceme; sepals less than 1 mm long; fruit 7–10 mm in diameter, black, not ribbed: **1. E. axillaris**

1. Leaves with midrib flat on the upper side; flowers solitary or umbellate; sepals c. 4 mm long; fruit 12–25 mm in diameter, red, distinctly ribbed: [**2. E. uniflora**]

1. Eugenia axillaris (Sw.) Willd. in L., Sp. Pl. ed. 4, 2:960. 1800. **FIG. 146.**

Eugenia baruensis Jacq., 1789.

Eugenia monticola of Griseb., 1860, not DC.

"Strawberry"

Shrub or small tree to 8 m tall, glabrous throughout; leaves rather leathery, elliptic or ovate, 3-7 cm long, obtusely acuminate, the glandular dots but faintly pellucid. Flowers 4 to 10 on very short racemes 2-4 mm long; pedicels 1-1.5 mm long; sepals mostly 0.8 mm long. Fruits black, more or less globose, 7-10 mm in diameter, alleged to be edible.

GRAND CAYMAN: *Brunt 2088, 2107, 2166, 2198; Correll & Correll 51009; Hitchcock; Kings GC 285; Proctor 15228, 27932; Sachet 424.* LITTLE CAYMAN: *Proctor 35099, 35167, 35204.* CAYMAN BRAC: *Millspaugh 1157; Proctor 28912.*
– Florida, West Indies, Mexico and northern Central America, in sandy or rocky thickets and woodlands.

[2. **Eugenia uniflora** L., the "Surinam Cherry", is sometimes cultivated in the Georgetown area. It is a small shrub with ovate leaves, the white flowers on very long slender pedicels. The red, ribbed fruits are edible.

GRAND CAYMAN: *Proctor 15308.*
– Cultivated sparingly in most tropical countries, and often becoming naturalized; probably native of South America.]

Genus 4. **PSIDIUM** L.

Shrubs or trees, often with strongly pinnate-veined leaves. Flowers axillary, solitary or in 3-flowered dichasia, or otherwise clustered; calyx closed or somewhat open in bud, usually splitting irregularly down to the ovary at anthesis, producing 4 or 5 distinct teeth or lobes; hypanthium usually prolonged beyond the summit of the ovary, the free portion splitting with the calyx-lobes. Petals 4 or 5, usually white. Stamens numerous in several series from a broad disk. Ovary usually 3-4-celled (rarely 2- or up to 7-celled), with numerous ovules on bilamellate placentas of parietal origin. Fruit a fleshy berry, sometimes large and edible; seeds more or less kidney-shaped, hard and bony; embryo curved or C-shaped, with small cotyledons and a long radicle.

A tropical American genus of more than 100 species.

1. **Psidium guajava** L., Sp. Pl. 1:470. 1753. **FIG. 147.**

"Guava"

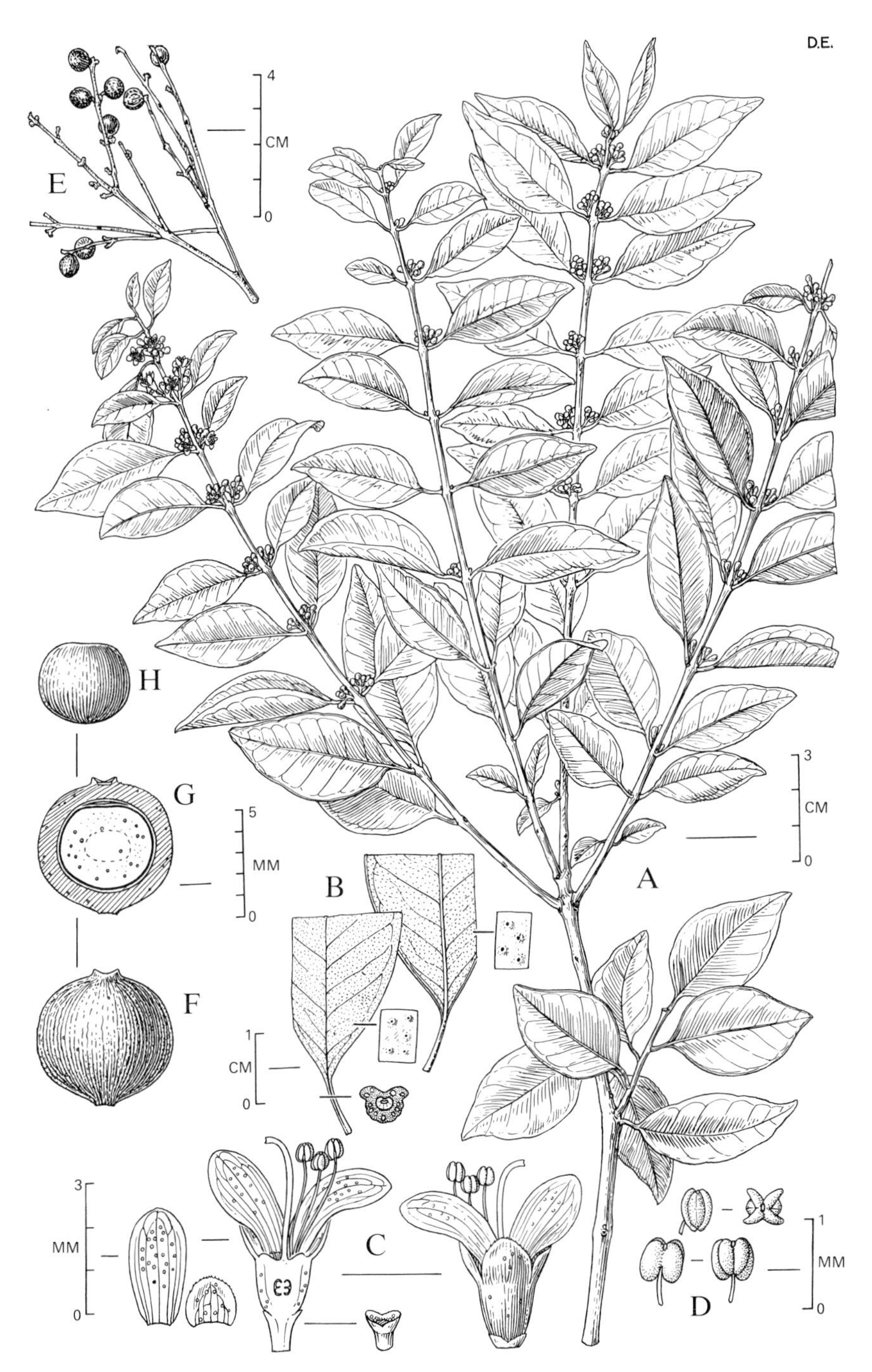

FIG. 146 **Eugenia axillaris.** A, flowering branch. B, details of leaf. C, details of flower. D, four views of anther. E, fruiting branchlets. F, fruit, G, same, long. section through fruit. H, seed. (D.E.)

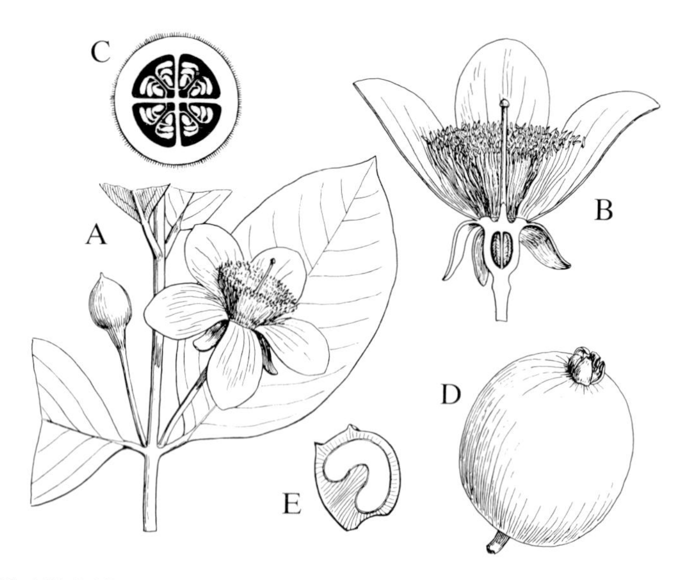

FIG. 147 **Psidium guajava.** A, portion of branch with leaf, bud and flower, x$^2/_3$. B, flower cut lengthwise, x1$^1/_3$. C, x-section of ovary, x6. D, fruit, x$^2/_3$. E, seed cut to show embryo, x7. (F. & R.)

Shrub or small tree rarely over 7 m tall, with 4-angled branchlets; leaves elliptic or oblong, mostly 7–14 cm long, prominently veined, the apex obtuse or acute, beneath clothed with soft, grayish, mostly appressed hairs; tissue with numerous pellucid dots. Flowers usually solitary in the leaf-axils, the peduncles 1–2 cm long; calyx completely closed in bud, the mature buds 1.3–1.6 cm long, puberulous toward the base. Petals white, 1–1.2 cm long. Style about as long as the petals, with flat, peltate stigma. Berry globose or pear-shaped, 2–6 cm long, with pinkish or yellowish edible flesh and numerous small seeds.

GRAND CAYMAN: *Brunt 1827, 1980; Hitchcock; Kings GC 385; Millspaugh 1378; Proctor 11985.* CAYMAN BRAC: *Proctor 28981.*
– Widely distributed in the American tropics, often planted and naturalized, its true natural range not known; introduced in the Old World tropics.

[PUNICACEAE

Punica granatum L., the "Pomegranate", is often cultivated around houses, and may persist after cultivation. GRAND CAYMAN: *Kings GC 295.*]

Family 60

ONAGRACEAE

Annual or perennial herbs, sometimes aquatic, rarely shrubs or trees; leaves alternate, opposite, or whorled, entire to dentate or pinnatifid; stipules absent. Flowers usually perfect and regular, mostly axillary and solitary but sometimes in spikes, racemes or panicles; calyx often prolonged above the ovary into a slender tube, cleft at apex into 2-6 (often 4) valvate lobes; petals 2-5 (often 4), rarely none, often inserted at base of a disk, soon falling. Stamens 1-8, inserted with the petals, the anthers 2-celled and attached dorsally to slender filaments. Ovary inferior, 1-6- (often 4-) celled; stigma capitate and entire or 4-lobed; ovules numerous or rarely solitary. Fruit a capsule, or sometimes nut-like or berry-like, often elongate and splitting into 4 valves, the valves separating from the seed-bearing axis; seeds usually numerous and small, with little or no endosperm.

A primarily temperate-climate family of about 20 genera and 650 species, widely distributed in both hemispheres.

Genus 1. **LUDWIGIA** L.

Annual or perennial herbs, often aquatic or growing in wet places, rarely shrubs or small trees. Leaves opposite or alternate, entire in our species. Flowers axillary, solitary; calyx not prolonged above the ovary, the lobes usually 4, rarely more; petals usually 4, sometimes lacking, yellow or rarely white. Stamens mostly 4 or 8. Ovary 4-6-celled; ovules numerous on prominent placentas. Fruit a terete, ribbed or angled capsule crowned by the calyx-lobes and disk; seeds numerous, minute.

A cosmopolitan genus of about 75 species, the majority in the American tropics.

KEY TO SPECIES

1. Capsule 4-angled, c. 1.5 cm long; calyx 4-5 mm long, scarcely exceeded by the petals: **1. L. erecta**

1. Capsule cylindric, 3-4.5 cm long; calyx 5-12 mm long, much exceeded by the petals: **2. L. octovalvis**

1. Ludwigia erecta (L.) H. Hara in Jour. Jap. Bot. 28:292. 1953.

Jussiaea erecta L., 1753.

Erect, glabrous herb to 1 m tall, much-branched, the branches obscurely angled by the decurrent petioles. Leaves almost linear to narrowly or broadly lance-acuminate, 5-8 cm long, mostly 0.8-1.5 cm broad, slightly rough on the margins, the petioles 2-5 mm long. Flowers small, sessile or subsessile, the yellow petals 4-5 mm long. Capsule 2-3 mm in diameter; seeds about 0.4 mm long.

GRAND CAYMAN: *Brunt 1893; Proctor 15303, 27954.*
– Nearly pantropical in distribution, chiefly in wet ditches and moist fields.

2. **Ludwigia octovalvis** (Jacq.) Raven in Kew bull; 15:476. 1962. **FIG. 148.**

Jussiaea suffruticosa L., 1753.

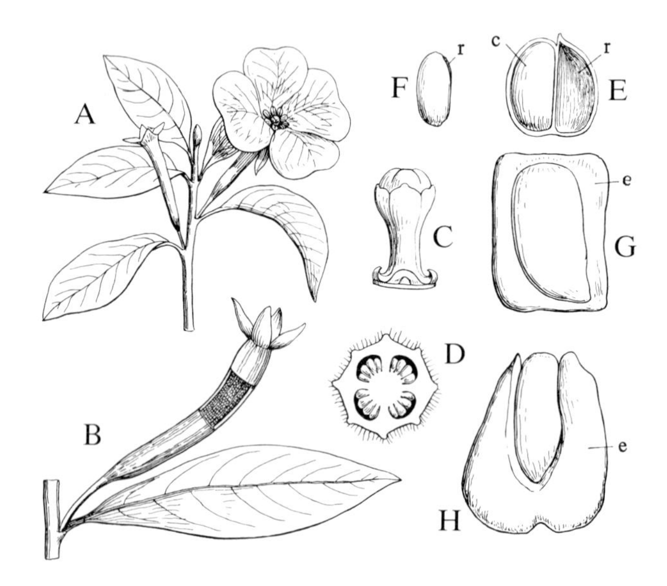

FIG. 148 **Ludwigia octovalvis.** A, branchlet with leaves, flower and young capsule, x2/3. B, leaf and ripe capsule with part of pericarp removed, x2/3. C, style, much enlarged. D, x-section of ovary enlarged. E, seed, x20; c, embryo; r, the hollow enlarged raphe. F, **L. erecta,** seed with inconspicuous raphe (r), x20. G, H seeds of two other species *(L. repens, L. leptocarpa),* x20; e, endocarp. (F. & R.)

Erect herb often subwoody at base, up to 1 m tall or more, almost glabrous to more or less villous-pubescent, the branches angled. Leaves usually lanceolate to oblong or lance-ovate, 3-11 cm long, up to 2.5 cm broad, narrowed at both ends. Flowers petiolate, with showy yellow petals up to 2 cm long. Capsule 2.5-3.5 mm in diameter; seeds about 0.6 mm long.

GRAND CAYMAN: *Brunt 1928; Proctor 15278.*
– Pantropical, frequently a weed of low moist ground.

Family 61

COMBRETACEAE

Trees or shrubs, often climbing, unarmed or spiny; leaves alternate, opposite, or rarely whorled, petiolate, simple, entire, and without stipules. Flowers usually perfect, sometimes polygamo-dioecious or unisexual, usually in spikes, racemes or heads, rarely paniculate. Calyx with tube adnate to the ovary, divided above into usually 4 or 5 lobes, these valvate in bud, persistent or deciduous. Petals none or 4–5. Stamens 4–5, or 8–10, inserted on the limb or base of the calyx, the filaments inflexed in bud; anthers attached at the middle, opening by longitudinal slits. Ovary wholly adnate to the calyx, thus appearing inferior, 1-celled, the style simple or filiform, the stigma simple or lobed; ovules 2–6, hanging from apex of the cell by slender stalks (stalks lacking in *Laguncularia*). Fruit leathery or drupe-like, often angled or winged, 1-seeded, usually not opening; seed usually elongate and grooved, without endosperm, the cotyledons often fleshy and oily.

A pantropical family of about 15 genera and more than 500 species.

KEY TO GENERA

1. Leaves alternate; petals lacking:

 2. Flowers in spikes: **[1. Terminalia]**

 2. Flowers in dense globose heads: **2. Conocarpus**

1. Leaves opposite; petals present: **3. Laguncularia**

[Genus 1. **TERMINALIA** L.

Erect shrubs or trees without spines; leaves alternate or apparently subopposite, often crowded at ends of the branches, and often bearing glands at the base beneath. Flowers perfect or staminate, small, usually green or white, borne in spikes; calyx with limb bell-shaped, cut to the middle with 4 or 5 lobes, soon falling; petals lacking. Stamens usually 10 in 2 equal series, the lower opposite the calyx-lobes, the upper alternating; filaments exserted. Stigma simple; ovules usually 2. Fruit dry or drupe-like, often winged.

A widely-distributed tropical genus of about 200 species, many yielding valuable timber.

1. **Terminalia catappa** L., Mant. Pl. 128. 1767. **FIG. 149.**

"Almond" or "Indian Almond"

A fast-growing tree with conspicuously whorled, horizontal branches; leaves clustered near ends of the branches, obovate, 10–30 cm long, rounded and abruptly pointed at the apex, tapered downward to a minutely subcordate base, almost glabrous. Flower-spikes 5–15 cm long, the perfect (fruit-producing) flowers toward the base, the distal flowers all being staminate only. Fruit a bony, flattened-ellipsoid, 2-edged drupe, 4–7 cm long, with thin edible flesh.

GRAND CAYMAN: *Brunt 1759; Hitchcock; Kings GC 30; Millspaugh 1316; Proctor 15116.* LITTLE CAYMAN: *Kings LC 55; Proctor 35094.* CAYMAN BRAC: *Proctor (sight).*

– Native of tropical Asia, commonly planted for ornament and shade throughout the tropics, especially in localities near the sea, and often becoming naturalized. The wood is hard and close-grained; the bark and leaves are astringent and contain tannin, and also together with the fruits yield a black dye used in some countries for making ink or dyeing textiles. The seeds are edible; they are very rich in oil and have an almond-like flavour.]

Genus 2. **CONOCARPUS** L.

Shrubs or trees, the leaves alternate, of firm texture, and often biglandular at the base. Flowers perfect and staminate, minute, in dense globular heads, these rather numerous in terminal panicles. Calyx-limb 5-parted, soon falling; petals none. Stamens 5, with exserted filaments and small, cordate anthers. Ovary compressed, 1-celled, with short subulate style and simple stigma; ovules 2, pendulous from apex of the cell. Fruits scale-like, over-lapping, 1-seeded, aggregated in a cone-like head.

A small genus of 2 species, one occurring in tropical America, the other in West Africa.

1. **Conocarpus erectus** L., Sp. Pl. 1:176. 1753 (as *"erecta"*). **FIG. 150.**

"Buttonwood", "Button Mangrove"

A low, trailing shrub in exposed situations, or an erect shrub or tree in more favourable localities. Leaves alternate, lanceolate to elliptic, obtuse or acute at both ends, glabrous (or in var. *sericeus* DC. covered with silky whitish tomentum,

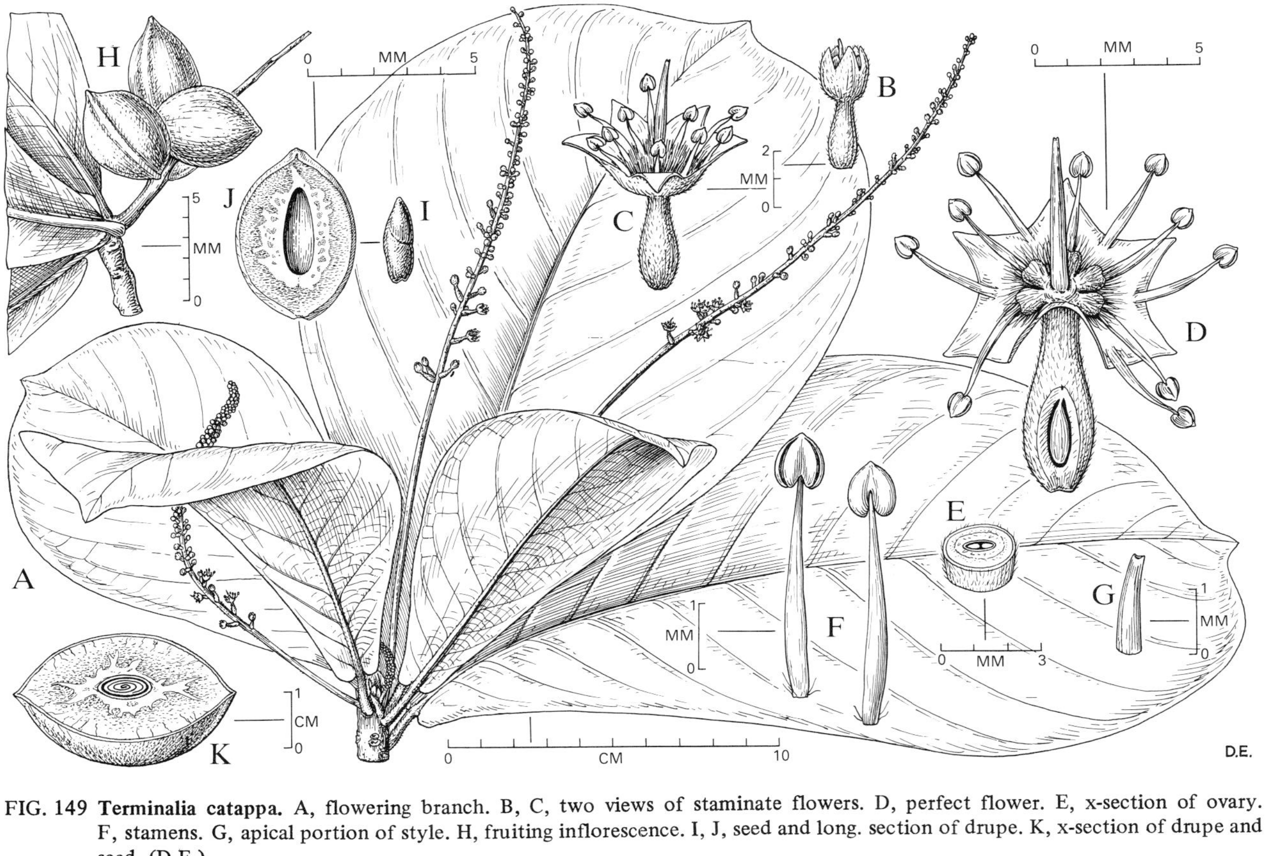

FIG. 149 **Terminalia catappa.** A, flowering branch. B, C, two views of staminate flowers. D, perfect flower. E, x-section of ovary. F, stamens. G, apical portion of style. H, fruiting inflorescence. I, J, seed and long. section of drupe. K, x-section of drupe and seed. (D.E.)

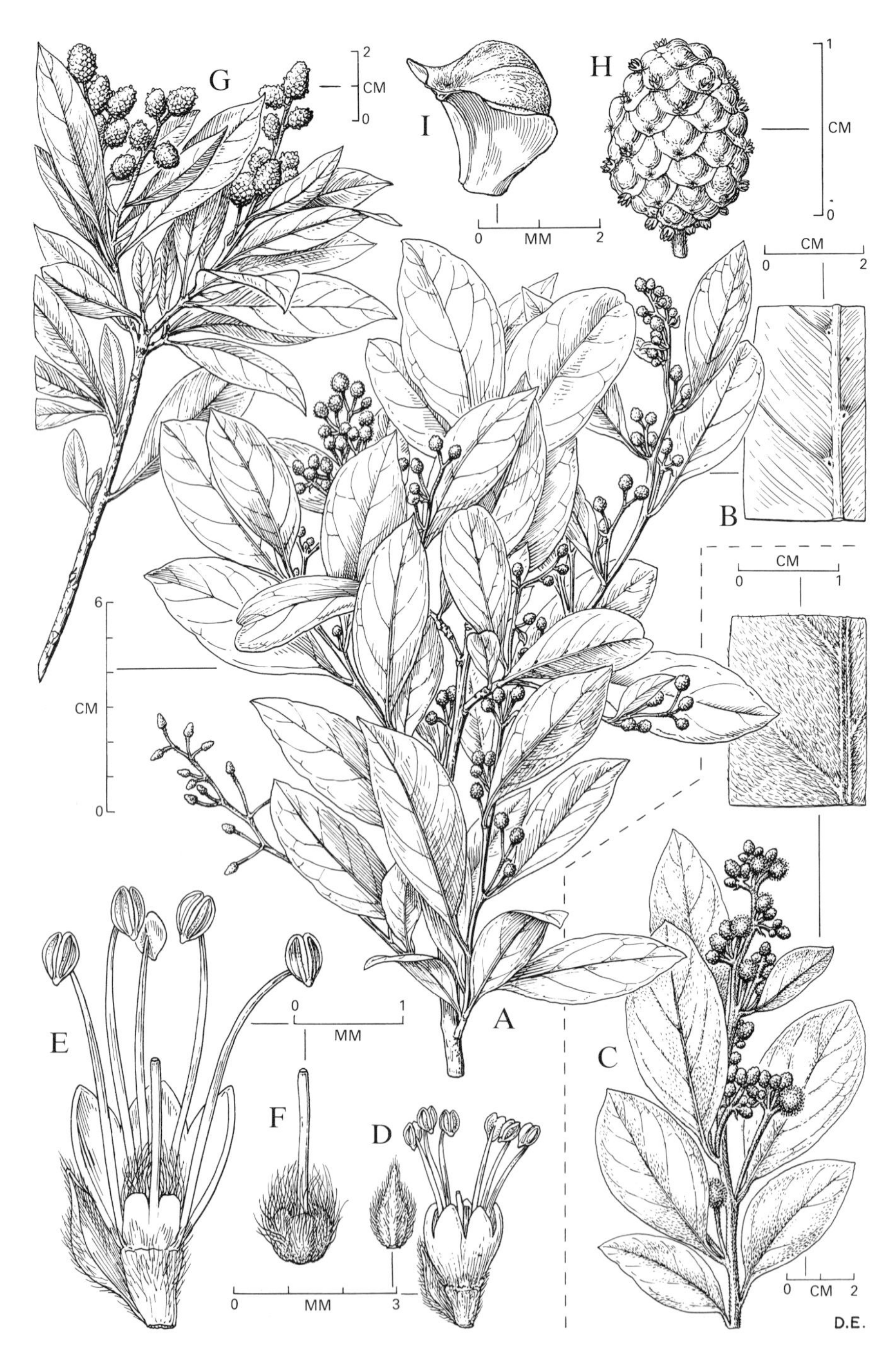

FIG. 150 **Conocarpus erectus.** A, flowering branch. B, detail of leaf. (C, flowering branch of var. *sericeus* and detail of leaf.) D, flower and bract. E, long. section of flower. F, pistil. G, fruiting branch. H, fruiting spike. I, single fruit. (D.E.)

Fig. 150 C), 2–10 cm long, the petiole with 2 glands on the upper (adaxial) surface at the base of the blade. Flower-heads c. 1 cm in diameter or less; calyx-limb c. 1 mm long.

GRAND CAYMAN: *Brunt 1628, 1971, 1972* (var. *sericeus*); *Hitchcock; Kings GC 27, GC 101, GC 137* (var. *sericeus*), *GC 138; Lewis GC 27; Millspaugh 1306; Proctor 15224; Sachet 390.* LITTLE CAYMAN: *Kings LC 21, LC 49* (var. *sericeus*); *Proctor 28103, 28174* (var. *sericeus*). CAYMAN BRAC: *Brunt 1677, 1678, 1679; Kings CB 11* (var. *sericeus*); *Millspaugh 1212* (var. *sericeus*); *Proctor 28925, 29048* (var. *sericeus*).

– Widespread in tropical America, a common element on the landward side of mangrove swamps but also occurring along rocky seashores. The wood is very fine-textured, hard and durable; it is highly valued as a fuel and for making charcoal. The bark is widely used for tanning leather. Var. *sericeus* is often planted as an ornamental shrub.

Genus 3. **LAGUNCULARIA** Gaertn. f.

A shrub or trees; leaves opposite, rather thick and leathery, the veins obscure, with minute glands in the tissue on both sides near the margins, and with 2 glands at apex of the petiole. Flowers mostly perfect with a few staminate ones intermixed, in pubescent axillary spikes and in a terminal panicle of usually 3 spikes, the central one often 3-branched. Bracteoles 2, scale-like, below the calyx. Calyx-limb cup-like, 5-cleft to the middle, persistent; petals 5, small and roundish, soon falling. Stamens 10 in 2 series, the filaments included, the anthers cordate. Ovary 1-celled, crowned by a disk; stigma 2-lobed; ovules 2, pendulous from apex of cell. Fruit leathery, angled and unequally-ribbed, the angles narrowly winged or marginate.

A small genus of 2 species occurring in tropical American and West Africa.

1. **Laguncularia racemosa** (L.) Gaertn. in Gaertn. f., Fruct. 3:209, t. 217. 1805. **FIG. 151.**

"White Mangrove"

A bushy small tree sometimes 15 m tall or more; leaves oblong to oval, 3–11 cm long, obtuse or rounded at both ends, often notched at the apex, glabrous, with petioles, 0.5–2 cm long. Calyx finely pubescent, 2–3 mm long; petals about equalling the calyx. fruit 1.5–2 cm long.

GRAND CAYMAN: *Kings GC 66, GC 171; Sachet 458.* LITTLE CAYMAN: *Kings LC 29, LC 30; Proctor 28075, 28173.* CAYMAN BRAC: *Brunt 1685; Kings CB 7; Proctor 29127.*

– Widespread in saline swamps of tropical America and West Africa, usually associated with other mangrove species. The wood is hard, heavy, strong and dense, but is little used except for fuel. The bark is sometimes used for tanning leather.

Family 62

RHIZOPHORACEAE

Trees or shrubs, usually glabrous; leaves mostly opposite and with stipules, rarely alternate and lacking stipules, simple and more or less leathery, entire or sometimes serrulate; stipules interpetiolar, united in pairs, soon falling. Flowers perfect, usually subtended by bracteoles, and arranged in small axillary cymes, panicles, spikes or racemes, rarely solitary. Calyx-tube more or less adnate to the ovary or rarely free, the limb extending beyond the ovary and cleft into 3–14 lobes, these valvate and persistent. Petals as many as the sepals and often shorter, mostly concave and embracing the stamens, sessile or clawed, usually more or less lacerate or fringed. Stamens usually 2–4-times as many as the petals, or rarely the same number, inserted on the margin of a perigynous or epigynous disk, the lobes of the disk sometimes elongated into staminodes; filaments short or elongate; anthers 2-celled, opening by longitudinal slits. Ovary usually inferior, mostly 2–5-celled, or the septa disappearing and 1-celled; style with simple or lobed stigma; ovules usually 2 in each cell, hanging side by side from the axis above the middle. Fruit leathery or fleshy, crowned by the persistent calyx-limb, mostly indehiscent or tardily splitting, 1-seeded or the cells 1-seeded; seeds pendulous, with or without an aril; endosperm fleshy or lacking.

A pantropical family of about 16 genera and 120 species.

Genus 1. **RHIZOPHORA** L.

Glabrous shrubs or trees supported by curved prop-roots and aerial roots, the branchlets thick and marked by leaf-scars; leaves opposite, petiolate, leathery, elliptic and entire. Peduncles 2- or 3-forked, bearing few flowers; flowers leathery; calyx mostly 4-parted; petals 4 or rarely 5. Stamens 8–12, inserted with the petals on a thick, cup-like disk; filaments very short; anthers elongate, acuminate, with numerous round pollen-sacs but eventually 2-valved. Ovary half-inferior, 2-celled, prolonged above the calyx as a fleshy cone; style awl-shaped, with a 2-toothed stigma; ovules 2 in each cell. Fruit leathery, surrounded above the base by the reflexed sepals, 1-celled, 1-seeded; seed without endosperm; radicle becoming elongate and perforating the apex of the fruit while still on the tree, eventually falling upright into the mud or water beneath.

Three or more species, widely distributed along tropical seashores.

1. **Rhizophora mangle** L., Sp. Pl. 1:443. 1753. **FIG. 152.**

"Red Mangrove"

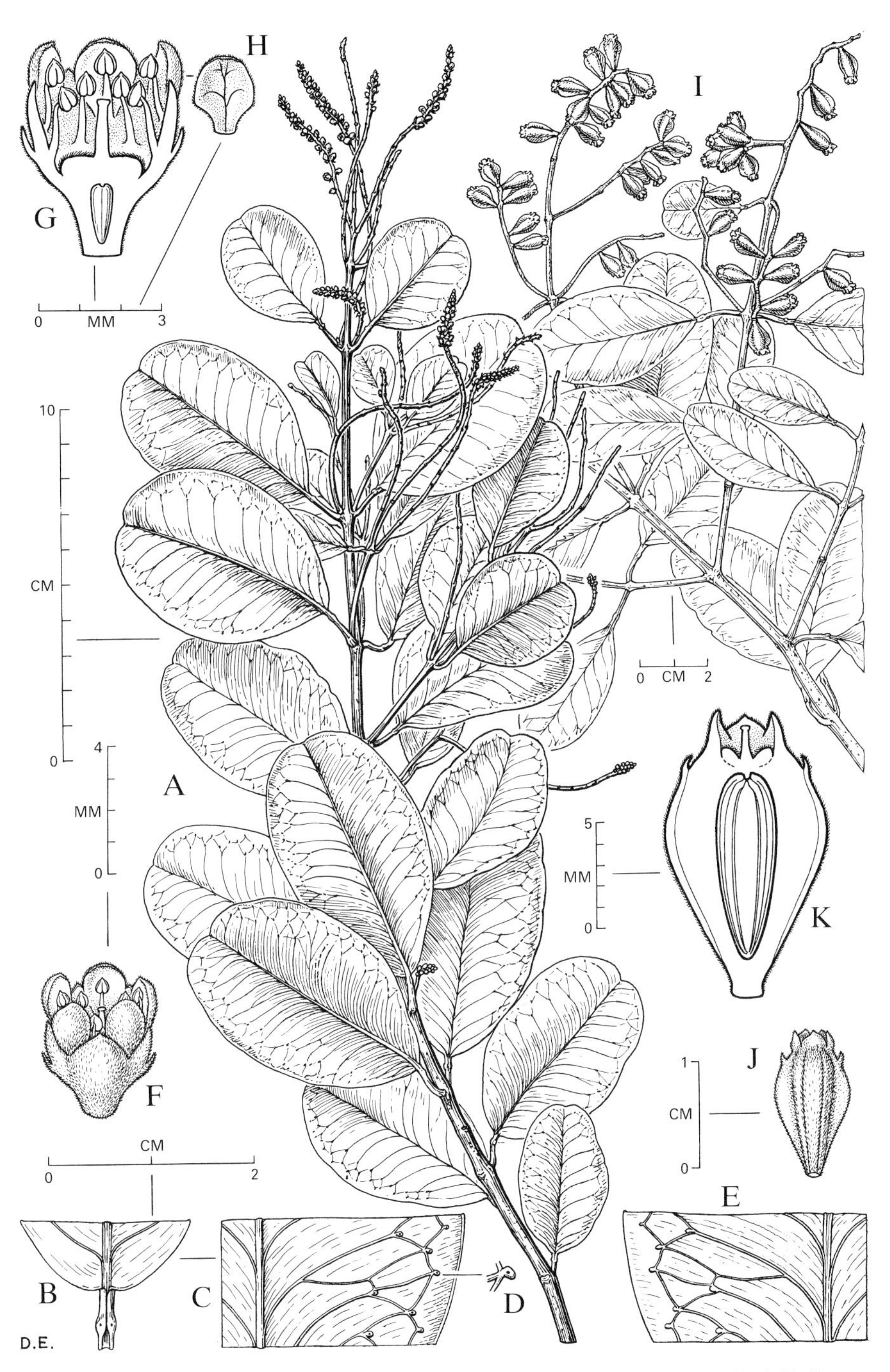

FIG. 151 **Laguncularia racemosa.** A, flowering branch. B, C, D, E, details of leaf. F, flower. G, long. section of flower. H, petal. I, fruiting branch. J, single fruit. K, long. section of fruit. (D.E.)

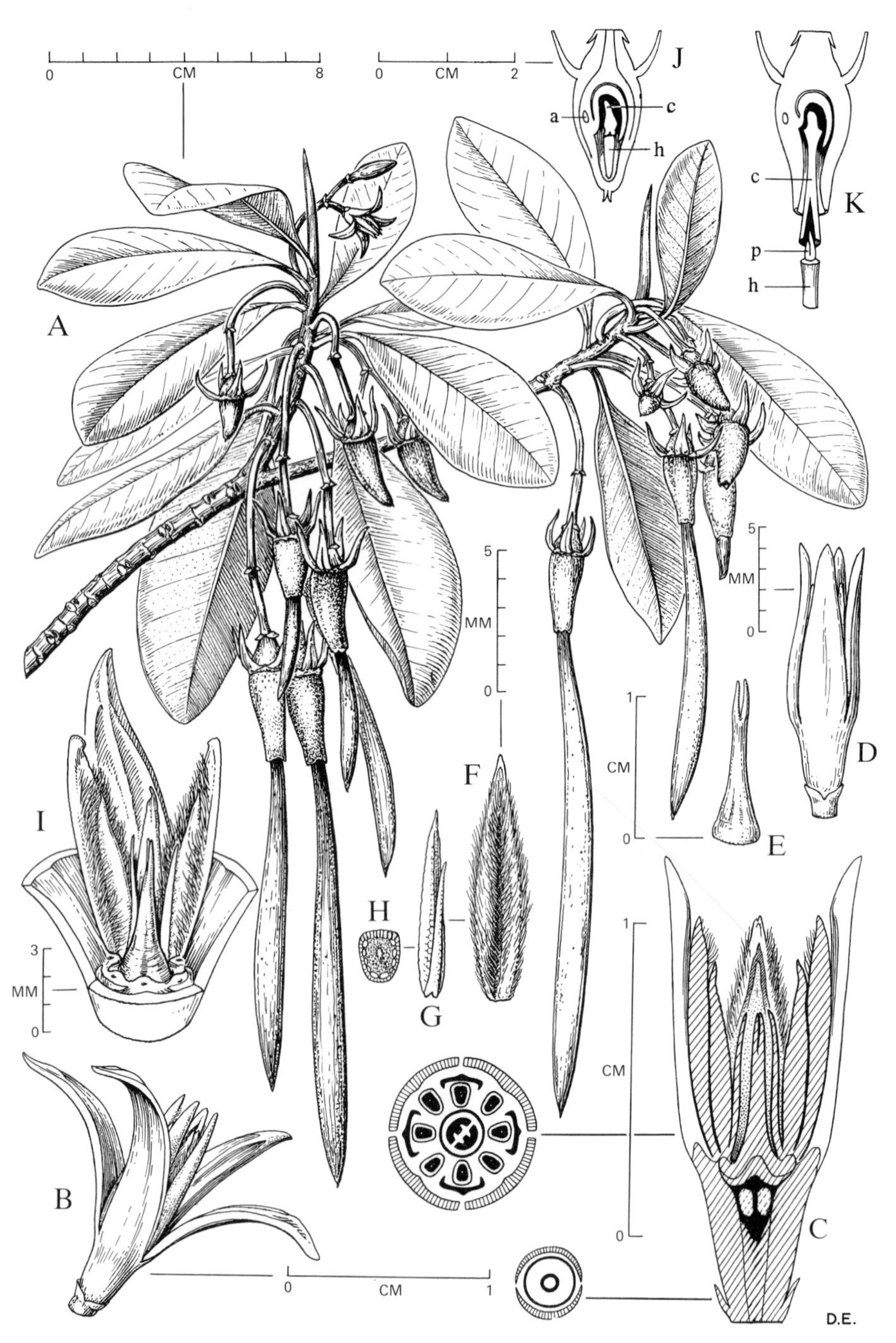

FIG. 152 **Rhizophora mangle.** A, branch with flower, fruits, and developing radicles. B, flower. C, long. section of flower, and x-sections at two levels. D, flower with sepals removed. E, pistil. F, inner face of petal. G, stamen. H, x-section of anther. I, dissection of flower. J, long. section of fruit; a, abortive cell; c, cotyledon; h, hypocotyl. K, long. section of fruit after germination of seed. (D.E.)

A tree, sometimes low, shrubby and diffuse, otherwise up to 18 m tall or more; leaves 5–15 cm long, obtuse, entire, deep green, the nerves obscure; stipules 2.5–4 cm long. Peduncles mostly 2-3-flowered; calyx 1 cm long, becoming reflexed; petals creamy yellow, 7–8 mm long, villous inside, chiefly below the apex. Stamens 8, c. 5 mm long. Fruit 2.5–3.5 cm long but soon appearing much longer because of the developing radicle.

GRAND CAYMAN: *Hitchcock; Kings GC 176A; Sachet 457.* LITTLE CAYMAN: *Kings LC 31, LC 31A.* CAYMAN BRAC: *Kings CB 8; Millspaugh 1211; Proctor 29126.*

– Muddy or sandy seashores of tropical America and islands of the Pacific. The red mangrove with associated species forms an important plant community of tropical coasts, and is abundant in the Cayman Islands. *Rhizophora* is especially adapted to saline aquatic habits partly by virtue of its numerous prop-roots; these roots catch and hold mud and silt so that coast-lines are not only protected from erosion but are actually extended into the sea, the mangrove thickets gradually filling with soil and eventually becoming more or less solid land. Mangrove seedlings also drift about and take root on sand-bars to form small islands. The prop-roots form a tangle that becomes the habitat for numerous marine and semi-marine organisms. *Rhizophora* has hard, heavy, strong, durable wood, used in some regions for many purposes, such as miscellaneous construction and for posts, pilings, and railway ties. It produces a superior kind of charcoal, and the bark has been much used for tanning leather. The young shoots are used to produce various dyes, the colours depending on the kind of salts used in preparing them. In spite of these various potential uses, mangroves should not be indiscriminately cut for any purpose without due consideration of their very real value in protecting shorelines from sea-erosion.

Family 63

OLACACEAE

Shrubs or trees, the leaves usually alternate, entire and without stipules. Flowers small, perfect or unisexual, solitary or in cymes or racemes; calyx cup-shaped, persistent, with 4–6 teeth or lobes, sometimes greatly enlarged in fruit. Petals 4–6, free or united below into a bell-shaped corolla, inserted on the receptacle or at the margin of a disk, valvate. Stamens 4–12, more or less adnate to the petals, all fertile or some of them sterile; anthers 2-celled. Ovary free, 1-celled or imperfectly 3-5-celled below, with a central placenta at the apex; ovules usually 3, pendulous. Fruit a drupe, 1-seeded; embryo minute, at the apex of the fleshy endosperm.

A widely-dispersed tropical family of 25 genera and about 250 species.

KEY TO GENERA

1. Corolla-lobes densely bearded within; plants armed with spines: **1. Ximenia**

1. Corolla-lobes not bearded; plants unarmed: **2. Schoepfia**

Genus 1. **XIMENIA** L.

Shrubs or trees, often armed with spines, these formed from abortive branchlets; leaves often clustered on short spurs. Flowers in short axillary cymes or sometimes solitary; calyx 4–5-toothed or -lobed; petals 4 or 5, valvate, densely white-bearded within. Stamens free, twice as many as the petals; anthers linear. Ovary partly 3-celled; style simple, with subcapitate stigma; ovules 3, linear. Fruit a drupe with abundant pulp and a hard stone.

A genus of at least 9 species, 5 of them in Mexico, I confined to Hispaniola, 1 in South Africa, 1 in certain Pacific islands, and another of pantropical distribution.

1. Ximenia americana L., Sp. Pl. 2:1193. 1753. **FIG. 153.**

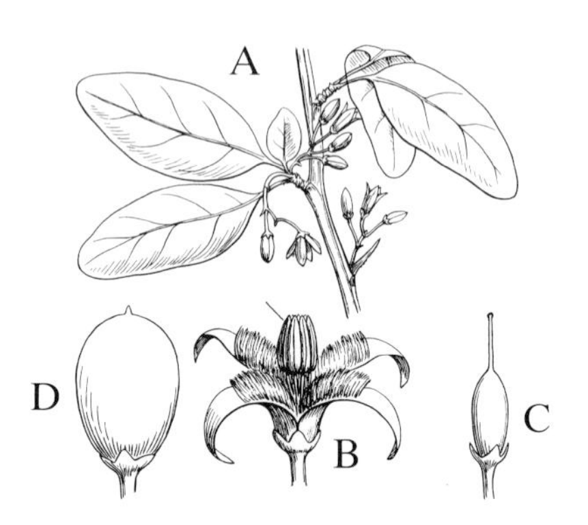

FIG. 153 **Ximenia americana.** A, portion of branch with leaves and flowers. B, flower; a, anthers. C, pistil. D, drupe. (F. & R.)

"Hog Plum"

A densely-branched shrub or small tree to 6 m tall or more, the branches usually spiny; leaves glabrous, oblong to elliptic, 3–7 cm long, rounded or obtuse at both ends, deciduous. Flowers fragrant; corolla yellowish-white, 4-lobed, the lobes reflexed, 6–9 mm long. Fruit yellow or reddish, 1.4–1.7 cm long, the flesh edible.

GRAND CAYMAN: *Brunt 2142; Kings GC 353; Proctor 15161.* LITTLE CAYMAN: *Proctor 28086, 35091.*

– Pantropical, chiefly in coastal thickets but in mountainous countries often with a very wide altitudinal distribution. This species is occasionally cultivated for its edible fruits, which can be eaten raw or cooked. The wood is fragrant, reddish-yellow, fine-textured, very hard and heavy, and has been used in India as a substitute for sandalwood. The bark is occasionally used for tanning leather. Although *Ximenia* is called "Hog Plum" in the Cayman Islands, this name is usually applied elsewhere (e.g., in Jamaica) to a species of *Spondias* (Anacardiaceae), while the general common name for *Ximenia* elsewhere is "Tallow Plum".

Genus 2. **SCHOEPFIA** Schreber

Glabrous shrubs or small trees with coriaceous leaves. Flowers small and inconspicuous, in short racemes which are solitary or clustered in the leaf-axils; calyx cup-shaped, obscurely toothed; disk entire, adnate to the ovary, enlarging as the ovary ripens, at length almost enveloping the fruit, forming an evident ring around it near the apex. Petals 4–6, inserted at margin of the disk, cohering to form a bell-shaped corolla with free lobes at the apex. Stamens equal in number to the petals, adnate to the corolla-tube. Ovary partly immersed in the disk, imperfectly 3-celled, with short or long style and 3-lobed stigma; ovules 3, pendulous from apex of the placenta. Fruit a drupe with thin flesh and hard seed.

A small genus of about 12 species in the American and Asiatic tropics. Some have been shown to have parasitic root connections to other plants.

1. Schoepfia chrysophylloides (A. Rich.) Planch. in Ann. Sci. Nat. ser. 4, 2:261. 1854.

A shrub or small tree up to 7 m tall but usually much lower; leaves petiolate, the blades ovate or elliptic, 2–7 cm long, blunt at both ends. Flowers dimorphic (long- or short-styled), paired on short peduncles, the minute calyces of each pair coherent; corolla cream, 3–3.5 mm long. Fruit red, narrowly obovoid, 6.5–7.5 mm long.

LITTLE CAYMAN: *Proctor 28121.* CAYMAN BRAC: *Proctor 29051.* Both of the cited specimens are sterile; flower and fruit details have been taken from material collected elsewhere, and from literature.

– Bahamas, Cuba, Jamaica, and Hispaniola, in dry rocky woodlands, apparently nowhere very common.

Family 64

LORANTHACEAE

Parisitic shrubs, vines, or sometimes small trees containing chlorophyll, attached to host plants by specialized roots called haustoria. Branches terete or angled, glabrous or pubescent; leaves usually present and well-developed, opposite, simple, often somewhat fleshy. Flowers small or more usually large, often showy, perfect or if unisexual then the plants dioecious, usually borne in groups of 3 (dichasia) on axillary or terminal racemes or panicles, rarely solitary. Calyx represented in rudimentary form by a circular rim or calyculus at apex of ovary. Petals 4–8, valvate, free or united into a tube, regular or (in some African species) zygomorphic. Stamens as many as the petals, adnate to their base or middle and usually shorter; anthers usually 4-celled (or 2-celled by coalescence), opening longitudinally. Disk present or lacking. Ovary inferior, 1-celled, capped by the calyculus; style simple (sometimes elongate) with minute terminal stigma; ovules in the strict sense lacking, the equivalent structure undeveloped until after pollination. Fruit berry-like, with small solitary seed surrounded by viscid pulp; endosperm compound and without chlorophyll.

A chiefly tropical family of about 57 genera and more than 200 species, a few occurring in temperate regions. Widely known as "Mistletoes", a term also (and primarily) applied to members of the related family Viscaceae.

Genus 1. **DENDROPEMON** (Blume) Reichb.

Small parasitic shrubs with coriaceous or somewhat fleshy, mostly flat leaves, and small bracteolate flowers in simple axillary racemes, the bracteoles connate, cupulate. Receptacle-rim produced into a truncate or 4–6-toothed calyculus. Corolla of usually 5 or 6 free petals, these usually small. Stamens short, borne at the base of the petals; anthers dorsifized. Style relatively short, with a terminal stigma which in some species is the same diameter as the style, in others more or less capitate.

A primarily West Indian genus of about 15 species.

1. **Dendropemon caymanensis** Proctor in Phytologia 35:403. 1977.

Stems c. 2 mm thick near the base, up to c. 20 cm long, terete, cinereous, branched; branches more or less compressed, smooth and glabrous, with internodes 0.7–1.5 cm long. Leaves with petioles 1–1.5 mm long, the blades oblong to oblanceolate-oblong, 0.5–2.5 cm long, 3–10 mm broad, subacute and apiculate at the apex, narrowed toward the base, the median nerve prominent especially beneath, the lateral nerves few, obscurely prominulous or obsolete, the margins flat and entire, the texture subcoriaceous. Inflorescences much shorter than most leaves, with peduncles 1.5–4.5 mm long and rhachis up to 1 cm long, rather stiff, 4–8-flowered, smooth and glabrous throughout, the rhachis more or less compressed. Pedicels 0.5–1 mm long; bracteoles with shortly triangular free apex. Flower-buds oblong, obtuse; calyculus cup-shaped and truncate; petals of open flowers ligulate, c. 1.8 mm long and 0.5–0.6 mm broad. Style 0.9–1.1 mm long, with subcapitate stigma. Berries obovoid, blackish-purple, up to 5.5 mm long.

LITTLE CAYMAN: *Proctor 35215 (type).*

– Endemic. This species is known only from the type collection, which was found growing as a parasite on *Capparis cynophallophora.* it appears to be related to *D. purpureus* (L.) Krug & Urban of the Bahamas, Cuba, Hispaniola and Puerto Rico, and to *D. rigidus* Urban & Ekman of the Dominican Republic.

Family 65

VISCACEAE

Parasitic shrubs (some taxa reduced to minute size) containing chlorophyll, attached to host plants by specialized roots called haustoria. Branches articulate and terete, angled or flattened, glabrous or pubescent. Leaves opposite, simple, often rather fleshy, or else reduced to scales and apparently lacking. Plants either monoecious or dioecious, the flowers very small and always unisexual, rarely solitary, usually in rows or series on the internodes of articulated axillary or terminal spikes, which in turn may be arranged in symmetric compound inflorescences. Perianth-segments usually 3 (rarely 4), valvate, or in very reduced genera sometimes lacking or indistinguishable from the apical part of the ovary, united at base. Stamens as many as the perianth-segments, adherent to them and much shorter, the anthers nearly sessile; anthers 1–4-celled, the cells splitting transversely. Ovary inferior, 1-celled, at apex with minute, sessile, knoblike stigma; calyculus absent. True ovules lacking, but embryo development different in many details from that of Loranthaceae. Fruit berry-like, crowned by the persistent perianth-segments, with solitary seed surrounded by viscid pulp. endosperm simple and green.

A chiefly tropical family of about 7 genera and more than 350 species, some occurring in North Temperate regions. The great majority of the species belong to the genus *Phoradendron,* which is confined to the Western Hemisphere. The European *Viscum album* L. was the original "Mistletoe", allegedly with magical properties, but all members of the family are now known by this name.

Genus 1. **PHORADENDRON** Nutt.

Small shrubs mostly parasitic on broad-leaved woody plants, rarely on conifers or on other species of *Phoradendron.* Stems terete, 4-angled, or flattened and 2-edged, rarely winged. Leaves usually evident or conspicuous, rarely apparently absent. Inflorescence of axillary jointed spikes with a pair of small fleshy or scarious bracts subtending each joint; flowers arranged in 2-6 (rarely 8) rows on each internode, each flower sessile and more or less immersed in a small pit; plants monoecious or dioecious. Staminate perianth bearing an almost sessile 2-celled anther at the base of each lobe. Ovary with capitate stigma. Berries white, yellow, orange, or red.

A widespread Western Hemisphere genus of more than 200 species, the majority occurring in tropical areas.

KEY TO SPECIES

1. Flowers at anthesis 0.7-0.8 mm in diameter, yellowish-brown when dried; leaves less than 1 cm broad: **1. P. quadrangulare**

1. Flowers at anthesis 1-1.2 mm in diameter, usually black when dried; leaves up to 2.8 cm broad: **2. P. rubrum**

1. Phoradendron quadrangulare (Kunth) Krug & Urban in Engl. Bot. Jahrb. 24:35. 1897. **FIG. 154.**

P. quadrangulare var. *gracile* Krug & Urban, 1897.

P. gracile (Krug & Urban) Trelease, 1916.

Stems up to 40 cm long, and to 5 mm thick near the base; branches with the younger internodes acutely and subequally 4-angled; cataphylls (sheathing scales) occurring only at the base of the lowest internode of a branch. Leaves narrowly oblanceolate, flaccid, mostly 1.5-4 cm long, 3-9 mm broad, obtuse at the apex, attenuate at base, very obscurely 3-nerved from the base. Spikes usually solitary in the leaf-axils, 1-2 cm long, 2-3-jointed; ripe fruits subglobose, smooth, 2.5-3 mm in diameter, pale orange, yellow, or nearly white, somewhat translucent.

GRAND CAYMAN: *Correll & Correll 51033; Kings GC 150, GC 365.* LITTLE CAYMAN: *Proctor 35194.* CAYMAN BRAC: *Kings CB 91.*
– In the broad sense widely distributed in tropical America, but many of the variants have been described as separate species. The Cayman collections resemble the population occurring in Jamaica, where it has been called *P. gracile.* The Grand Cayman plants grew as parasites on *Guapira discolor* (Nyctaginaceae), while that of Little Cayman was on *Croton linearis* (Euphorbiaceae). The Cayman Brac specimen was collected on *Bauhinia divaricata* (Leguminosae: Caesalpinioideae, *Kings CB 91 A*).

2. Phoradendron rubrum (L.) Griseb., Fl. Brit. W. I. 314. 1860. **FIG. 154.**

"Scorn-the-ground"

Stems up to 50 cm long (but often less), and to 6 mm thick near the base, the oldest terete, the intermediate internodes more or less 4-angled, and those nearest the apex flattened and rather sharply 2-angled, especially just below a node; cataphylls occurring only at the base of the lowest internode of a branch. Leaves oblanceolate to obovate or sometimes elliptic, rather thick and leathery, 2–8 cm long, (5-) 10–28 mm broad, rounded at the apex, more or less cuneate at the base, with a very short petiole, obscurely 3–5-nerved from the base. Spikes 1–3 in the leaf-axils, 1–3 cm long, 3–5-jointed; ripe fruits subglobose, smooth, c. 4 mm in diameter, reddish or orange.

GRAND CAYMAN: *Brunt 1652; GC 165, GC 388; Lewis GC 14; Proctor 15111, 15259.* LITTLE CAYMAN: *Kings LC 36, LC 43; Proctor 28122, 28185, 35075, 35100.* CAYMAN BRAC: *Proctor 28959, 35148.*

– Bahamas and Cuba. This common species is most often found as a parasite on *Guapira discolor* (Nyctaginaceae), but has been recorded twice on *Swietenia mahagoni* (Meliaceae) (*Brunt 1652, Proctor 15259*) and once on *Croton linearis* (Euphorbiaceae) (*Proctor 35148*). *Phoradendron* species are usually not host-specific, but nevertheless seem to show definite preferences which may vary from one area to another. As these plants are adapted to dispersal by birds, it is possible that what seem to be host-preferences of the *Phorandendron* species may actually reflect the perching preferences of the birds that eat the berries.

Family 66

CELASTRACEAE

Shrubs or trees, erect or sometimes climbing; leaves opposite or alternate, rarely whorled, simple and entire or toothed but never lobed; stipules absent or small and soon falling. Flowers usually in axillary cymes, greenish or white, perfect or else unisexual and monoecious or dioecious; pedicels frequently jointed. Calyx small, with 4–5 imbricate lobes, persistent. Petals 4–5, short, spreading, sessile below the margin of the disk, imbricate in bud. Stamens 4–5, inserted on or near the margin of the disk, with awl-shaped filaments; anthers dehiscing introrsely. Ovary 3–5-celled, the short thick style entire or sometimes 3–5-lobed; ovules 2 or sometimes 1 in each cell, erect or rarely pendulous. Fruit a drupe or capsule; if capsular, the carpels fused; seeds with or without an aril; endosperm usually present and fleshy; embryo usually rather large, with flat, foliaceous cotyledons.

A cosmopolitan family of more than 40 genera and 850 species.

KEY TO GENERA

1. Fruit a capsule; seed with an aril; flower-parts in 5's: **1. Maytenus**

1. Fruit a drupe; seed without an aril:

  2. Flower-parts in 5's: **2. Elaeodendron**

  2. Flower-parts in 4's:

    3. Flowers bisexual; some leaves in whorls: **3. Crossopetalum**

    3. Flowers unisexual; plants mostly dioecious:

      4. Leaves opposite: **4. Gyminda**

      4. Leaves alternate: **5. Schaefferia**

Genus 1. **MAYTENUS** Molina

Shrubs or small trees, mostly glabrous; leaves alternate, coriaceous, entire or serrate; stipules minute and soon falling. Flowers small, polygamous or rarely functionally unisexual and the plants dioecious, axillary and solitary, clustered, or in small cymes. Stamens inserted outside and below the disk, with ovate-cordate anthers; disk round with wavy margin. Ovary immersed in the disk and confluent with it, 2-4-celled; style absent or very short; ovules 1 or 2 in each cell, erect. Fruit a hard, 1-3-celled capsule, 2-3-valved; seeds more or less enclosed by a thin aril; endosperm fleshy or apparently none.

A tropical American genus of perhaps 70 species, the majority in South America.

1. **Maytenus buxifolia** (A. Rich.) Griseb., Cat. Pl. Cub. 53. 1866.

"Bastard Chelamella"

A shrub or rarely a small tree with gray twigs; leaves oblong to obovate or spatulate, mostly 1–3 cm long, of smooth hard texture, obtuse and slightly notched at the apex. Flowers few in clusters, on pedicels 1–4 mm long; calyx-lobes roundish, 0.5 mm long; petals greenish-yellow, ovate, nearly 2 mm long. Fruits red or orange, globose-obovoid, 5–9 mm long, apiculate.

LITTLE CAYMAN: *Kings LC 67; Proctor 28091.*
– Bahamas, Cuba, and Hispaniola, in dry rocky thickets and scrublands.

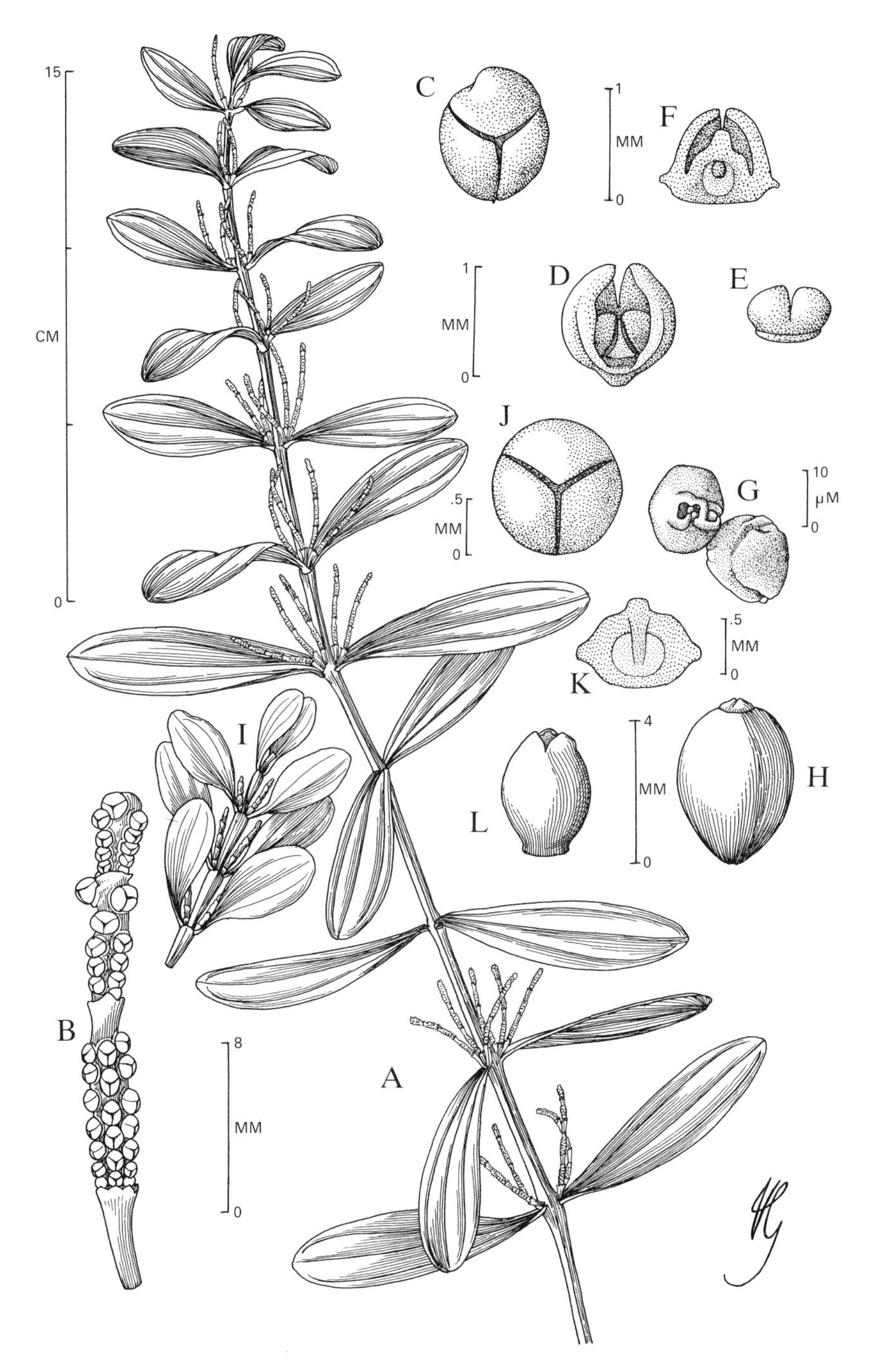

FIG. 154 **Phoradendron quadrangulare.** A, flowering branch. B, flower-spike. C, flower. D, staminate flower, opened. E, single stamen. F, long. section of pistillate flower. G, pollen-grains as shown by scanning electron microscope. H, fruit. **Phoradendron rubrum.** I, habit of small branch. J, flower. K, pistillate flower, x-section. L, fruit. (G.)

Genus 2. **ELAEODENDRON** Jacq. f. ex Jacq.

Shrubs or small trees; leaves opposite and alternate, entire or crenate; stipules minute, soon falling. Flowers perfect, or often unisexual and dioecious, in cymes on axillary peduncles, the parts in 5's. Stamens inserted under and outside the disk, with globose anthers; petaloid staminodes present in pistillate flowers. Ovary confluent with the disk, 2-5-celled; style very short, the stigma 2-5-lobed; ovules 2 in each cell, erect. Fruit a dry or pulpy drupe, the stone 1-3-celled with 1 or 2 seeds in each cell; seeds without an aril.

A pantropical genus of nearly 60 species.

1. **Elaeodendron xylocarpum** (Vent.) P.DC., Prodr. 2:11. 1825.
var. **attenuatum** (A. Rich.) Urban, Symb. Ant. 5:88. 1904.

A small tree up to 10 m tall, sometimes lower and shrubby; leaves opposite or some of them alternate, oblong to elliptic or obovate, mostly 4–10 cm long, the apex obtuse to acute, the margins nearly entire or sparingly crenulate. Inflorescence shorter than the leaves; flowers pale green on pedicels 1–3 mm long; sepals c. 1 mm long, petals 2-3 mm long. Fruits yellowish, globular to ellipsoid, 1.5-3 cm long, rounded and smooth or minutely apiculate.

GRAND CAYMAN: *Brunt 1698, 1829, 1979; Hitchcock; Proctor 15033, 15110.*
– As interpreted here, one of several intergrading and variable varieties of a species occurring throughout the West Indies. Var. *attenuatum* is best represented in the Bahamas, Cuba, and Hispaniola but does not occur in Jamaica. It grows in dry thickets and rocky woodlands.

This taxon is placed in the genus *Cassine* by Adams (1972) and several other modern writers. However, *Cassine* interpreted strictly is confined to Africa, and interpreted broadly causes confusion in delimiting other genera in the Celastraceae.

Genus 3. **CROSSOPETALUM** P. Br.

Shrubs or low trees; leaves opposite or in whorls of 3 (rarely alternate), short-petiolate, the margins entire, crenate, or spiny; stipules minute. Flowers small, perfect, in cymes or subsolitary at the ends of short or long peduncles, the parts mostly in 4's. Calyx lobed; petals reflexed. Stamens inserted between lobes of the disk; anthers subglobose. Ovary usually 4-celled, with short style and 4-lobed stigma; ovules solitary in each cell, erect. Fruit a small dry or fleshy drupe; seed with or without an aril and with fleshy endosperm.

A tropical American genus of about 20 species.

KEY TO SPECIES

1. Stems and underside of leaves glabrous; flowering pedicels 1-2.5 mm long: **1. C. rhacoma**

1. Stems and underside of leaves densely pubescent; flowering pedicels 4-6 mm long: **2. C. caymanense**

1. Crossopetalum rhacoma Crantz, Inst. Rei Herb. 2:321. 1766. **FIG. 155.**

Rhacoma crossopetalum L., 1759.

"Snake Berry", "Tobacco Berry", "Wild Tobacco"

Usually a shrub 1-3 m tall, the glabrous young twigs with 4 raised longitudinal lines; leaves opposite or sometimes in whorls of 3, in Cayman plants oblanceolate, oblong, or obovate, 1-3 cm long, the apex obtuse or rounded and often slightly notched, the margins crenulate. Inflorescence a 1-3-times-forked cyme with glabrous peduncle and bracts; pedicels 1-2.5 mm long; calyx 0.7 mm long, glabrous; petals obovate-elliptic, 1-1.2 mm long, green tinged with red. Style erect with 4 recurved linear stigmas at the apex. Fruit obliquely obovoid, 5-7 mm long, scarlet, crowned by the persistent style.

GRAND CAYMAN: *Brunt 2054; Hitchcock; Proctor 15091, p.p. (IJ).* LITTLE CAYMAN: *Proctor 28098.* CAYMAN: BRAC: *Proctor 29141.*
– Florida and throughout the West Indies to St. Lucia, also in the Belize cays, Bonaire and Colombia, mostly in coastal scrublands.

2. Crossopetalum caymanense Proctor in Sloanea 1:2. 1977. **FIG. 155.**

A shrub 1-2 m tall, the densely pubescent young branches with 4 prominent longitudinal ridges; leaves opposite or sometimes in whorls of 3, lanceolate to ovate or sometimes oblong, 1-5 cm long, the apex acute and sharply mucronate, the margins distantly serrulate. Inflorescence a 2-5-times-forked cyme with densely pubescent peduncle and bracts; pedicels glabrous, at anthesis and in fruit flexuous and 4-6 mm long; calyx 0.8-0.9 mm long, pubescent; petals roundish, 1.1-1.5 mm long, cream or pale salmon. Style nearly obsolete, the stigma minutely 4-lobed. Fruit globose or ellipsoid, 5-6 mm long, crimson.

GRAND CAYMAN: *Proctor 15026, 15091 (BM), 15184 (type), 27968.* CAYMAN BRAC: *Kings CB 12.*
– Endemic, occurring in rocky thickets and woodlands. The Cayman Brac specimens bear larger and more ovate leaves than those of Grand Cayman.

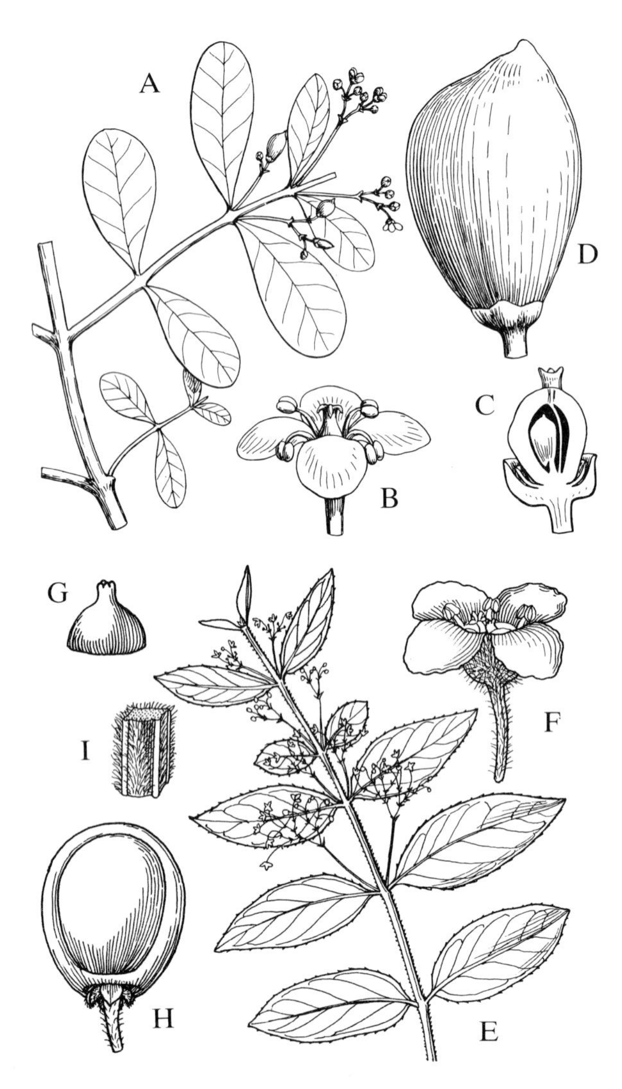

FIG. 155 **Crossopetalum rhacoma.** A, twig with leaves and flowers, x1. B, flower, x10. C, ripening ovary cut lengthwise, x10. D, fruit, x5. (F. & R.). **Crossopetalum caymanense.** E, twig with leaves and flowers, x1. F, flower, x10. G, ovary x20. H, fruit, x4. I, section of stem, x8. (G.)

Genus 4. **GYMINDA** Sarg.

Shrubs or small trees; leaves opposite, short-petiolate, the margins entire or crenulate-serrate above the middle; small stipules present. Flowers small, unisexual and dioecious, in small, stalked, axillary cymes, the parts in 4's. Calyx deeply lobed; petals reflexed. Staminate flower with stamens inserted in lobes of the disk; anthers oblong. Pistillate flower with 3-celled ovary and a 2-lobed stigma; ovule solitary and pendulous in each cell. Fruit a small drupe, 1- or sometimes 2-seeded; seeds with thin, fleshy endosperm.

A pan-Caribbean genus of 3 species.

1. **Gyminda latifolia** (Sw.) Urban, Symb. Ant. 5:80. 1904. **FIG. 156.**

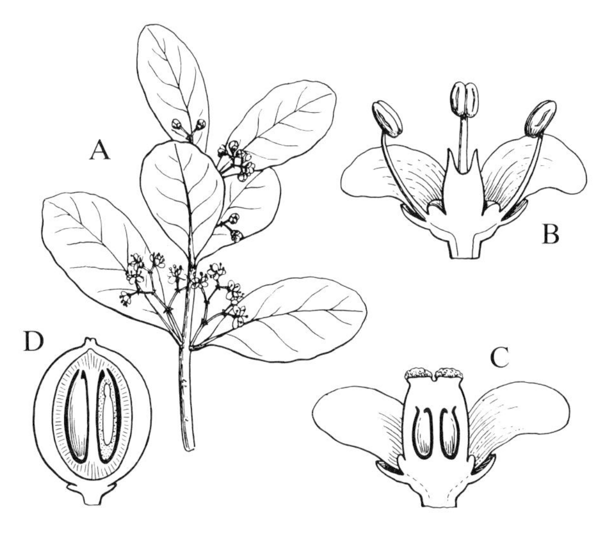

FIG. 156 **Gyminda latifolia.** A, branchlet with leaves and flowers, x$^2/_3$. B, staminate flower cut lengthwise, x8. C, pistillate flower cut lengthwise, x8. D, drupe cut lengthwise, x6. (F. & R.)

A glabrous shrub or small tree to 7 m tall, the young twigs marked with 4 raised lines. Leaves oblanceolate to obovate, 1–5 cm long, the apex rounded or obtuse. Flowers fragrant, whitish; sepals 0.6–0.8 mm long, petals 1.6–2.2 mm long. Fruit black, more or less ellipsoid, 4–8 mm long.

GRAND CAYMAN: *Brunt 2051; Correll & Correll 51028; Proctor 15066, 15196, 15230.* LITTLE CAYMAN: *Proctor 28093, 35135.* CAYMAN BRAC: *Proctor 29140.*
– Florida and the West Indies, in rocky thickets and woodlands.

Genus 5. **SCHAEFFERIA** Jacq.

Glabrous shrubs or small trees; leaves alternate or clustered on short, spur-like branches, the margins entire; stipules minute. Flowers unisexual and usually dioecious, solitary or clustered in the leaf-axils, with parts in 4's. Calyx deeply lobed; petals hypogynous. Stamens hypogynous or inserted below the margin of a small, inconspicuous disk. Ovary 2-celled; ovules solitary in each cell, erect; style short with prominent 2-lobed stigma. Fruit a small dry drupe, 2-seeded; seeds without aril and with scant endosperm.

A tropical American genus of 16 species.

1. **Schaefferia frutescens** Jacq., Enum. Pl. Carib. 33. 1760. **FIG. 157.**

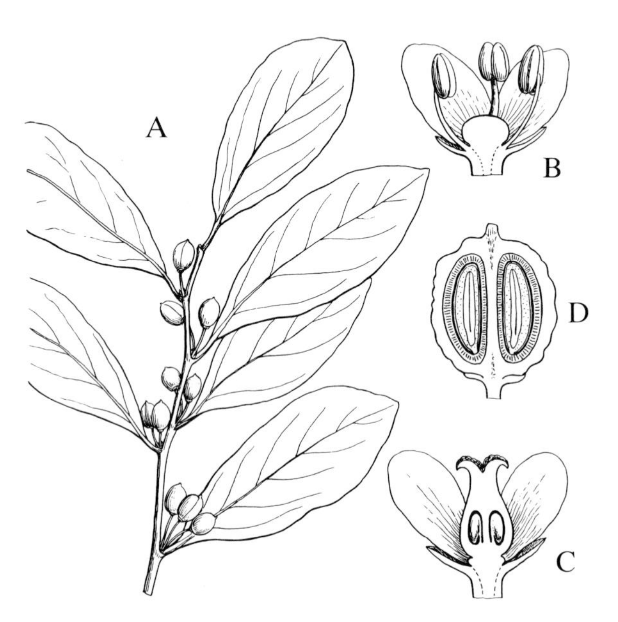

FIG. 157 **Schaefferia frutescens.** A, fruiting branch, x$^{2}/_{3}$. B, staminate flower cut lengthwise, x7. C, pistillate flower cut lengthwise, x7. D, drupe cut lengthwise, x4. (F. & R.)

Shrub or small tree to 6 m tall, the young twigs striate with raised lines; leaves narrowly to broadly elliptic, 2–5 cm long, sharply acute or subacuminate at apex, the veins prominulous on both sides. Flowers 1-several in a cluster, pale green, on pedicels 1-3 mm long; calyx 0.7-1 mm long; petals 3-4 mm long. Fruit spherical to ovoid, 4-6 mm long, orange or scarlet, crowned by the persistent stigmas.

GRAND CAYMAN: *Correll & Correll 51031; Proctor 11996, 15164.* LITTLE CAYMAN: *Proctor 28182.* CAYMAN BRAC: *Proctor 15330, 29028.*
– Florida and the West Indies; attributed also to Mexico and Ecuador. This species grows in dry, rocky woodlands.

Family 67

BUXACEAE

Shrubs or small trees, rarely herbs; leaves opposite or alternate, usually entire and of hard texture; stipules lacking. Flowers unisexual, monoecious or rarely dioecious, in axillary or supra-axillary, lax or dense spikes or racemes, the terminal flowers often pistillate and the others staminate, each one subtended by a bract. Perianth of 4-6 imbricate sepals, or lacking; petals none. In staminate flowers, the stamens free and opposite the sepals, or indefinite; anthers 2-celled. In pistillate flowers the ovary usually 3-celled, terminating in 3 simple styles; ovules 2 or rarely 1 in each cell, pendulous. Fruit a dehiscent capsule or sometimes drupe-like, usually crowned by the 3 persistent styles; seeds with more or less fleshy endosperm, or rarely none.

A small, scattered, cosmopolitan family of about 7 genera and 100 species.

Genus 1. **BUXUS** L.

Shrubs or small trees, usually densely branched; leaves opposite, sessile or short-petiolate. Bracts numerous, several often without flowers. Staminate flowers usually stalked, with 6 sepals in 2 series; stamens 4. Pistillate flowers sessile, with 6 sepals in 2 series. Fruit a 3-horned capsule, splitting loculicidally through the horns; seeds oblong, 3-cornered, with a small strophiole.

A widely-distributed genus of about 45 species, the majority West Indian.

1. Buxus bahamensis Baker in Hook., Ic. Pl. t. 1806. 1889. **FIG. 158.**

Usually a shrub 1-2 m tall, glabrous throughout; leaves elliptic or sometimes oblong, rigid, 1.5-3 cm long, sharply acute and mucronulate at the apex, the venation obscure. Flowers greenish-white, the staminate on pedicels 2-3 mm long; sepals 1-2 mm long. Capsule 5-7 mm long.

LITTLE CAYMAN: *Proctor 28138, 35136.*
– Bahamas, Cuba and Jamaica, in dry rocky scrublands.

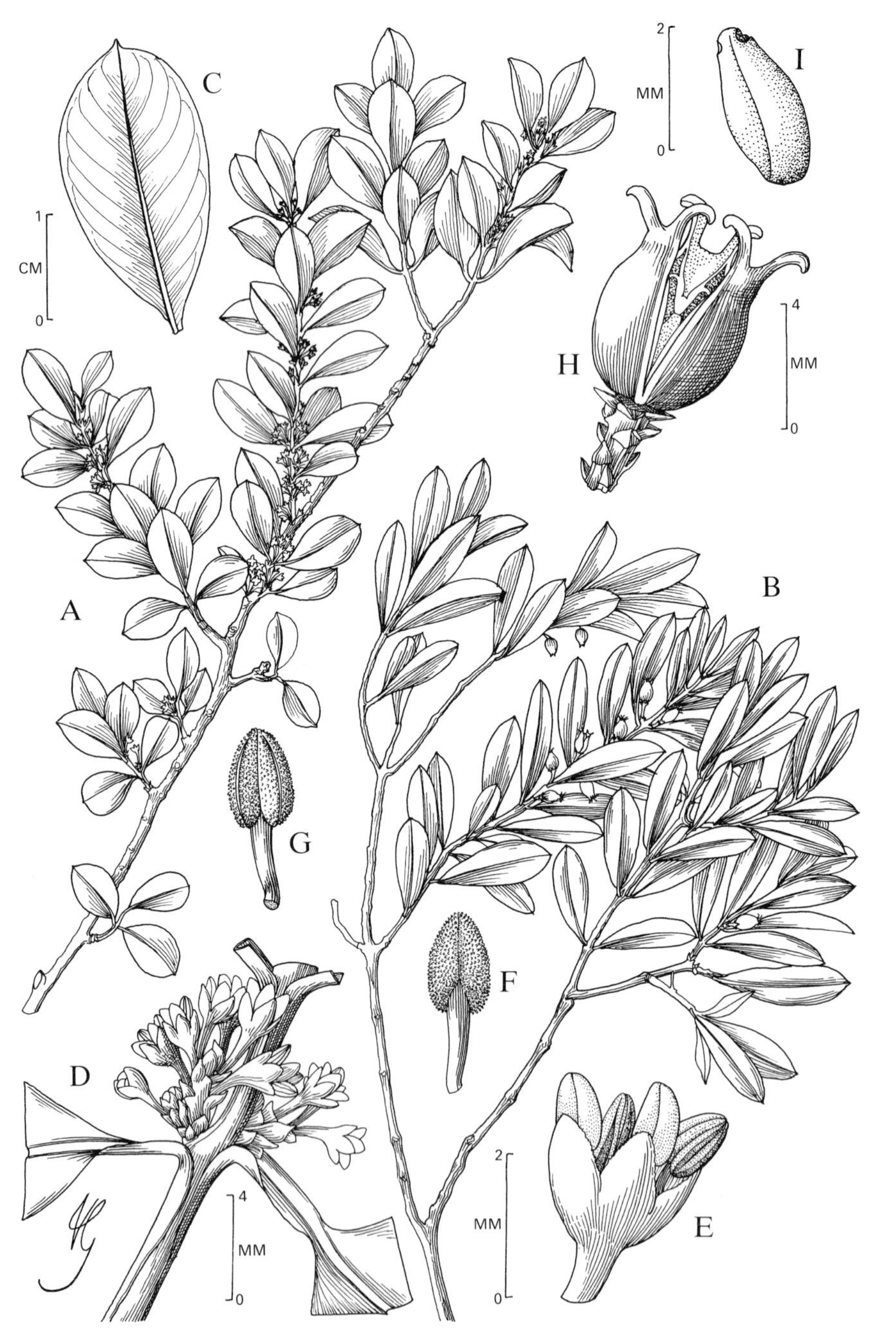

FIG. 158 **Buxus bahamensis.** A, flowering branch. B, fruiting branch. C, leaf. D, flower-clusters. E, staminate flower. F, G, two views of stamen. H, fruit. I, seed. (G.)

Family 68

EUPHORBIACEAE

Herbs, shrubs or trees, sometimes scandent or twining, often with milky sap; leaves usually alternate, sometimes opposite or whorled, simple or rarely digitately compound, palmate-lobed, or absent; stipules often present. Inflorescence various in form, the flowers usually small, unisexual, monoecious or dioecious, generally regular. Perianth sometimes lacking, often dissimilar in staminate and pistillate flowers, with either a calyx only, or a calyx and corolla, the segments free or united, imbricate or valvate. Staminate flowers with receptacle sometimes expanded into a disk within the stamens, or the glands or lobes of the disk alternating with stamens of the outer series; stamens sometimes indefinite, or often as many as the sepals or fewer, sometimes only 1; filaments free or united; rudimentary ovary present or absent. Pistillate flowers with disk ring-like or cup-like, or of separate glands, or absent; ovary usually 3-celled (rarely otherwise), styles usually 3, free or united, entire, cleft or fringed; ovules 1 in each cell, or 2 collateral ones, pendulous, anatropous, attached at inner angle of the cell, the raphe ventral; micropyle often covered with a cushion-like caruncle (obturator), this persisting on the seed. Fruit usually a "schizocarp" capsule splitting 3 ways into 2-valved cocci, these separating from the persistent axis, or sometimes drupe-like and indehiscent. Seeds as many as the ovules; endosperm usually copious and fleshy; cotyledons mostly broad and flat or folded.

One of the largest plant families, with more than 200 genera and 7,000 species. Many are easily recognizable by the combination of milky sap and dry 3-celled fruit, but there are numerous exceptions to these characters. The family includes a number of plants of great economic importance, of which *Hevea*, the source of most natural rubber, is the most valuable.

SYNOPSIS OF SUBFAMILIES

Ovules 2 in each cell of the 3-celled ovary; fruit a capsule; sap never milky; leaves entire, unlobed, and non-glandular: *SUBFAM. 1. PHYLLANTHOIDEAE*

Ovules 2 in each cell of the 2-celled ovary; fruit a drupe; sap never milky; leaves 3-foliolate, non-glandular: *SUBFAM. 2. OLDFIELDIOIDEAE*

Ovules 1 in each cell of the ovary; fruit a capsule or drupe; sap often coloured or milky; leaves often lobed and/or with petiolar glands or marginal teeth: *SUBFAM. 3. EUPHORBIOIDEAE*

SUBFAMILY. 1. PHYLLANTHOIDEAE

KEY TO GENERA

1. Flowers with petals; leaves pinnate-veined, usually deciduous at flowering time:

  2. Staminate flowers with free filaments; pistillate flowers subsessile, the fruits on stalks not longer than diameter of the fruit: **1. Savia**

  2. Staminate flowers with filaments united into a column, this expanded at apex into a disk bearing the anthers; pistillate flowers and fruits very long-stalked: **2. Astrocasia**

1. Flowers without petals; leaves variously veined or absent, but if present not deciduous at flowering time:

  3. Plants dioecious; staminate flowers with rudimentary ovary; leaves 3–5-veined from the base: **3. Chascotheca**

  3. Plants monoecious; staminate flowers without rudimentary ovary; leaves pinnate-veined or absent: **4. Phyllanthus**

Genus 1. **SAVIA** Willd.

Dioecious shrubs or small trees; leaves alternate, entire; stipules present. Flowers axillary, the parts in 5's; rudimentary petals present. Staminate flowers densely clustered, subsessile; disk ring-like, outside the stamens; filaments free, alternate with the petals; a rudimentary ovary present, with 3 short styles. Pistillate flowers with 3-celled ovary and separate styles, each with 2 awl-shaped acuminate branches; ovules 2 in each cell. Seeds ovoid or 3-edged, without a caruncle.

A West Indian genus of probably less than 10 species.

1. Savia erythroxyloides Griseb. in Mem. Amer. Acad. n.s. 8:157. 1860.

"Wild Coco-plum"

A shrub or small tree to 5 m tall; leaves glabrous, leathery, elliptic to obovate, 2–7 cm long, the apex obtuse or rounded, sometimes shallowly notched; veins prominulous especially on the upper side. Capsules glabrous, depressed-globose, c. 8 mm in diameter, slightly and obtusely 3-lobed.

GRAND CAYMAN: *Brunt 1874, 1895, 2003, 2188; Proctor 15023, 15030, 15167, 15177.* LITTLE CAYMAN: *Kings LC 64; Proctor 28094. 35108.* CAYMAN BRAC: *Proctor 15326, 28943.*

– Florida, Bahamas, Cuba, Jamaica, Hispaniola, and the Swan Islands, in dry rocky thickets and woodlands. As here interpreted, not specifically distinct from *S. bahamensis* Britton.

Genus 2. **ASTROCASIA** Robins. & Millsp.

Glabrous shrubs or small trees; leaves alternate, membranous, on long slender petioles, the margins entire. Flowers dioecious, clustered in the leaf-axils; petals present. Staminate flowers on short pedicels; sepals 5, imbricate; petals 5, erect; stamens 10, the filaments connate into a slender column, this expanded at the apex into a disk; anthers ellipsoid, sessile, horizontally dehiscent; ovary rudiment lacking. Pistillate flowers on very long slender pedicels; ovary 3-celled. Fruit an elastically dehiscent, 3-grooved capsule; seeds irregularly globose, without a caruncle.

A chiefly Central American genus of about 4 species.

1. **Astrocasia tremula** (Griseb.) Webster in Jour. Arnold Arb. 39:208. 1958.

Phyllanthus glabellus Fawc. & Rendle, 1919, not *Croton glabellus* L., 1759.

An arborescent shrub or small tree to 5 m tall; leaves with petioles 3-5 cm long or more, the blades elliptic or rhombic-ovate, mostly 4-8 cm long, blunt at apex, whitish-glaucous on the under surface, often deciduous before flowering; stipules lance-linear, soon falling. Staminate flowers on pedicels mostly 5-10 mm long; sepals 1.5 mm long, petals to 2.5 mm long; anthers 4, attached around margin of the peltate connective, opening transversely. Pistillate flowers on pedicels lengthening in fruit to as much as 4.5 cm; sepals 5, to 2 mm long; petals 5, to 3.5 mm long; style-branches 2-lobed, fleshy. Capsules depressed-globose, 8-10 mm in diameter; seedes 3 or 6, often all abortive, flattish-ellipsoidal, 4.5 mm long.

GRAND CAYMAN: *Proctor 11975 (♀), 15140 (♂).*

– Jamaica, in rocky woodlands.

Genus 3. **CHASCOTHECA** Urban

Dioecious, glabrous shrubs with terete branchlets; leaves alternate, distichous, entire, 3-5-nerved from the base; stipules small, biauriculate. Flowers in small clusters from minute, cushion-like spurs in the leaf-axils or often at defoliated nodes; sepals 5; petals lacking. Stamens 5, with filaments connate at the base, the rudimentary ovary columnar and terminating in 3 recurved styles. Pistillate flowers with ovary 3-celled and 2 ovules in each cell; styles 3, free, forked at the apex. Capsule of 3 two-valved cocci; seeds bullate-rugose, with abaxial hilum; caruncle absent; endosperm fleshy.

A small genus of 3 species confined otherwise to Cuba and Hispaniola.

KEY TO SPECIES

1. Leaves elliptic, with midrib whitish beneath and petiole 0.5–0.7 mm in diameter at the top; fruiting pedicels 4–10 mm long; capsules 2–3 mm in diameter: **1. C. neopeltandra**

1. Leaves broadly ovate to orbicular, with midrib dark beneath and petiole 0.3–0.4 mm in diameter at the top; fruiting pedicels 12–25 mm long; capsules 4–5 mm in diameter: **2. C. domingensis**

1. Chascotheca neopeltandra (Griseb.) Urb., Symb. Ant. 5:14. 1904.

A dense-crowned shrub to 3 m tall or more, with numerous wide-spreading branches; leaves with petioles 2–5 mm long, the blades 2–4 cm long, obtuse to acutish at the apex, pale or glaucous beneath with the fine venation faintly or scarcely evident. Sepals membranous, 1.5 mm long, the margins minutely ciliate. Staminate flowers subsessile. Pistillate flowers not seen.

GRAND CAYMAN: *Brunt 1817 (♂).*
– Cuba and Hispaniola, in thickets on limestone rock.

2. Chascotheca domingensis (Urb.) Urb., Symb. Ant. 5:14. 1904.

Shrub 1.5–2 m tall, the branchlets with horizontal frond-like habit; leaves with delicate, filiform petioles 2–6 mm long, the blades light green 1–2 cm long, obtuse or rounded at the apex, the fine, densely reticulate venation evident on the under side. Sepals membranous, 1.5 mm long, with margins minutely ciliate. Staminate flowers subsessile or on pedicels to 2 or 3 mm long. Pistillate flowers not seen, apparently undescribed.

GRAND CAYMAN: *Brunt 2187* (sterile); *Proctor 15165* (mixture, ♂ flowers and fruits). CAYMAN BRAC: *Proctor 35150* (sterile).
– Hispaniola, apparently rare in dry rocky woodlands.

Genus 4. **PHYLLANTHUS** L.

Herbs, shrubs, or trees of very diverse habit, the branching either unspecialised and either spiral or distichous, or else "phyllanthoid", i.e., the spiralled leaves on the main axes reduced to stipule-like "cataphylls" which subtend deciduous branchlets, these either bearing distichous leaves or else leafless and more or less broadly flattened (then called "phylloclades"). Leaves present or absent, always simple and entire, usually glabrous, the petiole always shorter than the blades; stipules deciduous or persistent. Plants usually monoecious, rarely dioecious. Inflorescences axillary; calyx 4-6-lobed, the lobes imbricate in bud; petals lacking; disk nearly always present, commonly divided into segments alternating with the calyx-lobes. Staminate flowers stalked; stamens 2-15 (mostly 3-6), the filaments free or connate, the anthers free or connate; rudimentary ovary absent. Pistillate flowers stalked or subsessile; staminodes absent (except in *P. acidus*); ovary usually 3-celled with 2 ovules in each cell; styles erect or spreading, free or united into a column, variously divided or else dilated into an entire or lacerate stigma. Fruit mostly a dehiscent capsule, rarely berry-like or drupe-like; seeds usually 2 in each cell, rarely only 1 developing; endosperm cartilaginous.

A nearly cosmopolitan genus of about 750 species, the majority of them Old World in distribution. *Phyllanthus acidus* (L.) Skeels, the "Jimbling", originally from Brazil, is commonly cultivated for its abundant edible (but sour!) fruits.

KEY TO SPECIES

1. Plants clothed with green leaves:

  2. Small annual herb; leaf-blades less than 11 mm long; capsules less than 2.2 mm in diameter: **1. P. amarus**

  2. Shrubs usually 1-3 m tall; leaf-blades more than 1 cm long (up to 8 cm); capsules more than 4 mm in diameter (up to 10 mm):

    3. Leaf-blades not over 2.5 cm long; pedicels of staminate flowers 5-7 mm long, the flowers themselves c. 1.5-2 mm in diameter; capsules 4-4.5 mm across: **2. P. caymanensis**

    3. Leaf-blades up to 8 cm long; pedicels of staminate flowers 8-15 cm long or more, the flowers themselves c. 2.5-4 mm in diameter; capsules up to 10 mm across: **3. P. nutans**

1. Plants (except seedlings) leafless, the ultimate branches being flattened, green and leaflike with nodes represented by marginal notches whence are produced the flowers and fruits: **4. P. angustifolius**

1. **Phyllanthus amarus** Schum. & Thonn. in Kongl. Danske Vidensk. Selsk. Skr. 4:195-196. 1829.

Phyllanthus niruri of many authors, not L., 1753.

Erect annual herb mostly 10–30 cm tall, with phyllanthoid branching, the main stem simple or branched; deciduous branchlets 4–12 cm long, with 15–30 leaves; leaves membranous, usually elliptic-oblong, mostly 5–11 mm long, obtuse or rounded and often apiculate, more or less glaucous beneath. Flowers paired beneath the leaf-axils, all but the lowest axils having one staminate and one pistillate flower, each minutely stalked. Capsules oblate, obtusely 3-angled, mostly 1.9–2.1 mm in diameter; seeds sharply 3-angled.

GRAND CAYMAN: *Hitchcock; Millspaugh 1339, 1363; Proctor 15103, 15244.* LITTLE CAYMAN: *Proctor 35108.* CAYMAN BRAC: *Proctor 29092.*
– A pantropical weed of open waste ground.

2. **Phyllanthus caymanensis** Webster & Proctor*

Glabrous shrub up to 2.5 m tall with simple (pinnatiform) deciduous branchlets; leaves membranous, short-petiolate, the blades ovate or rhombic-ovate, mostly 1.5–2.5 cm long, obtusely to acutely pointed at the apex, somewhat pale or sub-glaucous beneath; stipules 1.2–1.5 mm long. Flowers in axillary cymules, each cluster with one central female flower and several lateral male flowers; calyx-lobes of male flowers 1.3–1.7 mm long; stamens 3, the filaments connate into a column 0.4–0.5 mm high; female flowers with filiform pedicels becoming 8–12 mm long in fruit, the calyx-lobes 1.4–1.7 mm long; styles free, spreading, bifid. Capsules oblate, prominently veiny, greenish; seeds angled, 1.9–2 mm long, nearly smooth.

LITTLE CAYMAN: *Proctor 35145.* CAYMAN BRAC: *Proctor 35151 (type).*
– Endemic; related to *P. mocinianus* and *P. mcvaughii* of Mexico and Central America.

3. **Phyllanthus nutans** Sw., Prodr. Veg. Ind. Occ. 27. 1788.

An irregularly-branched glabrous to hirsutulous shrub 1–3 m tall, without deciduous branchlets; leaves short-petiolate, the blades usually elliptic or ovate, mostly 2–8 cm long, obtuse or rarely acute at the apex, often somewhat glaucous or purple-tinged beneath; stipules 3–5 mm long, yellowish. Inflorescences axillary or pseudoterminal, often racemiform; male flowers in small clusters (cymules) at intervals on the inflorescence-axis; filaments united into a column 0.7–1.1 mm high; female flowers solitary or 2–3 in small clusters mostly on pseudoterminal inflorescences, with pedicels mostly 10–27 mm long. Capsules oblate-spheroidal, obscurely 6-ribbed, c. 6 mm high and up to 10 mm broad; seeds 3-angled, 4.2–7 mm long, smooth.

**Phyllanthus caymanensis* Webster & Proctor, sp. nov., ab aliis speciebus sect. *Nothoclema* differt ramulis glabris simpliciter pinnatiformibus, foliis ovatis subacutis conspicue (cont.)

The Cayman representatives of this species have been assigned to two subspecies, as follows:

1. Leaves ovate, flat, obtuse or rounded at base; flowers in axillary and pseudoterminal inflorescences; calyx-lobes subentire; stylar column 0.5-2.3 mm high: **3a.** ssp. **nutans**

1. Leaves elliptic, with narrowly revolute margins, acute at both ends; flowers all axillary, solitary or the male and female paired at each axil; calyx-lobes denticulate; stylar column 0.5-0.7 mm high: **3b.** ssp. **grisebachianus**

3a. Phyllanthus nutans ssp. **nutans. FIG. 159.**

GRAND CAYMAN: *Brunt 1994, 2139, 2194; Proctor 11977, 15245.* LITTLE CAYMAN: *Proctor 28039.*
– Jamaica and the Swan Islands, in various habitats; the Cayman plants occur in dry rocky woodlands.

3b. Phyllanthus nutans ssp. **grisebachianus** (Muell. Arg.) Webster in Jour. Arnold Arb. 39:61. 1958.

LITTLE CAYMAN: *Kings LC 42.*
– Eastern Cuba; in habitats similar to those of ssp. *nutans.* It has been pointed out that the Cayman populations of *P. nutans* show transitional features between the two subspecies. Several collections from Little Cayman *(Proctor 35111, 35209, 35212)* have not been determined to subspecies.

4. Phyllanthus angustifolius (Sw.) Sw., Fl. Ind. Occ. 2:1111. 1800. **FIG. 159.**

Phyllanthus linearis of Hitchcock, 1893, not Swartz, 1788.

"Duppy Bush", "Duppy Basil"

A shrub mostly 1-3 m tall, or occasionally treelike and 6 m tall, with phyllanthoid branching, the axes glabrous; cataphylls of main stems clustered at apex in a scaly cone rounded in outline and 3-6 mm thick, light to dark brown. Leaves absent except on seedlings. Branchlets bipinnatiform, the ultimate axes broadened into phylloclades, these lance-linear to elliptic or obovate-lanceolate and mostly 3-8 cm long and 2-10 mm broad. Flowers in small staminate or bisexual

---

**Phyllanthus caymanensis* (cont.) venosis, antheris compressis subacutis pollinis grana striato-reticulata. Type: Cayman Islands, Cayman Brac, Foster Land Distr., rocky woodland c. 0.7 mi NW of Pollard Bay, alt. c. 100 ft., 7 Aug. 1975, *G. R. Proctor 35151* (IJ, holotype).

clusters at notches of the phylloclades; pedicels mostly 2-6 mm long; calyx greenish-cream, pale buff, or shades of red. Capsules oblate, 3-4 mm in diameter; seeds 3-angled, 1.4-2.6 mm long.

GRAND CAYMAN: *Brunt 1758, 1828; Correll & Correll 51002; Hitchcock; Howard & Wagenknecht 15030; Kings GC 116, GC 202; Proctor 11974, 11976; Sauer 4111.* LITTLE CAYMAN: *Proctor 28052, 28188, 35085, 35086, 35191; Sauer 4171.* CAYMAN BRAC: *Fawcett; Kings CB 70; Matley; Proctor 28915, 28916.*

– Jamaica and the Swan Islands, in rather dry rocky thickets and woodlands. The Little Cayman population is notably variable in phylloclade-shape and flower-colour.

[**Breynia disticha** J.R. & G. Forst., a *Phyllanthus*-like shrub with green and white variegated leaves, is cultivated in Grand Cayman *(Kings GC 308)* and Cayman Brac *(Proctor, sight record)*. It is said to originate from islands of the Pacific.]

SUBFAMILY 2. OLDFIELDIOIDEAE

Genus 5. **PICRODENDRON** Planch.

Small bushy trees; leaves deciduous, alternate, digitately 3-foliolate; stipules small, soon falling. Flowers dioecious, without petals. Staminate flowers lacking a calyx, the naked stamens in clusters of 3-54 on a convex receptacle subtended by usually several imbricate bracts, the clusters on stalked axillary spikes crowded near the ends of the branchlets and appearing with the young leaves; anthers 2-celled, opening lengthwise, rudiment of ovary lacking. Pistillate flowers solitary in leaf-axils, long-stalked, the pedicel expanded at apex into a concave receptacle; calyx of 4-5 unequal valvate sepals; ovary superior, 2-celled, with slender terminal style and 2 large spreading stigmas; ovules 2 in each cell, pendulous and inserted below an obturator. Fruit a thin-fleshed 1- or 2-seeded drupe; seeds without endosperm.

An Antillean endemic genus once thought to contain 3 species, but here considered probably to consist of but a single variable species. *Picrodendron* has been placed by some authors in a family of its own, of dubious affinity, but more recent consensus places it in the Euphorbiaceae.

1. Picrodendron baccatum (L.) Krug & Urban in Engl. Bot. Jahrb. 15:308. 1892. **FIG. 160.**

"Wild Plum", "Cherry", "Jamaica Walnut", "Black Ironwood"

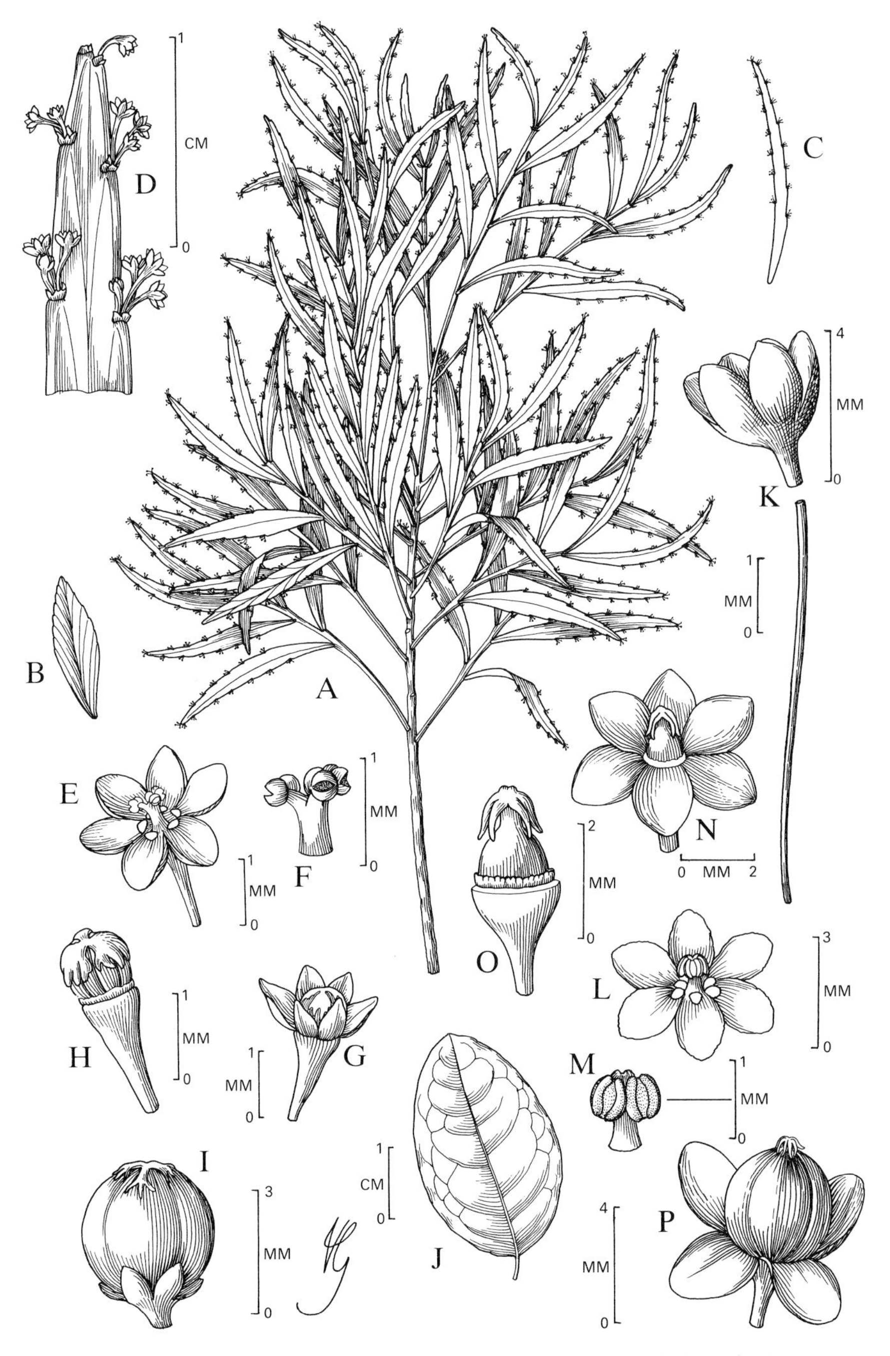

FIG. 159 **Phyllanthus angustifolius.** A, habit. B, C, two forms of phylloclade. D, apex of phylloclade with clusters of flowers. E, staminate flower. F, stamens (androecium). G, pistillate flower. H, gynoecium. I, fruit. **Phyllanthus nutans** ssp. **nutans.** J, leaf. K, flower showing long pedicel. L, staminate flower. M, stamens (androecium). N, pistillate flower. O, gynoecium. P, fruit. (G.)

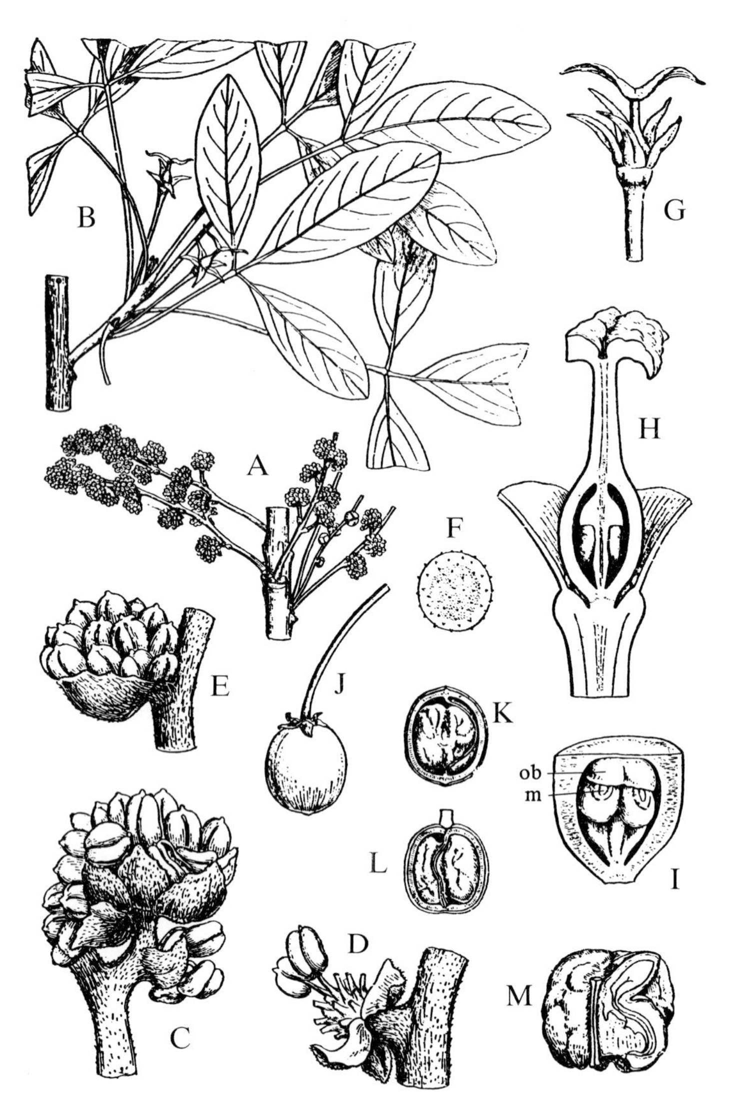

FIG. 160 **Picrodendron baccatum.** A, part of twig with staminate inflorescence, x$^2/_3$. B, same with pistillate flowers, x$^2/_3$. C, D, E, clusters of stamens with subtending bracts, x7. F, pollen grain, x400. G, pistillate flower, x2. H, ovary and style cut lengthwise, x5. I, ovary cut lengthwise through one cavity, x10; ob, obturator; m, micropyle. J, fruit, x$^2/_3$. K, same cut open, with one seed, x$^2/_3$. L, same cut open, with two seeds, x$^2/_3$. M, seed cut to show radicle and cotyledons x1½. (F. & R.)

A tree to 12 m tall, or in exposed situations low and shrubby; leaflets mostly narrowly to very broadly elliptic, rarely obovate, mostly 2.5-9 cm long, acutish to blunt or rounded at the apex, glabrous to rather densely puberulous chiefly on veins beneath; ultimate venation finely reticulate. Staminate inflorescences minutely puberulous; anthers c. 1 mm long. Pistillate flowers with denticulate sepals mostly 3-5 mm long, thickening in fruit. Fruits subglobose or broadly ellipsoid, c. 2 cm in diameter, greenish to orange-yellow, with bitter flesh; seeds alleged to be edible. "but should be eaten with caution" (Fawcett & Rendle, 1920, p. 275).

GRAND CAYMAN: *Brunt 1763; Kings GC 131; Proctor 15248, 27976.* LITTLE CAYMAN: *Kings LC 77; Proctor 28038.* CAYMAN BRAC: *Proctor 29125.*
– Bahamas, Cuba, Hispaniola, Jamaica and the Swan Islands, in rocky woodlands and thickets. The fruits are a favourite food of iguanoid lizards (genus *Cyclura*).

SUBFAMILY 3. EUPHORBIOIDEAE

KEY TO GENERA

1. Flowers solitary or in spikes, cymes, or panicles, not fused and aggregated into bisexual cyathia:

  2. Leaves palmately veined, lobed, or divided:

    3. Shrubs or trees; flowers in panicle-like inflorescences:

      4. Stamens branched and fasciculate, anthers many (up to 1000) per flower; leaves peltate: **11. Ricinus**

      4. Stamens relatively few and unbranched; leaves not peltate:

        5. Flowers with petals; sap not milky; roots not tuberous:

          6. Calyx-lobes free, imbricate; fruit a capsule; seeds with a caruncle: **6. Jatropha**

          6. Calyx-lobes valvate, fused into a spathe; fruit indehiscent; seeds without a caruncle: **[Aleurities**, p. 464]

        5. Flowers without petals; sap milky; roots with starchy tubers: **[Manihot**, p. 464]

    3. Herb; flowers in terminal spikes: **Croton lobatus**, (see **7. Croton**)

2. Leaves simple and pinnate-veined (rarely 3-nerved at base):

7. Petals present, at least in staminate flowers:

8. Stamens more or less inflexed in bud; plants bearing stellate hairs or scales: **7. Croton**

8. Stamens erect in bud; plants clothed with malpighiaceous hairs: **8. Argythamia**

7. Petals absent in both staminate and pistillate flowers:

9. Fruit a dry capsule; sap not milky:

10. Flowers clustered on small cushions in the leaf-axils; leaves without glands: **9. Adelia**

10. Flowers in spikes (the distal staminate portion often deciduous in fruit):

11. Leaves with conspicuous glands at top of petiole; pubescence stellate: **10. Bernardia**

11. Leaves without glands; pubescence if present not stellate:

12. Pistillate flowers sessile and subtended by conspicuous bracts; pubescent small herbs (or cultivated shrubs): **12. Acalypha**

12. Pistillate flowers with long-stalked ovary and fruit, the bracts if present minute; glabrous shrubs or small trees: **13. Ateramnus***

9. Fruit a fleshy drupe; sap milky, poisonous: **14. Hippomane**

1. Flowers fused into a usually bisexual cyathium, usually with one central female flower surrounded by 4 or 5 staminate flowers each represented by a single stamen; cyathium bordered by 1-several usually conspicuous glands, these often with petaloid appendages, the whole cyathium rather flower-like in general appearance:

13. Cyathium spurred, somewhat slipper-shaped; glands hidden within spur: **[Pedilanthus.** p. 489]

13. Cyathium not spurred, the glands on its rim:

14. Gland of cyathium usually 1 or rarely more, never with petaloid appendage; leaves subtending inflorescence often coloured or pale, at least at base: **16. Poinsettia**

---

*See p. 67, Recent Nomenclatural Changes

14. Glands of cyathium usually 4 or 5, with or without petaloid appendages:

15. Leaves alternate or apparently absent; stipules absent; stem-axis persisting: **15. Euphorbia**

15. Leaves opposite; stipules present and persistent; main stem-axis aborting within 2 nodes of cotyledons, the stem always forking at this point: **17. Chamaesyce**

Genus 6. JATROPHA L.

Herbs, shrubs, or small trees; leaves alternate with blades entire, toothed, or palmately lobed; stipules present, often glandular. Plants monoecious, rarely dioecious; flowers in terminal, often long-stalked dichasia, the lower flowers pistillate, the distal ones staminate; calyx 5-lobed; petals 5, imbricate to contorted, free or coherent. Staminate flowers with mostly 8–12 connate stamens; rudimentary ovary absent. Pistillate flowers with 2–3-celled ovary; styles 2 or 3, connate at base, the stigmas 2-forked; ovules solitary. Fruit a capsule with 1 seed in each locule; seeds with caruncle; endosperm copious.

A large tropical genus of perhaps 150 species, the majority in America and Africa. [*Jatropha multifida* L. is cultivated in Grand Cayman (*Proctor, sight record*), and doubtless other ornamental members of this genus.]

KEY TO SPECIES

1. Leaves palmately lobed below the middle; leafy nodes with conspicuous glandular stipules: **1. J. gossypifolia**

1. Leaves palmately veined but not or only slightly and broadly lobed; leafy nodes without glandular stipules: **2. J. curcas**

1. Jatropha gossypifolia L., Sp. Pl. 2:1006. 1753.

"Bitter Cassava", "Wild Cassava"

A low shrub often less than 1 m tall; leaves long-petiolate, the petioles with glandular hairs, the blades mostly 3–12 cm across, deeply divided into 3–5 lobes and narrowly cordate at base, more or less pubescent; stipules cut into thread-like glandular hairs 3–5 mm long. Flowers deep crimson or purple; sepals glandular-ciliate,

3-5 mm long; petals glabrous, c. 4 mm long. Stamens 10-12, the filaments much longer than the anthers and united above the middle. Capsule c. 1 cm in diameter, glabrous; seeds 7-8 mm long.

CAYMAN BRAC: *Proctor 28917.*
– Pantropical, often a weed of pastures and open waste ground, occasionally planted for ornament.

2. Jatropha curcas L., Sp. Pl. 2:1006. 1753.

"Physic Nut"

A shrub or small tree to 7 m tall, with milky sap; leaves long-petiolate, without glandular hairs, the blades roundish-ovate, mostly 4-12 cm across, widely cordate at the base, and entire, slightly angulate, or broadly lobed, glabrous. Flowers yellowish-green, in small corymbose cymes; sepals 3.5-4.5 mm long; petals c. 6.5 mm long, pubescent within. Stamens 9, the 5 inner filaments united halfway, the 4 outer united at base. Capsule fleshy on the outside at first, 2-3-celled, 2.5-4 cm long; seeds c. 2 cm long.

GRAND CAYMAN: *Hitchcock.*
– Widespread in tropical America, sometimes planted as a "quickstick" fence. There is some doubt whether it still occurs in the Cayman Islands. The seeds of this species are valued for their purgative properties, and for their oil, which can be used in the manufacture of soap.

[**Aleurites moluccana** (L.) Willd., the "Candlenut", is cultivated in Georgetown, GRAND CAYMAN *(Brunt 1989; Hitchcock; Millspaugh 1337).* It is a monoecious tree to 12 m tall or more, the young twigs and other parts densely clothed with minute stellate hairs, the leaves long-petioled, with blades 3-5-nerved at the base, entire or lobed, and up to 20 cm long or more. The small white flowers are borne in terminal panicle-like cymes. The fruits are 5-6 cm in diameter and contain 1 or 2 large, nut-like seeds. These seeds are edible "but should be eaten with caution" (Fawcett & Rendle); pickling in alcohol or roasting are said to reduce their purgative effect. Although locally called "Walnut", this species is not, of course, related to the true walnut. Another species of *Aleurites, A. fordii,* is widely cultivated in southeastern United States and elsewhere as a source of Tung Oil, used in the manufacture of varnish.]

[**Manihot esculenta** Crantz, the "Cassava" plant (**FIG. 161**), is cultivated rather widely in Grand Cayman *(Kings GC 309, Millspaugh 1293),* also to a lesser extent on Little Cayman and Cayman Brac, and may in the long run become naturalized. It is a shrubby plant with tuberous, starchy roots and alternate, palmately-lobed leaves. The unisexual flowers are monoecious in terminal panicles. This species is a native of Brazil, but is cultivated in many tropical countries for its starchy tubers. Cassava starch is low in protein, lacking in vitamins, and nutritionally is an inferior food. On the other hand, it has the advantage of being drought-resistant and easy to grow, and can be harvested at various seasons. The leaves can be eaten as a cooked vegetable.]

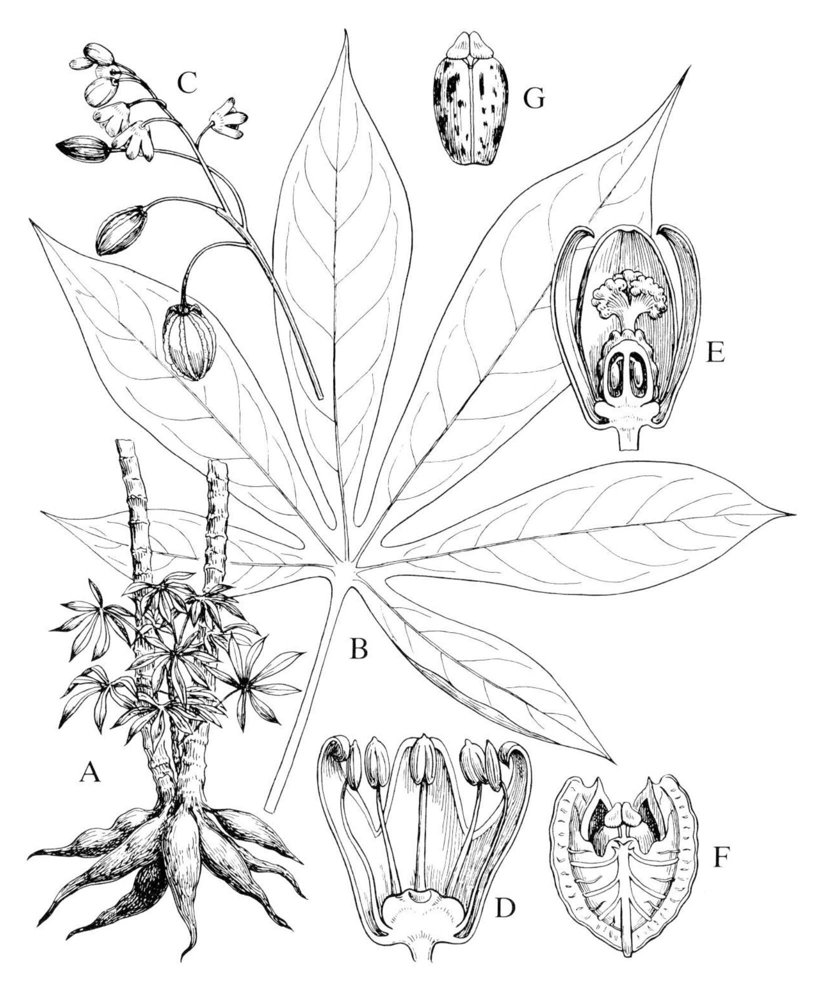

FIG. 161 **Manihot esculenta.** A, lower part of stem showing young shoot and tubers, much reduced. B, small leaf, x1. C, portion of inflorescence, x⅓. D, male flower cut lengthwise, x5. E, female flower cut lengthwise, x2. F, one carpel of ripe fruit showing seed, x1. G, seed, x1. (F. & R.)

Genus 7. **CROTON** L.

Herbs, shrubs or trees, the stems often containing coloured or resinous sap but not milky latex; indumentum at least in part of branched or scale-like trichomes. Leaves alternate, pinnately- or palmately-veined or sometimes lobed; petioles sometimes bearing glands at the junction of the leaf-blades. Plants monoecious or sometimes dioecious; flowers usually in spike-like, often terminal racemes, the staminate several on the apical portion, the pistillate often solitary near the base; petals usually present on staminate flowers and absent from pistillate ones. Staminate calyx 5-lobed; stamens mostly 8–20, free, the filaments inflexed in bud; rudimentary ovary absent. Pistillate calyx usually 5–7-lobes; ovary mostly 3-celled, each cell with 1 ovule; styles free or nearly so, 1-several-times-forked. Fruit a 3-seeded capsule; seeds with copious endosperm.

A very large, widely-distributed, but chiefly tropical American genus of between 700 and 1000 species (authorities disagree on the number), well represented in the West Indies. Very few have any economic importance, although the aromatic oils and alkaloids of a few have medicinal or flavouring value. The horticultural "Crotons" are properly called *Codiaeum* (see below, p. 538).

KEY TO SPECIES

1. Leaves simple and entire; shrubs:

  2. Leaves narrowly linear (less than 2 mm wide) and less than 2 cm long: **1. C. rosmarinoides**

  2. Leaves broader and longer:

    3. Leaves clothed with stellate hairs or non-silvery scales beneath, or rarely glabrous:

      4. Leaves linear-oblong, white or yellowish beneath; plants dioecious: **2. C. linearis**

      4. Leaves elliptic or ovate, green beneath; plants monoecious: **3. C. lucidus**

    3. Leaves clothed with minute silvery scales: **4. C. nitens**

1. Leaves 3–5-lobed, the margins crenate-serrate; an annual herb: **5. C. lobatus**

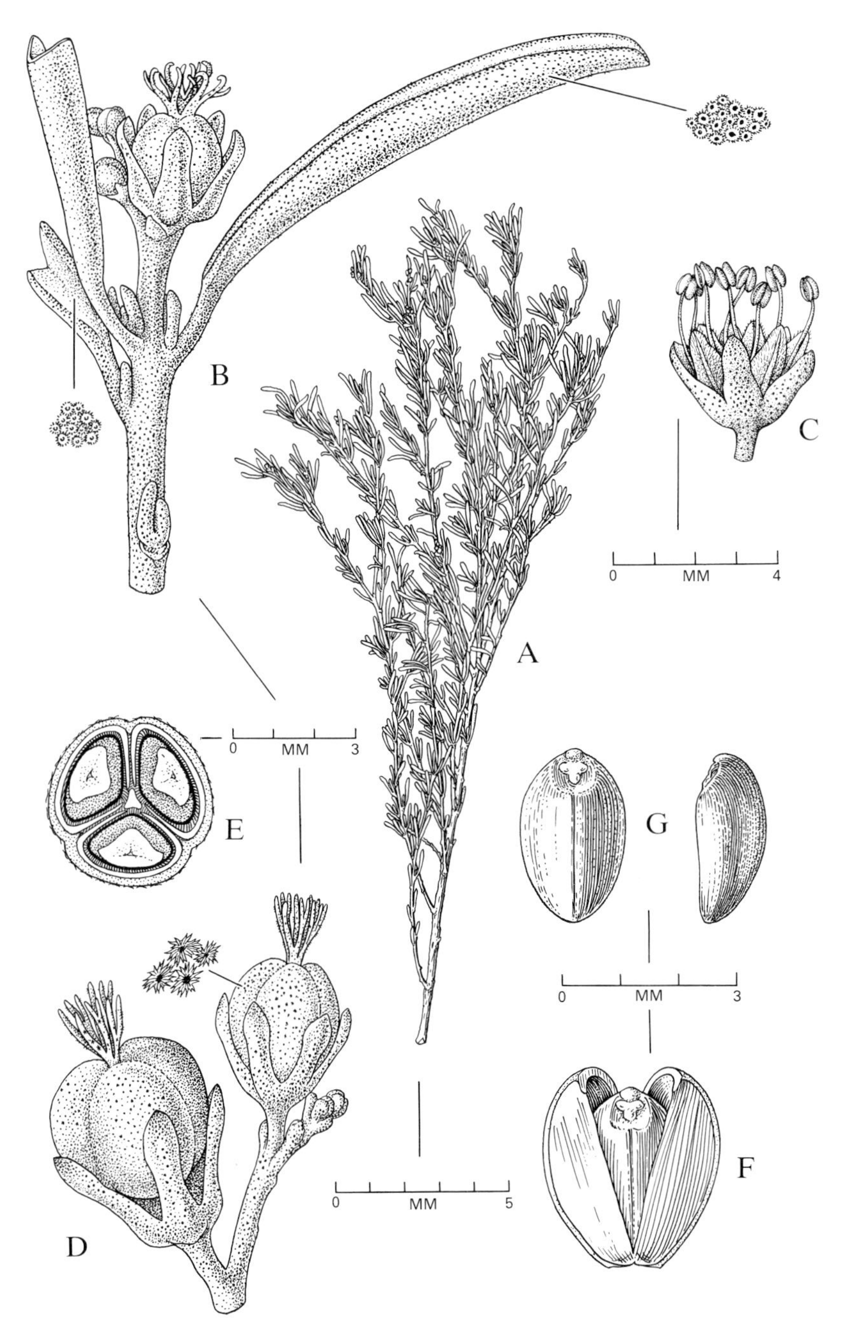

FIG. 162 **Croton rosmarinoides.** A, branch, habit. B, branchlet, much enlarged, with young raceme. C, staminate flower. D, pistillate flower and young fruit. E, x-section of fruit. F, portion of capsule, seed within. G, two views of a seed. (D.E.)

1. Croton rosmarinoides Millsp. in Britton & Millsp., Bahama Fl. 222. 1920. **FIG. 162**

A densely-branched monoecious shrub about 1 m tall or less; leaves subsessile, mostly 6–19 mm long and 0.6–2 mm broad, the margins strongly incurved beneath, the lower (abaxial) surface densely clothed with minute whitish, brown-centered stellate scales, the upper by very minute greenish punctate scales. Racemes terminal, short and densely-flowered, densely scaly throughout. Petals of staminate flowers woolly-ciliate; stamens 6, the filaments woolly at the base. Pistillate flowers 1–4 per inflorescence; styles 4-branched, stellate-scaly. Capsules ellipsoid-globose, densely stellate-scaly.

LITTLE CAYMAN: *Proctor 28137, 35073.* CAYMAN BRAC: *Proctor 29062.* – Bahamas and Cuba, in dry rocky scrublands.

2. Croton linearis Jacq., Enum. Syst. Pl. 32. 1760; Sel. Stirp. Amer. 256, t. 162, f. 4. 1763. **FIG. 163.**

Croton cascarilla L., 1753, in part, not as to type.

"Rosemary"

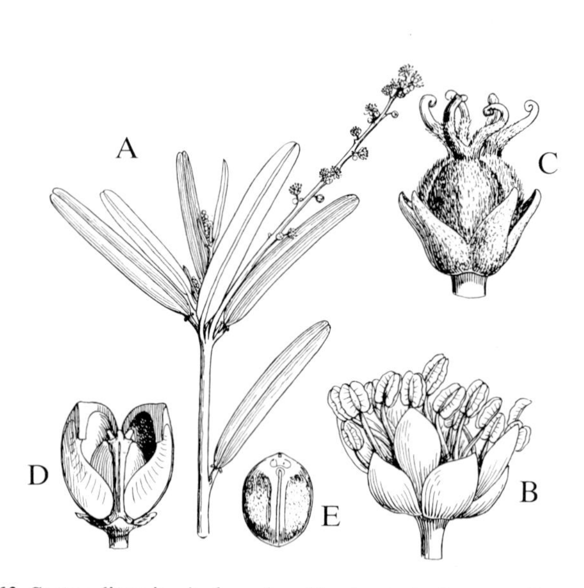

FIG. 163 **Croton linearis.** A, branch with staminate flower, x7. C, pistillate flower, x10. D, coccus with seed, x3. E, seed, x4. (F. & R.)

An aromatic shrub mostly 1-2 m tall, the branchlets and undersides of leaves densely clothed with white or yellowish stellate-hairy scales; leaves short-petiolate, linear-oblong, up to 5 cm long and mostly 2-9 mm broad, glabrous above; petioles with 2 cylindric glands at apex on the upper (adaxial) side. Plants dioecious; racemes usually longer than the leaves, with scattered subsessile flowers. Staminate calyx 1.5 mm long; stamens 13-15, with glabrous filaments. Pistillate calyx 1.8 mm long; ovary clothed with minute stellate scales; styles 2-branched. Capsules globose, c. 5 mm in diameter, 3-grooved, minutely stellate-scaly.

GRAND CAYMAN: *Brunt 1655; Correll & Correll 50995; Hitchcock; Kings GC 52, GC 392; Millspaugh 1312; Proctor 15058; Sachet 365.* LITTLE CAYMAN: *Kings LC 9; Proctor 28029.* CAYMAN BRAC: *Kings CB 49; Millspaugh 1217; Proctor 28969.*
– Florida, Bahamas, and the Greater Antilles except Puerto Rico, in rocky thickets and old pastures.

3. Croton lucidus L., Syst. Nat. ed. 10, 2:1275. 1759.

A monoecious shrub 1-2 m tall; leaves usually long-petiolate, glabrate or stellate-pubescent, the blades elliptic or ovate, 2-12 cm long, acute to acuminate at the apex, the tissue minutely pellucid-dotted. Racemes 3-8 cm long; staminate calyx and petals c. 2 mm long, densely pubescent; stamens 11-12; pistillate calyx 5-6.5 mm long, densely pubescent; styles short, deeply 4-branched. Capsules oblong-ellipsoidal, glabrate, 10-12 mm long.

GRAND CAYMAN: *Brunt 2135, 2148; Hitchcock; Proctor 15025, 27981; Rothrock 77, 172, 177.* LITTLE CAYMAN: *Kings LC 13, LC 53; Proctor 28139.* CAYMAN BRAC: *Millspaugh 1199, 1204, 1206, 1216; Proctor 29061.*
– Bahamas and Greater Antilles; frequent in rocky thickets and woodlands.

4. Croton nitens Sw., Nov. Gen. & Sp. Pl. 100. 1788.

Croton glabellus of Fawcett & Rendle, 1920, not L., 1763.

C. eluteria of Adams, 1972, not (L.) Sw., 1788.

"Wild Cinnamon"

An aromatic shrub or small tree to 10 m tall or more, the branchlets brownish-scaly; leaves oblong or elliptic to very broadly ovate, 2-10 cm long, blunt to acutish or acuminate at the apex, densely clothed beneath with minute pale silvery scales having brown-punctate centers, scattered similar scales also on the upper (adaxial) side; tissue also minutely pellucid-dotted. Plants monoecious. Racemes simple or branched, axillary and terminal, much shorter than the leaves; all flowers stellate-scaly and with woolly-margined petals. Staminate calyx 2–2.4 mm long; stamens 10–13, the long-exserted filaments glabrous except at the base. Pistillate flowers like the staminate; styles 4–6-branched. Capsules obovoid-globose, tuberculate and stellate-scaly, 7–9 mm long.

GRAND CAYMAN: *Brunt 2116, 2137; Hitchcock; Proctor 15239.* LITTLE CAYMAN: *Proctor 28030, 28110.* CAYMAN BRAC: *Proctor 28968.*

– Jamaica, Swan Islands, Mexico and Central America; related forms in Colombia and Ecuador. In the Cayman Islands, this species is frequent in dry rocky woodlands. *Croton nitens* is closely related to *C. eluteria* of the Bahamas, Cuba, and Hispaniola, and by some authors the two are united under the latter name. However, the two populations are distinctively different in appearance, and specimens placed side by side can be distinguished at a glance. Submerging them under one name would conceal the fact that the Cayman plants are allied to those of Jamaica and not to those of Cuba and the Bahamas. *C. eluteria* as it occurs in the Bahamas is the source of high-priced aromatic "Cascarilla Oil", used in medicines, alcoholic beverages, and perfumes. Similar oils occur in *C. nitens*, but have not been exploited.

5. **Croton lobatus** L., Sp. Pl. 2:1005. 1753.

A weedy monoecious herb up to 1 m tall, the branches pilose with simple hairs; leaves thin and glabrous to sparingly pilose, usually 3-lobed, rarely 5-lobed, the lobes mostly 2–7 cm long, constricted toward the base, the tips acuminate, the margins crenate-serrate; petioles often as long as the blades. Racemes terminal or axillary, up to 10 cm long. Staminate flowers glabrous; stamens 10–13; pistillate flowers with glandular-ciliate sepals; ovary stellate-pubescent and pilose; styles 3–8-branched. Capsule globose-ellipsoid, c. 8 mm long; seeds 5 mm long.

GRAND CAYMAN: *Hitchcock.*

– Florida, West Indies (except Jamaica) continental tropical America, and tropical Africa. This plant is usually a weed of fields and open waste places; its continued presence in the Cayman Islands requires confirmation.

[**Codiaeum variegatum** (L.) Blume is the correct name for the ornamental shrubs commonly called "Croton". This species is, in fact, not closely related to the botanical genus of that name. Although the leaves of *Codiaeum* are exceedingly variable in size, shape, and colouration, all the forms belong to a single botanical species, this native to the region of Malaya and islands of the Pacific; it is much cultivated throughout the tropics.]

Genus 8. **ARGYTHAMNIA** Sw.

Monoecious or very rarely dioecious perennial herbs, shrubs or rarely small trees, always bearing "malpighiaceous"[1] hairs at least when young, and often containing purplish pigment. Leaves alternate, petiolate, entire to serrulate; stipules present. Inflorescence an axillary raceme, with pistillate flowers at the base. Staminate flowers with 4 or 5 pubescent valvate sepals and 4 or 5 petals alternate with the sepals; glands 4 or 5; stamens 4–15 in 1 or 2 series, the filaments joined at the base; rudimentary ovary absent. Pistillate flowers with 5 pubescent valvate sepals, the petals 5, rudimentary or absent; glands 5, opposite the sepals, inserted on disk of the ovary; ovary superior, subglobose, 3-celled, each cell with 1 pendulous ovule; styles 3, 1-2-forked. Capsule splitting into 3 1-seeded cocci; seeds rough-coated, and with endosperm.

An American genus of about 50 tropical or warm-temperate species.

1. **Argythamnia proctorii** Ingram in Gentes Herb. 10:25. 1966.

A straggling monoecious shrub to 1.5 m tall, the young branchlets densely clothed with malpighiaceous hairs; leaves narrowly to broadly elliptic or rarely narrowly obovate, 1-6.5 cm long, acute and mucronate at the apex, the margins subentire to shallowly serrate with gland-tipped teeth, more or less woolly with soft malpighiaceous hairs on both sides. Racemes to 5 mm long. Staminate flowers with 4 sepals, these 1.75 mm long, bearing dense malpighiaceous hairs outside and simple hairs within; petals 4, c. 1.5 mm long, hairy like the sepals; stamens 4, exserted, hairy. Pistillate flowers with 5 sepals, these 2.5 mm long, increasing to 3.5 mm in fruit; petals 5, minute, hairy; glands 5, squarish, almost as thick as wide; styles woolly, 3-times-forked. Capsule 3 mm long and 4. 5 mm wide, villous; seeds ovoid-subglobose, c. 2 mm in diameter, lightly ridged.

GRAND CAYMAN: *Brunt 2158; Proctor 15043 (type), 27987.* LITTLE CAYMAN: *Proctor 28123, 35081.* CAYMAN BRAC: *Proctor 28941.*
– Endemic; occurs chiefly in rocky woodlands, or sometimes in sandy thickets. Closely related to *A. candicans*, a widespread West Indian species.

Genus 9. **ADELIA** L., nom. gen. cons.

Dioecious shrubs or trees, often with spinescent twigs, glabrous or bearing simple non-glandular hairs; leaves alternate or crowded at nodes, simple, entire, and

---

[1] Hairs attached at the middle, thus 2-armed or T-shaped. It should be noted that one arm is sometimes much longer than the other, the point of attachment being near, but not at, one end.

pinnately veined. Flowers small and without petals, clustered on small woolly cushions in leaf-axils. Staminate flowers short-stalked or sessile; calyx 4–5-lobed, valvate; disk ring-like, thick and fleshy; stamens 8–17, the filaments united at least near the base; rudimentary ovary minute or absent. Pistillate flowers long-stalked; calyx 5–6-lobed, the lobes reflexed; ovary 3-celled with 1 ovule per cell; styles more or less free, deeply lacerate. Capsule 3-valved, 3-lobed, dehiscing to leave a persistent columella; seeds without a caruncle; endosperm fleshy.

A tropical American genus of about 10 species.

1. **Adelia ricinella** L., Syst. Nat. ed. 10, 2:1298. 1759. **FIG. 164.**

A shrub to 3 m tall or more, rarely tree-like, with whitish bark, the ultimate branches often spine-tipped; leaves oblanceolate to obovate or elliptic, 1–6.5 cm long, rounded, blunt or acutish at the apex, glabrous above, and lightly pubescent

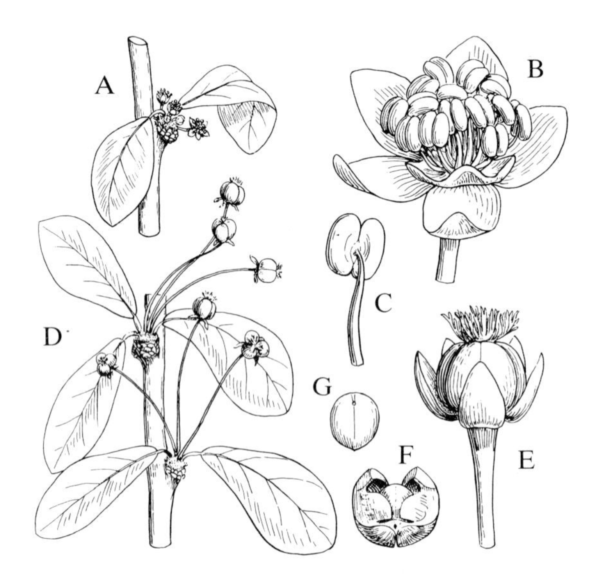

FIG. 164 **Adelia ricinella.** A, portion of stem with staminate flower. B, staminate flower, x5. C, stamen, x10. D, portion of stem with pistillate flowers, x$^{2}/_{3}$. E, pistillate flower with ovary ripening, x4. F, coccus with seed, x2. G, seed, x2. (F. & R.)

along the midvein beneath. Staminate flowers with calyx 2–3 mm long, on pedicels about the same length; stamens 8–15. Pistillate flowers on pedicels 1–5 cm long, with calyx c. 3 mm long; ovary tomentose. Capsule deeply 3-lobed, 6–8 mm in diameter; seeds brown, globose.

GRAND CAYMAN: *Brunt 2138; Proctor 15228.* LITTLE CAYMAN: *Proctor 28119.* CAYMAN BRAC: *Proctor 29115.*
– West Indies south to Tobago and Curacao, mostly in dry rocky thickets and woodlands.

Genus 10. **BERNARDIA** Houst. ex P. Browne

Herbs or shrubs, monoecious or dioecious, bearing simple or stellate hairs; leaves alternate, with toothed margins and often with basal glands; stipules present. Flowers in unisexual, axillary, bracteate spikes, the bracts conspicuous; petals absent. Staminate flowers minute, with calyx globose and entire in bud, splitting into 3 or 4 valvate lobes at anthesis; disk absent or vestigial; stamens 3–25, with free, erect filaments and 4-celled anthers; rudimentary ovary absent or vestigial. Pistillate flowers sessile, the calyx 4–9-parted with imbricate sepals; disk annular or dissected; ovary 3-celled with a single ovule in each cell; styles 3, variously forked. Capsule 3-valved, with a persistent columella after dehiscence; seeds without a caruncle; endosperm fleshy.

A tropical American genus of between 30 and 40 species.

1. **Bernardia dichotoma** (Willd.) Muell. Arg. in Linnaea 34:172. 1865.

Croton dichotomus Willd., 1805.

Bernardia carpinifolia Griseb., 1859.

A shrub up to 3 m tall, dioecious or monoecious, the branchlets tomentose with simple and stellate hairs; leaves deciduous at flowering time, more or less elliptic or rhombic-ovate, mostly 2–8 cm long, blunt to acute at the apex, the margins closely or distantly serrate or denticulate, tomentose with stellate hairs on both sides but especially beneath, and the tissue with two small round glands toward the base beneath. Spikes mostly 8–20 mm long, densely tomentose. Staminate flowers in clusters of 3–5, each cluster subtended by a bract c. 1.5 mm long; calyx c. 2 mm long; stamens 15–25, exserted. Pistillate flowers with calyx c. 2.5 mm long, the sepals unequal; ovary densely tomentose. Capsule deeply 3-lobed, 7–9 mm in diameter, densely tomentose.

GRAND CAYMAN: *Hitchcock; Millspaugh 1265; Proctor 15042.* LITTLE CAYMAN: *Proctor 35104.* CAYMAN BRAC: *Proctor 29030.*
– West Indies south to Grenada, in dry thickets and woodlands.

[Genus 11. **RICINUS** L.

A glabrous, monoecious, wide-branching shrub with watery sap; leaves alternate, long-petiolate, the blade peltate and palmately 7–11-lobed, with serrate margins; stipules sheathing, soon falling. Inflorescences terminal or appearing axillary due to sympodial growth, paniculate, the lower flowers staminate, the distal ones pistillate. Staminate flowers with calyx globose in bud, splitting into 3–5 valvate lobes at anthesis; stamens very numerous (up to 1,000), the filaments partially connate into fascicles at the base, irregularly branched; rudimentary ovary absent. Pistillate flowers with calyx like that of the staminate ones, soon falling; ovary 3-celled, a single ovule in each cell; styles joined below, forked upwardly, with papillate branches. Capsule spiny or smooth, splitting into three 2-valved cocci; seeds with conspicuous caruncle; endosperm copious.

A single species of African origin, now widely distributed in tropical and warm-temperate regions.

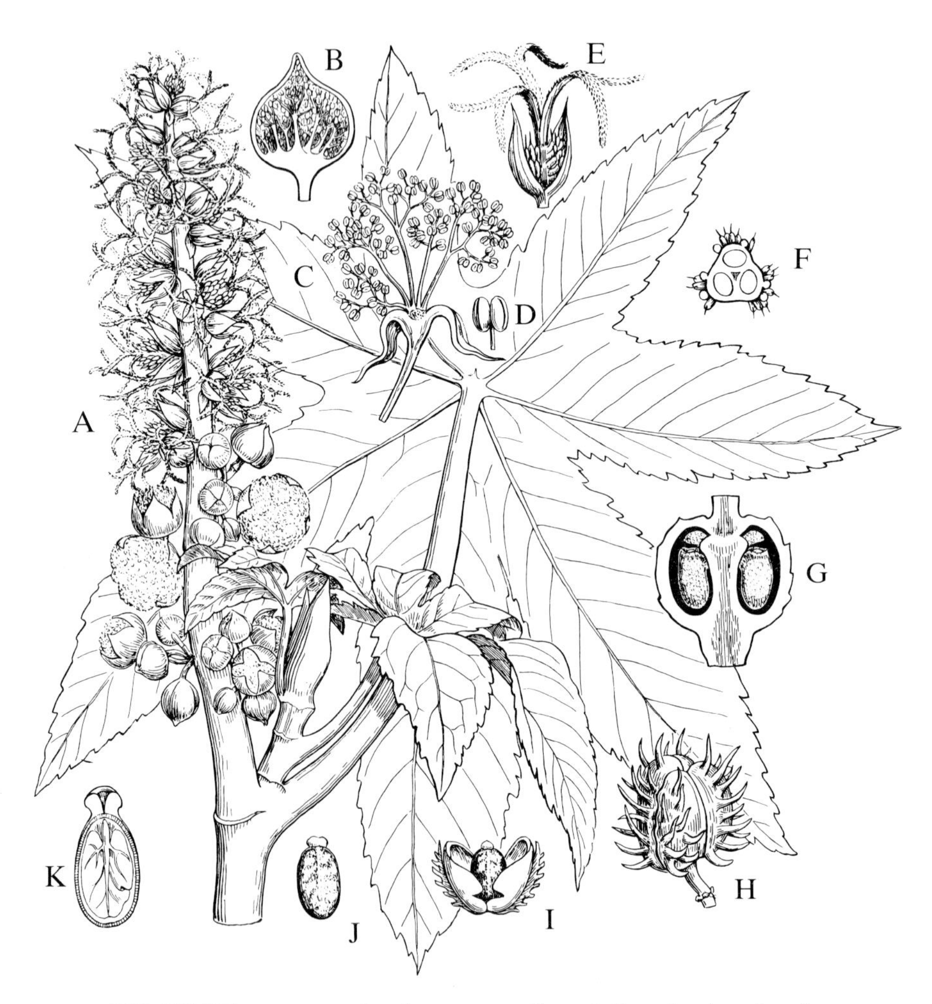

FIG. 165 **Ricinus communis**. A, upper portion of flowering branch, x$^{2}/_{3}$. B, staminate flower just before opening, x2. C, same open, x2. D, anther, x8. E, pistillate flower, x2. F, x-section of ovary, x4. G, ovary cut lengthwise to show ovule and obturator, x10. H, capsule, x1. I, coccus split open, x1. J, seed, x1. K, same, x1½. (F. & R.)

1. Ricinus communis L., Sp. Pl. 2:1007. 1753. **FIG. 165.**

"Castor-oil Plant"

With characters of the genus. Plants usually 2–5 m tall; leaf-blades from 10 to 100 cm across, the lobes acute and pinnately veined the marginal serrations more or less glandular. Capsules 12–21 mm in diameter; seeds ellipsoid, somewhat flattened, variously mottled, mostly 10–20 mm long.

GRAND CAYMAN: *Brunt 1983; Kings GC 128, GC 415; Millspaugh 1389; Proctor 15236. 15237.* LITTLE CAYMAN: *Kings LC 108.* CAYMAN BRAC: *Kings CB 80, CB 81.*
– Roadsides, old fields, and open waste ground. The seeds yield the castor oil of commerce.]

Genus 12. **ACALYPHA** L.

Herbs, shrubs or rarely trees, monoecious or dioecious, bearing simple hairs or glands; leaves alternate, petiolate, simple, entire or toothed, and with pinnate or palmate venation. Flowers in terminal or axillary unisexual or bisexual spikes, the staminate flowers several at each node subtended by a minute bract, the pistillate flowers 1–3 at each node, subtended by a usually large, foliaceous, lobed bract; petals and disk absent. Staminate flowers with calyx closed in bud, valvately splitting into 4 lobes at anthesis; stamens 4–8, the filaments free or joined at the base; anther-sacs 1-celled, pendent, elongate and worm-like, opening at the apex; rudimentary ovary absent. Pistillate flowers with 3–5 imbricate calyx-lobes; ovary usually 3-celled with solitary ovules; styles free, thread-like, usually much-branched. Fruit a 3-seeded capsule; seeds with or without a caruncle; endosperm present.

A large genus of about 400 species, the majority tropical American, with the greatest concentration of species in the Caribbean region. *Acalypha hispida* Burm. f., "Red Puss-tail", from the Indonesian area, and *A. amentacea* Roxb. subsp. *wilkesiana* (Muell. Arg.) Fosberg "Joseph's Coat", from the Pacific islands, are shrubs often cultivated in gardens.

KEY TO SPECIES

1. Inflorescences bisexual; leaves usually less than 1.5 cm long with very short petioles: **1. A. chamaedrifolia**

1. Inflorescences unisexual; leaves mostly 2–10 cm long, the petioles often longer than the blades: **2. A. alopecuroidea**

1. **Acalypha chamaedrifolia** (Lam.) Muell. Arg. in DC., Prodr. 15(2):879. 1866.

A small subwoody herb with woody taproot, the stems usually less than 15 cm long, erect or procumbent, the young branches pubescent. Leaves elliptic, oblong or ovate, 0.5-1.5 cm long or sometimes a little longer, blunt to acute at the apex, the margins crenate-dentate, pubescent or glabrate. Spikes 1-2 cm long, terminal or in upper axils, with 1-5 pistillate flowers near the base; staminate flowers red; pistillate flowers with bracts c. 3 mm long, divided at apex into 7-9 triangular lobes, each bract enclosing 2 flowers. Capsule 3-lobed, 1.6 mm long; seeds ellipsoidal.

CAYMAN BRAC: *Proctor 28919.*
– Florida and the West Indies, on sandy banks or exposed rocky hillsides.

2. **Acalypha alopecuroidea** Jacq., Collect. 3:196. 1789.

Erect annual herb to about 35 cm tall, the branches pubescent. Leaves long-petiolate, the blades ovate, 1.5-7 cm long, acuminate at the apex, the margins sharply serrate except toward base, glabrous or sparingly pubescent. Staminate spikes axillary, very short and inconspicuous, 2-9 mm long. Pistillate spikes terminal, 2-5 cm long, densely-flowered and plume-like, tipped with a bristle-like appendage terminating in an abortive flower; bracts hairy, each segment with a long apical bristle. Capsules solitary within each bract, 1.3-1.4 mm long.

GRAND CAYMAN: *Hitchcock; Kings GC 89; Proctor 15292.*
– Widely distributed in tropical America, a weed of fields, roadsides and open waste places.

Genus 13. **ATERAMNUS*** P. Browne

Glabrous shrubs or trees, usually monoecious; leaves alternate, entire or toothed, with short non-glandular petioles; stipules present. Inflorescences spike-like, axillary, protected by a conspicuous bud, bisexual, with usually only 1 basal pistillate flower per spike; calyx small or absent; petals absent; disk absent. Staminate flowers with rudimentary calyx of 1 or 2 small sepals, or absent; stamens 2 or 3 in lateral flowers of a cymule, 3-5 in central flowers; filaments free or joined at the base; rudimentary ovary absent. Pistillate flowers usually 1 at the lowest node of the inflorescence; calyx minute, of 2 or 3 reduced sepals; ovary sessile or stalked, 3-celled with a solitary ovule in each cell; styles free or joined at the base, simple, slender and recurved. Fruit a capsule with a persistent, 3-winged columella; seeds subglobose, smooth, and with a caruncle.

A small Caribbean genus of 12 species.

---

*Properly **GYMNANTHES**. See p. 67 Recent Nomenclatural Changes.

1. Ateramnus lucidus* (Sw.) Rothm. in Fedde, Rep. Sp. Nov. 53:5. 1944. **FIG. 166.**

Gymnanthes lucida Sw., 1788.

"Crab Bush"

An evergreen shrub or small tree; leaves leathery and glossy, more or less oblanceolate, 2.5-10 cm long, obtuse at the apex, the margins entire or with a few low teeth, the veins prominulous on both sides, and the tissue often with 1 or 2

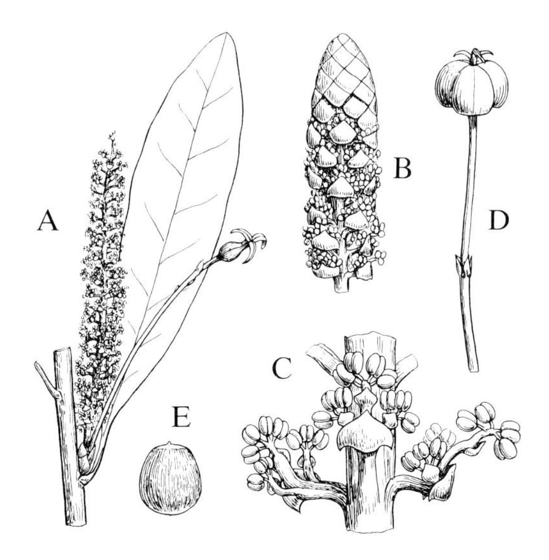

FIG. 166 **Ateramnus lucidus.** A, leaf and inflorescence, x1. B, upper part of young inflorescence, x2. C, part of inflorescence with three bracts and staminate flowers in their axils, x10. D, capsule, x1. E, seed, x2. (F. & R.)

glands near the base beneath. Spikes 1-3 cm long, densely-flowered. Staminate flowers 3 to each bract. Pistillate flower with short pedicel greatly elongating in fruit; sepals scale-like, minute, not all attached at the same level; ovary stalked above the sepals, the stalk short at first but elongating to 1.5-2 cm long in fruit, appearing as a continuation of the pedicel. Capsules deeply 3-lobed, nearly 1 cm in diameter.

GRAND CAYMAN: *Kings GC 243; Lewis 3863; Proctor 15031, 15032.* LITTLE CAYMAN: *Proctor 35188.* CAYMAN BRAC: *Proctor 29016.*
– Florida and the West Indies, in dry rocky woodlands. The wood is heavy, hard, close-grained, and capable of receiving a beautiful polish; it is occasionally made into canes.

---

*Properly renamed **Gymnanthes lucida**. See p. 67, Recent Nomenclatural Changes.

Genus 14. **HIPPOMANE** L.

Glabrous, monoecious shrubs or trees with poisonous milky latex; leaves alternate, long-petiolate, the petioles with a single gland at the apex, the blades cordate and pinnately-veined, the margins toothed or spiny. Stipules present, soon falling. Inflorescence a terminal, bisexual spike with thickened rhachis, the bracts biglandular; petals and disk absent. Staminate flowers several to numerous in dense glomerules at each of the distal nodes; calyx 2-3-lobed, the lobes imbricate; stamens 2 with cohering filaments; rudimentary ovary absent. Pistillate flowers 1 or few, sessile, solitary at lower nodes of the spike; calyx 3-lobed; ovary 6-9-celled (rarely less) with 1 ovule in each cell; styles 2 or 3, joined at base, recurved, undivided. Fruit a small apple-like drupe, with yellowish or reddish fleshy exocarp, the bony endocarp with numerous blunt projections; seeds flattened-elongate, without a caruncle; endosperm present.

A tropical American genus of 3 species, one of them widespread, the others confined to Hispaniola.

1. **Hippomane mancinella** L., Sp. Pl. 2:1191. 1753. **FIG. 167.**

"Manchineel"

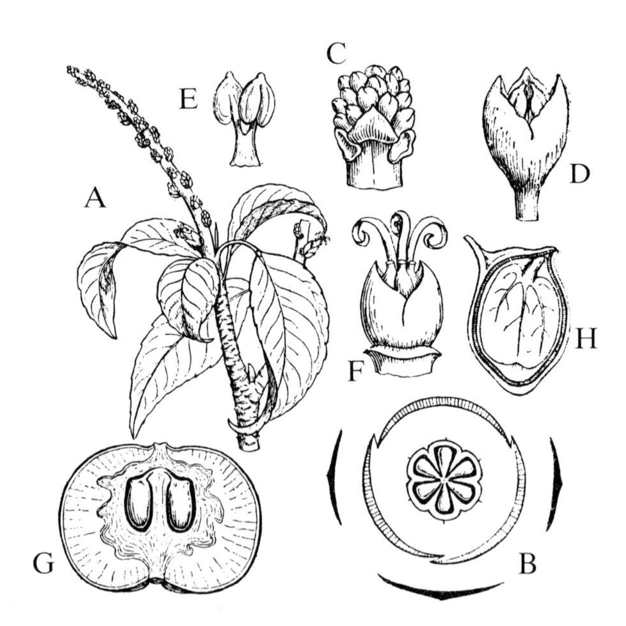

FIG. 167 **Hippomane mancinella.** A, branchlet with leaves and inflorescence, x$^2/_3$. B, diagram of pistillate flower. C, cluster of staminate flowers, x3. D, staminate flower just opening, x12. E, stamens, x14. F, pistillate flower with three styles cut off, x3. G, section of fruit, x1. H, section of seed, x4. (F. & R.)

An evergreen shrub, or tree to 10 m tall or more; leaves with cordate-ovate blades usually 2-7 cm long or sometimes longer, the apex sharply acute or subacuminate, the margins distantly serrate, the tissue shiny green above, the finely reticulate venation prominulous. Spikes usually 3-9 cm long, the rhachis dark red or purple. Fruit globular, 2.5-3.5 cm in diameter; seeds 4-6 mm long.

GRAND CAYMAN: *Brunt 1789, 1790; Hitchcock; Kings GC 114; Proctor 15064.* LITTLE CAYMAN: *Proctor 28100.* CAYMAN BRAC: *Proctor 29133.* – Florida, West Indies, and continental tropical America, in coastal thickets and woodlands. The wood of this tree is prized for cabinet-work and construction, but because of its poisonous sap is hazardous to saw unless dry. The fruits are likewise poisonous, and children should be warned to shun them.

Genus 15. **EUPHORBIA** L.[1]

Herbs, shrubs or trees, sometimes succulent, with copious milky latex, monoecious or rarely dioecious; leaves simple and alternate, opposite or whorled, in succulent forms soon-falling or sometimes apparently absent; stipules present or absent, sometimes glandular. Inflorescence a cyathium (aggregate flower), the 5 cup-like lobes alternating at their tips with 4 or 5 glands, these with or without appendages. Staminate flowers in 4-5-cymes, the subtending bracteoles partly fused to the involucre or reduced or absent; calyx and corolla absent; stamen 1. Pistillate flowers terminal, solitary; calyx of 3-6 united sepals or absent; ovary 3-celled with a single ovule in each cell; styles 3, free or joined at the base, usually forked. Fruit a capsule or rarely a drupe; seeds more or less ovoid, angled or terete, smooth or variously roughened; caruncle present or absent.

A world-wide genus of about 1,200 species, extremely diverse in appearance. The name "Euphorbia" is said to have been first applied to a succulent member of this genus by King Juba of Mauritania (reigned B.C. 25 -A.D. 18) in honour of his doctor, Euphorbus, the new plant and the worthy physician both being of notably fleshy build. "Euphorbos" is a Greek adjective meaning "well-fed".

KEY TO SPECIES

1. Low, rosette-like herb with a woody taproot; leaves numerous, crowded, alternate: **1. E. trichotoma**

1. Plants shrubby or tree-like, the stems leafless or with minute scattered leaves on new growth;

---

[1] The circumscription of this genus is taken from Burch's treatment in "Flora of Panama", Ann. Missouri Bot Gard. 54:332. 1967.

2. Straggling shrub to 1.5 m tall; branches angled; flowers yellow; seeds white: **2. E. cassythoides**

2. Stout erect shrub or small tree to 7 m tall or more; branches terete; flowers greenish; seeds dark brown: **[3. E. tirucalli]**

1. **Euphorbia trichotoma** Kunth in H.B.K., Nov. Gen. & Sp. Pl. 2:60. 1817.

Glabrous, pale green, perennial herb with a woody taproot, the branches not usually more than 20 cm long, spreading or ascending; leaves numerous and crowded, nearly sessile, mostly oblanceolate or narrowly ovate, 2–8 mm long, the margins entire to minutely erose or serrulate. Cyathia in terminal 3-branched clusters or solitary, the involucre 2 mm high, the glands obreniform, 1 mm broad, yellow. Capsule broadly 3-lobed, 3–4 mm in diameter; seeds whitish, 1.5 mm long, smooth.

GRAND CAYMAN: *Kings GC 23a; Proctor 11972, 15215. Sachet 442.* LITTLE CAYMAN: *Proctor 35178.* CAYMAN BRAC: *Millspaugh 1185, 1232; Proctor 15309, 28918.*
– Florida, Bahamas, Cuba, and Mexico, on sandy sea-beaches.

2. **Euphorbia cassythoides** Boiss., Cent. Euph. 20. 1860.

Arthrothamnus cassythoides (Boiss.) Millsp., 1909.

A straggling, pale-barked shrub up to 1.5 m tall or more, the pale green branches leafless, more or less whorled, and sharply 6–7-angled, with gummy-resinous nodes; vestigial leaf-scales few, opposite. Cymes terminal, few-flowered, the yellow cyathia 1.5 mm long; styles hairy. Capsules ovoid; seeds white, 3-angled, foveolate.

GRAND CAYMAN: *Brunt 2173; Kings GC 364.*
– Bahamas, Cuba, and Hispaniola, in dry rocky or sandy woodlands, rare. Herbarium specimens of this species have been mistaken for *Rhipsalis baccifera* (Cactaceae).

[3. **Euphorbia tirucalli** L., Sp. Pl. 1:452. 1753.

A stout shrub or small tree, the green, smooth, terete, mostly naked branches alternate or clustered in brush-like masses, the ultimate ones 5–7 mm thick and bearing scattered, alternate, narrow leaves 6–12 mm long, these soon falling.

Inflorescences of small, sessile, terminal clusters, the cyathia 3 mm long. Ovary sessile at first, becoming exserted on a pedicel-like stalk 8 mm long in fruit. Capsule obtusely 3-lobed; seeds dark brown, smooth, 4 mm long.

GRAND CAYMAN: *Brunt 2028.*
– Indigenous to tropical Africa but widely cultivated in warm countries, often becoming naturalized.]

[*Euphorbia milii* Ch. des Moulins (*E. splendens* Bojer) from Madagascar, the "Crown-of-thorns" or "Crucifixion Plant", is cultivated in Grand Cayman (*Kings GC 306*); also the red-leaved shrub *Euphorbia cotinifolia* L.; and probably other ornamental species as well.]

Genus 16. **POINSETTIA** Graham

Monoecious herbs or shrubs with milky latex; leaves alternate and opposite, petiolate, those subtending the inflorescence often brightly coloured or whitish at least toward the base; stipules minute or lacking. Inflorescence a cyathium, several or many of these clustered in terminal dichasia or pleiochasia; lobes 5, usually more or less fringed, the involucral glands cup-like, usually 1 but up to 5 in central cyathia. Staminate flowers few to many, naked, with a single stamen. Pistillate flowers terminal, solitary, naked; ovary 3-celled with 1 ovule per cell; styles 3, united at the base, bifid. Fruit a 3-lobed dehiscent capsule; seeds ovoid, the caruncle vestigial or absent.

An American genus of about 12 species. *P. pulcherrima* (Willd. ex Kl.) Graham, the ornamental "Poinsettia", is widely grown for its showy bracts.

KEY TO SPECIES

1. Cyathial gland circular at the apex; bracts green, often pale or purple-mottled at base but never red; seeds angled: **1. P. heterophylla**

1. Cyathial gland flattened and 2-lipped; bracts green or red at the base; seeds not or scarcely angled: **2. P. cyathophora**

1. Poinsettia heterophylla (L.) Kl. & Gke. in Monatsb. Akad. Berlin 1859:253. 1859. **FIG. 168.**

Euphorbia geniculata Ort., 1797.

An annual or sometimes persisting herb, usually less than 50 cm tall; leaves alternate below, opposite above, the membranous blades pandurate, obovate or elliptic, rarely lanceolate, the margins entire or with 1 to few broad teeth, glabrous or pubescent, those subtending the inflorescence-cluster often pale or purple-mottled at the base. Cyathia glabrous, the gland with a slightly flared round opening. Capsule c. 5 mm broad; seeds up to 2.5 mm long, tuberculate.

GRAND CAYMAN: *Kings GC 145; Proctor 15079;*
– Southeastern United States and tropical America, a widely-distributed and common weed of roadsides and fields.

2. Poinsettia cyathophora (Murray) Kl. & Gke. in Monatsb. Akad. Berlin 1859: 253. 1859.

Euphorbia heterophylla of authors, not L., 1753.

"Bleeding Heart", "Starlight", "Ground Dove Berry"

An annual or sometimes persisting herb up to 0.8 m tall or more; leaves alternate below, opposite above, the membranous blades pandurate, ovate, lanceolate or sublinear, the margins entire or broadly toothed, glabrous or sparingly pubescent, those subtending the inflorescence-cluster often bright red at base or throughout. Cyathia glabrous or sparingly short-pubescent, the gland flattened and 2-lipped. Capsules c. 5 mm broad; seeds up to 3 mm long, tuberculate.

GRAND CAYMAN: *Brunt 2033, 2058; Kings GC 17, GC 139; Maggs II 59; Proctor 15020.* LITTLE CAYMAN: *Kings LC 51.* CAYMAN BRAC: *Kings CB 45.*
– Eastern United States and tropical America, also in the Old World, a common weed of waste places and disturbed areas.

Genus 17. **CHAMAESYCE** S.F. Gray[1]

Monoecious herbs or subshrubs, annual or persisting, prostrate to ascending or erect, with milky latex throughout; stems with main axis aborting within 2 nodes of the cotyledons, the secondary axes few to many, rarely (in prostrate species) rooting at the nodes. Leaves opposite, simple, the blade oblique or unequal-sided at base and with entire or serrate margins; stipules present, often conspicuous. Inflorescence a cyathium, solitary, or several clustered in cymules at nodes; cyathia 5-lobed, with 4 glands (sometimes a 5th vestigial one) alternating with the lobes and

---

[1] Treatment adapted from that of Burch in Ann. Missouri Bot. Gard. 53:90-99. 1966.

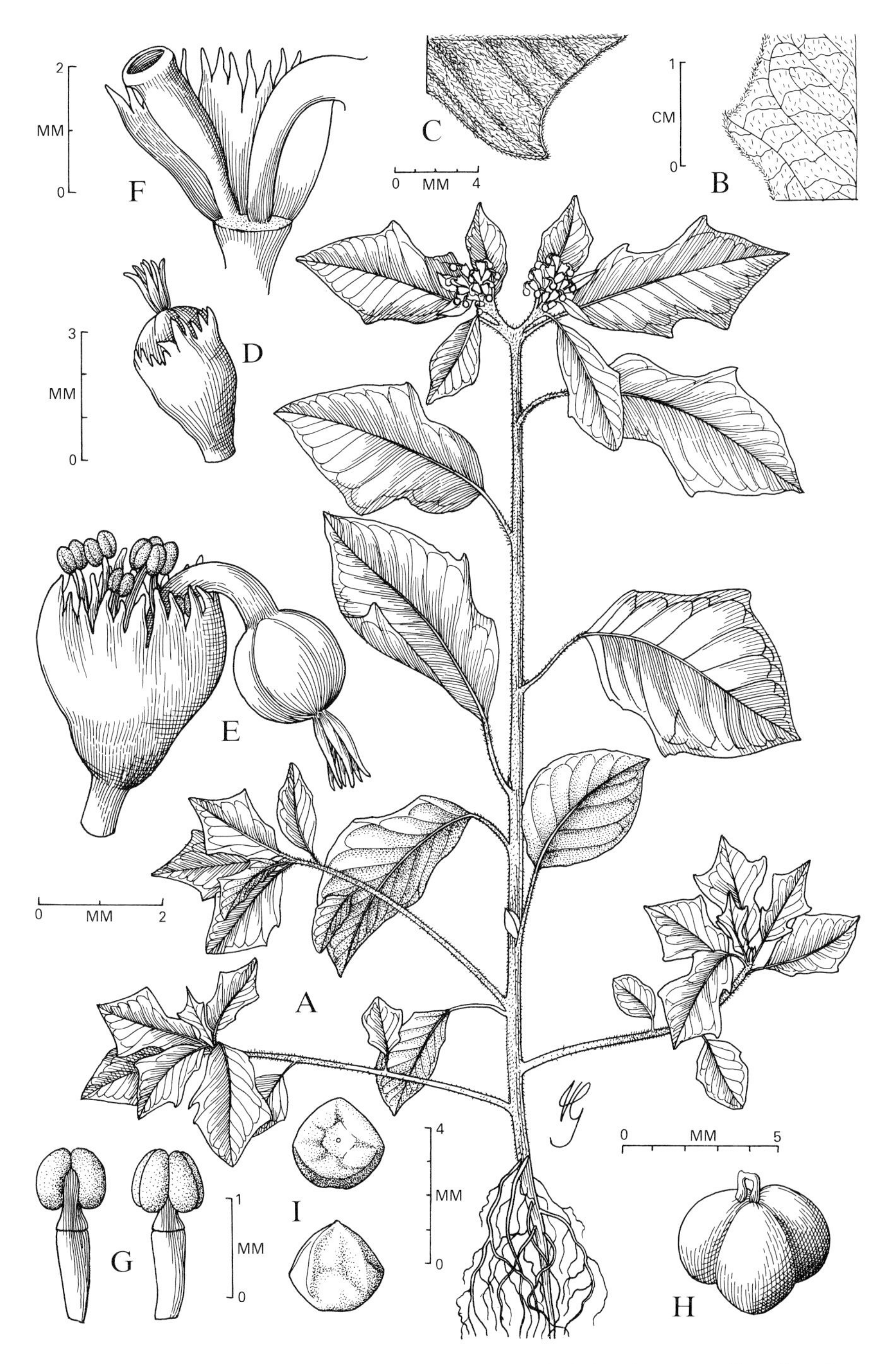

FIG. 168 **Poinsettia heterophylla.** A, habit, B, portion of leaf showing upper (dorsal) surface. C, portion of leaf showing lower (ventral) surface. D, immature cyathium. E, mature cyathium at anthesis. F, cyathium cut open to show gland. G, two views of stamen. H, capsule. I, two views of seed. (G).

each bearing a petal-like appendage (rarely the appendage obsolete). Staminate flowers few to many, naked, with 1 stamen. Pistillate flowers terminal, solitary, naked; ovary 3-celled with 1 ovule in each cell; styles 3, free or united at base, partly bifid. Fruit an exserted capsule; seeds ovoid, angled or terete, smooth or variously sculptured; caruncle absent.

A world-wide genus of about 250 species, the majority in tropical America.

KEY TO SPECIES

1. Capsules glabrous:

2. Plants erect or ascending:

3. Plants herbaceous; leaf-margins toothed; cyathia in dense, stalked, glomerate clusters; capsule less than 1.4 mm long: **1. C. hypericifolia**

3. Plants woody; leaf-margins entire; cyathia solitary in uppermost leaf-axils; capsule c. 2 mm long: **2. C. mesembrianthemifolia**

2. Plants prostrate or decumbent:

4. Stems numerous, fine and dense, rarely exceeding 0.5 mm diam. and 6 cm in length, from a woody rootstock; stipules less than 0.5 mm long; leaves serrulate throughout: **3. C. torralbasii**

4. Stems few, rather open, often up to 2 or 3 mm in diameter and usually 10-30 cm long, the rootstock not or scarcely woody; stipules often 1 mm long; leaves minutely serrulate at apex only: **4. C. blodgettii**

1. Capsules more or less pubescent:

5. Cyathia solitary at leafy nodes, never in stalked glomerate clusters:

6. Stems pubescent at least on one side; ovary and capsule pubescent on the angles: **5. C. prostrata**

6. Stems glabrous; ovary and capsule sparsely and minutely strigulose on the sides: **6. C. bruntii**

5. Cyathia in stalked, glomerate clusters:

7. Stems branching at base but seldom near the tip; inflorescences both lateral and terminal; plants mostly rather robust and ascending: **7. C. hirta**

7. Stems branching freely to near the tip; inflorescences always terminal on branches; plants mostly low and decumbent: **8. C. ophthalmica**

1. Chamaesyce hypericifolia (L.) Millsp. in Field Mus. Bot. 2:302. 1909; Burch in Rhodora 68:160-163. 1966. **FIG. 169.**

Euphorbia glomerifera (Millsp.) L.C. Wheeler, 1939; Adams, 1972.

"Chick Weed"

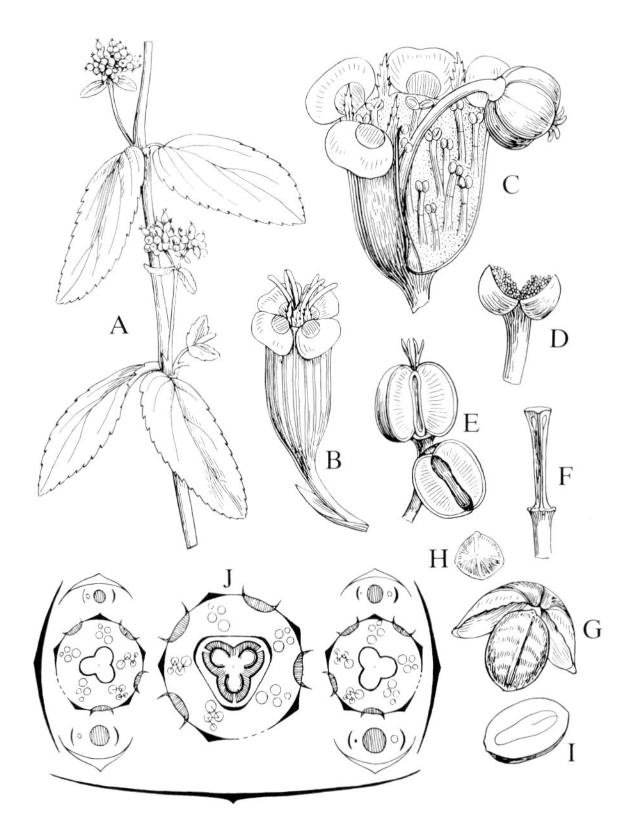

FIG. 169 **Chamaesyce hypericifolia.** A, part of flowering branch, x$^2/_3$. B, cyathium, x16. C, cyathium cut open, x24. D, stamen, much enlarged. E, capsule, x10. F, columella of capsule, much enlarged. G, coccus with seed, much enlarged. H, seed, x10. I, seed cut lengthwise, much enlarged. J, diagram of partial inflorescence. (F. & R.)

An annual herb, the ascending or erect stems to 40 cm tall and 3 mm in diameter at the base; leaves glabrous, ovate-elliptic to elliptic-obovate, somewhat falcate, 1.5-3.5 cm long, the apex acute, the margins serrulate. Cyathia glabrous, in short-stalked axillary and terminal glomerules. Capsules glabrous, subspherical, to 1.2 mm in diameter; seeds ovoid, 4-sided, to 0.8 mm long.

GRAND CAYMAN: *Brunt 1721, 1931; Hitchcock; Kings GC 296; Lewis GC 39; Maggs II 60; Millspaugh 1291, 1304; Proctor 15109.* LITTLE CAYMAN: *Proctor 35185.* CAYMAN BRAC: *Proctor 28994.*

– West Indies and the warmer parts of continental America, a common weed of fields, pastures, and clearings.

2. **Chamaesyce mesembrianthemifolia** (Jacq.) Dugand in Phytologia 13:385. 1966.

Chamaesyce buxifolia (Lam.) Small, 1903.

"Tittie Molly"

A subwoody herb or miniature shrub, erect or sometimes decumbent, often 30 cm tall or more; leaves more or less overlapping the stem, glabrous and often glaucous, ovate to elliptic, 5-12 mm long, with entire margins; stipules whitish, united, to 1 mm long, more or less fringed. Cyathia solitary at upper nodes, glabrous outside, densely bearded within; appendages white. Capsules glabrous, subspherical, 2-2.5 mm in diameter; seeds broadly ovoid, 1.3 mm long.

GRAND CAYMAN: *Brunt 1882, 2011, 2070; Hitchcock; Kings GC 23, GC 272; Lewis GC 23; Millspaugh 1262; Proctor 11973; Sachet 402, 441; Sauer 3319. 4107.* LITTLE CAYMAN: *Kings LC 93; Proctor 28070; Sauer 4166.* CAYMAN BRAC: *Kings CB 26; Millspaugh 1180, 1196, 1233; Proctor 29077.*

– Florida, West Indies, and coasts of Central and northern South America, frequent to abundant on sandy seashores or in pockets of coastal rocks. The latex of this species is reputed to be effective in removing warts or wart-like skin eruptions.

3. **Chamaesyce torralbasii** (Urban) Millsp. in Field Mus. Bot. 2:412. 1916.

A prostrate, glabrous, perennial herb, the wiry, densely-branched stems with noticeably thickened nodes; leaves short-petiolate, obliquely ovate or rotundate, 2-3.5 mm long or more (larger in Cuban specimens), obtuse at the apex and shallowly cordate at the base, with thickened denticulate margins. Cyathia few, terminal or solitary in the uppermost leaf-axils; appendages 4, minute, subequal, entire; styles forked below the middle. Capsules glabrous; seeds 4-angled, glabrous (or reported elsewhere to be minutely hairy).

CAYMAN BRAC: *Proctor 29042.*
– Cuba and possibly elsewhere. Cayman Brac plants were found "in pockets of exposed limestone pavement".

4. **Chamaesyce blodgettii** (Engelm.) Small, Fl. S.E. U.S. 712. 1903.

An annual glabrous herb, the branches prostrate, ascending, or rarely erect, to 30 cm long; leaves oblong or oblong-oval, 4–10 mm long, minutely serrulate at apex; stipules broadly triangular, often more than 1 mm long, ciliate. Cyathia solitary in upper leaf-axils, glabrous outside, bearded within; appendages entire or 2–3-crenate. Capsules glabrous, 3-cornered-globular, 1.5–2 mm in diameter; seeds 4-angled, c. 1 mm long.

GRAND CAYMAN: *Hitchcock; Kings GC 372; Millspaugh 1257, 1258, 1314, 1333; Proctor 15123, 15240; Sachet 385.* LITTLE CAYMAN: *Proctor 28145, 28168.* CAYMAN BRAC: *Proctor 29076.*
– Florida, West Indies and Central America, a weed of clearings, open waste ground, and second-growth woodlands.

5. **Chamaesyce prostrata** (Ait.) Small, Fl. S.E. U.S. 713. 1903.

Annual herb with prostrate, minutely pubescent, much-branched stems, often forming mats. Leaves oblong or broadly obovate, 4–7 mm long, puberulous or glabrate on both sides, the margins more or less serrulate; stipules broadly triangular, ciliate. Cyathia solitary in leaf-axils, glabrous or puberulous outside, bearded within; appendages crenulate. Capsules 3-angled, pilose on the angles, c. 1 mm in diameter; seeds pink, sharply 4-angled and with numerous transverse ridges.

GRAND CAYMAN: *Correll & Correll 51001; Proctor 11971.*
– Widespread in warm-temperate and tropical regions, a weed of open waste ground and sandy clearings.

6. Chamaesyce bruntii Proctor in Sloanea 1:2. 1977.

A prostrate perennial herb with woody taproot, the wiry glabrous stems densely branched. Leaves purplish, short-petiolate, obliquely ovate or broadly oblong, 2–5 mm long, glabrous, broadly obtuse at the obscurely toothed apex, subcordate at base; stipules inconspicuous, consisting of 2 or 3 linear segments 0.3–0.5 mm long. Cyathia solitary in the upper leaf-axils, the 4 subequal appendages lunate-orbicular, c. 0.3 mm wide. Capsules broadly 3-lobed, strigose-pubescent, c. 1 mm in diameter; seeds 4-angled, smooth, 0.8–0.9 mm long.

LITTLE CAYMAN: *Proctor 28146 (type), 35083.* (Another collection, *Proctor 35090,* differs from typical *C. bruntii* in possessing sparse, lax pubescence on stems and leaves).
– Endemic; occurs in sandy clearings and natural sandy glades.

7. Chamaesyce hirta (L.) Millsp. in Field Mus. Bot. 2:303. 1909.

"Dove Weed"

Annual pubescent herb, the mostly erect stems up to 35 cm tall, clothed with multicellular hairs. Leaves ovate to lanceolate or rhombic, 1–3.5 cm long, pubescent especially on the under surface, the margins serrate; stipules lacerate, to 1 mm long. Cyathia in stalked, dense, terminal and axillary clusters, strigulose throughout; appendages present or absent. Capsules strigulose, c. 1 mm in diameter; seeds reddish, wedge-shaped with sharp angles, c. 0.8 mm long.

GRAND CAYMAN: *Millspaugh 1278; Proctor 11970.* LITTLE CAYMAN: *Proctor 35181.* CAYMAN BRAC: *Kings CB 50; Millspaugh 1175; Proctor 29103.*
– A pantropical weed of open waste ground.

8. Chamaesyce ophthalmica (Pers.) Burch in Ann. Missouri Bot. Gard. 53:98. 1966.

An annual pubescent herb, the freely-branching stems mostly low and decumbent up to 20 cm long, clothed with multicellular hairs; leaves oblong-lanceolate, more or less rhombic, mostly 5–18 mm long, pubescent on both sides, the margins finely serrate; stipules bifid with narrow lobes, 0.5–0.8 mm long. Cyathia in stalked, dense clusters, these always terminal on leafy branches, strigulose throughout; appendages absent or minute. Capsules strigulose, c. 1 mm in diameter; seeds reddish, sharply quadrangular, 0.6–0.7 mm long.

GRAND CAYMAN: *Millspaugh 1292, 1298; Proctor 31052.* LITTLE CAYMAN: *Proctor 35138.* CAYMAN BRAC: *Millspaugh 1213; Proctor 28024, 29336.*
– Florida and the West indies, a common weed.

[**Pedilanthus tithymaloides** ssp. **parasiticus** Kl. & Garcke, the "Monkey Fiddle", is frequently cultivated. GRAND CAYMAN: *Kings GC 118.*]

Family 69

RHAMNACEAE

Trees or shrubs, rarely herbs, sometimes scandent, often with spines, rarely with tendrils; leaves simple, alternate or opposite, often 3-5-nerved from the base, the margins entire or serrate; stipules small and deciduous or sometimes modified into spines. Flowers small, usually perfect, mostly in small axillary cymes, green or yellowish. Calyx with more or less obconic tube and 4-5 triangular, valvate lobes; petals 4-5 or none, inserted in throat of the calyx and often smaller than its lobes, often hooded or infolded, at the base sessile or clawed. Stamens 4-5, inserted with the petals and often concealed by them; anthers oblong or 2-lobed, opening by slits. Disk perigynous. Ovary sessile, free or immersed in the disk, superior or somewhat connate with the calyx-tube, usually 3-celled, or cells sometimes 2 or 4; style erect with capitate or 3-lobed stigma; ovules usually 1 in each cell, erect from base of the cell. Fruit usually a 3-coccous capsule or drupe with 1-3-celled stone, the seeds solitary in the cells; seeds often with aril at the base; endosperm fleshy, scant, or rarely none, the embryo large, the cotyledons flat or plano-convex.

A cosmopolitan family of about 50 genera and 600 species.

Genus 1. **COLUBRINA** L.C. Rich.

Trees or shrubs, sometimes scrambling; leaves alternate, pinnate-nerved or sometimes 3-nerved from the base, the margins entire or toothed; stipules small, soon falling. Calyx 5-lobed, the lobes keeled on the inside, the tube forming a cupule confluent with the fruit. Petals 5, hooded and clawed. Stamens 5 with short, slender filaments. Disk thick, 5- or 10-lobed. Ovary immersed in the disk and confluent with it, 3-celled; style 3-lobed or 3-branched. Fruit a subglobose 3-coccous capsule; seeds flattish-ellipsoid with scant endosperm.

A genus of about 15 species, all but one confined to the American tropics.

KEY TO SPECIES

1. Leaves entire, pinnate-nerved:

2. Leaves thick, densely tomentose on both sides; capsule with cupule reaching halfway from base or nearly so: **1. C. cubensis**

2. Leaves thin, minutely puberulous beneath and glabrous on upper (adaxial) side; capsule with cupule reaching one-third from the base or less: **2. C. elliptica**

1. Leaves serrate, 3-nerved at base, glabrous: **3. C. asiatica**

1. Colubrina cubensis (Jacq.) Brongn. in Ann. Sci. Nat. sér. 1, 10:369. 1827.

"Cajon"

A shrub to 3 m tall or more, rarely a small tree, the young branches, leaves, and inflorescence densely velvety-tomentose; leaves oblong or elliptic 2-7 cm long, rounded or acute at the apex, the pinnate venation strongly raised beneath and channelled on the upper side. Cymes stalked, longer than the petioles; flowers yellow-green; calyx densely pubescent, the lobes c. 2 mm long; petals equalling the calyx. Fruit globose, c. 7 mm in diameter.

GRAND CAYMAN: *Brunt 1813; Correll & Correll 51046; Hitchcock; Howard & Wagenknecht 15027; Kings GC 390; Millspaugh 1256; Proctor 15016.* LITTLE CAYMAN: *Proctor 35186.* CAYMAN BRAC: *Millspaugh 1150, 1230; Proctor 15314.*
– Florida, Bahamas, Cuba, and Hispaniola, in dry rocky woodlands. Reports of *Colubrina arborescens* (Mill.) Sarg. (under other names) apparently were based on misidentifications of *C. cubensis*.

The leaves of this species are sometimes used to make tea.

2. Colubrina elliptica (Sw.) Briz. & Stern in Trop. Woods. no. 109:95. 1958. **FIG. 170.**

Colubrina reclinata (L'Her.) Brongn., 1827..

"Wild Guava"

Shrub or small tree 2-10 m tall or more, with slender, finely pubescent branches; leaves of thin texture, broadly elliptic or ovate, 2-10 cm long, blunt to sub-acuminate at the apex, glabrous above, minutely puberulous beneath, and often bearing a marginal gland on either or both sides near the base; petioles finely pubescent. Cymes short-stalked, few-flowered; flowers pale green; calyx

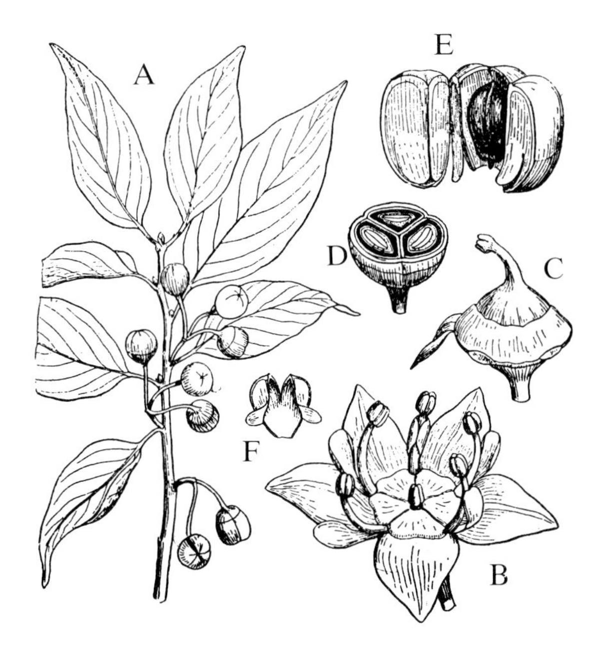

FIG. 170 **Colubrina elliptica.** A, branchlet with leaves and fruit, x⅔. B, flower, x7. C, unripe fruit with one calyx-lobe still attached, x4. D, ripe fruit cut across, x2. E, ripe fruit splitting open, x2. F, endocarp of one coccus after splitting, x1⅓. (F. & R.)

puberulous, the lobes 1.3 mm long; petals slightly shorter. Fruit glabrous, 6–8 mm in diameter.

GRAND CAYMAN: *Correll & Correll 50994; Hitchcock; Proctor 15162.* LITTLE CAYMAN: *Proctor 28114.* CAYMAN BRAC: *Proctor 29137.*

– Florida, West Indies, Yucatan, Guatemala, and Venezuela, in dry rocky woodlands.

3. **Colubrina asiatica** (L.) Brongn. in Ann. Sci. Nat. sér. 1, 10:369. 1827.

A bushy shrub with long, trailing glabrous branches; leaves with petioles mostly 1–1.5 cm long, these puberulous in one side, the glabrous blades glossy green, ovate-acuminate, 3–8 cm long. Cymes shorter than the petioles, glabrous; flowers greenish-cream. Capsules 7–9 mm in diameter.

GRAND CAYMAN: *Brunt 1633; Correll & Correll 51032; Kings GC 349; Proctor 15191; Sachet 399, 433.* LITTLE CAYMAN: *Proctor 28159; Sauer 4172.* CAYMAN BRAC: *Proctor 29105.*
– Pantropical in seaside thickets.

[**Ziziphus mauritiana** Lam., or "Cooly Plum", from southern Asia and Africa, has been planted on Grand Cayman and is likely to become naturalized. It is a small tree with spiny, tomentose branches; the oblong to oval leaves are mostly 4–5 cm long, strongly 3–5-nerved from the base, light green and glabrate on upper side and whitish-tomentose beneath. The fleshy drupes are edible.]

Family 70

VITACEAE

More or less woody vines with copious watery sap, the stems often swollen at the nodes, climbing by means of tendrils which are either sterile peduncles or sometimes simple branches of flowering peduncles. Leaves alternate, simple or digitately 3–5-foliolate or pedate, rarely bipinnate; stipules present. Flowers regular, perfect or unisexual, in racemes or cymose panicles usually borne opposite the leaves; calyx small, entire or with 4–6 teeth or lobes; petals 4–5, valvate, soon falling. Stamens 4–5, opposite the petals, inserted at base of the disk or between its lobes; anthers free or connate, short, 2-celled, opening inwardly. Disk various or none. Ovary usually immersed in the disk, 2–5-celled, the cells with 1 or 2 ascending anatropous ovules; style short or none, the stigma capitate or discoid. Fruit a berry, 1–6-celled, the cells 1–2-seeded; seeds with cartilaginous endosperm at the base of which is the short embryo.

A chiefly tropical and subtropical family of about 12 genera and 700 species. *Vitis vinifera* L. is the "Grape", cultivated in most warm and warm-temperate countries. Over 25,000,000 metric tons of wine are made throughout the world from grapes every year. Dried grapes are called "raisins".

Genus 1. **CISSUS** L.

Herbaceous or woody plants generally climbing by means of tendrils; leaves simple or 3-foliolate. Flowers in more or less umbellate cymes, usually opposite a leaf; calyx short, subentire; petals 4, often adnate to each other and falling away like a cap at anthesis; disk 4-lobed, adnate to base of the ovary. Ovary 2-celled, each cell with 2 ovules. Berries 1-4-seeded, not edible; seeds ovoid or obtusely 3-cornered.

A genus of about 200 species, widely distributed in tropical regions.

KEY TO SPECIES

1. Leaves simple: **1. C. sicyoides**

1. Leaves 3-foliolate:

   2. Flowers greenish, greenish-cream or whitish; leaflets usually 1–3 cm long, deeply and closely toothed above the middle: **2. C. trifoliata**

   2. Flowers bright red; leaflets mostly 2–6 cm long, subentire, with minute distant teeth: **3. C. microcarpa**

1. Cissus sicyoides L., Syst. Nat. ed. 19, 2:897. 1759.

Glabrous subwoody vine; leaves simple, long-petiolate, the blades narrowly to broadly ovate, mostly 4–9 cm long, subacuminate at the apex, broadly wedge-shaped, truncate or shallowly cordate at the base, the margins with distant minute teeth. Cymes shorter than the opposing leaves, 2-3-forked. Flowers small, greenish-cream. Berries obovoid-globose, black 8–10 mm in diameter.

GRAND CAYMAN: *Proctor 15297.* LITTLE CAYMAN: *Proctor 35221.* CAYMAN BRAC: *Proctor 29083.*
– Florida, West Indies, and continental tropical America, climbing on trees, rocks and fences.

2. Cissus trifoliata L., Syst. Nat. ed. 10, 2:897. 1759.

A glabrous, somewhat fleshy subwoody vine; leaves trifoliate, the leaflets obovate-wedge-shaped, deeply and closely toothed above the middle, usually

1-3 cm long. Cymes longer than the opposing leaves. Flowers greenish-cream, long-stalked, the pedicels up to 8 mm long. Berries ovoid-globose, black, 6-7 mm long, mucronate.

GRAND CAYMAN: *Brunt 2103; Kings GC 151; Proctor 15128, 27938; Sachet 432.* LITTLE CAYMAN: *Kings LC 99; Proctor 28031.* CAYMAN BRAC: *Proctor 28932.*

– Florida, West Indies, and northern South America, in dry rocky thickets and on fences and old stone walls. This species has been used in folk medicine for treating coughs and back-ache.

3. **Cissus microcarpa** Vahl, Ecolog. 1:16. 1796. **FIG. 171.**

"Pudding Withe", "Daffodil"

A subwoody scrambler or vine with few tendrils, the branches variously 4-winged or -angled; leaves glabrous, trifoliate, the terminal leaflet subrhombic-elliptic,

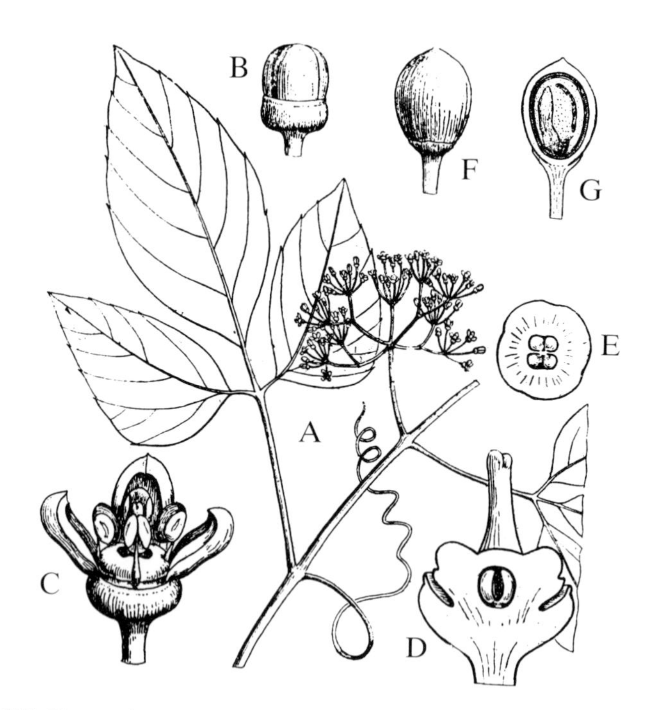

FIG. 171 **Cissus microcarpa.** A, leaf and inflorescence, x$^2/_3$. B, flower-bud, x4. C, flower, x5. D, flower with petals removed, cut lengthwise, x11. E, x-section of ovary, x11. F, fruit, x2. G, same cut lengthwise, x2. (F. & R.)

3-6 cm long or more, the lateral ones obliquely ovate and smaller, all blunt to subacuminate at the apex, the margins distantly and minutely mucronate-serrulate. Cymes red, many-flowered, shorter than the opposing leaves, the branches and pedicels puberulous. Flowers red, the pedicels 3-5 mm long. Berries ovoid-globose, 7-8 mm in diameter, purple.

GRAND CAYMAN: *Kings GC 119.* LITTLE CAYMAN: *Proctor 35117.* CAYMAN BRAC: *Kings CB 95; Proctor 28947.*

– Cuba, Jamaica, Central and South America, in dry or moist woodlands, climbing on rocks and trees. The Cayman plants as described above are somewhat transitional toward *C. caustica* Tussac, a species of Cuba and Hispaniola.

Family 71

SURIANACEAE

Shrubs; leaves alternate, simple, entire and usually of thick texture; stipules minute or absent. Flowers regular, perfect or polygamous, solitary or in small panicles, these axillary or subterminal; calyx 5-lobed with persistant imbricate lobes; petals 5 or absent. Stamens 10, with free slender filaments. Ovary 1-celled or of 5 distinct 1-celled carpels, with a free style springing from the base of each; ovules 2 in each cell, ascending from the base. Fruit a nut, dry drupe, or of indehiscent achene-like carpels, these 1-seeded; seeds without aril and with little or no endosperm.

A small family consisting of one monotypic pantropical genus and one other *(Stylobasium)* of 2 species occurring in southwestern Australia.

Genus 1. **SURIANA** L.

A seashore shrub clothed with mixed capitulate and acicular hairs; leaves densely alternate; stipules absent. Flowers perfect, subterminal, solitary or in small, few-flowered panicles; petals clawed, yellow. Stamens unequal in length, those opposite the petals shorter and sometimes without anthers. Ovary of 5 distinct 1-celled carpels. Fruiting carpels achene-like, surrounded by the persistent calyx; seeds with thick, horseshoe-shaped embryo.

A genus of 1 pantropical species.

1. **Suriana maritima** L., Sp. Pl. 1:284. 1753. **FIG. 172.**

"Juniper"

Characters of the genus. Bushy, 1–2 m tall, rarely more; leaves linear-spatulate, 1–3 cm long. Sepals ovate-lanceolate, 6–10 mm long; petals erose at the apex, shorter than the calyx. Ripe carpels 4–5 mm long, finely pubescent.

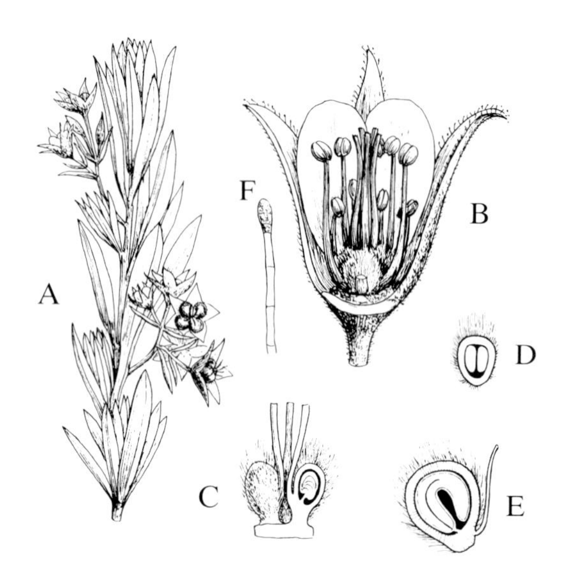

FIG. 172 **Suriana maritima.** A, branch with inflorescence, x²/₃. B, flower with part of the calyx and corolla and one stamen removed, x4. C, pistil cut lengthwise, x6. D, x-section of a carpel, x6. E, ripe achene (nutlet) cut lengthwise, x2. F, hair from calyx, much enlarged. (F. & R.)

GRAND CAYMAN: *Brunt 2009, 2064; Hitchcock; Kings GC 22, GC 264; Millspaugh 1253; Proctor 15122; Sachet 398.* LITTLE CAYMAN: *Kings LC 85; Proctor 28077.* CAYMAN BRAC: *Kings CB 57; Millspaugh 1151; Proctor 28923.* – Common on sandy or rocky tropical seashores.

Family 72

SAPINDACEAE

Shrubs or trees with watery sap, or sometimes woody or herbaceous vines climbing by means of tendrils; leaves alternate, mostly pinnately or bipinnately compound, sometimes 3- or 1-foliolate; stipules absent in most genera. Flowers usually small and white, mostly polygamo-dioecious, regular or irregular, variously arranged in racemes, panicles or corymbs, axillary or sometimes terminal. Sepals 4–5, rarely none, mostly free and often unequal, imbricate; petals 4–5 or none, equal or unequal, often scaly or bearded within; disk complete in regular flowers or represented by 2 or 4 glands in irregular flowers. Stamens usually 8, usually hypogynous and inserted within the disk; anthers oblong, 2-lobed, or linear. Ovary superior, mostly 3-celled, entire or lobed, with 1–2 ovules in each cell, attached to the axis; style terminal or basal between the ovary-lobes, simple or divided, with simple stigma. Fruit a capsule or indehiscent and drupe-like, berry-like, leathery, or consisting of 2–3 samaras; seeds usually with or sometimes without aril; endosperm none, the embryo thick and often folded or spirally twisted.

A large, chiefly tropical family of about 150 genera and 2000 species.

KEY TO GENERA

1. Plants climbing by tendrils: **1. Cardiospermum**

1. Plants erect, never with tendrils:

  2. Leaves compound:

    3. Leaves 3-foliolate, having a terminal leaflet; fruits drupe-like:

      4. Plant wholly glabrous; petioles marginate; leaves with closely parallel venation: **6. Hypelate**

      4. Plant puberulous at least on inflorescence; petioles not marginate; leaf-venation not closely parallel: **2. Allophylus**

    3. Leaves 2–4-foliolate or more, without a terminal leaflet:

      5. Sepals and petals 4; fruit a 1-seeded drupe: **[3. Melicoccus]**

      5. Sepals and petals 5; fruit a fleshy 3-celled capsule: **[4. Blighia]**

  2. Leaves simple; fruit a winged capsule: **5. Dodonaea**

Genus 1. **CARDIOSPERMUM** L.

Climbing herbs or shrubs with paired tendrils arising from the peduncles; branches slender, grooved or ribbed; leaves biternate, the leaflets crenate or serrate, often with pellucid dots or lines in the tissue. Flowers white, irregular, polygamo-dioecious, with jointed pedicels, in axillary racemes or corymbs; sepals 4–5, concave, the outer ones smaller; petals 4, two with crested scale which has a bearded appendage pointing downward, the other two with a scale which has a wing-like crest on the back. Disk-glands 2, opposite the petals with the appendage. Stamens 8, the filaments unequal. Style 3-lobed. Capsule of 3 inflated membranous lobes, splitting loculicidally; seeds globose, blue-black, often arillate at the base; cotyledons large, transversely folded.

A pantropical genus of 12 species.

KEY TO SPECIES

1. Capsule subglobose, puberulous, 3–4 cm in diameter; seed with a heart-shaped, distinctly bilobed hilum: **1. C. halicacabum**

1. Capsule top-shaped or subglobose, glabrous, less than 3 cm in greatest diameter; seed with a semicircular hilum: **2. C. corindum**

1. Cardiospermum halicacabum L., Sp. Pl. 1:366. 1753.

A mostly herbaceous vine, the stems glabrous or puberulous, 5–6-ribbed; leaves 3–10 cm long, with ovate or lanceolate leaflets, variously toothed or lobed, puberulous. Inflorescence umbel-like, long-stalked; flowers mostly 4–5 mm long. Seeds 4–5 mm in diameter, with white hilum nearly as broad as the seed.

GRAND CAYMAN: *Hitchcock; Kings GC 47.* CAYMAN BRAC: *Millspaugh 1165, 1200, 1201.*
– Pantropical, thickets and woodlands.

2. Cardiospermum corindum L., Sp. Pl. ed. 2, 1:526. 1762.

Herbaceous to subwoody vine, the stems puberulous, 5–7-ribbed; leaves 2–7 cm long, with lanceolate to ovate leaflets, incised or crenate, puberulous. Inflorescence umbel-like, long-stalked; flowers mostly 3–5 mm long. Seeds 2.5–4 mm in diameter, with white semicircular hilum.

GRAND CAYMAN: *Millspaugh 1309; Proctor 15181.* LITTLE CAYMAN: *Proctor 28125, 35096, 35180.* CAYMAN BRAC: *Proctor 29017, 29049.*
– Pantropical, in thickets and woodlands.

Genus 2. **ALLOPHYLUS** L.

Erect shrubs or small trees; leaves often long-petiolate, trifoliolate or 1-foliolate, the leaflets often with pellucid dots or lines in the tissue. Inflorescence an axillary raceme or panicle; flowers very small, polygamo-dioecious; sepals 4, opposite in pairs, concave, broadly imbricate, the 2 outer ones smaller; petals 4, each with a small 2-lobed scale within. Disk lobed or of 4 glands. Stamens 8. Ovary commonly

2-celled and deeply 2-lobed, united by the 2-3-lobed style; ovules solitary; only 1 ovary-lobe usually developing to form a fruit. Fruit a single indehiscent drupe-like coccus (rarely 2 cocci), dry or fleshy, usually obovoid or globose; seed erect, on a very short fleshy aril.

A pantropical genus of perhaps 175 species.

1. **Allophylus cominia** (L.) Sw., Nov. Gen. & Sp. Pl. 62. 1788.
var. **caymanensis** Proctor in Sloanea 1:2. 1977. **FIG. 173.**

A bushy shrub to 3 m tall; young branches puberulous or glabrate, with raised lenticels. Leaves long-petiolate, trifoliate, the leaflets elliptic or oblanceolate to narrowly obovate, the central one 5-10 cm long, short-acuminate at the apex, the margins distantly crenate-serrate, glabrous or nearly so on the upper side, minutely puberulous or glabrate beneath and with small tufts of white hairs (domatia) in the vein-axils. Inflorescence often equalling or exceeding the leaves, simple or with 1 or 2 short branches, puberulous. Flowers greenish-cream, c. 1 mm long, the stamens exserted. Fruit more or less globose, 4-4.5 mm in diameter.

GRAND CAYMAN: *Correll & Correll 51021; Hitchcock; Proctor27979, 31039, 31053 (type).* LITTLE CAYMAN: *Proctor 35219.* CAYMAN BRAC: *Fawcett; Proctor 28965, 29346.*
– The variety is endemic, growing in rocky thickets and woodlands. Var. *cominia*, which is widely distributed in the West Indies and Central America, is usually much more densely pubescent (a variable character), has larger leaves, inflorescences much more branched, and larger fruits.

[Genus 3. **MELICOCCUS** P.Br.

Glabrous trees; leaves abruptly pinnate, the leaflets in 2 or 3 subopposite pairs, subsessile, entire. Racemes simple or paniculate, terminal on lateral branchlets, bearing numerous flowers. Flowers polygamo-dioecious; calyx deeply 4-5-lobed; petals 4 or 5, roundish or obovate. Disk 4-5-lobed. Stamens 8. Ovary 2-3-celled; stigma peltate, 2-3-lobed. Fruit a drupe; seed enclosed in a large, pulpy, edible aril.

A tropical American genus of 2 species.

1. **Melicoccus bijugatus** Jacq., Enum. Syst. Pl. Carib. 19. 1760.

"Genip"

Tree to 12 m tall or more, with smooth gray bark; leaves deciduous annually, the young leaves appearing with the flowers in March or April; petiole and rhachis flat, often winged; leaflets elliptic or ovate-elliptic, 5–15 cm long, subacuminate at apex; margins often undulate. Staminate inflorescence often longer than the leaves, much-branched; pistillate inflorescence shorter and less branched. Open flowers 6–8 mm in diameter, scented. Fruits green, c. 3 cm in diameter, with edible pulp.

GRAND CAYMAN: *Brunt 2012, 2060; Kings GC 162; Proctor 15193.* CAYMAN BRAC: *Proctor 29109.*
– Native to continental tropical America from Nicaragua to Surinam, widely planted elsewhere and becoming naturalized. It is a handsome shade-tree; the wood is hard and heavy, suitable for general construction purposes.]

[Genus 4. **BLIGHIA** Koenig

Trees; leaves large, even-pinnate, with 3–5 pairs of entire, glabrous leaflets. Flowers polygamous in axillary racemes or panicles; calyx 5-parted.; petals 5, with a scale at the base of each about half as long as the petal. Disk ring-like, somewhat 8-lobed; stamens 8, inserted within the disk. Ovary 3-celled, with 1 ovule in each cell. Fruit a somewhat fleshy 3-valved capsule; seeds large, black, surrounded at base with a large, whitish or yellowish, fleshy aril.

A tropical African genus of 6 species.

1. **Blighia sapida** Koenig in Koenig & Sims in Ann. Bot. 2:571, t. 16, 17. 1806.

"Akee"

A tree to 10 m tall, the young branchlets yellowish-tomentose; leaves 15–30 cm long, the 3–5 pairs of leaflets opposite or subopposite, glabrous above, puberulous beneath, the midrib and veins prominent. Racemes shorter than the leaves, tomentose, the flowers greenish-white, stalked, fragrant; calyx 2.5–3 mm long; petals 4 mm long, each with wide basal scale. Disk tomentose. Fruits 7–10 cm long, red, glabrate outside, densely tomentose within; aril edible when ripe.

GRAND CAYMAN: *Kings GC 346.* CAYMAN BRAC: *Proctor 29110.*
– Introduced to Jamaica from West Africa in the 1780's, and since widely planted there and to a lesser extent in other West Indian islands and elsewhere, often becoming naturalized. The fleshy arils are poisonous unless the capsules are ripe (i.e., split open) when picked, but are wholesome and nutritious if properly prepared. They are usually served with salt codfish, or sometimes with bacon.]

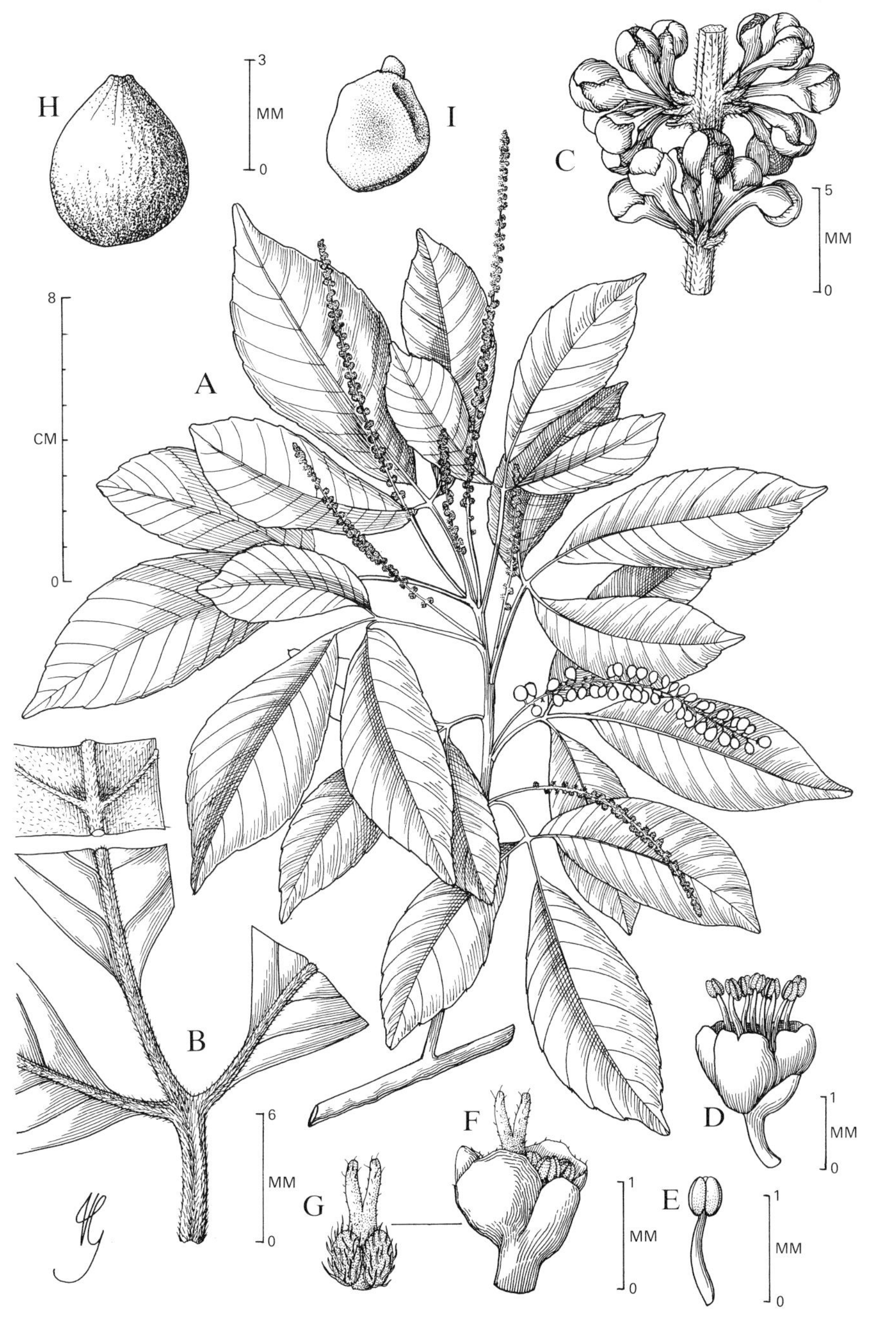

FIG. 173 **Allophylus cominia** var. **caymanensis.** A, habit. B, segments of leaf. C, section of inflorescence with flower-buds. D, staminate flower. E, stamen. F, perfect pistillate flower just opening. G, pistil, showing 2-lobed ovary and 2-branched style. H, fruit. I, seed. (G.)

Genus 5. **DODONAEA** L.

Shrubs or trees, usually viscid; leaves simple or rarely pinnate; stipules absent. Flowers unisexual or polygamo-dioecious, whitish, pale green or yellowish, in axillary or terminal racemes, corymbs or panicles. Sepals 2–5, valvate or narrowly imbricate; petals absent. Disk lacking in the staminate flower, short and stalk-like in the pistillate. Stamens 5–8, the anthers nearly sessile, linear-oblong, obtusely 4-angled. Ovary sessile, 3–6-angled, 3–6-celled; style 3–6-lobed at the apex; ovules 2 in each cell. Fruit a somewhat papery capsule, 2-6-angled, the angles acute to broadly winged, with 1–2-seeded cells; seeds without aril.

A chiefly Australian genus of more than 50 species.

1. **Dodonaea viscosa** (L.) Jacq., Enum. Pl. Carib. 19. 1760. **FIG. 174.**

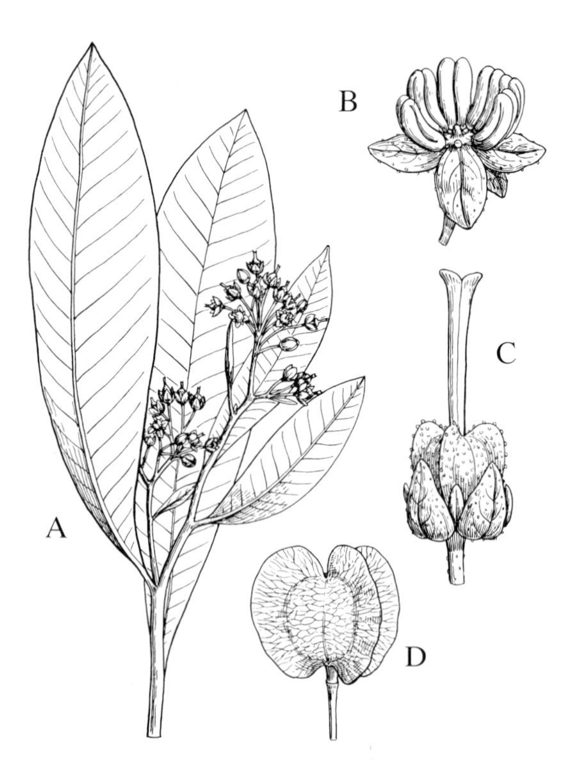

FIG. 174 **Dodonaea viscosa.** A, branchlet with leaves and flowers, x$^2/_3$. B, staminate flower with a sepal pressed down and a stamen removed, x4. C, pistillate flower, x4. D, fruit, x1. (F. & R.)

A shrub up to 3 m tall, with very viscid foliage, the branches reddish-brown, longitudinally ridged; leaves simple, mostly oblanceolate or narrowly oblong, usually 4–11 cm long and 1.5–4 cm broad above the middle, gradually narrowed to the base, a petiole virtually absent. Flowers pale green, dioecious, in small lateral corymbs; sepals 3 mm long. Capsules usually 3-celled and 3-winged, 1.5–2.5 cm wide, deeply notched at the apex.

GRAND CAYMAN: *Kings GC 386; Millspaugh 1264; Proctor 11964.* LITTLE CAYMAN: *Kings LC 69; Proctor 28078.*

– Pantropical, in various habitats; many variants have been described, but the Cayman plants seem to approximate the typical variety.

Genus 6. **HYPELATE** Sw.

A glabrous shrub or small tree; leaves alternate, trifoliate, with narrowly margined petioles. Flowers small, white, polygamo-monoecious, in axillary panicles; sepals 5, imbricate, soon falling; petals 5. Stamens 8, inserted on the disk, shorter and imperfect in the female flower. Ovary 3-celled, rudimentary in the male flowers; ovules 2 in each cell. Fruit a thin-fleshed, 1-seeded drupe; seed without endosperm, the cotyledons folded irregularly.

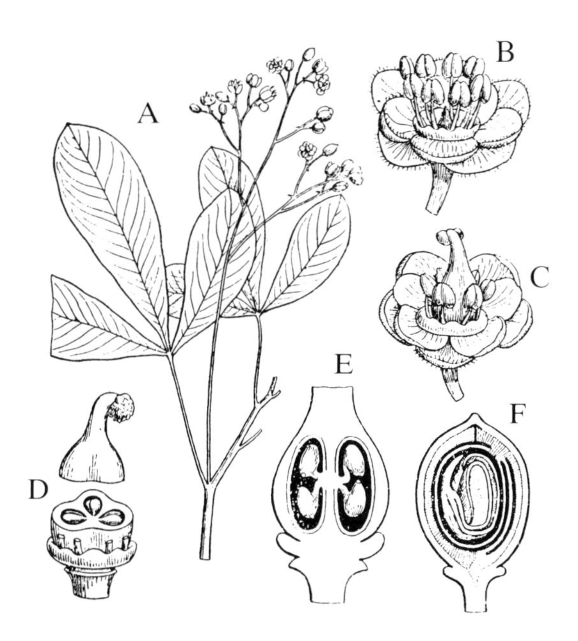

FIG. 175 **Hypelate trifoliata.** A, portion of branch with leaves and flowers, x$^2/_3$. B, staminate flower, x3. C, pistillate flower, x3. D, ovary cut across, x6. E, ovary cut lengthwise, x12. F, fruit cut lengthwise, x6. (F. & R.)

A monotypic genus of the Florida keys and West Indian islands.

1. **Hypelate trifoliata** Sw., Nov. Gen. & Sp. Pl. 61. 1788. **FIG. 175.**

"Pompero", "Plumperra", "Wild Cherry"

Shrub or small tree to 8 m tall or more; leaves with petioles 1-3 cm long, the oblanceolate to obovate sessile leaflets mostly 1-4 cm long; venation closely parallel and prominulous. Panicles exceeding the leaves; sepals 2.5-3 mm long, petals 2 mm long. Fruits ellipsoid, 6-8 mm long, brown or black.

GRAND CAYMAN: *Brunt 2195; Kings GC 377, GC 380; Proctor 15218.* LITTLE CAYMAN: *Proctor 28102, 35129.* CAYMAN BRAC: *Matley; Kings CB 92; Proctor 29025.*
– Distribution of the genus, frequent in dry rocky woodlands. The wood is said to be very heavy, hard, close-grained, and a rich dark brown; it is very durable in contact with the soil, and is used for posts, in ship-building, and for the handles of tools.

Family 73

BURSERACEAE

Shrubs or trees, more or less resinous or aromatic; leaves alternate, simple or usually odd-pinnate, sometimes 3-foliolate, usually not pellucid-dotted; stipules absent. Flowers small, perfect or polygamo-dioecious, in racemes or panicles. Calyx 3-5-lobed, the lobes imbricate or valvate; petals 3-5, free or rarely united toward the base, imbricate or valvate, alternating with the sepals. Disk ring- or cup-shaped, free or adnate to the calyx-tube. Stamens twice as many as the petals or rarely the same number, inserted at the base or margin of the disk; filaments free; anthers subglobose or oblong, 2-celled. Ovary superior and free, 2-5-celled, 3-angled, with short style and undivided or 2-5-lobed stigma; ovules 2 in each cell, attached to the axis above the middle, usually pendulous. Fruit capsule-like or drupe-like, dehiscent or indehiscent, containing 2-5 hard nutlets; seeds with membranous walls; endosperm lacking.

A pantropical family of about 17 genera and 500 species.

Genus 1. **BURSERA** L.

Shrubs or trees, often with thin papery bark, and with aromatic resinous sap; leaves deciduous, variously pinnate or sometimes 1-foliolate, the leaflets opposite, entire or serrate, and with rhachis naked or winged. Flowers polygamous or perfect;

calyx 3-4-lobed or -parted, imbricate; petals 3-4, valvate. Disk ring-like. Stamens 6-10, subequal, inserted at base of the disk. Ovary 3-5-celled, the stigma 3-5-lobed. Fruit an ovoid, more or less 3-angled capsule, 2-3-valved, with 1-several hard nutlets.

An American genus of about 80 species, the majority in Mexico; all are capable of yielding a fragrant resin called "copal", used (in Mexico and Central America chiefly) as incense, in domestic medicine, as an ingredient of varnish, and for numerous other purposes.

1. **Bursera simaruba** (L.) Sarg. in Gard. & For. 3:260. 1890. **FIG. 176.**

"Birch", "Red Birch"

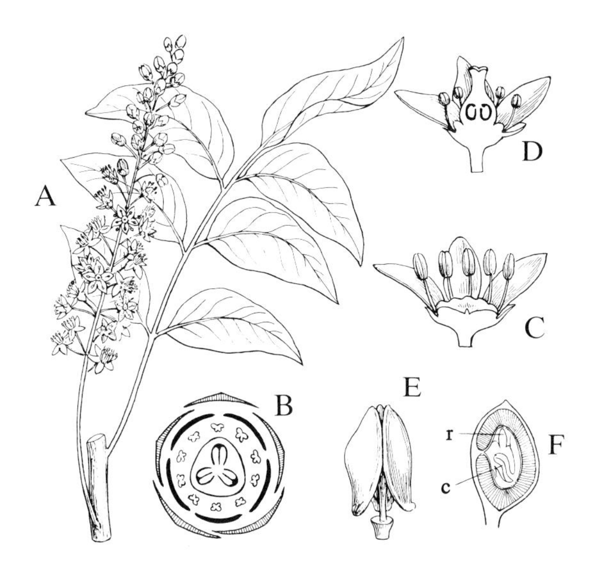

FIG. 176 **Bursera simaruba.** A, portion of branch with small leaf and inflorescence, x$^{2}/_{3}$. B, diagram of perfect flower. C, staminate flower, x4. D, fertile flower, x4. E, drupe, x1. F, stone cut lengthwise, x1½; c, cotyledons; r, radicle. (F. & R.)

A tree to 15 m tall or more, the young bark greenish, the old bark light reddish, peeling off in paper-like sheets; leaves petiolate, pinnate with usually 5-7 unequal-sided leaflets, these more or less pubescent when young, becoming glabrous with age, lance-oblong to broadly ovate, mostly 4–10 cm long, acuminate at the apex. Flowers white, in racemes appearing before or with the new leaves; staminate flowers 5-parted, with 10 stamens; perfect flowers 3-parted, with 6 stamens. Fruits 8–12 mm long, dark red.

GRAND CAYMAN: *Brunt 1816, 2023; Hitchcock; Proctor 15027, 15139; Sachet 324.* LITTLE CAYMAN: *Proctor 28046.* CAYMAN BRAC: *Proctor 28957.* – Florida, West Indies, Central and northern South America, common in dry woodlands and along roadsides. This tree is often used for living fenceposts, and to a partial extent its distribution is artificial for this reason. The wood is light and soft but firm, and is used somewhat for making crates, boxes and match-sticks.

Family 74

ANACARDIACEAE

Shrubs or trees, often containing more or less poisonous resin or sap; leaves alternate or very rarely opposite, and simple, 1–3-foliolate, or odd-pinnate; stipules absent or the lowest leaflet stipule-like. Flowers perfect or polygamo-dioecious, usually regular, in panicles or rarely racemes; calyx 3–7-lobed, rarely spathe-like or irregularly rupturing; petals usually 3–7, free, sometimes persistent. Disk usually ring-like. Stamens commonly twice as many as the petals, rarely otherwise, inserted at base of the disk; filaments free; anthers opening inwardly. Ovary 1-celled or rarely 2–5-celled; styles 1–3; ovules solitary, more or less pendulous. Fruit usually drupe-like and indehiscent, superior and free or surrounded either by the enlarged base of the calyx or disk, or sometimes located at the top of a fleshy receptacle formed from the base of the calyx and top of the pedicel, 1–5-celled, the flesh frequently oily or with caustic sap; seeds with little or no endosperm, the cotyledons fleshy.

A widely-distributed family of about 65 genera and 600 species. The "Cashew" tree, *Anacardium occidentale* L., belongs to this family; also "Poison Ivy" and "Poison Sumac" *(Toxicodendron)*, a group of two notorious North American plants.

KEY TO GENERA

1. Leaves simple: **[1. Mangifera]**

1. Leaves pinnate:

  2. Ovary 1-celled; fruits not over 1.5 cm long, not edible, the plants with poisonous sap:

3. Leaflets 3-7, entire; a glabrous tree to 10 m tall: **2. Metopium**

3. Leaflets usually 11-17, with toothed margins; often unbranched pubescent shrub to 2 m tall: **3. Comocladia**

2. Ovary 2-5-celled; fruits plum-like, 2.5 cm long or more, edible, the plants not poisonous: **4. Spondias**

[Genus 1. **MANGIFERA** L.

Trees; leaves alternate, simple and entire. Flowers polygamous, in terminal panicles; calyx 4-5-parted, deciduous, with imbricate sepals; petals 4-5, imbricate and spreading. Disk of 4-5 fleshy lobes, alternate with the petals. Stamens 1 or 4-5, inserted within or on the disk, only 1 (rarely more) fertile and larger than the others. Ovary free, sessile, 1-celled, with lateral style and simple stigma. Drupe fleshy, the stone compressed and sometimes 2-valved.

A genus of about 30 species chiefly in tropical Asia.

1. **Mangifera indica** L., Sp. Pl. 1:200. 1753.

"Mango"

A tree of 10-15 m with a dense, rounded or spreading crown; leaves glabrous, oblong-lanceolate, 10-20 cm long. Flowers greenish-white, in large panicles; sepals 2.5 mm long; petals 5 mm long; fertile stamens 1 or 2. Fruits edible, varying greatly in size, shape and colour, comprising many horticultural variants.

GRAND CAYMAN: *Brunt 1991; Hitchcock; Proctor 15185.* CAYMAN BRAC: *Proctor (sight record).*
– Originally from India, but now cultivated in various forms throughout the tropics for its edible fruits, and often becoming naturalized. In Grand Cayman trees can be found in areas remote from habitations wherever seeds have been casually tossed aside.]

Genus 2. **METOPIUM** P.Br.

Glabrous trees with caustic sap; leaves petiolate, odd-pinnate, the leaflets coriaceous, entire. Flowers small, greenish, in axillary panicles; sepals 5, imbricate; petals 5, imbricate. Disk ring-like. Stamens 5, with short, subulate filaments. Ovary 1-celled, the style short, the stigma 3-lobed. Fruit a small drupe.

A chiefly West Indian genus of 3 species.

1. Metopium toxiferum (L.) Krug & Urban in Urban, Bot. Jahrb. 21:612. 1896.

"Poison Tree"

A tree to 10 m tall or more, with wide-spreading branches and poisonous sap; leaves 15–30 cm long, with 3–7 stalked leaflets, these with blades ovate to suborbicular or obovate, 3–10 cm long, often obtuse or notched at the apex and narrowed or unequally cordate at the base, bright shining green especially on the upper side. Panicles many-flowered, sometimes exceeding the leaves. Drupe oblong, orange-yellow, 1–1.5 cm long.

GRAND CAYMAN: *Hitchcock.* LITTLE CAYMAN: *Proctor 28128.*
– Florida, Bahamas, Cuba, Hispaniola and Puerto Rico, in dry rocky or sandy woodlands. A different species occurs in Jamaica.

Genus 3. COMOCLADIA L.

Shrubs or small trees with poisonous sap, this turning blackish on exposure to air; trunk or main stem slender, often unbranched or few-branched, the leaves crowded in a palm-like rosette toward the top; leaves alternate, odd-pinnate, the leaflets rather numerous and opposite to subopposite, entire or toothed, reduced in size toward base of the leaf. Panicles axillary or appearing to be terminal, usually shorter than the leaves; flowers minute, crowded, polygamous, sessile or subsessile; calyx 3–4-cleft, persistent, with imbricate lobes; petals 3–4, red, imbricate. Disk 3-lobed. Stamens inserted in notches of the disk. Ovary 1-celled, with 3 stigmas. Fruit an oblong-ellipsoid or ovoid drupe.

A genus of about 20 species, occurring chiefly in the West Indies and Mexico.

1. Comocladia dentata Jacq., Enum. Pl. Carib. 16. 1760.

"Maiden Plum"

A slender shrub rarely more than 2 m tall, often unbranched, the younger parts puberulous; leaves with 11–17 leaflets, these ovate to oblong, 3–13 cm long, the margins sharply toothed, the under-surface puberulous and with prominently reticulate venation. Panicles 20–25 cm long; flowers 3-parted. Drupes ovoid, 7–8 mm long.

GRAND CAYMAN: *Kings GC 115; Lewis 3612, 3822; Proctor 15017, 27977.* LITTLE CAYMAN: *Proctor 35183.*
– Cuba and Hispaniola, in dry thickets, pastures and old fields, a noxious weed wherever it occurs.

Genus 4. **SPONDIAS** L.

Small to large trees; leaves alternate, odd-pinnate, the leaflets mostly opposite. Flowers small, polygamous, in axillary or terminal panicles; calyx small, 4-5-lobed, deciduous; petals 4-5, valvate in bud. Disk cup-shaped, crenate. Stamens 8-10, inserted below the disk. Ovary sessile, free, 3-5-celled; styles 3-5, connivent above. Fruit a fleshy drupe, the stone large and 1-5-celled.

A small genus of about 8 species, widely distributed in tropical regions at least in cultivation.

1. **Spondias purpurea** L., Sp. Pl. ed. 2, 1:613. 1762. **FIG. 177.**

"Plum"

A low, spreading, deciduous tree to 5 m tall or more, with arching lateral branches, the bark smooth and grayish; leaves glabrous, with 5-13 or more opposite

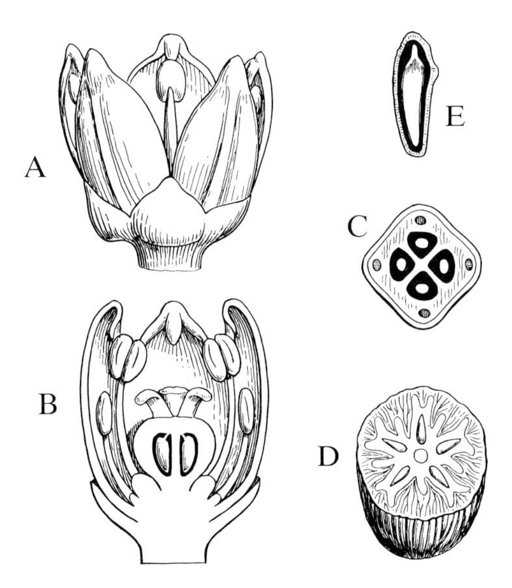

FIG. 177 **Spondias purpurea.** A, staminate flower, x6. B, pistillate flower cut lengthwise, x6. C, x-section of ovary, x9. D, endocarp of a related species *S. mangifera*, enlarged. E, cell of same with embryo, enlarged. (F. & R.)

or alternate leaflets, these elliptic-oblong to obovate, 1.5-4 cm long, acute and mucronate at the apex, and unequal-sided at the base. Panicles small and few-flowered, produced when the branches are leafless at nodes on the older wood. Flowers red-purple, the petals 3 mm long. Fruits red or yellow, 3–3.5 cm long, edible.

GRAND CAYMAN: *Brunt 2018.* CAYMAN BRAC: *Proctor 29111.*
– Widespread in tropical America; often planted as a "quickstick" fence or for its edible fruits, so that its true natural range would be difficult to ascertain. It is probably not indigenous in the Cayman Islands, but nevertheless can be found growing wild in rocky thickets.

[Other species of *Spondias*, such as *S. mombin* L., the "Hog Plum" and *S. dulcis* Parkinson, the "June Plum", may perhaps occur as cultivated trees in the Cayman Islands, but have not been encountered during the preparation of the present book.]

Family 75

SIMAROUBACEAE

Shrubs or trees, the bark usually very bitter and containing oil-sacs; leaves alternate or rarely opposite, pinnate or rarely 3- or 1-foliolate; stipules usually lacking. Flowers regular, dioecious or polygamous, sometimes perfect, in mostly axillary panicles or racemes, rarely in spikes or the flowers solitary; calyx 3–5-lobed; petals 3–5 (rarely none), imbricate or valvate. Disk ring-like, cup-like, or elongate to form a gynophore, entire or lobed. Stamens inserted at base of the disk, as many or twice as many as the petals; filaments free; anthers 2-celled, opening inwardly. Ovary 2–5-lobed, rarely entire, 1–5-celled; styles 2–5, free or sometimes fused at the apex into a cap-like stigma; ovules 1 or 2 in each cell, attached at the inner angle. Fruit a drupe, capsule or samara; seeds with or without endosperm.

A pantropical family of about 20 genera and 120 species.

Genus 1. **ALVARADOA** Liebm.

Shrubs or small trees with bitter juice; leaves unequally pinnate, with numerous small alternate leaflets. Flowers dioecious, small, numerous in narrow spreading or drooping racemes, these axillary and terminal; calyx of 5 valvate sepals; petals 5 or lacking. Staminate flowers with large, deeply 5-lobed disk; stamens 5, alternate with the sepals, inserted between the lobes of the disk. Pistillate flowers without stamens; ovary 2–3-celled with only 1 cell fertile; styles 2–3, free and recurved; ovules 2 in each cell. Fruit a 2–3-winged samara.

A tropical American genus of 5 or 6 species.

1. Alvaradoa amorphoides Liebm. in Nat. For. Kjoebenhavn Vid. Medd. 1853:100. **FIG. 178.**

"Wild Spanish Armada"

A deciduous shrub to 3 m tall or more, rarely a small tree; branchlets puberulous; leaves clustered near the ends of the branches, with 19-51 membranous leaflets, these oval or oblong, 1-2.5 cm long, rounded at the apex, puberulous on both sides or glabrous on the upper side, noticeably pale or whitish beneath. Flowers on slender pedicels, greenish, the staminate racemes up to 25 cm long, the pistillate ones much shorter, dense and plume-like. Samaras narrowly lance-oblong, dull reddish, 1-1.5 cm long, densely pilose and ciliate with long slender spreading hairs.

GRAND CAYMAN: *Brunt 2192; Hitchcock; Kings GC 320; Millspaugh 1282; Proctor 15198, 15229.* LITTLE CAYMAN: *Proctor 35093.*
– Florida, Bahamas, Cuba, Central and South America, in dry rocky woodlands. The wood is said to be valuable as fuel because it burns slowly and for a long time.

Family 76

RUTACEAE

More or less aromatic shrubs or trees, rarely herbs, occasionally scandent, sometimes armed with prickles; leaves alternate or opposite, usually compound, occasionally 1-foliolate or simple, nearly always more or less pellucid-dotted with oil-glands; stipules absent. Flowers perfect, polygamous or dioecious, in axillary or terminal cymes, panicles, racemes, spikes or clusters, rarely solitary, the parts usually in 4's or 5's, rarely 3's. Sepals imbricate or rarely none; petals imbricate or sometimes united. Stamens as many or twice as many as the petals, rarely more numerous; filaments free or united below, inserted on a hypogenous disk or sometimes adnate to the corolla-tube; anthers 2-celled, often gland-tipped, opening inwardly. Ovary usually 4-5-celled or the carpels free at the base and united in the styles or stigmas, or altogether free and 1-celled; ovules usually 2 (rarely 4) in each carpel. Fruit various, of follicles, or a samara, drupe, or berry; seeds solitary or several in each cell, with or without endosperm.

A widely-distributed family of about 120 genera and 900 species, chiefly of warm climates, especially numerous in South Africa and Australia.

KEY TO GENERA

1. Ovary 2-5-lobed; fruit of 1-5 cocci; leaves pinnately compound with 5 or more leaflets: **1. Zanthoxylum**

1. Ovary not lobed; fruit a drupe or berry; leaves simple or if compound not more than 3-foliolate:

 2. Leaves simple; stamens numerous (20–60): **[Citrus]**

 2. Leaves 3-foliolate; stamens 4–10:

  3. Flowers solitary in the leaf-axils, the parts in 3's; plants armed with spines: **[Triphasia]**

  3. Flowers in terminal and axillary panicles, the parts in 4's or 5's; plants unarmed: **2. Amyris**

Genus 1. **ZANTHOXYLUM** L.

Trees or shrubs, often more or less prickly; bark aromatic; leaves alternate, even- or odd-pinnate, or rarely 1-foliolate; leaflets opposite or alternate, frequently inequilateral, crenulate or entire, usually glandular-punctate (at least along the margin), and the rhachis often winged. Plants dioecious or polygamous; flowers small, white to greenish-yellow, unisexual and/or bisexual, in axillary short spikes or clusters or in terminal or axillary panicles. Calyx 3-5-cleft, deciduous or persistent, or rarely lacking; petals (3-5 (-10). Staminate flowers with 3-5 hypogynous stamens. Pistillate flowers without stamens or with scale-like staminodes; carpels 1–5, distinct or partially united, on an elevated fleshy gynophore, each 1-celled with 2 pendulous ovules in each cell. Fruits dry, of separate 2-valved follicles, each with 1 shining black seed; seeds with fleshy endosperm, at maturity often hanging from the carpels on slender funicles.

A pantropical genus of about 215 species, a few occurring in temperate regions. It is here considered to include the genus *Fagara* L., with which it intergrades. (See Brizicky in J. Arnold Arb. 43:80-83. 1962).

KEY TO SPECIES

1. Inflorescence and rhachis of the leaves puberulous with stellate hairs; floral parts in 4's or 5's; leaves densely pellucid-punctate; prickles absent: **1. Z. flavum**

1. Inflorescence and other parts glabrous; floral parts in 3's; leaves not pellucid-punctate; stems often prickly: **2. Z. coriaceum**

1. Zanthoxylum flavum Vahl, Eclog. 3:48. 1807. **FIG. 179.**

Fagara flava (Vahl) Krug & Urban, 1896; Adams, 1972.

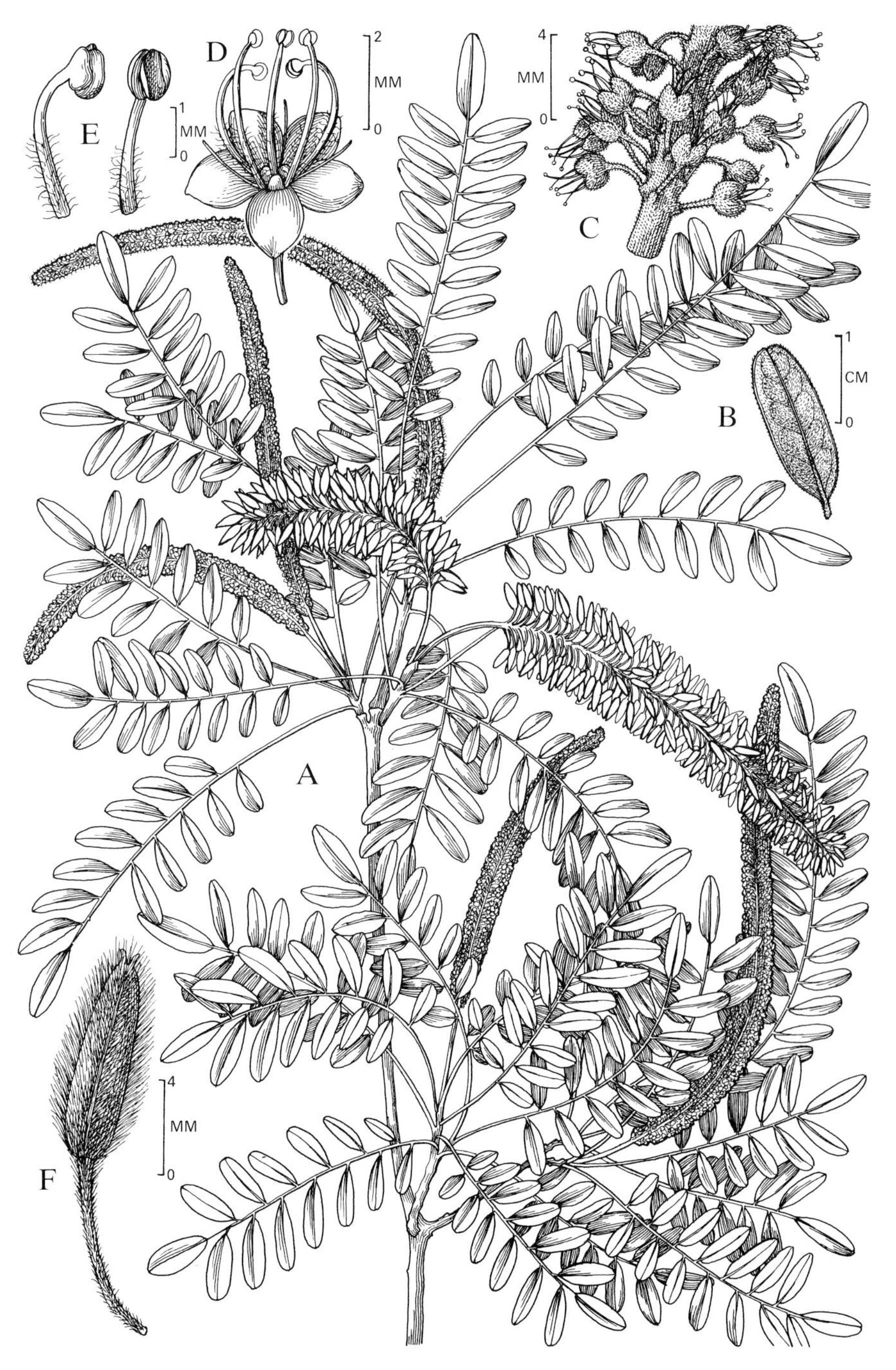

FIG. 178 **Alvaradoa amorphoides.** A, habit. B, single leaflet. C, portion of flowering raceme. D, flower. E, two stamens. F, fruit. (G.)

"Satinwood"

A small tree up to 7 m tall or more; leaves 10–30 cm long, usually with an odd leaflet; leaflets 5–11, opposite, elliptic-oblong to broadly ovate, 4–9 cm long, acute or somewhat acuminate at the apex, often inequilateral at the base, with

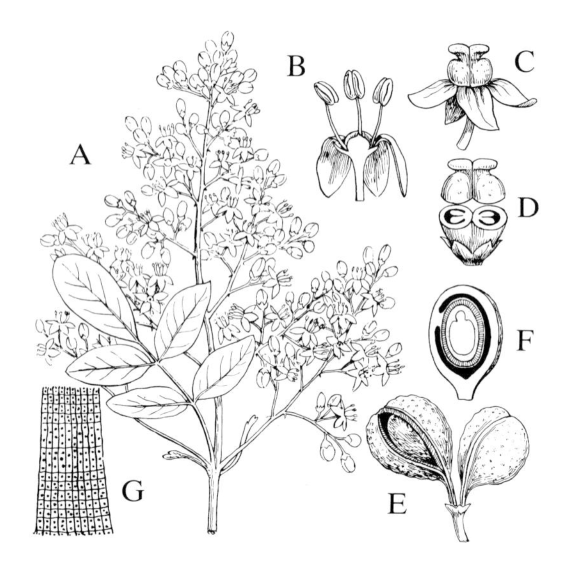

FIG. 179 **Zanthoxylum flavum.** A, inflorescence with a small leaf, x$^2/_3$. B, staminate flower cut lengthwise, x3. C, pistillate flower, x3. D, pistil cut across, x4. E, fruit of 2 cocci, one open showing the seed, x2. F, coccus cut lengthwise showing the embryo, x2. G, x-section of small portion of wood. (F. & R.)

lightly crenate margins. Sepals open in bud, stellate-puberulous, 0.5 mm long; petals greenish-white, glandular, 3–4 mm long. Fruiting cocci obovate-roundish, 4–6 mm long.

GRAND CAYMAN: *Proctor 15131.* LITTLE CAYMAN: *Proctor 35115.*
– Florida, Bermuda, and the West Indies, often in sandy woodlands. The wood is prized for cabinet-work and furniture, as it has a yellow, satiny luster, a rippled grain, and takes a high polish.

2. Zanthoxylum coriaceum, A.Rich., Ess. Fl. Cub. 326, t. 34. 1841.

"Shake Hand"

A glabrous shrub or small tree to 5 m tall, usually armed at least on young branches with sharp spines. Leaves with or more often without an odd leaflet, 7–15 cm long; leaflets 4–9 or rarely more, opposite and oblong, elliptic or obovate, mostly 3–6 cm long, blunt, rounded or emarginate at the apex, the margins entire, dark green and very shiny on the upper surface and with prominulous venation. Inflorescence a dense corymbose terminal panicle (rarely axillary). Flowers greenish cream; sepals 0.5-0.7 mm long; petals 2.5-4 mm long; ovary of 3 carpels. Fruiting cocci roundish-ellipsoid, rough, 5-6 mm long, minutely apiculate.

GRAND CAYMAN: *Kings GC 338; Millspaugh 1274; Proctor 11982.* LITTLE CAYMAN: *Proctor 35105, 35216.* CAYMAN BRAC: *Proctor 29013.*

– Florida, Bahamas, Cuba, and Hispaniola, in dry rocky woodlands at low elevations. Adams (1972, p. 386) ascribed two other species to the Cayman islands, namely *Z. cubensis* P. Wils. and *Z. spinosum* (L.) Sw. Both of these are very similar to *Z. coriaceum,* and these records may be due to misidentifications. *Z. cubense* can be distinguished by having larger leaves, more open panicles, and 1-carpellate flowers. *Z. spinosum,* which is a Jamaican endemic, has thinner leaf-tissue and smaller white flowers with hooded petals. Identifications are often difficult in this group of species.

[**Citrus aurantifolia** (Christm.) Swingle, the "Lime"; **C. aurantium** L., the "Sour Orange" or "Seville Orange"; **C. paradisi** Macf., the "Grapefruit"; **C. sinensis** (L.) Osbeck, the "Sweet Orange", and perhaps other citrus fruits, are quite often planted. All are indigenous to southeastern Asia. Only the first is likely to become naturalized in the Cayman Islands.]

[**Triphasia trifoliata** (Burm.f.) P.Wilson, a Chinese species, was reported by Hitchcock in 1893 as being naturalized on Grand Cayman. This record has not been substantiated by subsequent collectors.]

Genus 2. **AMYRIS** L.

Unarmed shrubs or trees, glabrous or pubescent; leaves opposite or alternate, 1-3-foliolate or pinnately compound; leaflets opposite, with numerous pellucid dots in the tissue. Flowers small, white, perfect, in terminal and axillary panicles; pedicels 2-bracteate; floral parts in 4's or 5's; calyx urn-shaped, toothed, persistant; petals imbricate. Stamens twice as many as the petals, inserted on the disk; filaments threadlike. Ovary 1-celled, the style short or lacking, the stigma capitate; ovules 2, pendulous. Fruit a 1-seeded drupe, usually black; seeds without endosperm; cotyledons glandular.

A tropical American genus of about 18 species.

1. Amyris elemifera L., Syst. Nat. ed. 10, 2:1000. 1759. **FIG. 180.**

"Candlewood", "White Candlewood", "Torchwood"

A shrub to 3 m tall or more, sometimes tree-like, glabrous throughout; leaves 3–5-foliolate, the leaflets stalked, lanceolate to broadly ovate, 2–6 cm long,

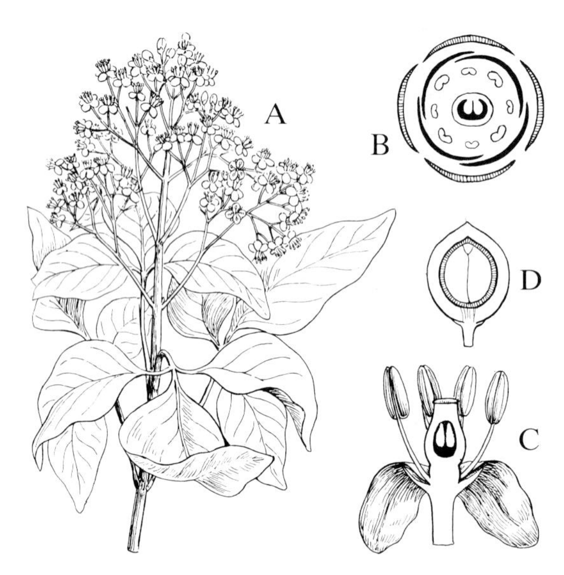

FIG. 180 **Amyris elemifera.** A, branchlet with leaves and staminate inflorescence, x$^2/_3$. B, diagram of perfect flower. C, perfect flower cut lengthwise, x6. D, drupe cut lengthwise, x2. (F. & R.)

acuminate, the margins entire or crenulate, the tissue densely dotted with resinous glands. Drupe globular, 5–8 mm in diameter.

GRAND CAYMAN: *Brunt 2001; Kings GC 342; Proctor 15163.* LITTLE CAYMAN: *Proctor 28101.* CAYMAN BRAC: *Kings CB 93; Proctor 29014.*
– Florida, West Indies and Central America, in rocky woodlands. The wood is hard, heavy, strong and close-grained; it is extremely durable and can take a brilliant polish. Most of the Cayman examples seen are too small to be of much use. Because of its high resin content, *Amyris* wood will burn when it is green.

Family 77

MELIACEAE

Trees or shrubs, the wood often scented; leaves alternate, usually pinnate, sometimes 2-pinnate, 3-pinnate, or 1–3-foliolate; leaflets entire or rarely serrate;

stipules absent. Inflorescence an axillary or terminal panicle; flowers regular, perfect or rarely polygamo-dioecious. Calyx 4-6-lobed, the lobes imbricate; petals 4-6, imbricate, convolute or valvate in bud, free or sometimes adnate to lower part of the stamen-tube. Stamens 8-10 (or 4-6 in *Cedrela*); filaments united to form an entire, toothed, or lobed tube, rarely free; anthers sessile or stalked, inserted within the mouth of the stamen-tube or on its margin, 2-celled. Disk ring-like or columnar, free or adnate to the stamen-tube or ovary. Ovary 2-5-celled, with usually 2 or more ovules in each cell; stigma disk-like or capitate. Fruit a dehiscent capsule, or rarely a drupe or berry; seeds solitary or numerous in each cell, sometimes winged, with or without endosperm.

A pantropical family of about 45 genera and 800 species, some of them valuable sources of timber.

KEY TO GENERA

1. Leaves pinnate; leaflets entire; fruit a capsule:

  2. Leaflets mostly 3 pairs; seeds not winged, partly enclosed by a red aril; capsules less than 1.5 cm long, not woody: **1. Trichilia**

  2. Leaflets 4-8 pairs; seeds winged and without aril; capsules more than 3 cm long, woody:

    3. Capsules 8-10 cm long, opening at the base; anthers sessile on the stamen-tube: **3. Swietenia**

    3. Capsules 3-4 cm long, opening at the apex; anthers on filaments free above their attachment to the column: **4. Cedrela**

1. Leaves 2-3-pinnate; leaflets serrate; fruit a drupe: [**2. Melia**]

Genus 1. **TRICHILIA** L.

Trees or shrubs; leaves odd-pinnate or rarely 1-3-foliolate; leaflets opposite or alternate, sometimes pellucid-dotted. Panicles axillary or terminal; flowers whitish or yellowish; calyx cup-like, 4-5-lobed; petals 4-5, rarely 3, free or nearly so, imbricate or valvate. Stamens usually 8 or 10; filaments broadly winged and united more or less into a tube; anthers inserted in notches at apex of the filaments, erect and exserted. Ovary 2-3-celled, more or less immersed in a disk, with usually (rarely 1) ovules in each cell. Fruit a 2-3-celled capsule, dehiscent from the apex; seeds 1 or 2 in each cell, subtended by a fleshy red aril; endosperm lacking.

A rather large genus of about 200 species occurring in tropical America and Africa.

KEY TO SPECIES

1. Panicles corymbose, at ends of the branches; petals 5-8 mm long; capsules tomentose: **1. T. glabra**

1. Panicles umbel-like, crowded, axillary; petals 3-3.5 mm long; capsules glabrous: **2. T. havanensis**

1. Trichilia glabra L., Syst. Nat. ed. 10, 2:1020. 1759.

A small tree to 8 m tall or more; leaves mostly with 3 pairs of leaflets and a terminal one, these elliptic or ovate, 3-7 cm long, bluntly acuminate at the apex, glabrous or nearly so, a membranous expansion occurring in the axils of the nerves beneath, this often minutely hairy. Inflorescence puberulous, usually 4-10 cm long; ovary pubescent. Capsules globular, greenish, tomentose, 1-1.5 cm long.

GRAND CAYMAN: *Proctor 11983.* LITTLE CAYMAN: *Proctor 28028, 35095, 35101.* CAYMAN BRAC: *Proctor 15327, 29019.*
– Jamaica, the Swan Islands and Cozumel, in rocky thickets and woodlands.

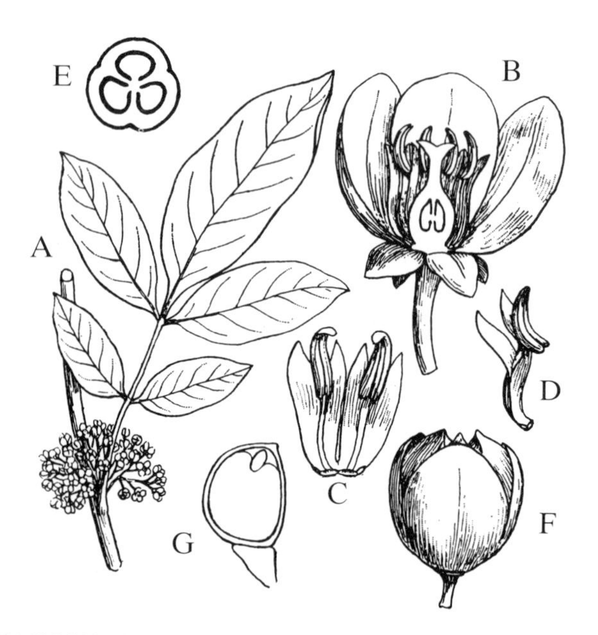

FIG. 181 **Trichilia havanensis.** A, leaf and inflorescence, x½. B, flower cut lengthwise, x7. C, two stamens, x10. D, stamen, side view, x10. E, x-section of ovary, x14. F, capsule, x2. G, long.section of seed, x2. (F. & R.)

2. **Trichilia havanensis** Jacq., Enum. Pl. Carib. 20. 1760. **FIG. 181.**

A shrub or small tree to 5 m tall or more; leaves mostly with 3 pairs of leaflets and a terminal one, these mostly elliptic, 3–10 cm long or more, bluntly acuminate at the apex, glabrous except near the point of attachment, the nerve-axils beneath without a membranous expansion. Inflorescence less than 3 cm long; ovary glabrous. Capsule globular, dark brown, glabrous, c. 1 cm long.

GRAND CAYMAN: *Brunt 1935; Hitchcock.*
– Cuba, Jamaica, and continental tropical America, in a variety of habitats. Cayman examples were found in moist thickets near Georgetown.

[Genus 2. **MELIA** L.

Trees or shrubs; leaves pinnate or 2–3-pinnate with a terminal leaflet. Panicles axillary, ample, with numerous flowers; flowers perfect, violet or purple. Calyx 5-6-lobed, imbricate in bud; petals 5–6, convolute, spreading. Stamen-tube long and slender, toothed at the apex; anthers 10–12, sessile within the apex of the tube. Disk ring-like. Ovary 5-6-celled with 2 ovules in each cell; style long and slender, the capitate stigma 5-6-lobed. Fruit a somewhat fleshy drupe, the stone 1–6-celled with 1 seed in each cell; seeds with fleshy endosperm.

A genus of 10 or 12 species indigenous to tropical Asia and Australia.

1. **Melia azederach** L., Sp. Pl. 1:384. 1753.

"Lilac"

A bushy ornamental shrub, sometimes arborescent, the young growing parts minutely stellate-puberulous; leaves 2-pinnate or nearly 3-pinnate, up to 40 cm long, the pinnae opposite in 2–5 pairs and a terminal one; ultimate leaflets lanceolate, 2–6 cm long, with serrate margins, glabrous. Sepals 1.5–2 mm long, puberulous; petals 7–9 mm long, pale violet or white; stamen-tube c. 6 mm long, deep purple. Drupe ellipsoid, c. 1.5 cm long, yellow, the stone deeply grooved.

GRAND CAYMAN: *Kings GC 297; Lewis 15; Maggs II 57; Millspaugh 1354.*
CAYMAN BRAC: *Proctor 29124.*
– Native of tropical Asia, but now cultivated and naturalized throughout the tropics. The fruits are said to be poisonous, and together with the dried leaves are used in India to protect clothing, etc., from insect attack. The "seeds" are used to make beads.]

Genus 3. **SWIETENIA** L.

Medium-sized to very large trees with dark red wood; leaves even-pinnate, glabrous, the leaflets opposite, entire, and strongly unequal-sided. Panicles axillary and subterminal, much shorter than the leaves; flowers small, white, polygamo-monoecious; calyx 5-lobed, the lobes imbricate; petals 5, convolute in bud. Stamen-tube urn-shaped and with 10 teeth, the anthers attached internally between the teeth. Disk saucer-shaped. Ovary 5-celled with 10–14 ovules in each cell, attached to the central axis; stigma disk-like. Fruit a woody capsule, 5-valved and dehiscent from the base, leaving a persistent 5-winged central axis; seeds numerous, each with a terminal oblong wing, overlapping downward in 2 rows in each cell; endosperm fleshy.

A tropical American genus of 3 species, all notable for their extremely valuable wood.

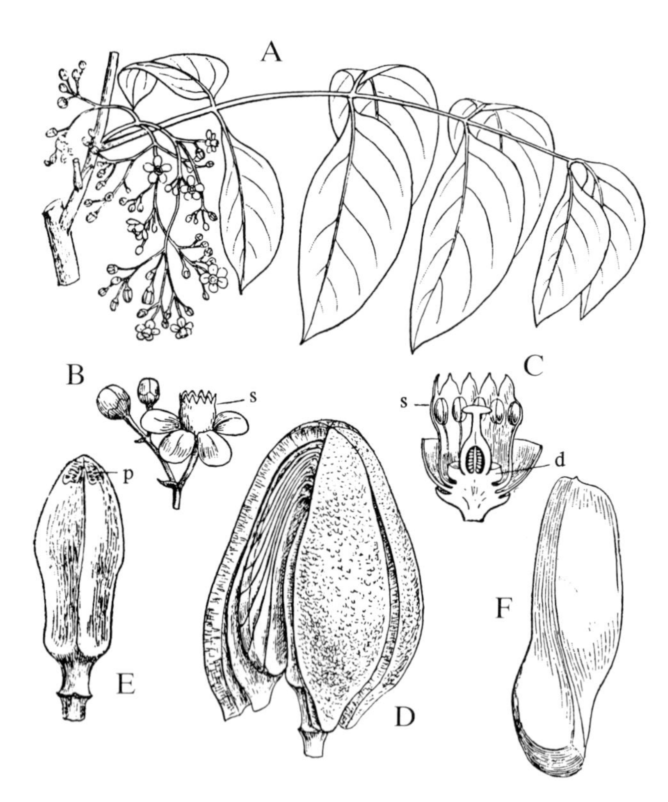

FIG. 182 **Swietenia mahagoni.** A, leaf and inflorescence, x½. B, buds and open flower, x1½; s, staminal tube. C, perfect flower cut lengthwise, x5; s, staminal tube; d, disk. D, fruit, ripe and splitting open (one valve removed), x½. E, central axis of fruit, x½; p, attachment-points of seeds. F, seed, x$^2/_3$. (F. & R.)

1. Swietenia mahagoni Jacq., Enum. Pl. Carib. 20. 1760. **FIG. 182.**

"Mahogany"

A tree to 10 m tall or more, with a rounded crown; leaves up to 15 cm long, the leaflets mostly in 4 pairs, recurved-lanceolate or -ovate, 3–6 cm long, sharply acuminate. Calyx c. 1 mm long or less, with rounded glabrous lobes; petals and stamen-tube c. 3 mm long. Seeds brown, c. 6 cm long.

GRAND CAYMAN: *Brunt 1650; Hitchcock; Kings GC 246; Proctor 15135, 27989.* LITTLE CAYMAN: *Kings LC 5; Proctor 28179.* CAYMAN BRAC: *Proctor (sight record).*
– Florida and the West Indies only; reports of this species from Central and South America pertain to *S. macrophylla* G. King. In the Cayman Islands *S. mahagoni* occurs chiefly in rocky woodlands, where it could formerly be found with massive trunks up to 1 m in diameter or more. Nearly all of these large, very old trees have long since been cut.

Genus 4. **CEDRELA** L.

Large trees with aromatic reddish wood; leaves even-pinnate, the leaflets in many pairs, opposite or subopposite, entire and more or less unequal-sided. Panicles large, terminal; calyx 4–5-lobed; petals 4–5, erect, keeled on the inner side toward the base, the keel adherent to the disk. Disk elevated and forming a column, 4–6-lobed. Stamens 4–6, adherent to the column and extending free beyond it. Ovary sessile at the apex of the column, 5-celled, with 8–12 ovules in each cell, in two series pendent from the axis; stigma disk-like. Fruit a woody capsule, 5-valved and dehiscent apically almost to the base, separating from a persistent 5-winged axis; seeds pendulous, each with a terminal wing, overlapping downward in 2 rows in each cell; endosperm scanty.

A tropical American genus of 15 species.

1. Cedrela odorata L., Syst. Nat. ed. 10, 2:940. 1759. **FIG. 183.**

Cedrela mexicana M.J.Roem., 1846.

"Cedar"

A tree up to 18 m tall or more, with a long straight trunk; leaves deciduous, renewed at time of flowering, usually 30–50 cm long, the leaflets in 7 or 8 pairs, narrowly oblong-ovate, 5–11 cm long or more, more or less long-acuminate. Panicles often 30–35 cm long, open and lax; flowers unpleasantly-scented, greenish-cream; calyx c. 1.5 mm long; petals c. 6 mm long, densely puberulous. Seeds light brown, 2 cm long.

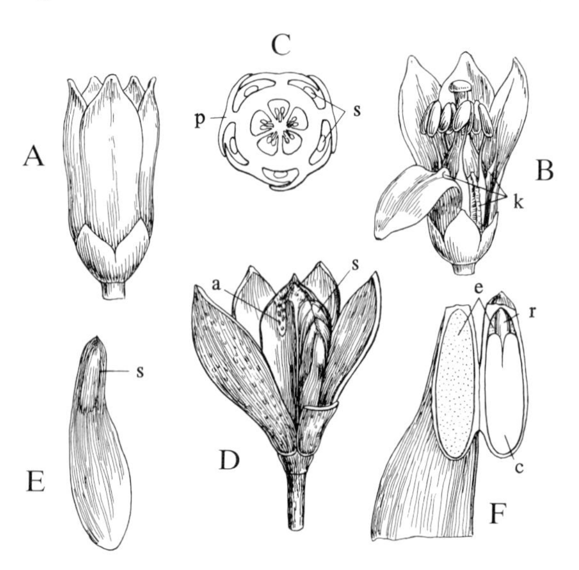

FIG. 183 **Cedrela odorata**. A, flower, x4. B, same with one petal removed; k, keel of petal attached to the ovary. C, same cut across, enlarged; s, filaments; p, petal. D, capsule with one valve cut away, x$^2/_3$; s, seeds; a, placenta with seeds removed. E, seed, x1; s, position of embryo. F, seed cut lengthwise, x2; c, cotyledon; e, endosperm; r, radicle. (F. & R.)

GRAND CAYMAN: *Kings GC 337; Proctor 31033.* CAYMAN BRAC: *Proctor 29368.*

– West Indies and continental tropical America, in various habitats. The rather soft but durable, handsome wood is used for many purposes, and is one of the most valuable of all tropical timbers. It is virtually impervious to attacks by insects.

Family 78

ZYGOPHYLLACEAE

Shrubs, trees, or annual or perennial herbs, frequently more or less resinous; leaves opposite, or alternate by abortion of one of each pair, petiolate, even-pinnate or in some cases simple or digitately compound; stipules small or minute. Flowers solitary or clustered in the leaf-axils, regular or nearly so, perfect; sepals usually 5 and free, usually imbricate; petals 5, free, imbricate, valvate or convolute. Stamens 10, free, in two series, the outer ones larger than the inner; anthers opening inwardly. Ovary of 2–5 united carpels, sessile or short-stalked, the styles united into one, terminated by a compound stigma; ovules 1-several in each carpel, pendulous or ascending. Fruit a capsule, or else separating into few or several often spiny nutlets, these 1- or few-seeded; seeds with or without endosperm, the cotyledons fleshy.

A widespread tropical and subtropical family of about 20 genera and 200 species.

KEY TO GENERA

1. Leaflets mostly in 6–9 pairs; sepals deciduous; fruit of 5 spiny nutlets: **1. Tribulus**

1. Leaflets mostly in 3–4 pairs; sepals persistent; fruit of 10–12 small tuberculate nutlets surrounding the persistent beak-like style-axis: **2. Kallstroemia**

Genus 1. **TRIBULUS** L.

Prostrate annual or perennial herbs; leaves opposite, one of each pair alternately smaller than the other or sometimes lacking; stipules obliquely lanceolate. Flowers solitary on long axillary peduncles; sepals soon falling; petals 5, yellow or orange, much larger than the sepals. Stamens 10, the 5 inner ones shorter and glandular at the base. Ovary sessile, 5-celled, surrounded at the base by an urn-shaped 10-lobed disk; ovary-cavities divided by transverse partitions into 3–5 compartments with 1 ovule in each. Fruit depressed, 5-angled, consisting of 5 bony nutlets which ultimately separate, each bearing spines, and each containing 3–5 seeds; seeds without endosperm.

A pantropical genus of about 10 species.

1. **Tribulus cistoides** L., Sp. Pl. 1:387. 1753. **FIG. 184.**

"Buttercup", "Jim Carter Weed"

Herb with prostrate stems up to 50 cm long or more, densely pubescent; leaves white-pubescent beneath; leaflets oblong, 4–17 mm long, unequal-sided; stipules 4–9 mm long. Peduncles arising from axils of the shorter leaves and longer than them; sepals 7–9 mm long; petals yellow, obovate, 1.5–2.5 cm long. Fruit 6–9 mm long, each nutlet with 2 longer spines at the top and 2 shorter ones near the base.

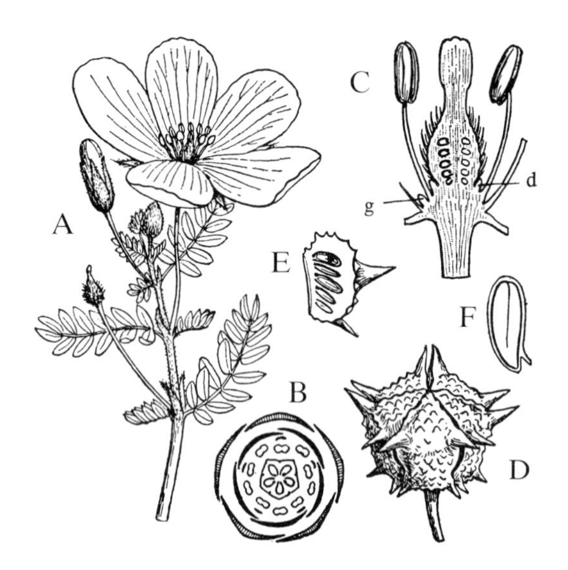

FIG. 184 **Tribulus cistoides.** A, flowering branch, x²/₃. B, floral diagram. C, flower cut lengthwise, calyx and corolla removed, x4, showing pistil with hypogynous disk (d), two stamens, and one staminal gland (g). D. fruit somewhat enlarged. E, coccus cut lengthwise, enlarged. F, seed cut lengthwise, enlarged. (F. & R.)

GRAND CAYMAN: *Brunt 2008; Kings GC 225; Proctor 15157.* LITTLE CAYMAN: *Kings LC 57; Proctor 28172.* CAYMAN BRAC: *Proctor 29120.* – Southern United States south to Guatemala, West Indies, and South America, usually in open sandy ground at low elevations.

Genus 2. **KALLSTROEMIA** Scop.

Prostrate or ascending, annual or perennial herbs; leaves opposite, one of a pair often alternately smaller than the other or sometimes lacking; stipules linear. Flowers solitary on axillary peduncles; sepals 5-6, persistent; petals 4-6, orange or yellow, soon falling; stamens 10–12, of unequal size, the smaller ones glandular at the base. Ovary sessile, 10–12-celled, with 1 pendulous ovule in each cell; styles united into a beak-like column persistent in fruit. Fruit of 10-12 hard, tuberculate, indehiscent, 1-seeded nutlets, at maturity these falling away from the persistent central axis; seeds without endosperm.

A tropical American genus of 12 or more species.

1. Kallstroemia maxima (L.) Torr. & Gray, Fl. N. Amer. 1:213. 1838. **FIG. 185.**

"Parsley"

A trailing herb, the branched stems up to 50 cm long or more, sparsely pubescent; leaves glabrate; leaflets elliptic or oblong-elliptic, somewhat unequal in size, the apical ones largest, 1-2 cm long, unequal-sided; stipules 4-5 mm long.

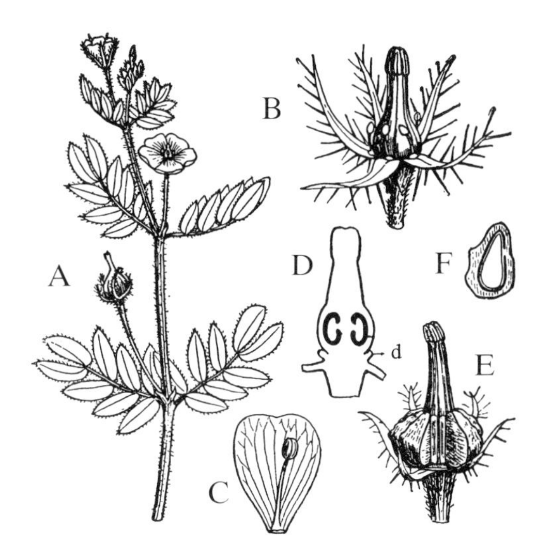

FIG. 185 **Kallstroemia maxima.** A, flowering branch, x$^{2}/_{3}$. B, flower with the petals and five larger stamens adhering to them removed, x5. C, petal with its adhering stamen, x2. D, pistil cut lengthwise, x5; d, hypogynous disk. E, fruit with calyx, two cocci detached, x3. F, coccus cut lengthwise, x2. (F. & R.)

Peduncles shorter or longer than the leaves; sepals 4-6 mm long; petals cream or pale yellowish, 6-8 mm long. Fruit 8-9 mm long, the nutlets glabrous, 4-5 mm long, tuberculate-ridged.

GRAND CAYMAN: *Brunt 2113; Kings GC 403; Millspaugh 1342; Proctor 27939.*
– Southern United States, West Indies, and continental tropical America, a common weed of roadsides, fields, and open waste ground.

[**Guaiacum officinale** L., the "Lignum Vitae", is occasionally seen as a cultivated plant in Grand Cayman. It is a small tree with dense rounded crown, blue flowers, and orange fruits. It has a wide natural distribution in the West Indies and Central America, and its absence from the Cayman Islands as a native species is surprising, since the ecological conditions in these islands would seem ideal for its growth.]

Family 79

ERYTHROXYLACEAE

Glabrous shrubs or trees; leaves alternate, simple, and entire; stipules present, often overlapping on young branches. Flowers small, solitary or clustered in the leaf-axils or at defoliated nodes, regular, perfect, often heterostylous; pedicels bracteate and usually angled. Calyx with 5 imbricate lobes, persistent; petals 5, free, deciduous, usually with a 2-lobed appendage (the ligule) on the inner side. Stamens 10 in 2 series, their filaments united below into a tube. Ovary superior, 3-celled, mostly with 2 of the cells sterile, the fertile cell with 1 (rarely 2) pendulous ovule; styles 3, usually each ending in a small capitate stigma. Fruit a drupe, usually 1-seeded; seed with or without farinaceous endosperm.

A pantropical family of 2 genera and over 200 species, all but one in the genus *Erythroxylum.*

Genus 1. **ERYTHROXYLUM** P. Browne

With characters of the family. The best-known member of this genus (and family) is *Erythroxylum coca* Lam. of the Andean region in South America, from whose leaves is obtained the drug cocaine.

KEY TO SPECIES

1. Leaves 3-7 cm long or more, with 2 delicate longitudinal lines demarcating a central zone on the underside; fruits 7-9 mm long: **1. E. areolatum**

1. Leaves 0.5-2.5 cm long, lacking longitudinal zones; fruits 5-6 mm long: **2. E. rotundifolium**

1. Erythroxylum areolatum L., Syst. Nat. ed. 10, 2:1035. 1759. **FIG. 186.**

"Smoke Wood"

A shrub or small tree to 8 m tall; leaves broadly elliptic, the apex often shallowly notched (retuse); venation finely reticulate and prominulous, especially on the upper side; deciduous, the flowers appearing with the new leaves in April and May.

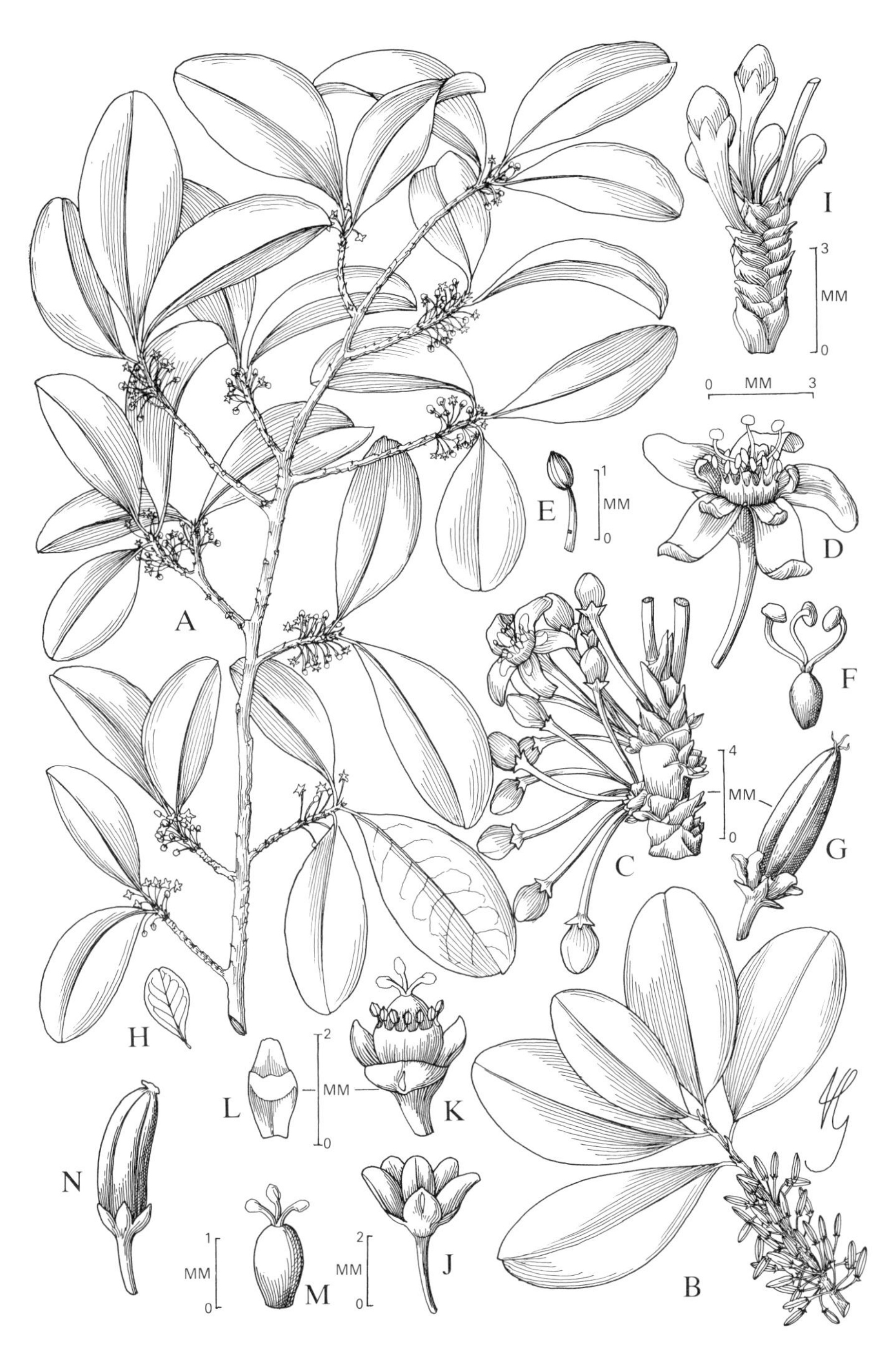

FIG. 186 **Erythroxylum areolatum.** A, habit, flowering branch. B, fruiting branch. C, portion of flowering branch. D, flower. E, stamen. F, ovary, styles and capitate stigmas. **G, fruit. Erythroxylum rotundifolium.** H, leaf. I, portion of flowering branch (buds only). J, flower. K, flower with petals removed. L, sepal. M, ovary, styles and stigmas. N, fruit. (G.)

Flowers white, numerous, fragrant, heterostylous; pedicels up to 7 mm long, enlarged toward the apex. Calyx 1.5–2 mm long; petals 2.5–3 mm long. Drupes red, subtended by the persistent staminal tube and calyx.

GRAND CAYMAN: *Brunt 1754, 2000, 2153; Kings GC 204a, GC 333; Proctor 11984, 15296.* LITTLE CAYMAN: *Proctor 35092.*
– Bahamas, Greater Antilles, Mexico and northern Central America, in rocky woodlands. The wood is reddish-brown, hard, heavy, fine-textured, and very durable.

2. Erythroxylum rotundifolium Lunan, Hort. Jam. 2:116. 1814. **FIG. 186.**

A shrub to 3 m tall or more; leaves thin, obovate to roundish, the apex often shallowly notched, the venation prominulous on upper side but not as finely reticulate as in *E. areolatum*, and apparently not deciduous all at one time, the flowers appearing among mature leaves at various seasons but especially in July. Flowers white, solitary or few together; pedicels usually not over 4 mm long; calyx c. 1 mm long; petals 1.5–2.5 mm long. Drupes red.

GRAND CAYMAN: *Brunt 2185; Correll & Correll 51029; Proctor 15175, 15205.* LITTLE CAYMAN: *Kings LC 63; Proctor 28106, 28187.* CAYMAN BRAC: *Proctor 29022.*
– Bahamas, Greater Antilles and Yucatan, in rocky thickets and woodlands. Neither of the Cayman species of *Erythroxylum* contains any cocaine.

Family 80

MALPIGHIACEAE

Mostly trees or shrubs, often scandent, the pubescence most often of "malpighiaceous" hairs, i.e., hairs attached at or near the middle, thus 2-armed or flattened T-shaped; these hairs in some species stiff and sharp, easily detached, and very irritating to the skin; leaves usually opposite and simple, often glandular either on the petiole, near the midrib, or on the leaf-margins; stipules present. Flowers mostly perfect and regular, in axillary or terminal racemes or panicles, rarely solitary; calyx 5-lobed, often bearing large, sessile glands; petals 5, usually concave, narrowed below into a claw. Stamens 5 or 10, the filaments more or less united at the base. Ovary superior, of 3 free carpels or these more or less united into a 3-celled ovary with 1 ovule in each cell; styles 3, free or united into a single style with 3-lobed stigma. Fruit a drupe containing one 3-celled stone or 1–3 separate 1-celled stones, or else a capsule or samara.

A pantropical family of perhaps 60 genera and about 850 species, best represented in the American tropics. Several South American species of this family are known to contain alkaloids with hallucinogenic properties.

KEY TO GENERA

1. Flowers pink or white; styles free; drupe containing crested stones; plant (in our single species) densely clothed with stinging hairs: **1. Malpighia**

1. Flowers yellow; styles united into 1; drupe containing smooth stones; plant without stinging hairs: **2. Bunchosia**

Genus 1. **MALPIGHIA** L.

Erect shrubs or small trees; leaves without glands, sometimes with stinging hairs, the margins entire or toothed; stipules minute, deciduous. Flowers in axillary or terminal umbellate clusters or corymbs, rarely solitary; calyx with 6–10 glands; petals pink, red or white, unequal, often fringed or ciliate at the apex. Filaments and ovary glabrous; filaments and styles free. Ovary 3-lobed. Fruit a red or orange 3-stoned drupe, each stone with 3–5 dorsal crests.

A tropical American genus of about 45 species.

1. Malpighia cubensis Kunth in H.B.K., Nov. Gen. & Sp. 5:145. 1821. **FIG. 187.**

Malpighia angustifolia of Hitchcock, 1893, not L., 1762.

"Lady Hair"

A shrub to 2 m tall; leaves narrowly oblong, 1.5–2.5 cm long or more, obtuse or acute, clothed on both sides with numerous yellowish, stiff, stinging malpighiaceous hairs. Flowers 1 or few in a glabrous umbellate cluster, the pedicels 7–8 mm long, enlarged toward the apex; calyx 2.5–3 mm long, completely covered by the glands; petals white or pale pink, 6–7 mm long, long-clawed. Drupes subglobose, 8–9 mm in diameter.

GRAND CAYMAN: *Brunt 2073; Hitchcock; Howard & Wagenknecht 15025; Kings GC 113; Proctor 15045.*
– Cuba, in rocky or sandy thickets.

[*Malpighia emarginata* Sessé & Moç. ex DC. (often called *M. punicifolia*) the

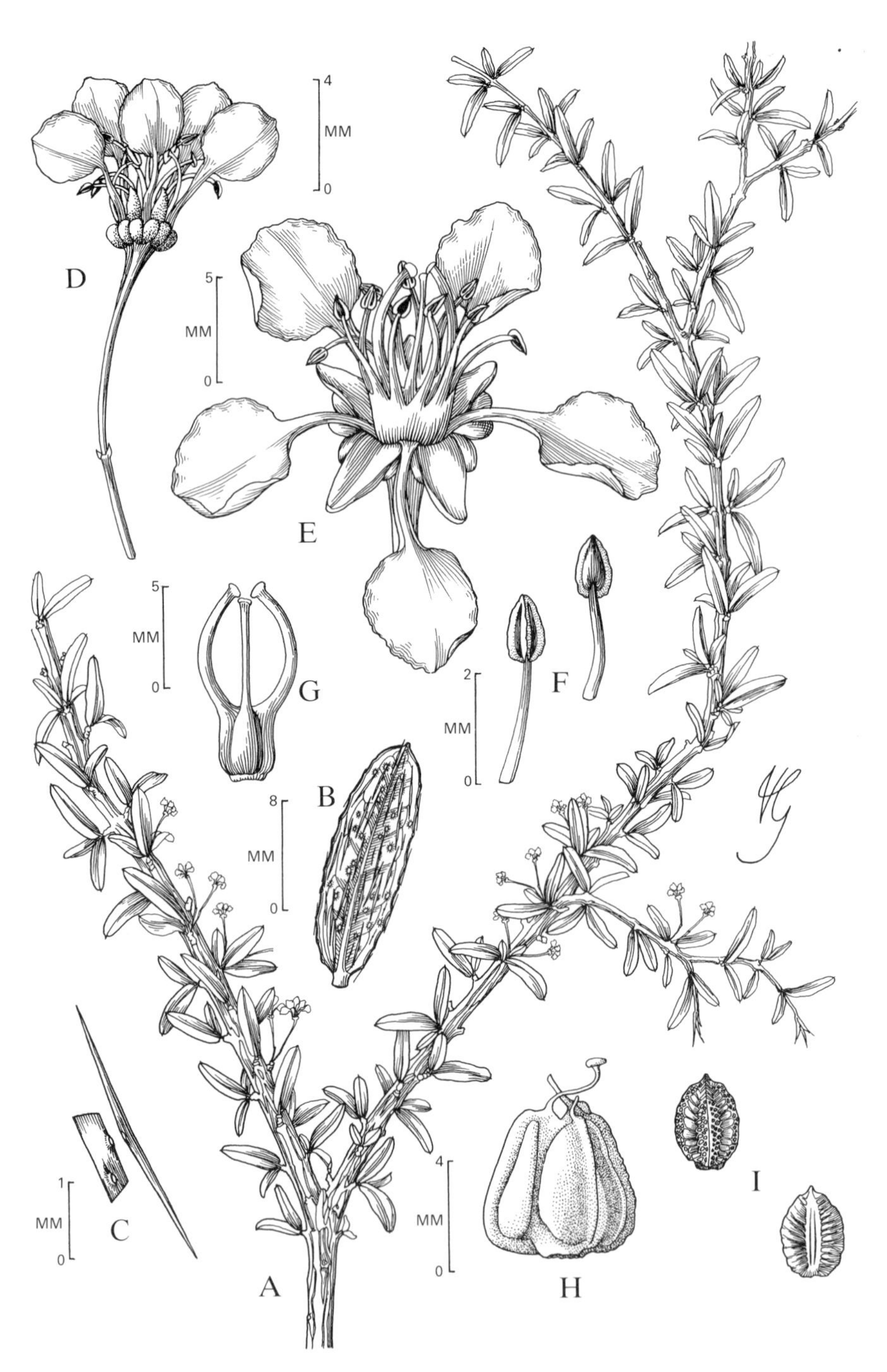

FIG. 187 **Malpighia cubensis**. A, habit. B, underside of single leaf, showing 2-armed hairs and scars where others have broken off. C, a single 2-armed hair and its point of attachment. D, habit of flower; the portion of stalk above the bracteoles is the pedicel. E, flower enlarged. F, two stamens. G, ovary, style and stigmas. H, fruit. I, two views of stone. (G.)

West Indian "Cherry", may occur as a cultivated plant, but was not seen during the preparation of the present book. It is a large, glabrous shrub with pink flowers. The red, edible fruits are one of the richest known sources of vitamin C.]

Genus 2. **BUNCHOSIA** L.C. Rich.

Shrubs or small trees, glabrate or pubescent; leaves usually with 2 glands on the lower surface a little above the base near the midrib; stipules minute, deciduous. Flowers in axillary racemes; calyx with 8-10 glands; petals yellow. Filaments glabrous, united at the base. Ovary 2-3-celled; styles united, terminating in a triangular stigma. Fruit a red, 2-3-stoned drupe, the stones smooth and uncrested.

A tropical American genus of about 40 species.

1. **Bunchosia media** (Ait.) DC., Prodr. 1:581. 1824.

Bunchosia swartziana Griseb., 1859.

A shrub to 4 m tall or more, rarely a small tree; leaves glabrous or nearly so, elliptic, 3.5-6 cm long or more, bluntly acuminate. Racemes usually shorter than the leaves; pedicels puberulous; calyx c. 2 mm long; petals 7-9 mm long, long-clawed. Drupes somewhat ellipsoid and 2-3-lobed, 8 mm long, orange.

GRAND CAYMAN: *Brunt 2183.*
– Cuba, Jamaica and Hispaniola, in dry rocky woodlands.

Family 81

POLYGALACEAE

Herbs, shrubs, or trees, sometimes woody vines, often with glands in the leaf-tissue, flowers and fruits; leaves alternate, opposite or whorled, simple and entire, with very short petioles; stipules absent. Flowers perfect, irregular (zygomorphic), in racemes or panicles; sepals 5, free or the lower 2 united, the two lateral ones much larger than the others and petaloid, forming the "wings"; petals 3 or rarely 5, the anterior one forming a boat-shaped "keel". Stamens usually 8, the filaments united for most of their length into a "sheath", this split on the upper side and usually united at the base with the keel or the upper petals or both; anthers mostly confluently 1-celled, opening by a subterminal pore. Ovary superior, usually 1-2-celled with 1 style and a 2-lobed often fringed stigma; ovules solitary or rarely 2-6, pendulous. Fruit a capsule, drupe, or samara; seeds usually pubescent, with an aril and with endosperm.

A cosmopolitan family of 12 genera and about 800 species.

Genus 1. **POLYGALA** L.

Erect herbs, shrubs, or trees; leaves alternate, opposite, or whorled. Flowers usually small and white, pink, or purple, in racemes or rarely umbellate; petals 2 upper ones and a boat-shaped keel, the latter clawed, sometimes 3-lobed and usually with an apical beak or crest. Ovary 2-celled with solitary ovules; style slender, bent, more or less excavate at the apex, the stigma 2-lobed. Fruit an equally or unequally 2-celled capsule, often winged or marginate, compressed contrary to the partition, splitting at the margin.

A very widely distributed variable genus of more than 500 species.

1. Polygala propinqua (Britton) Blake in Contr. Gray Herb. n.s. 47:16, t. 1, fig. 5. 1916.

A shrub to 2.5 m tall or more, with pale, brittle, densely puberulous branchlets; leaves broadly elliptic or oval, 2–3.5 cm long, glabrate above, minutely punctate-strigillose beneath. Inflorescences axillary, very short, few-flowered, puberulous; flowers whitish or greenish-white; sepals 1–1.3 mm long, the wings up to 1.5 mm long; keel c. 3 mm long. Capsule minutely pubescent or subglabrous, transversely oblong, lobed about 1/3 its length, the lobes divergent, 7–7.5 mm long, 9.5–11 mm wide; seeds 5.3 mm long.

LITTLE CAYMAN: *Proctor 28104, 35087.*
– Cuba, in rocky or sandy thickets or woodlands.

Family 82

UMBELLIFERAE

Annual or perennial herbs with hollow stems, rarely subwoody or arborescent; leaves alternate or rarely opposite, the petiole usually dilated at the base into a sheath; blades simple or usually compound, often more or less finely dissected; stipules usually lacking. Inflorescence umbellate or sometimes head-like; umbels simple or more often compound, terminal or lateral, sometimes numerous in a panicle, the individual umbellate clusters often subtended by an involucre of bracts. Flowers small, regular or sometimes irregular, perfect or often polygamo-monoecious, protandrous (staminate elements maturing before the pistillate); calyx represented by 5 small teeth around the upper edge of the ovary, or these absent; petals 5, inserted at the apex of the ovary, equal or the outer ones enlarged. Stamens 5, free, the filaments inflexed in bud. Ovary inferior, 2-celled, crowned by a conspicuous disk; styles 2, distinct; ovules solitary in each cell, pendulous, anatropous. Fruit dry, crowned by the persistent disk and styles, and consisting of 2 indehiscent 1-seeded "mericarps", these eventually separating; mericarps bearing longitudinal ribs sometimes extended into wings, with oil-tubes between or under them, rarely lacking; seeds with cartilaginous endosperm and small embryo.

A world-wide family of more than 200 genera and 2,850 species, best represented in temperate regions and on tropical mountains. The family includes numerous important cultivated plants, chiefly used for food or condiment. Among the best-known are Celery (*Apium graveolens* L.), Carrot (*Daucus carota* L.), Parsley (*Petroselinum crispum* (Mill.) Nyman), and Anise (*Pimpinella anisum* L.). Some other members of this family are poisonous, among the most notorious being *Conium maculatum*, the "Poison Hemlock" famous in the history of Greece and other Mediterranean countries; it is supposed to be the plant by which Socrates was put to death.

KEY TO GENERA

1. Leaves and umbels simple: **1. Centella**

1. Leaves and umbels compound: **[Anethum]**

Genus 1. **CENTELLA** L.

Perennial herbs with prostrate, rooting stems bearing clusters of long-petiolate leaves at the nodes; leaves with simple blades, palmately-veined, cordate at the base. Inflorescence of simple, subcapitate, few-flowered umbels on long, axillary peduncles; subtending bracts 2; calyx-teeth absent; styles short, hair-like. Fruits laterally compressed, prominently ribbed; oil-tubes lacking.

A widespread genus of about 20 species, the majority in Africa.

1. Centella asiatica (L.) Urban in Mart., Fl. Bras. 11, pt. 1:287. 1879. **FIG. 188.**

Hydrocotyle asiatica L., 1753.

Centella erecta (L.f.) Fernald, 1944.

A creeping, glabrous herb with somewhat fleshy stems and leaves; leaves with petioles up to 22 cm long or more, usually much longer than the blades; blades broadly ovate to orbicular, mostly 2–5 cm long, rounded at the apex, shallowly cordate at the base, the margins repand-dentate. Peduncles shorter than the leaves; umbels headlike, 2-4-flowered, the flowers sessile or subsessile. Fruits 3-4 mm long and slightly broader, ribbed and reticulate.

GRAND CAYMAN: *Brunt 1952, 1976; Proctor 27978.*
– Pantropical, usually in swampy meadows or areas of impeded drainage.

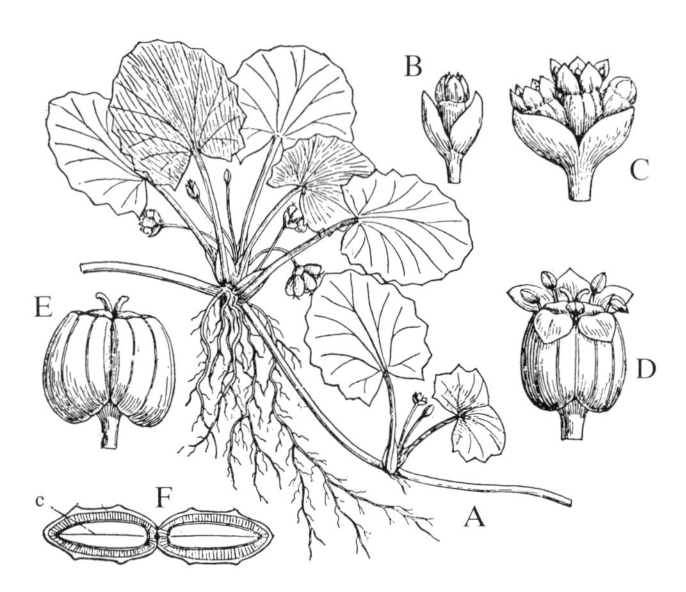

FIG. 188 **Centella asiatica.** A, portion of plant bearing flowers and fruit, x1. B, young inflorescence, x2½. C, inflorescence with lateral flowers developed, x2½. D, flower, x4. E, fruit, x4. F, x-section of fruit, x7; c, cotyledons. (F. & R.)

[**Anethum graveolens** L., the "Dill", has been recorded as cultivated and possibly escaped near Georgetown, Grand Cayman (*Kings GC 203, GC 207*). Both the foliage and the fruits of this species are much used for flavouring food, especially in the United States (e.g., "Dill pickles"). It probably cannot be expected to thrive in the climate of the Cayman Islands.]

Family 83

LOGANIACEAE

Herbs, shrubs, vines or trees; leaves opposite or rarely whorled, simple, entire or toothed; stipules present or the leaf-bases connected by a transverse line. Flowers regular and perfect, usually arranged in dichotomous cymes or in panicles; calyx 4-5-parted, usually short, with imbricate segments; corolla gamopetalous, 5-lobed.

Stamens 5, inserted in the corolla-tube and alternate with the lobes; filaments short; anthers 2-celled, opening inwardly. Ovary superior, usually 2-celled; ovules usually many in a cell, attached to the axis; style simple or forked. Fruit a 2-celled, many-seeded capsule, rarely berrylike or drupelike and indehiscent; seeds variable, sometimes winged; endosperm usually copious.

A rather heterogeneous family of about 32 genera and nearly 800 species, widespread in tropical and warm-temperate regions. By some authors it is divided into 6 smaller families. Among well-known plants of this complex are species of *Buddleja,* widely planted for ornament, and the numerous species of *Strychnos,* whose seeds yield the poisonous alkaloid strychnine.

Genus 1. **SPIGELIA** L.

Annual or perennial herbs, glabrous or pubescent, the stems terete or 4-angled. Leaves entire, decussate or in whorls at the top of the stems. Inflorescence usually terminal, consisting of 1-several 1-sided spikes or sometimes short spikes situated in the forks of branches, the individual flowers sessile or nearly so. Calyx 5-parted, with 2 or more linear glands at the inner base of each lobe; corolla funnel-shaped. Style filiform, simple, jointed near the middle, the upper portion quickly deciduous. Capsule 2-lobed, circumscissile above the persistent, cup-shaped base; seeds few, top-shaped, ellipsoid or ovoid, variously sculptured.

An American genus of about 50 species.

1. **Spigelia anthelmia** L., Sp. Pl. 1:149. 1753. **FIG. 189.**

A small erect annual herb, simple or few-branched, seldom more than 30 cm tall; lower leaves opposite, the upper ones larger and in a whorl of 4 (actually 2 closely decussate pairs) subtending the inflorescence, 3-9 cm long or more, minutely scabrid on the upper side and margins. Spikes 1-several in a terminal cluster, each spike usually 3-10 cm long, with 10-20 flowers, rarely less. Calyx-lobes 2-3 mm long; corolla usually pale pink, 5-10 mm long. Capsule 3-5 mm long, 4-6 mm broad, finely muricate; seeds 12-15 per capsule.

GRAND CAYMAN: *Brunt 1940, 1949; Kings GC 86.* CAYMAN BRAC: *Proctor 28949.*

– Florida, West Indies and continental tropical America; naturalized in tropical Africa and Indonesia, a weed of fields and open waste ground. The cut and wilted plant is extremely poisonous and may cause fatalities among livestock; the fresh plant is apparently less dangerous.

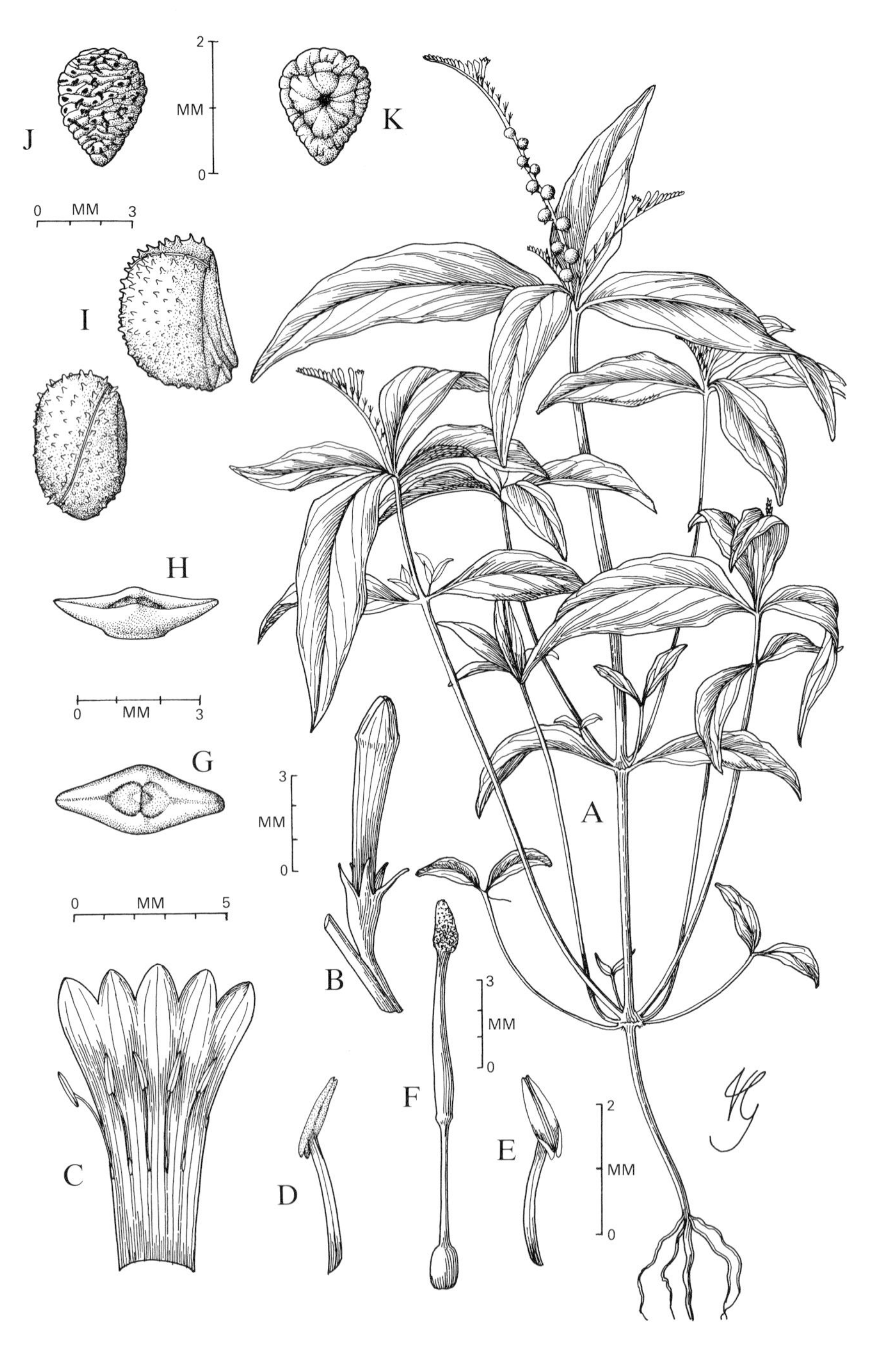

FIG. 189 **Spigelia anthelmia.** A, habit. B, flower-bud. C, corolla opened out to show stamens. D, E, two views of stamen. F, ovary, style and stigma. G, H, two views of persistent base of fruit. I, two views of fruit. J, K, two views of seed (G.)

Family 84

GENTIANACEAE

Annual or perennial herbs, glabrous throughout, the sap usually bitter. Leaves opposite and entire, often connate at the base or connected by a transverse line; true stipules lacking; rarely, the leaves reduced to scales and the whole plant lacking chlorophyll. Flowers regular and perfect, usually in a dichasial cyme; calyx 4-6-toothed or -lobed, imbricate or open in bud; corolla gamopetalous, 4–6-lobed, contorted in bud and persistent around the fruit. Stamens as many as the corolla-lobes and alternate with them, inserted in the throat or tube of the corolla. Ovary superior, usually 1-celled, with 2 parietal placentas each bearing numerous ovules; style simple. Fruit usually a 2-valved capsule, many-seeded; seeds various; endosperm usually copious.

A world-wide family of about 80 genera and 900 species.

Genus 1. **EUSTOMA** Salisb.

Erect annual herbs; leaves sessile and often clasping. Flowers few, rather large, long-stalked, and blue, purplish or white; calyx deeply 5–6-parted, the segments narrow and keeled; corolla bell-shaped with short tube, the limb deeply 5–6-lobed. Stamens with hair-like filaments and oblong, versatile anthers. Style persistent in fruit, the stigma 2-lamellate. Seeds small, very numerous, foveolate.

A small American genus of 3 species.

1. **Eustoma exaltatum** (L.) Salisb., Parad. Lond., t. 34. 1806. **FIG. 190.**

Erect herb, more or less glaucous throughout, usually 50 cm tall or less, simple or sparingly branched; stems terete; leaves oblong or oblanceolate, 2–6 cm long, rounded or obtuse and apiculate at the apex. Flowers light violet-blue, a deep violet blotch at the base within, 2–3.5 cm long; calyx-segments lance-linear, 1–1.5 cm long, united only at the base; corolla constricted below the lobes. Capsule oblong, 1.5 cm long, rounded at the apex.

GRAND CAYMAN: *Brunt 1831, 1916; Proctor 27927.*
– Southern United States, West Indies, and continental tropical America, most often in sandy, brackish or subsaline situations at low elevations.

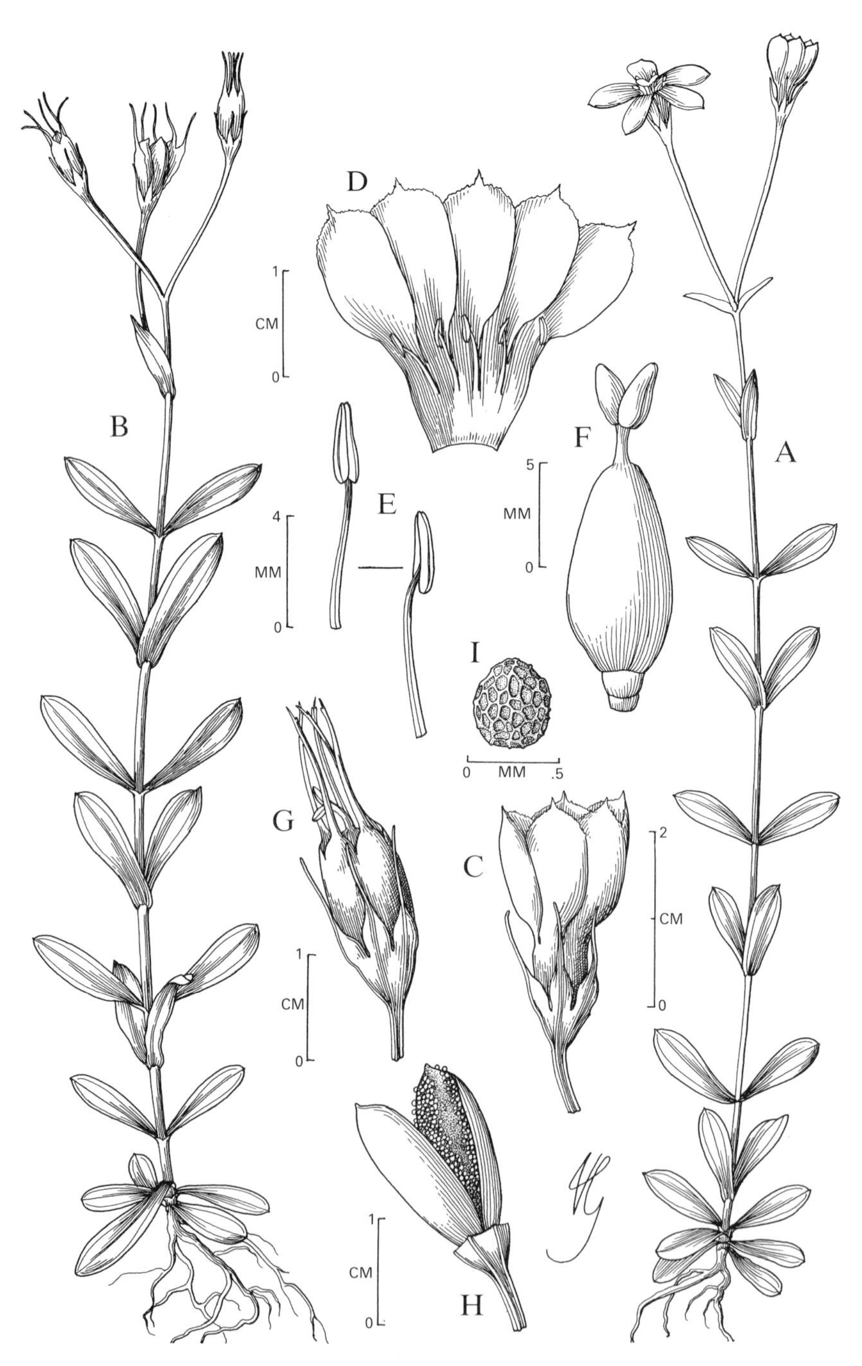

FIG. 190 **Eustoma exaltatum.** A, habit with flowers. B, habit with fruits. C, flower. D, corolla opened out. E, two stamens. F, ovary, style and 2-lamellate stigma. G, fruit. H, capsule after dehiscence. I, seed. (G.)

Family 85

APOCYNACEAE

Herbs, shrubs or trees, or often herbaceous or woody vines, usually containing milky latex; leaves opposite or whorled (rarely alternate), simple and entire; stipules present or absent. Flowers perfect, regular or nearly so, twisted in bud, in cymes or panicles, or sometimes solitary. Calyx usually 5-lobed, often with glandular appendages within; corolla more or less funnel-shaped, 5-lobed. Stamens 5, inserted on the corolla-tube, often connivent around the stigma but the filaments usually free; anthers 2-celled. Ovary superior, 2-celled, the carpels free or united; style simple with a large stigma; ovules 1 to many in each carpel. Fruit of 2 distinct follicles, or the carpels more or less united to form a capsule, drupe, or berry; seeds naked, or plumed at the apex, or with a papery wing, or sometimes with an aril; endosperm present.

A cosmopolitan family best represented in the tropics, with about 200 genera and more than 2000 species. Few are of any economic value, but some are grown for ornament. The white latex of several species is known to be poisonous, while in other cases it is edible.

Species commonly cultivated in the Cayman Islands include *Allamanda cathartica* L. and *A. hendersonii* Bull., both called "Yellow Allamanda"; *Nerium oleander* L., the "Oleander"; two species of *Plumeria* (see below); *Strophanthus gratus* (Wall. & Hook.) Baill., the "Blue Allamanda" (*Kings GC 300*); and perhaps more than one species of *Tabernaemontana.*

KEY TO GENERA

1. Plants erect, never vine-like; anthers free from the stigma, the cells without basal appendages:

  2. Leaves alternate, crowded; shrubs with thick branchlets; flowers in terminal cymes: **1. Plumeria**

  2. Leaves distinctly opposite; herbs or shrubs, the branchlets slender; flowers borne in leaf-axils, solitary or in cymes:

    3. Herbs with white or pink flowers; mature follicles less than 3 mm thick: **[2. Catharanthus]**

    3. Shrubs or small trees with small greenish-yellow flowers; mature follicles 10 mm thick or more: **3. Tabernaemontana**

1. Plants scandent or vine-like; anthers attached to the stigma, their cells with basal prolongations:

4. Calyx bearing glandular appendages inside the lobes; corolla cream or yellow:

5. Corolla cream and with a cylindric tube; sap not or scarcely milky: **4. Echites**

5. Corolla bright yellow and with a funnel-shaped tube; sap distinctly milky: **5. Urechites**

4. Calyx without glandular appendages inside the lobes; corolla white: **6. Rhabdadenia**

Genus 1. **PLUMERIA** L.

Shrubs or trees with stout, thick branches exuding abundant milky latex when cut; leaves crowded, alternate, petiolate; flowers in terminal, stalked cymes. Calyx 5-parted nearly to the base; corolla salver-shaped, without appendages within, equally 5-lobed. Stamens inserted near base of the corolla tube, included, the anthers free. Ovary of 2 distinct carpels, these with many ovules. Follicles 2, distinct, thick and wide-divergent; seeds many, winged at the base; endosperm fleshy.

A tropical American genus of about 40 described species.

1. Plumeria obtusa L., Sp. Pl. 1:210. 1753.

"Wild Jasmine"

A shrub or small tree to 4 m tall or more, glabrous throughout; leaves oblong to oblong-oblanceolate or obovate, 7–20 cm long including the petioles, and blunt, rounded or notched at the apex. Inflorescence of 4–5 branches in a dense sub-umbellate cluster at the apex of the long peduncle; calyx c. 3 mm long; corolla white with a yellow eye, the spreading obovate lobes 1.5–2 cm long. Follicles 10–16 cm long and to 1.4 cm thick.

GRAND CAYMAN: *Brunt 1764; Hitchcock; Lewis GC 1; Proctor 11988, 15194.* LITTLE CAYMAN: *Kings LC 8; Proctor 28043.* CAYMAN BRAC: *Millspaugh 1229; Proctor 15315, 28909, 29066.*
– Bahamas, Cuba, Hispaniola, Jamaica and the Swan Islands, mostly in dry, rocky, coastal thickets.

[*Plumeria rubra* L., the "Jasmine" or "Frangipani" in both pink and cream forms, and *P. pudica* Jacq., a species with large dense heads of white flowers, are traditionally planted in Cayman Island graveyards. The first-named is a native of Central America now cultivated in all tropical countries; the second is said to have been introduced from the Corn Islands at a fairly recent date.]

[Genus 2. **CATHARANTHUS** G.Don

Erect, subwoody perennial herbs or low shrubs. Flowers 1-4 in the leaf-axils; calyx 5-cleft; corolla with narrowly tubular throat and flat, salver-shaped limb, the throat pubescent and thickened in the apical part. Stamens inserted in the apical part of the corolla-tube, included; anthers free. Carpels 2, distinct but slightly cohering, alternating with 2 disk-glands (nectaries); ovules many in each carpel. Fruit of 2 narrowly cylindric, many-seeded follicles; seeds without appendages.

A small tropical genus of 8 species, all but one in Madagascar.

1. **Catharanthus roseus** (L.) G.Don, Gen. Syst. 4:95. 1837. **FIG. 191.**

Vinca rosea L., 1753.

Lochnera rosea (L.) Reichb., 1828.

"Periwinkle", "Burying-ground Flower", "Ramgoat Rose"

An erect herb usually less than 50 cm tall, more or less finely pubescent throughout. Leaves oblong or elliptic-oblong, mostly 2-7 cm long, obtuse and apiculate at the apex. Flowers in axillary pairs, subsessile; calyx-lobes 3-5 mm long; corolla white or pink, or white with a pink eye, the tube c. 2.5 cm long, the lobes broadly obovate, 1.5 cm long. Follicles pubescent, longitudinally ribbed, 1.5-3.5 cm long, 2-3 mm thick.

GRAND CAYMAN: *Brunt 2118; Kings GC 51, GC 221, GC 222; Millspaugh 1331; Proctor 15093.* LITTLE CAYMAN: *Proctor 35197.* CAYMAN BRAC: *Kings CB 60; Proctor 28988.*

Originally described from Madagascar, now cultivated and escaping in nearly all warm countries.]

Genus 3. **TABERNAEMONTANA** L.

Shrubs or trees, usually glabrous or nearly so. Flowers in axillary cymes; calyx 5-parted almost to the base, bearing numerous glandular appendages within; corolla with salver-shaped limb. Anthers not connivent, free from the stigma. Ovary of 2 distinct carpels, with or without a basal ring-like nectary; ovules many. Fruit of 2 rather broad and fleshy follicles; seeds embedded among the fleshy arils.

A tropical American genus of about 50 species

1. Tabernaemontana laurifolia L., Sp. Pl. 1:210. 1753.

A small tree to 7 m tall or more; leaves elliptic or elliptic-oblong, 7–17 cm long, indistinctly subacuminate, with petioles 1–2 cm long. Inflorescences congested, much shorter than the leaves, subumbellate; pedicels 2–3 mm long; calyx-lobes 3 mm long, imbricate; corolla greenish-yellow, the tube 15 mm long, the lobes narrow and 10 mm long. Ovary surrounded at the base by a low, ring-like nectary. Follicles 3–4.5 cm long.

GRAND CAYMAN: *Brunt 2147, 2184; Correll & Correll 51006; Howard & Wagenknecht 15031; Proctor 11990.*
– Jamaica; a doubtful record from Guatemala is probably not the same. This is normally a species of rocky coastal thickets and dry woodlands on limestone.

Genus 4. **ECHITES** P.Br.

Slender, twining, somewhat woody vines with colourless, watery sap; leaves opposite, without glands. Flowers in axillary cymes, rarely solitary; calyx 5-parted almost to the receptacle, each lobe bearing within at the base a single, often dissected glandular appendage; corolla salver-shaped, without appendages in the tube. Anthers connivent and adherent to the stigma, the cells appendaged at the base, 2-lobed. Ovary of 2 distinct carpels with many ovules, surrounded at the base by 5 more or less distinct nectar-glands; stigma spindle-shaped. Fruit of two distinct, divergent follicles; seeds numerous, plumed at the apex.

A tropical American genus of 7 species.

1. Echites umbellata Jacq. Enum. Pl. Carib. 13. 1760. **FIG. 192.**

"Nightshade"

A glabrous, suffrutescent vine, scrambling over bushes or trees to a height of 5 m; leaves elliptic to very broadly ovate, 4–10 cm long, acute or subacuminate at the apex and often more or less recurved or folded. Inflorescences 2–7-flowered,

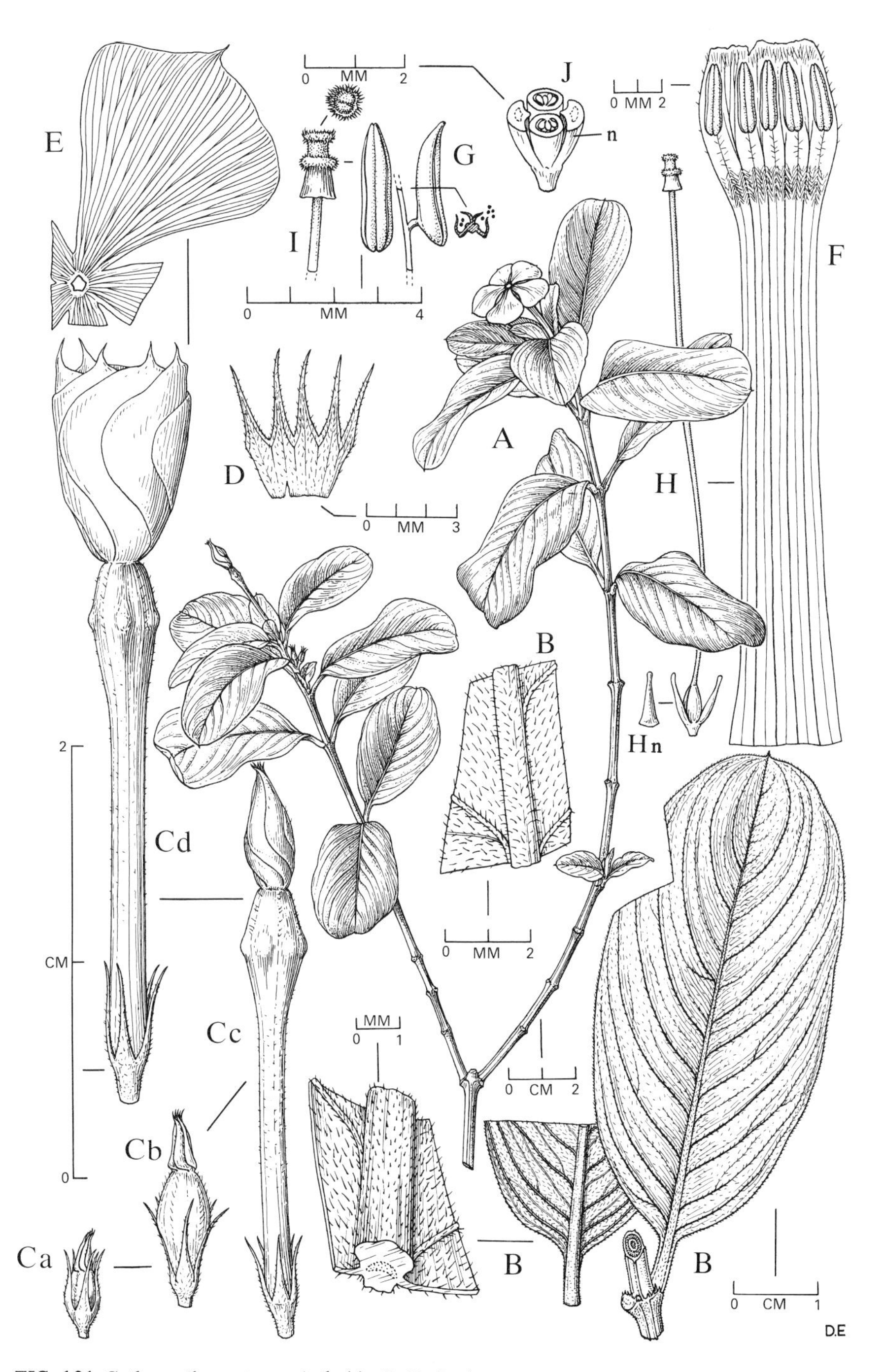

FIG. 191 **Catharanthus roseus.** A, habit. B, B, B, details of leaf. Ca, b, c, d, stages in development of flower. D, calyx spread out. E, portion of corolla-limb. F, inside of corolla-tube spread out. G, details of anther. H, pistil; Hn, nectary (disk-gland). I, details of stigma. J, x-section through carpels; n, nectary. (D.E.)

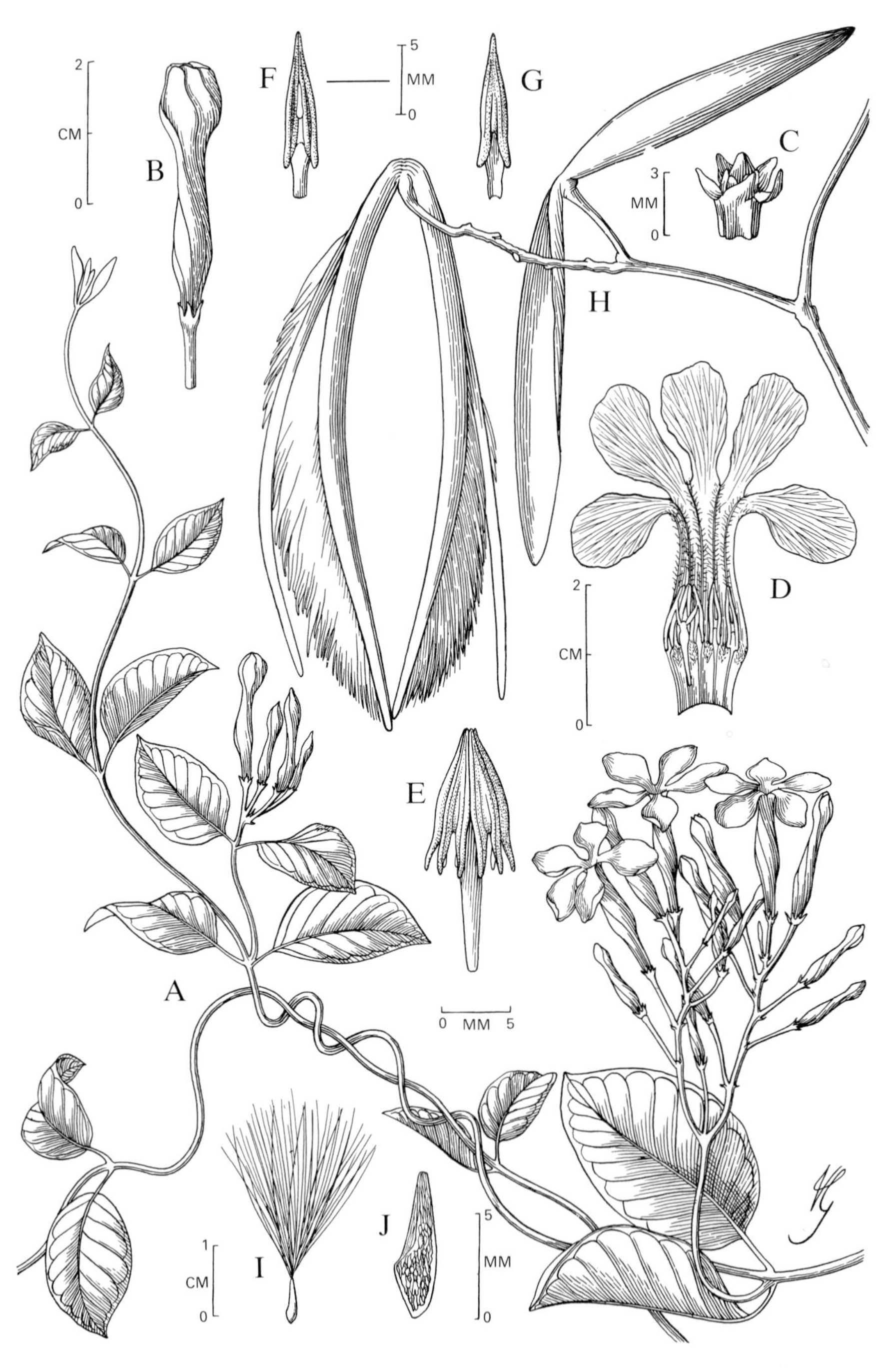

FIG. 192 **Echites umbellata.** A, branch with leaves and flowers. B, flower-bud. C, calyx showing appendages. D, corolla opened to show stamens. E, stamens showing connivent anthers. F, G, two views of anthers. H, fruiting branchlet. I, seed with plume. J, seed enlarged, with plume removed. (G.)

stalked; calyx-lobes scarious, sharply acute, 3-4 mm long; corolla-tube 3-4 cm long, dilated below the middle and above spirally twisted and gradually constricted; lobes obliquely obovate, c. 2 cm long. Follicles usually 10-20 cm long.

GRAND CAYMAN: *Brunt 2022; Kings GC 173; Lewis 3845; Millspaugh 1243; Proctor 15267; Sachet 427.* LITTLE CAYMAN: *Kings LC 1, LC 86.* CAYMAN BRAC: *Kings CB 24; Proctor 29021.*
– Florida, Bahamas, Greater Antilles except Puerto Rico, the Swan Islands, and the Caribbean coasts of continental America, in dry thickets and woodlands.

Genus 5. **URECHITES** Muell. Arg.

Woody or suffrutescent vines; leaves usually opposite and without glands. Flowers large, yellow, in axillary or subterminal cymes; calyx 5-parted nearly to the receptacle, the imbricate lobes bearing glandular appendages within at the base; corolla funnel-shaped, without appendages within. Anthers connivent and adherent to the stigma, the base enlarged and narrowly 2-lobed, appendaged at the apex. Ovary of 2 distinct carpels with many ovules, surrounded at the base by 5 more or less distinct nectar-glands; stigma spindle-shaped. Fruit of 2 distinct follicles; seeds numerous, dry, beaked and hairy at the apex.

A tropical American genus of 2 species.

1. **Urechites lutea** (L.) Britton in Bull. N.Y. Bot. Gard. 5:316. 1907. **FIG. 193.**

Echites andrewsii Chapm., 1860.

"Yellow Nightshade"

A usually pubescent slender vine; leaves oblong to broadly ovate, 1.5-6 cm long or more, obtuse at the apex, the margins often revolute, dark green on the upper side and pale green beneath. Cymes few-flowered; flowers on slender pedicels 0.8-1.5 cm long; calyx-lobes lance-acuminate, mostly 7-9 mm long; corolla-tube 8-12 mm in diameter, the obovate lobes 2-3 cm long. Follicles linear, 10-15 cm long.

GRAND CAYMAN: *Kings GC 91, GC 91B, GC 401; Lewis GC 4; Maggs II 67; Millspaugh 1373; Proctor 15089; Sachet 428.*
– Florida and the West Indies, in thickets, pastures and scrublands. This species is known to be poisonous to livestock.

Genus 6. **RHABDADENIA** Muell. Arg.

Slender, glabrous woody vines; leaves opposite, not glandular. Flowers solitary or few in axillary or subterminal umbellate cymes; calyx 5-parted almost to the receptacle, rather broad, lacking glandular appendages within; corolla funnel-shaped, lacking appendages within. Anthers connivent and adherent to the stigma, their bases enlarged and narrowly 2-lobed. Ovary of 2 distinct carpels with many ovules, surrounded at the base by 5 more or less distinct nectar-glands; stigma spindle-shaped. Fruit of 2 distinct, terete follicles; seeds numerous, beaked and hairy at the apex.

A small genus of 3 tropical American species.

1. Rhabdadenia biflora (Jacq.) Muell. Arg. in Mart., Fl. Bras. 6(1):175. 1860. **FIG. 194.**

Echites paludosa Vahl, 1798.

An often high-climbing woody vine with slender branches; leaves oblong or elliptic-oblong, 4–11 cm long, obtuse and apiculate at the apex, noticeably pale beneath. Cymes 1-5-flowered, on peduncles equalling or exceeding the leaves; pedicels 7–13 mm long; calyx-lobes ovate-oblong, usually 4–6 mm long; corolla pure white, the tube 1.5–2 cm long, the lobes obovate, 2–2.5 cm long. Filaments densely pubescent. Follicles slender, terete, 9–14 cm long.

GRAND CAYMAN: *Brunt 1773, 1801; Hitchcock; Kings GC 199, GC 348; Lewis 3846; Maggs II 65; Proctor 15046; Sachet 449.* LITTLE CAYMAN: *Kings LC 34; Proctor 35116.*
– Florida, Bahamas, Greater Antilles, and the Caribbean coasts of Central and South America, a characteristic species of mangrove swamps.

Family 86

ASCLEPIADACEAE

Mostly perennial herbs, erect, scandent or twining, or sometimes woody shrubs (rarely arborescent), usually with milky sap; leaves opposite or rarely whorled; stipules poorly developed or lacking. Inflorescence a usually umbel-like or raceme-like axillary cyme; flowers of very complicated structure, perfect, regular; calyx with very short or no tube, the 5 lobes imbricate or open in bud; corolla 5-lobed, the lobes twisted or valvate; usually a corona present, this simple or consisting of 5 or more scales or lobes, adnate to the corolla-tube or to the staminal column, highly varied in structure. Stamens 5, inserted at or near the base of the corolla, the filaments flat, short, and usually joined to form a tube, united with the stigma to

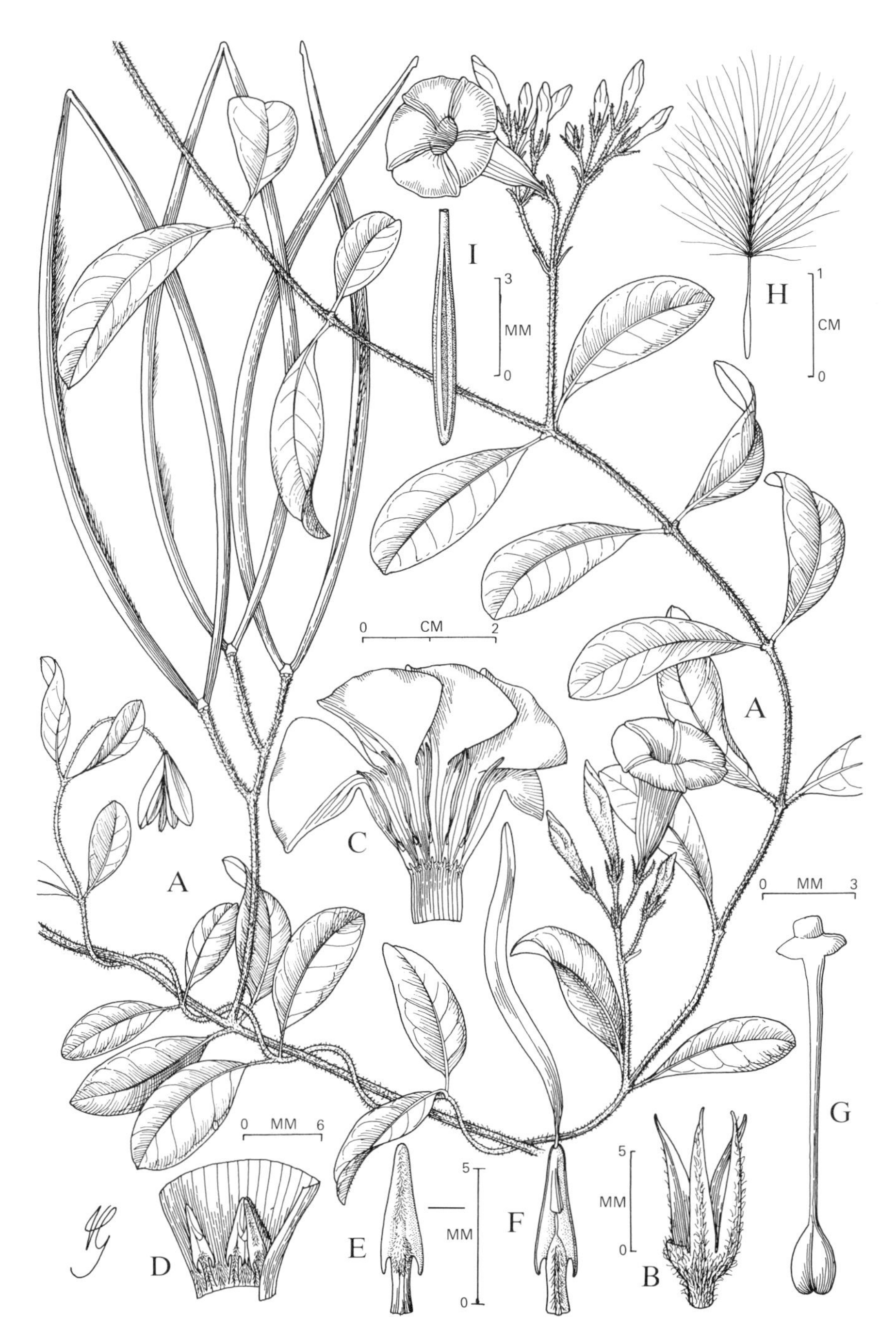

FIG. 193 **Urechites lutea.** A, branch with leaves, flowers and fruits. B, calyx with one lobe removed. C, corolla opened out to show stamens. D, enlarged portion of same. E, stamen, dorsal view. F, stamen with appendage, ventral view; a, appendage. G, ovary, style and stigma. H, seed with plume. I, seed enlarged, with plume removed. (G.)

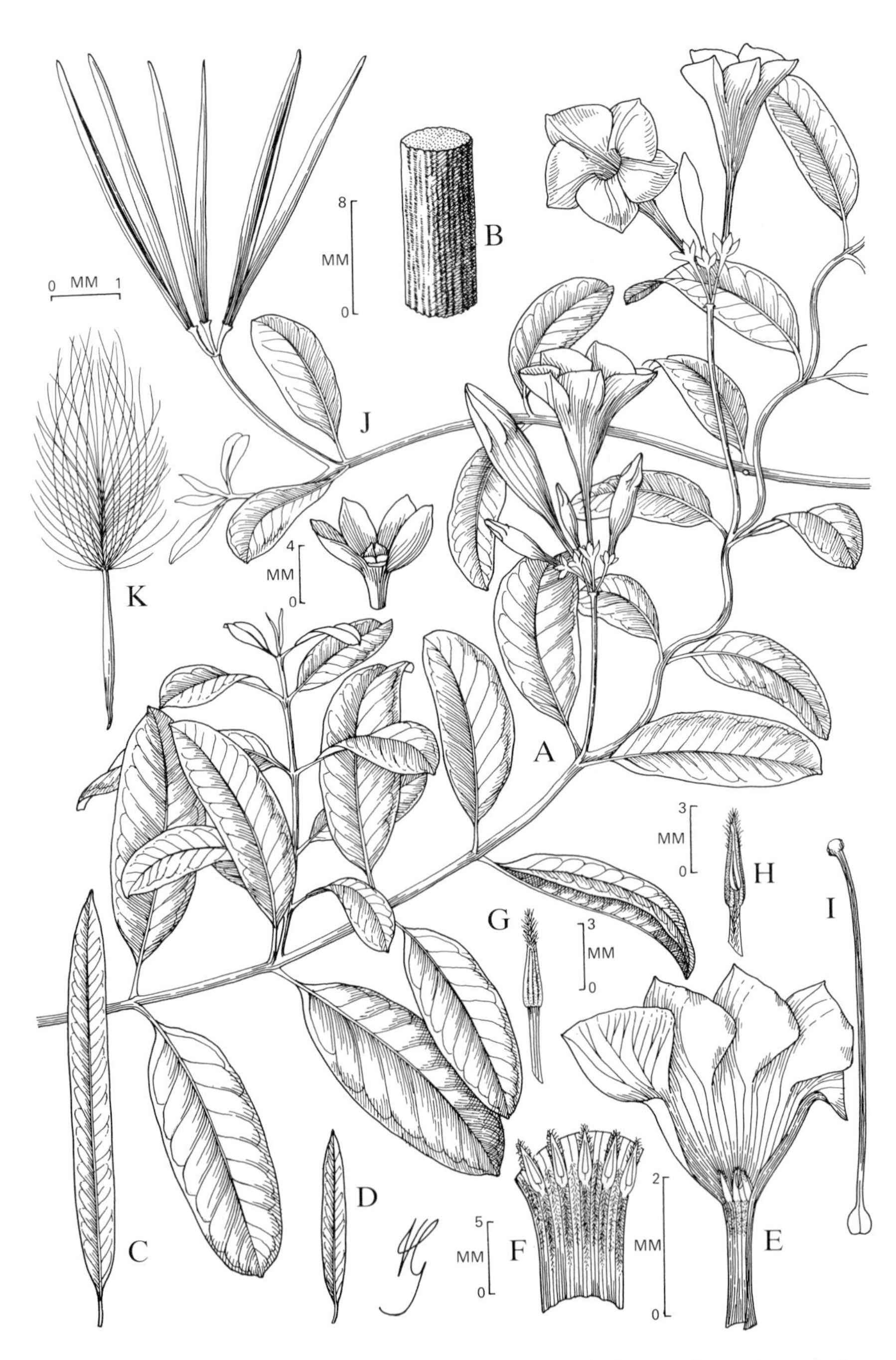

FIG. 194 **Rhabdadenia biflora.** A, branch with leaves and flowers. B, section of stem. C, D, variations of leaf-shape. E, corolla opened to show stamens. F, portion of same, enlarged. G, H, two views of anther. I, ovary and style. J, fruiting branch. K, seed with plume. (G.)

form a "gynostegium"; anthers 2-celled, opening inwardly, often extended at the base and tipped with a membranous appendage; pollen aggregated in waxy or granular masses called "pollinia", these usually solitary in each anther-cell, and agglutinated with the stigma by 5 glandular "corpuscles" which extrude the pollinia after dehiscence of the anthers. Ovary superior, of 2 distinct carpels, surrounded by the stamen-tube; styles 2, distinct below the stigma; stigma 1, peltately dilated and forming a more or less pentagonal disk; ovules numerous in each carpel, pendulous and overlapping on the placentas. Fruit of 2 follicles, or often only 1 developing; seeds numerous, compressed, and usually bearing a tuft of long soft white silky hairs; endosperm thin and cartilaginous; embryo large, with flat cotyledons.

A family of perhaps 130 genera and 2,000 species, widely distributed in temperate and tropical regions. There has been much disagreement about the classification of this group, as the almost incredible complexity of the flowers has led to the proposal of numerous genera whose validity is difficult to evaluate. Except perhaps for the orchids, the flowers of Asclepiadaceae are the most highly specialized of any group of plants.

KEY TO GENERA

1. Plants erect, not vine-like:

  2. Herb with lanceolate, petiolate leaves; corona-segments with a horn-like process within: **1. Asclepias**

  2. Shrub, sometimes of tree-like habit, the very broad leaves almost sessile with clasping bases; corona-segments spurred near the base: [**2. Calotropis**]

1. Trailing or twining vines:

  3. Flowers greenish or white; calyx-lobes less than 4 mm long:

    4. Corolla glabrous, less than 6 mm broad when expanded; corona simple: **3. Cynanchum**

    4. Corolla pubescent, 10 mm broad or more when expanded; corona double: **4. Sarcostemma**

  3. Flowers purple or violet, large; calyx-lobes more than 10 mm long: [**5. Cryptostegia**]

Genus 1. **ASCLEPIAS** L.

Usually erect, perennial herbs; leaves opposite or whorled. Flowers in umbels on axillary or terminal peduncles; calyx 5–10-glandular within at the base; corolla deeply 5-lobed, the lobes reflexed; scales of the corona 5, attached to the stamen-tube, erect and concave-hooded, ligulate within. Pollinia pendulous from apex of the anther-cell. Stigma flat, 5-angled or obtusely 5-lobed. Follicles acuminate, smooth or rarely spiny-tuberculate.

A chiefly North American genus of about 120 species.

1. **Asclepias curassavica** L., Sp. Pl. 1:215. 1753. **FIG. 195.**

"Red Top", "Hippa Casini"

Erect herb usually 30–60 cm tall with mostly simple, somewhat pubescent stems; leaves opposite or sometimes in whorls of 3, lanceolate, mostly 6–12 cm long, long-acuminate at the apex and attenuate to the short-petioled base, glabrous or nearly so. Umbels 1-several in the upper axils; pedicels 1–2 cm long, pubescent. Corolla scarlet or bright red, the reflexed segments 6–7 mm long; corona yellow, the hoods erect, 4–5 mm long, shorter than the horns. Follicles narrowly spindle-shaped, 6–9 cm long, glabrous.

GRAND CAYMAN: *Hitchcock; Kings GC 152, GC 328; Millspaugh 1323; Proctor 15142.*

– Florida, West Indies and continental tropical America, a common weed of roadsides and pastures. This species is alleged to be poisonous to livestock; it has been used in folk medicine to induce vomiting.

[Genus 2. **CALOTROPIS** R.Br.

Shrubs or small trees; leaves broad, opposite and nearly sessile. Flowers in terminal or axillary umbellate cymes; calyx with several to many glands within at the base; corolla 5-cleft with broad lobes not reflexed; scales of the corona 5, fleshy, attached to the stamen-tube, lobed or toothed above and short-spurred at the base. Pollinia pendulous from the anther-cells. Stigma sharply 5-angled. Follicles thick, blunt, and rather rough.

A genus of 6 species occurring in tropical Africa and Asia.

1. **Calotropis procera** (Ait.) R.Br. in Ait.f., Hort. Kew. ed. 2, 2:78. 1811. **FIG. 196.**

"French Cotton"

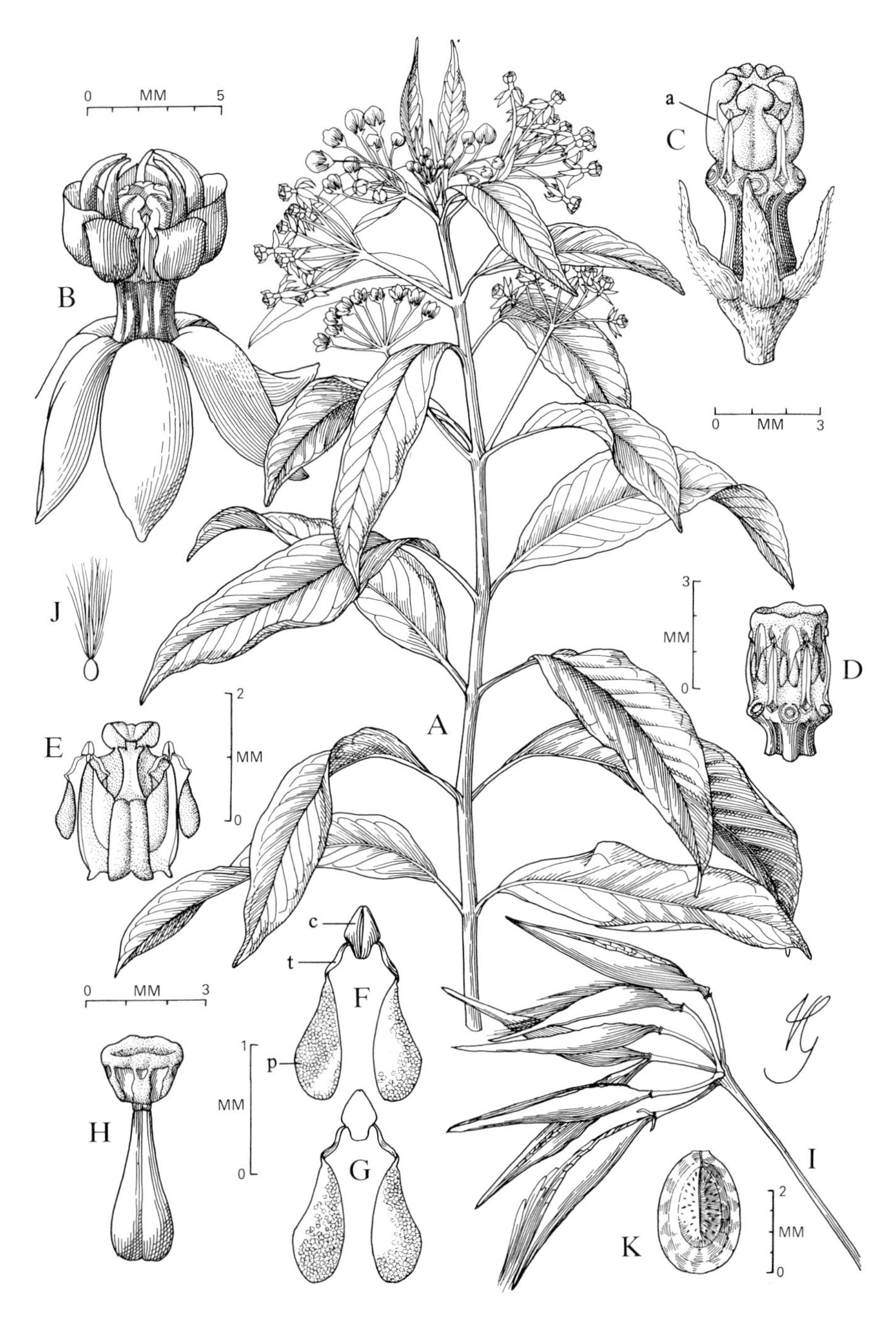

FIG. 195 **Asclepias curassavica.** A, upper portion of stem with leaves and flowers. B, flower. C, flower with corolla and corona removed; a, anther. D, same with anthers removed to show pollinia. E, part of same enlarged. F, G, two views of pollinia; c, corpuscle; t, translator; p, pollinium. H, pistil. I, fruiting inflorescence. J, seed with plume. K, seed enlarged, the plume removed. (G.)

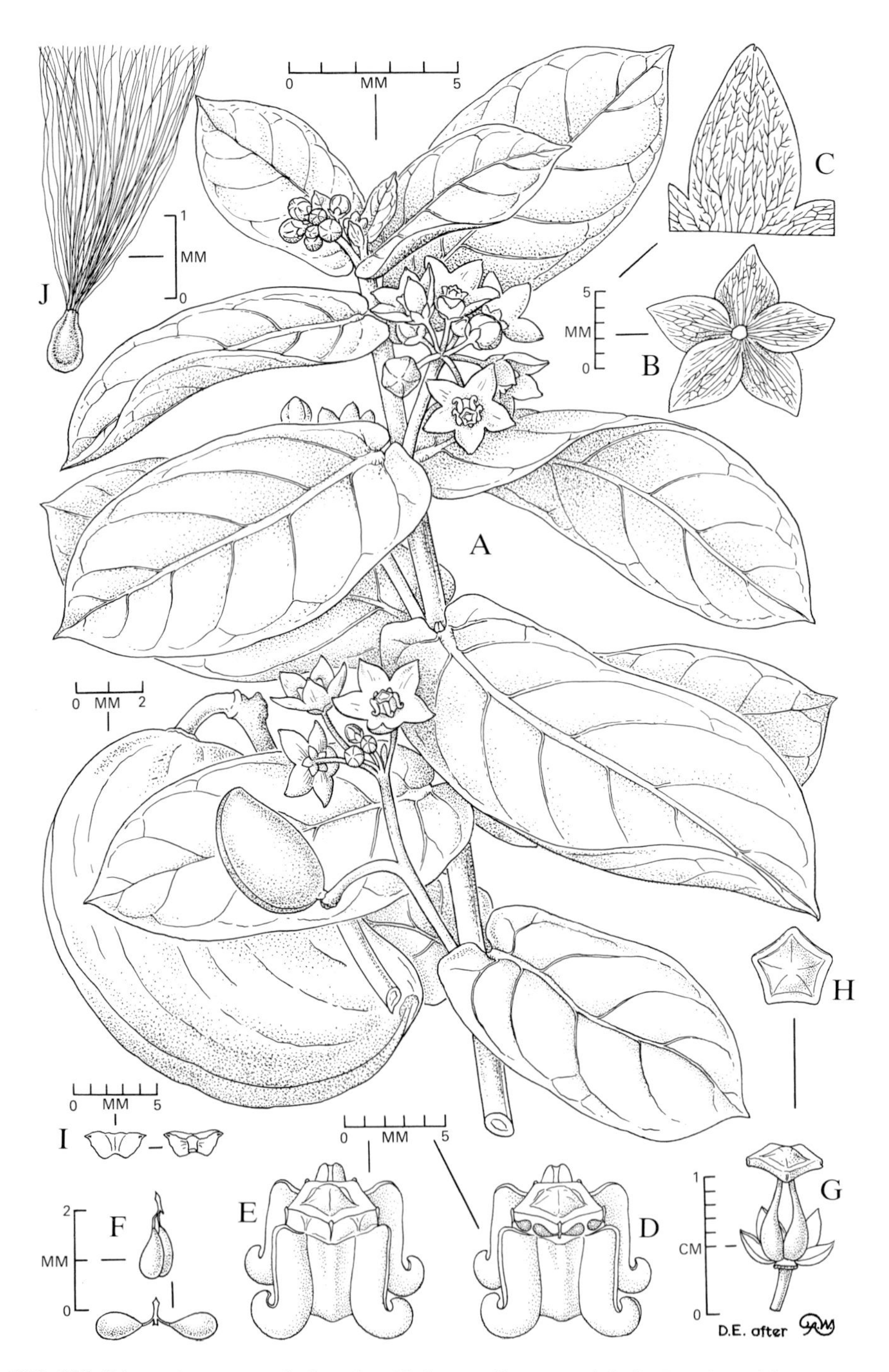

FIG. 196 **Calotropis procera.** A, branch with leaves, flowers and fruit. B, calyx with corolla removed. C, portion of corolla. D, corona (c) and gynostegium with pollinia in position after dehiscence of the anthers. E, same with pollinia removed. F, pollinia before and after dehiscence. G, carpels capped by peltate stigma. H, stigma see from above. I, long.-section through stigma. J, seed with plume.(St.)

A shrub or small tree to 5 m tall, the younger parts deciduously white-felted, becoming glabrous with age; leaves oblong-obovate or rotund, 8–18 cm long, up to 12 cm broad, cuspidate at the apex, narrowly cordate-clasping at the base, with very short, stout petiole. Cymes stalked, few-many-flowered; pedicels 1–3 cm long; calyx woolly; corolla bell-shaped, 2–3 cm broad when expanded, white, the fleshy triangular lobes violet within; corona deep violet; stigma pale green. Follicles greenish, swollen, 8–12 cm long, minutely woolly-puberulous.

GRAND CAYMAN: *Proctor 15121*. CAYMAN BRAC: *Proctor (sight record)*. – Apparently native of tropical Africa, but planted and naturalized in most warm countries, especially in dry places. The latex contains a bitter substance and is somewhat poisonous; the silky floss from the seeds can be used for stuffing pillows or for weaving cloth.]

Genus 3. **CYNANCHUM** L.

Slender perennial vines, herbaceous or somewhat woody; leaves opposite, small or sometimes reduced to scales, the stem green. Flowers very small, in sessile or stalked axillary umbel-like cymes; calyx 5-glandular within at the base, or sometimes without glands; corolla bell-shaped or subrotate, the lobes valvate, often bearded within; corona of 5 scales, these distinct or united, sometimes variously dissected or with internal projections, rarely lacking. Pollinia pendulous, often compressed. Stigma flat at the apex or apiculate. Follicles narrowly acuminate, usually smooth.

A pantropical genus of perhaps 200 species, many of them poorly understood because the complex flowers are too small to be studied without a microscope.

KEY TO SPECIES

1. Leaves subsessile, linear, up to 6 cm long or more; corolla-lobes acuminate-attenuate, 4 mm long: **1. C. angustifolium**

1. Leaves petiolate, oblong to ovate, 2 cm long or less; corolla-lobes obtuse, 2 mm long: **2. C. picardae**

1. Cynanchum angustifolium Pers., Syn. Pl. 1:274. 1805.

Metastelma palustre (Pursh) Schltr., 1899.

Cynanchum salinarum (Wr. ex Griseb.) Alain, 1955.

A trailing glabrous vine; leaves narrowly linear, 2-8 cm long, usually less than 3 mm broad, sharply acute at the apex. Peduncles 1.5-3 cm long; cymes up to 14-flowered, forming an umbel-like head; calyx-lobes narrowly lanceolate, c. 2 mm long; corolla greenish often tinged dull rose or purplish, the lobes ovate with attenuate tips; corona-lobes 1.5-2 mm long, notched. Anther-wings 1 mm long. Follicles 4.5-7 cm long and c. 5 mm thick.

GRAND CAYMAN: *Proctor 31044.*
– Southeastern United States, Bahamas, and Cuba, in various habitats, but most often in somewhat saline situations near the sea. The Cayman plants were growing on bare sand beside the sea, southwest of Rum Point.

2. **Cynanchum picardae** (Schltr.) Jiménez in Rhodora 62:238. 1960.

Metastelma schlechtendalii of Millspaugh, 1900, not Dcne. in DC., 1844.

A very slender, rather high-climbing vine, glabrous or nearly so, forming tangles; leaves oblong to ovate or lance-ovate, mostly 0.8-2 cm long, minutely apiculate at the apex, rounded at the base, the delicate petioles 1.5-4 mm long. Cymes subsessile, 2-5-flowered; calyx-lobes oblong, obtuse, 0.5-0.8 mm long; corolla greenish-cream, the lobes oblong, obtuse, c. 2 mm long, very minutely puberulous within; corona-lobes spatulate; gynostegium sessile. Follicles narrowly spindle-shaped, 3-5 cm long and c. 4 mm thick.

LITTLE CAYMAN: *Proctor 35125.* CAYMAN BRAC: *Millspaugh 1197.*
– Hispaniola, chiefly in dry thickets and scrublands near the sea. The Cayman Brac plants were collected "climbing over shrubbery at southwest point".

Genus 4. **SARCOSTEMMA** R.Br.

Herbaceous or suffrutescent vines, often glaucous; leaves opposite. Flowers in axillary, stalked umbels, usually white or greenish-white; calyx minutely 5-glandular within; corolla broadly bell-shaped or subrotate, shallowly or deeply 5-lobed, the lobes often contorted; outer corona ring-like, entire, adnate to the base of the corolla; inner corona of 5 scales adnate to the base of the stamen-tube, their blades free, broad and flat to concave or sac-like. Pollinia oblong or elongate, pendulous. Stigma flat, protuberant, or bearing a short forked beak. Follicles acuminate, smooth.

A pantropical genus of about 35 species.

1. Sarcostemma clausum (Jacq.) Schult. in L., Syst. Veg. ed. nov. 6:114. 1820. **FIG. 197.**

A twining, often high-climbing vine, the stems green and glabrous; leaves glabrous, somewhat fleshy, almost linear to elliptic, mostly 2–4 cm long, acuminate or cuspidate at the apex. Umbels long-stalked, the peduncles stout, 3.5–7.5 cm long, few- to many-flowered; flowers on slender pubescent pedicels; calyx densely pubescent; corolla white, the lobes 6–8 mm long, pubescent outside. Follicles 5–6.5 cm long, pubescent or glabrate.

GRAND CAYMAN: *Brunt 1767, 1857, 1960; Hutchings (IJ); Kings GC 211; Proctor 15176.* LITTLE CAYMAN: *Proctor 35121.* CAYMAN BRAC: *Proctor 29370.*
– West Indies and continental tropical America, common in thickets and woodlands especially along the borders of mostly brackish or subsaline swamps.

[Genus 5. **CRYPTOSTEGIA** R.Br.

Glabrous, high-climbing vines; leaves broad, opposite. Flowers large, in terminal cymes; calyx deeply 5-parted; corolla funnel-shaped with short tube, the lobes twisted; corona-scales 5, pointed, entire or 2-lobed. Stamens with filaments short and joined only at the base, the anthers connivent around the convex stigma; pollen granular, the grains cohering in small masses. Follicles thick, woody, divergent, ribbed and 3-winged.

A genus of 2 known species, the following and one from Madagascar. Some authors place this and some other genera with granular pollen and free filaments in a separate family, the Periplocaceae.

1. Cryptostegia grandiflora R.Br., Bot. Reg. t. 435. 1820.

A woody, trailing or climbing vine; leaves oblong-elliptic, 5–10 cm long, short-acuminate at the apex, on rather stout petioles c. 1 cm long. Inflorescence few-flowered, puberulous; calyx-lobes lance-acuminate; corolla light violet outside, whitish within, 5–6 cm long. Follicles 9–12 cm long.

GRAND CAYMAN: *Brunt 1853.* CAYMAN BRAC: *Proctor (sight record).*
– Said to be a native of India, but widely cultivated for ornament in many tropical countries, often escaping and becoming naturalized. The latex is capable of yielding a considerable amount of rubber, and has been used for this purpose. It is reported to be poisonous to livestock.]

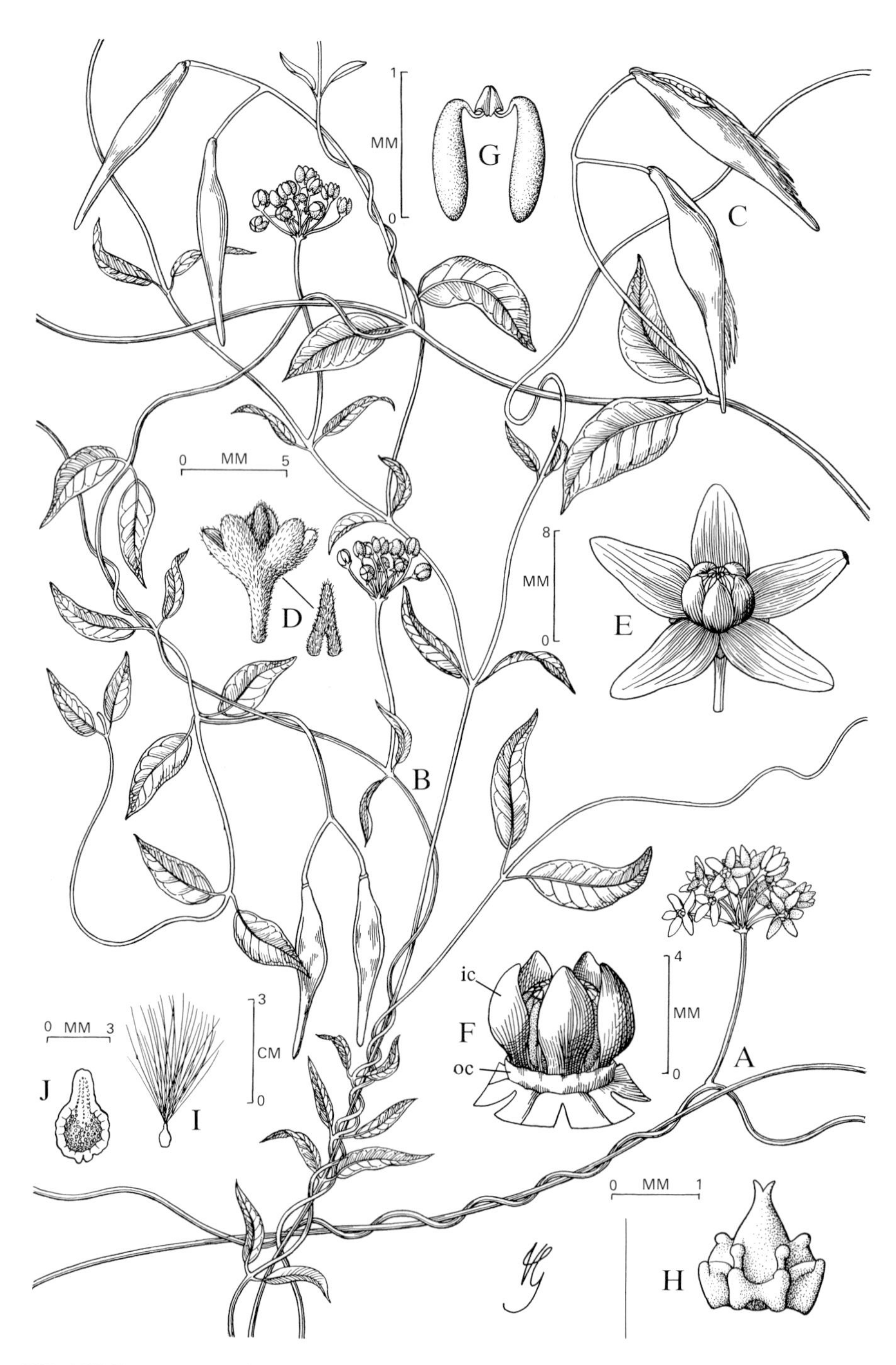

FIG. 197 **Sarcostemma clausum.** A, B, C, portions of stem with leaves, buds, flowers, and fruits. D, calyx. E, flower. F, central part of flower, enlarged; oc, outer corona; ic, scale of inner corona. G, pollinia. H, gynostegium; s, stigma. I, seed with plume. J, seed enlarged, plume removed. (G.)

[**Stephanotis floribunda** Brongn. is or has been cultivated in Grand Cayman (*Kings GC 241*). It is a vine with fragrant white tubular flowers in umbels.]

Family 87

SOLANACEAE

Herbs, shrubs, vines, or sometimes trees; leaves alternate or sometimes in unequal pairs, simple or pinnately compound; stipules lacking. Flowers perfect, regular or nearly so, solitary or in cymes, borne outside the axils or between the paired leaves; calyx usually 5-lobed or -cleft; corolla gamopetalous, mostly 5-lobed, the lobes more or less folded in bud. Stamens usually 5, inserted inside the corolla-tube and alternate with the lobes; anthers 2-celled, variously dehiscent. Ovary superior, usually 2-celled, with numerous ovules on thick axile placentas; style slender and simple. Fruit a capsule or berry; seeds numerous, with fleshy endosperm.

A world-wide, chiefly tropical family of about 90 genera and 2,200 species. A few are economically important, e.g., the "Irish" Potato, *Solanum tuberosum* L., (native of the Andean region of South America), the Tomato, *S. lycopersicum* L., Tobacco, *Nicotiana tabacum* L., and others.

KEY TO GENERA

1. Flowers more than 15 cm long; ovary 4-celled; unarmed scrambling woody vine: **5. Solandra**

1. Flowers less than 5 cm long; ovary 2-celled; erect herbs or shrubs, or if scrambling or vine-like then armed with spines:

 2. Calyx inflated around the fruit: **1. Physalis**

 2. Calyx not inflated:

  3. Flowers solitary, axillary or in forks of the stem; corolla not folded in bud: **2. Capsicum**

  3. Flowers in racemes or cymose panicles:

   4. Corolla spreading, the tube rotate or else shorter than the lobes: **3. Solanum**

   4. Corolla tubular, the very short lobes much shorter than the tube: **4. Cestrum**

Genus 1. **PHYSALIS** L.

Annual or perennial herbs, rarely somewhat woody, glabrous or with simple, branched or stellate hairs; leaves simple, alternate or clustered, petiolate, the margins entire to wavy or toothed. Flowers usually solitary at forks of the stem or arising laterally, remote from the leaf-axils; calyx 5-lobed, in fruit inflated like a papery bladder around the fruit; corolla bell-shaped, usually more or less yellow, and often with dark spots at the base. Stamens 5, inserted near the base of the corolla-tube; anthers splitting lengthwise. Style very slender with truncate or capitate stigma. Fruit a few- to many-seeded berry; seeds roundish or kidney-shaped, more or less flattened.

A cosmopolitan genus of about 100 species, most abundant in the American tropics.

1. Physalis angulata L., Sp. Pl. 1:183. 1753.

A tender, branched herb to 50 cm tall or more, glabrous or with scattered minute hairs on the young parts; leaves long-petiolate, the blades lance-ovate to broadly ovate, mostly 2.5–9 cm long, shortly-acuminate, the margins nearly entire to irregularly incised- or undulate-toothed. Corolla pale yellow, indistinctly spotted at the base within, or unspotted, 6–12 mm long. Fruiting calyces 10-angled or 10-ribbed, 2-3.5 cm long, on slender stalks 1-2.5 long; berry 10–12 mm in diameter.

GRAND CAYMAN: *Brunt 2032; Hitchcock; Kings GC 208.*
– Southern United States, West Indies, and continental tropical America, a frequent weed of fields, clearings and waste places.

Genus 2. **CAPSICUM** L.

Annual or perennial herbs or shrubs with forking stems; leaves often in unequal pairs, simple, entire or nearly so. Flowers solitary or sometimes in small clusters, the peduncles arising from forks of the stem or from leaf-axils; calyx 5-toothed or subentire; corolla usually white or cream, the tube very short, the 5 lobes spreading, imbricate. Stamens 5, inserted in the throat of the corolla-tube; anthers opening lengthwise. Ovary 2- or rarely 3-celled; stigma club-shaped. Berries red, yellow or green, erect or nodding, usually pungent-flavoured.

A genus of an uncertain number of species, perhaps 30, all native to tropical America. *Capsicum annuum* L. is cultivated in many forms, including the condiment "Red Pepper" and such non-pungent varieties as "Sweet Pepper" and "Pimiento".

1. **Capsicum baccatum** L., Mant. Pl. 47. 1767. **FIG. 198.**

"Bird Pepper", "Wild Pepper"

Bushy herb or shrub to 2 m tall or more, glabrous or minutely pubescent, leaves narrowly ovate, acute or acuminate, 1.5–5 cm long or more, petiolate. Pedicels solitary or sometimes paired, 1–2 cm long, somewhat thickened toward the apex. Calyx 1–2 mm long, almost truncate or very shallowly 5-lobed. Corolla cream, c. 6 mm across when expanded. Anthers greenish-blue. Berries usually red, ovoid or globose, 5–10 mm long or sometimes longer.

GRAND CAYMAN: *Kings GC 236.*
– Throughout tropical America, and naturalized in the Old World tropics; frequent in thickets and along roadsides; sometimes cultivated. The fruits are very hot to the taste.

Genus 3. **SOLANUM** L.

Herbs or shrubs, sometimes scandent or climbing, unarmed or spiny, and often bearing stellate hairs; leaves alternate, simple and entire or variously lobed or pinnatifid, sometimes bearing prickles. Flowers white to violet, solitary or in simple or compound cymes, or sometimes in umbel-like clusters, the peduncles arising laterally more less remote from the leaves. Calyx bell-shaped, 4–5-lobed; corolla rotate, the tube very short, the 4 or usually 5 lobes spreading or recurved, folded in bud. Stamens 5, inserted near the base of the corolla-tube, the filaments usually short, the anthers relatively long and bright yellow, connivent around the style, opening by more or less terminal pores or longitudinal slits. Ovary usually 2-celled; stigma small, capitate or obscurely 2-lobed. Fruit a globose berry, fleshy or leathery; seeds numerous and more or less flattened.

A very large tropical and subtropical genus of more than 1,000 species.

KEY TO SPECIES

1. Plants glabrous or bearing simple hairs; spines lacking:

  2. Herbs; corolla white, not over 6 mm across when expanded; berries black: **1. S. americanum**

  2. Shrubs; corolla pale blue-violet, more than 15 mm across when expanded; berries blue: **2. S. havanense**

1. Plants more or less clothed with stellate hairs; spines nearly always present on stems and leaves:

3. An erect shrub with straight spines: **3. S. bahamense**

3. A trailing or scrambling shrub with recurved spines: **4. S. lanceifolium**

1. Solanum americanum Mill., Gard. Dict. ed. 8. 1768.

Solanum nodiflorum Jacq., 1789.

An erect annual herb to 1 m tall, minutely puberulous throughout with simple hairs; leaves long-petiolate, the blades membranous, narrowly to broadly ovate, 2-8 cm long, acuminate, the margins entire or with 1-few short lobes or teeth near the base; petioles up to 2 cm long. Peduncles lateral, 1-2 cm long; flowers 3-10 in an umbel-like cluster; pedicels 7-10 mm long; calyx 1 mm long; corolla-lobes 2 mm long, longer than the tube, spreading or recurved. Berries black, 5-10 mm in diameter.

GRAND CAYMAN: *Brunt 2056; Hitchcock; Millspaugh 1351; Proctor 15277.* CAYMAN BRAC: *Kings CB 65; Proctor 29079.*
– A pantropical weed chiefly of open waste ground.

2. Solanum havanense Jacq., Enum. Pl. Carib. 15. 1760.

An erect or somewhat straggling shrub 1-2 m tall, of smooth texture, minutely puberulous with simple hairs on the young parts; leaves elliptic, 2-5 cm long or more, obtuse at the apex and acuminate at the base. Inflorescence lateral or sub-terminal, few-flowered on peduncles less than 1 cm long; pedicels 5-6 mm long; calyx deeply 5-lobed, c. 4 mm long; corolla-limb broadly spreading, the shallow lobes much shorter than the united portion. Berries c. 15 mm in diameter.

GRAND CAYMAN: *Brunt 2200.*
– Cuba, Jamaica and Hispaniola, in rocky woodlands most frequently near the sea.

3. Solanum bahamense L., Sp. Pl. 1:188. 1753. **FIG. 199.**

A more or less spiny shrub usually 1-2 m tall, minutely stellate-puberulous throughout; spines straight and slender, 3-8 mm long, scattered on stems and leaves; leaves lanceolate to oblong, 3-11 cm long, acute or obtuse at the apex, the margins entire or somewhat wavy. Cymes simple and raceme-like, shorter than or

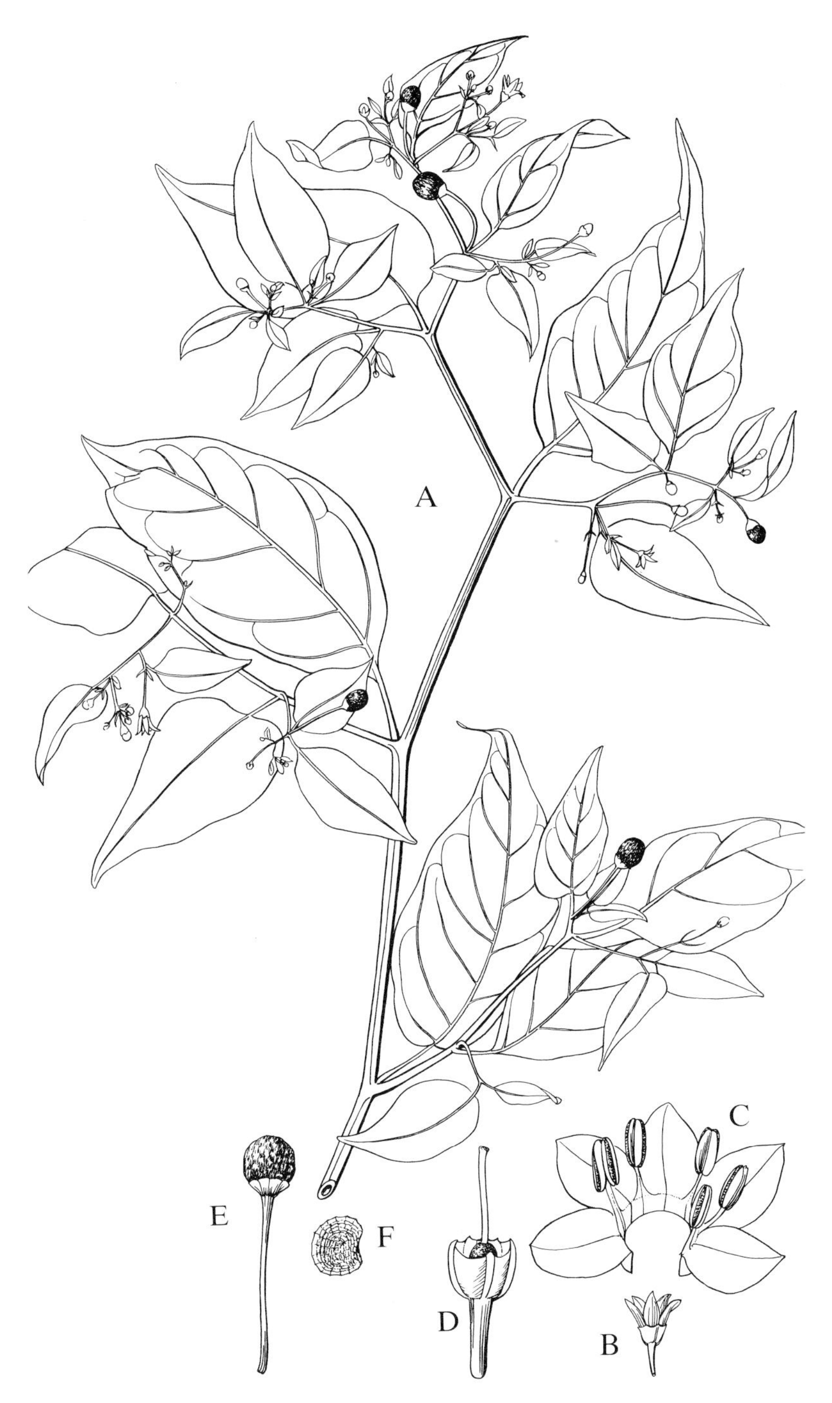

FIG. 198 **Capsicum baccatum.** A, habit, x1½. B, flower, x3. C, corolla cut open, showing stamens, x9. D, calyx and pistil, x9. E, fruit, x3. F, seed, x6. (St.)

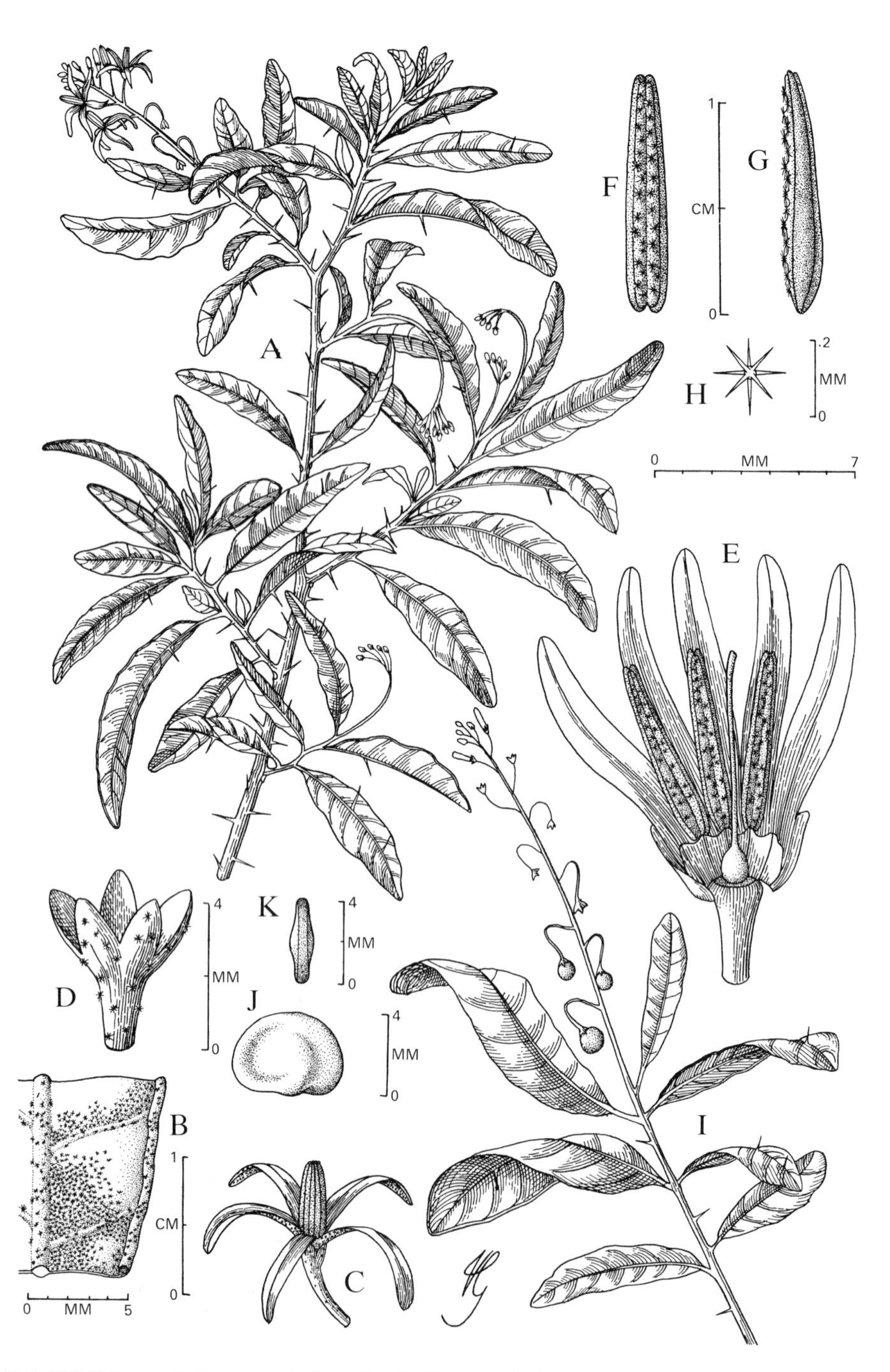

FIG. 199 **Solanum bahamense.** A, branch with leaves and flowers. B, enlarged portion of leaf, showing stellate hairs. C, flower. D, calyx. E, flower cut open to show stamens and pistil. F, G, two views of anther. H, stellate hair. I, branch with fruits. J, K, two views of seed. (G.)

sometimes exceeding the leaves; pedicels 6–12 mm long; calyx c. 1.5 mm long; corolla blue-violet or rarely white, with very short tube and narrow strap-like lobes 6–8 mm long. Berries red, 6–8 mm in diameter.

GRAND CAYMAN: *Brunt 1779, 1997; Correll & Correll 51014; Kings GC 317; Proctor 11978.* LITTLE CAYMAN: *Kings LC 33.* CAYMAN BRAC: *Proctor 29035.*
– Florida, Bahamas and the Greater Antilles except Puerto Rico; doubtfully distinct from *S. racemosum* Jacq. of Puerto Rico and the Lesser Antilles. This species grows in rocky thickets and woodlands. Outside our area it tends to be very polymorphic, varying widely in the presence or absence of spines and other characters, but the Cayman population is quite uniform.

4. **Solanum lanceifolium** Jacq., Coll. 2:286. 1789.

Solanum scabrum Vahl, 1798, not Mill., 1768.

A very prickly scrambling or scandent shrub with stems up to 4 m long, more or less stellate-tomentose throughout; spines short, rather thick and recurved, 1-5 mm long, scattered on stems, petioles and midribs of the leaves beneath leaves lanceolate to ovate, 4–20 cm long, petiolate. Inflorescence a contracted cymose raceme usually less than 6 cm long; pedicels 6-12 mm long, elongating in fruit; calyx 1–2 mm long, enlarging in fruit to 3–5 mm; corolla white, with very short tube and narrow strap-like lobes 7–12 mm long, stellate-pubescent on the outside. Berries red or orange, 8–10 mm in diameter.

GRAND CAYMAN: *Hitchcock; Kings GC 317; Proctor 15276.*
– Cuba, Hispaniola, Tortola and the Lesser Antilles, appearing to intergrade with a series of other species in Central and South America, where it is a noxious weed. The Cayman plants were collected in moist swales near Georgetown.

Other species of *Solanum* may yet be recorded from the Cayman Islands, including *S. torvum* Sw., the “Susumber”, and *S. erianthum* D.Don. In view of their ranges, the apparent absence of these two from Cayman is surprising.

Genus 4. **CESTRUM** L.

Shrubs or small trees; leaves alternate, simple and entire. Flowers in paniculate cymes, these often contracted, usually axillary; calyx 5-lobed or -toothed; corolla salver-shaped or more often funnel-shaped with long slender tube and short, spreading lobes. Stamens usually 5, inserted in the corolla-tube, included; filaments often pilose below and sometimes with a tooth-like appendage; anthers small, splitting lengthwise. Ovary 2-celled, usually short-stalked; ovules few. Fruit a small berry; seeds oblong, smooth.

A tropical American genus of about 150 species.

1. Cestrum diurnum L., Sp. Pl. 1:191. 1753.

Shrub 2-5 m tall, or sometimes a small tree to 10 m tall, mostly glabrous; leaves oblong to ovate or elliptic, cuneate at base, blunt to acuminate at the apex, up to 15 cm long and 6.5 cm broad, the petiole up to 2.5 cm long. Flowers fragrant at night; calyx varying from less than 3 mm long to 8 mm long; corolla greenish-white to cream. Berries purplish to deep blue or black.

A number of varieties have been described from various parts of the American tropics. Two of these occur in the Cayman Islands as follows:

1. Peduncles 0.5-1 cm long, shorter than or not much exceeding the petioles; flowers pedicellate in loose simple (rarely branched) racemes; calyx distinctly lobed; berries black: **1a. C. diurnum** var. **venenatum**

1. Peduncles 2.5-5 cm long, much longer than the petioles; flowers sessile in tight heads or short spikes; calyx truncate or nearly so; berries purple: **1b. C. diurnum** var. **marcianum**

1a. Cestrum diurnum var. **venenatum** (Mill.) O.E. Schulz in Urban, Symb. Ant. 6:263. 1909.

A shrub or small tree to 7 m tall or more, glabrous or minutely puberulous on the young parts; leaves thin, oblong to elliptic, 5-11 cm long, acute or sub-acuminate at the apex, often somewhat folded. Calyx 3.5-4.5 mm long, the rounded lobes ciliolate; corolla cream or white with cream lobes, the slender tube c. 10 mm long, the obtuse lobes minutely puberulous around the margin. Berries ellipsoid, 6-7 mm long.

GRAND CAYMAN: *Proctor 15013.* CAYMAN BRAC: *Kings CB 17; Millspaugh 1192; Proctor 15320, 29033.*
– Jamaica and the Swan Islands, in rocky woodlands.

1b. Cestrum diurnum var. **marcianum** Proctor in Sloanea 1:3. 1977.

A shrub 1.5 m tall, minutely woolly-puberulous on the youngest parts; leaves pale green, narrowly ovate-oblong, 5-8 cm long, blunt to acutish at the apex; petioles 3-5 mm long. Calyx 2.5-3 mm long, truncate or nearly so, the limb ciliolate; corolla white, the tube 10-11 mm long, the obtuse lobes minutely puberulous. Berries subglobose, 5-6 mm in diameter.

GRAND CAYMAN: *Proctor 15294 (type).*
– Cuba. Cayman plants were found in thickets bordering a stony pasture. This differs from all other variants of *C. diurnum* in its long-stalked, spicate or capitate inflorescence and in some other details, but resembles *C. diurnum* var. *portoricense* in the small, nearly truncate calyx.

Cestrum laurifolium L'Her. was recorded from Grand Cayman by Fawcett, but his specimen (if any) has not been located; it seems very likely to have been a misidentification of *C. diurnum* var. *venenatum.* True *C. laurifolium* has sessile inflorescences, and therefore presumably could not be confused with *C. diurnum* var. *marcianum.*

Genus 5. **SOLANDRA** Sw.

Woody, high-climbing or scrambling vines; leaves alternate, simple and entire. Flowers very large, solitary, terminal; calyx tubular and somewhat inflated, membranous, 2-5-cleft; corolla funnel-shaped, the tube cylindric below and more or less abruptly much-expanded above, at the apex divided into 5 broad, imbricate, spreading lobes. Stamens 5, inserted at the base of the corolla-tube, included, the declined anthers oblong. Ovary imperfectly 4-celled; ovules many; stigma subcapitate. Fruit a large pulpy berry.

A tropical American genus of about 10 species.

1. **Solandra longiflora** Tussac, Fl. Antill. 2:49, t. 12. 1818.

A glabrous, scrambling woody vine; leaves petiolate, elliptic or obovate, 5–12 cm long, acute at the apex. Calyx 6–7 cm long; corolla opening white, turning creamy yellow, 17–20 cm long or more, the narrow tubular part much longer than the calyx, the lobes finely undulate-toothed. Berry globose, c. 4 cm in diameter.

CAYMAN BRAC: *Proctor 29343.*
– Cuba and Hispaniola, in various habitats; the Cayman Brac plants were found in dense woodland along the brink of north-facing limestone cliffs.

Family 88

CONVOLVULACEAE

Annual or perennial herbs, twining vines, shrubs, or rarely trees, sometimes parasites with elongate slender stems and leaves reduced to scales; sap often milky; leaves alternate, simple and entire, lobed or variously dissected; stipules absent. Flowers axillary, solitary or in cymose heads or panicles, regular or nearly so, usually soon withering. Calyx of usually 5 free sepals, these imbricate, equal or unequal, usually persistent and often enlarging in fruit. Corolla gamopetalous and tubular, funnel-, bell- or salver-shaped, the more or less expanded limb 5-lobed or -toothed, or almost entire, twisted and often plaited or folded in bud, the folds indicated in the expanded corolla by longitudinal lines or stripes. Stamens 5, inserted in the base of the corolla-tube and alternating with its lobes; anthers

2-celled, splitting lengthwise. Disk ring-shaped or absent. Ovary superior, sessile, mostly 2-3-celled with solitary ovules or 1-celled with 4 ovules; style simple or forked, or styles 2; stigma capitate or 2-lobed. Fruit a usually 1-4-celled capsule, dehiscent or indehiscent, often more or less enclosed by the persistent calyx; seeds erect, glabrous or hairy, with scanty but hard endosperm.

A nearly cosmopolitan family with about 50 genera and 1,800 species, particularly well represented in tropical America and Asia.

KEY TO GENERA

1. Ovary and fruit deeply 2-lobed or the carpels distinct; small creeping plant with minute flowers: **1. Dichondra**

1. Ovary and fruit entire, not lobed:

 2. Styles 2, distinct, with elongate stigmas: **2. Evolvulus**

 2. Style 1, entire:

  3. Stigmas elliptic or oblong, flattened; flowers rather small and white: **3. Jacquemontia**

  3. Stigmas globose or capitate; flowers if white much larger:

   4. Filaments glandular at the base; anthers often spirally twisted at or soon after anthesis; pollen smooth; corolla white or yellow: **4. Merremia**

   4. Filaments glabrous or pilose at the base but not glandular; anthers never spirally twisted; pollen spiny; corolla not yellow: **5. Ipomoea**

Genus 1. **DICHONDRA** J.R. & G. Forst.

Slender perennial herbs, usually creeping and rooted at the nodes, more or less clothed with both simple and "malpighiaceous"[1] hairs; leaves small, roundish-cordate or reniform, entire and long-petiolate. Flowers minute, stalked, solitary in the leaf-axils; sepals subequal, distinct; corolla broadly bell-shaped, deeply 5-lobed. Stamens shorter than the corolla, inserted just below the sinuses of the lobes; anthers small, nearly globose. Ovary 2-lobed, each lobe a single carpel with 2 ovules; styles 2, arising between the ovary-lobes; stigmas capitate. Fruit of 1 or 2 capsules, each usually 1-seeded; seeds subglobose, smooth.

A nearly pantropical genus of about 12 species or less.

---

[1]Hairs attached at or near the middle, thus 2-armed or flattened T-shaped.

1. Dichondra repens J.R. & G. Forst., Char. Gen. Pl. 39, t. 20. 1776. **FIG. 200.**

A prostrate, creeping herb, forming mats, the delicate individual stems mostly 15-30 cm long, minutely appressed-hairy; leaves long-petiolate, the blades more or

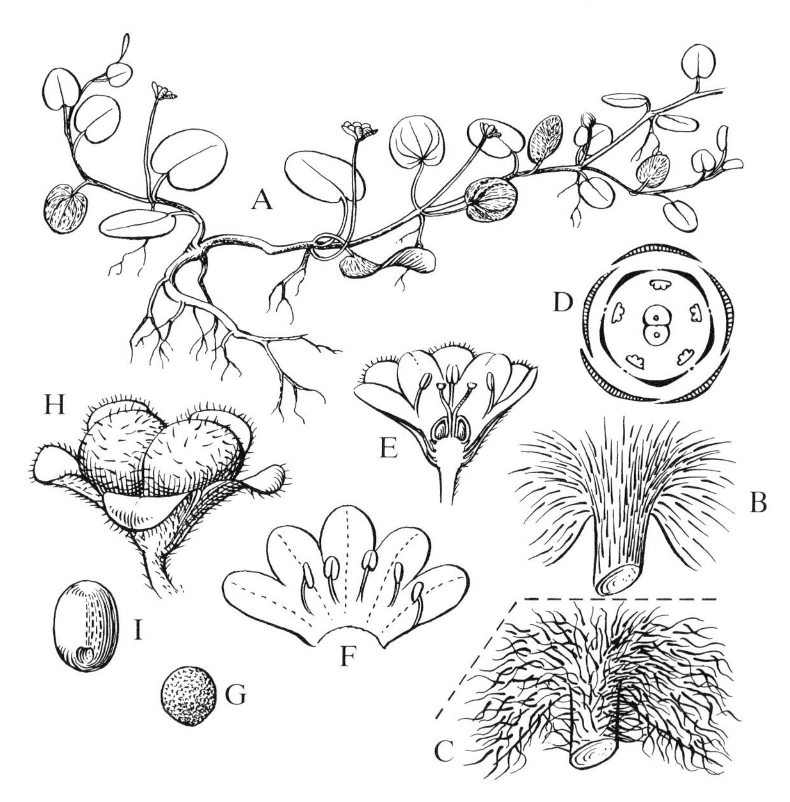

FIG. 200 **Dichondra repens.** A, habit x½. B, leaf-base enlarged to show arrangement of hairs, x8. C, same showing hairs on leaf from New Zealand, x8. D, floral diagram. E, long.section of flower, x4. F, corolla spread out to show stamens, x4. G, pollen grain, x800. H, capsule, x4. I, seed, x4. (St.)

less reniform, mostly 3-12 mm broad, sparsely and minutely hairy on both surfaces. Peduncles shorter than the petioles; sepals oblong-spatulate, c. 2 mm long; corolla yellowish, c. 2 mm long; capsules hairy.

LITTLE CAYMAN: *Proctor 28193.*

– Widespread in tropical and warm-temperate America, occurring in a number of varieties. The single Cayman specimen is not adequate for identification with any particular variety. In Little Cayman this species grows in sandy clearings and is rare.

Genus 2. **EVOLVULUS** L.

Annual or perennial herbs or small shrubs, the stems slender, creeping or erect, not twining; leaves small, simple and entire. Flowers axillary, solitary or in stalked, 1-several-flowered clusters, rarely in terminal spikes; sepals equal or subequal; corolla small, blue or white, usually rotate, funnel- or salver-shaped, the limb obscurely 5-angled but unlobed, folded in bud. Stamens 5, inserted at the mouth of the corolla-tube. Ovary 2-celled, each cell with 2 ovules, or rarely 1-celled with 4 ovules; styles 2, free or slightly joined at the base, each 2-forked at the apex. Capsule 2–4-valved, 1–4-seeded; seeds glabrous, smooth or minutely warty.

A genus of more than 100 species, widely distributed in tropical and warm-temperate regions.

KEY TO SPECIES

1. Small erect shrub; leaves few, minute, more or less scale-like: **1. E. arbuscula**

1. Trailing or creeping herbs; leaves numerous, conspicuous:

  2. Leaves orbicular or obovate, notched at the apex; peduncles very short or obsolete, the pedicel scarcely exceeding the petioles: **2. E. nummularius**

  2. Leaves narrowly elliptic or oblong, acute and mucronate at the apex; peduncles hair-like, often equalling or exceeding the leaves: **3. E. convolvuloides**

1. Evolvulus arbuscula Poir. in Lam., Encycl. Meth. Bot., Suppl. 3:459. 1813.

"Crab Bush"

A knee-high, brushy shrub with numerous stiff, slender, gray-green branchlets, these glabrate or sparsely pubescent; leaves minute, lance-linear, 1–2 mm long, sericeous. Flowers solitary or paired in the upper axils, the peduncles nearly obsolete, the sericeous pedicels 2–4 mm long. Sepals acuminate, c. 2 mm long; corolla white, 7–9 mm broad when expanded, puberulous in stripes to each lobe. Capsule glabrous, 2 mm in diameter.

LITTLE CAYMAN: *Kings LC 35, LC 76a; Proctor 28087, 28130.*
– Cuba, Jamaica and Hispaniola; closely-related forms also occur in the Bahamas. Our plants grow in dry sandy clearings or open glades of dry woodlands. This species is apparently the sole host of the diminutive land-snail *Cerion nanus*, endemic to Little Cayman.

2. Evolvulus nummularius (L.) L., Sp. Pl. ed. 2, 1:391. 1762.

Perennial creeping herb, rooting at most of the nodes; leaves distichous, 5-17 mm long. Flowers mostly solitary; peduncles nearly obsolete, the bracteoles located at base of the pedicel, the latter 2-6 mm long; sepals c. 2.5 mm long; corolla white or rarely pale blue, 5-7 mm broad when expanded. Capsule glabrous, 2-3 mm in diameter.

GRAND CAYMAN: *Kings GC 247 (not seen).*
– Pantropical, a frequent weed of grazed pastures, lawns, and open waste ground.

3. Evolvulus convolvuloides (Willd.) Stearn in Taxon 21:649. 1972.

Evolvulus glaber Spreng. in L., 1824.

Trailing or ascending herb, often suffrutescent at the base or growing from a woody taproot, the slender branches usually rooting at some of the nodes, finely sericeous when young; leaves 5-28 mm long, glabrate. Flowers 1 or 2 at the apex of a filiform peduncles up to 20 mm long; pedicels 1.5-4 mm long; sepals c. 4 mm long; corolla pale blue, 7-10 mm broad. Capsule glabrous, c. 2.5 mm in diameter.

GRAND CAYMAN: *Brunt 2154; Proctor 27967.*
– Florida, West Indies, and northern South America; the Cayman plants were found in open, brackish, seasonally flooded ground with *Salicornia.*

Genus 3. **JACQUEMONTIA** Choisy

Trailing or twining vines, herbaceous or subwoody; leaves usually entire, sometimes toothed or lobed. Flowers axillary in small, stalked, umbel-like cymes or heads; sepals 5, equal or unequal; corolla white, blue or violet, bell-shaped or somewhat rotate, the limb 5-angled. Stamens 5, mostly shorter than the corolla, the filaments sometimes dilated at the base. Ovary 2-celled, each cell with 2 ovules; style filiform, with 2 stigmas. Capsule globose, 2-celled, 4-seeded; seeds glabrous, tuberculate or hairy.

A pantropical genus of about 60 species.

1. Jacquemontia havanensis (Jacq.) Urb., Symb. Ant. 3:342. 1902.

Jacquemontia jamaicensis (Jacq.) Hall.f., 1899.

A slender, creeping or climbing, tangled, subwoody vine, the young stems and leaves puberulous; leaves sublinear to narrowly or broadly oblong, mostly 5-25 mm long, the apex acute or notched with a mucro. Inflorescence 1-few-flowered; sepals c. 2 mm long; corolla white, c. 1 cm long. Capsules 3-4 mm long; seeds ovoid, narrowly marginate, rough.

GRAND CAYMAN: *Brunt 1819; Kings GC 149, GC 354; Proctor 11969, 15171.* LITTLE CAYMAN: *Kings LC 96; Proctor 28088.* CAYMAN BRAC: *Kings CB 41; Millspaugh 1198, 1205; Proctor 29024.*

– Florida, West Indies, Yucatan, and the Turneffe Islands of Belize, frequent in dry rocky thickets and scrublands.

Genus 4. **MERREMIA** Dennst. ex. Endl., nom. cons.

Small or large vines; leaves simple and entire to variously lobed or palmately compound. Flowers axillary, solitary or in more or less condensed, stalked cymes; sepals subequal, often enlarging in fruit; corolla bell- or funnel-shaped, usually white or yellow. Stamens and style included. Ovary 2-4-celled with 2 ovules in each cell; style filiform with a globose or biglobose stigma. Capsule 2-4-celled, dehiscing lengthwise by valves, the pericarp thin and fragile; seeds glabrous or pubescent.

A pantropical genus of about 60 species.

KEY TO SPECIES

1. Leaves simple and entire, deeply cordate; inflorescence umbel-like, few- to many-flowered and dense; corolla yellow: **1. M. umbellata**

1. Leaves palmately divided into 5-7 bluntly-toothed segments; inflorescence few-flowered, loose; corolla white with a purplish eye: **2. M. dissecta**

1. Merremia umbellata (L.) Hall.f. in Bot. Jahrb. 16:552. 1893.

Ipomoea polyanthes Roem. & Schult., 1819.

I. mollicoma Miq., 1850.

A trailing or twining herbaceous vine with glabrate stems and milky sap; leaves long-petiolate, the blades narrowly triangular to broadly ovate, mostly 3-9 cm long, long-acuminate, glabrate or puberulous. Inflorescence 3-18-flowered, on stout

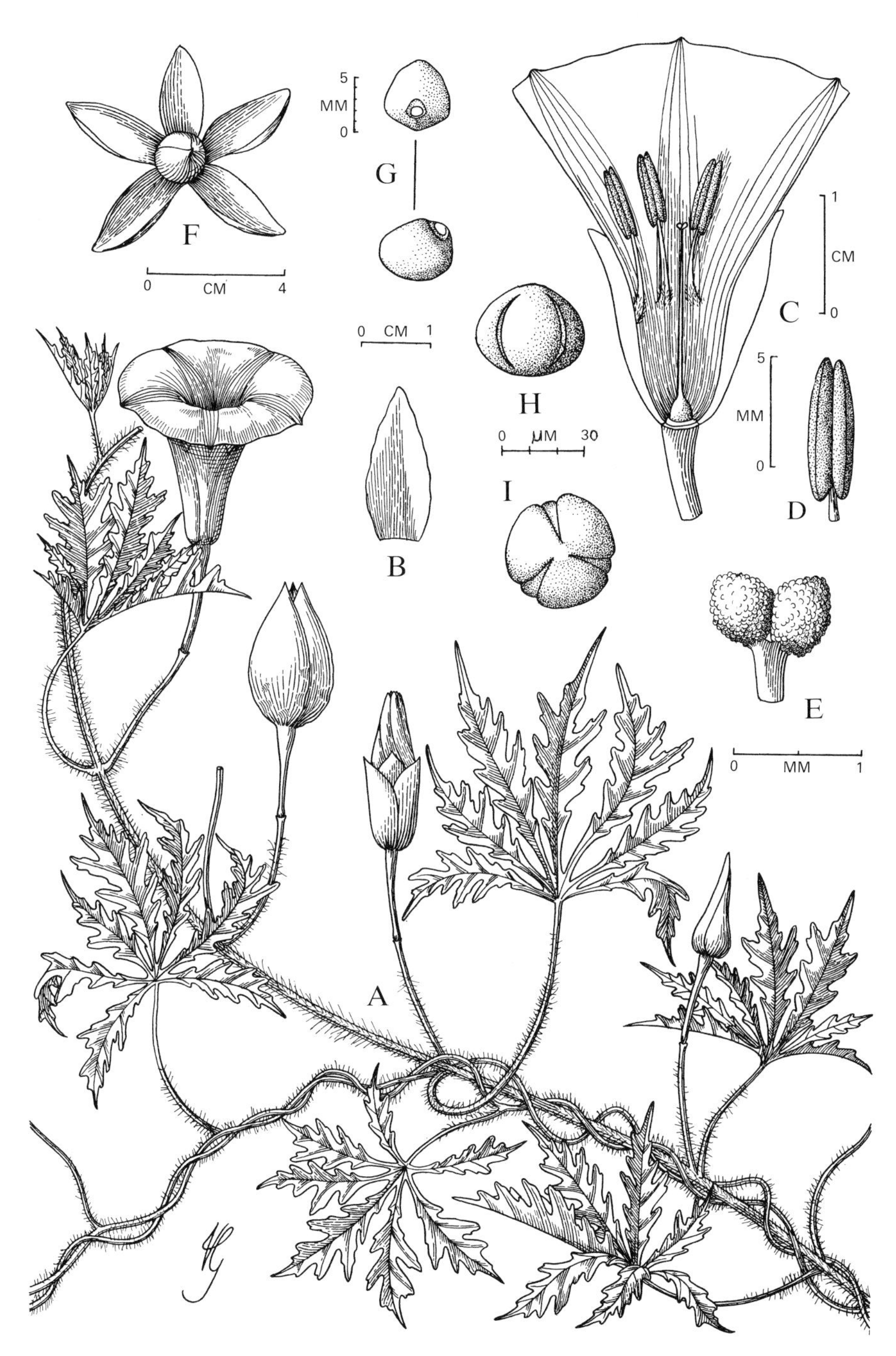

FIG. 201 **Merremia dissecta.** A, branches with leaves, flowers, and immature fruit. B, sepal. C, flower cut open to show stamens and pistil. D, anther. E, stigma. F, capsule with persistent calyx. G, two views of seed. H, I, two views of pollen grain. (G.)

peduncles 3–5 cm long; pedicels c. 1 cm long, thickened at the apex. Sepals 6–8 mm long; corolla bright yellow, c. 2.5 cm long. Anthers not twisted. Capsule 8–10 mm in diameter, partly enclosed by the persistent calyx; seeds velvety-pubescent.

GRAND CAYMAN: *Correll & Correll 51003; Lewis 3833; Millspaugh 1322; Proctor 15148.*
– Pantropical, frequent in thickets and old fields.

2. Merremia dissecta (Jacq.) Hall.f. in Bot. Jahrb. 16:552. 1893. **FIG. 201.**

A trailing or twining herbaceous vine with hirsute stems; leaves palmately divided almost to the base, the 7–9 segments coarsely sinuate-dentate, glabrous except for the long, hairy petioles. Flowers solitary or 2–3 together on long, stout, often hairy peduncles; pedicels 1–2 cm long, much thickened at the apex. Anthers spirally twisted. Capsules c. 15 mm in diameter, surrounded by the enlarged, fleshy calyx; seeds black, glabrous.

GRAND CAYMAN: *Brunt 1722, 2016; Proctor 15081.* CAYMAN BRAC: *Kings CB 84; Proctor 29097.*
– Southern United States, West Indies, and continental tropical America, along the borders of thickets, overgrown fields, and open waste places.

Genus 5. IPOMOEA L.

Creeping or twining, sometimes erect herbs or shrubs, rarely trees; leaves entire, angled, lobed, or palmately or pinnately divided. Flowers often large, soon-withering, solitary or in stalked axillary clusters, rarely aggregated in terminal panicles. Sepals 5, equal or unequal, persistent and often enlarged in fruit; corolla bell- or funnel-shaped, the limb entire or 5-angled, usually plaited. Stamens included or sometimes exserted; pollen spiny. Ovary 2–4-celled; style simple; stigma capitate. Capsule 2–4-celled, 2–4-seeded; seeds glabrous or often pubescent.

A genus of between 400 and 500 species, widely distributed in tropical and warm-temperate regions. *I. batatas* (L.) Lam., the "Sweet Potato", is one of the most important food-crops of tropical countries. Other species are often grown for ornament.

KEY TO SPECIES

1. Corolla bright red or scarlet, with exserted stamens and style:

  2. Leaves pinnate with narrow linear segments: **1. I. quamoclit**

2. Leaves entire or usually somewhat angled or lobed: **2. I. hederifolia**

1. Corolla various colours or white, but never bright red; stamens included (except in *I. macrantha,* with large white flowers):

3. Leaf-blades deeply 3-lobed or else digitately divided into 3–7 leaflets, cut nearly to junction of the petiole:

4. Corolla 1.5–2 cm long: **3. I. triloba**

4. Corolla 4–7 cm long:

5. Sepals and ovary glabrous; wild plants with twining or climbing stems:

6. Sepals c. 6 mm long: [**4. I. cairica**]

6. Sepals 10–20 mm long: **5. I. indica** var. **acuminata**

5. Sepals ciliate and ovary hairy; cultivated plant with trailing stems and large edible tubers: [**I. batatas**, p. 649]

3. Leaf-blades entire or somewhat lobed but not divided nearly to junction of the petiole:

7. Sepals 15–20 mm long:

8. Corolla pink, mauve, or bluish-purple, with funnel-shaped tube not over 5 cm long; sepals attenuate at the apex: **5. I. indica** var. **acuminata**

8. Corolla white, with cylindric tube 6–7 cm long; sepals broad at the apex; flowers opening at night: **6. I. violacea**

7. Sepals less than 15 mm long:

9. Sepals and ovary glabrous; wild plants:

10. Stems trailing on sand, rooting at nodes; leaves rather thick or leathery, often notched at apex:

11. Leaves broadly oblong to suborbicular, rarely twice as long as broad; corolla pink or purple: **7. I. pes-caprae**

11. Leaves oblong or narrowly triangular, 2–3 times as long as broad; corolla white: **8. I. stolonifera***

---

*See p. 67, Recent Nomenclatural Changes.

10. Stems climbing or twining over bushes or fences, not rooting at nodes; leaves thin, acuminate at the apex: **9. I. tiliacea**

9. Sepals ciliate and ovary hairy; plants cultivated for edible tubers: **[I. batatas,** p. 649]

1. **Ipomoea quamoclit** L., Sp. Pl. 1:159. 1753.

A slender, glabrous, annual twiner; leaves oblong to ovate in outline, 4–9 cm long, pinnately divided into linear segments mostly 1 mm wide or less, a small finely-divided accessory leaf resembling a stipule arising at the base of each regular leaf. Peduncles 1-6-flowered; sepals 4–6 mm long, mucronulate; corolla 2.5–4 cm long with flat, lobate limb. Ovary 4-celled. Capsules ovoid, c. 10 mm long.

CAYMAN BRAC: *Proctor 29367.*
– Southeastern United States, West Indies, and continental tropical America, introduced in the Old World tropics; the Cayman Brac specimens were found in roadside thickets.

2. **Ipomoea hederifolia** L., Syst. Nat. ed. 10, 2:925. 1759. **FIG. 202.**

Ipomoea coccinea of Griseb., 1862, not L., 1753.

A slender, annual twiner, glabrous or nearly so; leaves long-petiolate, the simple, cordate blades very broadly ovate, up to about 7 cm long and broad, acuminate at the apex and with margins entire or somewhat angular and with a few broad, lobe-like teeth. Flowers few in elongate dichasial cymes; sepals c. 3 mm long with a subulate tooth of about the same length or longer arising from the back of each; corolla 2–4 cm long, the limb obscurely 5-lobed. Ovary 4-celled. Capsules globose, 6–7 mm in diameter.

GRAND CAYMAN: *Brunt 1688.*
– Eastern and southern United States, West Indies and continental tropical America, in open waste ground and moist thickets.

3. **Ipomoea triloba** L., Sp. Pl. 1:161. 1753. **FIG. 203.**

A slender annual twiner, the usually glabrous stems mostly trailing; leaves long-petiolate and with blades cordate and usually deeply 3–5-lobed or rarely entire, more or less ovate or sagittate in outline, mostly 2–5 cm long, the apex acuminate.

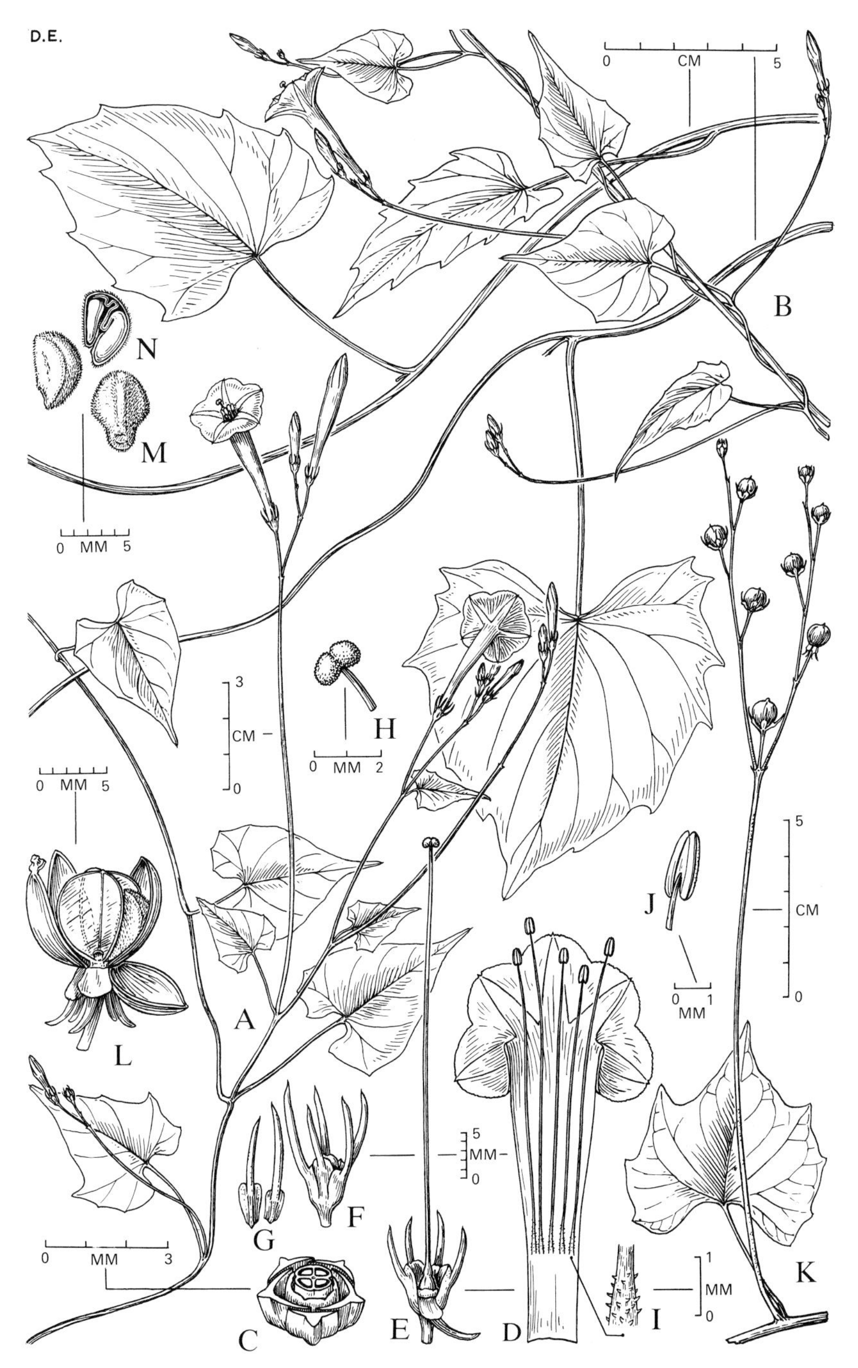

FIG. 202 **Ipomoea hederifolia.** A, B, habit. C, x-section through base of flower. D, corolla spread out to show stamens. E, calyx with pistil. F, calyx alone. G, separated sepals. H, stigma. I, base of filament. J, anther. K, fruiting inflorescence. L, capsule after dehiscence. M, two views of seed. N, section through seed. (D.E.)

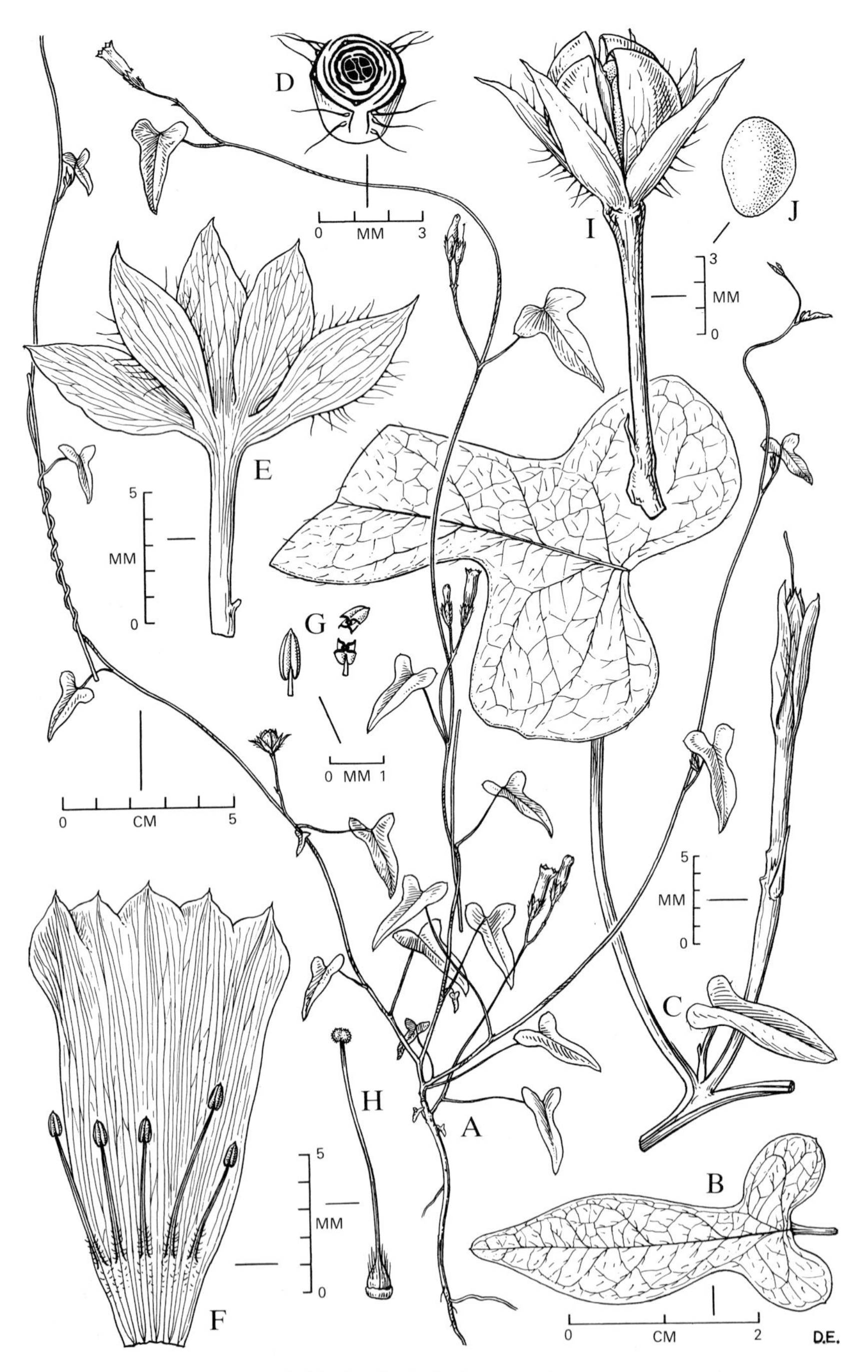

FIG. 203 **Ipomoea triloba.** A, habit. B, single leaf. C, portion of stem showing insertion of inflorescence. D, x-section through base of flower. E, calyx spread out. F, corolla spread out, showing stamens. G, anther, whole and cut crosswise. H, pistil. I, capsule. J, seed. (D.E.)

Flowers solitary or few, clustered on a rather long peduncles; sepals c. 6 mm long, acuminate, hairy; corolla pink or rose. Ovary 2-celled. Capsules subglobose, c. 4 mm in diameter.

GRAND CAYMAN: *Brunt 1941, 2109; Hitchcock; Kings GC 85, GC 330; Millspaugh 1385; Proctor 27936, 27955; Sachet 373, 419.* LITTLE CAYMAN: *Proctor 35141, 35184.* CAYMAN BRAC: *Proctor 28977.*
– Southeastern United States, West Indies, and continental tropical America; naturalized in the Old World tropics; a weed of sandy ground, moist pastures, clearings and thickets.

[4. **Ipomoea cairica** (L.) Sweet, Hort. Brit. 287. 1827.

A slender, nearly glabrous trailing or climbing vine, the stems becoming subwoody. Leaves palmately divided into 5 lanceolate to ovate segments, the middle one longest and 1.5–4 cm long. Flowers 1–10 in umbel-like cymes on a short peduncle; sepals ovate, acute, c. 6 mm long; corolla lavender or light mauve, 4–5 cm long. Capsules c. 10 mm in diameter; seeds hairy.

GRAND CAYMAN: *Proctor 11967.*
– Native of tropical Asia and Africa, now introduced and naturalized in most warm countries.]

5. **Ipomoea indica** (Burm.) Merrill, Interpr. Rumph. Herb. Amboin. 445. 1917. var. **acuminata** (Vahl) Fosberg in Bot. Notiser 129:38. 1976.

Ipomoea acuminata (Vahl) Roem. & Schult. in L., 1819.

I. cathartica Poir. in Lam., 1816.

I. jamaicensis of Hitchcock, 1893, not G. Don, 1837.

Slender perennial vine climbing to 5 m or more, the young stems more or less puberulous. Leaves long-petiolate, the blades cordate, ovate, mostly 4–8 cm long, the margins varying from entire to deeply 3-lobed, acuminate at the apex. Flowers solitary or few in a small, loose cluster; peduncles rather stout, 2–4 cm long, appressed-puberulous; sepals lance-acuminate, unequal, mostly 15–20 mm long; corolla mostly 5.5–7 cm long. Capsules subglobose, 10–12 mm in diameter; seeds puberulous.

GRAND CAYMAN: *Brunt 2140; Hitchcock; Lewis 3830, 3844a; Millspaugh 1244, 1246, 1372, 1381, 1403; Proctor 15107, 15108, 27980; Sachet 420.* LITTLE CAYMAN: *Kings LC 19; Proctor 35112.* CAYMAN BRAC: *Kings CB 74; Millspaugh 1227, 1235.*
– Pantropical; common in roadside thickets and along the borders of moist woodlands.

6. Ipomoea violacea L.Sp. P1.1: 161. 1753. **FIG. 204.**

Ipomoea macrantha Roem. & Schult. in L. 1819.

Ipomoea tuba (Schlecht.) G. Don, 1837.

"Cowslip"

Perennial, glabrous, creeping or high-climbing vine (to 12 m or more), with white latex, the older stems becoming subwoody. Leaves rather fleshy, long-petiolate, the blades deeply cordate, very broadly ovate, 4–15 cm long, abruptly acuminate at the apex, the margins entire or with a short, acuminate lobe on either side. Flowers nocturnal, solitary or sometimes 2 together on a stout peduncle; sepals obtuse, 20 mm long; corolla white, the limb 7–8 cm across. Capsules subglobose, 2–2.5 cm in diameter, enclosed by the enlarged fleshy sepals; seeds densely hairy on the angles.

GRAND CAYMAN: *Brunt 1654, 1746, 1877; Fawcett; Hitchcock; Proctor 11968.* LITTLE CAYMAN: *Kings LC 45; Proctor 28045.* CAYMAN BRAC: *Kings CB 103; Millspaugh 1234; Proctor 28930.*
– Pantropical, chiefly in sandy, swampy, or rocky coastal thickets, often in exposed situations. The cliffs toward the eastern end of Cayman Brac are in many places festooned with this vine.

7. Ipomoea pes-caprae (L.) R. Br. in Tuckey, Narrat. Exped. Zaire 477. 1818. ssp. **brasiliensis** (L.) Ooststr. in Blumea 3:533. 1940. **FIG. 205.**

"Bay Vine"

A perennial, subsucculent trailing vine, the glabrous stems reaching 6 m or more in length; leaves on petioles up to 15 cm long, the blades entire, mostly 5-10 cm long, broadly notched at the apex. Flowers 1-few on stout peduncles; pedicels 1-3 cm long or more; sepals ovate-oblong, apiculate, the outer 6–10 mm long, the inner 8–15 mm long; corolla 4–5 cm long. Ovary 2-celled. Capsules subglobose, c. 1.5 cm in diameter; seeds brownish-pubescent.

GRAND CAYMAN: *Kings GC 67; Lewis GC 24; Maggs II 58; Proctor 15062.* LITTLE CAYMAN: *Kings LC 18; Proctor 28165.* CAYMAN BRAC: *Kings CB 19; Millspaugh 1228; Proctor 28997.*
– Pantropical on sandy seashores. The subspecies *pes-caprae* is restricted to certain Old World localities.

8. Ipomoea stolonifera* (Cyrillo) J.F.Gmel. in L., Syst. Nat. ed. 13, 2:345. 1791.

Ipomoea carnosa R.Br., 1810.

---

*Properly **I. imperati**. See p.67, Recent Nomenclatural Changes.

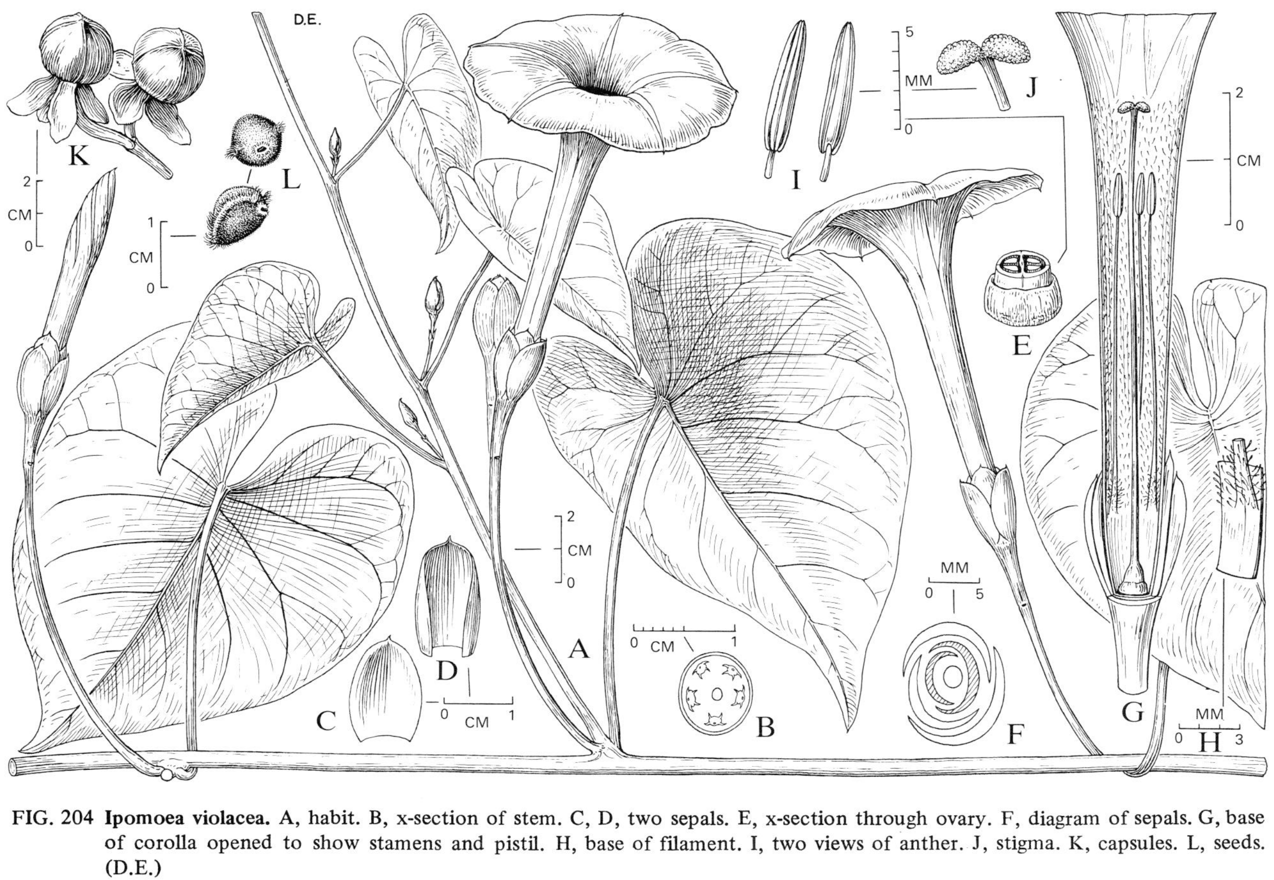

FIG. 204 **Ipomoea violacea.** A, habit. B, x-section of stem. C, D, two sepals. E, x-section through ovary. F, diagram of sepals. G, base of corolla opened to show stamens and pistil. H, base of filament. I, two views of anther. J, stigma. K, capsules. L, seeds. (D.E.)

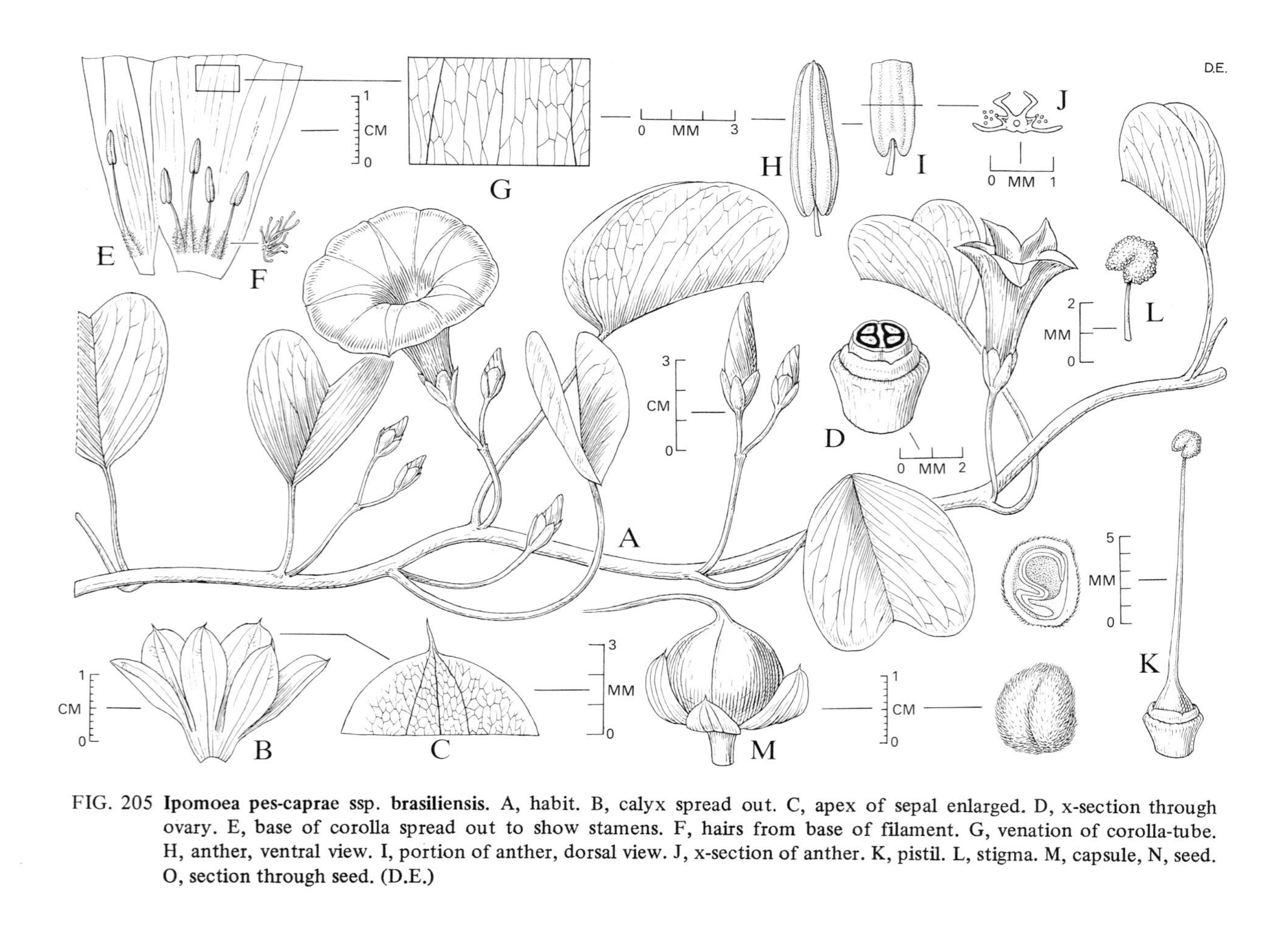

FIG. 205 **Ipomoea pes-caprae** ssp. **brasiliensis.** A, habit. B, calyx spread out. C, apex of sepal enlarged. D, x-section through ovary. E, base of corolla spread out to show stamens. F, hairs from base of filament. G, venation of corolla-tube. H, anther, ventral view. I, portion of anther, dorsal view. J, x-section of anther. K, pistil. L, stigma. M, capsule, N, seed. O, section through seed. (D.E.)

Trailing herb, the glabrous stems extensively creeping on or under sand and rooting at most nodes; leaves rather long-petiolate, the leathery blades narrowly oblong or triangular, mostly 2.5–6 cm long, often slightly notched at the apex, the margins entire or sometimes irregularly lobed. Flowers solitary on short or long peduncles; sepals oblong, acute, 10–15 mm long; corolla 3.5–5 cm long, white, the tube often yellowish within and with a purplish spot at the base. Capsules globose, 10–15 mm in diameter; seeds smooth.

GRAND CAYMAN: *Millspaugh 1310.* LITTLE CAYMAN: *Proctor 28053.* CAYMAN BRAC: *Millspaugh 1222.*
– Pantropical on sandy beaches near the sea, apparently never very common.

9. **Ipomoea tiliacea** (Willd.) Choisy in DC., Prodr. 9:375. 1845.

Ipomoea fastigiata (Roxb.) Sweet, 1827.

"Hog Slip", "Wild Slip"

Herbaceous twining vine, glabrous or nearly so, the roots sometimes with tubers, the stems climbing to 5 m high or more; leaves long-petiolate, the blades shallowly cordate, broadly ovate, 4–7 cm long or more, acuminate and with entire margins. Flowers few to many in cymes on long peduncles; sepals lanceolate, 7–10 mm long, smooth and scarious, sharply pointed; corolla light mauve to pink or purplish with paler or whitish tube and darker spot at the base within, 3.5–5 cm long. Ovary 2-celled. Capsules depressed-globose, c. 6 mm long and 8 mm broad; seeds nearly black, glabrous.

GRAND CAYMAN: *Brunt 1716, 1936; Hitchcock; Kings GC 75, GC 299; Lewis 3844b.*
– Pantropical, frequent in roadside thickets, along borders of woodlands, and in overgrown pastures.

[**Ipomoea batatas** (L.) Lam., the "Sweet Potato", is or has been commonly cultivated in Grand Cayman (*Millspaugh 1290*) and Cayman Brac.]

Family 89

MENYANTHACEAE

Perennial herbs, mostly aquatic; leaves alternate, simple or trifoliate, and entire or toothed; stipules absent. Flowers regular, perfect, and solitary or in umbel-like cymes; calyx 5–6-cleft, persistent; corolla gamopetalous, rotate or funnel-shaped, the 5 or 6 lobes induplicate-valvate in bud. Stamens equal in number to the corolla-lobes and alternate with them, inserted in the corolla-tube; filaments free; anthers 2-celled, opening lengthwise. Ovary superior or partly inferior, 1-celled, with numerous ovules on usually 2 parietal placentas; style simple, with 2-3-lobed stigma. Fruit a capsule, indehiscent, bursting irregularly, or 2-valved; seeds with copious endosperm.

A small family of world-wide distribution, consisting of 5 genera and about 40 species.

Genus 1. **NYMPHOIDES** Séguier

Aquatic perennial herbs with fleshy rootstocks and elongate leafy flowering stems; leaves floating, simple and deeply cordate with entire margins. Flowers in clusters closely subtended by a leaf-blade, thus appearing to arise from the top of an elongate petiole, yellow or white, the parts in 5's; corolla deeply lobed, the lobes with fringed margins and often hairy within. Stamens inserted near the base of the corolla, with appendages on the corolla between them; anthers sagittate. Capsule rupturing irregularly.

A cosmopolitan genus of about 20 species.

1. **Nymphoides indica** (L.) Ktze., Révis. Gen. Pl. 2:429. 1891. **FIG. 206.**

Nymphoides humboldtiana (Kunth) Ktze., 1891.

"Water Snowflake"

Leaves glabrous, broadly ovate to nearly orbicular, 3–15 cm wide, often purplish beneath. Pedicels 2–5 cm long, erect when flowering and deflexed in fruit; calyx-lobes c. 6 mm long; corolla 2–2.5 cm across when expanded, white, yellow at the base. Flowers heterostylous, the yellow filaments 1 or 4 mm long, the ovary and style 10 or 6 mm long. Capsules ovoid, c. 6 mm long, many-seeded; seeds c. 1.5 mm in diameter, smooth.

GRAND CAYMAN: *Brunt 1793, 2165; Kings GC 210; Proctor 27956; Sachet 450.*
– Pantropical, in ponds or small pools, sometimes on mud where water has receded. Presumably this species is tolerant of somewhat brackish conditions.

Family 90

HYDROPHYLLACEAE

Annual or perennial herbs, rarely shrubs, often hairy or scabrid, sometimes armed with spines; leaves alternate or rarely opposite, sometimes in basal rosettes,

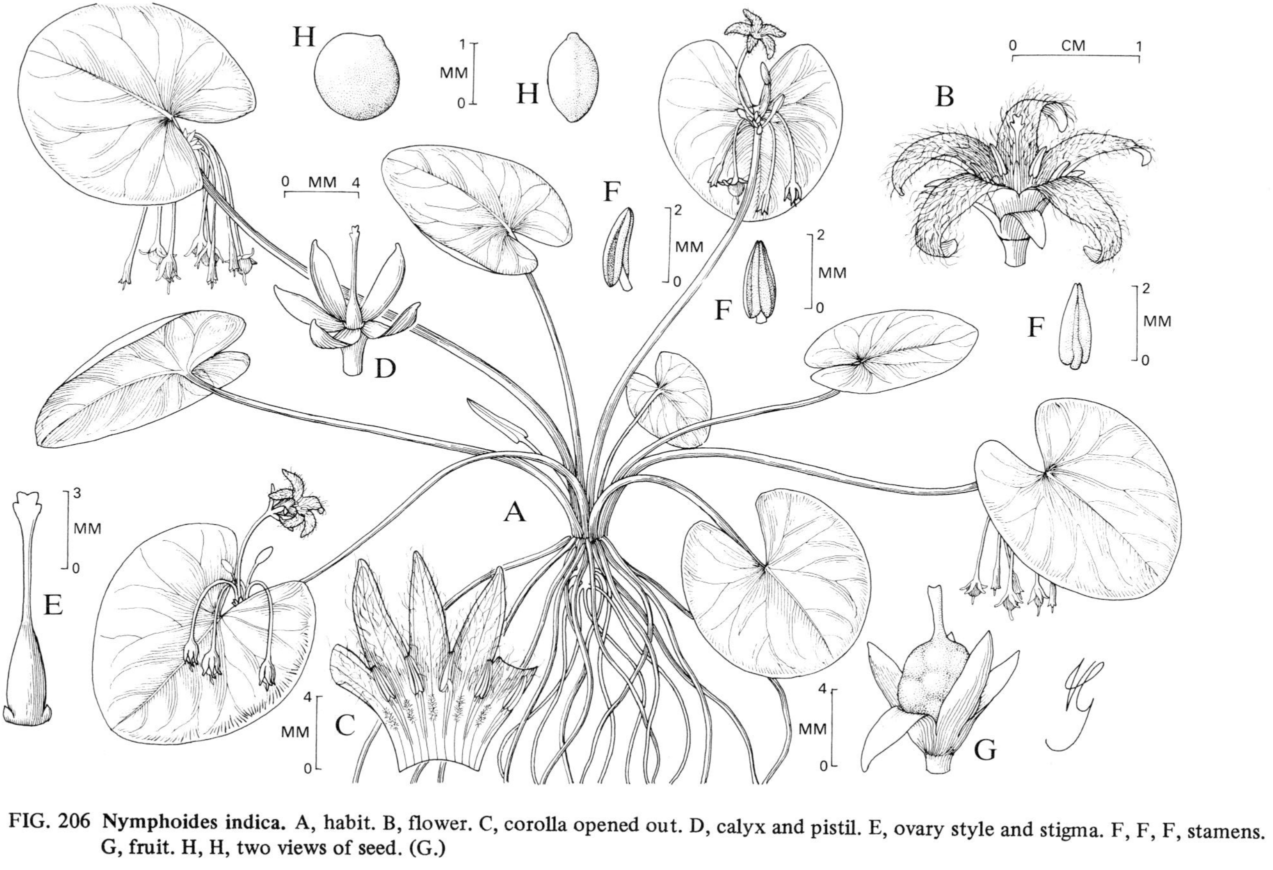

FIG. 206 **Nymphoides indica.** A, habit. B, flower. C, corolla opened out. D, calyx and pistil. E, ovary style and stigma. F, F, F, stamens. G, fruit. H, H, two views of seed. (G.)

entire to pinnately or palmately divided; stipules absent. Flowers perfect and regular, borne in dichasial, umbel-like, or helicoid cymes, or sometimes solitary in the leaf-axils. Calyx 5-lobed, imbricate, the lobes often with appendages between; corolla gamopetalous, 5-lobed, the lobes imbricate or contorted. Stamens usually 5, inserted on the corolla-tube near its base and alternate with the lobes; filaments often dilated at the base, or subtended by appendages; anthers 2-celled, opening lengthwise. Ovary superior or partly inferior, 2-carpellate but usually 1-celled with 2 parietal placentas meeting in the centre, the ovules 4 to numerous; style 1 or 2 (rarely more). Fruit a capsule, dehiscent or rarely indehiscent; seeds variously sculptured; endosperm copious or thin.

A widespread family of 20 genera and about 270 species, best represented in North America, absent from Australia.

Genus 1. **NAMA** L., nom. cons.

Prostrate to erect, annual or perennial herbs, sometimes partly woody; leaves alternate and mostly entire, rarely toothed. Flowers solitary or paired in the axils of the upper leaves, or several in reduced lateral or terminal cymes; calyx divided nearly to the base, with narrow, subequal lobes elongating in fruit; corolla white to purple, tubular or funnel-shaped. Stamens usually included, unequally inserted or unequal in length; filaments usually glabrous, dilated or appendaged at the base. Styles 2, usually free or sometimes partly united; stigmas small, capitate. Capsules dehiscent; seeds brown and variously pitted, reticulate or smooth.

A chiefly American genus of between 40 and 50 species.

1. **Nama jamaicense** L., Syst. Nat. ed. 10, 2:950. 1759. **FIG. 207.**

Annual herb with spreading prostrate branches, these pubescent and more or less winged by the decurrent petiole-margins; leaves obovate to spatulate, 1–6 cm long or more, obtuse at the apex. Flowers solitary or few, subsessile or stalked; sepals linear, c. 7 mm long in flower, longer in fruit; corolla white, 5–6 mm long, narrowly funnel-shaped with lobes 1–1.5 mm long. Capsules c. 5 mm long; seeds light brown, minutely pitted.

GRAND CAYMAN: *Hitchcock.*
– Florida, Texas, West Indies, and continental tropical America, a frequent weed of shady waste places.

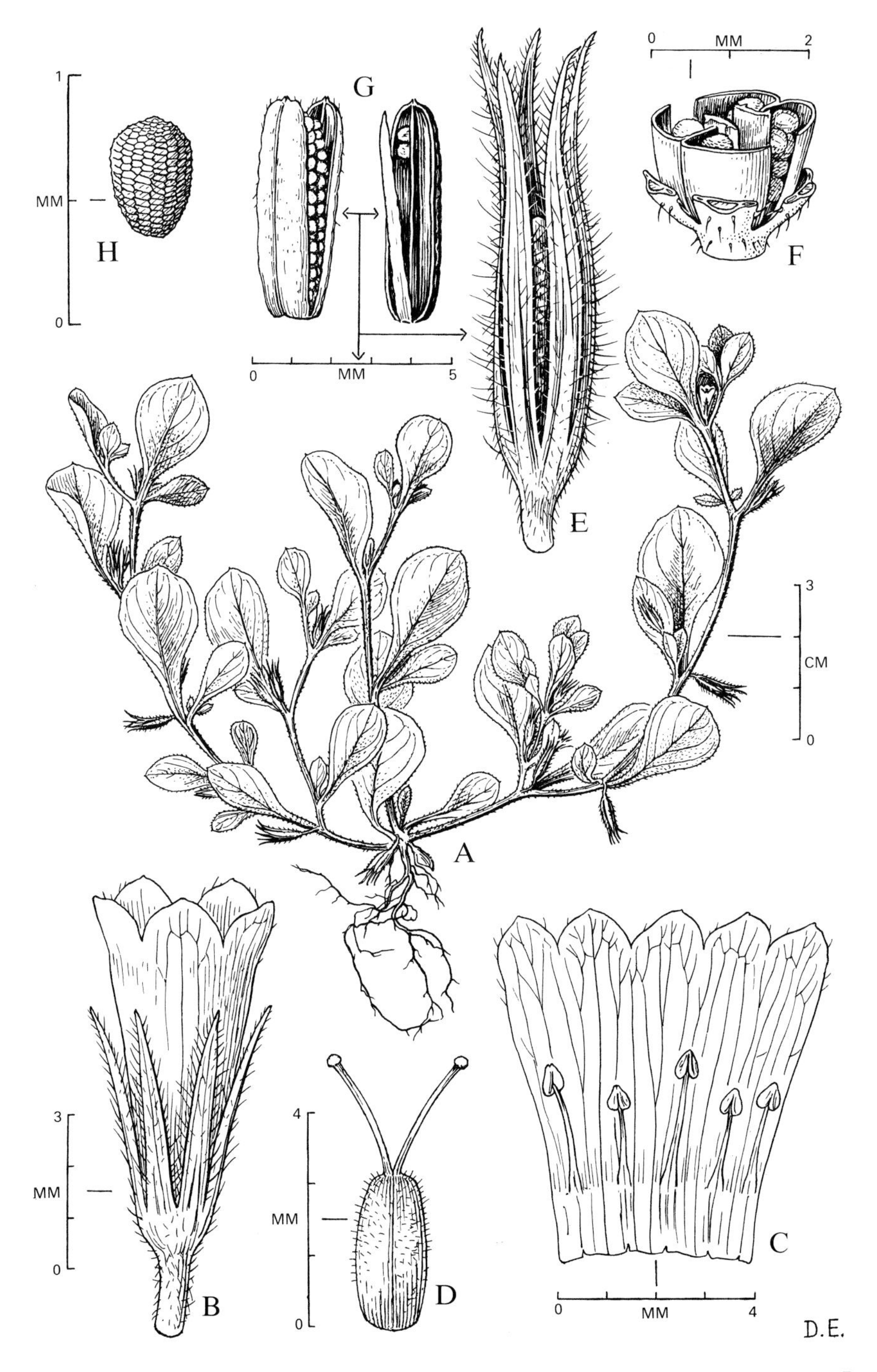

FIG. 207 **Nama jamaicensis.** A, habit, B, flower. C, corolla spread out to show stamens. D, ovary and styles. E, fruiting calyx. F, x-section through capsule. G, two views of capsule after dehiscence. H, seed. (D.E. ex St.)

Family 91

BORAGINACEAE

Annual or perennial herbs, or else shrubs, trees, or woody vines, usually with scabrid, setose, or hispid indument, rarely tomentose or glabrous; leaves alternate (rarely opposite or whorled), simple, entire or toothed; stipules none. Flowers usually regular, perfect or rarely polygamous by abortion, borne in usually dichotomous cymose spikes, racemes, heads or panicles. Calyx usually 5-toothed or -lobed, more or less bell-shaped, imbricate or often open in bud, usually persistent and often enlarging in fruit. Corolla gamopetalous, usually 5-lobed (rarely 6–8-lobed), tubular, funnel-, bell- or salver-shaped or subrotate, the lobes imbricate or twisted in bud, sometimes with pleats or appendages in the tube or partly closing the throat. Stamens as many as the corolla-lobes and alternating with them, inserted in the corolla-throat or -tube; filaments free; anthers 2-celled, opening lengthwise, attached above the base and more or less 2-lobed below the point of attachment. Disk ring-like, entire or 5-lobed, or apparently absent. Ovary superior, 2-celled and 2-carpellate, but often becoming falsely 4-celled at maturity, simple or deeply 4-lobed; ovules usually 4, 2 in each carpel; styles 1 or 2, terminal on simple ovaries, or sometimes arising from between the lobes on lobate ones, simple or 1–2-forked at the apex. Fruit of 4 1-seeded nutlets or a 1–4-seeded nut or drupe; seeds with or without endosperm.

A cosmopolitan family of about 100 genera and 2,000 species.

KEY TO GENERA

1. Style 1, with solitary simple or 2-lobed stigma, or stigma sessile, then usually large and peltate or conical; flowers in simple or forked one-sided spikes, uncoiling in development; herbs or shrubs:

 2. Shrubs, erect and up to 2 m tall or more, or vine-like; fruits drupe-like, fleshy at least at first, later drying into 2 or 4 nutlets:

  3. Leaves succulent, narrowly oblanceolate, covered with silky tomentum; fruits hollowed at the base: **1. Argusia**

  3. Leaves not as above; fruits not hollowed at the base: **2. Tournefortia**

 2. Annual herbs or low, much-branched or dense, small-leaved shrubs; fruits dry, breaking up into 2 or 4 nutlets: **3. Heliotropium**

1. Styles 2, free or united below, each simple or forked; flowers in cymose panicles or heads, these not one-sided or uncoiling in development; shrubs or trees:

4. Stigmas 2 (styles simple):

5. Expanded corolla 4-5 mm across, the lobes longer than the tube; ripe fruits yellow or ultimately black: **4. Ehretia**

5. Expanded corolla more than 8 mm across, the tube longer than the lobes; ripe fruits red: **5. Bourreria**

4. Stigmas 4 (styles forked): **6. Cordia**

Genus 1. **ARGUSIA** Amman ex Boehmer

Herbs, shrubs, or trees clothed with whitish-silky hairs. Leaves fleshy, alternate, more or less crowded, linear to oblong or obovate. Flowers in scorpioid or corymbose cymes; calyx sessile or pedicellate, the lobes cuneate to orbicular; corolla small, white, salver-shaped, the 5 valvate lobes shorter than the nearly cylindric tube; stamens included, with short filaments and elongate anthers; style simple. Fruit drupelike, dry and bony, hollowed at the base; carpels more or less embedded in hard, corky exocarp, at maturity the 2 nutlets ultimately separating.

A small genus of 3 species, all confined to seacoasts or saline shores, one (herbaceous) occurring in temperate Asia, another (a tree) widespread in the Old World tropics, and the third (a shrub) occurring in the West Indies.

1. **Argusia gnaphalodes** (L.) Heine, Fl. Nouv. Caléd. 7:108. 1976. **FIG. 208.**

Heliotropium gnaphalodes L., 1759.

Tournefortia gnaphalodes (L.) R. Br., 1819.

Mallotonia gnaphalodes (L.) Britton, 1915.

Messerschmidia gnaphalodes (L.) I. M. Johnston, 1935.

"Lavender", "Sea-lavender"

Dense, moundlike shrub to 2 m tall; leaves linear-oblanceolate, 3-6 cm long or more, 3-6 mm broad below the obtuse apex. Inflorescence stalked, scarcely exceeding the leaves; flowers in dense scorpioid cymes becoming headlike in fruit; calyx-lobes 2-3 mm long; corolla slightly exceeding the calyx, the tube pubescent outside. Fruits dark brown, glabrous, 5-6 mm long.

GRAND CAYMAN: *Fawcett; Hitchcock; Kings GC 263; Proctor 15047; Sachet 400.* LITTLE CAYMAN: *Kings LC 56; Proctor 28171.* CAYMAN BRAC: *Kings CB 102; Millspaugh 1177; Proctor 28931.*
– Bermuda, Florida, West Indies, and the coasts of Yucatan, Cozumel, Belize, and Venezuela, characteristically in sand at the top of sea-beaches.

Genus 2. **TOURNEFORTIA** L.

Shrubs, woody vines, or treelike; leaves alternate, entire. Flowers in scorpioid cymes usually forked 1-several times, sometimes elongate and flexuous. Calyx 5-lobed, one lobe often longer than the others. Corolla cylindric with 5 short spreading lobes. Stamens included; ovary 4-celled, sometimes 4-lobed; style terminal, usually forked at apex. Fruit a small entire or 4-lobed drupe, fleshy or dry, finally separating into 2-4 nutlets, these 1-2-seeded.

A diverse pantropical genus of perhaps 100 species or more.

KEY TO SPECIES

1. Leaves less than 7 cm long; a slender trailing or scrambling shrub; corolla-tube 2 mm long: **1. T. volubilis**

1. Leaves mostly 10-20 cm long; an erect shrub to 3 m tall or more; corolla-tube 5 mm long or more: **2. T. astrotricha**

1. **Tournefortia volubilis** L., Sp. Pl. 1:140. 1753.

"Aunt Eliza Bush"

Usually trailing or twining, rarely erect, the younger stems puberulous; leaves lanceolate or narrowly oblong to narrowly ovate, 2-7 cm long, acute or subacuminate at the apex, glabrous or sparsely puberulous. Inflorescence-branches very slender, curved or flexuous, up to 8 cm long but usually much shorter; calyx c. 1 mm long; corolla pale greenish, the lobes subulate, c. 1 mm long. Ripe fruits white, usually with black spots, glabrous, 2-3 mm in diameter.

GRAND CAYMAN: *Brunt 1996, 2132; Kings GC 140; Proctor 15155, 15170.* LITTLE CAYMAN: *Kings LC 12; Proctor 28112, 35102.* CAYMAN BRAC: *Proctor 29018.*
– Florida, Texas, West Indies and continental tropical America, chiefly in rocky thickets and woodlands; a variable species over much of its range, but the Cayman population is quite uniform.

2. **Tournefortia astrotricha** DC., Prodr. 9:520. 1845.

An erect or sometimes arborescent shrub 1.5-5 m tall, with stout branches, sometimes of open or straggling habit; leaves elliptic but often more or less asymmetric or folded, usually of rather harsh texture, acute or subacuminate at

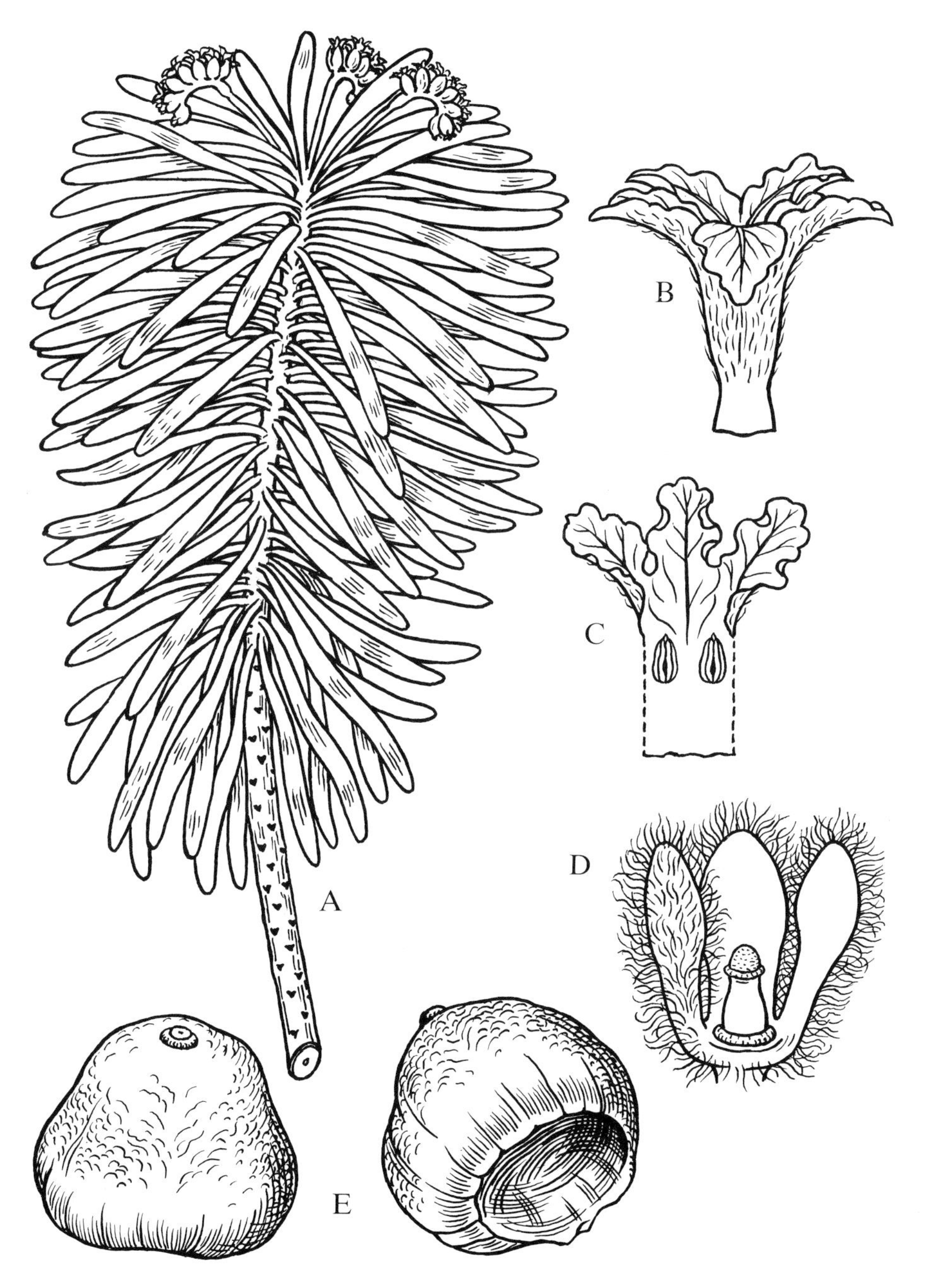

FIG. 208 **Argusia gnaphalodes** A, branch with leaves and inflorescence, x$^{3}/_{8}$. B, corolla, x6. C, corolla opened to show stamens, x6. D, calyx with two sepals removed to show pistil, x6. E, two views of fruit, x6. (St.)

the apex. Flowers odourless or somewhat fragrant, in helicoid one-sided cymes, these often becoming elongate and pendent; calyx c. 2.5 mm long; corolla cream or whitish turning pink. Ripe fruits pure white, unspotted, rather fleshy.

The Cayman plants of this species have been assigned to two varieties, as follows:

1. Leaves densely short-pubescent beneath; corolla c. 5 mm long; fruits to 10 mm in diameter: (a) var. **astrotricha**

1. Leaves glabrous beneath or with a few scattered hairs; corolla c. 6 mm long; fruits 6–9 mm in diameter: (b) var. **subglabra**

2a. Tournefortia astrotricha var. **astrotricha**

GRAND CAYMAN: *Kings GC 155; Proctor 11999, 15041.* CAYMAN BRAC: *Proctor 28946.*

– Jamaica and the Swan Islands, in dry rocky thickets and woodlands. The Swan Islands plants are intermediate in pubescence between the two varieties.

2b. Tournefortia astrotricha var. **subglabra** Stearn in Jour. Arnold Arb. 52:633. 1971.

GRAND CAYMAN: *Brunt 2080; ? Fawcett; ? Kings GC 343.* Fawcett's record of *T. cymosa* L. (= *T. glabra* L.) probably pertains to this variety, but his specimen has not been found for examination.

– Jamaica, in habitats like those of the preceding.

Genus 3. HELIOTROPIUM L.

Annual or perennial herbs or sometimes low shrubs; leaves alternate or sub-opposite, rarely whorled, always simple and entire. Flowers usually in one-sided scorpioid spikes, these simple or 1-2-forked, or sometimes the flowers solitary and internodal. Calyx of 5 teeth or lobes, often unequal; corolla small, with cylindric tube and spreading 5-lobed limb. Stamens included; anthers obtuse, mucronate or short-appendaged. Ovary 4-celled, the terminal stigma sessile or borne on a distinct style, the apex usually bearing a conic or elongate sterile appendage, this often bifid. Fruit dry, entire or lobed, at maturity separating into 2–4 bony nutlets, the nutlets 1-2-seeded; endosperm thin.

A widely distributed, diverse genus of about 150 species, sometimes not easily distinguishable from *Tournefortia.*

KEY TO SPECIES

1. Plant glabrous, succulent, and usually somewhat glaucous: **1. H. curassavicum**

1. Plant pubescent, the leaves not succulent:

2. Flowers borne on more or less elongate, distinctly scorpioid spikes; annual herbs, sometimes persisting and slightly woody at the base:

3. Corolla white, 2 mm long; fruits 2 mm long, the nutlets rounded and not ribbed: **2. H. angiospermum**

3. Corolla pale blue, 4-5 mm long; fruits 3-4 mm long, the nutlets pointed and strongly ribbed: **3. H. indicum**

2. Flowers solitary in the axils or on short, terminal, non-scorpioid spikes; depressed or erect shrubs:

4. Erect shrub to 1 m tall or more; leaves up to 10 mm long or more; flowers in short terminal spikes: **4. H. ternatum**

4. Prostrate, dense, matlike shrub not more than a few cm high and c. 15 cm across; leaves 2-3 mm long; flowers solitary in the upper axils: **5. H. humifusum**

1. Heliotropium curassavicum L., Sp. Pl. 1:130. 1753.

Annual succulent herb of pale blue-green colour, the branches decumbent-ascending; leaves sessile, linear-oblanceolate or narrowly spatulate, mostly 1-4 cm long and 2-6 mm broad. Flowers sessile or nearly so in simple or twin scorpioid cymes 1-4 cm long or more; calyx 1-2 mm long; corolla white, the tube c. 2 mm long. Anthers sagittate, sessile. Fruit 2 mm in diameter, separating into 4 nutlets.

GRAND CAYMAN: *Brunt 1830; Kings GC 322; Proctor 15137.*
– Widespread in tropical America in brackish or saline soils, mostly at or near sea-level.

2. Heliotropium angiospermum Murray, Prodr. Stirp. Goett. 217. 1770. **FIG. 209**, H-I.

Heliotropium parviflorum L., 1771.

"Scorpion Tail", "Bastard Chelamella"

Erect, loosely-branched herb, at length becoming somewhat woody toward the base, and up to 1 m tall but often lower, the stems sparsely clothed with appressed hairs. Leaves alternate or subopposite, petiolate, the blades lanceolate or narrowly ovate to elliptic, mostly 2-9 cm long, acute at the apex, the upper surface with scattered hairs, the lower pubescent on midrib and veins. Cymes simple or once-forked, up to 12 cm long in fruit. Stigma sessile. Fruit glabrous but covered with minute scales.

GRAND CAYMAN: *Brunt 2029; Hitchcock; Lewis 3856; Millspaugh 1287; Proctor 15159.* LITTLE CAYMAN: *Kings LC 41; Proctor 28058.* CAYMAN BRAC: *Kings CB 67; Proctor 29121.*
– Widespread in tropical America, frequent along roads and paths, in pastures, and in open waste ground.

3. **Heliotropium indicum** L., Sp. Pl. 1:130. 1753. **FIG. 209**, J-K.

"Scorpion Tail"

Erect, annual herb to 1 m tall or more, the stems and leaves more or less hispid; leaves petiolate, the blades broadly ovate, 4-6 cm long or often longer, acute at the apex and abruptly contracted at the base, the margins somewhat wavy. Flowers in simple scorpioid spikes up to 15 cm long in fruit, but much shorter while flowering. Stigma borne on an evident style. Fruit glabrous.

GRAND CAYMAN: *Kings GC 275.*
– A pantropical weed, probably of American origin.

4. **Heliotropium ternatum** Vahl, Symb. Bot. 3:21. 1794. **FIG. 209**, A-G.

A mostly erect, bushy shrub up to 1.5 m tall, densely white-pubescent throughout; leaves rigid, alternate, opposite or in whorls of 3, sharply acute at the apex, the hairs on the upper side with pustulate bases, the margin usually revolute. Flowers few in short terminal spikes; calyx 3-4 mm long, white-strigose; corolla white with yellow eye, the tube 4 mm long, the expanded limb mostly 3-4 mm across. Nutlets subglobose, black.

GRAND CAYMAN: *Brunt 2007, 2199; Proctor 15195.*
– West Indies and continental tropical America, in dry sandy or rocky thickets.

5. **Heliotropium humifusum** Kunth in H.B.K., Nov. Gen. & Sp. 3:85, t. 205. 1818.

A dense, matlike, suffrutescent shrub, diffusely branched, clothed throughout with stiff white strigose hairs mostly with pustulate bases; leaves densely overlapping,

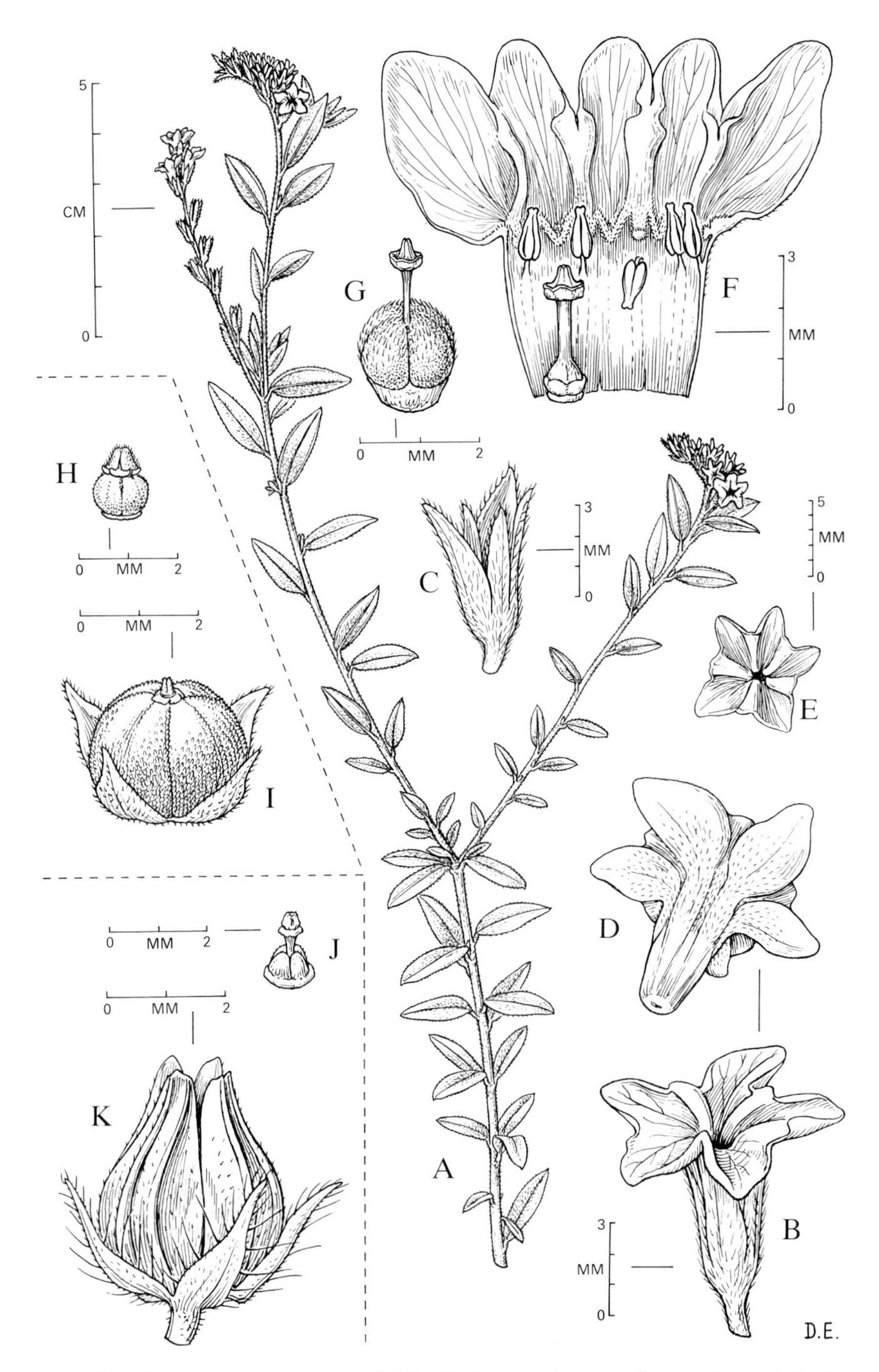

FIG. 209 **Heliotropium ternatum.** A, habit. B, flower. C, calyx. D, corolla, ventral view. E, corolla, view of limb. F, corolla spread out, showing stamens and pistil. G, fruit. H, **H. angiospermum**, pistil. I, fruit. J, **H. indicum**, pistil, K, fruit. (D.E. ex St.)

rigid, lance-oblong, acute at the apex and 1-nerved, attached to the stem by a broad base. Calyx strigose, 2–3 mm long; corolla white with yellow eye, the tube scarcely exceeding the calyx, the expanded limb 2–3 mm across or more. Stigma capitate on a short style. Fruit subglobose, enclosed by the calyx, the nutlets hispidulous.

GRAND CAYMAN: *Crosby, Hespenheide & Anderson 37 (GH, MICH); Kings GC 59, GC 123; Proctor 15159; Sauer 4214 (WIS).* LITTLE CAYMAN: *Kings LC 48, LC 78; Proctor 28129, 35137.* CAYMAN BRAC: *Proctor 35154.*
– Cuba and Hispaniola, usually in gravel-filled pockets of dry, exposed limestone, or in sandy clearings. Cuban plants of this affinity growing on serpentine appear to represent a different species, though usually identified as being the same.

Genus 4. **EHRETIA** P.Br.

Trees or shrubs; leaves alternate and petiolate, the blades entire, serrate or toothed. Flowers small, in terminal panicles or corymbose cymes; calyx of 5 segments imbricate or open in bud; corolla with short or cylindric tube, the 5 lobes spreading or recurved. Stamens usually exserted. Ovary imperfectly or completely 4-celled, with ovules attached laterally; style terminal, bifid, the 2 stigmas club-shaped or capitate. Fruit a globose or subglobose drupe, at maturity the stone separating into two 2-seeded or four 1-seeded nutlets; seeds with scant endosperm and ovate cotyledons.

A pantropical genus of about 50 species, the majority in the Old World. Some of them are valued for their timber.

1. **Ehretia tinifolia** L., Syst. Nat. ed. 10, 2:936. 1759. **FIG. 210.**

A glabrous arborescent shrub or small tree to 10 m tall or more; leaves broadly elliptic or oblong-elliptic, mostly 7–15 cm long, acutish at the apex, the midrib and primary veins prominent on the underside. Panicles many-flowered; calyx c. 1.5 mm long, the rounded lobes ciliate; corolla white or creamy-white, the tube 1 mm long, the reflexed or revolute lobes c. 2.5 mm long. Fruits c. 5 mm in diameter, with 2 nutlets.

GRAND CAYMAN: *Correll & Correll 51037, 51047; Proctor 15044, 31038.*
– Greater Antilles except Puerto Rico, the Swan Islands, Mexico and Honduras, in thickets and woodlands.

Genus 5. **BOURRERIA** P.Br., nom. cons.

Shrubs or small trees; leaves alternate, petiolate and entire. Flowers usually rather many in terminal corymbose cymes; calyx closed in bud, at anthesis

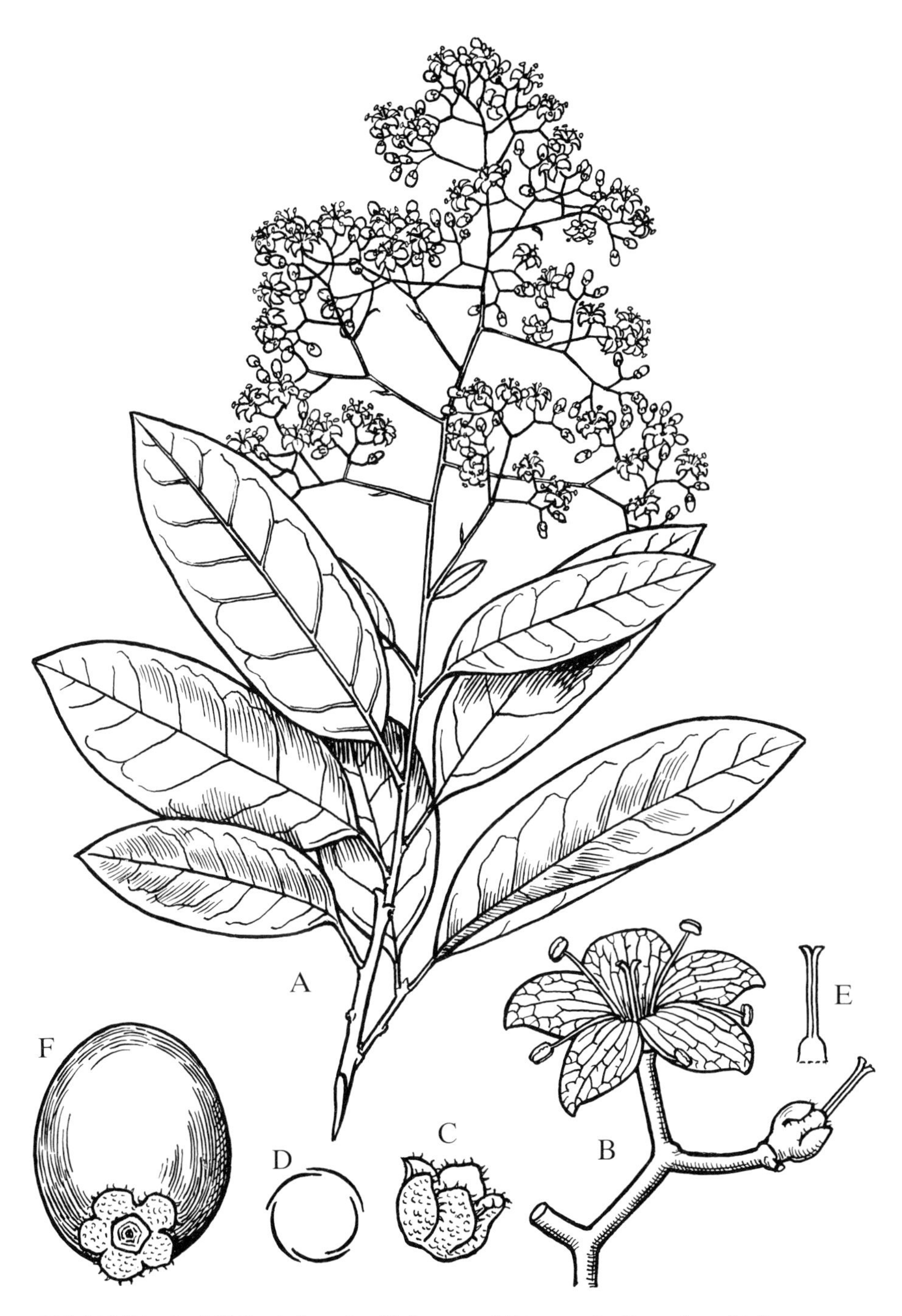

FIG. 210 **Ehretia tinifolia.** A, branch with leaves and flowers, x$^{3}/_{8}$. B, portion of inflorescence with flower, x6. C, calyx, x9. D, diagram of calyx. E, pistil, x6. F, fruit, x6. (St.)

splitting into 2–5 valvate teeth or lobes; corolla white, the 5 (or rarely 6) lobes imbricate in bud, the tube usually short, the throat widely dilated, and the lobes broad and spreading. Stamens borne in the corolla-tube, included or exserted, the filaments glabrous or villous near the base. Ovary 4-celled, the style terminal and more or less bifid at the apex, with truncate or subcapitate stigmas; ovules laterally attached. Fruit a thin-fleshed drupe enclosing 4 bony nutlets, these usually ridged on the back; endosperm fleshy.

A tropical American genus of between 15 and 20 species.

1. **Bourreria venosa** (Miers) Stearn in Jour. Arnold Arb. 52:625. 1971. **FIG. 211.**

A shrub or small tree to 6 m tall or more, glabrous or nearly so; leaves broadly elliptic to obovate or sometimes nearly rotund, 2–10 cm long. Inflorescence few- to many-branched, glabrous or puberulous; calyx 6–7 mm long, the lobes often splitting unequally; corolla c. 10 mm long, the lobes c. 4 mm long and broad. Style more or less distinctly 2-branched, the branches up to 2.5 mm long but often less or obsolete. Ripe fruits red, 5–8 mm in diameter.

GRAND CAYMAN: *Brunt 1715, 1804(?), 1814, 2069; Correll & Correll 51044; Fawcett; Hitchcock; Kings GC 310, GC 426; Proctor 11995. Sachet 438(?)* LITTLE CAYMAN: *Proctor 28149, 35144.* CAYMAN BRAC: *Proctor 29032.*
– Jamaica and the Swan Islands, in sandy or rocky thickets and woodlands. Perhaps not really distinct from *B. succulenta* Jacq., with undivided capitate stigma, which has a wide range from Florida and Mexico through the West Indies to Venezuela.

Genus 6. **CORDIA** L.

Shrubs or trees, often roughly pubescent, the hairs simple, branched or stellate; leaves alternate or mostly so, petiolate, the margins entire or toothed. Flowers in panicles, cymes, spikes or heads, sessile or stalked; calyx tubular to bell-shaped, and grooved, striate or smooth, usually 5-toothed or 3–10-lobed, usually persistent, often enlarging in fruit; corolla usually white or creamy, funnel-shaped to sub-rotate, with 5 or more lobes. Stamens equally or unequally inserted in the corolla-tube, included or shortly exserted. Ovary 4-celled, the style terminal and 2-lobed or 2-branched, the branches each 2-lobed or 2-branched; ovules 1–4, erect, laterally-attached. Fruit a drupe with a hard 1–4-celled stone; seeds without endosperm.

A pantropical genus of about 250 species.

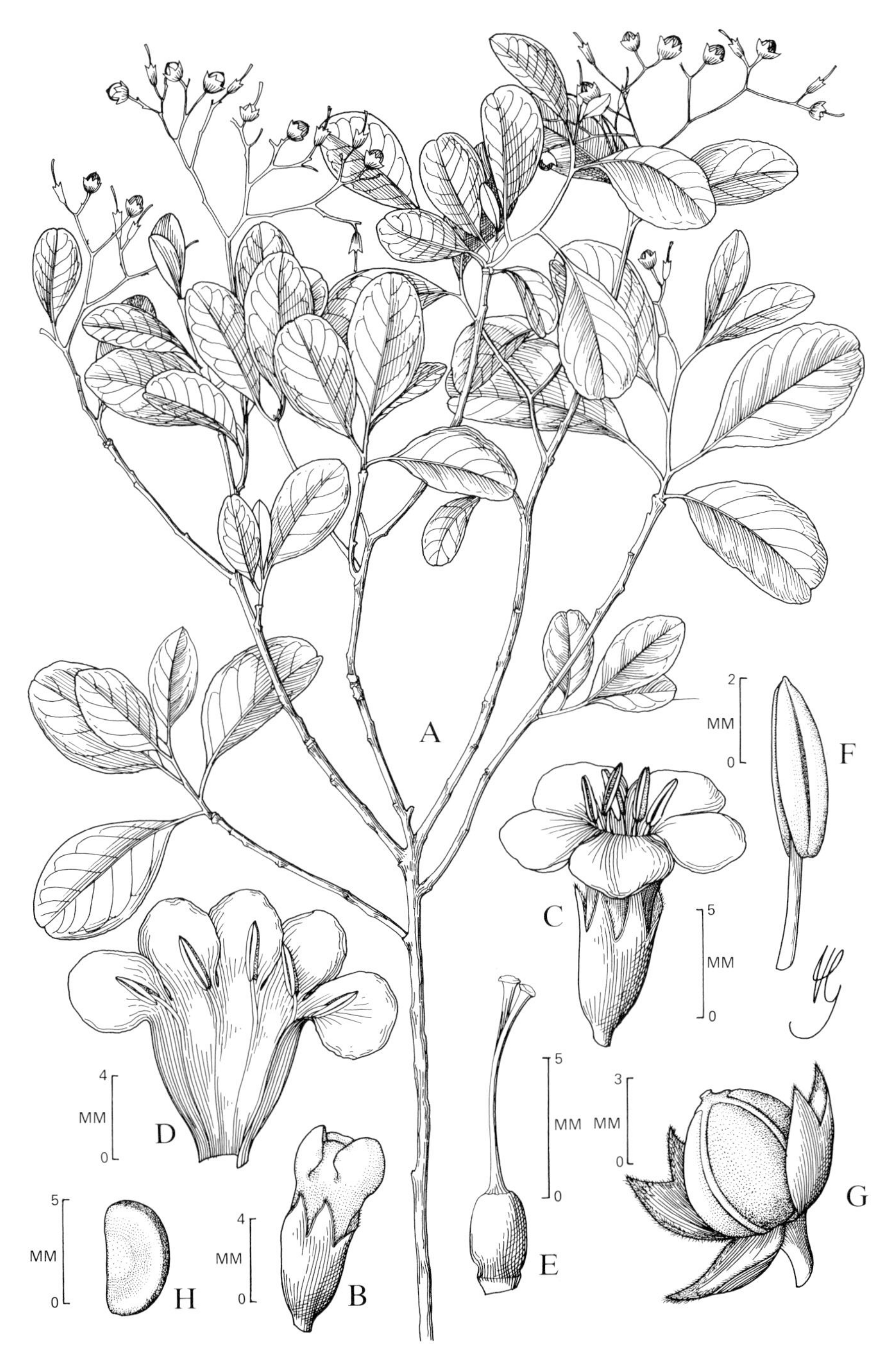

FIG. 211 **Bourreria venosa.** A, habit. B, flower-bud. C, flower. D, dissected corolla. E, pistil, showing ovary, style and stigmas. F, anther. G, fruit. H, seed. (G.)

KEY TO SPECIES

1. Inflorescence branched and open, corymbose or paniculate:

  2. Calyx 7-15 mm long in flower, enlarging in fruit; corolla 15-55 mm long; fruits not red:

    3. Leaves rough; corolla orange or scarlet: **1. C. sebestena**

    3. Leaves smooth; corolla white, persistent and turning brown, enclosing the fruit: **2. C. gerascanthus**

  2. Calyx c. 4 mm long; corolla c. 10 mm long; ripe fruits red, with viscid juice: **3. C. laevigata**

1. Inflorescence dense and unbranched:

    4. Inflorescence a cylindric spike; leaves densely pubescent beneath, on upper side scabrous but without hairs: **4. C. brownei**

    4. Inflorescence a globose head; leaves sparsely pubescent beneath and minutely glandular, on upper side with numerous pustulate hairs: **5. C. globosa**

1. Cordia sebestena L., Sp. Pl. 1:190. 1753.
var. **caymanensis** (Urb.) Proctor, in Sloanea 1:3. 1977.

Cordia caymanensis Urb., Symb. Ant. 7:344. 1912.

"Broadleaf"

Shrub or small tree to 10 m tall, very rough-hispid throughout; leaves very broadly elliptic or ovate to rotund, up to 30 cm long (but often much less), shortly subacuminate at the apex, the margins subentire or often the apical third sharply serrate, the base truncate or broadly cuneate; petioles up to 4 cm long. Inflorescence corymbose; calyx tubular, 10-12 mm long, with short, broad, unequal lobes; corolla salver-shaped, the tube 1.5-2 cm long, the limb 2-3 cm across. Stamens included. Flowers heterostylous, the styles once-forked if included or twice-forked if exserted. Ripe fruits white, acuminate, c. 3 cm long.

GRAND CAYMAN: *Brunt 1651; Hitchcock; Kings GC 63; Lewis 3823; Maggs II 55; Millspaugh 1261, 1358 (type of var.) Proctor 15169; Sachet 401, 406, 439.* LITTLE CAYMAN: *Kings LC4; Proctor 28047.* CAYMAN BRAC: *Kings CB 9; Millspaugh 1223; Proctor 29037.*
– The variety endemic; var. *sebestena* is widely distributed in coastal thickets

from Florida through the West Indies and along the eastern coasts of continental tropical America. Var. *caymanensis* differs from var. *sebestena* in a suite of overlapping characters, including the (on the average) larger, usually serrate leaves that are never cordate at the base, and in the smaller calyx and corolla. The rough, sandpaper-like leaves have traditionally been used to polish tortoise-shell. Var. *sebestena* is widely cultivated for ornament in tropical countries of both hemispheres.

2. **Cordia gerascanthus** L., Syst. Nat. ed. 10, 2:936. 1759.

"Spanish Elm"

An evergreen or sometimes deciduous tree to 10 m tall or more, the young branchlets glabrous; leaves narrowly to broadly elliptic or ovate, 5-15 cm long or more, acuminate at the apex; petioles of young leaves with scattered spreading hairs. Inflorescence rather densely corymbose, the branches puberulous; flowers fragrant; calyx 7-10 mm long, striate-grooved, hispidulous; corolla salver-shaped, the expanded limb 2-2.5 cm across, persistent in fruit and acting as a parachute for fruit-dispersal. Fruits oblong, enclosed by the calyx.

GRAND CAYMAN: *Brunt 1805; Howard & Wagenknecht 15029; Millspaugh 1273, 1300; Proctor 15299.* LITTLE CAYMAN: *Proctor 28126.* CAYMAN BRAC: *Kings CB 1; Proctor 15321.*
– Greater Antilles, Mexico, Central America and Colombia, in rather dry, rocky woodlands. The wood is useful for general construction.

3. **Cordia laevigata** Lam. in Tabl. Encycl. Méth Bot. 1:422. 1792.

Cordia nitida Vahl., 1793.

C. collococca of Hitchcock, 1893, not L., 1759.

A small tree to 12 m tall, the young branchlets and petioles finely rusty-pubescent; leaves elliptic to obovate, mostly 5-18 cm long, blunt or short-acuminate at the apex, the upper surface shining. Inflorescence corymbose, puberulous; calyx globose in bud, broadly bell-shaped after anthesis, 3-5-lobed; corolla creamy white. Fruits mostly 8-12 mm in diameter, red when ripe.

GRAND CAYMAN: *Correll & Correll 51016; Hitchcock.* CAYMAN BRAC: *Proctor 29082.*
– Greater Antilles and the Virgin Islands, the Swan Islands, and perhaps in Central America, in rocky woodlands usually on limestone. The Central American plants are alleged to have creamy-white instead of red fruits and might better be considered a different species. Lesser Antillean plants identified as *C. laevigata* differ in the tubular calyx and other details, and likewise are probably not conspecific.

4. Cordia brownei (Friesen) I.M.Johnst. in Jour. Arnold Arb. 31:177. 1950 **FIG. 212.**

A shrub usually 1-3 m tall with densely puberulous branchlets; leaves lanceolate to oblong or oblong-elliptic, rarely ovate, 2-8 cm long, blunt to acutish at the apex, the margins irregular. Spikes 1.5-3.5 cm long, on puberulous peduncles 2-4 cm long; flowers at apex of spike opening first, then in succession downwardly; calyx 2.5-3 mm long, very hairy (Cayman Brac) or hairs virtually lacking (Grand Cayman); corolla white, with small intermediate lobes between the main lobes. Ripe fruits subglobose, red, 3-4 mm in diameter.

GRAND CAYMAN: *Brunt 2136; Correll & Correll 50997; Kings GC 126; Proctor 11998, 15268, 27985; Sachet 382.* LITTLE CAYMAN: *Proctor 28142, 35126.* CAYMAN BRAC: *Proctor 29063, 35158.*
– Jamaica, in rocky thickets and woodlands.

5. Cordia globosa (Jacq.) Kunth in H.B.K., Nov. Gen. & Sp. 3:76. 1818. var. **humilis** (Jacq.) I.M.Johnst. in Jour. Arnold Arb. 30:98, 117. 1949.

"Black Sage"

A somewhat straggling shrub mostly 1-1.5 m tall, the young branchlets pustulate-puberulous and minutely glandular; leaves lance-elliptic or ovate, mostly 1-4 cm long, acute at the apex, the margins sharply toothed, the upper surface with numerous white pustulate hairs, the lower side with non-pustulate hairs on the nerves and veins and the whole surface densely clothed with minute glistening yellow glands. Inflorescence 0.6-1.5 cm in diameter on peduncles 0.5-2 cm long; calyx hispid, the lobes with hair-like tips 1-2 mm long; corolla white, 5-7 mm long. Ripe fruits ovoid, red, partly enclosed in the calyx.

GRAND CAYMAN: *Brunt 2013, 2101; Correll & Correll 51030; Fawcett; Hitchcock; Kings GC 126a; Proctor 12000, 15187; Sachet 414.*
– Var. *humilis* occurs in Florida, the Bahamas, Greater Antilles, and from Mexico to Panama, in sandy or rocky thickets and woodlands. Var. *globosa* occurs in the Lesser Antilles and northern South America. The latter differs especially in the non-glandular leaves and larger inflorescences, the individual flowers with longer, more hispid calyx-processes.

Family 92

VERBENACEAE

Herbs, shrubs or trees, or sometimes woody vines; branches often 4-angled, sometimes armed with thorns; leaves opposite or whorled, simple or palmately

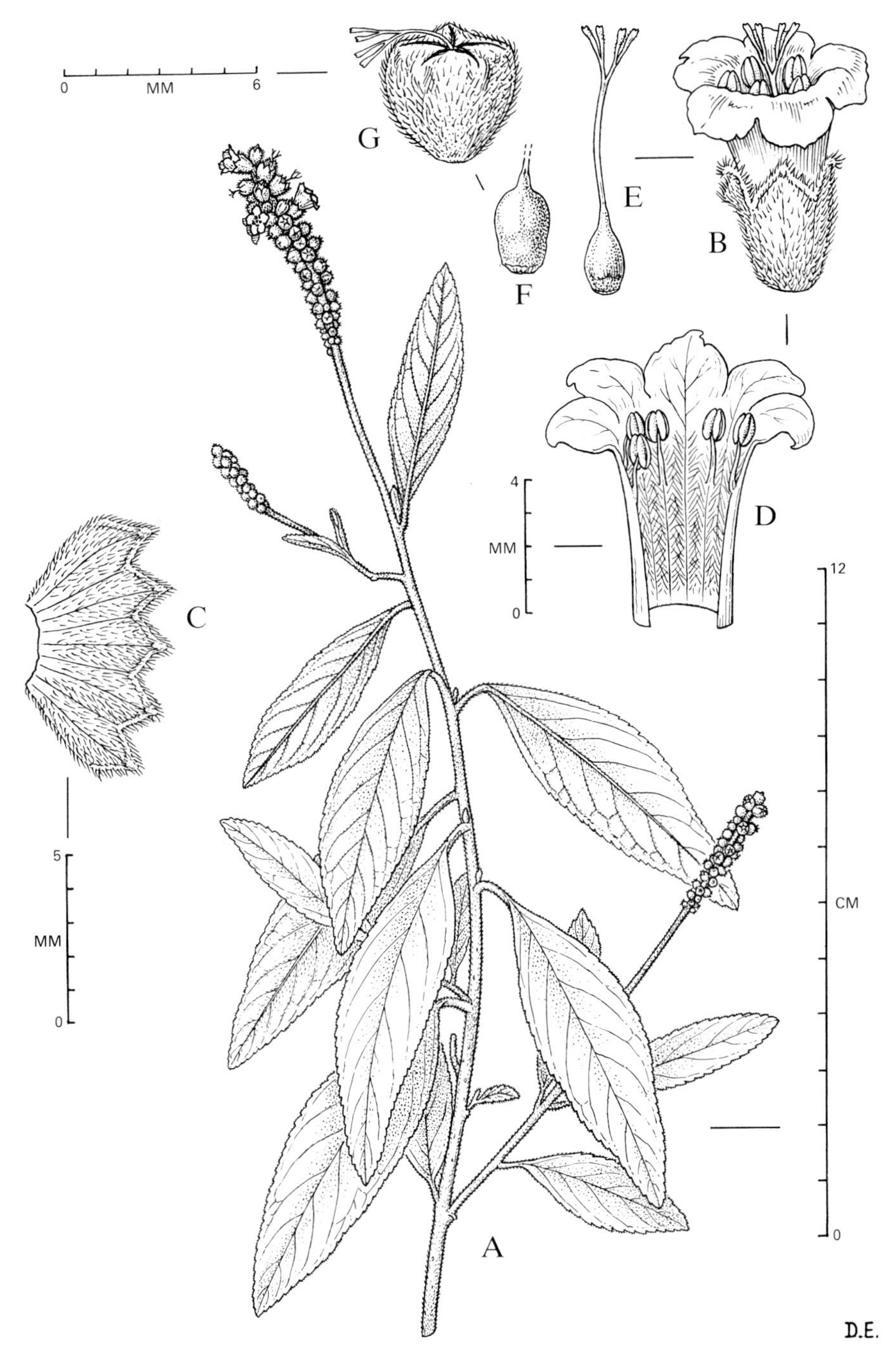

FIG. 212 **Cordia brownei. A,** habit. B, flower. C, calyx spread out. D, corolla opened to show stamens. E, pistil. F, ovary. G, fruit enclosed by calyx, with style protruding. (D.E. ex St.)

compound, the margins entire or variously toothed, lobed or incised; stipules absent. Flowers usually perfect, in axillary or terminal spikes, racemes, cymes or panicles, or sometimes in heads. Calyx 4–5-toothed or -lobed, persistent and often enlarging in fruit; corolla gamopetalous, regular or irregular, funnel-or salver-shaped, 4–5-lobed, often more or less 2-lipped. Stamens usually 4 (2 long and 2 short) or rarely only 2, the filaments inserted in the corolla-tube; anthers 2-celled, opening lengthwise; staminodes sometimes present. Ovary superior, usually 2-carpellate but often 4-lobed and usually 2–5-celled (rarely 1-celled), with 1 or 2 ovules in each cavity, basally or laterally attached to the placentas; style terminal and solitary. Fruit a drupe with 1 or more stones, or a dry schizocarp separating into 2 or 4 nutlets.

A rather large, widely-distributed family of about 100 genera and 2,600 or more species. Several are important sources of timber, e.g., *Tectona grandis* L.f., or "Teak". Among the species cultivated for ornament are *Holmskioldia sanguinea* Retz., or "Chinese Hat", several species of *Clerodendrum,* and *Petrea volubilis* L.

KEY TO GENERA

1. Inflorescence a spike, raceme, or head:

  2. Plants herbaceous:

    3. Fertile stamens 2; calyx closely appressed to the rhachis and more or less sunk in depressions: **1. Stachytarpheta**

    3. Fertile stamens 4; calyx not appressed to the rhachis:

      4. Inflorescence a loose spike with scattered flowers; fruiting calyx globose, bearing minute hooked hairs: **2. Priva**

      4. Inflorescence a dense head; fruiting calyx not globose: **3. Lippia**

  2. Plants woody:

    5. Plants erect or arching but not scrambling or vine-like; wild species:

      6. Inflorescence a more or less dense head:

        7. Fruit dry; calyx distinctly 2–4-toothed or -cleft: **3. Lippia**

        7. Fruit a small fleshy drupe; calyx with margin truncate or very obscurely toothed: **4. Lantana**

      6. Inflorescence a spike or raceme:

8. Calyx in flower smooth, in fruit dry and cup-shaped, subtending the fruit: **5. Citharexylum**

8. Calyx in flower ribbed, in fruit fleshy, contracted at the apex and enclosing the fruit: **6. Duranta**

5. Plant a scrambling woody vine, cultivated for its showy racemes of purple flowers with bluish calyces: [**Petrea** p. 682]

1. Inflorescence a cyme or panicle:

9. Corolla regular; plants not spiny:

10. Leaves of rather smooth texture on both sides, minutely gland-dotted beneath, glabrous or with unicellular hairs chiefly on the midrib; style-branches filamentous, 4 mm long or more; drupe with up to 4 1-seeded nutlets: **7. Aegiphila**

10. Leaves with raised densely reticulate venation beneath, densely pubescent with pluricellular hairs; style-branches c. 1 mm long; drupe with 1–4-celled stone: **8. Petitia**

9. Corolla irregular; plants spiny: **9. Clerodendrum**

Genus 1. **STACHYTARPHETA** Vahl, nom. cons.

Annual or perennial herbs or low shrubs; leaves opposite and toothed or serrate. Flowers in terminal spikes, sessile or partly immersed in the rhachis, each flower solitary in the axil of a bract. Calyx tubular, 5-toothed, the teeth equal or unequal; corolla with cylindric tube and 5 spreading lobes. Stamens 2, inserted above the middle of the corolla-tube and included; staminodes 2, small and inconspicuous; anther-cells divergent and opening in one continuous line. Ovary 2-celled, with 1 ovule attached laterally near the base of each cell; stigma subcapitate. Fruit dry, oblong, enclosed in the persistent calyx, eventually separating into 2 1-seeded nutlets; seeds without endosperm.

A tropical American genus of between 30 and 40 species.

1. **Stachytarpheta jamaicensis** (L.) Vahl, Enum. Pl. 1:206. 1804. **FIG. 213.**

"Vervine"

Annual herb, the stems decumbent or ascending to 1 m tall, glabrous throughout or with a few scattered hairs; leaves oblong-elliptic to ovate, mostly 2–9 cm long, obtuse at the apex, abruptly long-cuneate at the base, the margins coarsely crenate-serrate. Spikes glabrous, stiff, 10–50 cm long, up to 5 mm thick or more; bracts lanceolate or narrowly ovate, 4–6 mm long; calyx 5–7 mm long; corolla blue-violet or violet, 8–10 mm long.

GRAND CAYMAN: *Brunt 1901; Hitchcock; Kings GC 121, GC 290, GC 291; Lewis 3831, GC 36; Maggs II 53; Millspaugh 1340; Proctor 11981; Sachet 403;* LITTLE CAYMAN: *Proctor 28048.* CAYMAN BRAC: *Millspaugh 1173; Proctor 28990.*

– Florida, West Indies, and continental tropical America, introduced elsewhere, a weed of open waste places and dry sandy clearings and thickets.

Genus 2. **PRIVA** Adans.

Annual or perennial herbs, often with rough pubescence; leaves opposite, thin, the margins serrate or toothed. Flowers small, in terminal racemes, the stems often branching dichasially below the inflorescence so that the latter appears to arise from a fork. Calyx tubular at anthesis, 5-ribbed, the ribs ending in small teeth, persistent and enlarging to completely enclose the fruit except for a minute orifice. Corolla obliquely 5-lobed. Stamens 4, included, inserted in 2 pairs at different levels; anther-cells parallel or slightly divergent. Ovary 2-celled; stigma 2-lobed. Fruit dry and surrounded by the enlarged, bladdery calyx, consisting at maturity of 2 nutlets, these usually variously spined, ridged or roughened on the dorsal surface.

A pantropical genus of about 20 species.

1. **Priva lappulacea** (L.) Pers., Synops. Pl. 2:139. 1806. **FIG. 214.**

"Old Lady Coat Tail"

Ascending to erect annual herb often 50–75 cm tall, with quadrangular pubescent stems; leaves with petioles up to 2 cm long or more, the blades more or less ovate and mostly 3–8 cm long, acute to acuminate at the apex, the margins crenate-serrate, rough-hispidulous on the upper side. Racemes lax, up to 20 cm long; flowers short-stalked; calyx at anthesis 2–3 mm long, in fruit 5–6 mm long; corolla pale bluish or whitish, 4–5 mm long. Nutlets 3–3.5 mm long, curved-spiny on the dorsal side.

GRAND CAYMAN: *Brunt 1933; Hitchcock; Kings GC 87, GC 153; Millspaugh 1286.* CAYMAN BRAC: *Proctor 29366.*

– Florida, West Indies, and continental tropical America; introduced in the Old World tropics; a weed of open waste ground, fields, and roadsides.

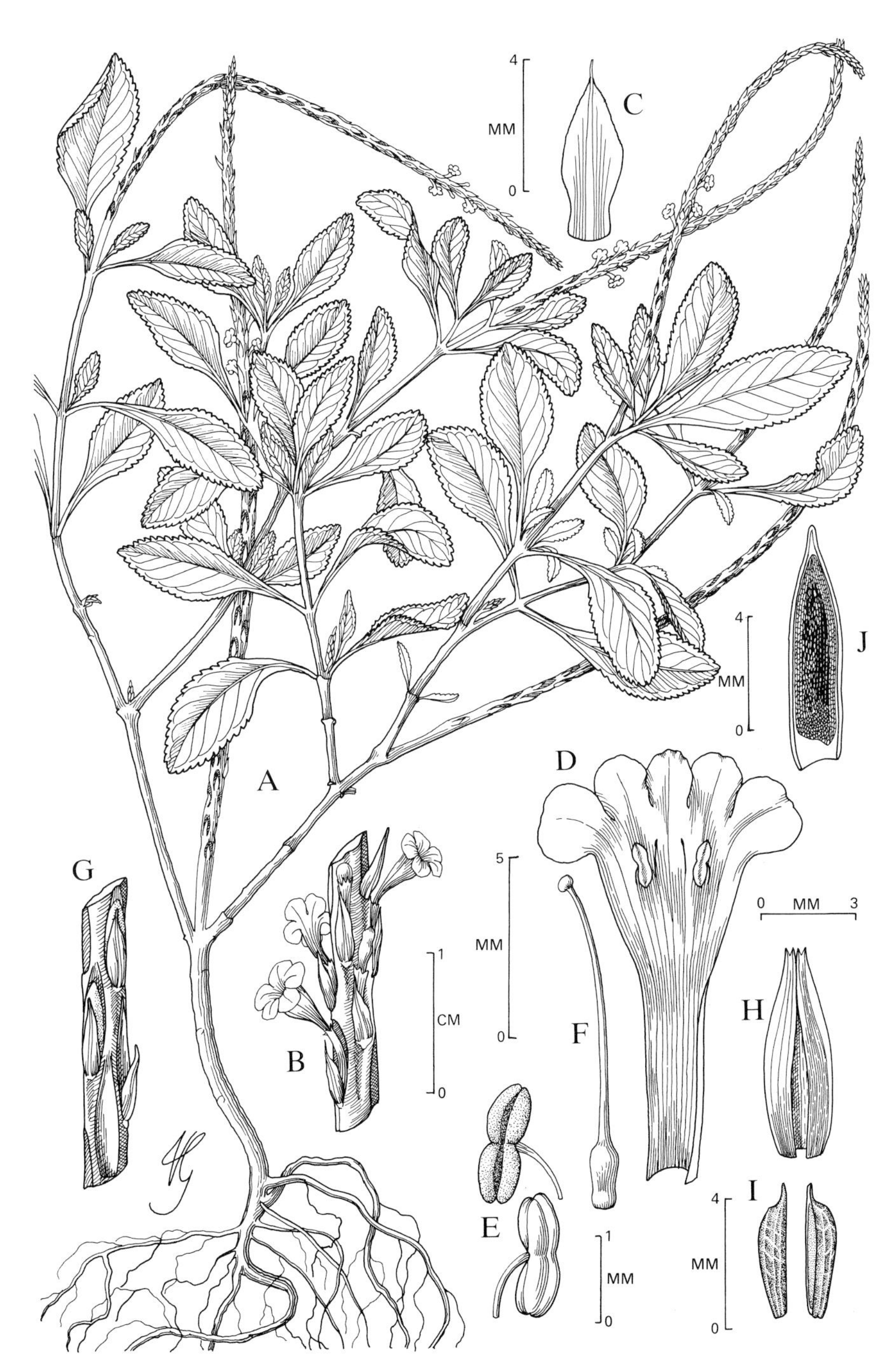

FIG. 213 **Stachytarpheta jamaicensis.** A, habit, B, portion of inflorescence showing bracts and flowers. C, bract. D, corolla cut open to show stamens. E, two views of anther. F, pistil. G, portion of fruiting spike. H, fruiting calyx. I, seeds. J, seed, enlarged. (G.)

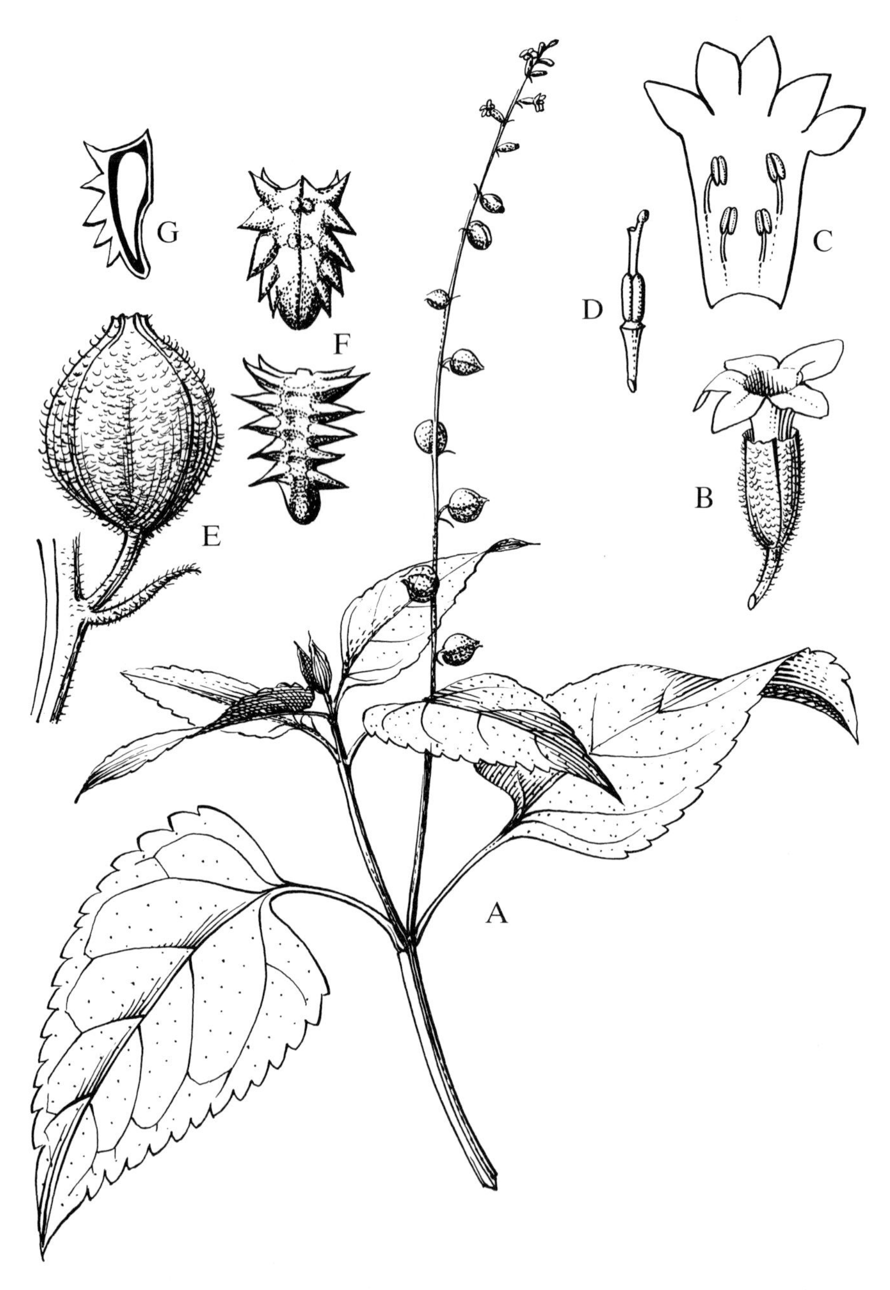

FIG. 214 **Priva lappulacea.** A, habit, x½. B, flower, x3. C, corolla spread out to show stamens, x3. D, pistil, x3. E, fruit, x3. F, two views of seed, x3. G, section through seed, x3. (St.)

Genus 3. **LIPPIA** L.

Herbs, shrubs, or small trees, sometimes aromatic; leaves opposite or in whorls of 3, entire or toothed. Flowers small, sessile, in axillary stalked heads or dense cylindric spikes, often with conspicuous bracts; calyx small, thin, 2-4-toothed; corolla with cylindric tube and 4 spreading lobes, one of them larger than the others. Stamens 4, in 2 pairs at different levels, usually included. Ovary 2-celled, each cavity with 1 basal ovule; stigma obscurely 2-lobed, oblique or recurved. Fruit small, dry, enclosed by the persistent calyx, at maturity usually separating into 2 hard nutlets; seeds without endosperm.

A genus of nearly 200 species, the majority of them in tropical America.

KEY TO SPECIES

1. Flower-heads nearly sessile or on peduncles usually 1 cm long or less; shrubby plant with strongly aromatic foliage: **1. L. alba**

1. Flower-heads on elongate peduncles much longer than the leaves and up to 9 cm long; trailing herb, sometimes with ascending or erect flowering branches, not aromatic: **2. L. nodiflora**

1. Lippia alba (Mill.) N.E.Br. in Britton & Wilson, Sci. Surv. Porto Rico & Virg. Is. 6:141. 1925.

"Sage"

Straggling or arching shrub up to 1 m tall or more, puberulous or tomentellous throughout; leaves opposite or in whorls of 3, oblong elliptic or ovate, mostly 1-5 cm long, blunt or broadly acute at the apex, the margins serrulate, the venation prominently raised on the lower surface and channeled on the upper. Flower-heads subglobose to shortly cylindric, 0.5-1.2 cm long; bracts ovate, 3-5 mm long; corolla pale rosy-lavender with tube 3-3.5 mm long.

GRAND CAYMAN: *Proctor 15189.* CAYMAN BRAC: *Kings CB 51.*
– West Indies, Texas and Mexico south to Argentina, usually in sandy or gravelly thickets near the sea. This species is occasionally cultivated for use in making tea.

2. Lippia nodiflora (L.) Michx., Fl. Bor. Amer. 2:15. 1803. **FIG. 215.**

Phyla nodiflora (L.) Greene, 1899.

"Match Head"

Creeping perennial herb, very minutely strigillose with malpighiaceous hairs throughout (appearing glabrous except under a strong lens); leaves oblanceolate to obovate or spatulate, mostly 2-4.5 cm long, blunt or acutish at the apex, long-cuneate at the base, the apical half with serrulate margins. Flower-heads globose at first, becoming cylindric and up to 1.5 cm long; bracts purplish and closely overlapping, 2-3 mm long; corolla pale pink or lavender-whitish with tube 2-3 mm long.

GRAND CAYMAN: *Brunt 1770; Hitchcock; Kings GC 61; Lewis GC 16; Millspaugh 1365; Proctor 15014; Sachet 423.* LITTLE CAYMAN: *Kings LC 72.* CAYMAN BRAC: *Proctor 29088.*

– Pantropical in moist pastures, ditches and sandy clearings.

Genus 4. **LANTANA** L.

Erect, scrambling or climbing shrubs, always more or less aromatic, the stems 4-angled, ribbed or subcylindric, sometimes armed with prickles; leaves opposite or whorled, more or less toothed. Flower in dense, stalked, axillary heads or contracted spikes, with or without a subtending involucre of bracts; inner bracts (subtending individual flowers) also present. Calyx membranous, truncate, sinuate or minutely 2-4-toothed, enlarging in fruit; corolla with slender, often curved tube and 4–5 lobes, one larger than the others. Stamens 4, in 2 pairs inserted at different levels, included; anthers ovate, parallel-celled. Ovary 2-celled with 1 ovule in each cavity. Fruit a small, globose, juicy drupe, the stone 2-celled or separating into 2 1-seeded nutlets; seeds without endosperm.

A mostly tropical American genus of perhaps 50 species, many of them not clearly differentiated, thus often difficult to identify. The following treatment of the Cayman species is not very satisfactory and cannot "work" outside the local geographic context. More field observations are needed on these plants.

KEY TO SPECIES

1. Flowers yellow or orange, changing to orange, red or mauve; inflorescence usually not subtended by a conspicuous involucre of bracts, the outer bracts seldom over 2 mm broad:

  2. Stems armed with numerous prickles; flowers changing to mauve, the corolla-tube c. 10 mm long, the limb 6–7 mm across: **1. L. aculeata**

  2. Stems unarmed or nearly so; corolla never mauve, the tube 4–8 mm long, the limb 3–6 mm across:

    3. Corolla tube 5–8 mm long:

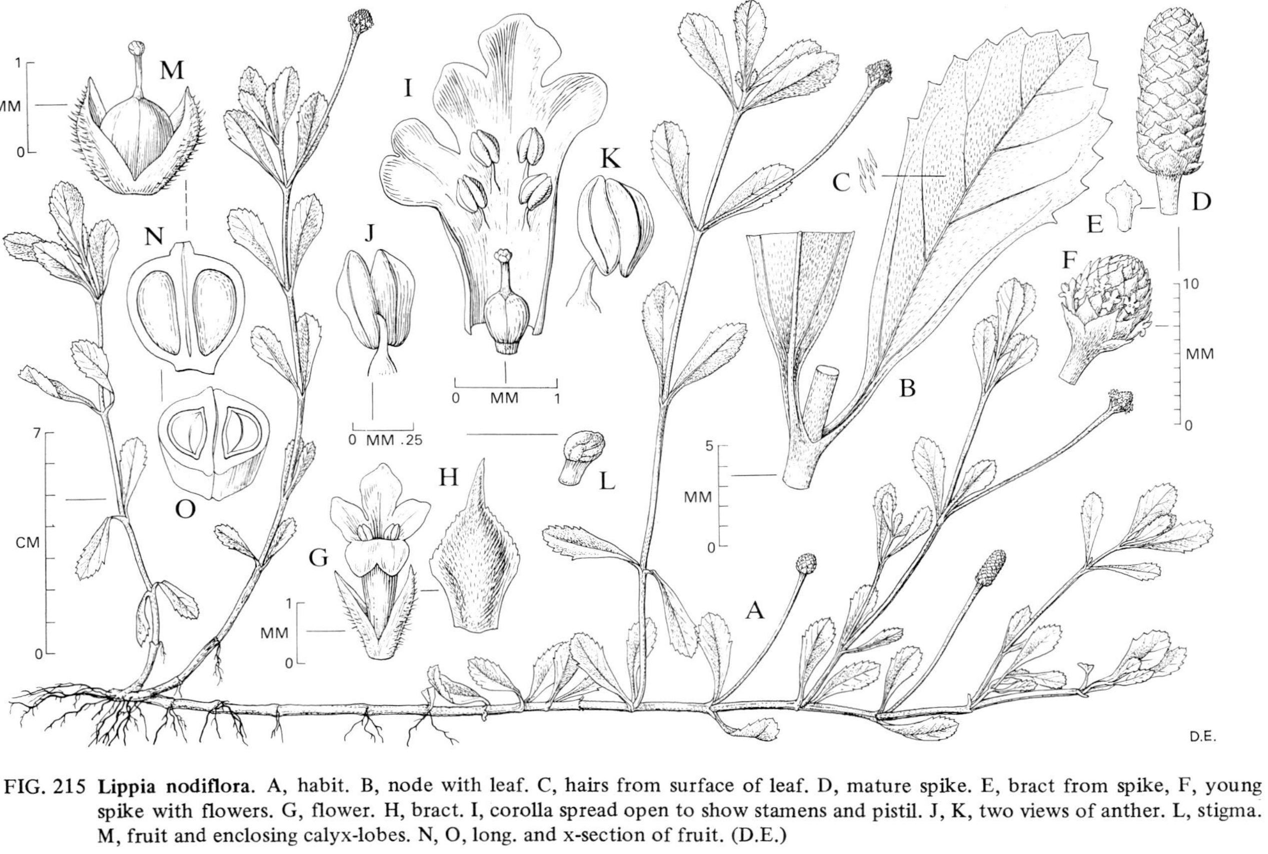

FIG. 215 **Lippia nodiflora**. A, habit. B, node with leaf. C, hairs from surface of leaf. D, mature spike. E, bract from spike, F, young spike with flowers. G, flower. H, bract. I, corolla spread open to show stamens and pistil. J, K, two views of anther. L, stigma. M, fruit and enclosing calyx-lobes. N, O, long. and x-section of fruit. (D.E.)

4. Lower surface of the leaves thinly pubescent or almost glabrous; calyx as broad as long: **2. L. camara**

4. Lower surface of the leaves with numerous short hairs; calyx longer than broad: **3. L. urticifolia**

3. Corolla tube c. 4 mm long: **4. L. bahamensis**

1. Flowers white, pale pink or lavender with yellow eye; inflorescence subtended by broad outer bracts 2–5 mm wide, forming an involucre:

5. Leaves rounded or blunt at the apex; tertiary venation irregular on the lower leaf-surface: **5. L. involucrata**

5. Leaves acute at the apex; tertiary venation demarcating regular oblong areoles on the lower surface: **6. L. reticulata**

1. Lantana aculeata L., Sp. Pl. 2:627. 1753.

A straggling shrub to 1 m tall but frequently less, with strong pungent odour; young branchlets glandular, sparingly pubescent, and beset with scattered sharp recurved prickles. Leaves ovate or broadly ovate, 3–7 cm long, acuminate at the apex, the margins coarsely crenate-serrate. Peduncles up to 6 cm long, thickened toward the apex. Flower-heads c. 2.5 cm in diameter. Drupes blue, c. 4 mm in diameter.

LITTLE CAYMAN: *Proctor 35179.* CAYMAN BRAC: *Proctor 28913.*
– With a wide but uncertain range in the tropics; in other regions said to intergrade with *L. camara* (see next), and by many recent authors not considered a distinct species or even a variety. However, the Cayman population of *L. aculeata* is so strikingly different from "normal" *L. camara* (as represented in the Cayman Islands) that it makes no sense in the local context to lump them together. No intergrades have been observed locally. Common on the north side of Cayman Brac in sandy thickets; rare on Little Cayman.

2. Lantana camara L., Sp. Pl. 2:627. 1753.

"White Sage"

An erect, mostly unarmed shrub to 2 m tall or more, with aromatic odour different from that of *L. aculeata;* young branchlets glabrate and glandular. Leaves narrowly to broadly ovate or oblong-ovate, mostly 2.5–6 cm long or more, blunt to acute at the apex, the margins crenate-serrate. Peduncles up to 5 cm long, slightly thickened at the top. Flower-heads 1.5-2 cm in diamter, corolla yellow changing to orange or red, the tube usually 6-8 mm long. Drupes blue-black, c. 3 mm in diameter.

GRAND CAYMAN: *Brunt 2057; Hitchcock; Howard & Wagenknecht 15028; Kings GC 314; Lewis GC 41; Millspaugh 1330, 1332; Sachet 368.* LITTLE CAYMAN: *Proctor 28035.* CAYMAN BRAC: *Millspaugh 1202, 1215.*
– Florida, West Indies, and continental tropical America; Cayman plants occur chiefly in rocky thickets and woodlands.

3. **Lantana urticifolia** Mill., Gard. Dict. ed. 8. 1768.

Lantana arida Britton, 1910.

"Sweet Sage"

Aromatic bushy shrub to 2 m tall; young branchlets scabrid and more or less pubescent. Leaves broadly ovate, mostly 2-6 cm long (rarely more), acute to acuminate at the apex, densely and softly pubescent on the under side, scabrid with short pustular hairs on the upper (adaxial) side. Peduncles mostly 2-4 cm long, thickened at the top. Flower-heads 1-1.5 cm in diameter (or more); corolla deep yellow, the tube usually 5-7 mm long. Drupes blue-black, 3-4 mm in diameter.

GRAND CAYMAN: *Kings GC 112; Proctor 15009.*
– Widely distributed in the West Indies, usually in rocky thickets or dry woodlands on limestone.

4. **Lantana bahamensis** Britton in Bull. N.Y. Bot. Gard. 3:450. 1905.

An erect unarmed shrub to 1 m tall or more, the young branchlets glandular and rather densely puberulous, the foliage very aromatic; leaves oblong-lanceolate or elliptic, blunt to acute at the apex, the margins crenulate-serrulate. Peduncles 0.5-2 cm long, very slender (almost filiform). Flower-heads 0.6-1.5 cm in diameter; corolla-tube c. 4 mm long. Drupes shining black, c. 3 mm in diameter.

GRAND CAYMAN: *Proctor 15230.*
– Bahamas and Cuba, in dry rocky thickets. This might be considered a much-reduced variant of *L. camara*, and more collections and field observations are necessary to establish its true relationship with that variable species.

5. **Lantana involucrata** L., Cent. Pl. 2:22. 1756.

"Roundleaf Sage", "Bitter Sage"

An erect, unarmed, aromatic shrub up to 1.5 m tall, glandular and puberulous throughout, the young branches stiff and nearly terete; leaves broadly elliptic or ovate to obovate or nearly orbicular, mostly 1-4.5 cm long, abruptly cuneate at the base, the margins finely crenulate, the petioles up to 10 mm long. Peduncles up to 6 cm long, rarely very short (3 mm or less). Flower-heads 0.8-1.7 cm in diameter (including the bracts); corolla-tube 2-3 mm long. Drupes light to deep purple or mauve, 3-4 mm in diameter.

GRAND CAYMAN: *Brunt 1638, 1752, 1811; Correll & Correll 51027; Kings GC 339; Lewis 3854; Millspaugh 1252; Proctor 15090; Sachet 413, 426.* LITTLE CAYMAN: *Kings LC 38; Proctor 28059.* CAYMAN BRAC: *Millspaugh 1218; Proctor 28914.*
– Florida, West Indies, Yucatan and Belize, and northern South America, chiefly in sandy coastal thickets.

6. **Lantana reticulata** Pers., Synops. Pl. 2:141. 1806.

Lantana stricta of Hitchcock, 1893, not Sw., 1788.

Similar to the preceding species and often scarcely distinguishable from it. The single Cayman record (GRAND CAYMAN: *Hitchcock*) needs further confirmation. Occurs elsewhere chiefly in the Greater and Lesser Antilles, in rocky thickets.

Genus 5. **CITHAREXYLUM** L.

Shrubs or trees with quadrangular or striate stems, usually unarmed or rarely spiny; leaves opposite or whorled, entire or toothed, usually with 1 or 2 glands at the base of the blade. Flowers in axillary and terminal, more or less elongate, spikes or racemes; calyx at anthesis tubular or bell-shaped, the margin truncate or 5-7-toothed or -lobed, becoming enlarged and cup-shaped in fruit; corolla funnel-shaped or salver-shaped, white or yellowish, the 5 lobes subequal, usually pubescent in the throat. Stamens 4, in 2 pairs at different levels, included, a fifth stamen represented by a rudimentary staminode. Ovary incompletely 4-celled, each cavity with a single lateral ovule; stigma 2-lobed. Fruit a juicy, berry-like drupe containing two 2-seeded nutlets.

A genus of more than 100 species in tropical and subtropical America.

1. Citharexylum fruticosum L., Syst. Nat. ed. 10, 2:1115. 1759.

A glabrous shrub or small tree to 5 m tall, the branchlets closely striate-grooved; leaves rather long-petiolate, the blades elliptic or oblong-elliptic to ovate, mostly 5-9 cm long or sometimes longer, rounded and often notched at the apex or sometimes abruptly short-acuminate, the upper surface shining and with prominulous venation. Spikes often curved or lax, mostly 4-10 cm long; flowers sessile, fragrant; calyx 3-4 mm long, with broadly triangular lobes; corolla white, with funnel-shaped tube 4-6 mm long. Drupes red at first, turning black, subglobose, 8-12 mm in diameter.

GRAND CAYMAN: *Correll & Correll 51000.* LITTLE CAYMAN: *Proctor 28135.* CAYMAN BRAC: *Proctor 29001.*
– Florida, West Indies and northern South America, in rocky thickets and woodlands.

Genus 6. **DURANTA** L.

Shrubs, sometimes arborescent, and in some species armed with spines; branches sometimes arching or pendent; leaves opposite or whorled, entire or toothed. Flowers in terminal and axillary racemes; calyx in flower tubular or bell-shaped, truncate and 5-ribbed, each rib ending in a small tooth, enlarging to enclose the fruit; corolla salver-shaped, the 5 lobes equal and spreading or oblique, usually pubescent in the throat. Stamens 4, in 2 pairs at different levels, included; anthers with parallel cells. Ovary imperfectly 8-celled, composed of four 2-celled carpels, each cavity with 1 ovule; stigma obliquely subcapitate. Fruit a fleshy drupe enclosed by the enlarged calyx and containing 4 2-seeded nutlets; seeds without endosperm.

A tropical American genus of about 35 species.

1. Duranta repens L., Sp. Pl. 2:637. 1753. **FIG. 216.**

A shrub to 3.5 m tall or more, the glabrate or puberulous branches often arching or trailing, unarmed or sometimes with short axillary spines; leaves glabrous or thinly puberulous, opposite or a few of them alternate, elliptic or obovate, 2-6 cm long, rounded, blunt or acute at the apex, the margins entire or obscurely crenate-serrate toward the apex. Racemes axillary and unbranched or terminal and branched, 2-6 cm long or sometimes much longer. Flowering calyx 3-4 mm long.

Corolla blue-violet or rarely white, the tube c. 6 mm long. Fruits orange, pear-shaped, 6–9 mm long.

GRAND CAYMAN: *Hitchcock; Kings GC 369; Proctor 15006.*
– Florida, West Indies and continental tropical America; introduced in the Old World tropics; sometimes cultivated as an ornamental. The Cayman plants occur wild in rocky woodlands.

[**Petrea volubilis** L., a woody ornamental vine, is cultivated in Grand Cayman.]

Genus 7. **AEGIPHILA** Jacq.

Shrubs or small trees, sometimes scandent, the stems terete or more or less 4-angled; leaves opposite or rarely whorled, simple, the petioles jointed at or near the base. Flowers usually polygamo-dioecious, in axillary or terminal cymes, these paniculate, corymbose, or head-like. Calyx bell-shaped or tubular, the apex truncate and entire or with 4–5 teeth or lobes, becoming enlarged and thickened in fruit. Corolla funnel- or salver-shaped with cylindric tube, the 4–5 lobes somewhat unequal or oblique. Stamens 4–5, equal or nearly so, exserted in the male flowers. Ovary incompletely 4-celled, with 1 laterally-attached ovule in each cavity; stigma of 2 linear branches, exserted in the female flowers. Fruit a 1–4-seeded drupe; seeds without endosperm.

A tropical American genus of perhaps 150 species.

KEY TO SPECIES

1. Stems, underside of leaves, and inflorescence densely clothed with velvety brownish hairs; calyx truncate with 4 minute teeth; corolla-tube 13 mm long or more: **1. A. caymanensis**

1. Stems, etc., glabrous or nearly so; calyx with 4 or 8 wavy lobes, rarely merely 4-toothed; corolla-tube 5–8 mm long: **2. A. elata**

1. Aegiphila caymanensis Moldenke in Fedde, Repert. Sp. Nov. 33:118. 1933.

A velvety-tomentose shrub; leaves oblong-lanceolate, 5–9 cm long, 2–3.8 cm broad, acute to acuminate at the apex, the under-surface (in addition to hairs) densely clothed with minute glistening yellow glands; fine venation obscure. Cymes

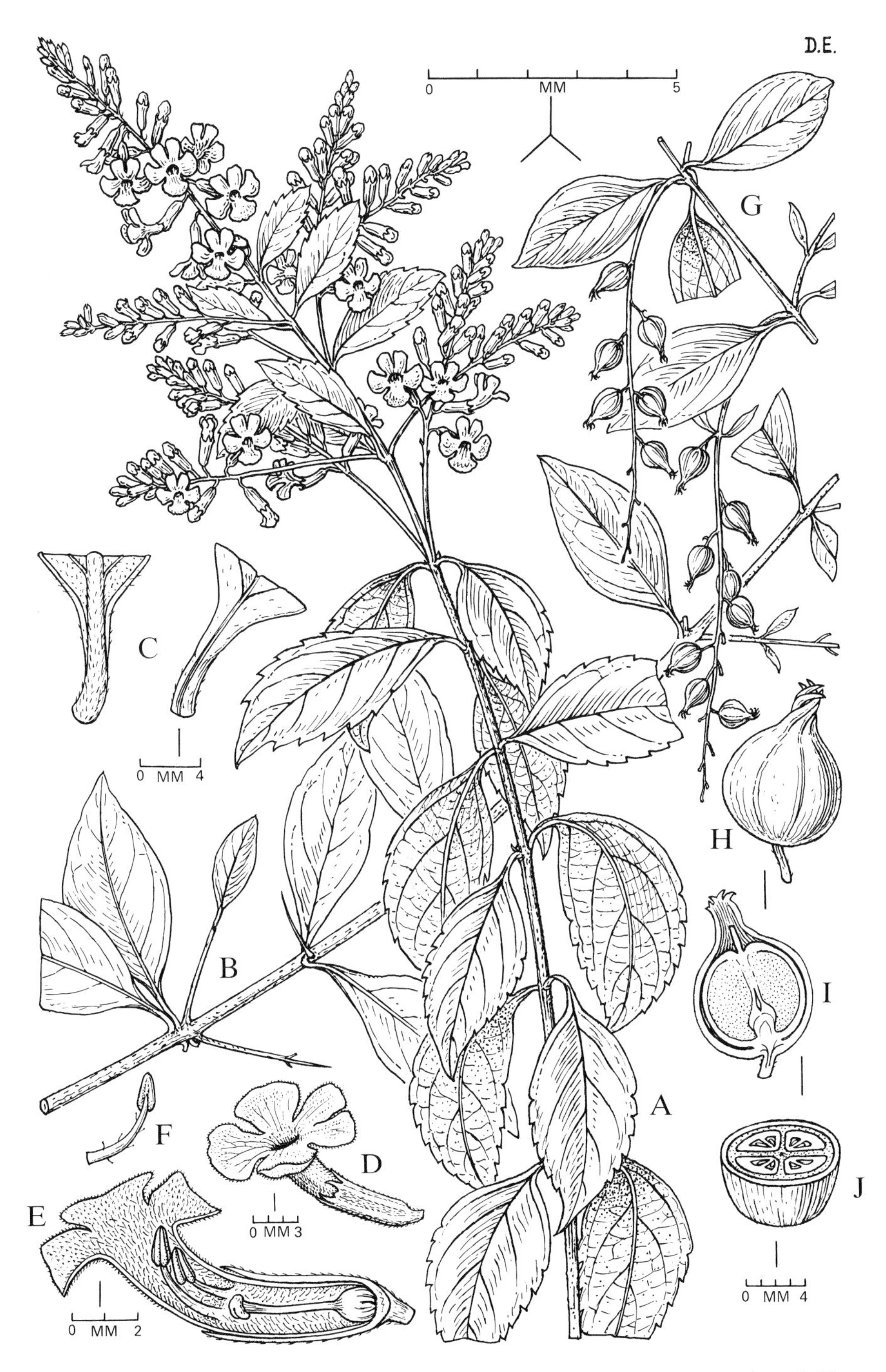

FIG. 216 **Duranta repens.** A, branch with leaves and terminal inflorescence. B, portion of older branch with spines. C, two views of leaf-base. D, flower. E, section through flower to show stamens and pistil. F, stamen. G, portion of fruiting branch. H, fruit. I, long-section of fruit. (St.)

terminal and in the upper axils, laxly few-flowered; pedicels 6-9 mm long, glandular; calyx c. 4 mm long, glandular; colour of corolla not known. Fruit oblong, c. 7 mm long.

GRAND CAYMAN: *Hitchcock (type)*; endemic and known only from this one collection of January 17, 1891. The species may be extinct.

2. Aegiphila elata Sw., Nov. Gen. & Sp. Pl. 31. 1788.

An erect or straggling glabrate shrub to 2 m tall or more; leaves oblong-ovate, 6–15 cm long, 3–6 cm broad, short-acuminate at the apex, more or less densely gland-dotted beneath but the dots not glistening yellow; fine venation evident or prominulous. Inflorescence a terminal cymose panicle, often densely-flowered; pedicels 2–6 mm long, densely puberulous; calyx 3–4 mm long, glandular and sparingly appressed-puberulous; corolla pale yellow. Fruit subglobose, yellow, 9 mm long or more.

GRAND CAYMAN: *Brunt 1751; Kings GC 60; Millspaugh 1281; Rothrock 158, 235.*
– West Indies and continental tropical America, along margins of sandy or rocky thickets and woodlands.

Aegiphila martinicensis Jacq. was attributed to the Cayman Islands by Moldenke in "The known geographic distribution of the members of the Verbenaceae . . .", p. 46, 1949, based upon a literature reference that was probably erroneous. *A. martinicensis* is similar to *A. elata* and the two have sometimes been confused.

Genus 8. **PETITIA** Jacq.

Unarmed shrubs or trees; leaves opposite, simple, entire and long-petiolate. Flowers small, rather numerous in axillary cymose panicles; calyx bell-shaped, subtruncate or 4-toothed, not enlarging in fruit; corolla short, salver-shaped, with 4 spreading equal imbricate lobes. Stamens 4, inserted near the top of the corolla-tube, the ovate anthers almost sessile. Ovary 2-celled, with 2 ovules in each cavity; style shortly forked at the apex. Fruit a small drupe containing a single 2–4-seeded stone.

A small genus of 2 species occurring in the Bahamas and the Greater Antilles.

1. **Petitia domingensis** Jacq., Enum. Pl. Carib. 12. 1760. **FIG. 217.**

"Fiddlewood"

A small tree to 10 m tall or more, sometimes smaller and shrubby, the younger parts densely and minutely brownish-puberulous; leaves with petioles 1.5–4 cm long, the blades lance-oblong or elliptic-oblong and mostly 7–14 cm long, acuminate at the apex, glabrate on the upper surface, beneath softly tomentellous and also densely coated with minute glistening yellow glands. Calyx puberulous, 2 mm long; corolla cream or greenish-white, the tube slightly longer than the calyx, the lobes 1 mm long. Fruits subglobose, blackish or red, 3.5–4 mm in diameter.

GRAND CAYMAN: *Brunt 1756, 1924-a; Correll & Correll 51012; Hitchcock; Kings GC 107, GC 409; Proctor 15087.* CAYMAN BRAC: *Millspaugh 1164; Proctor 15324.*
– Distribution of the genus, frequent or common in rocky woodlands. The wood is hard, heavy and strong; it is used for furniture and general construction.

Genus 9. **CLERODENDRUM** L.

Mostly shrubs or trees, sometimes herbs, the stems erect, arching or scandent; leaves opposite, subopposite or whorled, and simple, entire or toothed; petioles jointed at or near the base, when breaking off often leaving a raised or spine-like persistent base. Flowers more or less irregular, in cymose panicles, terminal and often in the upper axils; calyx usually bell-shaped, 5-toothed or -lobed, in fruit enlarging and subtending or enclosing the fruit. Corolla often with long straight or curved tube, the 5 subequal or unequal lobes spreading. Stamens 4, in 2 pairs inserted at different levels, long-exserted, with ovoid or oblong anthers. Ovary imperfectly 4-celled with one lateral ovule in each cavity; style shortly 2-branched. Fruit a drupe with thin, fleshy exocarp, the bony endocarp ultimately separating into 2 or 4 nutlets; seeds without endosperm.

A large pantropical genus of about 350 species, most abundant in Asia and Africa.

1. **Clerodendrum aculeatum** (L.) Schlecht. in Linnaea 6:750. 1831.

"Cat Claw"

A shrub up to 2 m tall or more, forming dense thickets, with puberulous and spiny arching or straggling branches; leaves opposite or occasionally a few alternate, or often in 3's or clustered, puberulous especially beneath. Cymes on peduncles 0.8–1.5 cm long, few-flowered, axillary or in dense terminal clusters; calyx 5-lobed, the lobes reflexed at anthesis; corolla white. Filaments purple, unequal. Fruit splitting into 2 hard nutlets at maturity.

Represented in the Cayman Islands by two varieties:

1. Leaves elliptic or lanceolate, up to 6 cm long or more and 2 cm broad; corolla-tube more than 15 mm long; fruits 5–7 mm in diameter: **1a.** var. **aculeatum**

1. Leaves narrowly lanceolate or lance-oblong, mostly less than 3 cm long and up to 0.8 cm broad; corolla-tube 10 mm long or less; fruits c. 4 mm in diameter: **1b.** var. **gracile**

1a. Clerodendrum aculeatum var. **aculeatum.**

GRAND CAYMAN: *Correll & Correll 50993; Hitchcock; Kings GC 133, GC 148; Millspaugh 1380; Proctor 15153.* LITTLE CAYMAN: *Proctor 28111.* CAYMAN BRAC: *Proctor 29135.*

– Bermuda, the West Indies, and Mexico to Venezuela and the Guianas, in dry rocky or gravelly thickets.

1b. Clerodendrum aculeatum
var. **gracile** Griseb. ex Moldenke in Carib. Forester 2:13. 1940.

GRAND CAYMAN: *Proctor 27941.*

– Cuba, in dry thickets. The peduncles of this variety are often 1-flowered, and all the parts are of a more slender appearance than in var. *aculeatum.*

Family 93

AVICENNIACEAE

Shrubs or small trees of mangrove swamps, the roots sending up erect pencil-like aerial pneumatophores; leaves opposite and simple; stipules absent. Flowers perfect, sessile in the leaf-axils or in terminal spikes or heads, each flower subtended by 3 imbricate bracteoles. Calyx of 5 imbricate sepals, almost free, persistent. Corolla gamopetalous, with short tube and 4 almost equal lobes. Stamens 4, inserted near the base of the corolla-tube, shortly exserted; anthers 2-celled, opening lengthwise inwardly. Ovary superior, incompletely 4-celled, with a free-based placenta having

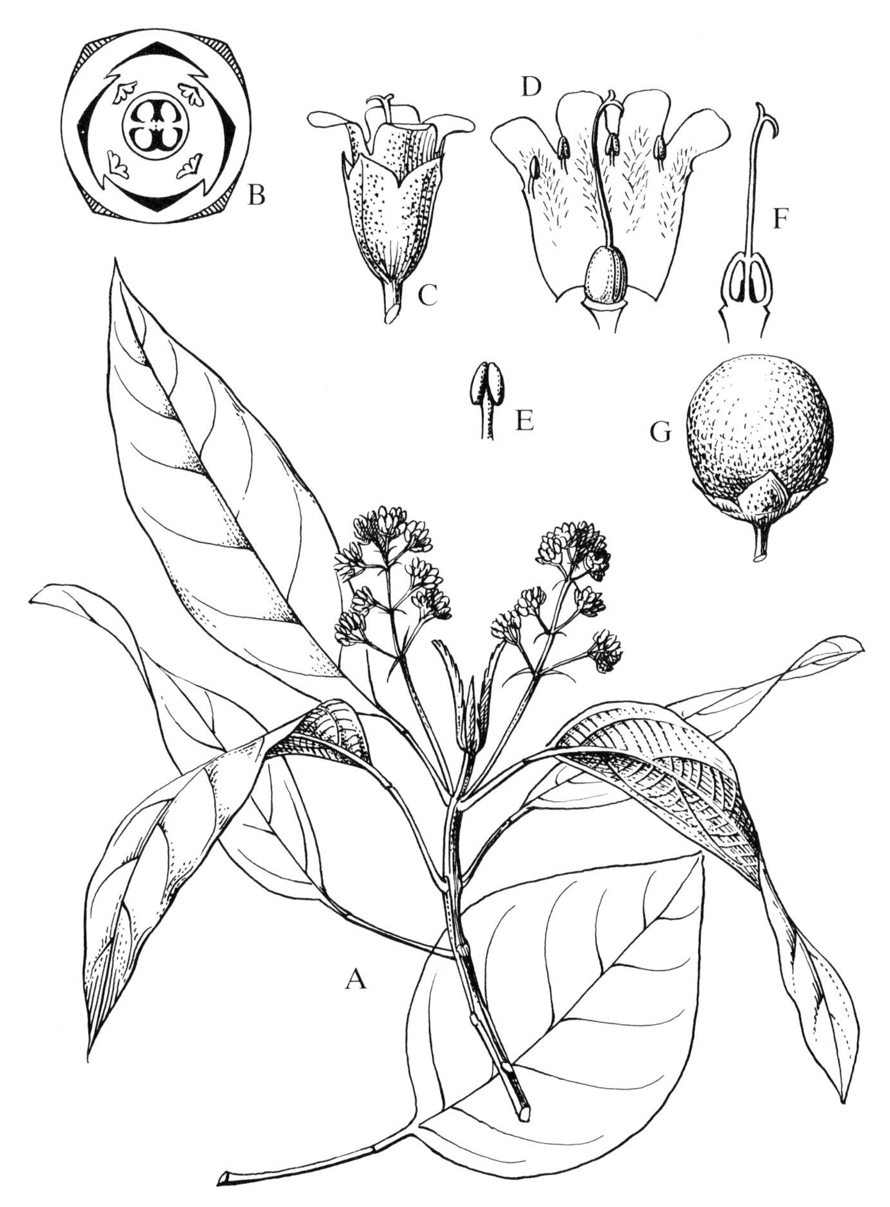

FIG. 217 **Petitia domingensis.** A, branch with leaves and flowers, x$^{2}/_{5}$. B, floral diagram. C, flower, x4. D, corolla spread open to show stamens and pistil, x4. E, anther, x8. F, long.section of pistil, x4. G, fruit, x4. (St.)

FIG. 218 **Avicennia germinans.** A, branch with leaves and flowers. B, C, details of leaf. D, flower. E, long.-section of flower, showing stamens and pistil. F, apex of style. G, fruiting branch. H, two views of fruit. I, fruit split open. J, K, stages in development of seedling. (D.E.)

4 pendulous ovules attached to the apex; style terminal, with 2 stigmatic lobes. Fruit a compressed, leathery, asymmetric, 1-seeded capsule, eventually splitting by 2 valves; seed with folded cotyledons, germinating in the fruit.

A single genus with about 11 species, widely distributed on tropical and subtropical seashores. By many authors this group is included in the Verbenaceae.

Genus 1. **AVICENNIA** L.

With the characters of the family.

1. **Avicennia germinans** (L.) L., Sp. Pl. ed. 3, 2:891. 1764. **FIG. 218.**

Avicennia nitida Jacq., 1760.

"Black Mangrove"

A shrub or tree to 15 m tall or more, the young branchlets and peduncles very minutely puberulous; pneumatophores up to 30 cm long; leaves oblong-elliptic or lance-oblong, 3-13 cm long, blunt or acute at the apex, the lower surface pale grayish-green. Spikes simple or 3-branched, densely-flowered toward the apex. Calyx puberulous and ciliate. Corolla white with yellow eye, the lobes densely puberulous on both sides. Style 3-4 mm long. Capsules 2.5-4 cm long.

GRAND CAYMAN: *Brunt 2043; Kings GC 255, GC 270; Proctor 15212; Sachet 459.* LITTLE CAYMAN: *Kings LC 28; Proctor 28073.*
– Seacoasts of subtropical and tropical America, also in West Africa.

Family 94

LABIATAE

Herbs or shrubs, rarely trees, often aromatic and usually with 4-angled stems; leaves opposite or whorled, simple, and entire, crenate, serrate, or lobed, often glandular; stipules absent. Flowers mostly perfect, rarely unisexual, solitary in the leaf-axils or more often in compact cymes, these axillary or terminal, often condensed into whorls spaced on the axes of racemes or panicles, or crowded into a spike or head; bracts small or large; bracteoles often present. Calyx tubular or bell-shaped, regular or 2-lipped, basically 5-lobed but the upper 3 teeth or lobes often united. Corolla gamopetalous, tubular at the base, the limb basically 5-lobed but usually 2-lipped, the 2 upper lobes usually to form an entire or notched hood, or rarely absent; the 3 lower lobes partly or wholly united to form the lower lip. Stamens usually 4 in 2 pairs inserted at different levels of the corolla-tube, or sometimes only 2; anthers 2-celled or sometimes 1-celled by abortion, opening lengthwise inwardly. A ring-shaped or unilateral disk or gland present. Ovary superior, of 2 bilobed carpels, becoming more or less 4-lobed and 4-celled, each cavity with 1 basal, erect ovule; style slender, central and usually gynobasic (i.e., arising basally between the lobes of the ovary); stigma 2-lobed or entire. Fruit of 4 free or paired, dry, 1-seeded nutlets, usually enclosed by the persistent calyx; seeds with little or no endosperm, the embryo straight and with flat cotyledons.

A large, cosmopolitan family of about 200 genera and 3,200 species.

KEY TO GENERA

1. Calyx regular or nearly so, truncate and with 5 equal teeth: **1. Hyptis**

1. Calyx distinctly lobed and 2-lipped or irregular:

  2. Stamens 2; corolla usually blue in Cayman species: **2. Salvia**

  2. Stamens 4; corolla whitish: **3. Ocimum**

Genus 1. **HYPTIS** Jacq., nom. cons.

Erect herbs, sometimes shrubby; leaves opposite, usually toothed. Flowers in short-stalked axillary clusters, terminal racemes, or panicles of contracted more or less whorled cymules. Calyx 10-nerved, truncate and bearing 5 subequal bristle-like teeth, or rarely equally 5-lobed; corolla 2-lipped, the upper lip 2-lobed, the lower one 3-lobed with the lateral lobes deflexed and the central lobe concave or sac-like. Stamens 4, slightly exserted; filaments glabrous; anthers 2-celled. Nutlets ovoid or ellipsoid, smooth or rough.

A large tropical American genus of about 400 species.

KEY TO SPECIES

1. Calyx-tube in fruit 5–7 mm long, with glandular and non-glandular hairs; foliage strongly aromatic: **1. H. suaveolens**

1. Calyx-tube in fruit 2.5–3 mm long, puberulous (the hairs not glandular) and with sessile glands; foliage not strongly aromatic: **2. H. pectinata**

1. Hyptis suaveolens (L.) Poit. in Ann. Mus. Hist. Nat. Paris 7:472, t.29, f.2. 1806.

Erect annual herb up to 1 m tall, clothed throughout with gland-tipped hairs, the stems becoming subwoody at the base; leaves long-petiolate, the blades broadly ovate to orbicular, 2-6 cm long or more, the margins crenate-serrate or doubly serrate; pubescent on both sides. Cymes short-stalked in the upper axils or forming a terminal raceme; corolla pale blue or whitish, the tube 4-6 mm long. Nutlets brown, ribbed, 3-4 mm long.

GRAND CAYMAN: *Proctor 15263.* CAYMAN BRAC: *Millspaugh 1154.*
– Widespread in the American tropics, mostly a weed of open waste ground; introduced into the Old World tropics.

2. Hyptis pectinata (L.) Poit. in Ann. Mus. Hist. Nat. Paris 7:474, t.30. 1806.

Erect shrub-like herb up to 2 m tall or more, puberulous throughout with minute non-glandular hairs interspersed with minute sessile glands; leaves long-petiolate, the blades lanceolate to broadly ovate and 2-6 cm long (longer toward base of the main stem), the margins irregularly crenate-dentate. Cymes subsessile, often forked, axillary among the upper leaves and also forming simple or branched terminal racemes; corolla whitish, the tube 1.5-2 mm long. Nutlets black, smooth, 0.5 mm long.

GRAND CAYMAN: *Hitchcock; Millspaugh 1341.*
– Pantropical, a weed of thickets and waste places.

Genus 2. **SALVIA** L.

Herbs or sometimes shrubs, erect or decumbent; leaves opposite, serrate or subentire. Flowers sessile or shortly stalked in few-flowered whorls at the nodes of terminal spikes, racemes or panicles. Calyx 2-lipped, the lips subequal and not spiny, in our species clothed with spreading, gland-tipped hairs; corolla strongly 2-lipped, the upper lip entire to 2-lobed, the lower lip spreading and more or less 3-lobed. Stamens 2; anthers 1-celled, the connective very much elongated, on one branch ascending and bearing the anther-cell at the apex, the other descending and more or less flattened. Nutlets smooth, usually becoming mucilaginous when wet.

A very widely distributed genus of more than 700 species. Culinary "Sage" is *S. officinalis* L., a native of the Mediterranean region. *S. splendens* Ker-Gawl. or "Red Salvia" is a scarlet-flowered species often cultivated for ornament.

The peculiar structure of the anthers in this genus causes them to function as a sort of lever; a bee, pushing into the flower in search of nectar, comes into contact with the sterile arm of the anther, thus raising it and causing the other end to descend onto the bee's back and dust it with pollen. The flowers are protandrous, i.e., the staminate part matures before the pistillate, thus ensuring cross-pollination. When the pistil matures, the style bends down and places the stigma in position to be touched first by a visiting insect.

KEY TO SPECIES

1. Mature calyces 2.8–3.5 mm long, not gaping in fruit: **1. S. occidentalis**

1. Mature calyces more than 5 mm long, obviously open and gaping:

  2. Annual decumbent-ascending to erect herb with mostly pale blue or sometimes white flowers; leaf-blades broadly ovate, puberulous but not white-woolly beneath: **2. S. serotina**

  2. Erect slender shrub to 1 m tall with deep blue flowers; leaf-blades narrowly lance-oblong, densely white-puberulous beneath: **3. S. caymanensis**

1. **Salvia occidentalis** Sw., Nov. Gen. & Sp. Pl. 14. 1788.

A straggling, diffusely-branched herb with ascending flowering-branches, the young parts sparsely pubescent; leaf-blades rhombic-ovate, mostly 1-5 cm long, acute at the apex, cuneate at the base, the margins coarsely serrate. Racemes slender and elongate, up to 12 cm long or more. Calyx at anthesis c. 2 mm long; corolla-tube c. 2.5 mm long. Nutlets c. 2 mm long.

GRAND CAYMAN: *Hitchcock; Proctor 15039.* LITTLE CAYMAN: *Kings LC 61.* CAYMAN BRAC: *Kings CB 3; Millspaugh 1186.*
– Throughout tropical America, common along roadsides, in open waste places, and in various other disturbed habitats.

2. **Salvia serotina** L., Mant. Pl. 25. 1767.

A decumbent to erect herb usually less than 30 cm tall, more or less puberulous throughout; leaf-blades mostly 1.5–3 cm long and up to 2.5 cm broad, obtuse or subacute at the apex, truncate-subcordate at the base, the margins crenate. Racemes 5–12 cm long. Calyx at anthesis 3–4 mm long; corolla tube c. 3 mm long. Nutlets c. 2 mm long.

GRAND CAYMAN: *Hitchcock.*
– Florida, West Indies, and Central America, a weed of pastures and cultivated fields.

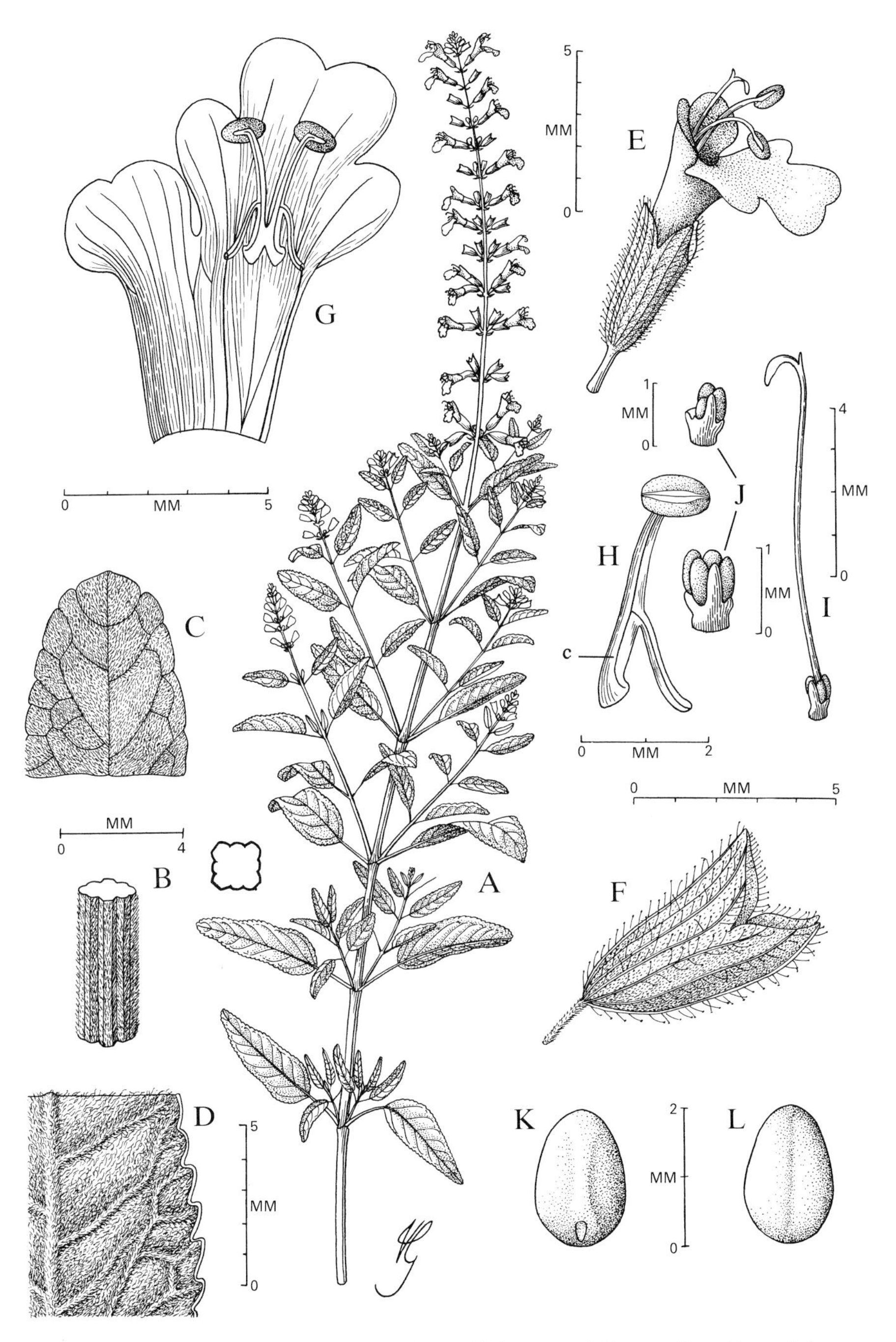

FIG. 219 **Salvia caymanensis.** A, portion of plant with leaves and flowers. B, sections of stem. C, upper side of leaf. D, lower side of leaf. E, flower. F, calyx. G, upper part of corolla cut open to show stamens. H, anther; c, connective. I, pistil. J, two ovaries. K, L, two views of nutlet. (G.)

3. Salvia caymanensis Millsp. & Uline ex Millsp. in Field Mus. Bot. 2:94. 1900. **FIG. 219.**

Salvia serotina var. *sagittaefolia* Millsp. 1900.

A small, stiffly erect shrub with canescent-puberulous branchlets; leaf-blades mostly 1-3.5 cm long and usually not over 1 cm broad, acutish at the apex, broadly short-cuneate at the base, the margins crenulate. Racemes 1.5–10 cm long. Calyx at anthesis 4-5 mm long; corolla-tube 4-6 mm long. Nutlets 1.9-2 mm long.

GRAND CAYMAN: *Brunt 2081; Kings GC 422; Millspaugh 1295 (type), 1391; Proctor 27969.*
– Endemic, in sandy thickets and clearings. Sometimes considered merely a local variant of *S. serotina*, but falls outside the normal variability of that species; in habit it is especially distinctive.

Genus 3. **OCIMUM** L.

Aromatic herbs or low shrubs; leaves opposite, usually serrate, the tissue finely dotted with pellucid glands. Flowers small, in 4–10-flowered whorls arranged in terminal racemes or panicles, the pedicels usually recurved. Calyx 2-lipped, the upper lip broad, the edges wing-like and decurrent along the tube, the lower lip 4-lobed, the lobes ending in bristly teeth. Corolla 2-lipped, the upper lip 4-lobed, the lower entire. Stamens 4, in 2 pairs, exserted. Nutlets smooth or wrinkled, often mucilaginous when moistened.

A genus of about 150 species, widely distributed in tropical and warm-temperate regions. *O. basilicum* L., or "Basil", is a well-known culinary herb.

KEY TO SPECIES

1. Plants annual; fruiting calyx 7–8 mm long: **1. O. micranthum**

1. Plants perennial; fruiting calyx 4–5 mm long: **[2. O. sanctum]**

1. Ocimum micranthum Willd., Enum. Hort. Berol. 630. 1809. **FIG. 220.**

"Pimento Basil"

A somewhat bushy herb usually less than 50 cm tall, thinly puberulous on the younger parts; leaf-blades broadly ovate-elliptic, mostly 2-6 cm long, short-acuminate at the apex, cuneate at the base, the margins subentire or obscurely serrate. Racemes compact, mostly under 6 cm long, the flowers in whorls of 6. Corolla whitish with pale violet mottling, c. 4 mm long. Nutlets black, 1-2 mm long.

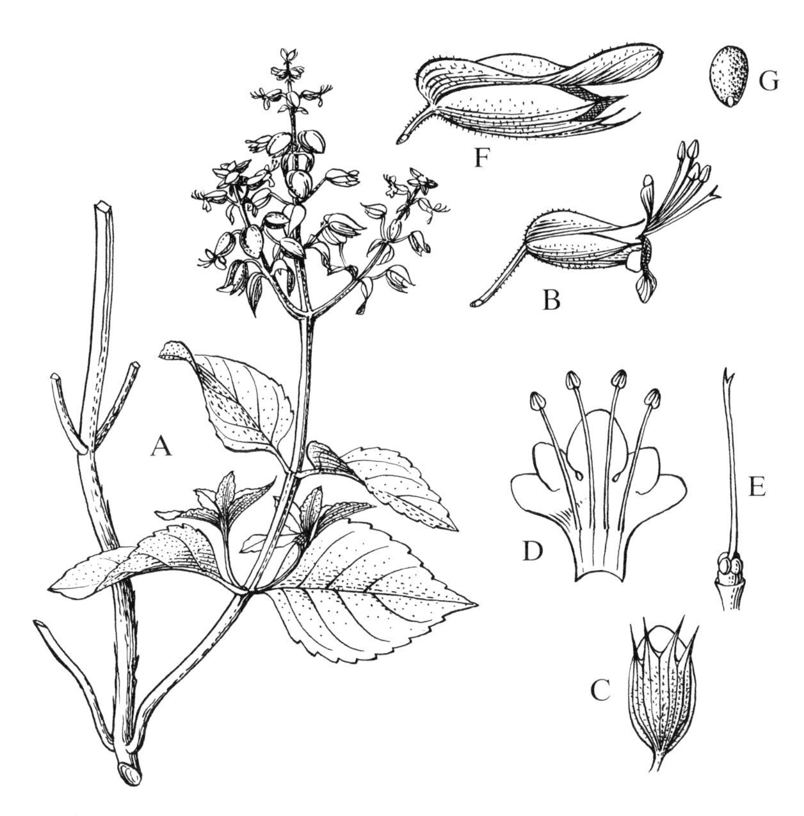

FIG. 220 **Ocimum micranthum.** A, habit, x½. B, flower, x4. C, underside of calyx, x4. D, corolla spread open to show stamens, x4. E, style, x4. F, fruiting calyx, x4. G, nutlet, x2½. (St.)

GRAND CAYMAN: *Correll & Correll 51042; Kings GC 213; Millspaugh 1266.* LITTLE CAYMAN: *Kings LC 6.*

– Florida, West Indies, and continental tropical America, common in fields, thickets, and open waste ground. The pleasant, aromatic odour of this species is due to the essential oil methyl cinnamate, which can be used like citronella as a mosquito-repellent.

[2. **Ocimum sanctum** L., Mant. Pl. 1:85. 1767.

A bushy herb or subshrub usually more than 70 cm tall, puberulous and thinly pilose on the younger parts; leaves long-petiolate, the blades mostly 2–5 cm long, acute at the apex, the margins coarsely serrate. Racemes simple or paniculate, up to 10 cm long, the flowers in whorls of 4–8. Corolla c. 3 mm long. Nutlets brown, 1–1.5 mm long.

GRAND CAYMAN: *Kings GC 135.*
– Native of the Old World tropics, now sparingly naturalized in the West Indies and continental tropical America.]

Family 95

OLEACEAE

Trees or shrubs, often scandent or vine-like; leaves usually opposite, rarely alternate or whorled, simple or pinnately compound; stipules absent. Flowers regular, perfect or unisexual, borne in terminal or axillary racemes, cymes or panicles. Calyx 4-many-lobed, valvate, or rarely absent; corolla gamopetalous or rarely with free petals, usually 4-lobed or rarely with more. Stamens 2 or rarely 4, hypogynous or inserted on the corolla; anthers 2-celled, opening lengthwise; disk absent. Ovary superior, 2-celled, usually with 2 ovules in each cavity, pendulous or ascending on axile placentas; style simple or absent; stigma capitate or bifid. Fruit a berry, drupe, capsule or samara; seeds with fleshy endosperm (rarely none) and straight embryo.

A widespread family of perhaps 30 genera and about 600 species, occurring in both temperate and tropical regions. The most important species in this family is *Olea europea* L., the "Olive", the fruit being eaten when pickled, and also the source of olive-oil.

KEY TO GENERA

1. Plants creeping or vine-like, with pinnate leaves; corolla-tube elongate; fruit of 2 berry-like lobes: **[1. Jasminum]**

1. Plants erect, with simple leaves; corolla-tube very short, or petals free or lacking; fruit a simple drupe:

 2. Flowers unisexual, dioecious, in small axillary clusters; calyx and corolla absent, or minute and soon falling: **2. Forestiera**

 2. Flowers perfect, in terminal and axillary panicles; calyx and corolla both present and well-developed: **3. Chionanthus**

[Genus 1. **JASMINUM** L.

Erect shrubs or often vines; leaves usually opposite and simple, 3-foliolate or odd-pinnate. Flowers solitary or usually in cymes, terminal or in the upper axils, often fragrant and showy; calyx 4–9-toothed or -lobed; corolla salver-shaped with cylindric tube and 4–5 or more imbricate lobes. Stamens 2, included. Ovary 2-celled with 2 (rarely 3–4) ovules in each cavity, these laterally attached near the base. Fruit a double berry, or one of the carpels sometimes abortive; seeds usually solitary, without endosperm.

A genus of about 200 species, in the tropical and warm-temperate regions of the Old World. Several are widely cultivated for their showy or fragrant flowers.

1. **Jasminum fluminense** Vell., Fl. Flumin. 10. 1825.

"Star of Bethlehem"

A scrambling shrub or vine, the younger parts pubescent; leaves 3-foliolate; leaflets broadly ovate, 1.5–5 cm long, acute or acuminate at the apex, the margins entire, and with tufts of woolly hairs in the nerve-axils beneath. Cymes on pubescent peduncles; flowers fragrant at night; corolla white, 5–9-lobed, the tube c. 1.5 cm long. Berries black, 5–8 mm in diameter.

GRAND CAYMAN: *Brunt 1888; Proctor 15076.*
– Native of tropical Africa, widely naturalized in the American tropics. In Grand Cayman this species is grown in gardens, has escaped and become naturalized in roadside thickets.]

Genus 2. **FORESTIERA** Poir.

Dioecious trees or shrubs; leaves mostly opposite, simple, entire or toothed, usually deciduous during dry seasons. Flowers small, 1 -few in small fascicles in the leaf-axils or at nodes on old wood; calyx deeply 4–6-lobed or absent; corolla none or rarely 1 or 2 small free petals present. Stamens 2 or 4. Ovary 2-celled, each cavity with 2 pendulous ovules. Fruit a 1-seeded drupe.

A genus of about 15 species occurring in southern United States, Mexico, northern Central America and the West Indies.

1. **Forestiera segregata** (Jacq.) Krug & Urb. in Engl., Bot. Jahrb. 15:339. 1892.

A diffusely-branched shrub to 2 m tall or more, rarely tree-like; branchlets stiff and minutely puberulous; leaves glabrous, narrowly elliptic, oblanceolate or obovate, mostly 2-6.5 cm long, obtuse at the apex, minutely gland-dotted beneath. Inflorescence subtended by sessile and clawed, ciliate bracteoles 2-2.5 mm long. Flowers yellowish-green, fragrant. Staminate flowers pedicellate, lacking calyx and corolla, consisting of 2-4 naked stamens with filaments 4-6 mm long. Pistillate flowers undescribed. Drupe obliquely spindle-shaped, acute, bluish-purple, c. 7 mm long.

GRAND CAYMAN: *Brunt 2152; Proctor 15063, 15066, 15138.* LITTLE CAYMAN: *Proctor 35134.*

– Florida, Bermuda, Bahamas, the Greater Antilles, Virgin Islands and Antigua, in dry rocky thickets and woodlands.

Genus 3. **CHIONANTHUS** L.

Trees or sometimes shrubs; leaves opposite, entire and more or less coriaceous. Flowers in terminal and axillary panicles, the branches racemose, thyrsoid, or cymose, or the inflorescence contracted to an umbel or head. Calyx small, 4-parted or -toothed; corolla of 4 free or nearly free valvate petals, these linear or oblong. Stamens 2 or rarely 4, borne on the base of the petals. Ovary 2-celled with 2 ovules in each cavity, attached laterally near the apex and pendulous. Fruit an ovoid, oblong or subglobose drupe with thin flesh and hard endocarp, usually containing a solitary seed.

A pantropical and warm-temperate taxon of more than 150 species.

1. Chionanthus caymanensis Stearn in Bot. Notiser 132:58. 1979. **FIG. 221.**

"Ironwood"

A shrub or small tree to 7 m tall, with ashy-gray bark, glabrous throughout, the young shoots clothed with numerous minute white waxy scales. Leaves with marginate petioles 0.5-1 cm long; blades obovate, 2-5.5 cm long, 1-2.5 (-3) cm broad above the middle, abruptly short-acuminate at the apex, long-cuneate at the base and decurrent on the petiole, very minutely gland-dotted on both surfaces; domatia absent. Panicles 3-6 cm long, many-flowered; flowers white, fragrant; calyx rugose, c. 1 mm long with triangular lobes; petals oblong-obovate, 2 mm long, joined at the base. Anthers ellipsoid, 1-1.4 mm long. Style 1 mm long; stigma bifid. Drupes obliquely ellipsoid, 7-8 mm long.

LITTLE CAYMAN: *Proctor 28116, 28183 (type).*

– Endemic; a similar plant with larger, elliptic leaves (up to 8 cm long or more) and smaller fruits, occurs on Grand Cayman (*Proctor 27958*) but has not yet been collected in flower; it may or may not be conspecific.

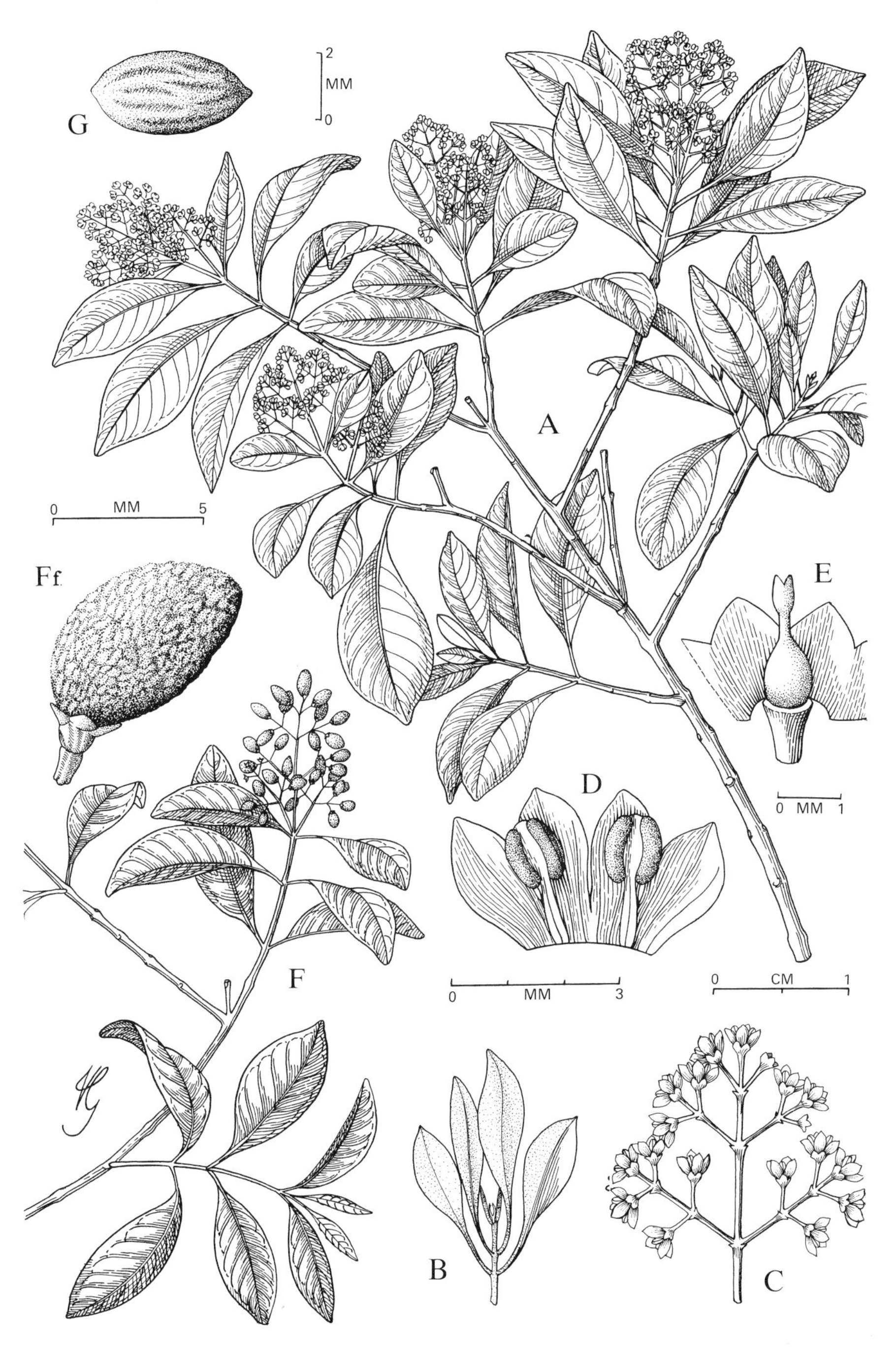

FIG. 221 **Chionanthus caymanensis.** A, habit, branch with leaves and flowers. B, young leafy branchlet. C, portion of flowering inflorescence. D, corolla opened out to show stamens. E, calyx (opened out) and pistil. F, fruiting branch; Ff, fruit. G, seed. (G.)

Family 96

SCROPHULARIACEAE

Herbs, shrubs or vines, rarely trees; leaves alternate, opposite or whorled, always simple, margins entire, toothed or lobed; stipules absent. Flowers perfect, more or less zygomorphic, solitary, clustered, or in racemes or panicles. Calyx nearly truncate or 4–5-toothed or -divided, persistent; corolla gamopetalous, tubular to broadly bell-shaped, often 2-lipped, with 4, 5 or more imbricate lobes. Stamens usually 4 (rarely 2 or 5), in 2 pairs inserted at different levels in the corolla-tube; anthers usually 2-celled, opening lengthwise. Disk usually present. Ovary superior, typically 2-celled; ovules on large axile placentas; style solitary, often persistent, the stigma usually more or less 2-lobed. Fruit usually a capsule variously dehiscent, or rarely a berry; seeds usually numerous, small, and smooth, angled or winged; endosperm fleshy, the embryo small.

A large, cosmopolitan family of about 200 genera and 3,000 species.

KEY TO GENERA

1. Leaves alternate: **3. Capraria**

1. Leaves opposite or whorled:

  2. flowers in lax, peduncled cymes; corolla red: **[1. Russelia]**

  2. Flowers solitary or clustered in the leaf-axils; corolla not red:

    3. Leaves (at least the larger ones) serrate or toothed; corolla less than 6 mm long:

      4. Flowers subsessile; corolla tubular, the lobes shorter than the tube: **2. Stemodia**

      4. Flowers on slender pedicels longer than the calyx; corolla rotate, the lobes longer than the tube: **4. Scoparia**

    3. Leaves entire, linear; corolla more than 10 mm long: **5. Agalinis**

[Genus 1. **RUSSELIA** Jacq.

Shrubs with striate or angled stems; leaves opposite or whorled, sometimes reduced to more scales. Flowers in simple or compound cymes; calyx deeply

5-lobed, the lobes ovate; corolla tubular or narrowly funnel-shaped, the 5 lobes much shorter than the tube, somewhat unequal. Stamens 4, the anther-cells divergent. Capsule glabrous, ovoid or globose, septicidally dehiscent; seeds small, ellipsoid, brown or black, and variously reticulate, pitted or ridged.

A genus of more than 50 species occurring naturally in Mexico and Central America; several are widely cultivated in warm countries.

1. **Russelia equisetiformis** Cham. & Schlecht. in Linnaea 6:377. 1831. **FIG. 222.**

Russelia juncea Zucc., 1832.

A lax, arching, much-branched shrub, the slender green branches striate and glabrous; leaves up to 7 in a whorl or obsolete, when present lanceolate to ovate, up to 2 cm long, with toothed margins, and gland-dotted beneath. Pedicels c. 1 cm long; sepals 2 mm long. Corolla crimson with tube up to 2.5 cm long. Capsules broadly ovoid, c. 5 mm long.

GRAND CAYMAN: *Brunt 1714; Hitchcock; Kings GC 80; Proctor 15188.* LITTLE CAYMAN: *Proctor 35201* (cult.) CAYMAN BRAC: *Proctor (sight record).* – Mexico and Central America, cultivated and naturalized elsewhere. In Grand Cayman this species now grows wild along sandy roadsides.]

Genus 2. **STEMODIA** L., nom. cons.

Herbs or low shrubs, mostly glandular-pubescent and aromatic; leaves opposite or whorled, serrate or toothed. Flowers solitary in the axils or in terminal, often leafy-bracted spikes or racemes; calyx 5-parted, the segments imbricate, equal and nearly free. Corolla with nearly cylindric tube, the limb 2-lipped, the upper lip notched or entire, the lower 3-lobed. Stamens 4; anthers-cells distinct, stalked. Stigma usually 2-lobed. Capsule 2-valved, the valves 2-cleft; seeds striate or reticulate.

A pantropical genus of about 30 species.

1. **Stemodia maritima** L., Syst. Nat. ed. 10, 2:1118. 1759.

Perennial sprawling herb with erect branches, the older stems becoming somewhat woody; leaves lanceolate, mostly 0.6–2 cm long, acute at the apex, sessile and clasping-subcordate at the base, the margins sharply serrate. Flowers solitary; calyx 2–3 mm long; corolla 5.5 mm long, pale blue or white. Capsules elongate-ovoid, 2.5 mm long; seeds minutely punctate, apiculate.

GRAND CAYMAN: *Brunt 1915, 2076, 2077; Hitchcock; Kings GC 65; Proctor 15036.*
– Bahamas, Greater Antilles and South America, in low moist ground, sandy clearings, or seasonally-flooded grasslands near the sea.

Genus 3. **CAPRARIA** L.

Perennial shrubby herbs; leaves alternate, toothed. Flowers solitary or paired (rarely 4 together) in the leaf-axils, usually long-stalked; calyx of 5 narrow, almost equal sepals; corolla white, bell-shaped, 5-lobed, the lobes subequal. Stamens 4 (rarely 5); anther-cells divergent, confluent. Stigma dilated. Capsule longitudinally grooved and 4-valved; seeds reticulate.

A tropical American genus of 4 species.

1. **Capraria biflora** L., Sp. Pl. 2:628. 1753.

Stems erect, to 1 m tall or more, usually hairy; leaves oblanceolate or narrowly elliptic, mostly 1.5–4 cm long, acute at the apex, narrowed to the point of attachment, sharply serrate on the distal half. Pedicels slender and flexuous, up to 15 mm long; sepals 4–6 mm long; corolla c. 1 cm long. Capsule oblong-ovate, 4–6 mm long; seeds light brown, 0.5 mm long.

GRAND CAYMAN: *Brunt 2075, 2130; Hitchcock; Kings GC 65a; Millspaugh 1364; Proctor 27940.*
– Throughout tropical and subtropical America, common along roadsides and ditches, in open waste ground, and in damp pastures.

Genus 4. **SCOPARIA** L.

Erect, branched herbs or low shrubs; leaves opposite or whorled, gland-dotted and with entire or serrate margins. Flowers solitary or paired (rarely 4) in the leaf-axils, usually long-stalked; calyx 4–5-parted, the segments imbricate, nearly free; corolla white, nearly rotate, 4-lobed with subequal lobes, densely bearded in the throat. Stamens 4; anther-cells distinct, parallel or divergent. Style pubescent, club-shaped with truncate or notched stigma. Capsule 2-valved, membranous; seeds numerous, angular.

A tropical American genus of 20 species, one of them (ours) also in the Old World tropics and subtropics.

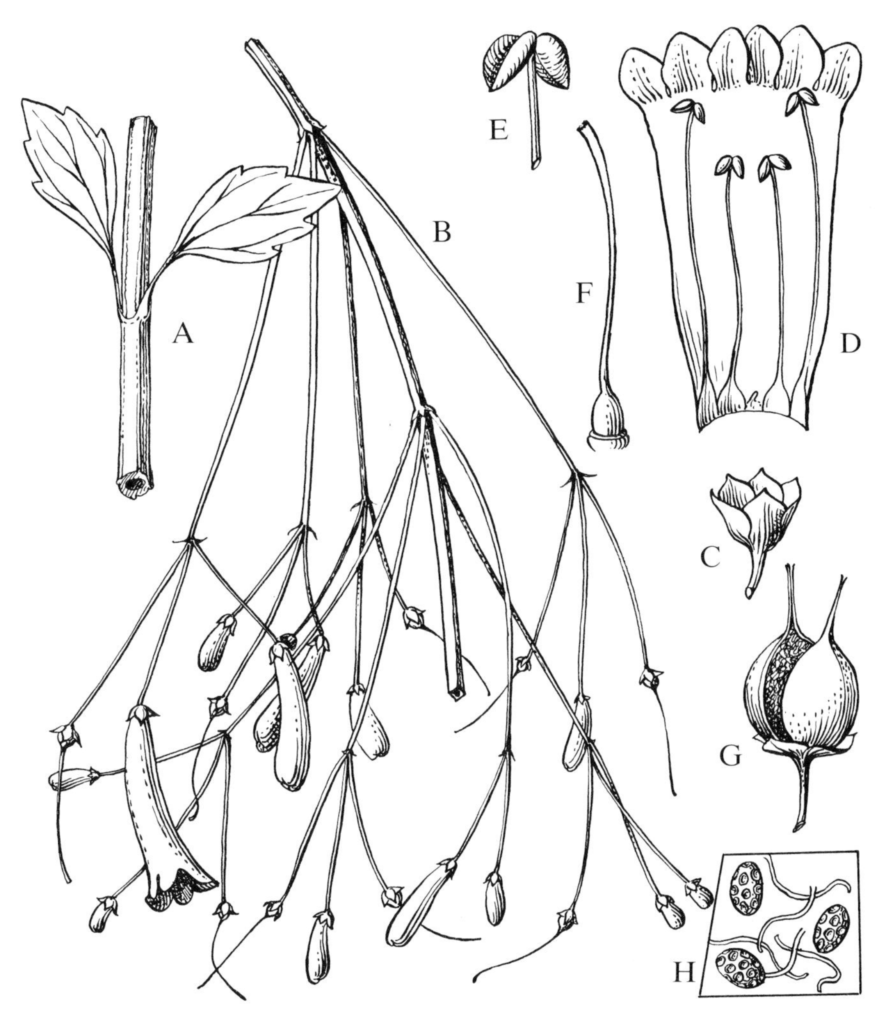

FIG. 222 **Russelia equisetiformis.** A, portion of stem with pair of leaves, x⁴/₅. B, portion of inflorescence, x⁴/₅. C, calyx, x1½. D, corolla spread open to show stamens. E, anther, x4. F, pistil, x1½. G, capsule, x1½. H, seeds, x6. (St.)

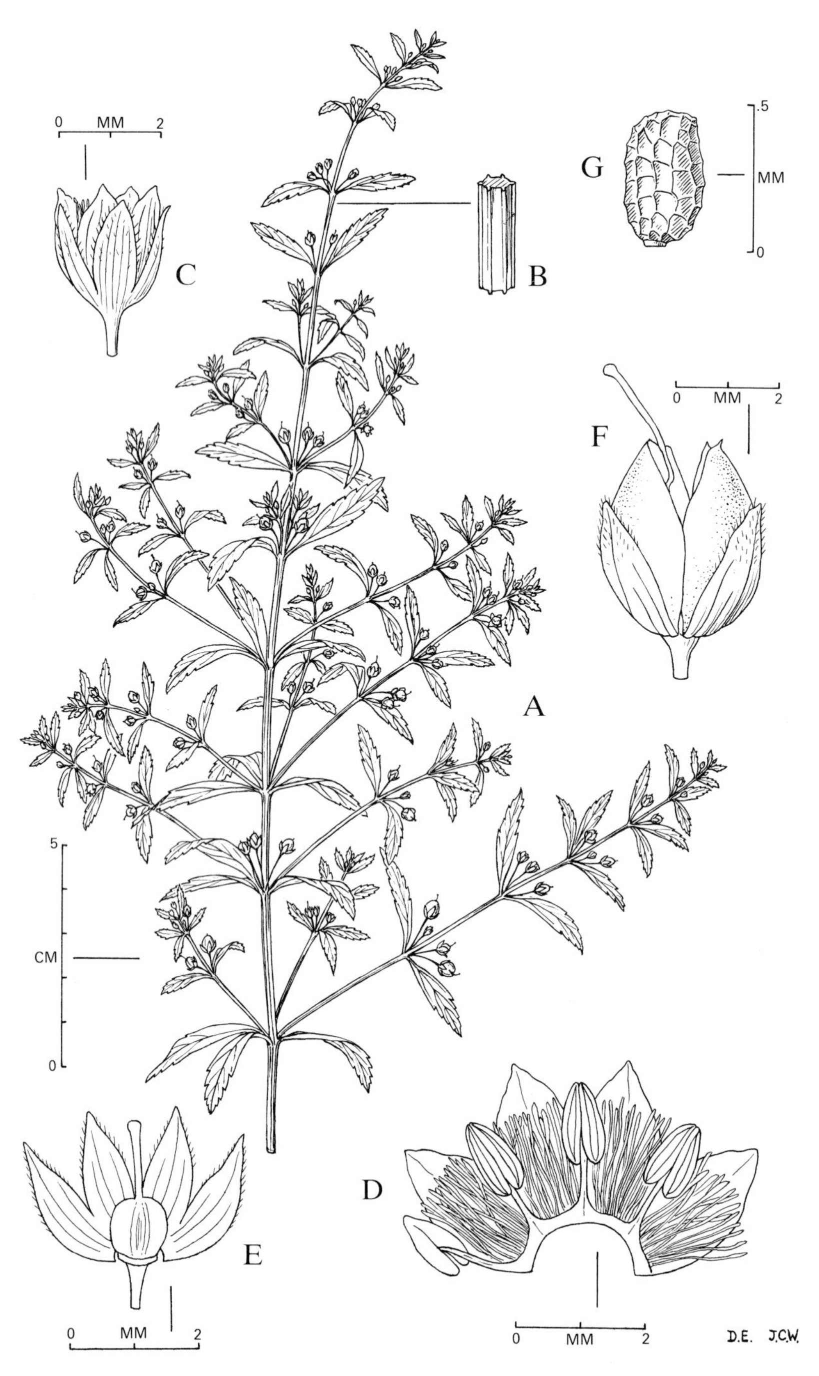

FIG. 223 **Scoparia dulcis.** A, habit. B, section of stem. C, flower. D, corolla spread open to show stamens and hairs. E, calyx with pistil. F, capsule. G, seed. (D.E. & J.C.W. ex St.)

1. Scoparia dulcis L., Sp. Pl. 1:116. 1753. **FIG. 223.**

A bushy annual herb with a taproot, up to 50 cm tall or more, glabrous throughout; leaves opposite or whorled, linear-oblanceolate or narrowly elliptic, mostly 0.5–4 cm long, at least the larger ones serrate in the distal half. Pedicels filiform, 5–8 mm long; calyx-lobes oblong, 1.5–2 mm long; corolla 3–4 mm across when expanded, with reflexed lobes. Capsules ovoid-globose, c. 3 mm long; seeds brown, 0.3–0.4 mm long.

GRAND CAYMAN: *Proctor 27970.*
– Widespread in the tropics and subtropics; Cayman plants were found in sandy thickets and clearings.

Genus 5. **AGALINIS** Raf., nom. cons.

Erect hemiparasitic terrestrial herbs; leaves opposite, narrow, and sessile. Flowers solitary in the bract-axils of loose racemes; calyx campanulate, 5-toothed or -lobed; corolla campanulate or funnel-shaped, the tube broad, the limb 5-lobed and somewhat 2-lipped, the posterior (upper) lobes inflexed in bud. Stamens 4, included, the filaments pubescent; anthers 2-celled. Capsule ovoid or globose, 2-valved; seeds numerous, angled.

An American genus of about 60 species.

1. Agalinis kingsii Proctor in Sloanea 1:3. 1977. **FIG. 224.**

Slender annual subglabrous herb to 50 cm tall or more; leaves linear-subulate, up to 4 cm long and 2 mm broad, often incurved; margins rough-edged. Racemes very lax and few-flowered; pedicels ascending, 6–10 mm long; calyx-tube c. 2.5 mm long, the lobes triangular; corolla pink, campanulate, 10–15 mm long, the lobes ciliate. Capsules globose, c. 4 mm long; seeds wedge-shaped, 0.7–0.9 mm long.

GRAND CAYMAN: *Kings GC 257 (type MO, isotype BM).*
– Endemic, collected "in mangrove swamps on the drier land" at Forest Glen, near North Side. The species has not been seen again by subsequent collectors. *A. kingsii* differs from *A. albida* Britton & Pennell of Cuba and Jamaica in its very much larger leaves, longer pedicels, and pink flowers. It differs from the related *A. purpurea* (L.) Pennell of the United States and Cuba in its glabrous leaves, longer pedicels, smaller corollas and smaller capsules.

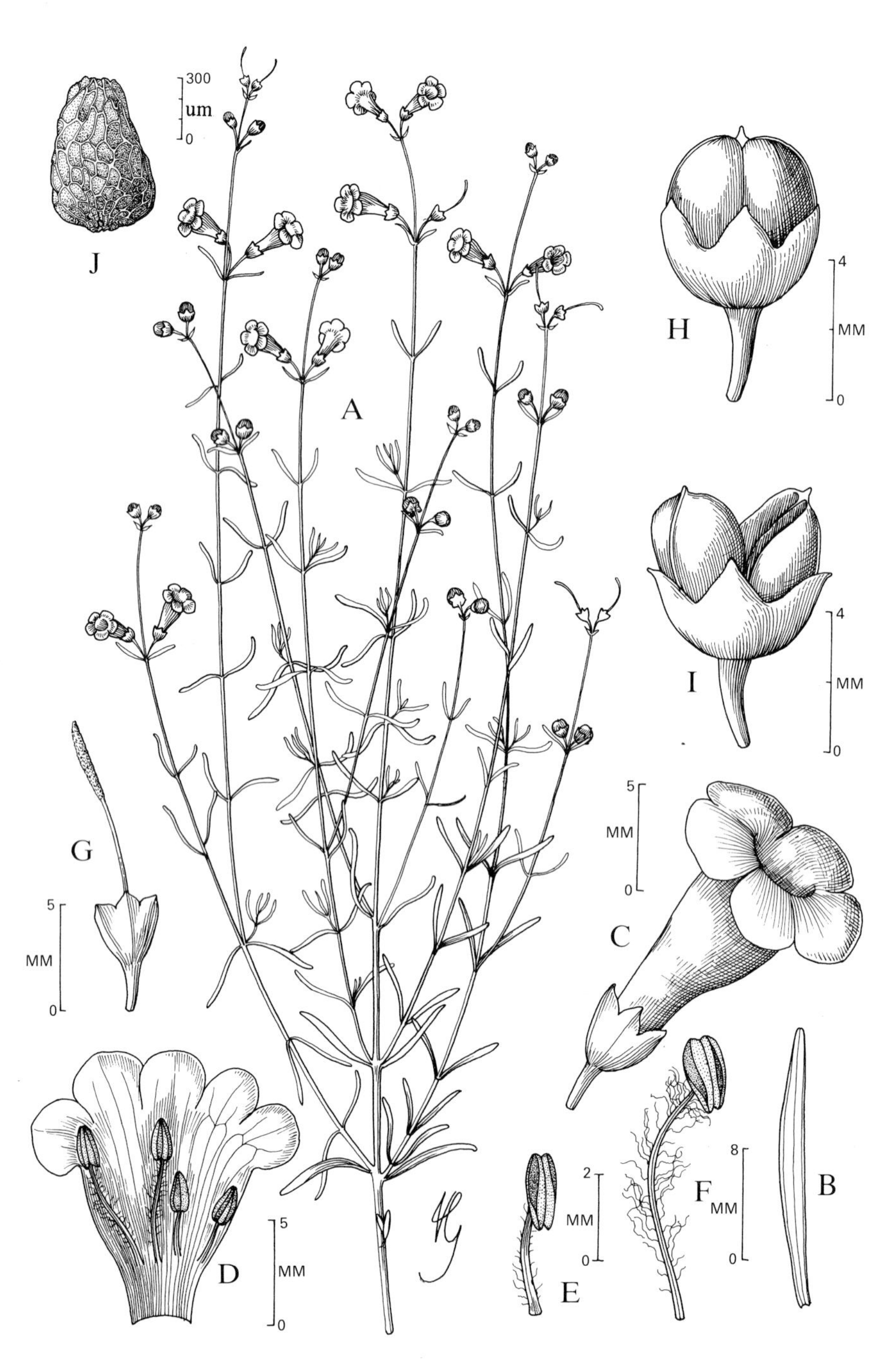

FIG. 224 **Agalinis kingsii.** A, habit. B, leaf. C, flower. D, corolla opened out. E, F, stamens. G, flower with corolla and stamens removed, showing calyx, style and stigma. H, fruit. I, fruit after dehiscence. J, seed. (G.)

Family 97

MYOPORACEAE

Shrubs or trees; leaves alternate or opposite, simple and entire, often glandular; stipules absent. Flowers perfect, zygomorphic, solitary or in axillary cymose clusters; calyx 5-lobed, persistent; corolla gamopetalous, 5-lobed, the limb oblique or 2-lipped. Stamens 4 (rarely 5), in 2 pairs inserted at different levels of the corolla-tube, sometimes accompanied by a staminode; anthers 2-celled, opening lengthwise. Ovary superior, usually 2-celled (rarely 3–10-celled); ovules 2–8 in each cavity, paired and pendulous on axile placentas; style terminal with simple stigma. Fruit a drupe; seeds small, with scant endosperm.

A chiefly Australasian family of 5 genera and about 180 species, a single representative occurring in the West Indies.

Genus 1. **BONTIA** L.

A glabrous, bushy, often arborescent shrub; leaves alternate, somewhat fleshy. Flowers solitary in the leaf-axils; calyx-segments imbricate; corolla with cylindric tube and 2-lipped limb, the upper lip 2-lobed, the lower 3-lobed and recurved, the middle lobe densely bearded. Ovary 2-celled with 4 ovules in each cavity. Drupe ovoid-acuminate.

A monotypic genus of the Caribbean area.

1. **Bontia daphnoides** L., Sp. Pl. 2:638. 8153. **FIG. 225.**

Plants up to 3 m tall or more; leaves narrowly lanceolate, mostly 4–12 cm long, acuminate at the apex, finely gland-dotted, the venation (except the midvein) obscure. Peduncles 1-2.5 cm long; calyx-lobes ovate-acuminate, 3-5 mm long, with hairlike tip; corolla dull yellow blotched with purple. c. 2 cm long, gland-dotted. Drupes 1-1.5 cm long, yellowish when ripe, crowned by the persistent elongate style.

GRAND CAYMAN: *Brunt 2055, 2177; Hitchcock.* LITTLE CAYMAN: *Proctor 28175.* CAYMAN BRAC: *Proctor 29136.*
– West Indies and north coast of South America, in subsaline coastal thickets and dry woodlands on limestone.

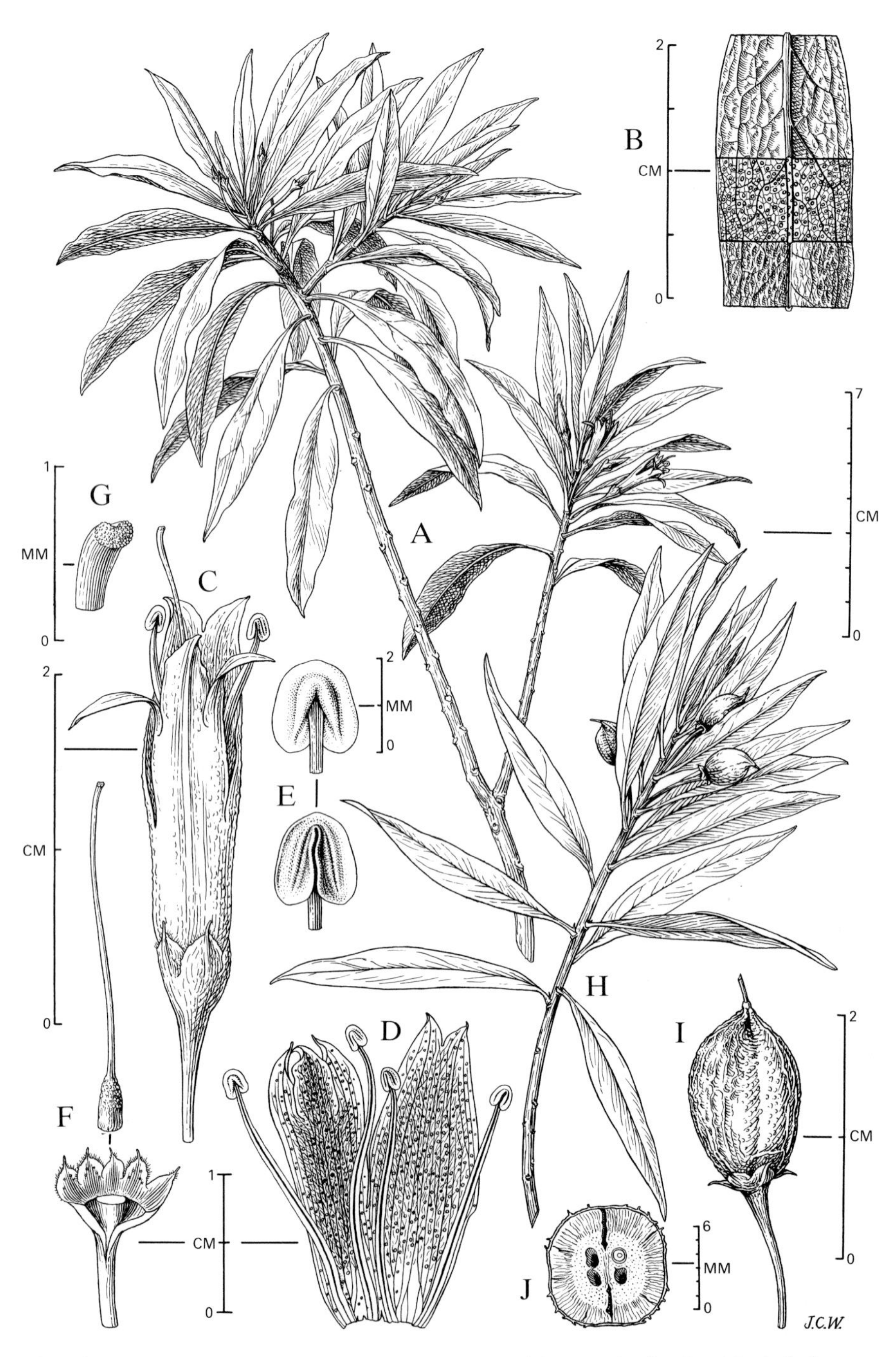

FIG. 225 **Bontia daphnoides.** A, branch with leaves and flowers. B, details of leaf. C, flower. D, upper part of corolla spread open to show stamens. E, two views of anther, F, calyx and pistil. G, stigma. H, fruiting branch. I, fruit. J, x-section of fruit. (J.C.W.)

Family 98

BIGNONIACEAE

Trees, shrubs, or vines, rarely herbs; leaves opposite or rarely alternate, simple, trifoliolate, pinnately compound, or sometimes digitate; if trifoliolate, the terminal leaflet often modified to a tendril; stipules absent. Flowers perfect, zygomorphic, often large, usually in cymes or racemes, sometimes solitary or clustered. Calyx bell-shaped, usually 5-toothed or -lobed, sometimes truncate or spathe-like; corolla gamopetalous, 5-lobed, sometimes 2-lipped. Stamens 4 (rarely 2), inserted in the corolla-tube; anthers 2-celled, opening lengthwise, the cells often widely divergent; 1-3 staminodes sometimes present. A hypogynous, ring-like or cup-shaped disk present. Ovary superior, 2-celled, with numerous ovules borne on 2 axile placentas in each cavity, or the ovary 1-celled with 2 bifid parietal placentas; style terminal, simple with 2-lobed stigma. Fruit indehiscent or else a 2-valved capsule; seeds without endosperm, those from capsular fruits often winged.

A pantropical family of about 120 genera and 800 species, the great majority in tropical America.

KEY TO GENERA

1. Leaves simple; fruits indehiscent; seeds not winged: **1. Crescentia**

1. Leaves compound; fruit an elongate dehiscent capsule; seeds winged:

  2. Leaves digitately 3-5-foliolate, the leaflets clothed with very minute peltate scales: **2. Tabebuia**

  2. Leaves pinnate:

    3. Leaflets entire; calyx spathe-like; corolla scarlet, very large: [**Spathodea**, p. 712]

    3. Leaflets serrate; calyx regular; corolla yellow: **3. Tecoma**

Genus 1. CRESCENTIA L.

Small trees, the branches with prominent nodes; leaves alternate or mostly fascicled on dwarf shoots in the axils of older, fallen leaves. Flowers solitary or several in a cluster from nodes on the old wood (or occasionally axillary); calyx leathery, closed in bud, splitting into 2 or 5 lobes at anthesis; corolla broadly bell-shaped with swollen tube and an oblique, 5-lobed limb. Stamens 4, in 2 pairs at different levels, included or slightly exserted. Ovary 1-celled. Fruit globose or ovoid, often very large, with a hard shell; seeds numerous, flattened, borne in pulp on spongy placentas.

A tropical American genus of 5 species.

1. **Crescentia cujete** L., Sp. Pl. 2:626. 1753. **FIG. 226.**

"Wild Calabash"

A tree up to 10 m tall (but often less), with stout branchlets; leaves oblanceolate or narrowly obovate, mostly 2–15 cm long, variable in length in the same fascicle

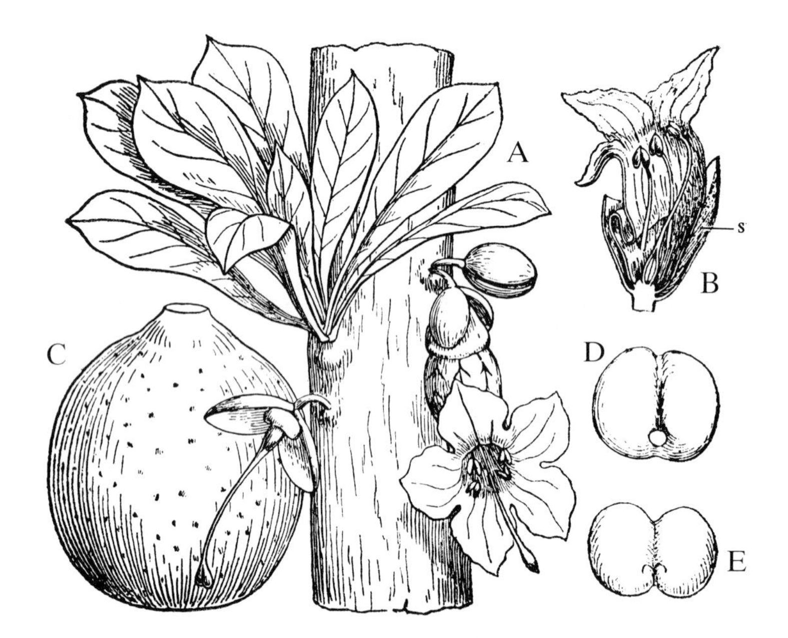

FIG. 226 **Crescentia cujete.** A, portion of woody stem bearing flowers and a cluster of leaves, x½. B, flower cut lengthwise, x½; s, staminode. C, fruit, very much reduced. D, seed, x¾. E, embryo, x¾; the radicle is concealed by the auricle of the cotyledon. (St.)

or on the same branch. Calyx up to 2 cm long, deeply 2-lobed; corolla 4–6 cm long, greenish-cream, often purple-veined. Fruits up to 25 cm in diameter (often less).

GRAND CAYMAN: *Brunt 2100; Kings GC 204; Proctor 15295.*
– Florida, West Indies and continental tropical America, in thickets, savannas, pastures, and along roadsides. The fruits, called "calabashes" (from the Spanish "calabazo", meaning a gourd), have traditionally been used in rural areas throughout its range as containers for water and other liquids; the shell is so tough that water can even be boiled in it. Less well known is the fact that the wood of this tree is exceedingly tough and durable, suitable for making furniture and the handles of tools.

Genus 2. **TABEBUIA** DC.

Trees or shrubs; leaves opposite or nearly so, usually 3–5-foliolate, sometimes 1-foliolate or simple. Flowers in terminal or axillary cymes or cymose panicles; calyx tubular or bell-shaped, toothed or shallowly lobed, sometimes more or less 2-lipped; corolla more or less funnel-shaped, the 5 spreading, slightly unequal lobes rounded and variously toothed, undulate or ruffled. Stamens 4, in 2 pairs inserted at different levels, a short staminode also usually present. Ovary 2-celled. Fruit an elongate, linear, lengthwise-dehiscent capsule; seeds with 2 membranous whitish wings.

A tropical American (chiefly West Indian) genus of more than 100 species. Several are frequently planted for ornament, e.g., *T. rosea* (Bertol.) DC., the "Pink Poui" and *T. rufescens* J.R.Johnst., the "Yellow Poui" (Jamaica).

1. **Tabebuia heterophylla** (DC.) Britton in Ann. Missouri Bot. Gard. 2:48. 1915.

Tabebuia pentaphylla of Hitchcock, 1893, not Hemsl., 1882.

T. riparia (Raf.) Sandwith, 1944.

"Whitewood"

A shrub or small tree to 5 m tall or more; leaflets usually 5- (sometimes 3-), stalked, more or less coriaceous and variable in shape, elliptic or oblong-elliptic to oblanceolate or obovate, mostly 2–8 cm long, the apex rounded or blunt to acutish, the venation finely reticulate. Flowers petiolate, solitary or in small cymose clusters on peduncles 1–3 cm long; calyx 9–12 mm long, slightly 2-lipped; corolla 4–7 cm long, light pink with yellow throat; stigma spatulate. Capsules 7–12 cm long, rarely longer, minutely lepidote like the leaves, beaked at the apex.

GRAND CAYMAN: *Brunt 1955, 2133; Correll & Correll 50999; Hitchcock; Howard & Wagenknecht 15023; Kings GC 132; Lewis GC 11; GC 57; Proctor 15055, 15227; Sachet 430.* LITTLE CAYMAN: *Kings LC 14; Proctor 28060.* CAYMAN BRAC: *Kings CB 73; Millspaugh 1214; Proctor 29054; Sauer 4139.*
– Greater Antilles, the Virgin Islands and northern Lesser Antilles, in dry rocky thickets and woodland on limestone. The wood is used for building cat-boats and schooners.

[**Spathodea campanulata** Beauv., the "African Tulip Tree", is planted as an ornamental in Grand Cayman (*Brunt 2019; Kings GC 95*).]

Genus 3. **TECOMA** Juss.

Erect shrubs or trees; leaves opposite, odd-pinnate or rarely simple; flowers in terminal racemes or panicles; calyx more or less bell-shaped, 5-toothed or -lobed; corolla funnel- or bell-shaped, the limb 5-lobed and slightly 2-lipped, the lobes nearly equal. Stamens 4, in 2 pairs inserted at different levels, included or exserted. Ovary 2-celled. Capsule linear-elongate, lengthwise-dehiscent, many-seeded, the seeds winged.

A tropical American genus of about 16 species.

1. **Tecoma stans** (L.) Kunth in H.B.K., Nov. Gen. & Sp. 3:144. 1819. **FIG. 227.**

"Shamrock", "Cow-stick", or "Hemlock" (Cayman Brac).

A shrub to 3 m tall or more; leaves with mostly 5 or 7 leaflets (rarely simple or 3-foliolate), these lanceolate or narrowly ovate, mostly 3–10 cm long, the terminal leaflet the largest, all acuminate and sharply serrate, puberulous beneath chiefly along the midrib and principal veins. Flowers in a short raceme; corolla 4–5 cm long. Capsules 11–20 cm long, glabrous.

GRAND CAYMAN: *Brunt 2072; Kings GC 154; Lewis 3851; Proctor 15085.* CAYMAN BRAC: *Proctor 28987.*
– Florida, West Indies, and continental tropical America, introduced and naturalized in the Old World tropics. The Cayman plants are common in sandy or rocky thickets and along roadsides.

FIG. 227 **Tecoma stans.** A, branch with leaves, flowers and fruits. B, calyx. C, corolla cut open to show stamens. D, two views of anther. E, pistil. F, same, enlarged. G, one valve of fruit, inside surface. H, seed. I, seed with wings removed. J, K, two views of pollen grain. (G.)

Family 99

ACANTHACEAE

Herbs, shrubs or trees, sometimes vines, the shoots often angled and swollen above the nodes; leaves opposite or rarely in whorls of 3, simple, entire or toothed, the epidermis often with cystoliths. Flowers perfect, zygomorphic or nearly regular; solitary, clustered or cymose in the leaf-axils, or in terminal spikes, racemes, cymes or panicles, often with large bracts. Calyx 4–5-lobed, imbricate or valvate, or reduced to a ring; corolla gamopetalous, 5-lobed and nearly regular or 2-lipped, sometimes 1-lipped. Stamens mostly 4, in 2 pairs inserted at different levels of the corolla-tube, sometimes reduced to 2; 1 or more staminodes sometimes present; anthers 1-2-celled, opening lengthwise. Disk present. Ovary superior, 2-celled, with 2 to many ovules in each cavity on axile placentas; style simple, with capitate or lobed stigma. Fruit usually a capsule, often elastically dehiscent, the valves rupturing and recurving explosively, flinging the seeds out, the action aided by the hook-like and hardened ovule-stalks; seeds usually flat, with little or no endosperm.

A pantropical and subtropical family of about 250 genera and 2,500 species. Many ornamental shrubs and vines belong to this family, which otherwise has little economic importance.

KEY TO GENERA

1. Flowers in spikes or racemes:

  2. Bracts large and conspicuous, completely concealing the calyx: **1. Blechum**

  2. Bracts minute, shorter than the pedicels: **[2. Asystasia]**

1. Flowers in panicles: **3. Ruellia**

Genus 1. **BLECHUM** P.Br.

Herbs with 4-angled, often straggling stems; leaves entire, wavy or crenate, usually pubescent. Inflorescence a dense terminal spike, the flowers partly concealed by large, overlapping bracts; flowers solitary or paired in the bract-axils, each subtended by 2 narrow bracteoles longer than the calyx. Calyx 5-parted, with linear segments; corolla narrowly tubular, with 5 spreading, nearly equal lobes. Stamens 4, inserted in the upper part of the corolla-tube; anthers oblong, with parallel cells. Capsule ovoid, pointed 6–16-seeded.

A tropical American genus of 10 species.

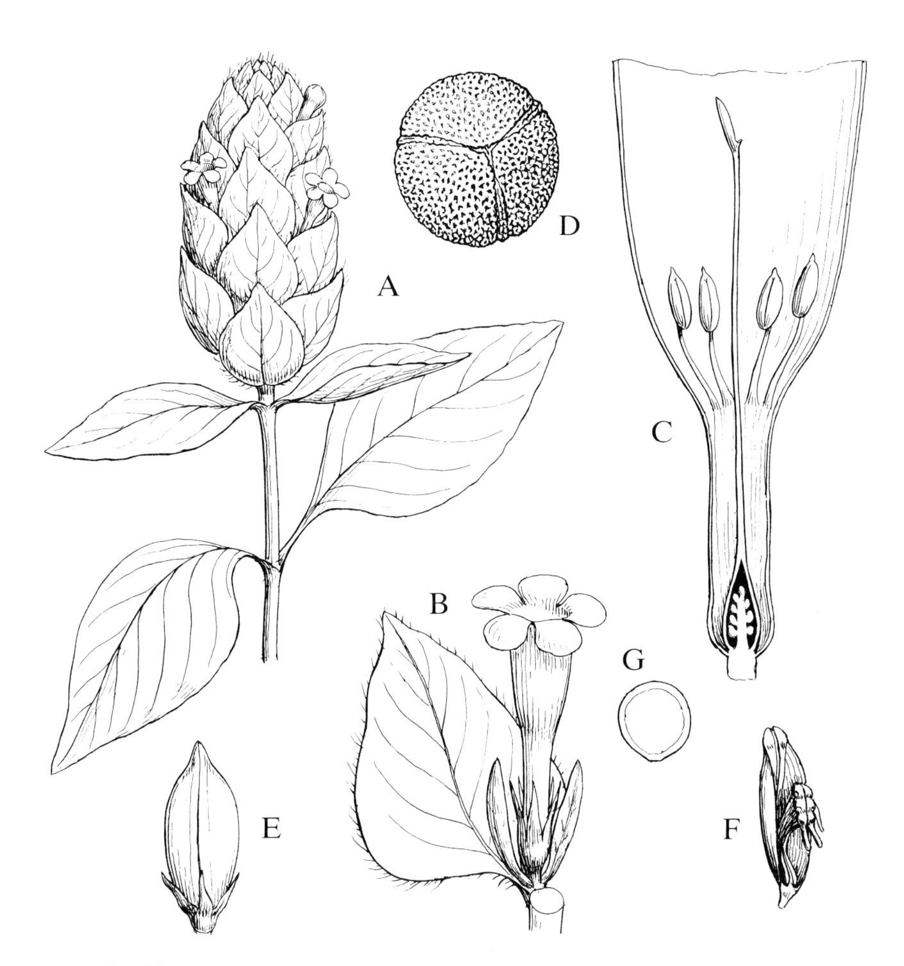

FIG. 228 **Blechum brownei.** A, habit, x1. B, single flower and bract, x3. C, corolla cut open to show stamens and pistil, x6. D, pollen grain greatly enlarged. E, capsule, x3. F, capsule after dehiscence, x3. G, seed, x6. (St.)

1. **Blechum brownei** Juss. in Ann. Mus. Paris 9:270. 1807. **FIG. 228.**

Blechum pyramidatum (Lam.) Urb., 1918.

An annual decumbent herb with ascending branches up to 50 cm tall or more, often rooting at the lower nodes; leaves narrowly ovate or elliptic, mostly 1.5-5 cm long, sparingly pubescent and with numerous linear cystoliths. Spikes 3-10 cm long, or rarely some flowers solitary in the leaf-axils; corolla pale violet or whitish, c. 15 mm long. Capsules 6–7 mm long, 12–16-seeded.

GRAND CAYMAN: *Brunt 1900; Hitchcock; Kings GC 88; Lewis 3860; Millspaugh 1328; Proctor 15075.* CAYMAN BRAC: *Millspaugh 1174; Proctor 29112.*

– West Indies and continental tropical America.

[Genus 2. **ASYSTASIA** Blume

Asystasia gangetica (L.) T. Anders. has been found becoming naturalized along roadsides and in clearings on Cayman Brac *(Proctor 29339)*. This is a trailing or straggling herb, sometimes scrambling over bushes to a height of 2 or 3 m, with ovate leaves and elongate, one-sided racemes of pale yellow or dull purplish flowers, the corolla 3.5–4 cm long. A native of tropical Asia introduced as a garden plant, this species is obviously a recent escape from cultivation.]

Genus 3. **RUELLIA** L.

Perennial herbs or shrubs with 4-angled stems; leaves entire or sometimes crenate. Flowers rather large, solitary or clustered in the leaf-axils, or in axillary and terminal cymose panicles; bracts narrow, and the bracteoles minute. Calyx deeply 5-lobed, the segments long and narrow; corolla funnel-shaped with the tube narrow at the base, the 5 spreading lobes nearly equal. Stamens 4, in 2 pairs, included. Style with recurved apex, the stigma simple or of 2 unequal lobes. Capsule cylindric, pointed at the ends, usually many-seeded; seeds flattened, ovate or orbicular, attached by their edges.

A large pantropical and subtropical genus of about 250 species.

KEY TO SPECIES

1. Leaves linear or narrowly lanceolate, less than 1 cm broad; calyx-lobes 6–9 mm long, usually with minute gland-tipped hairs; roots fibrous: **[1. R. brittoniana]**

1. Leaves elliptic or ovate, 1.5–4 cm broad; calyx-lobes 10–13 mm long, glabrous; roots tuberous-thickened: **2. R. tuberosa**

[**1. Ruellia brittoniana** Leonard in Jour. Wash. Acad. Sci. 31:96, f.1. 1941.

An erect herb with purplish stems; leaves 5–12 cm long or more, narrowly acuminate, the cystoliths aggregated beneath to form an irregular zig-zag pattern.

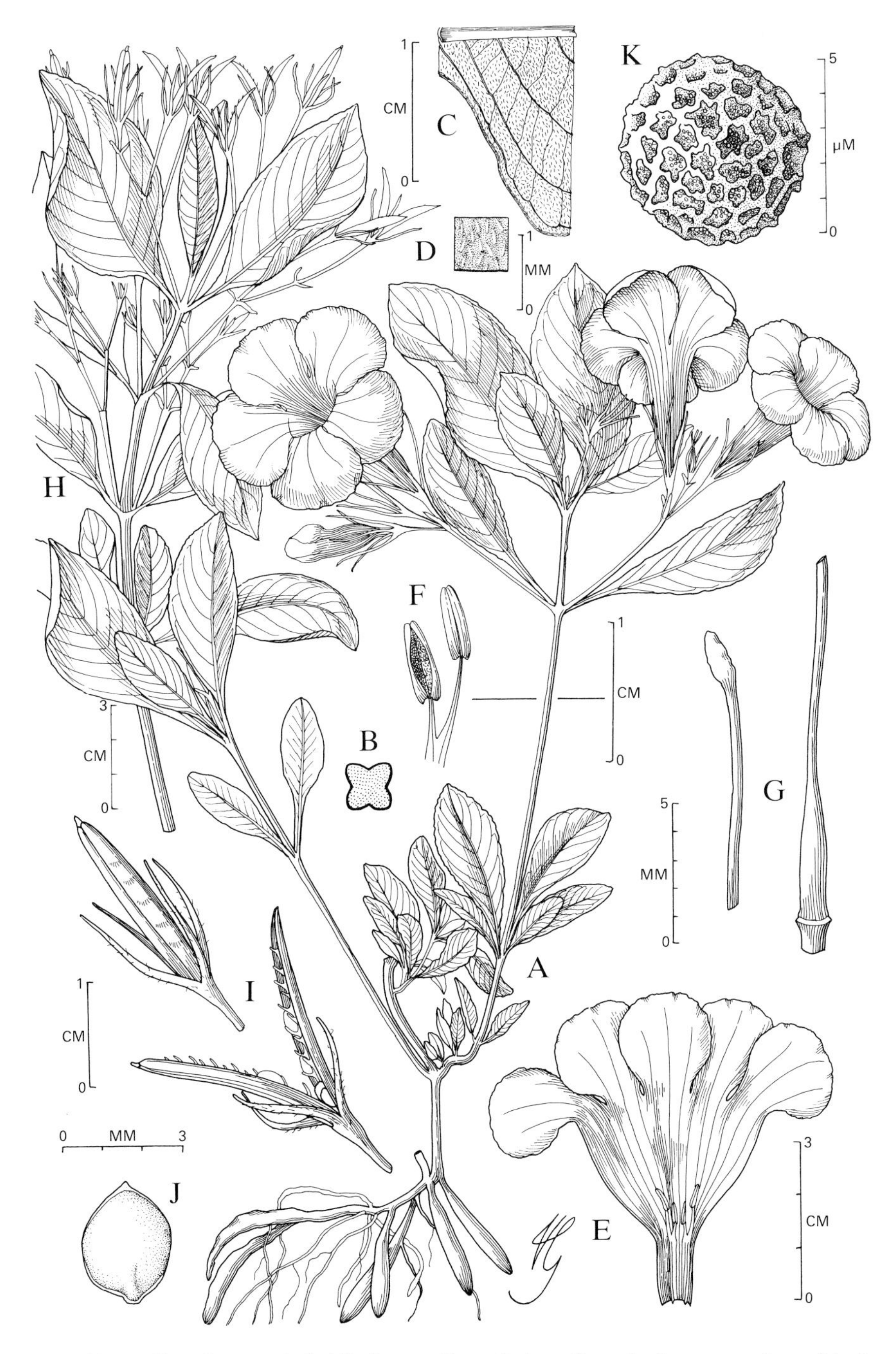

FIG. 229 **Ruellia tuberosa.** A, habit. B, x-section of stem. C, part of upper surface of leaf. D, enlarged portion of same, showing cystoliths. E, corolla cut open to show stamens. F, two anthers. G, pistil, cut through style. H, fruiting branch. I, fruit, closed and open. J, seed. K, pollen grain. (G.)

Inflorescences few-flowered or the flowers sometimes solitary; corolla violet-blue, 4–5 cm long, the tube puberulous. Capsules 2–2.5 cm long, glabrous except for a patch of hairs on each side of the beak, about 20-seeded; seeds nearly orbicular, 2–2.5 mm in diameter.

GRAND CAYMAN: *Brunt 1944.*
– Native of Mexico, cultivated and naturalized in many warm countries. The Cayman plants were found along sandy roadsides at West Bay.]

2. Ruellia tuberosa L., Sp. Pl. 2:635. 1753. **FIG. 229.**

Ruellia clandestina L., 1753.

"Heart Bush", "Duppy Gun"

Erect herb with numerous thickened roots; stems more or less pilose; leaves mostly 3–10 cm long, obtuse or acute, the minute linear cystoliths all scattered in a random pattern. Inflorescences few-flowered; corolla violet-blue, 4.5–6 cm long, the tube puberulous. Capsules c. 2 mm long, glabrous throughout, 10–15-seeded, dehiscing explosively when wet; seeds nearly orbicular, c. 2 mm in diameter.

GRAND CAYMAN: *Brunt 1970; Hitchcock; Kings GC 103; Maggs II 56; Millspaugh 1388; Proctor 15094; Sachet 408.* CAYMAN BRAC: *Proctor 28976, 35223.*
– West Indies and northern South America, usually in pastures and open waste ground or along roadsides. The local name "Heart Bush" refers to the use of its roots as a medicine for heart disease.

Family 100

GOODENIACEAE

Perennial herbs or shrubs with watery sap; leaves alternate, rarely opposite or basal only, simple and entire or toothed, rarely pinnatifid; stipules absent. Flowers perfect, usually zygomorphic, solitary or in cymes, racemes or heads arising from the leaf-axils; Calyx shortly 5-lobed, truncate or sometimes obsolete; corolla 5-lobed and 1–2-lipped, split down one side. Stamens 5, free or shortly adnate to the base of the corolla; anthers free or coherent around the style, 2-celled, opening lengthwise inwardly. Ovary inferior to superior, mostly 1–2-celled, with 1 or more erect or ascending ovules in each cavity, basal or on axile placentas; stigma simple or 2–3-branched, surrounded by a cup. Fruit a drupe, nut, or capsule; seeds with fleshy endosperm.

A chiefly Australasian family of about 12 genera and 300 species.

Genus 1. **SCAEVOLA** L., nom. cons.

Rather fleshy herbs or shrubs; leaves alternate or subopposite, usually entire. Flowers in axillary cymes or dichasia; calyx 5-lobed or the lobes nearly obsolete; corolla white or blue, bearded within, and with winged lobes. Stamens free. Ovary inferior, 2-celled, with 1 ovule in each cavity. Fruit a drupe with 2-seeded stone.

A genus of more than 80 species chiefly found in Polynesia and the Australian region. Two species occur in the West Indies, one widespread, the other Cuban.

1. **Scaevola plumieri** (L.) Vahl., Symb. Bot. 2:36. 1791. **FIG. 230.**

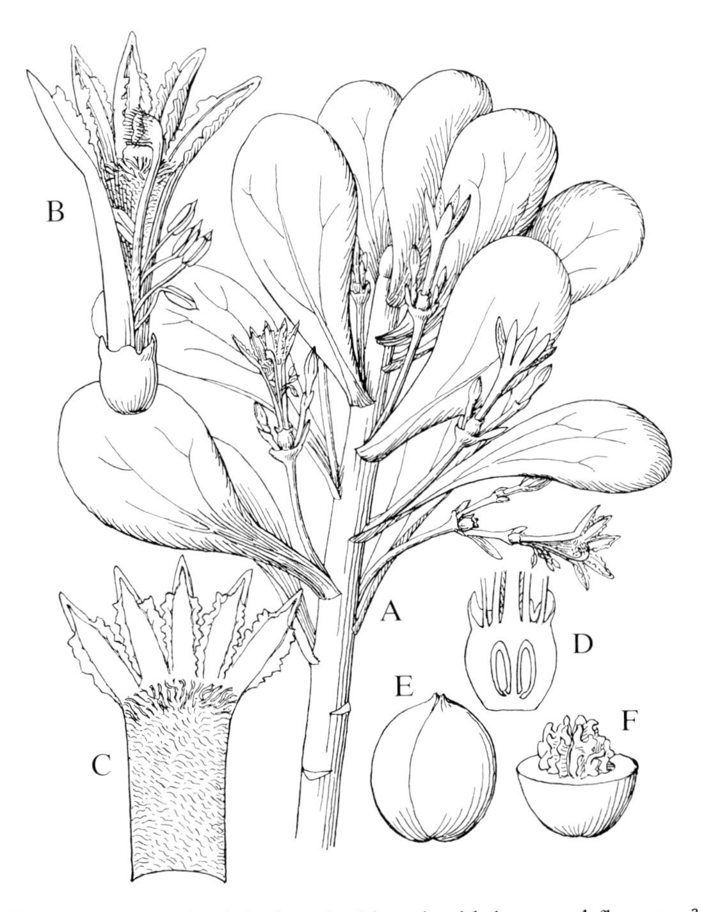

FIG. 230 **Scaevola plumieri.** A, end of branch with leaves and flowers, x$^2/_3$. B, flower, x2. C, corolla cut open, x2. D, ovary with calyx, cut lengthwise, x3. E, drupe, x1. F, drupe with upper half cut away exposing the stone, x1. (F. & R.)

"Bay Balsam"

A nearly glabrous, much-branched shrub to 1.5 m tall; leaves obovate, 4–8 cm long, rounded at the apex and of fleshy texture. Flowers in stalked dichasia; corolla c. 2.5 cm long, greenish and glabrous on the outside, the lobes white within. Drupes ellipsoid or subglobose, black, mostly 10–15 mm long, with rugose endocarp.

GRAND CAYMAN: *Brunt 2063; Proctor 15214.* LITTLE CAYMAN: *Kings LC 66; Proctor 28025.*

– Florida, West Indies, and the Caribbean coast of Central America, also along the coasts of tropical Africa, always at the top of sandy sea-beaches.

Family 101

RUBIACEAE

Herbs, shrubs or trees; leaves opposite or whorled, simple and entire or rarely toothed; stipules often sheathing the stem, sometimes divided into linear segments or reduced to glandular hairs, sometimes expanded and leaf-like. Flowers regular, usually perfect, solitary or in cymes or panicles, sometimes condensed to heads or glomerules, or rarely in spikes, with or without bracts. Calyx of 4–6 free or united sepals, or rarely the limb truncate. Corolla gamopetalous, mostly 4–6-lobed (rarely more), the lobes valvate, imbricate or twisted. Stamens as many as the corolla-lobes and alternate with them, inserted in the corolla-tube; anthers 2-celled, opening lengthwise. Ovary mostly inferior, crowned by a disk, usually 2-celled, the placentation various, rarely 1- or several-celled; ovules 1–many in each cavity; style usually slender, often forked toward the apex. Heterostyly is present in many genera. Fruit a capsule, berry, drupe, or schizocarp; seeds usually with plentiful endosperm.

A very large cosmopolitan family of more than 450 genera and about 6,000 species, most numerous in the tropics. Among the many economically important species of this family, the most valuable are those producing coffee (*Coffea* spp.) and quinine (*Cinchona* spp.). Ornamental species planted in Cayman gardens include *Ixora coccinea* L., and species of *Pentas.*

KEY TO GENERA

1. Plants woody; shrubs or (rarely) trees:

 2. Fruit a capsule; flowers solitary:

  3. Leaves thin, 3–5 cm long; corolla 5-lobed, white or cream with tube 25–30 mm long: **1. Exostema**

3. Leaves thick and fleshy, less than 1 cm long; corolla 4-lobed, yellow, with tube 5–6 mm long: **2. Rhachicallis**

2. Fruit a drupe or berry:

4. Flowers solitary in the leaf-axils:

5. Plants erect, more or less spiny; fruits berry-like, the seeds immersed in pulp:

6. Corolla 5-lobed; fruits black with hard outer rind, 8–12 mm in diameter: **3. Randia**

6. Corolla 4-lobed; fruits white, soft, c. 4 mm in diameter: **4. Catesbaea**

5. Plants arching or trailing, devoid of spines; fruits leathery, consisting of 2 indehiscent pyrenes: **14. Ernodea**

4. Flowers several or many in a cyme, panicle, raceme, spike or head:

7. Leaves (or many of them) in whorls of 3:

8. Low, dense shrub; leaves linear and sessile, 1–2 cm long and not over 2 mm broad; corolla pink, the tube c. 1.5 mm long: **5. Strumpfia**

8. Tall shrub; leaves elliptic or ovate and long-petiolate, 3–10 cm long and up to 4 cm broad; corolla yellow, the tube c. 25 mm long: **10. Hamelia**

7. Leaves opposite, never in 3's:

9. Inflorescence a panicle:

10. Panicles axillary; fruits 5–10-seeded: **6. Erithalis**

10. Panicles terminal; fruits 2-seeded: **12. Psychotria**

9. Inflorescence not a panicle:

11. Plants climbing or scrambling; individual flowers stalked: **7. Chiococca**

11. Plants erect and shrubby or tree-like; individual flowers sessile:

12. Flowers on the branches of a forked cyme; fruit a drupe:

13. Leaves and inflorescence glabrous; calyx-teeth 5, persistent; fruit 2-seeded: **8. Antirhea**

13. Leaves and inflorescence pubescent; calyx-teeth 2, deciduous; fruit 3-6-seeded: **9. Guettarda**

12. Flowers in dense globose heads; fruit a fleshy compound berry: **11. Morinda**

1. Plants herbaceous:

14. Stems delicate, creeping, rooting at the nodes, the stipules minute or obsolete; fruit a many-seeded capsule; corolla 4-5-lobed: **13. Hedyotis**

14. Stems usually erect, not rooting at the nodes, the stipules evident; fruit 2-seeded; corolla 4-lobed:

15. Fruit a schizocarp, separating into 2 nutlets: **15. Hemidiodia**

15. Fruit a dehiscent capsule: **16. Spermacoce**

Genus 1. **EXOSTEMA** (Pers.) L.C. Rich.

Shrubs or trees; leaves opposite, thin-textured; stipules small and soon falling. Flowers solitary in the leaf-axils, or in terminal corymbs or panicles; calyx cylindric or top-shaped, 5-lobed; corolla more or less salver-shaped with long slender tube and 5 linear reflexed lobes. Stamens 5, borne near the base of the corolla-tube; anthers linear, attached by the base, exserted. Ovary 2-celled, with numerous ovules in each cavity; stigma club-shaped or capitate. Fruit a 2-valved capsule, somewhat leathery or woody, splitting lengthwise; seeds numerous, broadly winged.

A tropical American genus of about 50 species.

1. Exostema caribaeum (Jacq.) Schult. in L., Syst. Veg. ed. nov. 5:18. 1819. **FIG. 231.**

"Bastard Ironwood"

A shrub to 3 m tall or slender tree to 6 m, the branchlets glabrous and with enlarged nodes; leaves elliptic or broadly elliptic, mostly 3–7 cm long, up to 3 cm broad, acute, glabrous or puberulous in the vein-axils beneath. Pedicels 4–8 mm long; calyx-lobes triangular, 0.5–1 mm long; corolla greenish-white, turning cream or pale yellow, the lobes narrowly linear and c. 3.5 cm long. Anthers 1.5–2 cm long. Capsules woody, 1.2–1.7 cm long; seeds suborbicular, 2.5 mm in diameter, winged all around.

GRAND CAYMAN: *Correll & Correll 51045; Proctor 15249.* LITTLE CAYMAN: *Proctor 28140.*
– Florida, West Indies, Mexico and Central America, in dry rocky woodlands.

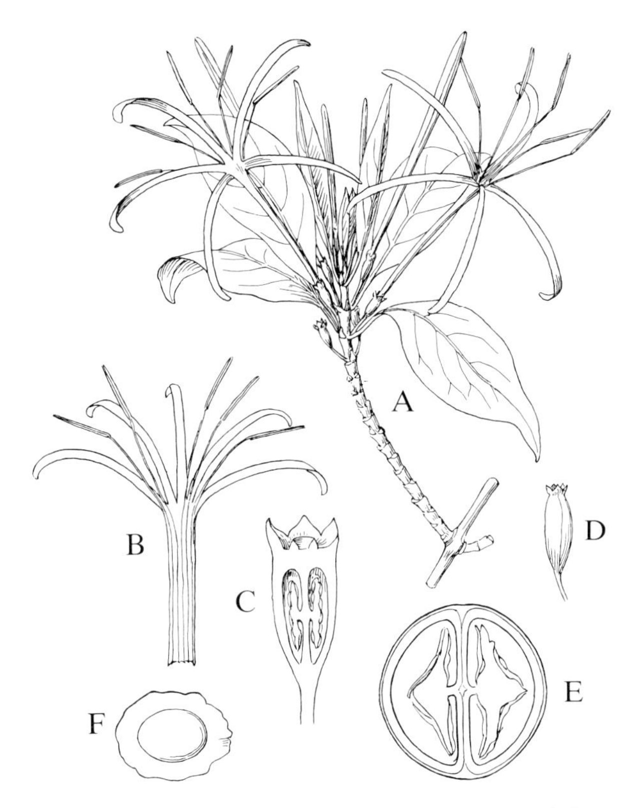

FIG. 231 **Exostema caribaeum.** A, branchlet with leaves and flowers, x$^2/_3$. B, upper part of corolla cut open, x$^2/_3$. C, ovary with calyx cut lengthwise, x4. D, capsule, x$^2/_3$. E, x-section of capsule, x4. F, seed x4. (F. & R.)

Genus 2. **RHACHICALLIS** DC.

A small, intricately-branched shrub; leaves opposite, densely overlapping; stipules sheathing, persistent. Flowers sessile, the base enclosed by the stipular sheath; calyx 4-toothed with smaller accessory teeth between the main ones; corolla salver-shaped with slender cylindric tube. Stamens 4, short, included. Ovary half-superior, 2-celled, each cavity with numerous ovules on a peltate placenta; style slightly 2-lobed. Capsule subglobose, 2-valved, septicidally dehiscent; seeds angular, pitted.

A monotypic West Indies genus.

1. Rhachicallis americana (Jacq.) Ktze., Revis. Gen. Pl. 1:281. 1891. **FIG. 232.**

"Juniper"

Stems flexible, canescent, mostly less than 1 m tall; leaves linear to oblong or ellipsoid, mostly 2-8 mm long; stipules mucronate and ciliate. Corolla deep yellow

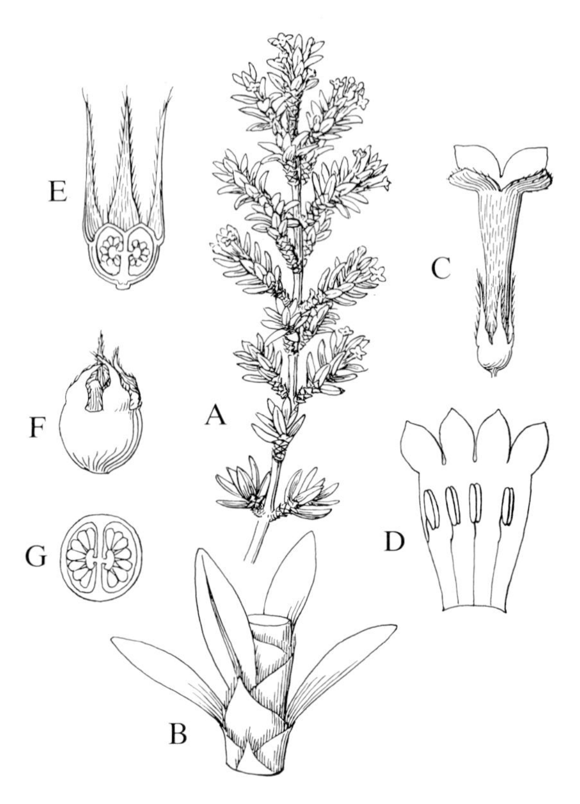

FIG. 232 **Rhachicallis americana.** A, branch with leaves and flowers, x$^2/_3$. B, small portion of branch showing leaves and stipules, x4. C, flower, x4. D, corolla cut open, x4. E, ovary with calyx cut lengthwise, x8. F, fruit, x4. G, x-section of fruit, x4. (F. & R.)

with a red eye, the tube 5-6 mm long, puberulous, the lobes densely pubescent on the outside. Capsules 3 mm long.

GRAND CAYMAN: *Correll & Correll 51023; Hitchcock; Kings GC 396; Lewis GC 37; Proctor 15199; Sachet 447.* LITTLE CAYMAN: *Proctor 28055.* CAYMAN BRAC: *Kings CB 58; Millspaugh 1178; Proctor 28926.*

– Bahamas, Greater Antilles (except Puerto Rico), and Yucatan, on exposed limestone rocks beside the sea.

Genus 3. **RANDIA** L.

Shrubs or trees, erect or scandent, sometimes with axillary or extra-axillary spines; leaves opposite; stipules interpetiolar, often sheathing. Flowers axillary or lateral, solitary or in clusters or corymbs; calyx-tube ovoid, obovoid or top-shaped, the limb tubular, cup- or bell-shaped, and truncate or 4-toothed or -lobed, the lobes often foliaceous. Corolla bell-, funnel-, or salver-shaped, 5-lobed (rarely 4–6-lobed), twisted in bud. Stamens usually 5, inserted in the throat or mouth of the corolla, included or exserted. Ovary usually 2-celled, with few to many ovules; style simple or bifid. Fruit a 2-celled berry with many compressed seeds immersed in pulp.

A pantropical genus of perhaps 300 species, mostly in the Old World.

1. **Randia aculeata** L., Sp. Pl. 2:1192. 1753. **FIG. 233.**

"Lancewood"

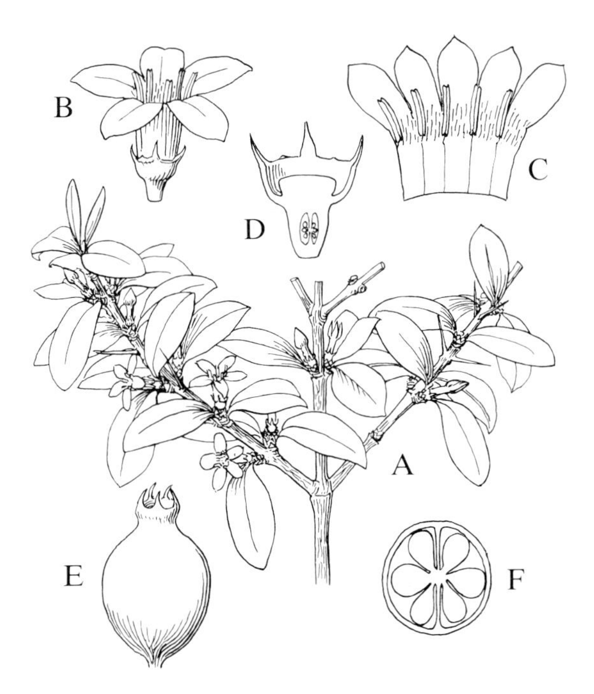

FIG. 233 **Randia aculeata.** A, portion of plant, x$^2/_3$. B, flower, x2. C, corolla cut open, x2. D, ovary with calyx, cut lengthwise, x4. E, berry, x2. F, x-section of berry, x2. (F. & R.)

A shrub to 2.5 m tall or more, glabrous or nearly so, with stiff horizontal or ascending branches occasionally armed with a few stout spines; leaves deciduous, narrowly to broadly obovate or nearly orbicular, mostly 1–5 cm long, acutish to rounded at the apex. Flowers fragrant, sessile or nearly so; corolla-tube green, 4–7 mm long, hairy within; lobes 4–6, white, 3.5–4 mm long. Berries ellipsoid or globose, 8–12 mm long.

GRAND CAYMAN: *Brunt 2071; Correll & Correll 51015; Kings GC 105; Millspaugh 1319; Proctor 11986, 15156; Sachet 429; Sauer 4104.* LITTLE CAYMAN: *Proctor 28041, 35120, 35172.* CAYMAN BRAC: *Millspaugh 1203; Proctor 28952.*

– Florida, West Indies, and Mexico to Venezuela, in dry rocky thickets and woodlands.

Genus 4. **CATESBAEA** L.

Shrubs with more or less spiny stems; leaves opposite or fascicled on short lateral spurs; stipules small, soon falling. Flowers small or large, solitary in the axils, short-stalked; calyx with 4 awl-shaped lobes; corolla white, funnel- or bell-shaped, glabrous within, the 4 lobes broad, valvate in bud. Stamens 4, inserted at the base of the corolla-tube, included or shortly exserted. Ovary 2-celled with few

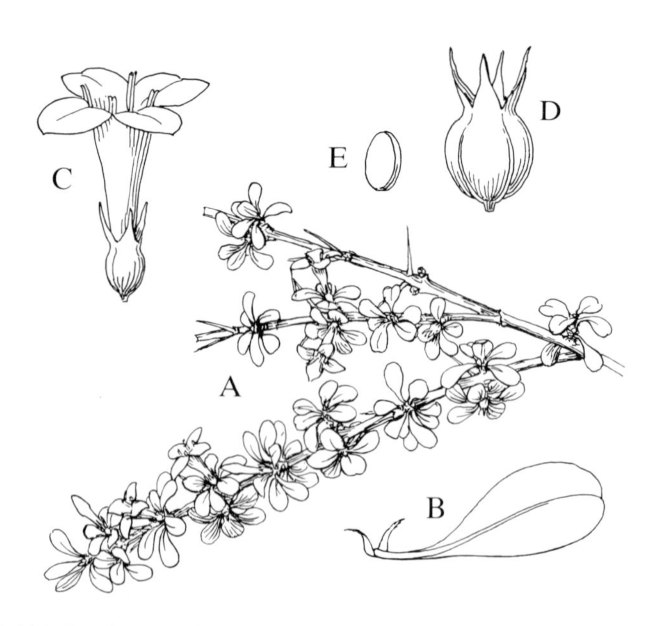

FIG. 234 **Catesbaea parviflora.** A, portion of plant with flowers, x$^2/_3$. B, leaf with stipules, x4. C, flower, x2. D, berry, x4. E, seed, x4. (F. & R.)

or many ovules; stigma bifid. Fruit a white (rarely black) berry, crowned by the persistent calyx-lobes; seeds compressed, angled or round; endosperm fleshy.

A West Indian genus of about 10 species.

1. **Catesbaea parviflora** Sw., Nov. Gen. & Sp. Pl. 30. 1788. **FIG. 234.**

A shrub usually 1 m tall or less (rarely up to 2.5 m), densely branched, the young branches arching, puberulous, the spines rather few or sometimes apparently lacking; leaves densely clustered, glabrous, rigidly leathery, obovate to orbicular, 4–10 mm long. Flower nearly sessile; corolla 5–6 mm long. Berries white, ovoid, c. 2 mm long.

GRAND CAYMAN: *Brunt 2005; Kings GC 254; Proctor 15168, 15217.* LITTLE CAYMAN: *Proctor 35107.* CAYMAN BRAC: *Proctor 29065.*
– Florida, Bahamas, Cuba and Jamaica, in dry rocky scrublands and thickets. The Cayman plants differ from those of Jamaica in their smaller fruits, more resembling those of Florida and the Bahamas in this respect. However, this size differential is overlapping and not wholly consistent. Black-fruited plants from Puerto Rico and the northern Lesser Antilles (sometimes included in a broad concept of *C. parviflora*) are here considered to represent a different species, *C. melanocarpa* Krug & Urb.

Genus 5. STRUMPFIA

A small, densely-branched shrub, the branches with very short nodes and rough with the rigid, persistent, spreading stipules; leaves in whorls of 3, rigidly leathery; stipules sheathing. Flowers in short axillary racemes; calyx 5-lobed, the lobes persistent; corolla deeply 5-parted, the tube very short. Stamens 5, borne near the base of the corolla-tube; anthers subsessile, erect, joined by their connectives to form a column surrounding the style. Ovary 2-celled with 1 ovule in each cavity; style glabrous but surrounded by a ring of hairs at the base. Fruit a small fleshy drupe containing a 1–2-celled stone; seeds oblong, with fleshy endosperm.

A monotypic West Indian genus.

1. **Strumpfia maritima** Jacq., Enum. Syst. Pl. Carib. 28. 1760. **FIG. 235.**

Low, flat-topped or mound-like shrub mostly less than 1 m tall, the young branches finely and densely white-tomentose; leaves sessile, linear, 1–2.5 cm long, 1–3 mm broad with revolute margins, puberulous on the upper side and densely white-tomentellous beneath. Racemes shorter than the leaves, few-flowered; corolla pink, the tube c. 1.5 mm long. Drupes globose, white, c. 3 mm in diameter.

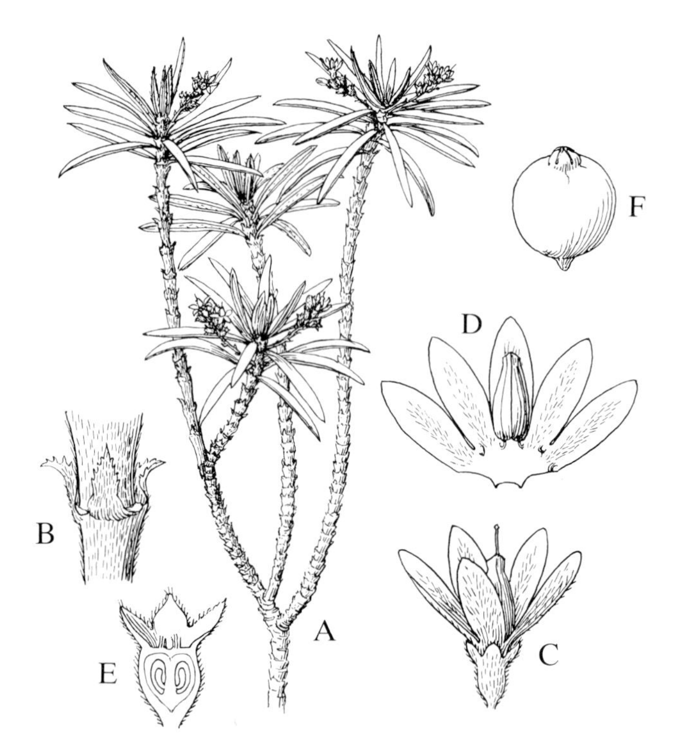

FIG. 235 **Strumpfia maritima.** A, portion of plant, x$^2/_3$. B, stipules, x4. C, flower, x4. D, corolla cut open, showing connate anthers. E, ovary with calyx cut lengthwise, x8. F, drupe, x4. (F. & R.)

GRAND CAYMAN: *Brunt 2010; Correll & Correll 51022; Fawcett; Hitchcock; Kings GC 124; Lewis GC 2; Proctor 15200; Sachet 393; Sauer 3312.* LITTLE CAYMAN: *Proctor 28095; Sauer 4190.* CAYMAN BRAC: *Kings CB 98; Millspaugh 1169; Proctor 29038.*

– Florida, West Indies and Yucatan, in dry rocky scrublands usually near the sea.

Genus 6. **ERITHALIS** P.Br.

Shrubs or rarely small trees, mostly glabrous; leaves opposite; stipules short, sheathing, persistent. Flowers in axillary or rarely terminal corymbose panicles; calyx-tube globose or ovoid, the limb more or less cup-shaped and truncate or 5-10-toothed, persistent on the fruit. Corolla with short tube and 5-10 lobes, these narrow and recurved. Stamens 5-10, inserted at the base of the corolla; filaments united at the base. Ovary 5-22-celled, with 1 pendulous ovule in each cavity. Fruit a globose drupe of 5-22 bony nutlets; seeds oblong, compressed.

A West Indian genus of about 10 species.

1. **Erithalis fruticosa** L., Syst. Nat. ed. 10, 2:930. 1759. **FIG. 236.**

"Black Candlewood"

A shrub usually 1-2 m tall but sometimes up to 5 m, nearly glabrous throughout; leaves leathery, elliptic or narrowly obovate to rotund, mostly 2.5-6.5 cm

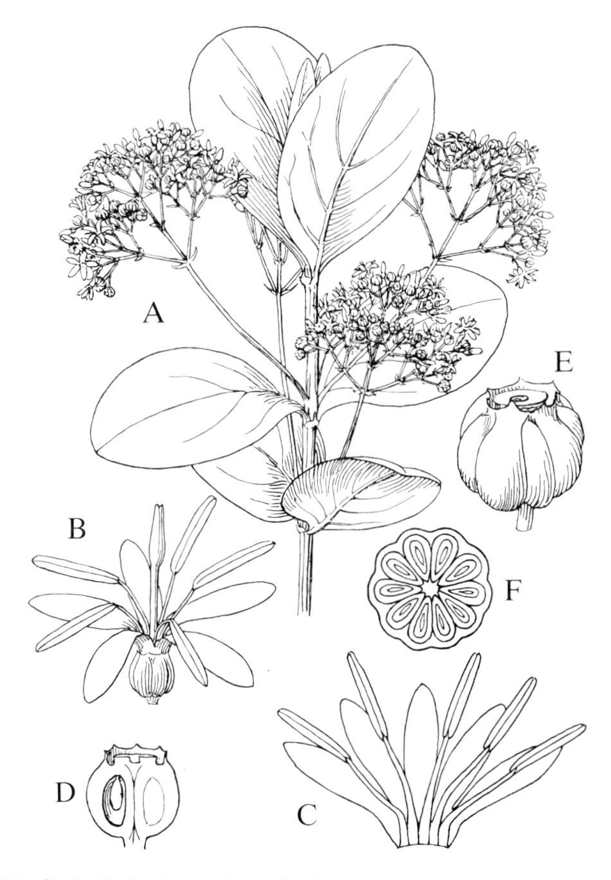

FIG. 236 **Erithalis fruticosa.** A, end of branch with leaves and flowers, x$^2/_3$. B, flower, x6. C, corolla cut open, x6. D, ovary with calyx cut lengthwise, x12. E, drupe, x6. F, x-section of drupe, x6. (F. & R.)

long, obtuse at the apex, the margins often minutely revolute, the venation somewhat obscure. Panicles exceeding the leaves. Flowers fragrant, corolla white or cream, with usually 4–5 lobes, these c. 4 mm long. Drupes purple-black, 2–2.5 mm in diameter.

GRAND CAYMAN: *Brunt 1763, 1818, 1917, 2059, 2068; Correll & Correll 51034; Fawcett; Hitchcock; Kings GC 335; Millspaugh 1251; Proctor 15092; Sachet 407.* LITTLE CAYMAN: *Kings LC 100; Proctor 28097.* CAYMAN BRAC: *Proctor 29064.*

– Florida, West Indies, and the east coast of Central America, in sandy or rocky thickets and scrublands, or sometimes on limestone cliffs.

Genus 7. **CHIOCOCCA** P.Br.

Glabrous shrubs, often trailing or climbing; leaves opposite, somewhat leathery and shining; stipules broad, with a sharp point, persistent. Flowers in axillary simple or compound racemes; calyx 5-toothed or -lobed, persistent; corolla funnel-shaped with glabrous throat, 5-lobed, the lobes valvate in bud. Stamens 5, inserted near the base of the corolla-tube, the filaments connate at the base, often pubescent; anthers linear, attached at the base, usually included. Ovary 2- (rarely 3-) celled, with solitary pendulous ovules; stigma entire or shortly 2-lobed. Fruit a flattened, leathery white drupe containing 2 nutlets; seeds compressed; endosperm fleshy.

A tropical American genus of perhaps 20 species.

KEY TO SPECIES

1. Leaves mostly 3–6 mm long or more; racemes often longer than the leaves: **1. C. alba**

1. Leaves mostly 1–3 cm long; racemes shorter than the leaves: **2. C. parvifolia**

1. Chiococca alba (L.) Hitchc. in Rep. Missouri Bot. Gard. 4:94. 1893. **FIG. 237.**

Chiococca parvifolia of Hitchcock, 1893, as to Cayman specimen, not Wullshl. ex Griseb., 1861.

Trailing, scrambling or climbing shrub with elongate stems to 6 m long or more; leaves ovate, lanceolate or narrowly elliptic, more or less acuminate at the apex.

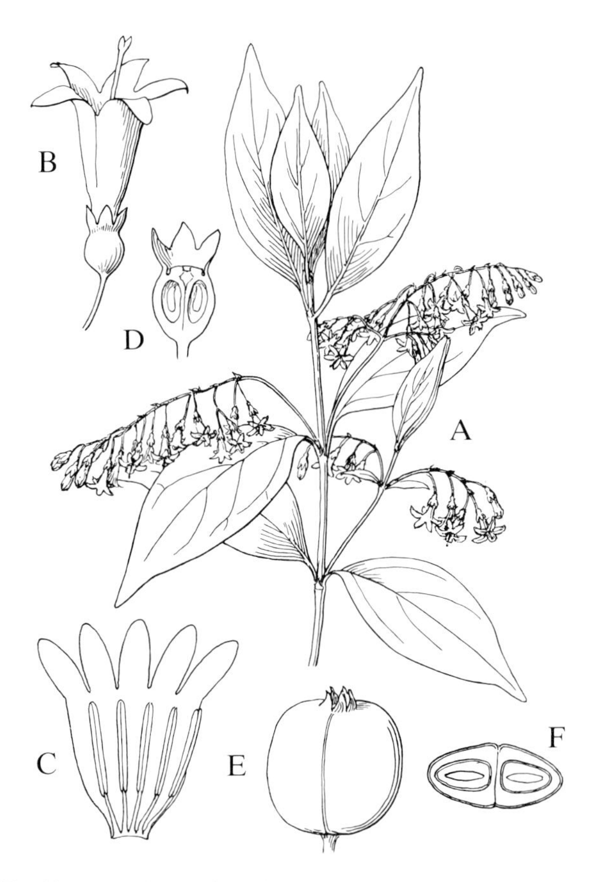

FIG. 237 **Chiococca alba.** A, branch with leaves and flowers, x$^2/_3$. B, flower, x4. C, corolla cut open, x4. D, ovary with calyx cut lengthwise. E, drupe, x4. F, x-section of drupe, x4. (F. & R.)

Racemes up to 7 cm long, often branched; corolla yellow, the tube 3–5 mm long, the lobes reflexed. Drupes 4–5 mm in diameter.

GRAND CAYMAN: *Brunt 1701, 1780, 2197; Hitchcock; Sachet 378.* LITTLE CAYMAN: *Proctor 28118.* CAYMAN BRAC: *Proctor 28953.*

– Florida, West Indies, Mexico and Central America, in rocky thickets and woodlands.

2. Chiococca parvifolia Wullschl. ex Griseb., Fl. Brit. W. I. 337. 1861.

A climbing shrub like the last; leaves narrowly to broadly elliptic or ovate, acute or blunt at the apex. Racemes up to 2 cm long, seldom branched but sometimes paired in the axils; corolla pale yellow, the tube c. 3 mm long, the lobes mostly erect. Drupe 4–5 mm in diameter.

CAYMAN BRAC: *Proctor 29000.*
– Florida, West Indies, Trinidad and Tobago, in rocky thickets and woodlands.

Genus 8. **ANTIRHEA** Commers.

Shrubs or trees, glabrous or pubescent; leaves opposite; stipules interpetiolar, soon falling. Flowers perfect or polygamous, usually sessile along the upper side of the branches of a 1–2-forked cyme, or rarely solitary. Calyx truncate or irregularly 4–5-toothed or -lobed, persistent. Corolla with cylindric or funnel-shaped tube and 4–5 short lobes, these imbricate in bud. Stamens 4–5, inserted in the throat of the corolla-tube, the filaments short, the anthers included or shortly exserted. Ovary 2–10-celled with solitary pendulous ovules; stigma capitate or 2–3-lobed. Fruit a small, fleshy drupe with 2–10-celled stone; seeds elongate, without endosperm.

A rather widespread tropical genus of about 40 species, occurring in the West Indies, Madagascar, tropical Asia and Australia. Some species yield valuable timber.

1. Antirhea lucida (Sw.) Hook.f. in Benth. & Hook.f., Gen. Pl. 2:100. 1873. **FIG. 238.**

A glabrous tree 5–10 m tall with light gray bark; leaves oblong-elliptic or broadly elliptic, 4–8 cm long or more, obtuse or acutish at the apex, shining light green on the upper side; stipules acuminate, 6–8 mm long. Inflorescence 1-forked with curved branches; flowers fragrant; calyx c. 3 mm long with 5 oblong, minutely ciliate lobes; corolla cream-white, bell-shaped, with tube 3.5–5 mm long. Drupes c. 1 cm long.

GRAND CAYMAN: *Brunt 1807.* LITTLE CAYMAN: *Proctor 35080, 35211.*
– Bahamas, Greater Antilles, St. Croix and the Swan Islands, in rocky thickets and woodlands.

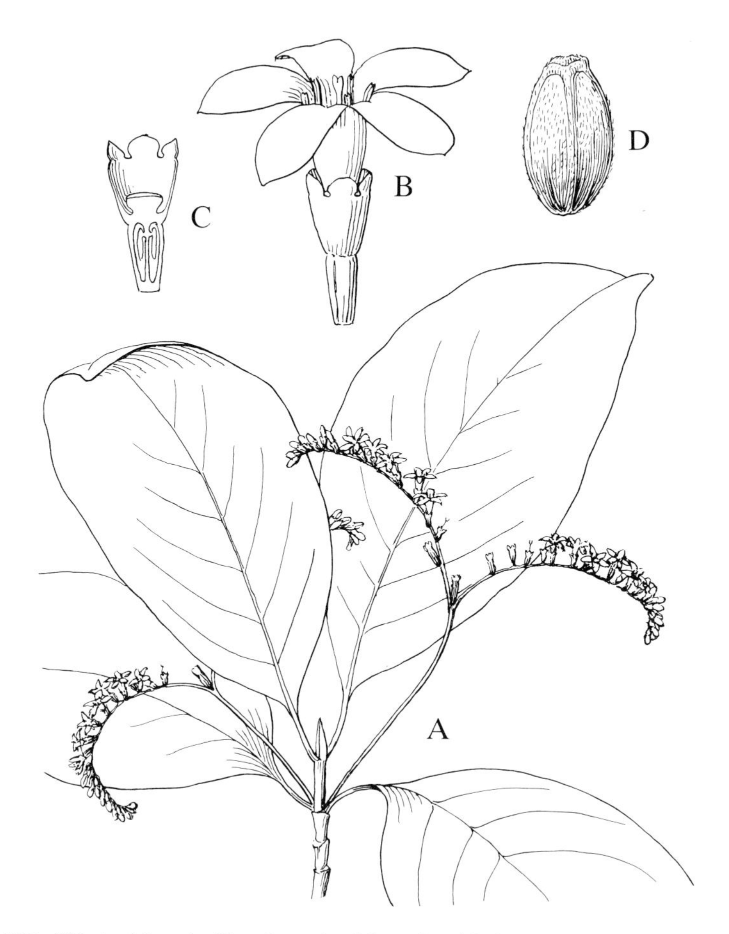

FIG. 238 **Antirhea lucida.** A, end of branch with leaves and flowers, x$^{2}/_{3}$. B, flower, x6. C, ovary with calyx cut lengthwise, x6. D, drupe, x2. (F. & R.)

Genus 9. **GUETTARDA** L.

Shrubs or small trees; leaves opposite or sometimes in whorls of 3; stipules interpetiolar, soon falling. Flowers perfect or polygamo-dioecious, sessile or subsessile on the branches of axillary forked cymes. Calyx with tubular limb truncate at the apex or irregularly toothed, deciduous. Corolla salver-shaped with long, cylindric tube sometimes curved, and 4–9 obtuse lobes, these imbricate in bud. Anthers sessile or subsessile in the upper part of the corolla-tube, included. Ovary 4–9-celled, with solitary pendulous ovules; stigma capitate, entire or short-lobed. Fruit a usually globose drupe with thin flesh and hard, 4–9-celled stone; seeds without endosperm.

A genus of more than 80 species, of which 60 occur in tropical America, 20 in New Caledonia, and 1 is widespread on tropical coasts.

1. Guettarda elliptica Sw., Nov. Gen. & Sp. Pl. 59. 1788.

A shrub to 3 m tall, or rarely a small tree; leaves ovate, more or less elliptic or narrowly obovate, mostly 1-7 cm long, obtuse or acute at the apex, puberulous on both sides. Peduncles slender, 0.5–2.5 cm long; inflorescences subcapitate, the cyme-branches short and few-flowered; corolla whitish, the tube 5 mm long or more, puberulous. Drupes dark red turning blackish, 4–8 mm in diameter, 3-6-seeded.

GRAND CAYMAN: *Brunt 2004; Proctor 15173, 15210.* LITTLE CAYMAN: *Proctor 28136, 35106, 35114, 35119.* CAYMAN BRAC: *Proctor 28954, 29027.*
– Florida, West Indies (except the Lesser Antilles), and Mexico to Venezuela, in dry, rocky thickets and woodlands. The Grand Cayman plants have much smaller leaves and shorter peduncles than those of Little Cayman and Cayman Brac.

Genus 10. **HAMELIA** Jacq.

Shrubs or trees; leaves thin, opposite or whorled and rather long-petiolate; stipules interpetiolar, soon falling. Flowers sessile or short-stalked on the branches of 2–3-forked or compound cymes. Calyx with 5 persistent lobes. Corolla yellow or red, tubular or narrowly bell-shaped, 5-angled at least in bud. Stamens 5, inserted near the base of the corolla-tube; filaments short; anthers linear, attached at the base, included. Disk prominent, persistent, forming a protuberance on the apex of the fruit. Ovary 5-celled, with numerous ovules on axile placentas; stigma narrow, entire. Fruit a 5-lobed, 5-celled berry; seeds numerous, small, variously angled or tuberculate.

A tropical American genus of about 40 species.

1. Hamelia cuprea Griseb., Fl. Brit. W. I. 320. 1861.

A shrub to 3 m tall or more, glabrous throughout or nearly so; leaves in whorls of 3, ovate or elliptic, 4–11 cm long (including petioles to 2.5 cm), acuminate at the apex, often recurved-plicate; stipules triangular-attenuate, to 2 mm long. Inflorescence laxly few-flowered; pedicels 2–7 mm long; corolla bright yellow streaked with orange, c. 2.5 cm long, constricted above the base. Berries ovoid, 5–7 mm long, ripening reddish black.

GRAND CAYMAN: *Kings GC 345; Proctor 15140.* LITTLE CAYMAN: *Proctor 35123.* CAYMAN BRAC: *Proctor 29011, 29072.*
– Cuba, Jamaica and Hispaniola, in rocky thickets and woodlands.

Genus 11. **MORINDA** L.

Usually glabrous shrubs or trees, or sometimes woody vines; leaves opposite or rarely in whorls of 3; stipules more or less sheathing. Flowers in dense globose heads, these stalked or subsessile, axillary or terminal, solitary or several in an umbel. Calyx-limb short and truncate or minutely toothed, persistent. Corolla usually white, salver-shaped, with usually 5 lobes (rarely 3-7), valvate in bud. Stamens as many as the corolla-lobes, inserted near the top of the tube, the anthers included or exserted. Ovary 2-4-celled with solitary ascending ovules; style with 2 linear arms. Fruit a fleshy compound berry ("syncarp") formed by union of the enlarged calyx-tubes, and containing numerous 1-seeded nutlets; seeds with fleshy endosperm.

A pantropical genus of about 80 species, the majority Indian and Malayan.

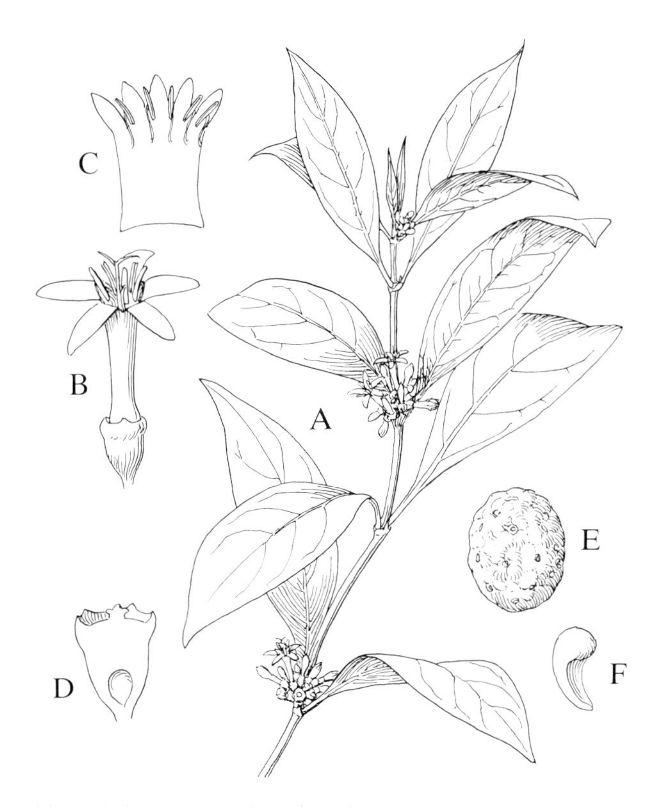

FIG. 239 **Morinda royoc.** A, branch with leaves and flowers, x$^2/_3$. B, flower, x3. C, corolla cut open, x2. D, ovary with calyx cut lengthwise, x6. E, fruit, x$^2/_3$. F, nutlet, x2. (F. & R.)

KEY TO SPECIES

1. Leaves less than 10 cm long and 3 cm broad; peduncles less than 5 mm long; syncarps not over 2 cm in diameter: **1. M. royoc**

1. Leaves up to 30 cm long or more and 6–15 cm broad; peduncles 10–25 mm long; syncarps to 10 cm long: **[2. M. citrifolia]**

1. Morinda royoc L., Sp. Pl. 1:176. 1753. **FIG. 239.**

"Yellow Root", "Rhubarb Root"

An erect or straggling shrub to 2 m tall; leaves lanceolate or narrowly elliptic, mostly 3–10 cm long, subacuminate at the apex; stipules cuspidate, inconspicuous. Corolla-tube c. 5 mm long, the 5 lobes shorter. Syncarps nearly spherical, yellowish when ripe.

GRAND CAYMAN: *Brunt 1649, 1812, 1926; Hitchcock; Kings GC 64; Millspaugh 1280, 1359; Proctor 15050.*
– Florida, Bahamas, Cuba, Jamaica, Hispaniola, Mexico to Venezuela, Curacao and Aruba, in sandy or rocky thickets or along the borders of pastures.

[2. **Morinda citrifolia** L., Sp. Pl. 1:176. 1753.

"Mulberry", "Hog Apple"

An erect shrub or small tree to 10 m tall; leaves broadly elliptic, acute or cuspidate at the apex, shining bright green; stipules membranous, up to 1.5 cm long. Corolla-tube 9–10 mm long, the lobes c. 4 mm long. Syncarps ovoid-ellipsoid, somewhat asymmetric or irregular, creamy-translucent when ripe and of foetid odour.

GRAND CAYMAN: *Brunt 1982; Proctor 15117, 31050.* LITTLE CAYMAN: *Kings LC 112.* CAYMAN BRAC: *Proctor 29094.*
– Native of tropical Asia and Australia, now widely naturalized in the American tropics. It frequently occurs in coastal thickets, as the fruits are apparently dispersed by floating in the sea. In India this species has been extensively cultivated for the production of a dye called "al", used for dyeing cloth various shades of red. This substance is obtained chiefly from the bark of the roots. The wood is said to be hard and durable.]

Genus 12. **PSYCHOTRIA** L., nom. cons.

Shrubs or sometimes small trees, rarely herbs or climbers; leaves usually opposite, rarely in whorls of 3 or 4; stipules free or more or less united, persistent or deciduous. Flowers often dimorphic (heterostylous), usually in terminal corymbs, cymes or panicles; calyx 5-toothed; corolla cylindric or funnel-shaped, straight, glabrous or hairy within, usually 5-lobed (rarely 4-6-lobed), the lobes valvate in bud. Stamens the same number as the corolla-lobes, inserted in the upper part of the tube; anthers included or exserted. Ovary 2-celled with solitary ovules erect from the base; style 2-armed. Fruit a berry, or a drupe with 2 nutlets, these often dehiscing longitudinally on the ventral side; seeds convex and smooth or ribbed on one side, and flat or concave and smooth on the other; endosperm fleshy or cartilaginous.

A very large, pantropical genus of more than 700 species.

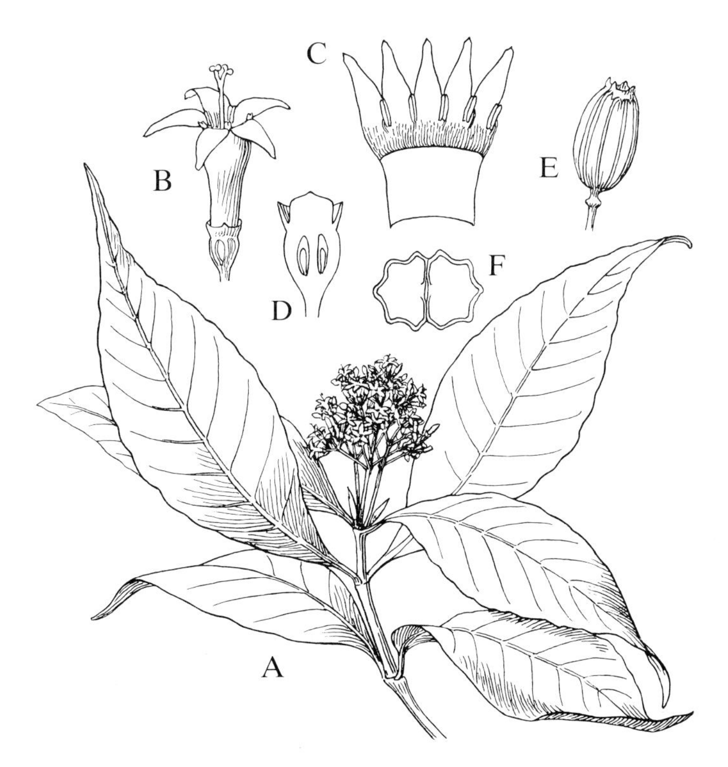

FIG. 240 **Psychotria nervosa.** A, end of branch with leaves and flowers, x$^2/_3$. B, flower, x4. C, corolla cut open, x4. D, ovary with calyx cut lengthwise, x8. E, fruit, x2. F, x-section of fruit, x8. (F. & R.)

1. Psychotria nervosa Sw., Nov. Gen. & Sp. Pl. 43. 1788. **FIG. 240.**

Psychotria undata Jacq., 1798.

"Strong Back", "Kidney Bush"

A shrub up to 2.5 m tall, rarely arborescent and up to 6 m, the young branchlets glabrate or reddish-pubescent; leaves narrowly elliptic or elliptic, mostly 5-16 cm long, acuminate at the apex, glabrate or puberulous beneath, the 9-16 pairs of side-nerves prominent; stipules fused and sheathing in bud, splitting on one side and soon falling. Panicles sessile, pubescent or glabrous; corolla white, the tube 2-3 mm long. Drupes ellipsoid, red, 6-7 mm long.

GRAND CAYMAN: *Brunt 1978; Correll & Correll 51043* (pubescent form); *Hitchcock; Kings GC 316; Lewis GC 33a; Proctor 11987, 15001, 15007; Sachet 376.*
– Florida, West Indies, and continental tropical America, variable; Cayman plants grow in rocky woodlands. Two rather different forms occur: (1), shrubby plants with pubescent stems, leaves and inflorescences, the latter rather compact; and (2), taller, more arborescent plants (e.g., *Brunt 1978*) which are nearly glabrous and have a more open type of inflorescence. Form 1 has been called *P. undata*, while form 2 resembles typical *P. nervosa* as it occurs in Jamaica. However, the distinctions are unstable and break down when the whole range of plants in this complex is considered.

Genus 13. **HEDYOTIS** L.

Small herbs or subshrubs, the stems erect, creeping or diffuse; leaves opposite; stipules entire or variously toothed or cut, rarely minute or apparently absent. Flowers solitary or cymose in the leaf-axils, often dimorphic (heterostylous); calyx 4-toothed or -lobed; corolla funnel- or salver-shaped, mostly 4-lobed, the lobes valvate in bud. Stamens usually 4, inserted above the middle of the corolla-tube; anthers included or exserted. Ovary 2-celled with many ovules on axile placentas; style 2-armed. Fruit a several- to many-seeded globose or 2-lobed capsule, dehiscing at the top or all the way down; seeds flattened or angled.

A pantropical subtropical or warm-temperate genus of about 300 species. This concept includes the species often separated as the genera *Houstonia* and *Oldenlandia.*

1. Hedyotis callitrichoides (Griseb.) W.H.Lewis in Rhodora 63:222. 1961.

Oldenlandia callitrichoides Griseb., 1862.

A delicate, creeping herb forming small mats, the glabrous stems 0.1-0.2 mm in diameter; leaves thin, petiolate, very broadly ovate or orbicular, 1-3mm long, obtuse at the apex, sparsely puberulous on the upper side; stipules minute or obsolete. Flowers solitary on delicate peduncles 4-5 mm long; calyx 4-5-lobed, the acuminate lobes hairy. Corolla white, 1.5-2 mm long, 4-5-lobed. Capsules oblong, 1.5-2 mm long; seeds flattened.

GRAND CAYMAN: *Hitchcock*.
– Bahamas, Cuba, Jamaica, Hispaniola, Guadeloupe, Yucatan, Trinidad and northern South America; reported from tropical Africa. The Cayman plants were found "growing on stone in a shallow well".

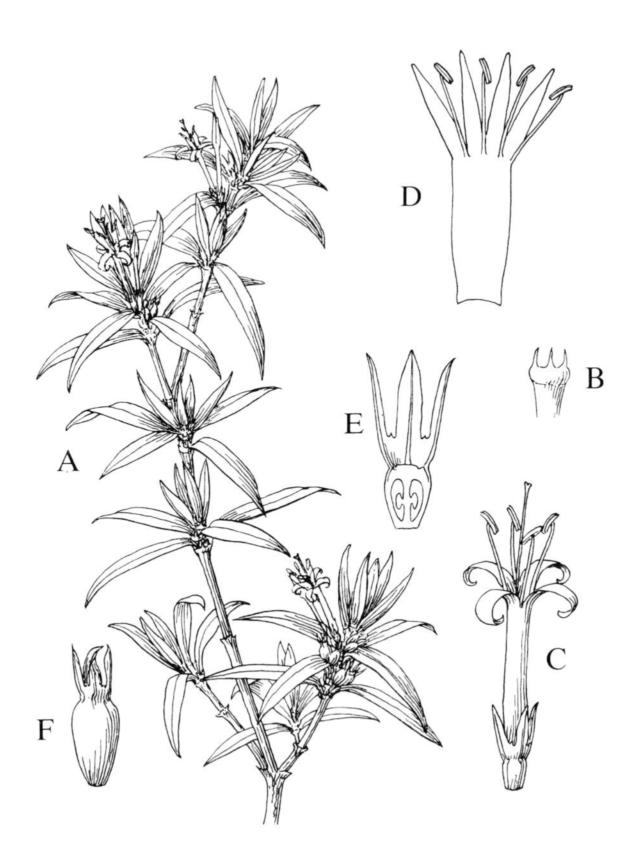

FIG. 241 **Ernodea littoralis.** A, branch with leaves, flowers and fruits, x²/₃. B, stipule, x2. C, flower, x2. D, corolla cut open, x2. E, ovary with calyx cut lengthwise, x4. F, fruit, x2. (F. & R.)

Genus 14. **ERNODEA** Sw.

Suberect or trailing slender shrubs; leaves opposite, crowded, and narrow; stipules sheathing, persistent. Flowers solitary and sessile in the leaf-axils; calyx with short tube, the limb 4–6-parted with narrow persistent segments. Corolla with long cylindric tube and 4–6 narrow reflexed lobes, these valvate in bud. Stamens as many as the corolla-lobes, inserted at the top of the tube and long-exserted. Disk fleshy. Ovary 2-celled with solitary axile ovules; style long-exserted and with capitate stigma. Fruit a dry-fleshed drupe with two 1-seeded, plano-convex nutlets.

A West Indian genus of 7 known species, 6 of them endemic to the Bahamas.

1. **Ernodea littoralis** Sw., Nov. Gen. & Sp. Pl. 29. 1788. **FIG. 241.**

"Guana Berry"

Stems 4-angled, trailing or ascending to 1 m long or more; leaves rigid, glabrous, linear to lanceolate, mostly 1–3 cm long, spine-tipped. Corolla pale pink, the tube 8–10 mm long. Drupes yellow, ellipsoid, 4–5 mm long, crowned by the persistent calyx-lobes.

GRAND CAYMAN: *Brunt 1635, 1815; Hitchcock; Kings GC 373; Millspaugh 1254; Proctor 15051.* LITTLE CAYMAN: *Kings LC 17; Proctor 28096.* CAYMAN BRAC: *Millspaugh 1194; Proctor 28999.*
– Florida, West Indies, Yucatan and Honduras, mostly in dry, rocky scrublands or thickets near the sea.

Genus 15. **HEMIDIODIA** K.Schum.

A perennial subwoody herb, the stems sometimes elongate; leaves opposite, nearly sessile; stipules sheathing, lacerate-bristly, adnate to the leaf-bases. Flowers in sessile glomerate clusters at the nodes; calyx obconic, 4-lobed; corolla funnel-shaped, 4-lobed, the lobes valvate in bud. Ovary 2-celled with solitary ovules; stigma capitate. Fruit dry, consisting of 2 nutlets dehiscent on the inner side near the base.

A monotypic tropical American genus.

1. **Hemidiodia ocymifolia** (Willd.) K.Schum. in Mart., Fl. Bras. 6(6):29, t.72. 1888.

Stems erect or straggling, up to 1 m tall or more (but often less), 4-angled, puberulous on the angles when young; leaves very narrowly elliptic, 2–4 cm long or more, acuminate scabrid on the margins and nerves; stipular bristles 2–3 mm long. Corolla white, 2–3 mm long. Fruits 3.5 mm long, puberulous.

GRAND CAYMAN: *Kings GC 303.*
– West Indies and continental tropical America, in thickets and pastures or along roadsides, chiefly in moist areas, not common.

Genus 16. **SPERMACOCE** L.

Annual or perennial herbs with 4-angled stems; leaves opposite; stipules united with the petioles to form a bristly sheath. Flowers several to many, sessile and glomerate in the leaf-axils. Calyx 4-lobed; corolla funnel-shaped, 4-lobed, the lobes valvate in bud. Stamens 4, inserted near the base of the corolla-tube; anthers included. Ovary 2-celled with solitary ovules; stigma 2-lobed. Fruit a hard-shelled capsule, one or both valves opening from the top; seeds oblong, with horny endosperm.

A pantropical genus of more than 150 species, now construed to include the plants long separated in the genus *Borreria*.

KEY TO SPECIES

1. One valve of the capsule always remaining closed:

 2. Stems and leaves minutely scabrid; capsules setulose: **1. S. confusa**

 2. Stems and leaves hirsute with white hairs; capsules densely long-hairy: **2. S. tetraquetra**

1. Both valves of the capsule opening: **3. S. assurgens**

1. Spermacoce confusa Rendle in J. Bot. 74:12. 1936; Nicolson in J. Arnold Arb. 58:446–447. 1977. **FIG. 242.**

Annual herb with erect stems to 45 cm tall or more (often less), scabrid on the angles; leaves narrowly lanceolate, lanceolate, or rarely narrowly ovate, mostly 2–5 cm long. Corolla white, c. 2 mm long. Fruits subglobose or ellipsoid, 2.5 mm long.

GRAND CAYMAN: *Hitchcock; Millspaugh 1302, 1338; Proctor 15260.*
– Southeastern United States, West Indies, and continental tropical America, common in open waste places, also frequently a weed of cultivated fields.

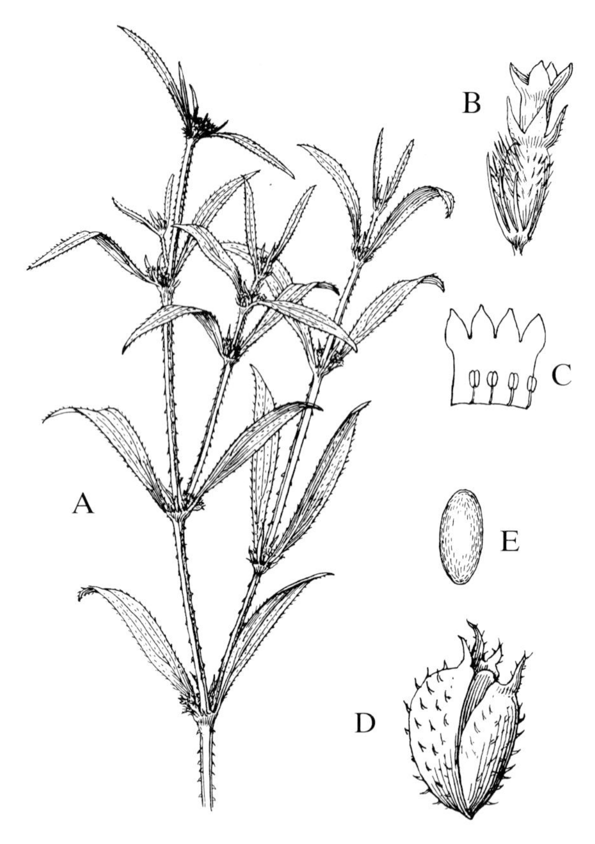

FIG. 242 **Spermacoce confusa.** A, portion of plant with flowers and fruits, x$^2/_3$. B, flower subtended by bract, x8. C, corolla cut open, x8. D, fruit, x8. E, seed, x8. (F. & R.)

2. Spermacoce tetraquetra A.Rich. in Sagra, Hist. Cub. 11:29. 1850.

Annual erect or sprawling herb like the last but not scabrid, and clothed nearly all over with long whitish hairs; leaves lanceolate or oblong-lanceolate, mostly 1.5-3 cm long. Corolla white, c. 2 mm long. Fruits globose, 2 mm long.

CAYMAN BRAC: *Proctor 29335.*

– Bahamas, Cuba, and Jamaica, in sandy clearings and open waste ground, appearing after rains.

3. **Spermacoce assurgens** Ruiz & Pavon, Fl. Peruv. 1:60, t.92, fig. C. 1798.[1]

Borreria laevis of all modern authors, not *Spermacoce laevis* Lam.,

Decumbent or erect annual or persisting herb, the stems glabrous or nearly so and up to 50 cm tall or more; leaves lanceolate or narrowly elliptic to elliptic, mostly 1-4 cm long, glabrate and rather strongly-nerved. Corolla white, hairy, c. 4 mm long. Capsules puberulous, 2.5-3 mm long.

GRAND CAYMAN: *Brunt 1892, 1932, 2096; Hitchcock; Proctor 15105; Sachet 418.* LITTLE CAYMAN: *Proctor 35199.* CAYMAN BRAC: *Proctor 28989.* – Florida, West Indies, and continental tropical America, introduced in Hawaii, a weed of pastures, roadsides, and open waste ground.

[CAPRIFOLIACEAE

Lonicera japonica Thunb., the "Honeysuckle" *(Kings GC 336)*, and *Sambucus simpsonii* Rehder, the "Elder" (*Hitchcock*) have been recorded from Grand Cayman, presumably on the basis of cultivated plants. The writer has seen no evidence that they still occur, or that they ever escaped or persisted outside cultivation.]

Family 102

COMPOSITAE

Herbs, shrubs, vines, or small trees; leaves alternate, opposite or whorled, simple or variously toothed, lobed or divided; stipules absent. Flowers (florets) small and crowded into heads (capitula) surrounded by an involucre of one or more series of free or connate bracts (phyllaries); rarely the heads compound, consisting of several 1-few-flowered capitula; heads solitary or arranged in spikes, racemes, cymes, corymbs, or panicles (just as if they were individual flowers). The receptacle from which the florets arise usually convex, sometimes conic or cylindric, or occasionally concave, and may be pitted or smooth and naked, scaly or hairy. Florets of 1 or 2 kinds in each head: perfect, unisexual or neuter, or rarely the heads dioecious; outer florets often ligulate (ray-florets), the inner ones tubular and without ligules (disk-florets), or all the florets may be ligulate or all tubular. Calyx superior, represented by a pappus of persistent or sometimes soon-falling hairs, bristles, awns or scales crowning the ovary, or sometimes reduced to a ring or absent. Corolla

---

[1] Thanks are due to Dr. F. R. Fosberg of the Smithsonian Institution for drawing the writer's attention to this name and citation.

gamopetalous, with a long or short tube, 4–5-lobed and regular (disk-florets) or ligulate (rarely 2-lipped) and irregular (ray-florets). Stamens 5 (rarely 4), inserted in the corolla-tube; filaments free; anthers connate into a tube surrounding the style, rarely free, 2-celled and opening lengthwise, often appendaged at each end. Ovary inferior, 1-celled with a solitary basal ovule; style of the perfect or pistillate flowers usually 2-armed, the arms smooth, papillose or hairy and of various shapes. Fruit a 1-celled, 1-seeded achene, indehiscent and usually dry, sometimes beaked, often crowned by the persistent pappus; seeds without endosperm.

A cosmopolitan family with an estimated 950 genera and more than 20,000 species, one of the largest plant families. All of the Cayman genera have the disk-florets (or all the florets) regular and 4–5-lobed; none have 2-lipped florets or all the florets ligulate. For convenience in so large a group, the genera of Compositae are arranged in series called Tribes, of which 7 occur in the Cayman Islands. These can be characterized as follows:

TRIBE I. VERNONIEAE. Leaves usually alternate. Heads discoid. Corolla white or purple. Anthers sagittate at the base. Style-arms more or less cylindric, hairy on the back. Pappus usually of bristles or scales. **Genus 1.**

TRIBE II. EUPATORIEAE. Leaves usually opposite, the upper sometimes alternate. Heads discoid. Corolla purple, white or whitish. Anthers blunt at the base. Style-arms long, subcylindric, hairy. Pappus usually of bristles. **Genera 2–3.**

TRIBE III. ASTEREAE. Leaves usually alternate. Heads radiate or disciform (heterogamous), or sometimes by suppression of the ray-florets homogamous. Receptacle usually naked. Corolla of the disk usually yellow, of the ray the same colour or different. Anthers blunt at the base. Style-arms flattened, appendaged. **Genera 4–6.**

TRIBE IV. INULEAE. As in the preceding, but anthers tailed at the base and the style-arms linear, without appendages. **Genus 7.**

TRIBE V. HELIANTHEAE. Leaves opposite or alternate. Heads usually radiate, sometimes disciform, or by suppression of the rays homogamous and discoid. Receptacle scaly. Corolla of the disk usually yellow, of the rays often of the same colour, sometimes different. Anthers entire at the base or with two very short points. Style-arms truncate or appendaged. Pappus usually of awns or scales, often lacking, rarely bristly. **Genera 8–22.**

TRIBE VI. HELENIEAE. Leaves opposite or alternate. Heads usually radiate, sometimes disciform, or by suppression of the rays discoid; phyllaries in one series. Receptacle naked. Corolla of the disk usually yellow. Anthers obtuse at the base. Style-arms truncate or appendaged. Pappus usually of scales. **Genera 23–24.**

TRIBE VII. SENECIONEAE. Leaves usually alternate. Heads usually radiate, less often disciform or discoid. Receptacle usually naked, sometimes scaly. Disk corollas red to yellow. Anthers with an apical appendage, often sagittate at the base. Style-arms usually truncate, with or without an appendage. Pappus of bristles. **Genera 25–26.**

Other Tribes occur in various parts of the world, but are not represented in the Cayman Islands.

KEY TO GENERA

1. Heads discoid, the florets all tubular and regular:

  2. Herbs:

    3. Leaves alternate (in *Porophyllum* both alternate and opposite):

      4. Phyllaries in 2-several series, distinctly imbricate:

        5. Heads in loose, open cymose panicles: **1. Vernonia**

        5. Heads in dense corymbs: **7. Pluchea**

      4. Phyllaries in 1 main series, essentially valvate:

        6. Phyllaries glabrous: **23. Porophyllum**

        6. Phyllaries pubescent:

          7. Florets cream or greenish-yellow; disk-florets perfect; marginal florets pistillate: **25. Erechtites**

          7. Florets crimson; all florets perfect: **[26. Emilia]**

    3. Leaves opposite:

      8. Leaves simple:

        9. Heads several or many in long-stalked corymbs: **2. Ageratum**

9. Heads solitary (rarely 2 or 3 on a common peduncle):

10. Peduncles and phyllaries glabrous; some of the leaves alternate: **23. Porophyllum**

10. Peduncles pubescent; phyllaries pubescent or at least ciliate:

11. Leaves more or less toothed; receptacle flat or slightly convex:

12. Phyllaries shorter than the florets; pappus of 2–4 awn-like bristles: **13. Melanthera**

12. Phyllaries foliaceous and much longer than the florets; pappus a ciliate cupule: **16. Eleutheranthera**

11. Leaves entire; receptacle conic: **18. Spilanthes**

8. Leaves compound (or at least some of them):

13. Heads unisexual, less than 3 mm in diameter, the pistillate ones 1-flowered, the staminate ones in more or less elongate racemes: **8. Ambrosia**

13. Heads bisexual, up to 10 mm or more in diameter always multiple-flowered, never in racemes: **21. Bidens**

2. Shrubs, or plants woody at least near the base:

14. Leaves alternate:

15. Heads on one-sided, simple or forked, spike-like cymes: **1. Vernonia**

15. Heads in corymbs or glomerate clusters:

16. Leaves less than 3 cm long, viscid-resinous and gland-dotted: **6. Baccharis**

16. Leaves mostly 4–10 cm long or more, not viscid-glandular:

17. Leaves entire, densely tomentose beneath; pappus of numerous white bristles: **7. Pluchea**

17. Leaves toothed, glabrate or sparingly puberulous beneath; pappus of 2 stiff awns: **17. Verbesina**

14. Leaves opposite or mostly so (a few alternate in *Isocarpha*);

18. Heads in leafy racemes: **9. Iva**

18. Heads in corymbs or glomerules:

19. Pappus of numerous bristles; receptacle without scales: **3. Eupatorium**

19. Pappus absent or of 2 awns; receptacle scaly:

20. Achenes oblong and 5-angled; pappus absent; plants woody only at the base: **10. Isocarpha**

20. Achenes flat; pappus of 2 awns; plants woody throughout: **19. Salmea**

1. Heads with at least some outer florets ligulate (the rays sometimes very inconspicuous):

21. Leaves alternate:

22. Leaves simple; pappus of numerous bristles:

23. Leaves rather few, narrowly oblanceolate, up to 10 mm broad or more; heads relatively few in open cymose panicles: **4. Aster**

23. Leaves very numerous, linear, those on the stem less than 3 mm broad except near the base; heads very numerous in rather dense thyrsoid panicles: **5. Conyza**

22. Leaves bipinnatifid; pappus of 2 slender recurved awns: **12. Parthenium**

21. Leaves opposite:

24. Erect shrub; leaves thick and leathery, often silvery-tomentose: **11. Borrichia**

24. Erect, decumbent or matlike herbs; leaves neither leathery nor silvery-tomentose:

25. Achenes without a pappus: **14. Eclipta**

25. Achenes with a pappus:

26. Pappus cup-like: **15. Wedelia**

26. Pappus of bristles, awns or scales:

27. Heads sessile in the leaf-axils and stem-forks; achenes of 2 kinds, the outer flat and with comb-like wings, the inner angular and crowned with a pappus of 2 or 3 bristles: **20. Synedrella**

27. Heads stalked; achenes all alike:

28. Leaves or leaflets broad with toothed margins; heads more than 5 mm across; plants not aromatic:

29. Pappus of 2 awns: **21. Bidens**

29. Pappus of many bristles: **22. Tridax**

28. Leaves linear or very small, entire or bristle-margined; heads less than 5 mm across, longer than broad; plants often aromatic: **24. Pectis**

TRIBE I. VERNONIEAE

Genus 1. **VERNONIA** Schreb., nom. cons.

Herbs or shrubs; leaves alternate, rarely opposite or whorled, entire or toothed, pinnate-veined. Heads discoid, solitary or in cymes, corymbs or panicles; phyllaries imbricate in several series; receptacle naked and smooth or minutely pitted. Florets 5-lobed. Anthers sagittate at the base. Style-arms linear or awl-shaped and usually terete, papillose or short-hairy. Achenes cylindric or top-shaped, angled or ribbed; pappus usually 2-seriate, the inner series of bristles, the outer usually of shorter bristles or scales.

A large pantropical to warm-temperate genus of more than 1,000 species.

KEY TO SPECIES

1. Annual herb with stalked heads in loose cymose panicles: **1. V. cinerea**

1. Shrub with inflorescence of one-sided cymose spikes, simple or branched; heads sessile: **2. V. divaricata**

1. Vernonia cinerea (L.) Less. in Linnaea 4:291. 1829.

Erect herb up to 50 cm tall or more, the stems puberulous; leaves rhombic-ovate or obovate, mostly 1.5–4 cm long, obtuse at the apex, long-cuneate at the base, the margins sinuate-toothed or subentire. Phyllaries in 5 series, pubescent, linear-oblanceolate with sharply acuminate tips. Achenes short-bristly, 1.5 mm long; pappus white, the inner bristles c. 3 mm long.

GRAND CAYMAN: *Brunt 1930; Kings GC 370; Proctor 15280.* CAYMAN BRAC: *Proctor 29091.*
– A pantropical weed of pastures, roadsides, and open waste places.

2. Vernonia divaricata Sw., Fl. Ind. Occ. 3:1319. 1806. **FIG. 243.**

"Christmas Blossom"

An erect shrub to 2 m tall or more, the young branches finely white-tomentose; leaves narrowly to broadly ovate, mostly 3–11 cm long, acuminate at the apex and

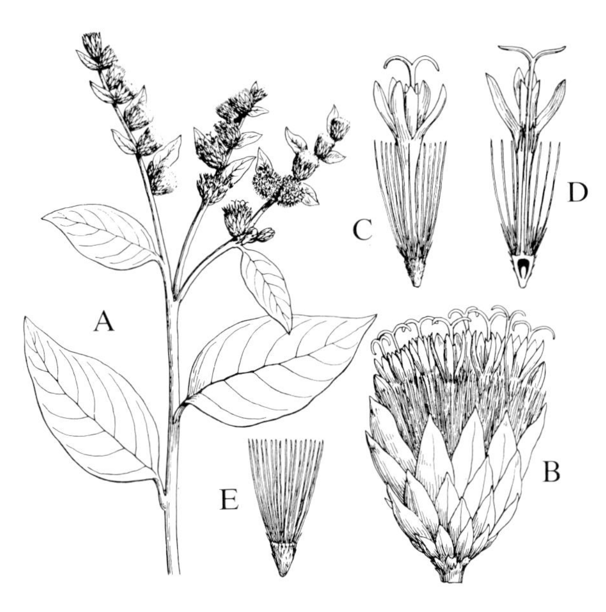

FIG. 243 **Vernonia divaricata.** A, branch with inflorescence, x$^{2}/_{3}$. B, flowering head, x5. C, floret, x6. D, floret cut lengthwise, x6. E, achene and pappus, x5. (F. & R.)

rounded at the base, the margins entire, slightly scabrid-roughened on the upper side and puberulous beneath. Inflorescence usually of 2-several curved, spike-like branches; involucres 5–6 mm long, the phyllaries in 5 series. Florets light purple, c. 6 mm long. Achenes appressed-silky, c. 1 mm long; pappus white, the inner bristles c. 3 mm long.

GRAND CAYMAN: *Brunt 1924; Correll & Correll 51020; Hitchcock; Proctor 15018.* LITTLE CAYMAN: *Proctor 28124, 35146.* CAYMAN BRAC: *Millspaugh 1161; Proctor 29023, 29352.*

– Jamaica; frequent in thickets, along the edges of clearings, and in dry rocky woodlands.

TRIBE II. EUPATORIEAE

Genus 2. **AGERATUM** L.

Annual or perennial herbs, or sometimes shrubs; leaves opposite or sometimes a few upper ones alternate. Heads discoid, many-flowered, in corymbs or panicles; phyllaries in 2–3 series, narrow and equal or nearly so; receptacle naked or scaly. Florets 5-lobed. Anthers obtuse at the base and with a terminal appendage. Style-arms elongate, exserted, terete. Achenes 5-angled and ribbed; pappus of 5 or more scales, these awned or not, free or united at the base, rarely reduced to a cup.

A mostly tropical American genus of about 35 species.

1. **Ageratum littorale** A.Gray in Proc. Amer. Acad. 16:78. 1881.

A decumbent-ascending herb with glabrous stems; leaves rhombic-ovate or elliptic, mostly 1.5–4 cm long including the slender petioles, acute at the apex and long-cuneate at the base, the margins crenate-serrate. Heads in small, dense corymbs at the apex of long, erect peduncles; involucres 3–4 mm long, the phyllaries minutely ciliate. Florets bright lavender-blue, c. 2.5 mm long. Achenes glabrous, c. 2 mm long; pappus a very short, crown-like cup.

GRAND CAYMAN: *Kings GC 176; Proctor 15037, 15145, 31049.*
– Florida Keys and Cape Sable, also on the Turneffe Islands, Belize; grows on beaches and coastal sands.

Genus 3. **EUPATORIUM** L.

Herbs or shrubs, sometimes scandent or vine-like; leaves opposite or whorled (rarely the upper ones alternate), pinnate-nerved, 3-nerved (from the base), or triplinerved (from above the base). Heads discoid, in corymbs or panicles, rarely solitary; involucres of various forms, the scarious phyllaries in 2–4 or more series; receptacle flat, convex or conic, usually naked or sometimes hairy or scaly. Florets 5-lobed or -toothed. Anthers obtuse at the base and with an apical appendage. Style-arms elongate and exserted, often expanded at the tip. Achenes 5-angled or -ribbed; pappus of numerous fine, usually scabrous bristles in a single series.

A very large, widespread genus of perhaps 1,200 species, the majority in the Western Hemisphere, a few in Africa and Asia.

KEY TO SPECIES

1. Heads cylindric, c. 10 mm long in flower; phyllaries strongly striate-nerved; leaves triplinerved: **1. E. odoratum**

1. Heads campanulate, c. 5 mm long in flower; phyllaries not striate; leaves 3-nerved: **2. E. villosum**

1. **Eupatorium odoratum** L., Syst. Nat. ed. 10, 2:1205. 1759.

A mostly erect, short-lived shrub to 2 m tall, the younger stems puberulous or scabrid; leaves petiolate, ovate or broadly ovate, mostly 2.5–8 cm long, acuminate at the apex and abruptly cuneate at the base, the margins broadly crenate-toothed, puberulous or glabrate on the upper side, densely puberulous beneath. Corymbs 2–3-branched; phyllaries in about 5 series, up to 7 mm long, green-tipped. Florets lavender-whitish, c. 5 mm long. Achenes scabrous on the angles, 3–4 mm long.

GRAND CAYMAN: *Hitchcock; Proctor 15293; Rothrock 166.* CAYMAN BRAC: *Millspaugh 1187; Proctor 29108.*
– Southern United States, West Indies, and continental tropical America, common along roadsides, in pastures, clearings, and in rocky thickets. This species has become an invasive weed in parts of Africa and Malaysia.

2. **Eupatorium villosum** Sw., Nov. Gen. & Sp. Pl. 111. 1788.

A bushy shrub to 2 m tall , more or less velvety-pubescent throughout; leaves petiolate, deltate-ovate, mostly 2–7 cm long, blunt to acute at the apex, truncate at the base, the margins entire or obscurely toothed. Heads numerous in dense corymbs; phyllaries 2–2.5 mm long. Florets whitish, 2.5–3 mm long, with glandular tube. Achenes sparsely puberulous, c. 1.5 mm long.

GRAND CAYMAN: *Brunt 1809, 1889; Correll & Correll 51010; Hitchcock; Kings GC 350; Millspaugh 1401; Proctor 11992; Rothrock 169; Sachet 366.* LITTLE CAYMAN: *Proctor 35143.* CAYMAN BRAC: *Kings CB 71; Proctor 28963.*
– Florida and sporadically throughout the West Indies, including the Swan Islands; common in clearings and in thickets on limestone.

TRIBE III. ASTEREAE

Genus 4. **ASTER** L.

Mostly perennial herbs (rarely annual); leaves alternate, and entire, toothed or divided. Heads radiate, usually in corymbs or panicles, rarely solitary or in racemes;

involucre bell-shaped, the phyllaries imbricate in 3–4 series; receptacle flat or convex. Ray-florets in 1 or 2 series, usually pistillate and fertile; disk-florets 5-lobed, perfect, and usually fertile. Anthers obtuse at the base. Style-arms of the disk-florets flattened, crowned with a short or long papillose appendage. Achenes usually compressed, 1–4-ribbed or ribless; pappus of bristles in 2 or 3 series.

A cosmopolitan genus of more than 500 species.

1. **Aster exilis** Ell., Sketch Bot. S. Carol. & Georgia 2:344. 1823. **FIG. 244.**

Erect annual herb up to 50 cm tall or more, glabrous or minutely puberulous; leaves elongate, narrowly oblanceolate, mostly 5–15 cm long, acuminate at the

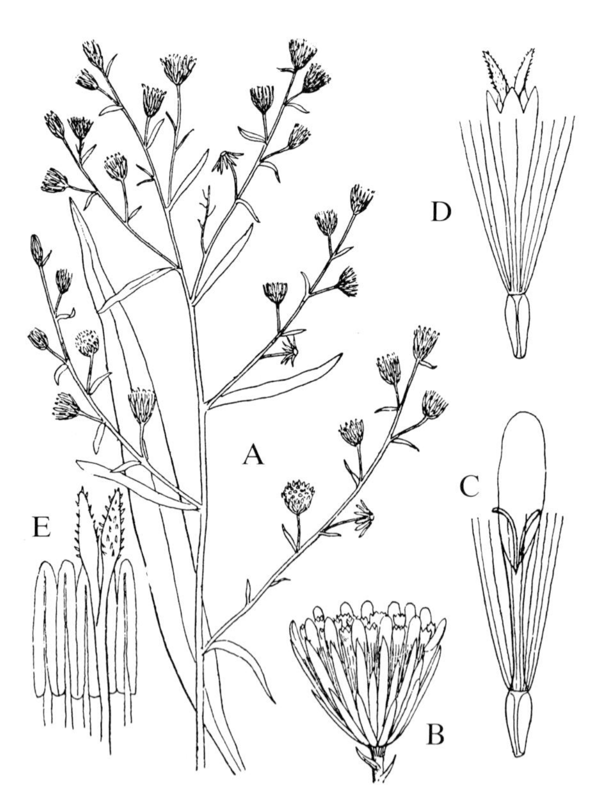

FIG. 244 **Aster exilis.** A, portion of plant, x$^2/_3$. B, head, x4. C, ray-floret with achene, x12. D, disk-floret with achene, x8. E, anthers and apex of style, x32. (F. & R.)

apex and long-tapered to the semi-clasping base, the margins entire or minutely serrulate. Heads 5-8 mm broad in flower; phyllaries 2-5 mm long, glabrous or nearly so. Ray-florets white or very pale mauve; disk-florets light yellow, c. 3.5 mm long. Achenes thinly pubescent, 1-1.5 mm long; pappus of white bristles c. 3 mm long.

GRAND CAYMAN: *Kings GC 419.*
– Southeastern United States, Bahamas, Cuba, Jamaica, and Mexico to Belize, chiefly along roadsides and on moist disturbed soil.

Genus 5. **CONYZA** Less., nom. cons.

Erect annual or perennial herbs; leaves usually in a basal rosette and alternate and numerous on the stems, entire, toothed, or deeply divided. Heads radiate, solitary or in corymbs or panicles; involucre bell-shaped, with phyllaries in 1-3 series; receptacle flat and naked, pitted. Ray-florets pistillate, in 2 or more series, with filiform corollas and very short ligules; disk-florets 4-5-toothed. Achenes narrow, with pappus of 2-3 rows of bristles.

A chiefly temperate and subtropical genus of about 60 species.

1. **Conyza canadensis** (L.) Cronquist in Bull. Torr. Bot. Club 70:632. 1943.

Erigeron canadensis L., 1753.
Conyza canadensis var. *pusilla* (Nutt.) Cronquist, 1943.

Erect weedy herb to 1 m tall or more; leaves on the stem linear or very narrowly oblanceolate and up to 5 cm long, the margins usually sparsely ciliate, the surfaces sometimes puberulous. Branches of the panicle raceme-like; heads mostly 3-5 mm across when in flower; phyllaries mostly 2-3 mm long, glabrous except for the minutely ciliolate tips. Ray-florets c. 2.5 mm long, the minute ligule white; disk-florets yellow, 4-toothed, c. 2.5 long. Achenes 1 mm long; pappus-bristles 3 mm long.

GRAND CAYMAN: *Brunt 1720; Kings GC 261, GC 351; Millspaugh 1259.* LITTLE CAYMAN: *Proctor 28162.* CAYMAN BRAC: *Proctor 28921.*
– Temperate North America and the West Indies, a common weed of roadsides, pastures, and open waste ground. Also widely naturalized in the Old World.

Genus 6. **BACCHARIS** L.

Dioecious shrubs; leaves alternate, occasionally reduced to scales, entire, toothed or lobed. Heads discoid with oblong to hemispheric involucre; phyllaries in several series; receptacle flat or convex, naked, sometimes minutely pitted. Staminate florets tubular with 5-lobed limb and rudimentary styles, their arms tipped with an ovoid pubescent appendage; anthers obtuse and entire at the base. Pistillate florets slender with 5-toothed or entire limb; style-arms smooth, exserted, truncate or club-shaped. Achenes somewhat compressed, 5–10-ribbed, developed only in pistillate heads; pappus of the staminate florets consisting of a few short bristles, of the pistillate flowers numerous in 1 or more series.

A New World genus of about 400 species, the majority tropical.

1. **Baccharis dioica** Vahl, Symb. Bot. 3:98, t.74. 1794.

A densely-branched, round-topped shrub 1 m tall or more, the young branchlets angled and resinous; leaves crowded, obovate, mostly 1.5–2 cm long or more, stiff, glabrous, and viscid, the apex minutely cuspidate. Heads sessile in dense clusters, the staminate c. 5 mm long and 20-flowered, the pistillate 7 mm long and 50-flowered; florets white. Achenes glabrous, 1–2 mm long; pappus dull whitish.

LITTLE CAYMAN: *Proctor 35190.* CAYMAN BRAC: *Proctor 29046.*
– Florida, and the West Indies to Montserrat, in rocky thickets and dry rocky scrublands.

TRIBE IV. INULEAE

Genus 7. **PLUCHEA** Cass.

Erect herbs or shrubs; leaves alternate, entire or toothed, rarely pinnatifid. Heads discoid, heterogamous, usually in terminal corymbose cymes; involucre ovoid or bell-shaped, the phyllaries in few or several series; receptacle flat, naked. Outer florets in several rows, pistillate and fertile; central florets usually few, perfect but sterile (functionally staminate). Corolla of the perfect florets tubular, 5-lobed, their anthers sagittate at the base with tailed auricles, their styles entire or forked, papillose. Corolla of the pistillate flowers filiform, 3-toothed or -lobed; style with 2 linear arms. Achenes 4–5-angled; pappus a single series of several or many scabrid bristles.

A pantropical genus of about 50 species.

KEY TO SPECIES

1. Annual herb with lanceolate leaves mostly less than 6 cm long (rarely longer); florets bright red-purple; achenes 1 mm long: **1. P. odorata**

1. Coarse shrub with elliptic leaves usually 6–15 cm long; florets dull light pink; achenes 0.5 mm long: **2. P. symphytifolia**

1. Pluchea odorata (L.) Cass. in Dict. Sci. Nat. 42:3. 1826

Pluchea purpurascens (Sw.) DC., 1836.

Aromatic herb to 1 m tall (rarely more), the stems puberulous; leaves mostly 2–6 cm long and 0.5–2 cm broad, at least the upper ones subsessile, acute or obtuse at the apex, the margins obscurely toothed with glandular serrations, puberulous and glandular on both sides. Corymbs mostly 2–4 cm across; heads 4–5 mm long. Achenes glandular; pappus-bristles c. 3 mm long.

GRAND CAYMAN: *Brunt 1662; Kings GC 357; Maggs II 51; Proctor 15261; Sachet 460.*
– Southeastern United States, West Indies, Mexico and Central America, chiefly in marshes and moist sandy ground.

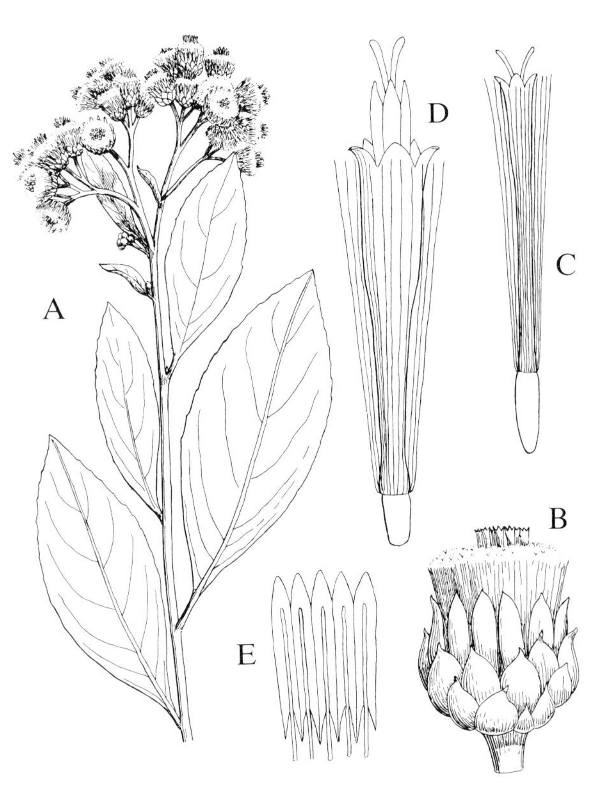

FIG. 245 **Pluchea symphytifolia.** A, end of branch with inflorescence, x$^2/_3$. B, head, x4. C, pistillate floret, x12. D, perfect floret, x12. E, anthers x12. (F, & R.)

2. Pluchea symphytifolia (Miller) Gillis in Taxon 26(5/6):588. 1977. **FIG. 245.**

Pluchea odorata of many authors, not (L.) Cass., 1826.

P. carolinensis (Jacq.) G.Don in Sweet, 1839; Adams, 1972.

A bushy shrub up to 3 m tall, the stems whitish-tomentose; leaves all petiolate, the blades elliptic, acute or blunt at the apex, the margins entire or with a few obscure glandular teeth, minutely puberulous and glandular on the upper (adaxial) surface, densely woolly-tomentose and glandular beneath. Corymbs mostly 8–15 cm across; heads 5–6 mm long. Achenes strigillose; pappus-bristles c. 3 mm long.

GRAND CAYMAN: *Hitchcock; Kings GC 357.*
– Florida, West Indies, Mexico to Venezuela, and islands of the Pacific; occurs in brackish thickets and open waste ground.

TRIBE V. HELIANTHEAE

Genus 8. **AMBROSIA** L.

Monoecious (rarely dioecious) annual or perennial herbs, sometimes woody at the base; leaves opposite or alternate (rarely in whorls of 3), entire to pinnately cut or dissected. Staminate heads in spikes or racemes, the pistillate solitary or in glomerules at the base of the staminate raceme. Staminate involucre hemispheric or saucer-shaped, 5–12-lobed; receptacle flat, naked or scaly; florets funnel-shaped, 5-toothed; anthers scarcely coherent, mucronate-tipped; style undivided, capitate-penicillate. Pistillate involucre closed, usually with small tubercles or prickles near the top and contracted above into a short beak surrounding the style; corolla absent; style deeply divided into 2 long, narrow arms. Achenes broad and thick, without a pappus, closely surrounded by the persistent, hardened involucre.

A mostly American genus of between 35 and 40 species. The wind-borne pollen of some species is a principal cause of "Hay-fever".

1. Ambrosia hispida Pursh, Fl. Amer. Sept. 2:743. 1814. **FIG. 246.**

"Geranium", "Running Wormwood"

A perennial, bitter-aromatic, trailing herb with ascending branches to about 35 cm tall or more, canescent-tomentose throughout; leaves ovate or broadly triangular in outline, mostly 2.5-5 cm long, bipinnatifid or the lower ones tripinnatifid. Racemes up to 8 cm long; staminate heads 6-20-flowered, 2-4 mm across, the involucres clothed with short hairs having pustulate bases; florets cream, densely glandular. Pistillate heads 3-10 in a cluster; achenes black, 2.5-3 mm long.

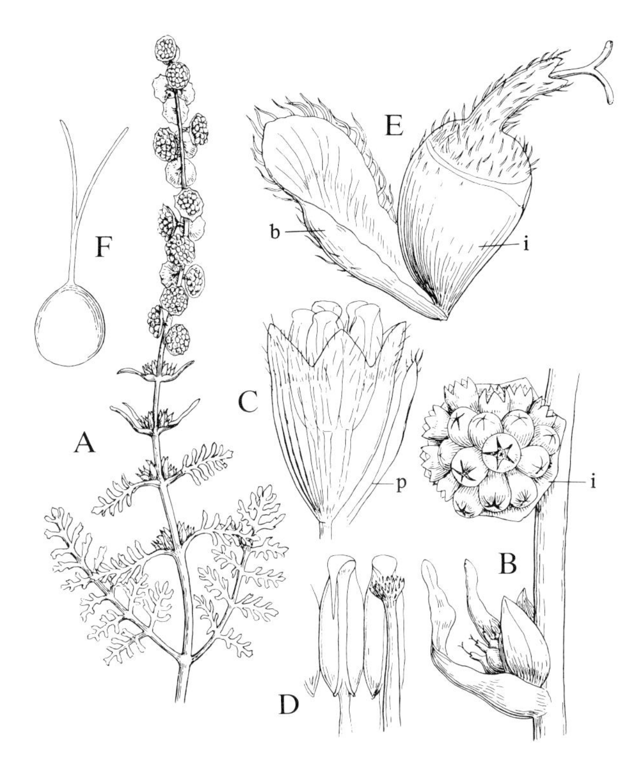

FIG. 246 **Ambrosia hispida.** A, branch with leaves and inflorescence, x$^2/_3$. B, a glomerule of pistillate heads (below) and a staminate head, x4; i, involucre. C, staminate floret detached with scale (p), x12. D, anthers and style from staminate floret, x12. E, pistillate head with bract (b), x12; i, involucre. F, ovary and style, x8. (F. & R.)

GRAND CAYMAN: *Brunt 1963; Hitchcock; Kings GC 260, GC 319; Proctor 15154; Sachet 443.* LITTLE CAYMAN: *Kings LC 25, LC 26; Proctor 28068.* CAYMAN BRAC: *Proctor (sight record).*

– Florida, West Indies, and Central America, mostly in sandy clearings near the sea; sometimes cultivated.

Genus 9. **IVA** L.

Annual or perennial herbs, or shrubs; leaves opposite or the upper ones alternate, simple and often of thick texture. Heads nodding, in spikes, racemes, or panicles, rarely solitary in the leaf-axils. Involucre hemispheric or cup-shaped, composed of few, rounded bracts; receptacle scaly, the chaff-like linear or spatulate scales enveloping the florets. Marginal florets 1–6, pistillate, fertile, their corollas short and tubular, or none; styles 2-armed. Disk-florets perfect in structure but functionally staminate only, their corollas funnel-shaped and 5-lobed; anthers entire at the base, scarcely coherent, tipped with mucronate appendages; styles undivided, dilated at the apex. Achenes compressed, obovoid, glabrous; pappus lacking.

An American genus of about 15 species.

1. **Iva cheiranthifolia** Kunth in H.B.K., Nov. Gen. & Sp. 4:276. 1820.

An aromatic shrub to 2 m tall, with arching canescent branches; leaves opposite and alternate, very narrowly elliptic or narrowly oblong-elliptic, mostly 1–6 cm long or more, 0.3–1.5 cm broad, sharp-pointed at the apex, 3-nerved from the base, the margins entire, finely puberulous and gland-dotted on both sides. Heads numerous, short-stalked, 3–5 mm across; phyllaries 3–5 nearly orbicular, puberulous. Achenes c. 2.5 mm long.

GRAND CAYMAN: *Brunt 1769, 2082; Hitchcock; Lewis 3829; Proctor 27928, 27953.*
– Bahamas and Cuba; common in brackish thickets and along the margins of swamps.

Genus 10. **ISOCARPHA** R.Br.

Perennial herbs, sometimes woody or shrublike at the base; leaves alternate or opposite, simple and entire or toothed. Heads discoid, solitary or clustered at the ends of long peduncles; involucre of rather few scales in 2–3 series; receptacle conic, with chaffy scales enveloping the florets. Florets all perfect and fertile, 5-lobed; anthers obtuse at the base; style-arms terete, slender, hairy. Achenes 4–5-angled, truncate at the apex and without pappus.

A genus of 10 tropical and warm-temperate American species.

1. **Isocarpha oppositifolia** (L.) Cass. in Dict. Sci. Nat. 24:19. 1822. **FIG. 247.**

A suffrutescent herb, the older stems woody and often decumbent, the young stems ascending and pubescent, up to 30 cm tall or more; leaves mostly opposite, often with smaller ones clustered in the axils, narrowly oblanceolate, mostly

1.5-4.5 cm long, obtuse at the apex, the margins mostly entire, pubescent and glandular on both sides, triplinerved. Heads 6-10 mm long, few in a small, dense cluster; phyllaries 2-nerved and mucronate, 3-4 mm long. Florets white, 2.5 mm long. Achenes 5-angled, dark, 2 mm long.

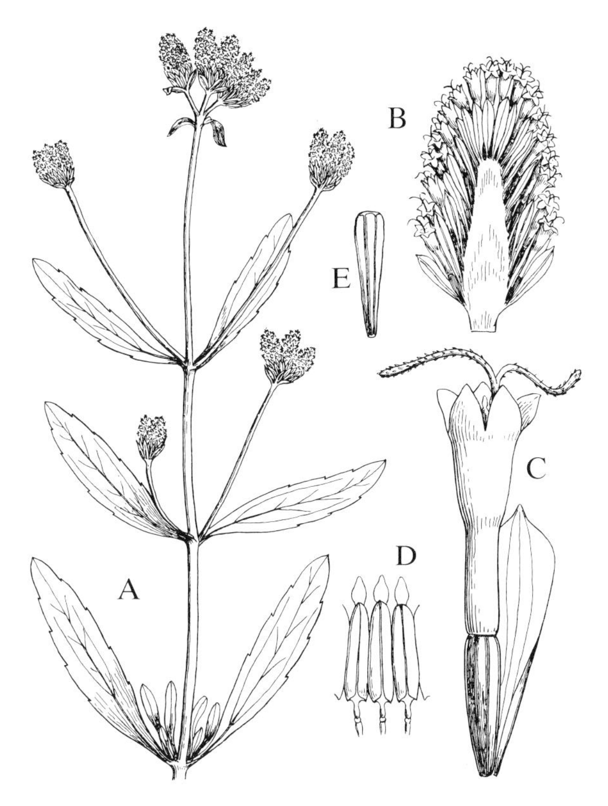

FIG. 247 **Isocarpha oppositifolia.** A, portion of plant, x2/3. B, head cut lengthwise, x4. C, floret, x12. D, anthers, x14. E, ripe achene, x8. (F. & R.)

CAYMAN BRAC: *Kings CB 54; Proctor 29044.*

– Bahamas, Cuba, Jamaica, and Texas to Venezuela, Trinidad and Tobago; occurs in dry, rocky thickets and scrublands.

Genus 11. **BORRICHIA** Adans.

Shrubs, often of fleshy texture; leaves opposite, simple and entire or toothed, 1-3-nerved. Heads solitary, terminal or in forks of the branches, radiate. Involucre hemispherical, the phyllaries leathery and in 2–3 series; receptacle slightly raised, with rigid concave scales surrounding the florets. Ray-florets pistillate, the ligules subentire or 2–3-toothed. Disk-florets perfect and fertile, 5-toothed; anthers entire at the base or nearly so; style-arms swollen and hairy above. Achenes oblong, 3–5-angled; pappus a short, toothed cup.

A small tropical American genus of 7 species.

1. **Borrichia arborescens** (L.) DC., Prodr. 5:489. 1836. **FIG. 248.**

"Bay Candlewood"

A rather succulent shrub to 1.5 m tall, occurring in two forms, one whitish-canescent, the other nearly glabrous throughout; leaves linear-oblanceolate to

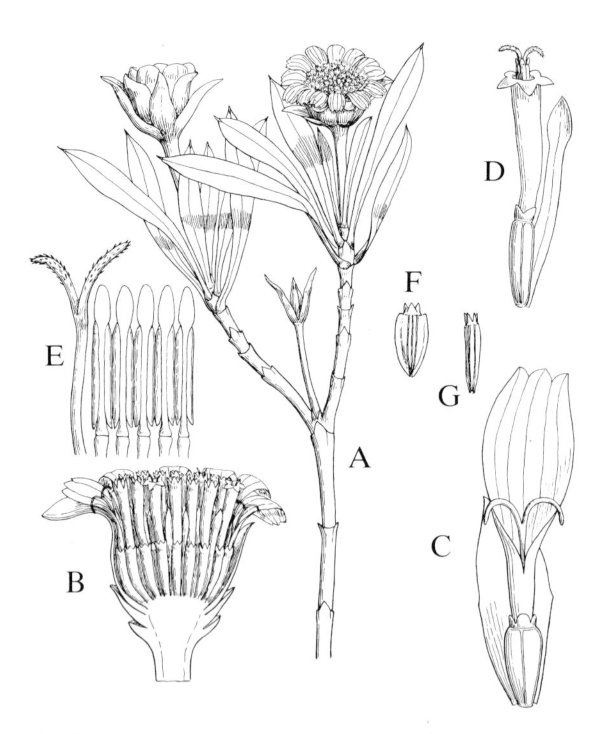

FIG. 248 **Borrichia arborescens.** A, portion of plant, x$^2/_3$. B, head cut lengthwise, x2. C, ray-floret, x4. D, disk-floret, x4. E, anthers and style, x8. F, achene of ray-floret, x3. G, achene of disk-floret, x3 (F. & R.)

narrowly obovate, 2–7.5 cm long, sharply mucronate at the apex, the margins entire. Peduncles mostly 1–2.5 cm long; heads of heavy, leathery texture, 1–2 cm in diameter or more; ligules yellow, 6–9 mm long; disk-florets 6 mm long. Achenes 3.5–4 mm long; pappus-cup c. 1 mm long.

GRAND CAYMAN: *Brunt 1687, 1850, 1884, 1973; Hitchcock; Kings GC 35, GC 38, GC 268, GC 269, GC 395; Millspaugh 1239, 1242, 1247; Proctor 11989; Sachet 392.* LITTLE CAYMAN: *Kings LC 40; Proctor 28026.* CAYMAN BRAC: *Kings CB 86; Proctor 28927.*

– Florida, West Indies and Yucatan; common on rocky seacoasts, sometimes in sandy or brackish thickets.

Genus 12. **PARTHENIUM** L.

Herbs or shrubs; leaves alternate, entire to pinnately-cut. Heads radiate, small, in terminal panicles; involucre bell-shaped, of few broad phyllaries in 2 or 3 series; receptacle more or less raised, bearing scales surrounding the disk-florets.

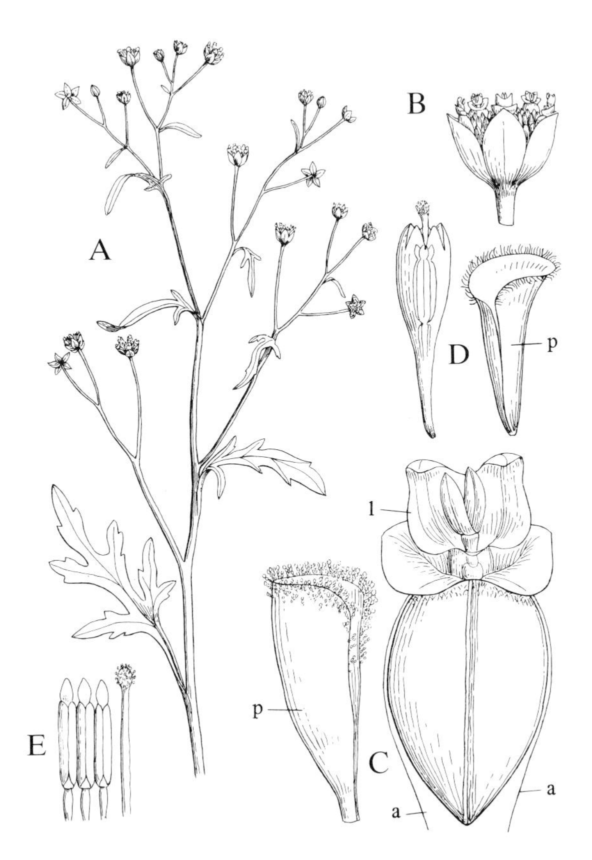

FIG. 249 **Parthenium hysterophorus.** A, portion of plant, x$^2/_3$. B, head, x6. C, ray-floret with achene and receptacular scale, x18; p, scale; l, ligule; a, a, awns. D, disk-floret with scale (p), x18. E, anthers and style of disk-floret, x24. (F. & R.)

Ray-florets few, pistillate, their ligules short, broad, 2-toothed and persistent. Disk-florets perfect but only staminate in function, funnel-shaped and 5-toothed; anthers entire at the base; styles unbranched. Achenes compressed, keeled on the inner face each surrounded by a phyllary and 2 concave receptacular scales, crowned by the persistent ligule and a pappus of 2–3 recurved awns.

A chiefly Mexican genus of 15 species.

1. Parthenium hysterophorus L., Sp. Pl. 2:988. 1753. **FIG. 249.**

An annual, diffusely-branched herb up to 50 cm tall or more, the stems more or less pubescent; leaves elliptic or obovate in outline, mostly 2–12 cm long, bipinnatifid, pubescent on both sides and densely gland-dotted beneath. Heads stalked, 3–5 mm in diameter, in loose open panicles; phyllaries in 2 series, ovate, clothed with minute, appressed, yellowish hairs. Scales of the receptacle with glandular margins. Florets and ligules white. Achenes black, 2 mm long, hairy at the top; pappus of 2 slender recurved awns.

GRAND CAYMAN: *Brunt 2020; Kings GC 45, GC 302; Lewis 3838, 3850; Proctor 15290.*
– Throughout the American tropics and subtropics, and introduced into the Old World; a weed of roadsides and open waste ground.

Genus 13. **MELANTHERA** Rohr

Perennial herbs or undershrubs; leaves opposite, simple but toothed or sometimes hastate-lobed. Heads solitary or rarely paired on long terminal or axillary peduncles, discoid (in our species) with all the florets perfect and fertile, or radiate, the ray-florets pistillate or neuter and the disk-florets perfect. Involucre hemispheric, the phyllaries in 2–3 series; receptacle convex, scaly, the scales concave. Disk-florets tubular, 5-lobed; anthers black, obtuse or minutely sagittulate at the base; style-arms with an acute, hairy appendage. Achenes more or less 3-angled; pappus of 2–4 stiff, ciliate, awn-like bristles, these soon falling, absent from the achenes of ray-florets.

A genus of about 50 species in tropical America and Africa.

1. Melanthera aspera (Jacq.) L.C.Rich. ex Spreng., Neue Entdeck. 3:40. 1822. **FIG. 250.**

"Soft Leaf" (so-called in jest, as the leaves are harsh, not soft)

A stiff bushy or scrambling herb to 1 m tall or more, the stems and leaves of rough, harsh texture; leaves long-petiolate, the blades broadly ovate or triangular-hastate, 3–12 cm long, acuminate at the apex, the margins coarsely crenate- or serrate-toothed, often with a sharp, short lobe on either side at the base. Peduncles up to 8 cm long; heads up to 1 cm in diameter. Achenes 3 mm long; pappus-bristles 1-3 mm long.

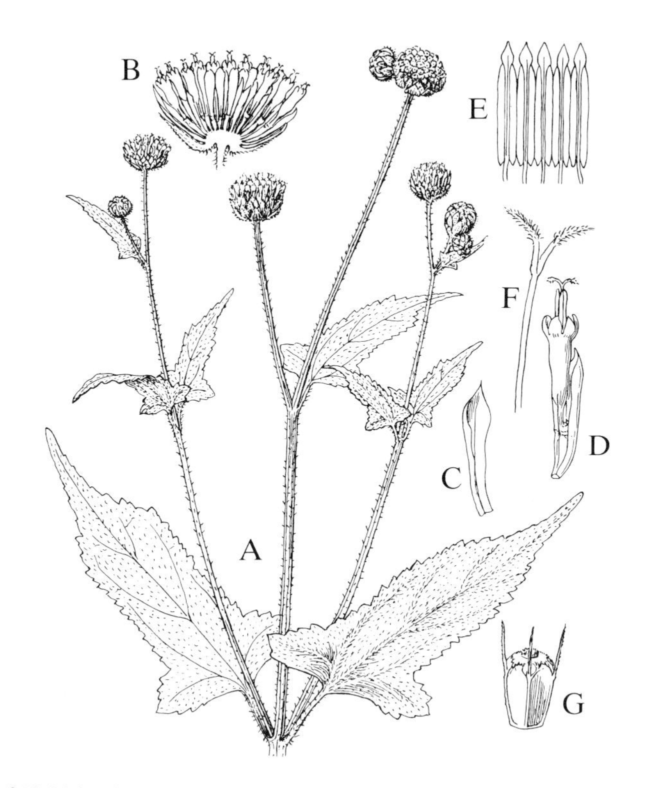

FIG. 250 **Melanthera aspera.** A, portion of plant, x$^{2}/_{3}$. B, head cut lengthwise, x2. C, receptacular scale, x4. D, floret with scale, x4. E, anthers, x8. F, style-arms, x8. G, achene, x4. (F. & R.)

LITTLE CAYMAN: *Kings LC 22; Proctor 28050, 28054.* CAYMAN BRAC: *Proctor 29123.*

– Florida, Bahamas, Cuba, Jamaica and the Yucatan Peninsula; in sandy thickets and clearings at low elevations, mostly near the sea.

Genus 14. **ECLIPTA** L., nom. cons.

Herbs; leaves opposite, simple, entire or toothed. Heads radiate, mostly solitary on axillary peduncles; peduncles 1-several in an axil; involucre campanulate, of few herbaceous phyllaries in 2 series; receptacle flat or slightly raised, scaly. Ray-florets pistillate or sterile, the ligule small and inconspicuous. Disk-florets perfect and fertile, the corolla tubular and 4-5-toothed; anthers obtuse at the base; style-arms with a short, obtuse appendage. Achenes cylindric or slightly compressed, rugose; pappus absent or of 2 short awns.

A small but widespread tropical genus of 3 or 4 species.

1. **Eclipta alba** (L.) Hassk., Pl. Jav. Rar. 528. 1848; Miq., 1856; Cronquist in Rhodora 47:398-399. 1945.

An erect or decumbent herb up to 50 cm tall or more, the stems strigose with basally-swollen hairs. Leaves sessile, very narrowly elliptic or narrowly lanceolate, mostly 2-7 cm long, bluntly acuminate at the apex and long-cuneate at the base, the margins entire or minutely toothed, the surfaces bearing strigose hairs. Peduncles 1-6 cm long; heads 6-9 mm across; phyllaries ovate, acute, 3.5-4 mm long; ray-florets with linear ligules c. 2 mm long. Achenes glabrous, rugose, black, 2 mm long; pappus of 2 very short awns joined by a fimbriate flange.

GRAND CAYMAN: *Brunt 1843; Kings GC 276; Proctor 15275.* CAYMAN BRAC: *Proctor 28975.*
– A pantropical weed of roadside ditches and low moist ground.

Genus 15. **WEDELIA** Jacq., nom. cons.

Herbs or shrubs; leaves opposite, usually toothed or lobed. Heads radiate, solitary on axillary or terminal peduncles; involucre campanulate; phyllaries few, in 2 series, those of the outer series firmly herbaceous, of the inner membranous at least toward base; receptacle flat or convex with scales enveloping the disk-flowers. Ray-florets pistillate and fertile, with 2-3-toothed ligules. Disk-florets perfect and fertile, tubular, 5-toothed or -lobed; anthers entire at their base or shortly sagittulate; style-arms linear, hairy toward the apex. Achenes thick and angled, or rarely narrowly winged; pappus cup-shaped, toothed, or of small scales, sometimes with 1 or 2 short awns, these soon-falling, or sometimes the pappus merely a ring or lacking altogether.

A pantropical and subtropical genus of about 70 species.

1. **Wedelia trilobata** (L.) Hitchc. in Rep. Missouri Bot. Gard. 4:99. 1893. **FIG. 251.**

"Marigold"

Herb with trailing stems rooting at the nodes, the branches and tips ascending, sparsely hairy or nearly glabrous throughout; leaves sessile or nearly so, obovate or oblong-obovate, 3–12 cm long, often strongly 3-lobed but sometimes merely

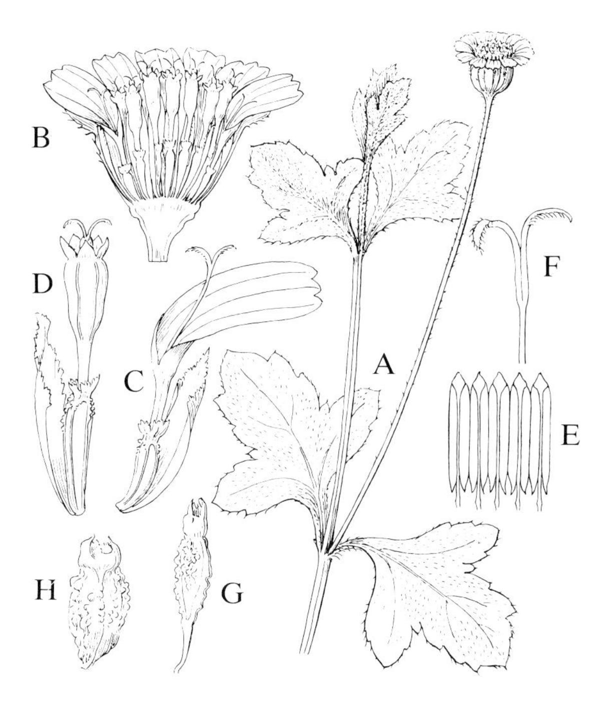

FIG. 251 **Wedelia trilobata.** A, branch with leaves and flower-head, x$^2/_3$. B, head cut lengthwise, x2. C, ray-floret with scale, x4. D, disk-floret with scale, x4. E, stamens x8. F, style-arms, x8. G, achene of ray-floret, x4. H, achene of disk-floret, x4. (F. & R.)

toothed or subentire. Peduncles up to 5 cm long or more; heads mostly 1.5–2 cm broad; outer phyllaries c. 12 mm long. Ligules yellow, 7–12 mm long, 3-toothed at the apex. Disk-florets 5 mm long. Achenes rugose, 4–5 mm long; pappus of irregular, short, united scales.

GRAND CAYMAN: *Brunt 1908; Hitchcock; Kings GC 315; Lewis GC 46; Maggs II 50; Millspaugh 1245; Proctor 15151; Sachet 440.* LITTLE CAYMAN: *Kings LC 50.*

– Florida, West Indies, continental tropical America, Africa and Hawaii; common along roadsides, in damp pastures, and in sandy thickets and clearings near the sea.

Genus 16. **ELEUTHERANTHERA** Poit. ex. Bosc.

Annual herbs; leaves opposite, simple, subentire or toothed. Heads discoid or rarely with a few minute, sterile ray-florets, solitary on short axillary or terminal peduncles, 1 or 2 peduncles from an axil. Involucre bell-shaped, of few foliaceous scales; receptacle convex, scaly, the scales surrounding the florets. Florets few, perfect, 5-toothed; anthers free, sagittulate at the base; style-arms linear, hairy on the back. Achenes thick, smooth or rough; pappus a small, ciliate cup and sometimes 2 or 3 short awns.

A genus consisting of 1 pantropical species and 1 in Madagascar.

1. Eleutheranthera ruderalis (Sw.) Sch. Bip. in Bot. Zeit. 24:165. 1866. **FIG. 252.**

An erect bushy herb to about 50 cm tall, the stems pubescent; leaves narrowly ovate or ovate, mostly 1.5–4 cm long, acute at the apex, the margins crenate-

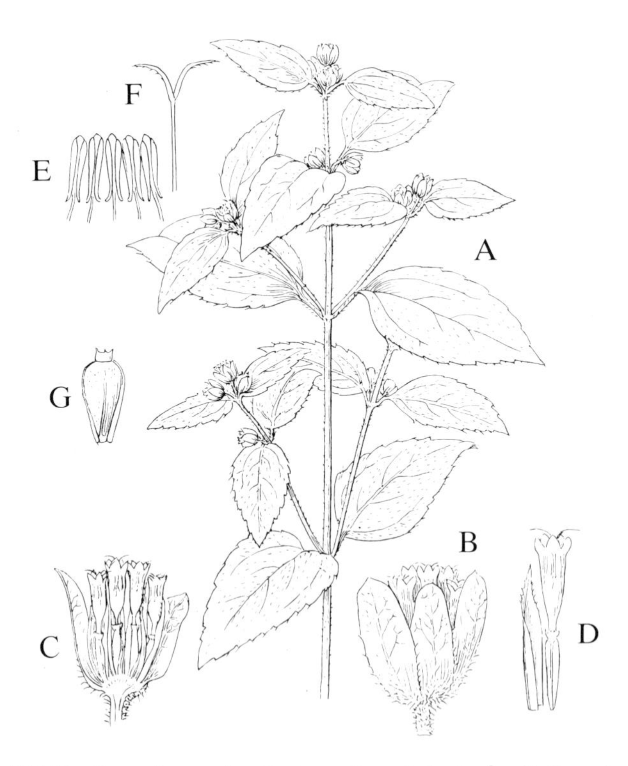

FIG. 252 **Eleutheranthera ruderalis.** A, portion of plant, x$^2/_3$. B, flowering head, x4. C, same cut lengthwise, x4. D, floret with scale, x6. E, anthers, nearly free, x16. F, style-arms, x16. G, ripe achene, x4. (F. & R.)

serrate, pubescent on both sides and gland-dotted beneath. Peduncles to 1 cm long or more; phyllaries to 7 mm long; receptacle scales c. 4 mm long, folded around the florets. Florets few, yellow, 5-toothed. Achenes obovoid, puberulous, 3 mm long.

GRAND CAYMAN: *Kings GC 304.*
– A pantropical weed of moist fields and pastures.

Genus 17. **VERBESINA** L.

Herbs or shrubs, sometimes arborescent; leaves opposite or alternate, usually toothed or lobed. Heads discoid (in our species) with all the florets perfect; or radiate, the ray-florets pistillate or sterile and the disk-florets perfect; solitary or few on long peduncles, or numerous in corymbs or corymbose panicles. Involucre hemispheric or bell-shaped, the scales in 2–6 series or sometimes apparently 1-seriate; receptacle convex or conic, bearing concave or folded scales. Disk-florets tubular and 5-lobed; anthers obtuse or minutely sagittulate at the base; style-arms with an acute, hairy appendage. Achenes compressed, with marginate or winged edges; pappus of 2 awns.

A tropical American genus of about 150 species.

1. **Verbesina caymanensis** Proctor in Sloanea 1:4. 1977. **FIG. 253.**

Shrub to 1 m tall or more, the branchlets striate and sparsely puberulous; leaves alternate, sessile, oblanceolate to narrowly obovate, 3–12 cm long, up to 4 cm broad, obtuse or acute at the apex, the margins broadly and shallowly serrate in the upper part, the serrations sharply gland-tipped; the surface thinly scabridulous on the upper side and minutely hispid or glabrescent beneath. Heads in small terminal corymbose panicles; phyllaries pubescent, acute, 4–5 mm long; receptacle-scales mucronate; florets 4.5–5 mm long. Achenes narrowly flattened-obovoid, 3.5–4.5 (-5) mm long, blackish and hispidulous, with a broad, whitish, notched and rugose wing along each side (or in some achenes the wings lacking); awns 1 or 2 (rarely 3), c. 3 mm long, often unequal, ciliate.

CAYMAN BRAC: *Proctor 29073, 29361 (type).*
– Endemic; confined to limestone cliffs and rocky thickets near Spot Bay. *V. caymanensis* differs from *V. propinqua,* its closest Jamaican counterpart, in various details of the leaves, florets and achenes. These species belong to a subgroup of *Verbesina* that by some authors has been treated as a separate genus *Chaenocephalus.* This subgroup has a remarkable distribution, with 7 species occurring in Jamaica and 6 others found in Ecuador, Peru and Argentina; the phytogeographic significance of such a distribution is not clear.

Genus 18. **SPILANTHES** Jacq.

Annual or sometimes perennial herbs; leaves opposite, entire or toothed. Heads long-stalked, solitary or in loose, open cymes, discoid (in our species) with all the florets perfect and fertile, or radiate, the ray-flowers pistillate and the disk-florets perfect. Involucre short, bell-shaped; phyllaries in 1 or 2 series; receptacle convex, conic or subcolumnar, bearing folded scales. Ray-florets, when present, with short inconspicuous ligules. Disk-florets tubular, with enlarged 4–5-lobed limb; anthers obtuse at the base; style-arms truncate. Achenes compressed, wingless pappus of 1–3 bristle-like awns, or absent.

A pantropical genus of about 60 species.

1. **Spilanthes urens** Jacq., Enum. Syst. Pl. Carib. 28. 1760.

Herb with trailing, nearly glabrous stems, often rooting at the nodes, the flowering-branches ascending to 20 or 30 cm (often less); leaves sessile, very narrowly to narrowly elliptic, mostly 2–9 cm long, acutish at the apex, triplinerved, glabrate or puberulous beneath. Heads discoid, solitary on long terminal peduncles, 0.8–1.3 cm in diameter; phyllaries 4–5 mm long; florets numerous, white, 2 mm long. Achenes obovate-oblong, 2.5 mm long, ribbed and thinly pubescent; pappus of 2 incurved, broad-based awns.

GRAND CAYMAN: *Brunt 1883, 1923; Correll & Correll 51051; Howard & Wagenknecht 15024; Kings GC 321; Lewis 3836; Proctor 15152; Sachet 452.* LITTLE CAYMAN: *Proctor 28069.*

– West Indies (except the Bahamas) and continental tropical America; frequent along moist roadsides, in seasonally-flooded pastures, and in damp sandy clearings near the sea.

Genus 19. **SALMEA** DC., nom. cons.

Erect or climbing shrubs; leaves opposite and entire, toothed or lobed. Heads discoid, in terminal or axillary cymose panicles or corymbs; involucre bell- or top-shaped; phyllaries in 2–5 series, membranous or dry and papery; receptacle conic or columnar, bearing folded scales. Florets tubular, enlarged above, 5-toothed;

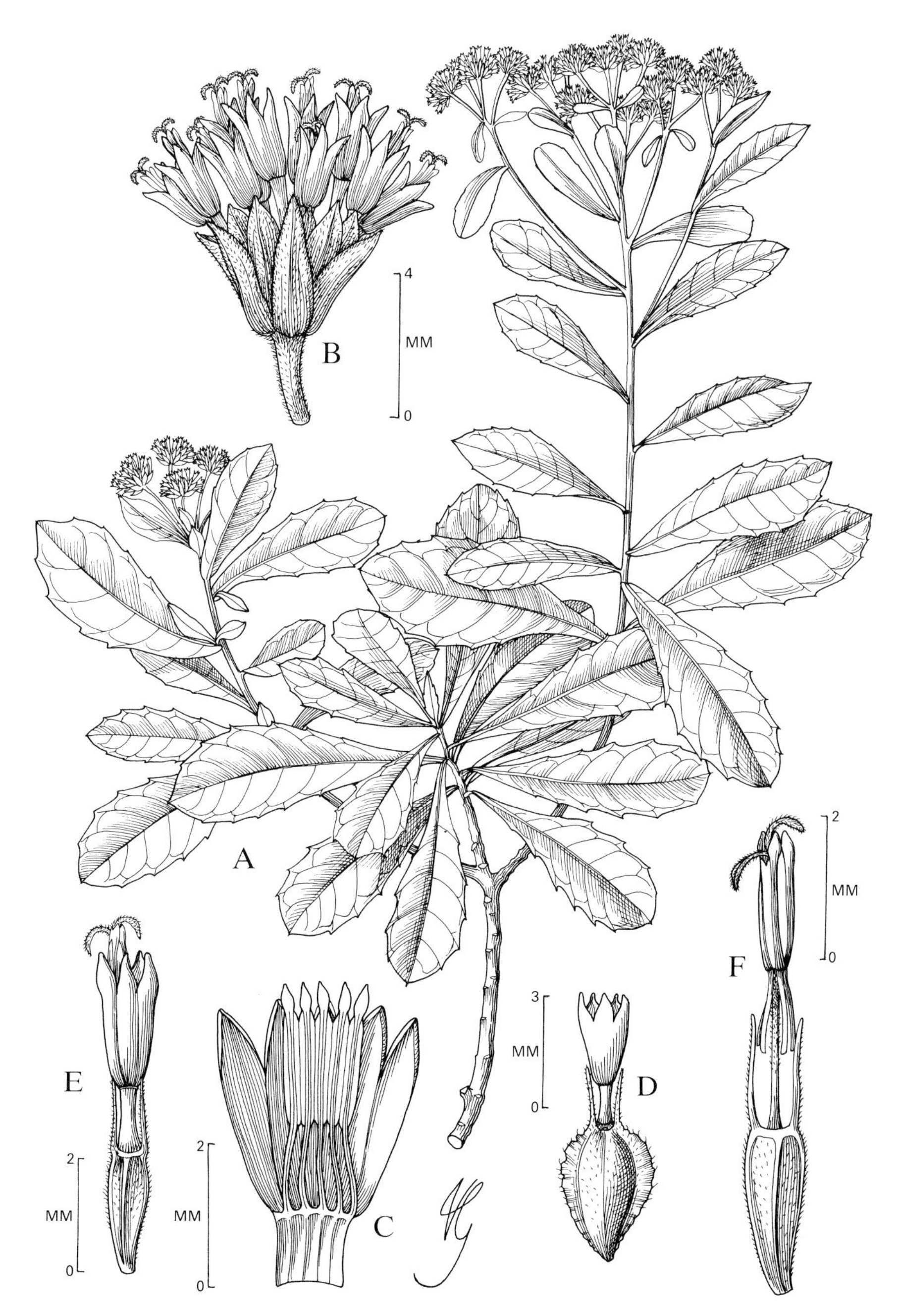

FIG. 253 **Verbesina caymanensis.** A, branch, habit. B, flowering head. C, single floret cut open to show the stamens. D, floret with winged achene. E, floret with unwinged achene. F, same, corolla partly removed. (G.)

anthers shortly sagittulate at the base; style-arms papillose at the top. Achenes flattened, dark, and with or without a pale border, usually ciliate; pappus of 2 subequal awns.

A genus of 12 species in the West Indies, Mexico and Central America.

1. Salmea petrobioides Griseb., Fl. Brit. W. I. 375. 1861.

A densely-branched glabrous shrub to 1.5 m tall or more; leaves somewhat fleshy, narrowly to broadly obovate, mostly 2–5 cm long, rounded, acutish or shortly apiculate at the apex, the margins entire, the lateral venation obscure. Heads in dense terminal corymbs; involucre glutinous, narrowly bell-shaped, c. 4 mm long, the phyllaries dry and long-persistent. Florets white, 2.5 mm long. Achenes black, c. 2 mm long, glabrous.

GRAND CAYMAN: *Brunt 1995; Correll & Correll 51035; Hitchcock; Lewis 3826; Millspaugh 1404; Proctor 11991.* LITTLE CAYMAN: *Proctor 35222.* CAYMAN BRAC: *Millspaugh 1231.*
– Bahamas and Cuba, in sandy or rocky thickets mostly near the sea.

Genus 20. **SYNEDRELLA** Gaertn., nom. cons.

Annual herbs; leaves usually opposite, serrate or subentire. Heads sessile, radiate; involucre oblong, of few bracts, the outer leaf-like; receptacle small, bearing flat chaffy scales among the florets. Ray-florets pistillate, fertile, each with a short, broad ligule. Disk-florets perfect and fertile, tubular and 4-toothed; anthers minutely sagittulate; style-arms with hairy appendages. Achenes compressed, dimorphic, those of the ray-florets crested-winged, those of the disk-florets with narrow entire wings and 2 awns.

A tropical American genus of 2 species, one of wide distribution, the other a native of Ecuador.

1. Synedrella nodiflora (L.) Gaertn., Fruct. & Sem. Pl. 2:456, t. 171, f. 7. 1791. **FIG. 254.**

Ucacou nodiflorum (L.) Hitchc., 1893.

An erect herb to 50 cm tall or more, sometimes flowering when very small; stems appressed-pubescent; leaves ovate, 1.5–6 cm long or more, acute at the apex, the margins obscurely serrate, appressed-pubescent on both sides and triplinerved.

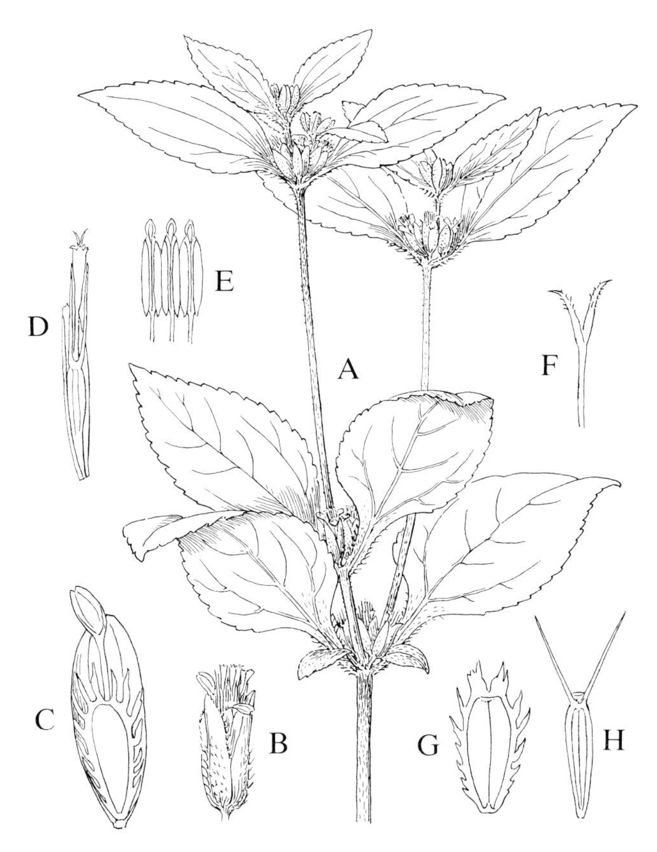

FIG. 254 **Synedrella nodiflora.** A, upper part of plant, x$^2/_3$. B, head in flower, x2. C, ray-floret with scale, x4. D, disk-floret with scale, x4. E, stamens, x14. F, style-arms, x14. G, achene of ray-floret, x4. H, achene of disk-floret, x4. (F. & R.)

Heads 1-several crowded together in the leaf-axils, each with 2 green phyllaries 7.5–10 mm long and several scarious inner ones. Ray-florets few, with yellow ligules c. 2 mm long; disk-florets 6–10, 2 mm long. Achenes 3.5–4 mm long.

GRAND CAYMAN: *Brunt 1934; Hitchcock; Kings GC 54, GC 305; Millspaugh 1276.*

– A common weed throughout tropical America; naturalized in the Old World tropics; frequent along roadsides, in pastures, and in open waste places.

Genus 21. **BIDENS** L.

Herbs, shrubs, or sometimes vines; leaves mostly alternate, simple and toothed, or variously divided. Heads usually radiate (rarely discoid only), solitary or few at ends of branches, sometimes in panicles; involucre more or less bell-shaped, of rather few phyllaries in 1 or 2 series; receptacle flat or slightly raised, bearing flat scales subtending the florets. Ray-florets in 1 series, usually neuter and sterile. Disk-florets perfect and fertile, tubular and 5-toothed; anthers entire at the base or minutely sagittulate; style-arms hairy and appendaged at the top. Achenes compressed or 4-angled, usually oblong or linear-oblong; pappus of 2–4 (rarely 6) awns usually upwardly- or downwardly-barbed, rarely smooth or lacking.

A cosmopolitan genus of about 230 species.

KEY TO SPECIES

1. Leaves bipinnate; outer phyllaries linear-subulate, soon reflexed, not ciliate; achenes mostly 4-awned; rays small, yellow: **1. B. cynapiifolia**

1. Leaves pinnate or simple; outer phyllaries spatulate, ciliate; achenes mostly 2–3-awned; rays conspicuous, white (sometimes absent): **2. B. alba** var. **radiata**

1. Bidens cynapiifolia Kunth in H.B.K., Nov. Gen. & Sp. 4:235. 1820.

Bidens bipinnata of Hitchcock, 1893, not L., 1753.

"Spanish Needle"

An erect or somewhat diffuse herb to 1 m tall or more, the spreading branches glabrate and striate-angled; leaves pinnate to bipinnate, the leaflets thin, ovate or lanceolate, the terminal one the largest and up to 5 cm long, caudate-acuminate at the apex, the margins serrate. Peduncles mostly 3–6 cm long; ray-florets 4 or 5. Achenes narrowed at the apex; awns 2–2.5 mm long, retrorsely barbed.

GRAND CAYMAN: *Hitchcock; Kings GC 362; Proctor 15074.* CAYMAN BRAC: *Proctor 29348.*
– West Indies and continental tropical America; a common weed of roadsides, fields, and open waste ground.

2. Bidens alba (L.) DC., Prodr. 5:605. 1836.
var. **radiata** (Sch.Bip.) Ballard & Melchert in Phytologia 32:291–298. 1975.

Bidens pilosa L. var. *radiata* of most recent authors; Adams, 1972.

"Spanish Needle"

Erect herb to 1 m tall (usually less), the young stems angled and puberulous. Leaves simple, or pinnate with 1-3 pairs of lateral leaflets; leaflets ovate or lanceolate, the terminal one up to 7 cm long or more (but often much smaller), acuminate at the apex, the margins coarsely serrate. Peduncles mostly 1-4.5 cm long; ray-flowers usually 5 or 6, the ligules 8-nerved; disk-florets light yellow. Achenes black, thinly hispidulous; awns up to 3 mm long, retrorsely barbed.

GRAND CAYMAN: *Hitchcock; Kings GC 73; Lewis GC 50; Millspaugh 1276; Proctor 15073; Sachet 492.* LITTLE CAYMAN: *Kings LC 60.*
– A pantropical weed of roadsides, fields, and waste places.

[**Cosmos caudatus** Kunth, an herb with dissected leaves and showy pink-rayed heads, is grown in gardens and may occasionally escape. (GRAND CAYMAN: *Kings GC418; Millspaugh 1353*). *Cosmos* differs from *Bidens* in its beaked achenes.]

Genus 22. **TRIDAX** L.

Perennial herbs with decumbent stems; leaves opposite and toothed or incised. Heads radiate, solitary on long terminal or axillary peduncles; involucre ovoid or hemispheric, its bracts subequal in 2-3 series; receptacle flat or convex, bearing chaffy scales subtending the disk-florets. Ray-florets pistillate, with 3-lobed or -toothed ligules. Disk-florets perfect and fertile, tubular and 5-lobed; anthers auricled or sagittate at the base; style-arms subulate-appendaged. Achenes compressed, silky-hairy; pappus of many plumose bristles or scales.

A tropical American genus of more than 25 species.

1. **Tridax procumbens** L., Sp. P. 2:900. 1753.

Stems decumbent-ascending, hairy; leaves lanceolate to ovate, mostly 2.5-6.5 cm long, acute to acuminate, coarsely serrate or toothed, rough-hairy. Peduncles up to 20 cm long; heads 7-12 mm in diameter, the phyllaries pubescent. Ray-florets 3-6, the ligules cream or pale yellow and 2.5-5 mm long; disk-florets light yellow. Achenes of the ray-florets c. 2.5 mm long with pappus 3 mm long; of disk-florets 2 mm long with pappus 6 mm long.

GRAND CAYMAN: *Brunt 1896a, 2015; Proctor 27942.* CAYMAN BRAC: *Kings CB 64; Proctor 15313.*
– Florida, West Indies, and continental tropical America, a weed of roadsides, pastures, and open waste places. Introduced in Africa and elsewhere in the Old World tropics.

TRIBE VI. HELENIEAE

Genus 23. **POROPHYLLUM** Guett.

Glabrous, glandular herbs; leaves opposite and alternate, entire or crenate, often glaucous. Heads discoid, solitary or in open corymbs on terminal peduncles; involucres subcylindric or narrowly bell-shaped, of few bracts in 1 series; receptacle naked. Florets with slender, elongate tube and a narrowly bell-shaped, 5-lobed limb; anthers obtuse at the base; style-arms linear, hairy at the top. Achenes linear, striate; pappus of rough or barbed bristles in 1 or 2 series.

A tropical to warm-temperate American genus of about 50 species.

1. **Porophyllum ruderale** (Jacq.) Cass. in Dict. Sci. Nat. 43:56. 1826.

"Stinking Bush"

A bitterly aromatic, erect, glaucous herb up to 70 cm tall or more; leaves long-petiolate, the blades thin and broadly elliptic to rotund, 1–3 cm long or more. Peduncles 2–4 cm long, swollen at the top; phyllaries 5, 17–20 mm long and 2-3.5 mm broad; florets with slender tube 12.5 mm long, green, the somewhat deflexed limb purplish-red. Achenes 8–8.5 mm long, hispidulous; pappus of numerous upwardly-barbed bristles c. 10 mm long.

GRAND CAYMAN: *Brunt 1627; Proctor 15305.*
– West Indies and continental tropical America; an occasional weed of sandy roadsides and open waste ground.

Genus 24. **PECTIS** L.

Annual or perennial glandular herbs, usually aromatic; leaves opposite, usually linear and often with stiff marginal hairs (cilia) below the middle. Heads radiate, small, solitary or in corymbs, sessile or on axillary or terminal peduncles; involucre cylindric or bell-shaped, the bracts glandular and few in 1 series; receptacle naked. Ray-florets pistillate and fertile, with entire or 3-toothed ligules. Disk-florets perfect and fertile, tubular with enlarged, regular or irregular, 5-lobed limb; anthers obtuse at the base; style-arms short and obtuse. Achenes linear, striate; pappus of few to many bristles or scales.

A tropical to temperate American genus of about 70 species.

KEY TO SPECIES

1. Plants erect, not or scarcely aromatic; leaves up to 7 cm long with scattered small glands beneath; pappus of 2-4 spreading, stiff, glabrous spines: **1. P. linifolia**

1. Plants prostrate, forming mats, strongly aromatic; leaves less than 1.5 cm long, the relatively large glands mostly in 2 rows beneath; pappus of erect setulose bristles: **2. P. caymanensis**

1. Pectis linifolia L., Syst. Nat. ed. 10, 2:1221. 1759.

Erect, diffusely-branched, annual, glabrous herb up to 70 cm tall or more; leaves lance-linear, 2-7 cm long, attenuate at both ends, and with 1-2 pairs of basal bristles. Heads 6-9-flowered on filiform, bracteolate peduncles 1-2.5 cm long; phyllaries purplish, 4.5-5.5 mm long. Ligules yellow with dark veins, c. 1 mm long. Achenes 4.5 mm long, hispidulous toward the top.

CAYMAN BRAC: *Proctor 29036, 29354.*
– West Indies, southwestern United States to northern South America, and the Galapagos Islands. The Cayman Brac plants were found scattered in rocky woodlands, not common.

2. Pectis caymanensis (Urb.) Rydb. in N. Amer. Fl. 34:204. 1916.

Pectis cubensis of Hitchcock, 1893, not Griseb., 1866.

P. cubensis var. *caymanensis* Urb., 1907.

"Tea Banker", "Mint"

Matlike perennial herb, subwoody at the base and with a woody taproot, the stems often pinkish; leaves oblong-linear or very narrowly lanceolate, 4-12 mm long, minutely scabrid toward the apex and sharply mucronate, and with 4-6 pairs of long cilia near the base. Peduncles mostly 5-10 mm long; ligules yellow, more or less longitudinally nerved. Achenes dark brown, minutely striate.

Occurs in two varieties, which can be distinguished as follows:

1. Stems glabrous, seldom more than 12 cm long; phyllaries ciliolate, c. 3 mm long; ligules c. 3 mm long; achenes 2-2.5 mm long, strigose with reddish hairs: **2a.** var. **caymanensis**

1. Stems sparingly hispidulous in lines, up to 25 cm long or more; phyllaries glabrous, c. 6 mm long; ligules c. 5 mm long; achenes 3-3.2 mm long, glabrous or minutely white-strigillose toward the base: **2b.** var. **robusta**

2a. Pectis caymanensis var. **caymanensis. FIG. 255.**

GRAND CAYMAN: *Brunt 1761; Correll & Correll 51054; Hitchcock; Howard & Wagenknecht 15026; Kings GC 58; Millspaugh 1279 (type); Proctor 11993.* LITTLE CAYMAN: *Proctor 28186, 35193.* CAYMAN BRAC: *Proctor 29080, 35155.*

– Cuba; occurs in sandy clearings or soil-filled pockets of exposed limestone. Frequently used to make a pleasantly aromatic tea.

2b. Pectis caymanensis var. **robusta** Proctor in Sloanea 1:4. 1977.

GRAND CAYMAN: *Proctor 31023 (type); Sachet 386.*

– Endemic; found growing in gravelly sand near the sea. This variety is generally larger and coarser in appearance than var. *caymanensis.*

[**Tagetes erecta** L., the "African Marigold", and **T. patula** L., the "French Marigold", are grown in gardens. In spite of their common names, both are natives of Mexico.]

TRIBE VII. SENECIONEAE

Genus 25. **ERECHTITES** Raf.

Annual or perennial herbs; leaves alternate, simple and entire to pinnatifid. Heads discoid but heterogamous, the outer florets (of 2 or more series) pistillate, the inner ones perfect, all fertile. Involucre cylindric, of numerous narrow bracts in 1 series, but often with a few very small accessory ones (bracteoles) outside; receptacle flat, naked. Outer florets filiform. Inner florets tubular and 5-toothed; anthers obtuse at base; style-arms of the perfect florets truncate. Achenes oblong; pappus of numerous long, silky bristles.

A genus of about 15 species, chiefly occurring in North America, Australia, and New Zealand.

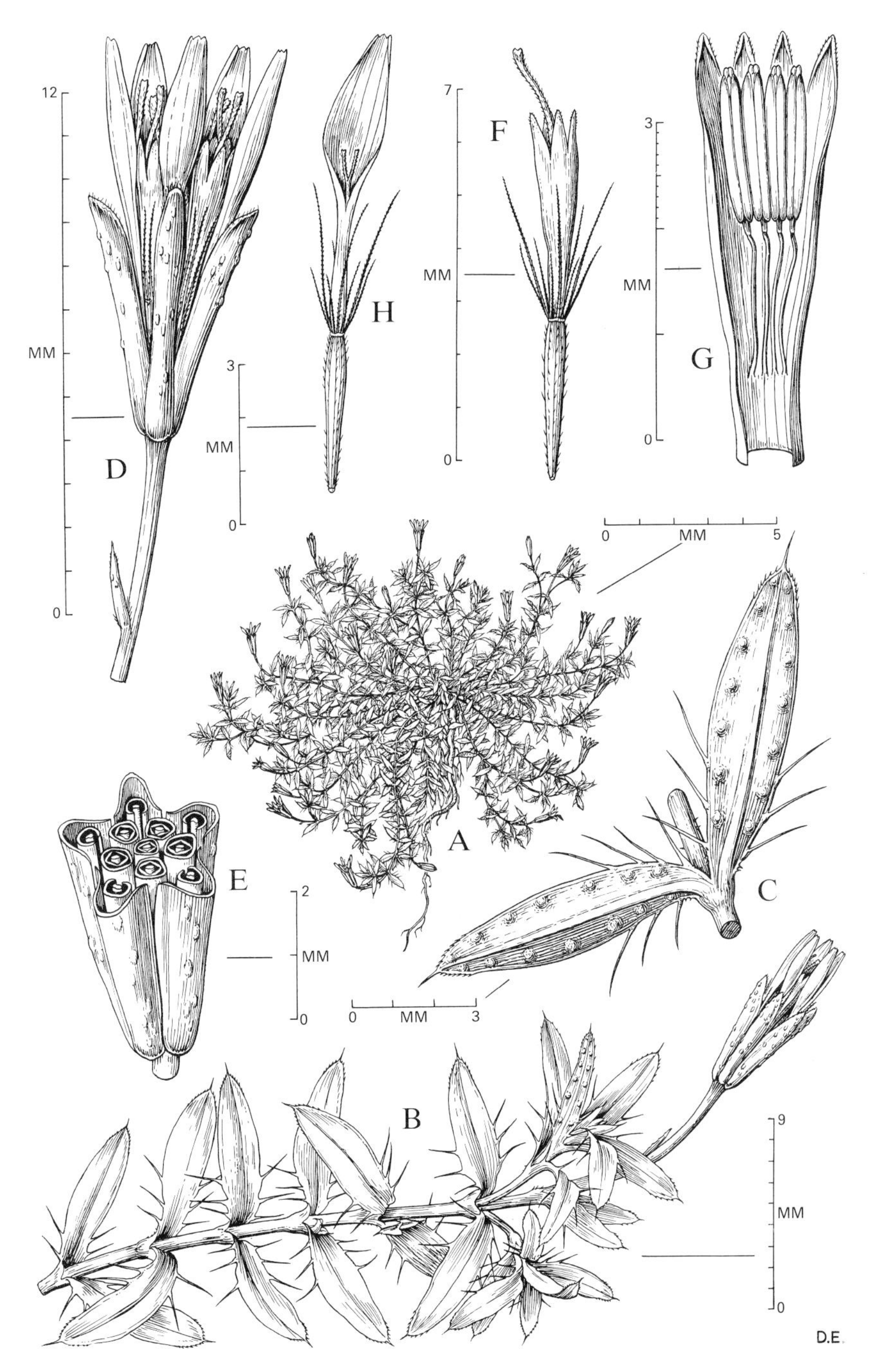

FIG. 255 **Pectis caymanensis** var. **caymanensis.** A, habit. B, flowering branch. C, pair of leaves. D, head. E, section through head. F, disk-floret with achene. G, corolla of disk-floret cut open to show stamens. H, ray-floret with achene. (D.E.)

1. Erechtites hieracifolia (L.) Raf. ex DC., Prodr. 6:294. 1838. **FIG. 256.**

Erect, glabrate or somewhat hairy herb up to 1 m tall; leaves sessile, linear-lanceolate to narrowly oblong, 5–17 cm long, often auricled at the base, and with

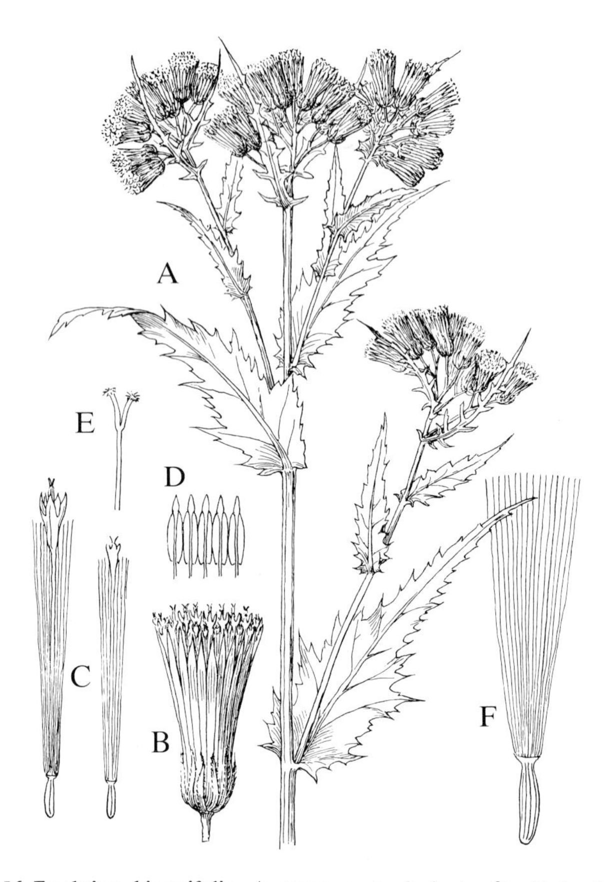

FIG. 256 **Erechtites hieracifolia.** A, upper part of plant, x$^2/_3$. B, head, x2¼. C, perfect (left) and pistillate florets, x4. D, anthers, x10. E, style-arms, x10. F, achene with pappus, x4. (F. & R.)

toothed or coarsely serrate-lobed margins. Heads clustered or in open corymbs; phyllaries 9–12 mm long. Achenes ribbed, slightly hairy, c.3 mm long; pappus of white, soft, smooth hairs 10–12 mm long.

GRAND CAYMAN: *Kings GC 72.*

– North America, Greater Antilles, northern South America, Hawaii, and central Europe; a weed of roadside banks and pastures. The distribution given is that of var. *hieracifolia;* other varieties occur in various regions but have not been found in the Cayman Islands.

[Genus 26. **EMILIA** Cass.

Annual or perennial herbs; leaves alternate and entire; toothed or pinnatifid. Heads discoid and with all florets perfect, on long peduncles, solitary or few in a lax corymb; involucre cylindric or narrowly campanulate, the narrow phyllaries in 1 series with no bracteoles outside; receptacle flat, naked, tuberculate. Florets tubular with an elongate, cylindric, 5-lobed limb; anthers obtuse at the base; style-arms with a short appendage. Achenes oblong, sub-terete or ribbed; pappus of numerous silky bristles.

A genus of about 30 species in the tropics of Asia and Africa, two being naturalized in the New World.

1. **Emilia fosbergii** Nicolson in Phytologia 32:34. 1975.

Emilia coccinea of many authors, not (Sims) Sweet, 1839.

E. javanica of Adams, 1972, not (Burm.f.) Robins., 1908.

A short-lived annual herb up to 60 cm tall or more (but often much less), nearly glabrous or thinly pubescent. Leaves (except the lowest) sessile and clasping, oblong, oblanceolate or obovate, 4–8 cm long, acute, and with irregularly-toothed or serrate margins. Heads few in loose corymbs; involucre 10–14 mm long; florets crimson or scarlet, 11–14.5 mm long, mostly numbering 40–65 in a head. Achenes 3.5–4 mm long, hispidulous on the angles; pappus of soft white hairs c. 7 mm long.

GRAND CAYMAN: *Brunt 2117; Kings GC 205.*
– Now pantropical, a common weed of fields and waste places.]

GLOSSARY AND INDEX

GLOSSARY

Abaxial On the side (of an organ) away from the axis, as the lower surface of a leaf.

Abortive Imperfectly developed, not fully developed at maturity, as abortive stamens with filaments only.

Abruptly pinnate Said of a compound leaf without a terminal leaflet.

Acaulescent Stemless; having the stem very short and often underground, thus appearing stemless; opposite of *caulescent.*

Achene A small, dry, indehiscent, 1-seeded fruit with a thin tight pericarp, as the individual fruit of a composite flower or of the Amaranthaceae.

Acicular Needle-shaped.

Acid Characterized by an excess of free hydrogen ions; opposite of *alkaline.*

Acidulous Slightly acid.

Acroscopic Facing or directed toward apex, e.g., of a fern-frond.

Acuminate Tapering to an apex, the sides concave along the taper.

Acute Tapering to a sharp apex, the sides straight along the taper.

Adaxial On the side (of an organ) toward the axis, as the upper surface of a leaf.

Adnate United with another (unlike) part, as the stamens to the corolla.

Adpressed Same as *appressed.*

Adventitious Said of plants recently introduced to an area where they are not indigenous; organs arising from abnormal positions, as buds near a wound or roots from a stem or leaf.

Aerial Living above the surface of the ground or water.

Aggregate fruit A cluster of ripened ovaries traceable to separate pistils of the same flower and inserted on a common receptacle.

Albumen An obsolete name for endosperm.

Alkaline Having the ability to neutralize acids; having a pH of more than 7; opposite of *acid.*

Alternate Any arrangement of leaves or other parts not opposite or whorled; placed singly at different heights along an axis or stem.

Alternation of generations 1. The growth of reproductive bodies into structures which differ from those from which they are reproduced. 2. The occurrence in one life-cycle of two or more modes of reproduction which are differently produced and which differ morphologically. Usually the sexual forms (gametophytes) and the asexual forms (sporophytes) alternate, the sporophytes containing double the number of chromosomes found in the gametophytic generation.

Anatropous Applied to ovules or seeds which grow in an inverted position; having an ovule bent over in growth so that the micropyle is near the base of the funicle, with the body of the ovule united with the funicle.

Angiosperm A plant with seeds enclosed in an ovary or pericarp.

Angulate More or less angular (not curved or rounded), having angles or corners usually of a determinate number.

Annual A plant of one year's (or one growing-season's) duration, completing its life-cycle in that period.

Annular In the form of a ring; arranged in a circle.

Annulus A rim of thickened cells around a fern sporangium, functioning as an elastic mechanism in shedding spores. (Plural: *annuli*)

Anther The pollen-producing part of a stamen.

Antherozoid One of the motile male cells produced in an antheridium; the sperm cell of a fern.

Anthesis The process or time of opening of a flower-bud and the expansion of the flower-parts.

Anthocarp A fruit formed by the fusion of part or all of the floral parts with the fruit itself.

Apex The tip of an organ, the extreme end or point farthest from the point of attachment; the growing point of a stem or root. (Plural: *apices*)

Apical Pertaining to the apex.

Apiculate Terminating in a short, sharp, somewhat flexible point, but not a spine.

Apogamous Developed without fertilization; development of an embryo without the fusion of gametes.
Appendage An attached subsidiary or secondary part, as hairs, prickles, or leaves of a stem.
Appressed Flattened against underlying or adjacent tissues; pressed down or against.
Approximate Close together but not united, as leaves along a stem; opposite of *distant*.
Aquatic Living in water; growing naturally in water or under water.
Arboreous Treelike or pertaining to trees.
Arborescent Attaining the size or character of a tree.
Archegone The egg cell of a fern, produced in an archegonium.
Arcuate Curved; bent in an arc.
Areolate Marked out into small spaces enclosed by anastomosing veins; reticulate.
Areole A space surrounded by anastomosing veins; a pit, spot or small raised area bearing hairs, glochids, or spines (or all three), as in many cacti.
Arid Dry, having little rainfall.
Aril An extra covering of part or all of a seed, which is an outgrowth from the hilum, and which may be more or less soft and fleshy, or else dry and bony.
Arillate Bearing an aril.
Aristate Bearing a terminal bristle or awn.
Articulate Jointed.
Ascending Directed obliquely upward, but not truly erect.
Asymmetric Not equally bilateral.
Attenuate Tapering gradually and narrowly.
Auricle An ear-shaped appendage or part, as earlike lobes at the base of leaves or small lobes at the summit of the sheath in many grasses.
Auriculate Having an earlike lobe or appendage.
Awl-shaped Narrow and tapering to a point; sharp-pointed from a broader base; subulate.
Awn A bristle-like appendage, especially on the glumes of grasses.
Axil The angle between a leaf-petiole and the stem on the upper (acroscopic) side; any angle between an organ and its axis.
Axile placentation A condition in which the ovules are borne on a central axis in the ovary.
Axillary Pertaining to an axil, or located in or attached in an axil.
Axis The main or central line of development of a plant or part of a plant; the main stem. (Plural: *axes*).

Barbed Beset with rigid points or short bristles, these usually reflexed; pertaining to a hair or other straight process which is armed with one or more teeth that point backward.
Basiscopic Facing or directed toward the base, e.g., of a fern-frond.
Bearded Bearing or furnished with long or stiff hairs.
Berry Any simple fruit having a pulpy or fleshy pericarp, usually with several or many seeds embedded in the pulp.
Bi- In combination, 2, twice, or doubled, as *biauriculate* (having two earlike appendages), *bicrenate* (having two rounded teeth), or *bilobed* (having 2 lobes).
Biennial A plant which completes its life-cycle in two growing seasons, usually fruiting during the second season.
Bifid Forked at least halfway to the base.
Bifurcate Forked at the tip.
Bilamellate Having two flat lobes or divisions.
Bilateral Having two sides which are equal.
Binomial nomenclature The system of applying two Latin names, a generic name and a specific name, to designate kinds of plants or animals.
Bipinnate Twice pinnately compound; doubly pinnate.
Bipinnatifid Twice pinnatifid, with the divisions of a pinnatifid leaf themselves pinnatifid.
Bipinnatiform Pertaining to a leaflike structure having bipinnatifid architecture.

Biseriate In two rows or series.
Bisexual Having both stamens and pistils in the same flower or inflorescence.
Biternate Pertaining to a leaf with three main divisions each having three leaflets.
Bract A reduced, scale-like leaf subtending a flower or branch of an inflorescence; a leaf or scale in whose axil an inflorescence, flower, or floral organ is produced. Sometimes applied to any leaf subtending a flower.
Bracteole A small bract or bractlet; a secondary bract, i.e., borne on a petiole, pedicel, or other secondary axis.
Bristle A rigid hair.
Bud An unopened flower; an undeveloped shoot or stem; an undeveloped axis covered with the rudiments of leaves.
Bulb A large bud, usually subterranean, made up of a short, thick stem with roots at the base and having a number of membranous or fleshy overlapping scale-like leaves.
Bulbil A deciduous bud usually formed on an aerial part of a plant, as on old inflorescences of *Agave* species. The term may also be applied to small subterranean offsets of bulbs.
Bulblet A little bulb, esp. the small deciduous buds formed in the axils of leaves, on the fronds of certain ferns, and in the inflorescence of a few grasses.
Bulbous Bulb-shaped.
Bullate Blistered or puckered; pertaining to a leaf with inflated convexities on the upper surface and corresponding cavities beneath; swollen or bubble-like.
Bur A fruit, seed, or head bearing hooked or barbed appendages.
Buttress A knee-like or plank-like outgrowth developing from the trunk-base of some kinds of trees.

Caespitose Growing or aggregated in tufts or small clumps.
Callose Hard and thick in texture.
Callosity A thickened or raised area; an area of tissue firmer than the surrounding tissue.
Callus A hard protuberance; in grasses, the swelling at the point of insertion of the lemma or palea; a roll of new tissue developed around a wound.
Calyx The outer whorl of the perianth or floral envelope, usually green in colour and composed of sepals; the sepals considered collectively. (Plural: *calyces*)
Calyx lobe One of the free, projecting parts of a united calyx.
Calyx tube The tube or cup of a calyx in which the sepals are united.
Cambium The thin layer of formative tissue beneath the bark of gymnospermous and dicotyledonous trees and shrubs, from which new wood and bark originate; a sheath of generative tissue usually located between the xylem and the phloem; the tissue from which secondary growth arises in stems and roots.
Campanulate Bell-shaped.
Canescent Covered with grayish downy pubescence.
Capitate Borne in heads; head-shaped.
Capitulate Borne in very small heads.
Capitulum A dense inflorescence consisting of sessile flowers.
Capsule A dry, dehiscent fruit resulting from the maturation of a compound (multi-carpellate) ovary.
Carpel A simple pistil or one unit of a compound pistil. A carpel usually consists of three parts: the *ovary* (a swollen basal portion containing the ovules, which after fertilization develop into seeds), the *style* (a usually narrow prolongation of the apex of the ovary), and the *stigma* (the specialized tip of the style on which pollen lodges and germinates).
Carpellate Pertaining to, or bearing carpels.
Cartilaginous Tough and firm but not bony, like gristle.
Caruncle A protuberance or appendage at or near the hilum of a seed.
Caryopsis The achene of a grass, differing from the achene of Compositae in being derived from a superior ovary.

Cataphyll Any rudimentary scale-like leaf which precedes the foliage-leaves, or (in *Phyllanthus*) subtends a deciduous branchlet; a bud-scale or rhizome-scale.
Catkin A pendulous, scaly spike with many simple, usually unisexual flowers; it usually falls as a unit after the pollen or seeds have been shed.
Caudate Bearing a tail-like appendage.
Caudex The stem or trunk of a tree-fern; the perennial base of an otherwise herbaceous, short-lived plant; the main axis of any woody plant. (Plural: *caudices*)
Caulescent Having an obvious main stem above the ground.
Cauliflorous Bearing flowers on old stems; flowering on the trunk of a woody plant, or on specialized spurs from it, or on the larger branches.
Cauline Pertaining to a stem or axis (as contrasted with *basal*).
Cell A microscopic unit of protoplasm, surrounded by a membrane, and in plants usually by a more or less rigid cell wall, the fundamental unit of structure and function of all living organisms. The term is also applied to the pollen-bearing cavity of an anther and to the ovule-bearing cavity of an ovary.
Cellular Composed of cells; containing small, enclosed spaces of similar shape and size.
Chartaceous Papery or firmly tissue-like in texture; having the texture of writing-paper.
Chlorophyll The green colouring matter in the cells of plants, which enables them to synthesize carbohydrates from carbon dioxide and water. It is a mixture of two green and two yellow pigments (chlorophyll a, chlorophyll b, carotin, and xanthophyll).
Chromosome One of the minute, rod-like bodies, usually definite in number in the cells of a given species, into which the chromatin (deeply staining material) of a cell nucleus becomes condensed during the process of cell division. The chromosomes carry the genes (units of heredity) in linear sequence, and by splitting lengthwise provide the mechanical basis for equal allocation of hereditary material to the daughter cells.
Cilia Minute protoplasmic threads acting as organs of motion which propel gametes and many unicellular organisms from place to place. The term is also applied to marginal hairs on leaves and other parts of plants.
Ciliate Fringed with hairs on the margins.
Ciliolate Fringed with minute hairs; minutely ciliate.
Cinereous Ash-coloured; light gray.
Circumscissile Opening or dehiscing by a transverse line around a fruit or an anther, the apical part usually being shed like a lid.
Clathrate Like a lattice; net-like.
Clavate Club-shaped, slender at the base and gradually thickening upward.
Claw The slender, tapered base or stalk-like part of a petal or sepal.
Cleft Divided into lobes separated by narrow or acute sinuses which extend more than half-way to the midrib.
Clinandrium In orchids, a cavity in the column between the anther sacs which often contains the stigmatic surface. (Plural: *clinandria*)
Clonal Pertaining to a clone.
Clone The vegetatively produced progeny of a single individual.
Coalescent United by growth, the fusing or organic cohesion of similar parts or organs.
Coccus A 1-seeded carpel into which compound fruits split when ripe. The term is also used for a bacterium having spherical or nearly spherical form, and for the spore mother cell of certain hepatics. (Plural: *cocci*)
Coherent Having two or more similar parts touching but not fused.
Colony A collection of organisms of the same kind growing together in close association.
Columella A small column or central axis; the persistent sterile central axis in some fruits, around which the carpels are arranged.
Column The structure formed by the fusion of stamen, style and stigma in the flowers of orchids; the lower undivided part of the awns of certain species of *Aristida*; the axis or central pillar of a capsule; the structure formed by the fusion of stamen filaments in the flowers of Malvaceae.
Columnar Shaped like a pillar.

Compound Consisting of 2 or more similar parts, as a leaf divided into leaflets, or a fruit or pistil made up to several carpels.

Concolorous Uniform in colour, having 1 colour only.

Cone An inflorescence or fruit covered with overlapping scales.

Confluent Blended together, as in ferns, those sori which spread so that they join those adjacent to them, or in flowers, those with anther-lobes which are united at the summit of the filament and diverge from that point.

Congeneric Belonging to the same genus.

Congeners Two or more species belonging to the same genus.

Conifers A general term used for cone-bearing trees; a tree or shrub belonging to the Coniferae.

Conjugate Coupled, as a pinnate leaf with two leaflets; joined or arranged in pairs.

Connate United with like structures during the process of formation; joined together, as confluent filaments or opposite leaves which are united at the base.

Connective The part of a filament which serves to connect the anthers; the stalk connecting the separated lobes of an anther.

Connivent Coming together or converging, but not organically connected.

Conspecific Belonging to the same species.

Contorted Twisted together; convoluted.

Convoluted Same as contorted.

Cordate Heart-shaped, with the notch at the base.

Coriaceous Of leathery texture.

Corm A solid, fleshy, underground base of a stem, usually bulb-like in shape, covered with thin membranes.

Corolla A collective term for the petals of a flower, whether separate or fused. The corolla usually more or less surrounds the stamens and pistil, and is subtended or encircled by the calyx.

Corona A more or less petaloid structure which develops between the corolla and the stamens, as in Asclepiadaceae.

Corpuscle A small body connecting the pollen masses in Asclepiadaceae, by means of which the pollen becomes attached to insects to accomplish cross-pollination. The term is also used for any minute body within a cell that has definite form and function.

Corymb A more or less flat-topped indeterminate flower-cluster in which the outer flowers have the longest pedicels and open first.

Corymbose With flowers arranged in corymbs.

Costa A ridge or midrib of a frond or leaf.

Costules The midribs of the subdivisions (segments or pinnules) of a fern frond, or of other compound leaves.

Cotyledon A seed leaf; a leaf-like organ, developed within the seed and in which nourishment for the young plant is usually stored. The number of cotyledons produced may be 1 (monocotyledonous plants), 2 (dicotyledonous plants), or several (many gymnosperms).

Crateriform Cup-shaped; crater-like.

Crenate Pertaining to leaf-margins with broad rounded teeth, these separated by narrow sinuses.

Crenulate Minutely crenate.

Cristate Bearing an elevated appendage resembling a crest; bearing elevated and toothed ridges. The term is often applied to leaves whose apices are abnormally forked, expanded, and twisted.

Crownshaft The smooth green sheath, formed by the connate expanded petioles of certain species of palm (as in *Roystonea*), which encloses the terminal bud. Palms with a crownshaft produce their inflorescences below the leaves.

Crustaceous Crustlike, or forming a brittle crust.

Cryptogams Plants which reproduce by methods other than by flowers and seeds.

Culm The stem of a grass or sedge, in grasses usually hollow except at the nodes. The term is not applied to the rhizome or other underground stems.

Cuneate Wedge-shaped, with the narrow point at the base of the leaf-blade, petal, or other structure.

Cupule A cup-shaped involucre of bracts adherent at least by their bases.

Cuspidate Tapering abruptly to a sharp point.

Cyathium An inflorescence reduced to look like a single flower; in the Euphorbiaceae, a cup-shaped involucre containing individually free male (staminate) flowers and usually a single female flower (gynoecium). (Plural: *cyathia*)

Cyme A flat-topped, usually few-flowered inflorescence in which the central terminal flowers open in advance of the outer ones. A *helicoid cyme* has the lateral branches always on the same side. A *scorpioid cyme* has the lateral branches occurring alternately on opposite sides.

Cymose Bearing or flowering in cymes.

Cymule A reduced or very small cyme; a cymose cluster.

Cystolith A small concretion of calcium carbonate (lime), usually acicular or elliptic in shape, occurring in the epidermal cells of certain plants, and usually visible as a minute raised whitish line.

Cytological Pertaining to *cytology*, the study of the internal structure, function, and life-history of cells.

Deciduous Falling off at certain seasons or stages of growth, such as leaves, petals, sepals, flowers, etc. This term is commonly applied to trees which are not evergreen.

Declinate Bent downward or forward, as the stamens in many flowers; directed downward from the base.

Decompound Pertaining to a leaf that is twice or more compound; repeatedly divided.

Decumbent Reclining at the base, but curved or bent upward toward the apex, said of stems lying on the ground but rising at the tip.

Decurrent Said of a leaf whose margins extend downward as ridges or narrow wings along the stem; any organ extending along the side of another.

Decussate A term applied to leaves or branches arranged in pairs, with each successive pair at right angles to the next pair, or at least growing out at a different angle.

Deflexed Bent sharply downward or outward.

Defoliated Pertaining to a node, stem, or branch from which the leaves have fallen.

Dehiscent Pertaining to an anther or fruit which opens naturally by valves, slits or pores.

Deltate Triangular; term used in preference to "deltoid" for flat structures such as leaves. (See also *ovate* versus *ovoid*, etc.)

Dentate Having a toothed margin, the teeth usually rather coarse and pointing outward.

Denticulate Finely-toothed, or minutely dentate.

Depauperate Reduced, undeveloped, impoverished, or dwarfed.

Dibrachiate Branched twice with widely spreading arms.

Dichasium A cyme with 2 lateral axes; a determinate inflorescence in which the first flower to open lies between 2 lateral flowers, the pedicels of the lateral flowers often elongating to form a false dichotomy, the central flower remaining sessile. (Plural: *dichasia*)

Dichasial branching Pertains to a compound dichasium in which the lateral flowers are replaced by branchlets that either terminate in secondary dichasia, or else in turn (rarely) give rise to tertiary dichasia.

Dichotomous Successively branched into more or less equal pairs; branching by repeatedly forking in pairs.

Dicot An abbreviation for "dicotyledonous plant".

Dicotyledon A plant with two seed-leaves or cotyledons.

Diffuse Loosely branching or spreading; widely or loosely spreading.

Digitate Having parts which diverge from the same point like the fingers of a hand; palmately divided.

Dimerous Having each whorl of two parts.

Dimidiate Unequally divided or lop-sided, as a leaf or leaflet with the principal vein ("mid-vein") much nearer to one margin than the other.
Dimorphic Occurring in two forms.
Dioecious Having staminate and pistillate flowers on different plants; the opposite of *monoecious.*
Diploid Having the number of chromosomes normally occurring in the somatic or vegetative cells of a given species, which is twice the number found in the gametes or the haploid generation of the same species. In many groups of plants the basic diploid chromosome number may be multiplied through hybridization or other factors, producing *polyploid* plants of various degrees (*triploids, tetraploids,* etc.).
Disc (disk) 1. Any flat, round growth. 2. A more or less fleshy torus developing from the receptacle within the calyx, or within the points of attachment of the corolla and stamens, and surrounding the base of the pistil; this structure is sometimes interpreted as representing coalesced nectaries or staminodes. 3. The flattened or conical receptacle in the heads of Compositae. 4. The flattened, clinging tip of some tendrils. 5. The base of a pollinium. 6. The expanded base of the style in Umbelliferae. 7. In a bulb, the solid base of the stem around which the scales are arranged.
Disciform (discoid) Shaped like a disc; round and flat.
Disk Alternative spelling for *disc.*
Disk-floret One of the tubular flowers making up the central part of the heads of Compositae, as distinguished from the *ray-florets* which occur around the margin of the heads in many species.
Dispersal The act of dispersing or scattering.
Dissected Deeply divided, or cut into many segments.
Distal Remote from the point of attachment; farthest from the axis; toward the apex.
Distant Widely scattered, said of leaves not closely attached or aggregated; remote; the opposite of *approximate.*
Distichous In two vertical ranks along an axis, producing leaves or flowers in two opposite rows.
Diurnal Functioning during the day; said of flowers that open by day and close at night; the opposite of *nocturnal.*
Divergent Said of two organs or structures that incline away from each other from the same point of attachment.
Domatium A small cavity or other form of structure on a plant, in which live minute insects or mites, apparently in symbiosis with the plant. (Plural: *domatia*).
Dormant Said of organisms or parts of organisms that are not actively functioning, as for example buds which do not develop unless stimulated to growth by special conditions.
Dorsal Pertaining to the outer or back surface of an organ; the surface away from the axis. See also *abaxial.*
Dorsiventral With a distinct upper and lower surface.
Drupe A fleshy, 1-seeded, usually indehiscent fruit; a stone-fruit.
Drupelet A little drupe.

Ecological Pertaining to the relation of organisms to their environment.
Ecologist A student of ecology.
Ecology The study of organisms in relation to their environment.
Egg-cell The female gamete.
Ellipsoid A solid or 3-dimensional body which is elliptic in section.
Elliptic A figure about twice as long as wide, widest in the middle, and rounded at both ends.
Emarginate With a shallow notch at the apex.
Embryo The rudimentary plant within a seed.
Endemic Occurring in one limited locality or region only; confined to a particular area and found nowhere else.

Endocarp The inner layer of a pericarp or fruit-wall.

Endosperm A multicellular, usually starchy or oily, nutritive tissue formed inside the seeds of many flowering plants, separate from the embryo; in gymnosperms, the term is applied to the prothallus of the female gametophyte.

Entire Pertaining to leaf-margins without teeth or marginal divisions; having a continuous, even margin.

Epicalyx A secondary calyx, as a group of bracts forming a calyx-like structure beneath the true calyx; an involucre formed of bracts.

Epicarp The outer layer of a pericarp or fruit-wall; sometimes called exocarp.

Epidermis The outermost layer of cells of a plant leaf or stem.

Epigynous Having the calyx, corolla, and stamens growing from the top of an inferior ovary.

Epipetric Growing on rocks: *epipetrous.*

Epipetrous Alternative spelling of *epipetric.*

Epiphytic Growing on other plants, but not parasitically.

Equitant Folded over as if astride; said of leaves that are folded together lengthwise in two ranks.

Erose With an irregular, uneven margin; gnawed or jagged.

Evolution The natural process through which organisms have acquired their characteristic structure and functions; the orderly development of species from pre-existing ones through differential natural selection of random variations.

Excentric One-sided.

Excurrent Directed outward. The term is also used to denote a leaf-vein that projects beyond the margin, or an axis that remains central and undivided, the other parts being regularly disposed around it.

Exine The integral wall of a spore, which may or may not be surrounded by a *perispore.*

Exocarp The outer layer of the pericarp or fruit-coat.

Exserted Projecting beyond, as stamens from a perianth; opposite of *included.*

Extrorse Facing or opening outward, away from the axis.

Falcate Sickle-shaped; curved like a sickle.

Farinaceous 1. Having a surface with a mealy coating. 2. Containing starch.

Farinose Covered with a mealy powder.

Fascicle A close cluster or bundle of flowers, leaves, stems, or roots.

Fasciculate In close clusters or bundles; arranged in fascicles.

Fermented Referring to organic substances transformed through the action of enzymes (catalysts produced by living cells), as the production of alcohol from sugar by yeasts.

Fertile Said of spore-bearing fern-fronds, pollen-bearing stamens, and seed-bearing fruits; opposite of *sterile.*

Fertilization The union of male and female reproductive cells.

Fiber (fibre) A thread, or thread-like structure; elongate, thick-walled cells many times longer than wide.

Fibrillose Covered with minute fibers; having a lined appearance as if composed of fine fibers.

Fibrous Composed or covered with tough, string-like fibers.

Filament The stalk of a stamen.

Filament-tube A tubular structure formed from coalesced filaments.

Filamentous Formed of fine fibers.

Filiform Thread-like; hair-like.

Fimbriate Having a fringed margin, the fringe composed of processes hairlike at the apex but more than one cell wide at the base.

Flabellate Fan-shaped; narrow at the base and much broader at the apex.

Flaccid Limp; lacking rigidity.

Flexuous Bent alternately in different directions; zigzag.

Floccose Bearing or covered with tufts of woolly hairs.

Flora 1. An aggregate term referring to all the plants occurring in a country or particular area. 2. A catalogue or descriptive account of the plants growing in a country or particular area.
Floral Pertaining to flowers or the parts of flowers.
Floret An individual flower of diminutive size, as in the grasses and Compositae.
Floristic Pertaining to a flora.
Flower A structure bearing stamens or pistils, or both, and usually having a calyx and corolla; the structure concerned with the production of seeds in the angiosperms.
Fluted Marked by alternating ridges and groovelike depressions.
Foetid Having an offensive smell.
Foliaceous Leaf-like; bearing leaves.
Foliage The leaves of a plant referred to collectively.
Foliolate Having leaflets; usually used as a suffix, with such prefixes as bi-, tri-, etc., or 2-, 3-, etc.
Follicle A dry, unilocular, capsular fruit that dehisces longitudinally by a *suture* (slit) on one side only, as in fruits of Asclepiadaceae.
Foveolate Marked with a regular uniform pattern like a honeycomb, or having numerous shallow pits in such a pattern; also said of a membrane having small perforations in such a pattern.
Free central placenta A placenta running through the center of a unilocular ovary and attached at the ends only.
Frond The leaf of a fern, which differs from a typical leaf in bearing reproductive organs (sporangia) on its surface or margins.
Fruit The structure which develops from the ovary of an angiosperm after fertilization, with or without additional structures formed from other parts of the flower.
Fungus A thallophyte lacking chlorophyll.
Funicle The stalk on which an ovule is borne.
Fusiform Spindle-shaped; swollen in the middle and narrowing gradually toward each end.

Galea A helmet-shaped petal, corolla-lip, calyx, or sepal.
Gamete A reproductive cell of either sex prior to fertilization.
Gametophyte The phase in the life-cycle of plants which bears the sex organs and gives rise to the gametes.
Gamopetalous Having the corolla-parts (petals) fused by their edges, at least at the base, to form a tube.
Gamosepalous Having a calyx composed of connate sepals.
Gelatinous Jelly-like.
Genera Plural of *genus.*
Generation The individuals of a species equally remote from a common ancestor.
Generic Pertaining to genera.
Genetic Pertaining to the inheritance of characters by means of genes.
Geniculate Bent abruptly like a knee.
Genus In classification, the principal subdivision of a family, a more or less closely related and definable group of plants comprising one or more species. The generic name is the first word of a binomial used to designate a particular kind of plant or animal. Large genera are frequently divided for convenience into subgenera, but in such cases the same generic name is still used for all the species. (Plural: *genera*)
Germination The beginning of growth from a spore or seed.
Gibbous Swollen on one side, often at the base.
Glabrate Nearly glabrous, bearing only a few scattered hairs.
Glabrescent Becoming glabrous with age.
Glabrous Lacking hairs, bristles, or scales.
Gland A small structure, prominence, pit, or appendage which usually secretes such substances as mucilage, oil, or resin. Similar non-secretory structures are often also called glands.

Glandular Pertaining to or having glands.
Glaucous Covered with an extremely fine whitish or bluish substance that is easily rubbed off.
Globose Spherical; globular.
Glochid A small barbed hair or bristle.
Glochidiate Beset with glochids; bearing glochids.
Glomerate Densely clustered; in a dense, compact cluster or head.
Glomerule A headlike cyme; a cyme of almost sessile (usually small) flowers condensed to form a head or capitate cluster.
Glume A chaff-like bract; specifically, one of the two empty bracts at the base of a grass spikelet.
Glutinous Covered with a sticky exudation.
Gonophore A stalk elevating the stamens and pistils; see also *gynophore.*
Granulate Finely roughened as if with grains of sand.
Granulose Same as granulate.
Gregarious Pertaining to a species whose individuals tend to occur together in groups.
Gymnosperm A plant whose seeds are not enclosed in an ovary or pericarp.
Gynobasic style A style which arises from the base of the ovary.
Gynophore A stalk or prolonged axis bearing the ovary at its apex.
Gynostegium The staminal crown in an *Asclepias* flower.

Habitat The kind of locality in which a plant grows.
Halophyte A plant that grows in a saline habitat or that tolerates salt.
Halophytic Growing in saline soil or in salt water.
Hastate Shaped like an arrowhead, but with the basal lobes diverging nearly at right angles to the midline of the axis; halberd-shaped.
Hastula The terminal (apical) part of the petiole of a palm leaf; also called a *ligule.*
Haustorium The food-absorbing sucker of a parasitic plant.
Head A dense cluster of sessile or nearly sessile flowers terminating an axis *(peduncle)*; a shortened spike reduced to a globular form; a *capitulum.*
Helicoid cyme A scorpioid inflorescence produced by the suppression of successive axes on the same side causing the sympodium to be spirally twisted.
Herb Any seed plant without a woody stem.
Herbaceous Not woody; dying at the end of a growing season.
Herbarium A collection of pressed, dried, identified plant specimens systematically arranged.
Heterogamous 1. Bearing two or more kinds of flowers (staminate, pistillate, hermaphroditic, and/or neutral) in a single inflorescence, as in the heads of Compositae. 2. Producing two kinds of gametes.
Heterosporous Producing two kinds of spores, *microspores* and *megaspores.*
Heterostylous Having styles of two or more distinct forms, or of different lengths.
Hilum The scar on a seed where the funicle was attached.
Hippocrepiform Horseshoe-shaped.
Hirsute Bearing long, rather coarse hairs.
Hispid Having bristly or stiff hairs.
Hispidulous Minutely hispid.
Holotype The particular single specimen cited by an author as the basis of a new species or other taxon. The identity of a holotype precisely fixes the application of a particular name.
Homogamous 1. Having all the flowers of an inflorescence alike, whether all staminate, pistillate, hermaphroditic, or neuter. 2. Having the anthers and stigmas mature at the same time.
Homosporous Producing spores that are all alike.
Humus Decaying organic matter in the soil; the mold or soil formed by the decomposition of vegetable matter.
Hyaline Translucent or transparent.

Hybrid The offspring of two different varieties, species, or genera; a cross between two different kinds of related plants.
Hygroscopic Readily absorbing moisture from the atmosphere; altering form, size, or position through changes in humidity.
Hypanthium The tubelike extension of a receptacle on which the calyx, corolla, and stamens are borne; a cuplike or shortly tubular structure formed by the fusion of the calyx-tube with the ovary wall, bearing the calyx-lobes, corolla, and stamens at the apex.
Hypogynium The swollen or perianth-like structure subtending the ovary in *Scleria* and some other Cyperaceae.
Hypogynous Having the calyx, corolla, and stamens at the base or below the free superior ovary.

Imbricate Overlapping like shingles on a roof.
Incumbent 1. Said of an anther attached to the inner face of its filament. 2. Said of a cotyledon with its back lying against the radicle.
Indehiscent Said of a fruit that remains closed after it is ripe; not opening by any regular process; opposite of *dehiscent.*
Indeterminate 1. The growth of a stem, branch or shoot not limited or stopped by the development of a terminal bud. 2. the indefinite branching of a floral axis because of the absence of a terminal bud. 3. An inflorescence with axillary buds which continue to develop indefinitely.
Indigenous Native to a country or particular area; not introduced.
Indument Any hairy covering or pubescence.
Indumentum Same as indument.
Induplicate With edges folded in, turned inward, or folded cross-wise.
Indurate Hardened.
Indusium A scale-like shield or covering overarching the sori of many ferns, originating as an outgrowth of the epidermis. (Plural: *indusia*). In some ferns, as *Adiantum*, the sori are protected by indusioid organs that structurally are not true indusia; these may be designated by such terms as *indusial flap, indusial flange,* or *false indusium.*
Inferior Said of one organ when it is below another, such as an *inferior ovary,* which has the perianth located on top.
Inflexed Turned in at the margins.
Inflorescence An arrangement of flowers on a stem or axis; a cluster of flowers or a single flower.
Infra-areolar Within an areole.
Inframarginal Near but not at the margin; submarginal.
Inframedial Located nearer to the midvein than to the margin.
Interfoliar Situated between two opposite leaves, as the stipules of many Rubiaceae.
Internode The part of a stem between two nodes.
Interpetiolar Between the petioles.
Intrapetiolar Inside or beneath the petiole (= *subpetiolar*), or between the petiole and the stem (= *intrafoliaceous*).
Involucre A cluster of modified leaves or bracts at the base of a flower cluster or capitulum.
Involute Having the upper surface of a leaf rolled inward.
Irregular Said of flowers that are asymmetrical, having parts of the same whorl (sepals or petals, or both) different in size and/or shape.

Keel A central dorsal ridge; the two united petals of a papilionoid flower, which together form a boat-shaped structure.

Lacerate Irregularly cleft or cut; a margin having a torn appearance.

Laciniate Cut into deep narrow segments; cut into pointed lobes separated by deep, narrow, irregular incisions.

Lamellate Composed of thin plates or scales; thinly stratified.

Lamina The expanded part of a leaf or frond; a leaf-blade.

Lanceolate Lance-shaped; usually applied to rather narrow leaves broadest below the middle and about 3 times longer than wide. The term is used, often in abbreviated form, to denote shapes intermediate between lanceolate and some other shape, as *lance-oblong, lance-linear,* etc.

Lanate Clothed with soft woolly hairs.

Lanose Covered with long and loosely tangled hairs.

Lateral On or at the side.

Latex Milky sap, as in Asclepiadaceae or some Euphorbiaceae.

Lax Loose; not rigid.

Leaf A lateral appendage arising from the node of a stem and subtending a bud; it is usually of expanded shape and green, being the chief organ of photosynthesis in most flowering plants.

Leaflet One of the subdivisions of a compound leaf.

Legume 1. A plant of the family Leguminosae. 2. A fruit formed from a single carpel opening by two sutures.

Lemma The lower of the two bracts immediately subtending or enclosing the floret of a grass.

Lenticel A corky spot on young bark that serves as a path of gas exchange between the atmosphere and internal tissues of the stem.

Lenticular Lens-shaped; shaped like a biconvex lens.

Lepidote Covered with minute scurfy scales.

Ligulate Strap-shaped; also said of organs that possess a ligule.

Ligule 1. The strap-shaped corolla-lobe of the ray-flowers in Compositae. 2. A membranous appendage projecting from the top of the leaf-sheath in grasses. The term also has other meanings not relevant to this book.

Limb The expanded part of a gamopetalous corolla, as distinct from the tube or throat.

Linear Line-like; long and narrow with parallel or nearly parallel margins.

Lingulate Tongue-shaped.

Lobate Divided into, or bearing lobes.

Lobe A rounded division of a plant organ; one of the segments of a leaf, leaflet, calyx, corolla, etc., that is divided to about the middle.

Locule One of the compartments or "cells" of an ovary or anther.

Loculicidal Said of a fruit that dehisces along a suture about midway between the partitions separating the carpels.

Lodicule A small scale outside the stamens in the flowers of grasses.

Lunate In the shape of a half-moon; crescent-shaped.

Malpighiaceous Pertaining to hairs attached at or near the middle, with two horizontal points oriented in opposite directions, especially characteristic of the Malpighiaceae.

Mangrove A salt-tolerant tree or shrub, especially of such genera as *Rhizophora, Avicennia,* and *Laguncularia.*

Marginate Having a margin of distinctive structure or colour, forming a well-defined border.

Medial Located at or near the middle.

Megasporangium A sporangium containing only megaspores. (Plural: *megasporangia*).

Megasporophyll The carpel of an angiosperm.

Mericarp One of the seed-like ripened carpels in the fruit of Umbelliferae.

Micropyle 1. A minute opening at the apex of an ovule through which the pollen tube often grows to reach the egg cell. 2. The corresponding opening in the integument of a seed, through which water may enter when the embryo plant starts to develop.

Microsporangium A sporangium containing microspores; a pollen sac. (Plural: *microsporangia*).
Midrib, midvein The main rib or central vein of a leaf or leaflike structure.
Miocene The geological period between the Oligocene and the Pliocene.
Monocot An abbreviation for "monocotyledonous plant".
Monoecious Having separate staminate and pistillate flowers on the same plant.
Monograph An exhaustive systematic account of a particular genus, family, or group of organisms.
Monolete Said of a fern spore having a single surface line or minute ridge separating two more or less flat areas, the line marking the narrow zone where 4 spore faces end, the adjoining flat areas being where two neighboring spores were in contact; occurring in cases where 4 spores are radially arranged about the zone-line in a disc-like uniaxial tetrad.
Monotypic Said of a taxon containing or comprised of but a single element, as a genus with but one species or a family with but one genus.
Montane Pertaining to mountains.
Mordant A chemical solution used to fix (render permanent) a dye, as in tissues prepared for microscopic study.
Morphological Pertaining to morphology.
Morphology The study of form and structure; also used to designate the structure of an organism as contrasted with its physiology or classification.
Motile Capable of self-movement.
Mottled Marked with spots or irregular patches of different colours or shades.
Mucilage A gummy or gelatinous plant secretion, usually belonging to the amylose group of hydrocarbons.
Mucilaginous Composed of, or covered with mucilage; slimy.
Mucro A short, sharp terminal point or tip.
Mucronate Having a relatively blunt apex ending abruptly in a mucro.
Mucronulate Slightly or minutely mucronate.
Multifid Cleft into several or many lobes or segments.
Multiple fruit A fruit derived from a cluster of flowers, i.e., the ripened ovaries are traceable to the pistils of separate flowers, as in the pineapple.
Multiseriate Arranged in many series, whorls, or rows.
Muricate Rough with many short, sharp points on the surface.
Muriculate Minutely or finely muricate.
Mycorrhiza A specialized rootlet or root-like structure in which the cells are permeated with or coated by a symbiotic fungus. (Plural: *mycorrhizae*).

Naked 1. Lacking a covering, such as a perianth, scales, or pubescence. 2. Said of an ovary that is not enclosed in a pericarp, or of a seed that develops when such a naked ovary ripens.
Natural history The study of plant and animal life.
Naturalist A student of plant and animal life.
Naturalized Said of a plant species introduced from another region that becomes established, maintains itself, and reproduces successfully in competition with the indigenous vegetation.
Nectar A sweet secretion produced by flowers.
Nectary A nectar-secreting gland.
Neotropical Pertaining to the tropical parts of the New World, i.e., the West Indies, Central America, and the tropical parts of South America.
Neuter Pertaining to flowers that lack both stamens and pistil.
Nocturnal Occurring or functioning during the night; the opposite of *diurnal.*
Node The point on a stem where a leaf or leaves are normally borne; a joint.
Nodose Having prominent or swollen nodes.
Nodule A small hard knot or rounded body.
Non-motile Not capable of self-movement.

Nucleus A dense, complex spheroidal body, surrounded by a membrane, occurring in the cytoplasm of a living cell; it contains substances that determine the transmission of hereditary characters and that become organized in minute bodies called chromosomes just prior to cell division. (Plural: *nuclei*).

Nut A hard or bony, dry, indehiscent fruit derived from two or more carpels enclosed in a dense pericarp and usually containing one seed. The term is loosely used for any hard, dry, one-seeded fruit.

Nutlet 1. A small nut. 2. A one-seeded portion of a fruit which fragments as it matures, as in some Boraginaceae.

Ob- In combination, inversely or oppositely, as *obovate*, reversed ovate, i.e., having the broader portion toward the apex, or *obcordate*, inversely heart-shaped, i.e., having the notch at the apex.

Oblate Having the form of a compressed sphere.

Oblong Longer than broad, with the margins nearly parallel.

Obsolete Not evident; rudimentary.

Obturator A plug or special structure closing an opening, as the small body accompanying the pollen masses of orchids and asclepiads which close the opening of the anthers, or the cushion-like structure (called a *caruncle*) closing the micropyle in Euphorbiaceae and often persisting as a protuberance on the seed.

Obtuse Blunt-pointed.

Ocrea A nodal sheath formed by the fusion of two stipules, as in many Polygonaceae. (Plural: *ocreae*).

Ocreola A small ocrea. (Plural: *ocreolae*).

Odd-pinnate Said of a pinnate leaf with an odd number of leaflets, i.e., pinnate with a single terminal leaflet; imparipinnate.

Oedema Accumulation of excessive fluid in tissues.

Oligocene The geological period between the Eocene and the Miocene.

Operculate Having a cap or lid.

Orbicular Circular in outline; round and flat; orbiculate.

Oval Broadly elliptic with the width greater than half the length.

Ovary The ovule-bearing part of the pistil.

Ovate Egg-shaped and flat, with the broader end at the base.

Ovule The structure that becomes a seed after fertilization.

Palea The upper of the two bracts immediately subtending or enclosing the floret of a grass.

Palmate With veins, lobes, or divisions radiating from a common point.

Pandurate Fiddle-shaped; constricted at or near the middle.

Panicle A branched or compound raceme; in more casual meaning, an irregular compound inflorescence with pedicillate flowers.

Paniculate With flowers arranged in panicles.

Pantropical Distributed generally throughout tropical regions.

Papilla A minute nipple-shaped projection. (Plural: *papillae*).

Papillate, papillose Having papillae on the surface.

Pappus The modified calyx in the florets of Compositae, consisting of scales, bristles, or plumes of various shape attached to the apex of the achene.

Paraphyses Sterile hairs occurring among the sporangia in ferns.

Parasitic Deriving nourishment at the expense of another organism.

Parietal Attached to or lying near and more or less parallel with a wall, as the placenta when it arises from the peripheral wall of a carpel.

Partite A suffix meaning divided or cleft, as 2-partite, 3-partite, etc.

Pedicel The stalk of a flower.

Pedicellate Having a pedicel; borne on a pedicel.

Peduncle A primary flower-stalk supporting two or more flowers, or a solitary flower if it is the remnant of a cluster.

Pedunculate Pertaining to or borne upon a peduncle.

Pellucid Wholly or partly transparent; translucent. Applied especially to various dots or lines in leaves which contain internal oil-glands that allow the passage of light.

Peltate Pertaining to a leaf-blade which is attached to its petiole somewhere on the lower surface instead of by the margin; derived from the latin 'pelta' meaning a small shield, hence shield-shaped.

Pendulous Hanging downward; drooping.

Penicillate Resembling a little brush; having a terminal tuft of hairs.

Perennial Continuing to live from year to year, as contrasted with *annual.*

Perfect Said of flowers having both stamens and pistil in functioning condition.

Perianth A collective term designating both the calyx and corolla considered together, especially if they are of similar colour and texture.

Pericarp The wall of a mature ovary or fruit.

Perigynous Having the perianth parts and stamens borne on or arising from the periphery of the ovary (not beneath it), with the calyx lobes, corolla, and stamens arising on or near the rim of the hypanthium. To be contrasted with *epigynous* and *hypogynous.*

Perisperm That part of the endosperm (stored nutrient material) of a seed that lies outside the embryo, or which surrounds the embryo.

Perispore The husklike outer covering which surrounds the *exine* of a spore; however, some fern spores wholly lack a perispore.

Persistent Retaining its place, shape, or structure; remaining attached after the growing period; not deciduous.

Petal A unit of the inner floral envelope or corolla of a polypetalous flower, usually white or variously coloured, seldom green.

Petaloid Resembling a petal.

Petiole The stalk of a leaf.

Petiolule The stalk of a leaflet in a compound leaf.

Phyllanthoid A type of branching, found in the genus *Phyllanthus* (Euphorbiaceae), in which the penultimate axes have spiral phyllotaxy with leaves modified as cataphylls which subtend deciduous, floriferous, distichous-leaved ultimate axes (or in some cases leafless and flattened phylloclades). The stems, in other words, are differentiated into persistent, flowerless, "leafless" long-shoots, and deciduous, floriferous, "leafy" short-shoots.

Phyllary An involucral bract, usually several to many in number, subtending the flower-head in Compositae.

Phylloclade A green flattened or rounded stem functioning as a leaf.

Phyllotaxy The system of leaf arrangement on an axis or stem.

Phytogeography The science or study of plant distribution.

Pilose Clothed with long, soft hairs.

Pinna One of the primary divisions of a compound leaf, or (especially) of a compound fern frond. (Plural: *pinnae*).

Pinnate Pertaining to a compound leaf or frond with a single primary axis, as contrasted with *palmate*. The term is sometimes used as a suffix preceded by a number indicating the degrees of division, as *2-pinnate, 3-pinnate,* etc.; the same meaning may also be conveyed by a prefix of latin derivation, as *bipinnate* or *tripinnate*, etc. If the term is used alone, without any prefix, it may mean 1-pinnate. A form such as "1-4-pinnate" indicates the extent of variation in the degrees of cutting of a leaf or frond.

Pinnatifid Pinnately cut more than halfway to the axis into *segments.*

Pinnatisect Cut to the midrib or rhachis in a pinnate pattern.

Pinnule A secondary (or tertiary, etc.) leaflet or pinna of a decompound leaf or frond. Usually distinguished from a *segment* by being attached to the next higher axis only by the base of its own axis, the tissue margin being entirely free.

Pistil The female organ of a flower, consisting when complete of an ovary, style and stigma. A simple pistil consists of a single carpel; a compound pistil consists of 2 or more carpels, these usually fused together.

Pistillate Pertaining to a flower having a pistil but no stamens.

Pistillode A rudimentary pistil in a staminate flower.

Pith The spongy tissue often occurring in the center of a dicotyledonous stem; also found in some gymnosperms.

Pitted Marked with small or minute depressions.

Placenta The part of the ovary or carpel where the ovules are attached.

Pleiochasium A cymose inflorescence in which each branch bears more than two lateral branches. (Plural: *pleiochasia*).

Plicate Folded or plaited.

Plumose Resembling a plume or feather.

Plumule The primary leaf-bud of an embryo.

Pluricellular Composed of two or more cells.

Pneumatophore An aerial structure which grows vertically upward from roots embedded in mud, composed of spongy tissue presumed to function as an aerating organ.

Pollen The dusty or sticky grains produced in the stamens of flowers. Each pollen grain is in fact a male gametophyte which, on contact with a suitable stigma, germinates to extend a "pollen tube" ultimately to an ovule. Fertilization is effected when one of the male nuclei enters the ovule and unites with the female nucleus.

Pollen-sac The cavity of an anther containing the pollen.

Pollen-tube See above under *pollen.*

Pollination The process by which pollen travels from an anther to a stigma.

Pollinium A coherent mass of pollen, as in orchids and asclepiads.

Polygamous Bearing perfect and unisexual flowers on the same plant.

Polymorphic With several or various forms; variable in structural characters.

Polyploid Having more than the standard number of chromosomes in its nuclei, usually expressed in terms of a definite multiple of the base number, as *tetraploid* meaning with four times the number of chromosomes found in the gametic cells of a normal "diploid" population.

Polystichous Having leaves borne in many rows or series and spreading in many directions.

Pore A small or minute aperture.

Prismatic 1. Shaped like a prism. 2. With many colours, like a rainbow.

Procumbent Prostrate but not taking root at the nodes; trailing.

Proliferous Bearing offshoots; often used to denote leaves or fronds which produce small root-bearing buds or bulbils that afford a means of vegetative reproduction.

Prominulous Minutely raised, as the veins of a leaf.

Prop root A root produced above the ground which serves as a prop or support to the plant, as in mangroves.

Prostrate Lying flat on the ground.

Protandrous With anthers maturing before the pistil in the same flower.

Prothallus The gametophyte of a fern. (Plural: *prothalli*).

Pruinose Bearing a waxy-powdery secretion on the surface.

Pseud-, pseudo- A prefix meaning false or spurious.

Pseudobulb The thickened or bulblike stems of certain orchids, usually borne above the ground or other substrate, and usually bearing one or more leaves at the apex.

Pseudocarp A false fruit; a fruit derived from parts other than the ovary, as in some Rosaceae, in which the "fruit" is chiefly composed of the greatly enlarged receptacle.

Pteridophyte One of the four main divisions of the plant kingdom according to some systems of classification, including the ferns and several other more or less remotely related groups which have a somewhat similar lifecycle.

Puberulous Minutely pubescent.

Pubescent Covered with soft, straight, short hairs.

Pulvinate Having a pulvinus.

Pulvinus A swelling at the base of a petiole or petiolule which frequently acts as a center of sensitivity, irritability, or movement in a leaf.
Punctate Marked with dots, translucent or otherwise.
Punctulate Minutely punctate.
Pustular, pustulate. Blister-like; covered with small blister-like prominences.
Pyrene A small, hard, stone-like seed in a drupe or similar fruit.

Quandrangular Having four sides.
Quadrinomial A 4-unit epithet applied to an organism or population, designating *genus, species, variety* or *subspecies*, and *form*, as *Pteridium aquilinum* var. *caudatum* forma *glabratum*, or *Pteridium aquilinum caudatum glabratum.*

Raceme A simple, elongate, indeterminate inflorescence bearing a number of pedicellate flowers, the pedicels of approximately equal length.
Racemose Having flowers in racemes.
Rachilla Same as *Rhachilla.* In grasses and sedges, the axis that bears the florets. (Plural: *rachillae*).
Radiate Spreading from or arranged around a common center.
Radicle The embryonic root of a germinating seed.
Raphe The continuation of a funicle adnate to the side of an ovule, usually evident as a raised line or ridge.
Rank A row, especially a vertical row, often used as a suffix, as *2-ranked, 3-ranked,* etc.
Ray-floret A ligulate flower in the flower-head of Compositae.
Receptacle In flowering plants, the more or less enlarged or elongated apex of the pedicel; in Compositae, the enlarged apex of the peduncle.
Reflexed Abruptly curved or bent downward or backward.
Regular Having flower parts of the same kind all alike in size and shape; radially symmetrical.
Remote Widely spaced; scattered; not close together.
Reniform Kidney-shaped.
Repand Undulate or wavy; having a slightly undulating or sinuous margin.
Replum A thin wall dividing a fruit into two chambers, formed by an ingrowth from the placenta and thus not a true part of the carpel walls.
Reticulate Netted; forming a network.
Retrorse Bent or turned backward or downward.
Retuse Having a shallow notch at an obtuse apex.
Revolute With the margins rolled under toward the lower (abaxial) side, as the margin of a leaf.
Rhachilla Same as *rachilla*. (Plural: *rhachillae*)
Rhachis 1. Any axis bearing lateral appendages or organs. 2. In compound leaves or fronds, the portion of the main axis to which the primary divisions or pinnae are attached. Sometimes spelled "rachis".
Rhizome An underground stem that gives rise to roots and aerial stems, distinguished from a true root by the presence of nodes, buds, or leaves, the latter sometimes reduced to scales.
Rhizophore A naked branch which grows down into the soil and develops roots from the apex.
Rind A tough outer layer, as on some fleshy fruits; sometimes used to designate any outer skin.
Root The usually underground part of a plant which supplies it with water and dissolved mineral nutrients, in structure always lacking nodes and leaves; the absorbtive, anchoring, and storage organ of vascular plants.
Rootlet A little root; a small branch of a root.
Rootstock An underground stem or rhizome; sometimes applied especially to an erect rhizome, as in some ferns.
Rosette A cluster of spreading or radiating basal leaves.

Rostellum A small beak; especially, a narrow extension of the upper edge of the stigma in some orchid flowers.
Rosulate In the form of a rosette.
Rotate Wheel-shaped with flat and spreading parts.
Rotund Rounded in outline; nearly orbicular but slightly inclined toward the oblong.
Rugose Wrinkled in appearance, as a leaf-surface with sunken veins.
Rugulose Minutely rugose.
Ruminate Having a crumpled appearance, or appearing as if chewed.
Russet Reddish or dull red.

Saccate Shaped like a pouch or little bag; sac-shaped.
Sagittate Shaped like an arrowhead, with prominent basal lobes pointing or curving downward.
Sagittulate Minutely sagittate.
Salverform, salver-shaped Said of a corolla composed of a slender tube abruptly expanding into a flat, rotate limb.
Samara A dry, indehiscent, one-seeded, winged fruit.
Saprophyte A plant which derives all its nourishment from dead animal or vegetable matter.
Scabrid Referring to a surface with scattered or intermittent roughness.
Scabridulous Minutely roughened.
Scabrous Referring to a surface that is rough with a covering of very short, stiff, bristly hairs, scales or points.
Scale 1. A usually small, dry, thin or membranous epidermal appendage, differing from a hair in being two or more cells in width. 2. A small, thin, semitransparent bract or leaflike structure, usually appressed, found on apical buds, bulbs, at the base of shoots, and on rhizomes and other organs; see also *cataphyll.*
Scandent Climbing or scrambling over rocks or other plants without the aid of tendrils, but often with adherent roots or rootlets.
Scape A leafless peduncle arising from the ground, either entirely naked or else at the most bearing scales or bracts.
Scarious Thin, dry, and membranous, and usually not green.
Schizocarp A dry compound fruit which splits apart at maturity into two or more single-seeded segments known as mericarps.
Scorpioid Coiled like a scorpion's tail, said of a coiled (*circinnate*) inflorescence in which the flowers are 2-ranked, being borne alternately on opposite sides of the axis.
Scrambling Said of a plant with weak, elongate stems that grow over other plants or any kind of support, but do not twine or have the aid of tendrils or aerial roots.
Scrub A type of vegetation composed of low and often densely-packed bushes.
Scurfy Covered with minute bran-like scales.
Secondary axis A branch of a main axis.
Secondary thicket; secondary vegetation Vegetation which comes up naturally after cutting, fire, or other disturbing factors.
Seed A mature, fertilized ovule, containing an embryo and some form of stored food material, and protected by a seed-coat or testa.
Semi- A prefix meaning partly, to some extent, incompletely, etc.
Sepal One of the parts of the calyx.
Sepaloid Sepal-like, resembling a sepal.
Septate Divided by partitions; having *septa*.
Septicidal Said of a capsule that dehisces along the partitions.
Septum A partition; a cross-wall; a dividing membrane. (Plural: *septa*).
Seriate Arranged in one or more rows or series.
Sericeous Clothed with a silky pubescence; covered with closely appressed fine soft hairs.
Serrate Having a saw-toothed margin, the teeth inclined toward the apex.
Serration A saw-like notch; a tooth of a serrate margin.

Serrulate Minutely serrate.
Sessile Lacking a stalk, as a leaf without a petiole.
Seta A bristle or bristle-like structure; a needle-shaped process. (Plural: *setae*).
Setaceous Bearing setae; covered with bristles.
Setiform Bristle-shaped.
Setose Covered with bristles.
Setulose Minutely setose.
Sheath Any more or less tubular structure surrounding a part, as the basal part of a grass leaf surrounding the culm.
Shrub A relatively low, usually several-stemmed, woody plant; a bush.
Silique A 2-valved capsular fruit in which the valves dehisce from a frame (the *replum*) on which the seeds grow, and across which a false partition is formed.
Sinuate Having a deeply wavy margin.
Sinus The notch between two lobes or segments. (Plural: *sinuses*).
Solitary Single; only one in the same place.
Sorus In ferns, a cluster of sporangia. (Plural: *sori*).
Spadix The thick or fleshy flower-spike of certain plants, as in the Araceae, surrounded or subtended by a *spathe.*
Spathe The bract or pair of bracts surrounding or subtending a flower cluster or spadix.
Spathulate, spatulate Shaped like a spatula, oblong with an attenuated base.
Species A term used in classification to denote a group or population of closely similar, mutually fertile individuals which show constant differences from allied groups that are more or less reproductively isolated. (Plural: *species*).
Specific name The name of a species; the second part of a binomial name.
Specimen A plant, or portion of a plant, prepared and preserved for study; a preserved sample intended to show the characteristics of a species or other taxon.
Spermatozoid A motile or free-swimming male gamete.
Spher- A prefix meaning round.
Spicate Spikelike; arranged in or having spikes.
Spike An unbranched, simple, elongate inflorescence bearing sessile or subsessile flowers.
Spikelet 1. A secondary spike, i.e., the part of a compound inflorescence which is itself a spike. 2. The ultimate unit in the inflorescence of grasses, consisting of two *glumes* and one or more florets each of which is subtended by a *lemma* and *palea.*
Spine Any sharp, rigid process or outgrowth, usually a modified branch, but also sometimes a modified stipule, petiole, or other part; a thorn.
Spinose Bearing a spine or spines.
Spinulose Minutely spinose.
Sporangium A minute capsule or sac containing spores. (Plural: *sporangia*).
Spore A minute, unicellular, reproductive body, analogous in function to a seed but lacking an embryo. In ferns, the spores of many or most species are produced through the process of *meiosis*, by which the number of chromosomes in the nucleus is reduced by half; germination of a fern spore gives rise to the gametophyte generation in the life-cycle of these plants.
Sporophyll A spore-bearing leaf or frond.
Sporophyte The asexual spore-producing plant or generation in plants having an alternation of generations.
Spur 1. A tubular elongation of the base of a petal or of a gamopetalous corolla. 2. An extension of the base of a leaf beyond its point of attachment. 3. A short branch in many trees on which flowers and fruits are borne. 4. A short branch borne from the axil of a scale-leaf or cataphyll and bearing the true foliage leaves of the plant.
Stamen The pollen-bearing organ of a flower, typically consisting of a filament and an anther, or the anther sometimes sessile.
Staminal cup, column, or *tube* Structures formed by various degrees of fusion of the filaments of stamens, as the staminal column of Malvaceae.

Staminate Having, producing, or consisting of stamens; having stamens and no pistil, as the staminate flowers of dioecious plants.
Staminode A sterile stamen or stamen-like structure; a structure occupying the position of, and often shaped like, a stamen, but lacking a fertile anther.
Standard The upper, broad, often erect to recurved petal in many leguminous flowers.
Stellate Star-shaped; having hairs or scales with branches or points radiating from a center.
Stem The main axis of a plant, bearing leaves and flowers, as contrasted with a root, which bears neither of these structures.
Sterile Barren or non-reproductive, as a fern-frond without sporangia or a flower lacking a pistil.
Stigma The part of a pistil (usually the apex) which is receptive to pollen grains and on which they germinate. (Plural: *stigmas* or *stigmata*).
Stigmatic Pertaining to the stigma.
Stipe The stalk or petiole of a fern-frond.
Stipel The stipule of a leaflet in a compound leaf.
Stipitate Having a stipe.
Stipular Having stipules, or relating to them.
Stipular spine A spine representing a modified stipule, or having the position of a stipule.
Stipule A more or less leafy appendage at the base of the petiole in many plants.
Stolon An elongate creeping stem (usually representing a modified basal branch) which roots at the nodes and often gives rise to new plants at some nodes, or at its tip, or both.
Stoloniferous Producing stolons.
Strand 1. A thread or fiber. 2. The zone of exposed beach above the high tide line.
Striate Marked with fine, linear, parallel lines.
Striation A fine, linear marking; a minute elongate ridge or furrow; one of a pattern of fine lines.
Strigillose Minutely strigose.
Strigose Bearing appressed, sharp, straight, stiff hairs.
Strobilus 1. A cone-like cluster of sporophylls. 2. An inflorescence or ovule-bearing structure made up largely of more or less imbricated scales, and usually of a somewhat conical shape. (Plural: *strobili*). The term is also written as *strobile.*
Strophiole An appendage at the hilum of some seeds; a caruncle.
Strychnine A poisonous alkaloid used medicinally in small doses as a nerve stimulant.
Stylar column The column in orchid flowers.
Style The elongate part of a pistil, bearing the stigma at its apex; lacking in flowers having a sessile stigma.
Sub- A prefix meaning less than, almost, approaching, etc., as *subcordate*, meaning slightly cordate, *subcylindric*, meaning not quite circular in cross-section, or *submarginal*, meaning almost at the margin.
Subfamily In classification, a category beneath that of family and above that of genus.
Subgenus In classification, a category beneath that of genus and above that of species; a subgeneric name is often used for one or more species that form a natural grouping within a genus, but does not become part of a binomial epithet.
Substrate The material in or on which a plant is rooted.
Subtend To be attached beneath and close to, as a bract below a flower.
Subterranean Underground; occurring beneath the surface of the soil.
Subtropical Inhabiting or characterizing regions bordering the tropics.
Subulate Awl-shaped.
Succulent Juicy or fleshy; having tissues thickened to conserve moisture.
Suffrutescent A stem which is woody and perennial at the base, but has an upper herbaceous portion that dies back at the end of a growing season. The term is also sometimes used to mean "slightly shrubby".
Sulcate Grooved lengthwise.
Superior Said of one organ when it is above another, such as a *superior ovary*, which has the perianth attached beneath it.

Supra-axillary Borne above the axil.
Supramedial Said of a sorus which is slightly closer to the margin than to the midvein.
Suture 1. A line of fusion or union. 2. A line along which dehiscence may occur.
Symbiosis The living together of dissimilar organisms, with benefit to one or both.
Sympodial Having a stem made up of a series of superposed branches arranged so as to appear like a simple axis; this is caused by the self-pruning and withering of the terminal bud at the end of each period of growth, so that successive growth-episodes must originate from lateral buds.
Syncarp A multiple or fleshy aggregate fruit.
Syncarpous 1. Composed of two or more united carpels. 2. Bearing an aggregate fruit.
Synonym One of two or more scientific names applied to the same taxon, one of which is correct and the others incorrect under the International Rules of Nomenclature. The term is commonly used to designate only the incorrect names.
Syntype One of two or more specimens (not duplicates of the same number) of equal rank on which a species, subspecies, variety, or form is based, in the absence of a designated holotype.

Taproot A central or leading root which penetrates deeply into the ground without dividing; a prolonged and relatively thick primary root.
Tawny Dull brownish-yellow.
Taxon In classification, a term applied to any coherent element, population, or grouping, regardless of level, as *order, family, genus, species*, etc. (Plural: *taxa*).
Taxonomic Pertaining to the systematic classification of organisms.
Taxonomy The science of classification.
Tendril A threadlike process or extension by which a plant grasps an object and clings to it for support; this structure may represent a modified stem, leaf, leaflet, or stipule.
Terete Circular in cross-section.
Terminal Pertaining to the end or apex.
Terrestrial Growing on the ground.
Tertiary Of the third order or rank, as *tertiary veins*, which are branches of secondary veins, which in turn arise from the primary vein or midvein.
Testa The outer coat or integument of a seed, usually hard and brittle.
Tetrahedral Having or made up of four faces.
Tetraploid Said of an organism whose cell nuclei contain twice the normal (diploid) number of chromosomes expected in members of its particular taxon; it is called tetraploid because it has four times the gametic number.
Thallophyte A plant lacking differentiation into roots, stems, leaves, and without internal vascular tissues.
Thyrse A compact or contracted panicle with the main axis indeterminate but the secondary and ultimate axes cymose and determinate.
Thyrsoid Resembling a thyrse.
Tissue An aggregate of cells similar in structure or function.
Tomentellous Minutely tomentose.
Tomentose Covered with matted, soft, woolly hairs.
Tomentum A covering of matted, soft, woolly hairs; wool-like pubescence.
Tortuous Marked by repeated twists, bends, or turns.
Torose Knobby; having a cylindric body abruptly swollen at intervals.
Torulose The diminutive of torose.
Torus The receptacle of a flower; the apical portion of a floral axis on which the parts of a flower are inserted.
Trapezoidal Having four unequal sides.
Tri- A prefix meaning three, as *trichotomous* meaning three-forked or three-branched, *trifid* meaning a three-cleft apex, *trifoliate* meaning with three leaflets, *trigonous* meaning three-angled, etc.

Tribe A taxon consisting of one or more genera forming a natural group within a family, but not equivalent to a subfamily.

Trichome Any hairlike outgrowth of the epidermis.

Triplinerved Said of a leaf with three main nerves or ribs, the lateral two arising from the median one above the base, contrasting with a trinerved leaf in which the three nerves all arise from the same point at the base of the lamina.

Triploid Said of an organism whose cell nuclei contain three times the normal gametic number of chromosomes, or one and one half the normal diploid number; such organisms usually arise by hybridization between a diploid and a tetraploid.

Tropical Strictly, the life-zone on either side of the equator where the day and night are of about equal length and the temperature is rather uniform throughout the year with a mean daily temperature of about 27°C (= 80°F); in this sense, such climatic variation as occurs in the zone is a result of fluctuations in rainfall. In a broader sense, the tropical zone is the region of the Earth's surface immediately north and south of the equator, and bounded approximately by the Tropic of Cancer (23½°N) and the Tropic of Capricorn (23½°S); within this very large area many climatic differences occur and a large number of life-zones are represented. Therefore the term "tropical" can mean different things according to the context in which it used.

Truncate Ending abruptly; having a base or apex that is nearly straight across as if cut off.

Tuber A relatively short, thickened rhizome with numerous buds capable of vegetative propagation, as in the "Irish" potato.

Tubercle A wart-like or knob-like projection.

Tuberculate Bearing tubercles.

Tufted Clustered or bunched, the units or divisions originating from approximately the same point; *caespitose.*

Turbinate Top-shaped.

Turgid Swollen, as by pressure of liquid from within.

Twining Twisting together or about another part or object more or less spirally.

Ultimate Last, final, or utmost; situated at the end, as the ultimate branch of a stem; farthest from the base.

Umbel An indeterminate, often flat-topped inflorescence consisting of several or many pedicellate flowers arising from a common point of attachment.

Umbellate Arranged in umbels, or pertaining to umbels.

Umbelliform In the shape of an umbel.

Undershrub A low-growing woody plant.

Undulate Wavy.

Undulate-toothed A wavy margin with teeth.

Uni- A prefix meaning, one, as *uniseriate* meaning in one row, or *unisexual* meaning of one sex.

Urceolate Pitcher-shaped; hollow and contracted at the mouth.

Urn-shaped In the shape of a vase or pitcher; *urceolate.*

Utricle 1. A bladdery, 1-seeded, usually indehiscent fruit, as in the Urticaceae. 2. Any small membranous sac or bladder-like body.

Vaginate Sheathed; surrounded by a sheath.

Valvate Having or pertaining to valves; opening by valves, as a capsule. The term is also applied to flower-buds whose segments meet exactly without overlapping.

Valve One of the segments into which the walls of a fruit separate by dehiscence; one of the parts of a dehiscent pericarp. The term is also applied to a partially detached flap of an anther wall in Lauraceae.

Variegated Marked with different colours or tints in spots, streaks, or patches.

Variety A morphological variant or variant group within a species, differing from other variants of the same species by one or more minor characteristics.

Vascular Furnished with vessels or ducts through which fluid passes.
Vascular bundle A strand of conductive tissue composed of xylem and phloem cells, these sometimes separated by a cambium, and sometimes also accompanied by sclerenchymatous (fibrous) supporting tissue.
Vascular plants Plants containing vascular tissues, such as ferns and seed plants.
Vegetation The total aggregation of plant-life in a particular area.
Vegetative reproduction Non-sexual reproduction; reproduction by means of non-sexual structures such as bulbils, stolons, or tubers.
Vein A strand of vascular tissue in a leaf-blade.
Veinlet A minute vein.
Venation Pertaining to the arrangement or pattern of veins.
Ventricose Swollen on one side.
Vermifuge A medicine for eliminating worms.
Versatile anther An anther attached near the middle and turning freely on its support.
Verrucose Covered with small wart-like projections.
Vestigial Rudimentary; partly-formed, often minute structure located where normal, fully-formed parts ought to be, or once existed.
Villous Clothed with long, soft, weak but straight hairs.
Vine 1. The plant which bears grapes, *Vitis vinifera*. 2. Any trailing or climbing plant.
Viscid Having a sticky coating or secretion.

Whorl The cyclic arrangement of appendages at a node.
Wing 1. Any thin, often dry and membranous expansion attached to an organ, as on petioles or fruits. 2. A lateral petal of a leguminous flower.

Xylem A complex tissue in the vascular system of higher plants, functioning chiefly for the conduction of fluids upward, but also for support and storage, and typically making up the woody part of a plant stem.

Zoology The study of animal life.
Zygomorphic Bilaterally symmetrical; divisible into similar halves in only one plane.
Zygote The cell resulting from the fusion of two gametes; a fertilized egg cell before its nucleus has divided to initiate growth.

INDEX OF COMMON NAMES WITH THEIR BOTANICAL EQUIVALENTS

(Bold page numbers indicate illustrations)

INDEX OF COMMON NAMES AND BOTANICAL EQUIVALENTS

INDEX OF BOTANICAL NAMES

(Accepted names are in roman. Synonyms of accepted species names are in italics. Family names are in bold type. Page numbers: bold type indicates illustration; + indicates further entries in same key).

*See p 67, Recent Nomenclatural Changes

*See p. 67, Recent Nomenclatural Changes

Printed in the UK for HMSO Dd 736445 C5 9/84